A Course in Real Analysis

Now in its second edition, *A Course in Real Analysis* provides a rigorous treatment of the foundations of differential and integral calculus. It proceeds gradually from an axiomatic characterization of the real number system to the study of differentiation and integration on m-dimensional surfaces. Proofs of theorems are given in detail, and many examples are provided to illustrate the concepts expressed in the theorems.

The book consists of three parts. Part I treats the calculus of functions of one variable. Traditional topics such as sequences, continuity, differentiability, Riemann integrability, numerical series, and the convergence of sequences and series of functions are covered. Optional sections on Stirling's formula, Riemann–Stieltjes integration, and other topics are also included.

The second part focuses on functions of several variables. It introduces the topological ideas (such as compact and connected sets) needed to describe analytical properties of multivariable functions. This part also discusses differentiability and integrability of multivariable functions, and it develops the theory of differential forms on surfaces in $\mathbb{R}^n$.

Many proofs and explanations in the first edition have been revised, and details have been added to clarify the exposition. Part III contains the Appendices on set theory and linear algebra, as well as solutions to some of the exercises are offered, while a full Solutions Manual contains complete solutions to all exercises for qualifying instructors.

Hugo D. Junghenn is a professor emeritus of mathematics at The George Washington University. He has published numerous journal articles and is the author of several books, including *Option Valuation: A First Course in Financial Mathematics, Principles of Analysis,* and *Discrete Mathematics with Coding.* His research interests include functional analysis, semigroups, and probability.

A Course in Real Analysis

Second Edition

Hugo D. Junghenn

CRC Press
Taylor & Francis Group
Boca Raton London New York

CRC Press is an imprint of the
Taylor & Francis Group, an **informa** business
A CHAPMAN & HALL BOOK

Designed cover image: Hugo D. Junghenn

Second edition published 2026
by CRC Press
2385 NW Executive Center Drive, Suite 320, Boca Raton FL 33431

and by CRC Press
4 Park Square, Milton Park, Abingdon, Oxon, OX14 4RN

CRC Press is an imprint of Taylor & Francis Group, LLC

Library of Congress Cataloging-in-Publication Data
Names: Junghenn, Hugo D. (Hugo Dietrich), 1939- author
Title: A course in real analysis / Hugo D. Junghenn.
Description: Second edition. | Boca Raton, FL : CRC Press, 2026. | Series: Textbooks in mathematics | Previous edition: 2015. | Includes bibliographical references and index.
Identifiers: LCCN 2025041598 (print) | LCCN 2025041599 (ebook) | ISBN 9781041045632 hardback | ISBN 9781041046707 paperback | ISBN 9781003629382 ebook
Subjects: LCSH: Mathematical analysis
Classification: LCC QA300 .J86 2026 (print) | LCC QA300 (ebook)
LC record available at https://lccn.loc.gov/2025041598
LC ebook record available at https://lccn.loc.gov/2025041599

ISBN: 978-1-041-04563-2 (hbk)
ISBN: 978-1-041-04670-7 (pbk)
ISBN: 978-1-003-62938-2 (ebk)

DOI: 10.1201/9781003629382

Typeset in Latin Modern font
by KnowledgeWorks Global Ltd.

Access the Support Material: www.routledge.com\9781041045632

To the Memory of My Parents

Rita and Hugo

Contents

Preface

The purpose of this text is to provide a rigorous treatment of the foundations of differential and integral calculus at the advanced undergraduate level. It is assumed that the reader has had the traditional three-semester calculus sequence and some exposure to elementary set theory and linear algebra. Regarding the last two subjects, the Appendices give a summary of most of the results used in the text. Linear algebra will not be needed until Part II.

The book consists of three parts. Part I treats the calculus of functions of one variable. Here, one can find the traditional topics: sequences, continuity, differentiability, Riemann integrability, numerical series, and convergence of sequences and series of functions. Optional sections on Stirling's formula, Riemann–Stieltjes integration, and other topics are also included. As the ideas inherent in these subjects ultimately rest on properties of real numbers, the book begins with a careful treatment of the real number system. For this we take an axiomatic rather than a constructive approach, guided as much by the need for efficiency of exposition as by pedagogical preference. Of course, presenting the real number system in this way begs the excellent question as to whether such a system exists. It is a question we do not answer, but the interested reader may wish to consult a text on the construction of the real number system from the natural numbers, or even on the philosophy of mathematics.

Part II treats functions of several variables. Many of the results in Part I, such as the chain rule, the inverse function theorem, and the change of variables theorem, have counterparts in Part II. The reader's exposure to the one-variable results should make the multivariable versions more meaningful and accessible. However, as might be expected, some results in Part II, notably the implicit function theorem and the iterated integral (Fubini–Tonelli) theorem, have no counterparts in Part I. Part II begins with a chapter on metric spaces. Here, we introduce the elementary topological ideas needed to describe some of the analytical properties of multivariable functions. Primary among these are the fundamental notions of compact set and connected set, which, for example, enable the extension to higher dimensions of the extreme value and intermediate value theorems. The remainder of Part II covers differentiability and integrability of multivariable functions. In regard to integrability of functions of several variables, we have chosen to develop from scratch the Lebesgue integral rather than extend the Riemann integral to higher dimensions, the former approach being in some ways more natural. The additional time required for this approach is, in my view, more than offset by the enormous utility of the Lebesgue integral. The last chapter of Part II develops the theory of differential forms on surfaces in $\mathbb{R}^n$. The chapter culminates with proofs of Stokes's theorem and the divergence theorem for compact surfaces. It is the hope that exposure to these topics at the concrete level of surfaces in $\mathbb{R}^n$ will ease the transition to more advanced courses such as calculus on differentiable manifolds.

Part III consists of the aforementioned Appendices on set theory and linear algebra, as well as solutions to some of the over 1600 exercises found in the text. Exercises that will find important uses later are marked with a downward arrow $\Downarrow$.

The book is an outgrowth of notes developed over many years of teaching real analysis to undergraduates at George Washington University. The typical two-semester course closely followed the non-starred sections of this text: Chapters 1–7 for the first semester and 8–13 for the second. Given the wealth of material, it was necessary to leave some proofs for students to read on their own, a not wholly unfortunate compromise. Material in some starred sections was assigned as optional reading.

I would like to thank my students whose critical eyes caught errors before they made their way into the first edition. Thanks also to readers who pointed out errors that slipped into the first edition.

I would also like to express my thanks to CRC Press, and in particular to my editor, Robert Ross, for giving me the opportunity to write a second edition. Many proofs and explanations in the first edition have been revised and details have been added to clarify the exposition. Of course, errors in the first edition have been corrected.

Finally, to my wife Mary for her support and understanding during the writing of both editions: thank you!

Hugo D. Junghenn
Vienna, VA
September 2025

List of Symbols

Part I

Functions of One Variable

Chapter 1

The Real Number System

If the notion of limit is the cornerstone of real analysis, then the real number system is the bedrock. In this chapter we begin with an axiomatic description of the system and then use the axioms to give a careful development of limit.

While the axiomatic approach gets us to where we want to be in the quickest and most efficient way, there is an alternate approach that starts out with a simpler system of numbers and axioms and constructs the real numbers from these. The initial system is the set $\mathbb{N}$ of *natural numbers* or *positive integers*, together with the *Peano Axioms.* The latter are used to define operations of addition and multiplication in $\mathbb{N}$. The set $\mathbb{Z}$ of all integers is constructed from $\mathbb{N}$ by attaching zero and the symbols $-n$, $n \in \mathbb{N}$, and defining the new operation of subtraction. The symbols m/n are then attached to $\mathbb{Z}$, forming the so-called rational number system $\mathbb{Q}$. At this stage it is possible to define addition, subtraction, multiplication, and division in $\mathbb{Q}$, which have the familiar arithmetic properties. The final step, the construction of the real number system $\mathbb{R}$ from $\mathbb{Q}$, may be viewed as "filling in the gaps" of the rational number line, these gaps corresponding to the so-called *irrational numbers.* This approach is certainly more appealing from a philosophical point of view, as it makes the real number system more "concrete," especially if you believe in natural numbers. But the process of constructing the real number system is lengthy and is somewhat removed from the spirit and goals of this book. For the details of this "bottom up" approach, the interested reader is referred to [7] or [10].

1.1 Algebraic Properties of $\mathbb{R}$

The real number system is defined as a nonempty set $\mathbb{R}$ together with two algebraic operations, called *addition* and *multiplication*, and an ordering *less than* that collectively satisfy three sets of axioms: the *algebraic* or *field axioms*, the *order axioms*, and the *completeness axiom.* The latter two sets of axioms are discussed in Sections 1.2 and 1.3, respectively. In this section we give the axioms that govern how addition $(+)$ and multiplication $(\cdot)$ are used in the real

DOI: 10.1201/9781003629382-1

number system. As we shall see, the axioms lead to all the familiar algebraic properties of real numbers.

Axioms of the Real Number System

In the following, a, b, c denote arbitrary members of $\mathbb{R}$.

- Closure under addition: $a + b \in \mathbb{R}$.
- Associative law of addition: $(a + b) + c = a + (b + c)$.
- Commutative law of addition: $a + b = b + a$.
- Existence of an additive identity: There exists a member 0 of $\mathbb{R}$ such that $a + 0 = a$ for all $a \in \mathbb{R}$.
- Existence of additive inverses: For each $a \in \mathbb{R}$, there exists a member $-a$ of $\mathbb{R}$ such that $a + (-a) = 0$.
- Closure under multiplication: $a \cdot b \in \mathbb{R}$.
- Associative law of multiplication: $(a \cdot b) \cdot c = a \cdot (b \cdot c)$.
- Commutative law of multiplication: $a \cdot b = b \cdot a$.
- Existence of a multiplicative identity: There exists a real number $1 \neq 0$ such that $a \cdot 1 = a$ for all $a \in \mathbb{R}$.
- Existence of multiplicative inverses: For each $a \neq 0$, there exists a member a^{-1} of $\mathbb{R}$ such that $a \cdot a^{-1} = 1$.
- Distributive law: $a \cdot (b + c) = a \cdot b + a \cdot c$.

The above properties are called the *field axioms* of $\mathbb{R}$. We shall use the following standard notation:

$$a - b = a + (-b), \quad ab = a \cdot b, \quad \frac{a}{b} = a/b = ab^{-1},$$
$$a + b + c = (a + b) + c = a + (b + c), \quad abc = (ab)c = a(bc),$$
$$a^n = \underbrace{aa \cdots a}_{n}, \quad a^{-n} = 1/a^n \ (a \neq 0), \quad \text{and} \quad a^0 = 1.$$

We also use the *summation* and *product* symbols $\sum$ and $\prod$ defined by

$$\sum_{j=m}^{n} a_j = a_m + a_{m+1} + \cdots + a_n \quad \text{and} \quad \prod_{j=m}^{n} a_j = a_m a_{m+1} \cdots a_n.$$

The field axioms may be used to derive the standard rules of algebra. Some of these are given in the following proposition; others may be found in Exercise 1.

1.1.1 Proposition. *The following algebraic properties hold in* $\mathbb{R}$:

(a) *The additive identity is unique; that is, if* $0'$ *is a real number such that* $a + 0' = a$ *for all* $a \in \mathbb{R}$, *then* $0' = 0$.

(b) *The additive inverse of a real number is unique; that is, if* $a + b = 0$, *then* $b = -a$.

(c) *The multiplicative identity is unique; that is, if* $1'$ *is a real number such that* $a \cdot 1' = a$ *for all* $a \in \mathbb{R}$, *then* $1' = 1$.

(d) $a \cdot 0 = 0$ *for all* $a \in \mathbb{R}$.

(e) *The multiplicative inverse of a nonzero real number is unique; that is, if* $ab = 1$, *then* $b = 1/a$.

(f) *If* $ab = 0$, *then either* $a = 0$ *or* $b = 0$.

(g) *If* $ab = ac$ *and* $a \neq 0$, *then* $b = c$.

(h) *If* $b \neq 0$ *and* $d \neq 0$, *then* $a/b = c/d$ *if and only if* $ad = bc$.

(i) *If* $a \neq 0$ *and* $b \neq 0$, *then* $(ab)^{-1} = a^{-1}b^{-1}$, *or* $\frac{1}{ab} = \frac{1}{a}\frac{1}{b}$.

Proof. (a) If $a + 0' = a$ for all a, then, in particular, $0 + 0' = 0$. But, by definition of 0 and commutativity of addition, $0 + 0' = 0'$. Therefore, $0' = 0$.

(b) By associativity and commutativity of addition,

$$b = b + 0 = 0 + b = (-a + a) + b = -a + (a + b) = -a + 0 = -a.$$

(c) If $a \cdot 1' = a$ for all a, then, in particular, $1 \cdot 1' = 1$. But by the definition of the multiplicative identity and the commutativity of multiplication, $1 \cdot 1' = 1'$. Therefore $1' = 1$.

(d) By the distributive property,

$$a \cdot 0 = a \cdot (0 + 0) = a \cdot 0 + a \cdot 0.$$

Adding $-(a \cdot 0)$ to both sides of this equation and using associativity of addition produces the desired equation.

(e) By associativity and commutativity of multiplication,

$$b = 1 \cdot b = (a^{-1}a)b = a^{-1}(ab) = a^{-1} \cdot 1 = a^{-1}.$$

(f) Assume $a \neq 0$. By (d) and commutativity and associativity of multiplication,

$$0 = a^{-1} \cdot 0 = (a^{-1})(ab) = (a^{-1}a)b = 1 \cdot b = b.$$

(g) By commutativity and associativity of multiplication,

$$b = 1 \cdot b = (a^{-1}a)b = a^{-1}(ab) = a^{-1}(ac) = (a^{-1}a)c = c.$$

(h) If $a/b = c/d$, then multiplying both sides by bd and using the commutativity and associativity of multiplication, we obtain $ad = bc$. Conversely, if $ad = bc$, then multiplying both sides by $1/(bd)$ yields $a/b = c/d$.

(i) By associativity and commutativity of multiplication,

$$(ab)(a^{-1}b^{-1}) = (aa^{-1})(bb^{-1}) = 1.$$

Now apply (e). □

The reader will notice that the assertions in the proposition are *implications*, that is, statements of the form p implies q, frequently written $p \Rightarrow q$. Such assertions may be proved directly by assuming p and then deducing q, or indirectly by assuming the negation of q and arguing to a contradiction or to the negation of p. Part (h) also contains an assertion of the form p if and only if q (hereafter, shortened to p iff q). Such an assertion is established by proving both $p \Rightarrow q$ and $q \Rightarrow p$. Throughout the text, we shall encounter many examples of such proofs. The reader is advised that a careful proof of an assertion requires that each (nontrivial) step be justified by citing hypotheses, appropriate axioms, or previously proved results.

One more point of logic: To prove that a general statement involving the universal quantifier "for every" (or "for all") is false, one must construct a *counterexample*. For example, the assertion that $xy = x + y$ for all real numbers x and y is clearly false. For a proof, one need only find a single pair of numbers x and y such that $xy \neq x + y$, for example, $x = y = 1$. On the other hand, to prove that $x^2 - y^2 = (x - y)(x + y)$ for all real numbers x and y, it is *not* sufficient to verify the statement for a specific pair of numbers; a general argument is needed here. For details on constructing proofs in mathematics, the reader is referred to [2].

The number systems described briefly in the introduction are now defined formally as the following subsets of the real number system:

- $\mathbb{N} = \{1, 2 := 1 + 1, 3 := 2 + 1, \ldots\}$ (positive integers),
- $\mathbb{Z} = \{0, \pm 1, \pm 2, \pm 3, \ldots\}$ (integers),
- $\mathbb{Q} = \{m/n : m, n \in \mathbb{Z}, n \neq 0\}$ (rational numbers),
- $\mathbb{I} = \mathbb{R} \setminus \mathbb{Q}$ (irrational numbers).

A precise definition of $\mathbb{N}$ is given in Section 1.4. From that definition it is possible to argue rigorously that $\mathbb{N}$ is closed under addition and multiplication. Assuming this, it follows that $\mathbb{Z}$ is closed under addition, subtraction, and multiplication, and $\mathbb{Q}$ is closed under addition, subtraction, multiplication, and division (Exercise 2).

For future reference we make the following definitions: An integer is said to be *even (odd)* if $n = 2k$ ($n = 2k + 1$) for some $k \in \mathbb{Z}$.

Exercises

1. Prove the following properties of addition and multiplication in $\mathbb{R}$:
   (a) $-(-a) = a$
   (b) $-(ab) = (-a)b = a(-b)$
   (c) $(-a)(-b) = ab$
   (d) $(-1)a = -a$
   (e) If b, c, $d \neq 0$, then $\dfrac{a/b}{c/d} = \dfrac{ad}{bc} = \dfrac{a}{b}\dfrac{d}{c}$
   (f) If $b \neq 0$ and $d \neq 0$, then $\dfrac{a}{b} + \dfrac{c}{d} = \dfrac{ad+bc}{bd}$.
2. Let $r, s \in \mathbb{Q}$. Assuming that $\mathbb{Z}$ is closed under addition and multiplication, prove that $r \pm s$, rs, $r/s \in \mathbb{Q}$, the last provided that $s \neq 0$.
3. If $r \neq 0 \in \mathbb{Q}$ and $x \in \mathbb{I}$, prove that $r \pm x$, rx, $r/x \in \mathbb{I}$.
4. Let $n \in \mathbb{N}$. Prove the following identities without using mathematical induction (discussed later):
   (a) ($\Downarrow$ 4.1.2.) $x^n - y^n = (x-y)\sum_{j=1}^{n} x^{n-j}y^{j-1}$
   (b) $x^n + y^n = (x+y)\sum_{j=1}^{n}(-1)^{j-1}x^{n-j}y^{j-1}$ if n is odd
   (c) $x^{-n} - y^{-n} = (y-x)\sum_{j=1}^{n} x^{j-n-1}y^{-j}$ if $x \neq 0$ and $y \neq 0$.
5. Define $0! = 1$ and for $n \in \mathbb{N}$ define $n! = n(n-1)\cdots 2\cdot 1$ (n *factorial*). Prove the following:
   (a) $\left(1-\frac{1}{n}\right)\left(1-\frac{2}{n}\right)\cdots\left(1-\frac{n-1}{n}\right) = \dfrac{n!}{n^n}$
   (b) $1\cdot 3\cdots(2n-1) = \dfrac{(2n)!}{2^n n!}$.
6. ($\Downarrow$ 1.4.5.) For $n \in \mathbb{Z}^+$ and $k = 0, 1, \ldots, n$, define the *binomial coefficient*

$$\binom{n}{k} = \frac{n!}{k!(n-k)!}$$

   (read "n choose k"). Prove that $\dbinom{n+1}{k} = \dbinom{n}{k-1} + \dbinom{n}{k}$.

7. Without using mathematical induction, prove that for any $n \in \mathbb{N}$,

   (a) $\displaystyle\sum_{k=0}^{n} \frac{1}{(k+1)(n-k+1)} = \frac{2}{n+2}\sum_{k=0}^{n}\frac{1}{k+1}$

   (b) $\displaystyle\sum_{k=0}^{n} \frac{1}{(2k+1)(2n-2k+1)} = \frac{1}{n+1}\sum_{k=0}^{n}\frac{1}{2k+1}$

   Suggestion. Use partial fractions.

8. Find a polynomial $f(x)$ of degree 2 such that $\sum_{k=1}^{n} f(k) = n^3$ for all n.

1.2 Order Properties of $\mathbb{R}$

The order relation on $\mathbb{R}$ is derived from the following:

Order Axiom

- There exists a nonempty subset $\mathbb{P}$ of $\mathbb{R}$, closed under addition and multiplication, such that for each $x \in \mathbb{R}$ exactly one of the following holds:

$$x \in \mathbb{P}, \quad -x \in \mathbb{P}, \quad \text{or} \quad x = 0.$$

  A real number x is called *positive* if $x \in \mathbb{P}$ and *negative* if $-x \in \mathbb{P}$.

1.2.1 Definition. Let a and b be real numbers. If $b - a \in \mathbb{P}$, we write $a < b$ or $b > a$ and say that *a is less than b* or that *b is greater than a*. ◊

1.2.2 Proposition. *The order relation $<$ on $\mathbb{R}$ has the following properties:*

(a) *$a < b$ iff $-a > -b$ (reflection property).*

(b) *If $a < b$ and $b < c$, then $a < c$ (transitive property).*

(c) *If $a < b$ and $c < d$, then $a + c < b + d$ (addition property).*

(d) *If $a < b$ and $c > 0$, then $ac < bc$ (multiplication property).*

(e) *For $a, b \in \mathbb{R}$, exactly one of the following is true: $a = b$, $a < b$, or $b < a$ (trichotomy property).*

(f) *If $x \neq 0$, then $x^2 > 0$. In particular, $1 > 0$.*

Proof. (a) $a < b$ iff $(-a) - (-b) = b - a \in \mathbb{P}$ iff $-a > -b$.

(b) By hypothesis, $b - a \in \mathbb{P}$ and $c - b \in \mathbb{P}$; hence, by closure under addition, $c - a = (b - a) + (c - b) \in \mathbb{P}$, that is, $a < c$.

(c) Similar to (b).

(d) Since $b - a,\ c \in \mathbb{P}$, $bc - ac = (b - a)c \in \mathbb{P}$, that is, $ac < bc$.

(e) This follows by applying the trichotomy property of $\mathbb{P}$ to $a - b$.

(f) If $x > 0$, then, by closure of $\mathbb{P}$ under multiplication, $x^2 > 0$. If $x < 0$, then $-x > 0$ so, by Exercise 1.1.1(c), $x^2 = (-x)(-x) > 0$. □

1.2.3 Definition. Let a and b be real numbers. If either $a < b$ or $a = b$, we write $a \le b$ or $b \ge a$ and say that a *is less than or equal to* b or that b *is greater than or equal to* a. If $A \subseteq \mathbb{R}$, we define $A^+ = \{x \in A : x \ge 0\}$. ◊

Note that by the trichotomy property,

$$a \le b \ \text{ and } \ b \le a \Rightarrow a = b. \tag{1.1}$$

The inequality $a \le b$ is sometimes called *weak inequality* in contrast to *strict inequality* $a < b$. The reader may check that parts (a)–(d) of the above proposition are valid if strict inequality is replaced by weak inequality.

The following definition gives a way of measuring the "distance" between two real numbers, a fundamental feature of limits.

1.2.4 Definition. The *absolute value* of a real number x is defined by

$$|x| = \begin{cases} x & \text{if } x \ge 0, \\ -x & \text{if } x < 0. \end{cases}$$ ◊

For example, $|0| = 0$ and $|2| = |-2| = 2$.

1.2.5 Proposition. *Absolute value has the following properties:*

(a) $|x| \ge 0$. (b) $|x| = 0$ *iff* $x = 0$. (c) $|-x| = |x|$.

(d) $-|x| \le x \le |x|$. (e) $|xy| = |x|\,|y|$. (f) $\left|\dfrac{x}{y}\right| = \dfrac{|x|}{|y|}$ $(y \ne 0)$.

(g) $|x + y| \le |x| + |y|$. (h) $\big|\,|x| - |y|\,\big| \le |x - y|$. (triangle inequalities)

Proof. Properties (a)–(e) are easily established by considering cases. For example, in (e), if $x \ge 0$ and $y \le 0$, then $xy \le 0$, hence

$$|xy| = -(xy) = x(-y) = |x|\,|y|.$$

For part (f), use (e) to obtain

$$|x| = \left|\frac{x}{y}\,y\right| = \left|\frac{x}{y}\right|\,|y|,$$

and then divide both sides by $|y|$.

For (g), we have $\pm x \le |x|$ and $\pm y \le |y|$ by (d), hence, $\pm(x+y) \le |x| + |y|$. Since one of the signed quantities on the left is $|x+y|$, the assertion follows.

From (g) we have

$$|x| = |(x-y)+y| \le |x-y| + |y|,$$

hence $|x| - |y| \le |x-y|$. Switching x and y and using (c) yields (h). □

1.2.6 Definition. Let S be a nonempty set of real numbers. The *largest element* or *maximum of* S is a member $\max S$ of S that satisfies

$$\max S \ge s \quad \text{for all } s \in S.$$

The *smallest element* or *minimum of* S, denoted by $\min S$, is defined analogously. ◊

A set need not have a largest or smallest member. The existence of $\max S$ and $\min S$ for a nonempty finite set may be established by mathematical induction (See Exercise 1.4.2.).

1.2.7 Definition. The *positive* and *negative parts of a real number* x are defined by

$$x^+ = \max\{x, 0\} \quad \text{and} \quad x^- = \max\{-x, 0\}.$$ ◊

Exercises

Prove the following:

1. (a) If $ab > 0$, then a and b have the same sign.
   (b) $a > 0$ iff $1/a > 0$.
   (c) Suppose either $b, d < 0$ or $b, d > 0$. Then $a/b > c/d$ iff $ad > bc$.
2. If $x > 1$, then $x^2 > x$. If $0 < x < 1$, then $x^2 < x$.
3. (a) If $0 < x < y$ and $0 < a < b$, then $0 < ax < by$.
   (b) If $x < y < 0$ and $a < b < 0$, then $0 < by < ax$.
   (c) Let $x, y > 0$. Then $x < y$ iff $x^2 < y^2$.
4. If either $0 < x < y$ or $x < y < 0$, then $1/y < 1/x$.
5. If $-1 < x < y$ or $x < y < -1$, then $x/(x+1) < y/(y+1)$. What if $x < -1 < y$?
6. If $0 < x < y$ and $n \in \mathbb{N}$, then

   (a) $0 < y^n - x^n \le n(y-x)y^{n-1}$ (b) $\dfrac{ny+1}{nx+1} < \dfrac{(n+1)y+1}{(n+1)x+1}$.

7. If $x > 1$, m, $n \in \mathbb{N}$, and $\dfrac{x-1}{x} < \dfrac{m}{n} < 1$, then $n > x$.

8. If $a < b$ and $0 < t < 1$, then $a < ta + (1-t)b < b$. In particular, $a < (a+b)/2 < b$.

9. $x^2 + y^2 + axy \geq 0$ for all x, $y \in \mathbb{R}$ iff $|a| \leq 2$.

10. If $a \leq b + x$ for every $x > 0$, then $a \leq b$.

11. If $0 < a \leq bx$ for every $x > 1$, then $a \leq b$.

12. If $a/x \leq x + 1$ for every $x > 0$, then $a \leq 0$.

13. For all x, y, z, $w \in \mathbb{R}$,
    (a) $2xy \leq x^2 + y^2$
    (b) $xy + yz + xz \leq x^2 + y^2 + z^2$
    (c) $(xy + zw)^2 \leq (x^2 + z^2)(y^2 + w^2)$
    (d) $(x + y)^2 \leq 2(x^2 + y^2)$

14. If x, $a > 0$, then $x + a^2/x \geq 2a$. Equality holds iff $x = a$.

15. (a) $|x - y| \leq |x - z| + |z - y|$
    (b) $|x - L| < \varepsilon$ iff $L - \varepsilon < x < L + \varepsilon$.

16. Let S, $T \subseteq \mathbb{R}$ be finite and nonempty. Define $-S := \{-s : s \in S\}$. Then
    (a) $\max(-S) = -\min S$
    (b) $\min(-S) = -\max S$
    (c) $\max(S \cup T) = \max\{\max S, \max T\}$
    (d) $\min(S \cup T) = \min\{\min S, \min T\}$.

17. For any x, $y \in \mathbb{R}$,
    (a) $x^+ \geq 0$, $x^- \geq 0$, $x = x^+ - x^-$, and $|x| = x^+ + x^-$
    (b) $x^+ = \big(|x| + x\big)/2$ and $x^- = \big(|x| - x\big)/2$
    (c) $x = y - z$ and $|x| = y + z$ imply $y = x^+$ and $z = x^-$
    (d) $(x + y)^+ \leq x^+ + y^+$ and $(x + y)^- \leq x^- + y^-$
    (e) $(x - y)^- \leq y$, if x, $y \geq 0$.

18. If $a \leq x \leq b$, then $|x| \leq \max\{|a|, |b|\}$.

19. (a) $\max\{x, y\} = \frac{1}{2}\big(x + y + |x - y|\big)$
    (b) $\min\{x, y\} = \frac{1}{2}\big(x + y - |x - y|\big)$.

20. (a) $\max\{a, b, c\} = \frac{1}{4}\left(a + b + 2c + |a - b| + \big|\, a + b - 2c + |a - b|\, \big|\right)$
    (b) $\min\{a, b, c\} = \frac{1}{4}\left(a + b + 2c - |a - b| - \big|\, a + b - 2c - |a - b|\, \big|\right)$.

21. Let $S = \{a_1, \ldots, a_n\}$, where $a_1 < \cdots < a_n$. Let $1 \le k < n$ and denote by $S_1, \ldots, S_m$ the subsets obtained by removing exactly k members from S, where $m = \binom{n}{k}$ is the binomial coefficient (see Theorem 1.4.5). Then

$$\max\{\min S_1, \ldots, \min S_m\} = a_{k+1}.$$

1.3 Completeness Property of $\mathbb{R}$

A system $(\mathbb{F}, +, \cdot, <)$ with the algebraic and order properties described in Sections 1.1 and 1.2 is called an *ordered field.* For example, $\mathbb{Q}$ is an ordered field under the algebraic operations and order relation inherited from $\mathbb{R}$ (Exercise 1.1.2). The same is true for the set

$$\mathbb{Q}(\sqrt{2}) := \{x + \sqrt{2}\,y : x, y \in \mathbb{Q}\}$$

(Exercise 19). This suggests that there are infinitely many ordered subfields of $\mathbb{R}$. The property that distinguishes $\mathbb{R}$ from all other ordered fields is *completeness,* described in this section.

1.3.1 Definition. A nonempty subset A of an ordered field $\mathbb{F}$ is said to be *bounded above* if A has an *upper bound,* that is, a member $u \in \mathbb{F}$ such that $a \le u$ for all $a \in A$. The notions of *bounded below* and *lower bound* are defined analogously. The set A is said to be *bounded* if it is bounded above and below. Any set that is not bounded is said to be *unbounded* (either above or below).◊

The subsets $\mathbb{Q}$ and $\mathbb{Z}$ of $\mathbb{R}$ are neither bounded above nor bounded below; $\mathbb{N}$ is bounded below but not above. The set $\{n/(n+1) : n \in \mathbb{N}\}$ is bounded above by 1 and below by 1/2.

1.3.2 Definition. Let A be a nonempty subset of an ordered field $\mathbb{F}$. An upper bound u_0 of A with the property that $u_0 \le u$ for all upper bounds u of A is called a *least upper bound* or *supremum* of A and is denoted by $\sup A$. A lower bound ℓ_0 of A such that $\ell \le \ell_0$ for all lower bounds ℓ of A is called a *greatest lower bound,* or *infimum,* of A and is denoted by $\inf A$. ◊

Note that if $\sup A \in A$, then $\sup A$ is the maximum of A, as defined in Section 1.2. Similarly, if $\inf A \in A$, then $\inf A$ is the minimum of A.

A nonempty, bounded subset of an ordered field need not have an infimum or a supremum. For example, because $\sqrt{2}$ is not rational (1.3.11 below), the bounded set $\{x \in \mathbb{Q} : x^2 < 2\}$ has neither an infimum nor a supremum in $\mathbb{Q}$. However, if the supremum or infimum of a set exists, then it is unique. This follows directly from (1.1).

The next proposition will be used frequently in ensuing discussions involving suprema and infima.

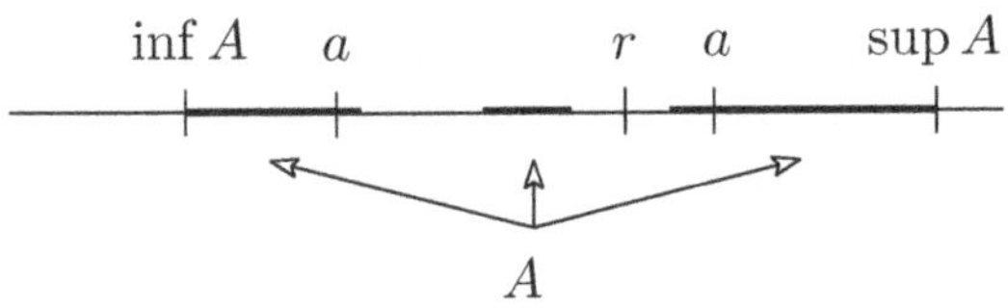

FIGURE 1.1: Supremum and infimum of A.

1.3.3 Theorem. *Let A be a nonempty subset of an ordered field $\mathbb{F}$.*

(a) *If* $\sup A$ *exists, then for each r with $r < \sup A$, there exists $a \in A$ such that* $r < a \leq \sup A$.

(b) *If* $\inf A$ *exists, then for each r with* $\inf A < r$, *there exists $a \in A$ such that* $\inf A \leq a < r$.

Proof. If $r < \sup A$, then r cannot be an upper bound for A, hence there exists $a \in A$ with $r < a$. The proof of (b) is similar. □

We may now state the property that distinguishes the real number system from all other ordered fields.

Every nonempty subset of $\mathbb{R}$ that is bounded above has a supremum.

This axiom is known as the *completeness property of* $\mathbb{R}$. It is the key ingredient needed for the formulation of a useful and robust theory of limits. From the completeness property, one may deduce (Exercise 1) the symmetric property

Every nonempty subset of $\mathbb{R}$ that is bounded below has an infimum.

The real number system may now be described as a *complete ordered field.* It may be shown that (up to isomorphism) there is exactly one such structure.

The following important consequence of completeness is sometimes useful in determining the infimum or supremum of certain sets. It asserts that positive integer multiples of a positive real number may be made arbitrarily large.

1.3.4 Theorem (Archimedean principle). *For any real numbers a and b with $a > 0$ there exists $n \in \mathbb{N}$ such that $na > b$.*

Proof. Suppose, for a contradiction, that $na \leq b$ for all $n \in \mathbb{N}$. The set $S = \{na : n \in \mathbb{N}\}$ is then bounded above and hence has a least upper bound, say u. Since $u - a < u$, the approximation property for suprema implies that $u - a < na$ for some $n \in \mathbb{N}$. But then $u < (n+1)a \in S$, contradicting that u is an upper bound for S. □

1.3.5 Example. Let

$$A = \left\{(-1)^n \frac{n}{n+1} : n \in \mathbb{N}\right\} = \left\{-\frac{1}{2}, \frac{2}{3}, -\frac{3}{4}, \dots\right\}.$$

Since A is bounded above by 1 and below by -1, $-1 \le \inf A \le \sup A \le 1$. Let $0 < r < 1$. By the Archimedean principle, we may choose an integer n such that $n > r/(1-r)$, and we may take n to be even. Then $r < n/(n+1) \in A$, which shows that r cannot be an upper bound of A. Therefore, $\sup A = 1$. Similarly, $\inf A = -1$. ◊

1.3.6 Theorem (Well-ordering principle). *Every nonempty subset A of $\mathbb{N}$ has a smallest member.*

Proof. Since A is bounded below (by 1), it has a greatest lower bound ℓ. The theorem will follow if we show that $\ell \in A$. Suppose, for a contradiction, that $\ell \notin A$. By the approximation property for infima, there exists $a \in A$ such that $\ell < a < \ell + 1$. Choose any real number r with $\ell < r < a$, for example, $r = (a+\ell)/2$. By the approximation property again, there exists $a' \in A$ such that $\ell < a' < r$. We now have $\ell < a' < a < \ell + 1$, which implies that $a - a'$ is an integer strictly between 0 and 1. As this is impossible,[1] it follows that ℓ must be a member of A. □

1.3.7 Theorem (Greatest integer function). *For each $x \in \mathbb{R}$ there exists a unique integer $\lfloor x \rfloor$ such that $x - 1 < \lfloor x \rfloor \le x$.*

Proof. The uniqueness is clear. To prove existence, apply the Archimedean principle twice: first to obtain an integer k such that $x + k > 1$ and then to conclude that the set $A := \{n \in \mathbb{N} : n > x + k\}$ is nonempty. By the well-ordering principle, A has a least member a. Then $1 < x + k < a$, so $a - 1$ is positive and $x - 1 < a - 1 - k$. Since $a - 1 < a$, $a - 1$ cannot be in A, so $x + k \ge a - 1$ or $a - 1 - k \le x$. Therefore, $x - 1 < a - 1 - k \le x$, hence the integer $a - 1 - k$ has the required property. □

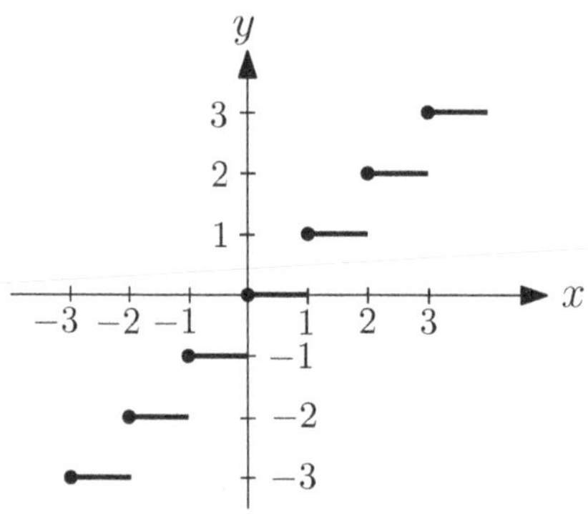

FIGURE 1.2: The greatest integer function.

[1]This is intuitively clear. The abstract definition of $\mathbb{N}$ given in Section 1.4 may be used to construct a rigorous proof.

The integer $\lfloor x \rfloor$ is called the *greatest integer in* x or the *floor of* x. The greatest integer function allows a simple proof of the following important result:

1.3.8 Corollary. *Between any pair of distinct real numbers lies a rational number.*

Proof. Let $a < b$. By the Archimedean principle, $n(b-a) > 1$ for some $n \in \mathbb{N}$, hence $na + 1 > nb$. Let $m := \lfloor na \rfloor + 1$. Since $na - 1 < \lfloor na \rfloor \le na$, we have $na < m \le na + 1 < nb$, hence $a < m/n < b$. □

1.3.9 Definition. (nth roots). Let n be a positive integer and let $b > 0$. The unique positive solution of the equation $x^n = b$ is called the *positive nth root of* b and is denoted by $b^{1/n}$. For $m \in \mathbb{Z}$ we define $b^{m/n} = (b^{1/n})^m$. As usual, we write $\sqrt{b}$ for $b^{1/2}$. ◇

The existence of $b^{1/n}$ is an easy consequence of the intermediate value theorem, proved in Chapter 3. Uniqueness follows from Exercise 1.1.4(a). We omit the straightforward (but admittedly tedious) proof of the following theorem, which summarizes the familiar rules of rational exponentiation.

1.3.10 Theorem. *For* $r, s \in \mathbb{Q}$ *and positive real numbers* a, b,

$$b^r b^s = b^{r+s}, \quad \frac{b^r}{b^s} = b^{r-s}, \quad (b^r)^s = b^{rs}, \text{ and } (ab)^r = a^r b^r.$$

The following proposition gives a simple way to generate irrational numbers.

1.3.11 Proposition. *If* n *is a positive integer that is not a perfect square, then* $\sqrt{n}$ *is irrational.*

Proof. By definition of the greatest integer function, $\sqrt{n} - 1 < \lfloor \sqrt{n} \rfloor \le \sqrt{n}$. Since $\sqrt{n}$ is assumed not to be an integer, the second inequality is strict. Thus

$$0 < \sqrt{n} - \lfloor \sqrt{n} \rfloor < 1.$$

Now suppose, for a contradiction, that $\sqrt{n}$ is rational. Then the set

$$A := \{m \in \mathbb{N} : m\sqrt{n} \in \mathbb{N}\}$$

is nonempty. By the well-ordering principle, A has a least member m_0. In particular, $m_0\sqrt{n} \in \mathbb{N}$, hence both of the quantities

$$m := m_0\big(\sqrt{n} - \lfloor \sqrt{n} \rfloor\big) \text{ and } m\sqrt{n} = m_0\big(n - \sqrt{n}\lfloor \sqrt{n} \rfloor\big)$$

are positive integers. But then $m \in A$, which is impossible since $m < m_0$. Therefore, $\sqrt{n}$ must be irrational. □

The same proof may be used to show that if $k \in \mathbb{N}$, $k \ge 2$, and n is not a perfect kth power, then $n^{1/k}$ is irrational. In later chapters, we shall see other important examples of irrational numbers, notably π and the base e of the natural logarithm.

1.3.12 Definition. The *extended real number system* is the set

$$\overline{\mathbb{R}} := \mathbb{R} \cup \{-\infty, +\infty\},$$

where $+\infty$, $-\infty$ are symbols with the following prescribed properties:

$$\begin{gathered}-\infty < x < \infty \quad \text{for all } x \in \mathbb{R},\\ x + \infty = +\infty \text{ if } -\infty < x \le +\infty, \quad x - \infty = -\infty \text{ if } -\infty \le x < +\infty,\\ x \cdot (+\infty) = +\infty \text{ if } 0 < x \le +\infty, \quad x \cdot (+\infty) = -\infty \text{ if } -\infty < x < 0,\\ x \cdot (-\infty) = -\infty \text{ if } 0 < x < +\infty, \quad x \cdot (-\infty) = +\infty \text{ if } -\infty \le x < 0,\\ \frac{x}{+\infty} = \frac{x}{-\infty} = 0 \text{ if } -\infty < x < +\infty.\end{gathered}$$ ◊

The above algebraic conventions are derived from limit considerations. Note that the operations

$$\pm\infty \mp \infty, \quad (\pm\infty)\cdot(\mp\infty), \quad \frac{\pm\infty}{\pm\infty}, \ \frac{\pm\infty}{\mp\infty}, \quad \text{and} \quad 0\cdot(\pm\infty) \tag{1.2}$$

are *not* defined.

1.3.13 Definition. If $A \ne \emptyset$ is not bounded above, we set $\sup A = +\infty$. Similarly, if A is not bounded below, we set $\inf A = -\infty$. We also define $\sup \emptyset = -\infty$ and $\inf \emptyset = +\infty$. ◊

The reader may verify that the approximation properties for suprema and infima given in 1.3.3 hold in the extended system $\overline{\mathbb{R}}$.

1.3.14 Definition. An *interval* in $\mathbb{R}$ is a nonempty set I with the property that $a, b \in I$ and $a < x < b$ imply that $x \in I$. An interval containing more than one point is said to be *nondegenerate*. ◊

Arguing cases, one may show that the definition of interval reduces to the following familiar subsets of $\mathbb{R}$:

$$\begin{aligned}(a, b) &:= \{x : a < x < b\}, & (a, b] &:= \{x : a < x \le b\},\\ [a, b) &:= \{x : a \le x < b\}, & [a, b] &:= \{x : a \le x \le b\}.\end{aligned}$$

For example, if I is unbounded below and bounded above and $b := \sup I \in I$, then $I = (-\infty, b]$. If, instead, I is bounded below and above such that $a := \inf I \in I$ and $b := \sup I \notin I$, then $I = [a, b)$. Intervals that contain their endpoints are said to be *closed*; those that don't contain *any* endpoints are called *open*. The *length* of a finite interval I with endpoints a, b is defined to be $|a - b|$ and will be denoted by $|I|$. Note that the length of a degenerate interval is zero.

Exercises

1. Prove that $\inf(-A) = -\sup A$, where $-A := \{-a : a \in A\}$. Conclude that every nonempty subset of $\mathbb{R}$ that is bounded below has a greatest lower bound.

2. Find the supremum and infimum of the following sets, where r_n denotes the remainder on division of $n \in \mathbb{N}$ by 3.[2]

   (a) $\{(-1)^n(r_n^2 + 3r_n + 2) : n \in \mathbb{N}\}$ (b) $\left\{\dfrac{(-1)^n n(r_n - 1)}{(n+1)(r_n+1)} : n \in \mathbb{N}\right\}$

   (c) $\left\{(-1)^n \dfrac{3n+2}{2n+3} : n \in \mathbb{N}\right\}$ (d) $\left\{\dfrac{\left[(-1)^{\lfloor n/3 \rfloor} - 1\right] n}{n+1} : n \in \mathbb{N}\right\}$.

3. Find the supremum and infimum of the following sets.

   (a) $\{x : x^2 - 5x + 6 < 0\}$ (b) $\{x : (x+3)(x-4) < -6\}$
   (c) $\{x : (x-4)/(x-3) < -2\}$ (d) $\{x : x - 2 < 1/(x-1)\}$
   (e) $\{x : (x-1)/x < 4\}$ (f) $\{x > 0 : x/(2-x) > 3\}$
   (g) $\{x : |x^2 - 3x + 2| \le 1/4\}$ (h) $\{x : |x-1| + |x-2| \le 3\}$
   (i) $\{x : \sqrt{x - 1/8} > x\}$ (j) $\{x : \sqrt{x + 1/8} > x\}$
   (k) $\{x : 2|x-1| + 3|y-2| < 6 \text{ for some } y \in \mathbb{R}\}$
   (l) $\{x : 2\left|x^2 - 1\right| + 3\left|y^2 - 2\right| < 6 \text{ for some } y \in \mathbb{R}\}$
   (m) $\{(-1)^n\left(\sin(n\pi/2) - n^{-1}\right) : n \in \mathbb{N}\}$
   (n) $\{(-1)^n\left(\sin(m\pi/2) - n^{-1}\right) : m, n \in \mathbb{N}\}$.

4. Let $A \subseteq B$ be nonempty subsets of $\mathbb{R}$ Prove that $\sup A \le \sup B$ and $\inf A \ge \inf B$.

5. ($\Downarrow$ 5.2.6.) For a nonempty bounded set A, define $|A| := \{|a| : a \in A\}$. Prove that
$$\sup|A| - \inf|A| \le \sup A - \inf A.$$
   Hint. Use $|x| - |y| \le |x - y|$.

6. For $r \in \mathbb{Q}$, $x \in \mathbb{R}$, and nonempty subsets A and B of $\mathbb{R}$, define
$$xA = \{xa : a \in A\} \qquad A + B = \{a + b : a \in A, b \in B\}$$
$$AB = \{ab : a \in A, b \in B\} \qquad A^r = \{a^r : a \in A\},\ A \subseteq (0, +\infty).$$
   Under the conventions described in 1.3.12, prove that

   (a) $\sup(A+B) \le \sup A + \sup B$, $\inf(A+B) \ge \inf A + \inf B$.

[2]For the existence of r_n, see Exercise 1.3.22.

(b) $\sup(xA) = x \sup A$, $\inf(xA) = x \inf A$ if $x \geq 0$.

(c) $\sup(AB) = (\sup A)(\sup B)$ and $\inf(AB) = (\inf A)(\inf B)$ if $A, B \subseteq (0, \infty)$.

(d) $\sup A^r = (\sup A)^r$, $\inf A^r = (\inf A)^r$ if $A \subseteq (0, \infty)$ and $r > 0$.

(e) $\sup A^{-1} = 1/\inf A$, $\inf A^{-1} = 1/\sup A$ if $A \subseteq (0, \infty)$.

7. Let $A \subseteq \mathbb{R}$ be nonempty such that $\inf\{|x - y| : x, y \in A, x \neq y\} > 0$ (for example, any set of integers). If A is bounded above, prove that $\sup A \in A$, that is, A has a maximum.

8. Let A be a nonempty bounded set and let $r \in \mathbb{R}$ such that $x - y < r$ for all $x, y \in A$. Show that $\sup A - \inf A \leq r$.

9. Prove that between any pair of distinct real numbers, there is an irrational number.

10. Prove that between any pair of real numbers $a < b$ there exist infinitely many rational numbers and infinitely many irrational numbers.

11. (Density of the dyadic rationals). Prove that for each pair of real numbers $a < b$ there exists $m \in \mathbb{Z}$ and $n \in \mathbb{N}$ such that $a < m/2^n < b$. (*Suggestion.* You might want to use the fact that $2^n > n$, a consequence of the binomial theorem, proved in the next section.) A number of the form $m/2^n$ is called a *dyadic rational.*

12. Let $m \in \mathbb{Z}$. Prove:

(a) $\lfloor x \rfloor = \lfloor -x \rfloor$ iff $x = 0$ (b) $\lfloor x \rfloor = -\lfloor -x \rfloor$ iff $x \in \mathbb{Z}$

(c) $-1 < x + \lfloor -x \rfloor \leq 0$ (d) $\lfloor x \rfloor + \lfloor m - x \rfloor = m$ or $m - 1$.

13. Let $m \in \mathbb{Z}$, $n \in \mathbb{N}$, $x_j \in \mathbb{R}$, and define

$$s := \sum_{j=0}^{n} x_j \quad \text{and} \quad t := \sum_{j=0}^{n} \lfloor x_j \rfloor.$$

Prove the following:

(a) $0 \leq \lfloor s \rfloor - t \leq n$ (b) $k \leq s - t < k + 1$ for some $k = 0, 1, \ldots, n$.

14. Let $b > 0$. Prove that $b^{m/n} = (b^m)^{1/n}$.

15. ($\Downarrow$ 4.1.2.) Prove that for $a, b > 0$ and $n \in \mathbb{N}$,

$$a^{1/n} - b^{1/n} = (a - b)\left(\sum_{j=1}^{n} a^{1-j/n} b^{(j-1)/n}\right)^{-1}.$$

16. Show that if $0 \leq a < b$ and $n \in \mathbb{N}$, then $a^{1/n} < b^{1/n}$.

17. Prove that if A is a bounded set, then there exists an integer N such that $|x| \leq N$ for all $x \in A$.

18. Let $a, b \in \mathbb{Q} \setminus \{0\}$ and $n \in \mathbb{N}$. Prove that $x := a + b\sqrt{n}$ is irrational iff n is not a perfect square.

19. Show that if $x, y \in \mathbb{Q}(\sqrt{2})$, then $x \pm y$, xy, $x/y \in \mathbb{Q}(\sqrt{2})$, the last provided that $y \neq 0$. Conclude that $\mathbb{Q}(\sqrt{2})$ is an ordered subfield of $\mathbb{R}$. Show that $\mathbb{Q}(\sqrt{2})$ is not complete.

20. (a) Find all $n \in \mathbb{N}$ such that $\sqrt{n+11} + \sqrt{n} \in \mathbb{Q}$.

    (b) Same question for $\sqrt{n+21} + \sqrt{n}$.

21. Let $p \in \mathbb{N}$ be prime, that is, divisible only by 1 and itself. Prove that

$$(\sqrt{n}+1)(\sqrt{n+p}+1)^{-1} \in \mathbb{Q} \quad \text{iff} \quad n = (p-1)^2/4.$$

22. (Division algorithm). Prove that for each pair of integers m and n with $n > 0$ there exist unique integers q (the *quotient*) and r (the *remainder*) such that

$$m = qn + r \quad \text{and} \quad 0 \leq r \leq n - 1.$$

    Suggestion. Consider the set of all nonnegative integers of the form $m - qn$, $q \in \mathbb{Z}$.

1.4 Mathematical Induction

In this section we give an abstract characterization of the natural number system. This will lead directly to the principle of mathematical induction.

1.4.1 Definition. A set S of real numbers is said to be *inductive* if $1 \in S$ and $x \in S$ implies that $x + 1 \in S$. The set $\mathbb{N}$ of *natural numbers* is then defined as the intersection of all inductive subsets of $\mathbb{R}$. ◊

For example, for $a < 1$ the sets $(a, +\infty)$ and $(a, +\infty) \cap \mathbb{Q}$ are inductive. More importantly, $\mathbb{N}$ itself is inductive. Indeed, since 1 is common to all inductive sets, $1 \in \mathbb{N}$, and if n is common to all inductive sets, then so is $n + 1$. We may therefore characterize $\mathbb{N}$ as the smallest inductive set (in the sense of set inclusion). The following is an important consequence:

1.4.2 Theorem (Principle of mathematical induction). *For each $n \in \mathbb{N}$, let $P(n)$ be a statement depending on n. Suppose that*

(a) *$P(1)$ is true,*

(b) *$P(n+1)$ is true whenever $P(n)$ is true.*

Then $P(n)$ is true for all n.

Proof. Let S denote the set of $n \in \mathbb{N}$ for which $P(n)$ is true. Then (a) and (b) imply that S is inductive and hence, as a subset of $\mathbb{N}$, must in fact equal $\mathbb{N}$. □

In applications part (a) is called the *base step* and part (b) the *inductive step.* The assumption in (b) that $P(n)$ is true is called the *induction hypothesis.*

The principle of mathematical induction has been loosely described as the "domino principle": If dominoes are lined up vertically in such a way that the $(n+1)$st domino will fall if the nth one falls, then, if the first domino is tipped, all the dominoes will fall.

Mathematical induction may be used to give a rigorous proof that $\mathbb{N}$ is closed under addition: Let $P(n)$ be the statement that $n + m \in \mathbb{N}$ for all $m \in \mathbb{N}$. Then $P(1)$ is true because $\mathbb{N}$ is inductive. Moreover, if for some n the statement $P(n)$ is true, that is, $n + m \in \mathbb{N}$ for all m, then clearly $P(n+1)$ is true. A similar argument shows that $\mathbb{N}$ is closed under multiplication.

Mathematical induction is indispensable for proving many important inequalities and formulas. We offer two examples; others may be found in the exercises.

1.4.3 Example. We prove by induction that $3^n n! > n^n$ for all $n \in \mathbb{N}$. This is obvious for $n = 1$. For the induction step, we need the fact (verified in Example 2.2.4) that $(1 + 1/n)^n < 3$, or equivalently, $(n+1)^n < 3n^n$, for all n. Assuming this, we see that if $3^n n! > n^n$, then

$$3^{n+1}(n+1)! = 3(n+1)3^n n! > 3(n+1)n^n > (n+1)^{n+1}. \qquad \diamond$$

1.4.4 Example. We derive a closed formula for $f(n) := \sum_{k=1}^{n}(3k-1)^2$ and then verify the result by induction. A little experimentation suggests that we should try a polynomial in n of degree 3, say $g(n) := An^3 + Bn^2 + Cn + D$. Then

$$\begin{aligned} g(n+1) - g(n) &= A\big[(n+1)^3 - n^3\big] + B\big[(n+1)^2 - n^2\big] + C\big[(n+1) - n\big] \\ &= 3An^2 + (3A + 2B)n + A + B + C \end{aligned}$$

and

$$\begin{aligned} f(n+1) - f(n) &= \sum_{k=1}^{n+1}(3k-1)^2 - \sum_{k=1}^{n}(3k-1)^2 \\ &= \big[3(n+1) - 1\big]^2 \\ &= 9n^2 + 12n + 4. \end{aligned}$$

Assuming that $f(n) = g(n)$ for all n, we may equate coefficients to obtain $A = 3$, $B = 3/2$, and $C = -1/2$. Since $f(1) = 4$, we see that $D = 0$. Thus, under the assumption that the sum has a closed form that is a cubic polynomial, we obtain the following formula

$$\sum_{k=1}^{n}(3k-1)^2 = 3n^3 + \tfrac{3}{2}n^2 - \tfrac{1}{2}n.$$

To prove the validity of the formula we use induction. When $n = 1$, each side equals 4. Assuming the formula holds for n, we have

$$\begin{aligned}\sum_{k=1}^{n+1}(3k-1)^2 &= \sum_{k=1}^{n}(3k-1)^2 + \big[3(n+1)-1\big]^2 \\ &= \big[3(n+1)-1\big]^2 + 3n^3 + \tfrac{3}{2}n^2 - \tfrac{1}{2}n.\end{aligned}$$

A little algebra shows that the last expression reduces to

$$3(n+1)^3 + \tfrac{3}{2}(n+1)^2 - \tfrac{1}{2}(n+1).$$

Thus the formula holds for $n+1$, completing the induction. ◊

The stalwart reader may wish to use the methods of the last example to derive and then verify by induction that

$$\sum_{k=1}^{n} k^4 = \frac{n}{30}\big(6n^4 + 15n^3 + 10n^2 - 1\big).$$

The following formula has important uses in combinatorics, probability theory, and infinite series.

1.4.5 Theorem (The binomial theorem). *For $a, b \in \mathbb{R}$ and $n \in \mathbb{N}$*

$$(a+b)^n = \sum_{k=0}^{n}\binom{n}{k}a^k b^{n-k},$$

where

$$\binom{n}{k} := \frac{n!}{k!(n-k)!}.$$

Proof. For $n = 1$ the formula asserts that

$$a + b = \binom{1}{0}a^0 b^1 + \binom{1}{1}a^1 b^0,$$

which follows from the convention $0! = 1$. Suppose that the formula holds for some $n \geq 1$. Writing $(a+b)^{n+1}$ as $(a+b)(a+b)^n$ and using the induction hypothesis, we have

$$\begin{aligned}(a+b)^{n+1} &= \sum_{k=0}^{n}\binom{n}{k}a^{k+1}b^{n-k} + \sum_{k=0}^{n}\binom{n}{k}a^{k}b^{n+1-k} \\ &= \sum_{k=1}^{n}\binom{n}{k-1}a^{k}b^{n+1-k} + \sum_{k=1}^{n}\binom{n}{k}a^{k}b^{n+1-k} + a^{n+1} + b^{n+1} \\ &= \sum_{k=1}^{n}\left[\binom{n}{k-1} + \binom{n}{k}\right]a^{k}b^{n+1-k} + a^{n+1} + b^{n+1} \\ &= \sum_{k=0}^{n+1}\binom{n+1}{k}a^{k}b^{n+1-k},\end{aligned}$$

where, for the last step, we used Exercise 1.1.6. By induction, the formula holds for all n. □

Exercises

1. ($\Downarrow$ 2.2.3.) Let $0 < a < x_1, y_1 < b := a + 1$ and define

$$x_{n+1} = a + \sqrt{|x_n - a|} \quad \text{and} \quad y_{n+1} = b - \sqrt{|b - y_n|}.$$

   Prove that $a < x_n < x_{n+1} < b$ and $a < y_{n+1} < y_n < b$ for all $n \in \mathbb{N}$.

2. Use induction to prove that a nonempty finite set has a maximum and a minimum.

3. ($\Downarrow$ 6.4.7.) Verify by induction that

$$\sum_{k=1}^{2n}\frac{(-1)^{k+1}}{k} = \sum_{k=n+1}^{2n}\frac{1}{k} \quad \text{for all } n \geq 1.$$

4. Use mathematical induction to prove the following formulas:

   (a) $\sum_{k=1}^{n} k = n(n+1)/2$

   (b) $\sum_{k=1}^{n} k^2 = n(n+1)(2n+1)/6$

   (c) $\sum_{k=1}^{n} k^3 = [n(n+1)/2]^2$

   (d) $\sum_{k=1}^{n} (2k-1)^2 = n(4n^2-1)/3$

   (e) $\sum_{k=1}^{n} (2k-1)^3 = n^2(2n^2-1)$

   (f) $\sum_{k=1}^{n} \frac{1}{\sqrt{k-1}+\sqrt{k}} = \sqrt{n}$

   (g) $\sum_{k=1}^{n} (4k^3 - 6k^2 + 4k - 1) = n^4$

   (h) $\sum_{k=1}^{n} \frac{2k+\sqrt{k(k-1)}-1}{\sqrt{k}+\sqrt{k-1}} = n\sqrt{n}.$

5. Use the methods of 1.4.4 to derive and verify a closed formula for $\sum_{k=1}^{n}(5k-4)^2$.

6. Use known formulas to calculate

   (a) $1 \cdot 2 + 2 \cdot 3 + 3 \cdot 4 + \cdots + 999 \cdot 1000$

   (b) $1 \cdot 3 + 3 \cdot 5 + 5 \cdot 7 + \cdots + 999 \cdot 1001$

   (c) $1 \cdot 3 + 5 \cdot 7 + 9 \cdot 11 + \cdots + 1001 \cdot 1003$

   (d) $1 \cdot 2 \cdot 3 + 4 \cdot 5 \cdot 6 + \cdots + 997 \cdot 998 \cdot 999$.

7. Use the principle of mathematical induction to prove the following variant: Let $n_0 \in \mathbb{Z}$ and let $P(n)$ be a statement depending on integers $n \geq n_0$ such that

   (a) $P(n_0)$ is true,

   (b) if $n \geq n_0$ and $P(n)$ is true, then $P(n+1)$ is true.

   Then $P(n)$ is true for every $n \geq n_0$.

8. Use the variant of mathematical induction in Exercise 7 to verify the following inequalities. (For (e) use $(1 + 1/n)^n > 2$, an easy consequence of the binomial theorem.)

   (a) $2n + 1 < 2^n, \ n \geq 3$

   (b) $n^2 < 2^n, \ n \geq 5$

   (c) $2^n < n!, \ n \geq 4$

   (d) $3^n < n!, \ n \geq 7$

   (e) $2^n n! < n^n, \ n \geq 6$

   (f) $8^n n! < (2n)!, \ n \geq 6$.

9. Use the variant of mathematical induction in Exercise 7 to prove that $n < \ln(n!), \ n \geq 6$.

10. Prove *Bernoulli's inequality*:

$$(1+x)^n \geq 1 + nx, \quad n \in \mathbb{Z}^+, \quad x \geq -1.$$

11. Use the principle of mathematical induction to prove the following variant: Let $n_0 \in \mathbb{Z}$ and let $P(n)$ be a statement depending on integers $n \geq n_0$ such that

    (a) $P(n_0)$ is true,

    (b) $P(n+1)$ is true whenever $P(j)$ is true for all $n_0 \leq j \leq n$.

    Then $P(n)$ is true for every $n \geq n_0$.

12. (Prime Factorization). Use the variant of induction in Exercise 11 to prove that every integer $n \geq 2$ may be written as a product of powers of prime numbers (for example, $72 = 2^3 \cdot 3^2$).

13. The *Fibonacci numbers* f_n are defined recursively by

$$f_0 = f_1 = 1 \quad \text{and} \quad f_{n+1} = f_n + f_{n-1}, \quad n \geq 1.$$

Use the variant of induction in Exercise 11 to prove that

$$f_n = \frac{1}{\sqrt{5}}\left(a^{n+1} - b^{n+1}\right), \quad a := \frac{1+\sqrt{5}}{2}, \quad b := \frac{1-\sqrt{5}}{2},$$

where a, b are the zeros of $x^2 - x - 1$.

14. Let a_0 and a_1 be arbitrary and define

$$a_{n+1} = \tfrac{1}{2}(a_n + a_{n-1}), \quad n \geq 1.$$

Use the variant of induction in Exercise 11 to prove that for all $n \geq 0$,

$$a_n = \frac{(-1)^n}{3 \cdot 2^{n-1}}(a_0 - a_1) + \frac{1}{3}(a_0 + 2a_1).$$

15. Use Exercise 1.3.22 and the variant of induction in Exercise 11 to prove that each $n \in \mathbb{N}$ may be uniquely expressed in the form $\sum_{k=0}^{p} d_k 10^k$ for some $p \in \mathbb{N}$ and $d_k \in \{0, 1, \ldots, 9\}$. The representation

$$n = d_p d_{p-1} \ldots d_0$$

is called the *decimal positional notation* for n. *Suggestion.* Use the preceding exercise.

1.5 Euclidean Space

The real number system may be used to construct other important mathematical systems, such as n-dimensional Euclidean space and the complex number system. In this section we construct the former. The reader may delay reading this section, as the material will not be needed until Chapter 8.

For $n \in \mathbb{N}$, let $\mathbb{R}^n$ denote the set of all n-tuples $\boldsymbol{x} := (x_1, x_2, \ldots, x_n)$, where $x_j \in \mathbb{R}$. Each such n-tuple is called a *point* or *vector*, depending on context. The distinction between points and vectors is important in physics and geometry, as it allows one to refer to a *vector at a point*, a notion useful in describing, say, forces or tangent vectors.

The set $\mathbb{R}^n$ has an algebraic structure defined as follows: Let

$$\boldsymbol{x} = (x_1, \ldots, x_n), \quad \boldsymbol{y} = (y_1, \ldots, y_n), \quad \text{and} \quad t \in \mathbb{R}.$$

The operations of *addition* $\boldsymbol{x}+\boldsymbol{y}$ and *scalar multiplication* $t\boldsymbol{x}$ in $\mathbb{R}^n$ are then defined, respectively, by

$$\begin{aligned}\boldsymbol{x}+\boldsymbol{y} &= (x_1,\ldots,x_n)+(y_1,\ldots,y_n) = (x_1+y_1,\ldots,x_n+y_n), \quad \text{and}\\ t\boldsymbol{x} &= t(x_1,\ldots,x_n) = (tx_1,\ldots,tx_n).\end{aligned}$$

We also define

$$-\boldsymbol{x} := (-x_1,\ldots,-x_n) \quad \text{and} \quad \mathbf{0} := (0,\ldots,0).$$

The following theorem asserts that $\mathbb{R}^n$ is a *vector space* under these operations (see Appendix B). The straightforward proof is left to the reader.

1.5.1 Theorem. *Addition and scalar multiplication on* $\mathbb{R}^n$ *have the following properties:*

- associativity of addition: $(\boldsymbol{x}+\boldsymbol{y})+\boldsymbol{z} = \boldsymbol{x}+(\boldsymbol{y}+\boldsymbol{z})$.
- commutativity of addition: $\boldsymbol{x}+\boldsymbol{y} = \boldsymbol{y}+\boldsymbol{x}$.
- existence of an additive identity: $\boldsymbol{x}+\mathbf{0} = \boldsymbol{x}$.
- existence of additive inverses: $\boldsymbol{x}+(-\boldsymbol{x}) = \mathbf{0}$.
- associativity of scalar multiplication: $(st)\boldsymbol{x} = s(t\boldsymbol{x})$.
- distributivity of a scalar over vector addition: $s(\boldsymbol{x}+\boldsymbol{y}) = s\boldsymbol{x}+s\boldsymbol{y}$.
- distributivity of a vector over scalar addition: $(s+t)\boldsymbol{x} = s\boldsymbol{x}+t\boldsymbol{x}$.
- existence of a scalar multiplicative identity: $1\boldsymbol{x} = \boldsymbol{x}$.

1.5.2 Definition. The *Euclidean inner product* $\boldsymbol{x}\cdot\boldsymbol{y}$ of $\boldsymbol{x} = (x_1,\ldots,x_n)$ and $\boldsymbol{y} = (y_1,\ldots,y_n)$ and the *Euclidean norm* $\|\boldsymbol{x}\|_2$ of $\boldsymbol{x}$ are defined by

$$\boldsymbol{x}\cdot\boldsymbol{y} = \sum_{j=1}^{n} x_j y_j \quad \text{and} \quad \|\boldsymbol{x}\|_2 = \left(\sum_{j=1}^{n} x_j^2\right)^{1/2} = \sqrt{\boldsymbol{x}\cdot\boldsymbol{x}}.$$

The set $\mathbb{R}^n$ with its vector space structure and the Euclidean inner product is called *n-dimensional Euclidean space.* ◊

The structure of Euclidean space allows one to define lines, planes, length, perpendicularity, angle between vectors, etc. These ideas will be useful in later chapters.

1.5.3 Theorem. *The inner product in* $\mathbb{R}^n$ *has the following properties:*

(a) $\boldsymbol{x}\cdot\boldsymbol{x} = \|\boldsymbol{x}\|_2^2$.

(b) $\boldsymbol{x}\cdot\boldsymbol{y} = \boldsymbol{y}\cdot\boldsymbol{x}$ (commutativity).

(c) $t(\boldsymbol{x} \cdot \boldsymbol{y}) = (t\boldsymbol{x}) \cdot \boldsymbol{y} = \boldsymbol{x} \cdot (t\boldsymbol{y})$ (associativity).

(d) $\boldsymbol{x} \cdot (\boldsymbol{y} + \boldsymbol{z}) = (\boldsymbol{x} \cdot \boldsymbol{y}) + (\boldsymbol{x} \cdot \boldsymbol{z})$ (additivity).

(e) $|\boldsymbol{x} \cdot \boldsymbol{y}| \le \|\boldsymbol{x}\|_2 \, \|\boldsymbol{y}\|_2$ (Cauchy–Schwartz inequality).

Proof. Properties (a) and (b) are immediate, and parts (c) and (d) follow, respectively, from the calculations

$$t\sum_{j=1}^{n} x_j y_j = \sum_{j=1}^{n} (tx_j)y_j = \sum_{j=1}^{n} x_j(ty_j) \text{ and } \sum_{j=1}^{n} x_j(y_j + z_j) = \sum_{j=1}^{n} x_j y_j + \sum_{j=1}^{n} x_j z_j.$$

The inequality in (e) holds trivially if $\boldsymbol{y} = \boldsymbol{0}$. Suppose $\boldsymbol{y} \ne \boldsymbol{0}$, so $\|\boldsymbol{y}\|_2 \ne 0$. By properties (a)–(d),

$$0 \le \|\boldsymbol{x} - t\boldsymbol{y}\|_2^2 = (\boldsymbol{x} - t\boldsymbol{y}) \cdot (\boldsymbol{x} - t\boldsymbol{y}) = \|\boldsymbol{x}\|_2^2 - 2t(\boldsymbol{x} \cdot \boldsymbol{y}) + t^2\|\boldsymbol{y}\|_2^2.$$

Setting $t = (\boldsymbol{x} \cdot \boldsymbol{y})/\|\boldsymbol{y}\|_2^2$, we obtain

$$0 \le \|\boldsymbol{x}\|_2^2 - 2(\boldsymbol{x} \cdot \boldsymbol{y})^2/\|\boldsymbol{y}\|_2^2 + (\boldsymbol{x} \cdot \boldsymbol{y})^2/\|\boldsymbol{y}\|_2^2 = \|\boldsymbol{x}\|_2^2 - (\boldsymbol{x} \cdot \boldsymbol{y})^2/\|\boldsymbol{y}\|_2^2,$$

which implies that $(\boldsymbol{x} \cdot \boldsymbol{y})^2 \le \|\boldsymbol{x}\|_2^2 \, \|\boldsymbol{y}\|_2^2$. Taking square roots yields (e). □

1.5.4 Theorem. *The Euclidean norm on* $\mathbb{R}^n$ *has the following properties:*

(a) $\|\boldsymbol{x}\|_2 \ge 0$ (nonnegativity).

(b) $\|\boldsymbol{x}\|_2 = 0$ *iff* $\boldsymbol{x} = \boldsymbol{0}$ (coincidence).

(c) $\|t\boldsymbol{x}\|_2 = |t| \, \|\boldsymbol{x}\|_2$ (absolute homogeneity).

(d) $\|\boldsymbol{x} + \boldsymbol{y}\|_2 \le \|\boldsymbol{x}\|_2 + \|\boldsymbol{y}\|_2$ (triangle inequality).

Proof. Parts (a) and (b) are clear, and (c) follows from

$$\|t\boldsymbol{x}\|_2^2 = \sum_{j=1}^{n} (tx_j)^2 = t^2 \sum_{j=1}^{n} x_j^2 = t^2\|\boldsymbol{x}\|_2^2.$$

For (d) we use 1.5.3:

$$\begin{aligned} \|\boldsymbol{x} + \boldsymbol{y}\|_2^2 &= (\boldsymbol{x} + \boldsymbol{y}) \cdot (\boldsymbol{x} + \boldsymbol{y}) \\ &= \|\boldsymbol{x}\|_2^2 + \|\boldsymbol{y}\|_2^2 + 2(\boldsymbol{x} \cdot \boldsymbol{y}) \\ &\le \|\boldsymbol{x}\|_2^2 + \|\boldsymbol{y}\|_2^2 + 2\|\boldsymbol{x}\|_2 \, \|\boldsymbol{y}\|_2 \\ &= (\|\boldsymbol{x}\|_2 + \|\boldsymbol{y}\|_2)^2. \end{aligned}$$

□

Exercises

1. Solve the following system of vector equations for $\boldsymbol{x}$ and $\boldsymbol{y}$ in terms of $\mathbf{a}$, $\mathbf{b}$, $\mathbf{c}$, $\mathbf{d}$, and $\mathbf{e}$, assuming that $(\mathbf{a}\cdot\mathbf{b})(\mathbf{d}\cdot\mathbf{b}) \neq 1$.

$$\boldsymbol{x} + (\boldsymbol{y}\cdot\boldsymbol{b})\mathbf{a} = \boldsymbol{c}$$
$$\boldsymbol{y} + (\boldsymbol{x}\cdot\boldsymbol{b})\mathbf{d} = \boldsymbol{e}.$$

2. Prove the following:
   (a) $\|\boldsymbol{x}+\boldsymbol{y}\|_2^2 - \|\boldsymbol{x}-\boldsymbol{y}\|_2^2 = 4(\boldsymbol{x}\cdot\boldsymbol{y})$ (polarization identity).
   (b) $\|\boldsymbol{x}+\boldsymbol{y}\|_2^2 + \|\boldsymbol{x}-\boldsymbol{y}\|_2^2 = 2\big(\|\boldsymbol{x}\|_2^2 + \|\boldsymbol{y}\|_2^2\big)$ (parallelogram rule).
   (c) $\big|\,\|\boldsymbol{x}\|_2 - \|\boldsymbol{y}\|_2\,\big| \le \|\boldsymbol{x}-\boldsymbol{y}\|_2$.
   (d) $\|\boldsymbol{x}_1 + \cdots + \boldsymbol{x}_n\|_2 \le \sum_{j=1}^n \|\boldsymbol{x}_j\|_2$ (generalized triangle inequality).

3. Suppose that $\boldsymbol{x}_i\cdot\boldsymbol{x}_j = 0$ for $i \neq j$. Prove that

$$\|\boldsymbol{x}_1 + \cdots + \boldsymbol{x}_k\|_2^2 = \sum_{j=1}^k \|\boldsymbol{x}_j\|_2^2.$$

4. ($\Downarrow$ Section 8.1.) For $\boldsymbol{x} = (x_1, \ldots, x_n)$ define

$$\|\boldsymbol{x}\|_1 = \sum_{j=1}^n |x_j| \quad\text{and}\quad \|\boldsymbol{x}\|_\infty = \max\{|x_1|, \ldots, |x_n|\}.$$

   Verify that $\|\cdot\|_1$ and $\|\cdot\|_\infty$ have the properties (a)–(d) of 1.5.4.

5. A nonempty subset C of $\mathbb{R}^n$ is said to be *convex* if $\boldsymbol{x}$, $\boldsymbol{y} \in C$ and $t \in [0,1]$ imply that $t\boldsymbol{x} + (1-t)\boldsymbol{y} \in C$. Let $r > 0$. Prove that $\{\boldsymbol{x} \in \mathbb{R}^n : \|\boldsymbol{x}\|_2 \le r\}$ is convex. Is the set $\{\boldsymbol{x} \in \mathbb{R}^n : \|\boldsymbol{x}\|_2 = r\}$ convex? What about the sets $\{\boldsymbol{x} \in \mathbb{R}^n : \|\boldsymbol{x}\|_1 \le r\}$ and $\{\boldsymbol{x} \in \mathbb{R}^n : \|\boldsymbol{x}\|_\infty \le r\}$?

6. Find positive constants a, b, c such that for all $\boldsymbol{x} \in \mathbb{R}^n$,

   $\|\boldsymbol{x}\|_2 \le a\|\boldsymbol{x}\|_1$, $\quad\|\boldsymbol{x}\|_1 \le b\|\boldsymbol{x}\|_\infty$, and $\quad\|\boldsymbol{x}\|_\infty \le c\|\boldsymbol{x}\|_2$.

7. Prove that if $\|\boldsymbol{x}\|_2 = \|\boldsymbol{y}\|_2 = \|(\boldsymbol{x}+\boldsymbol{y})/2\|_2 = 1$, then $\boldsymbol{x} = \boldsymbol{y}$. Is the same true for $\|\cdot\|_\infty$ or $\|\cdot\|_1$?

8. Show that in $\mathbb{R}^3$, $\boldsymbol{a}\cdot\boldsymbol{b} = \|\boldsymbol{a}\|\,\|\boldsymbol{b}\|\cos\theta$, where θ is the (smaller) angle between $\boldsymbol{a}$ and $\boldsymbol{b}$.

9. The *cross product* of vectors $\boldsymbol{a} = (a_1, a_2, a_3)$ and $\boldsymbol{b} = (b_1, b_2, b_3)$ in $\mathbb{R}^3$ is defined by

$$\begin{aligned}\boldsymbol{a}\times\boldsymbol{b} &= \left\langle \begin{vmatrix} a_2 & a_3 \\ b_2 & b_3 \end{vmatrix}, -\begin{vmatrix} a_1 & a_3 \\ b_1 & b_3 \end{vmatrix}, \begin{vmatrix} a_1 & a_2 \\ b_1 & b_2 \end{vmatrix} \right\rangle \\ &= \langle a_2b_3 - a_3b_2, a_3b_1 - a_1b_3, a_1b_2 - a_2b_1 \rangle.\end{aligned}$$

   Let θ be the (smaller) angle between $\boldsymbol{a}$ and $\boldsymbol{b}$. Verify the following:

(a) $(\boldsymbol{a} \times \boldsymbol{b}) \cdot \boldsymbol{a} = (\boldsymbol{a} \times \boldsymbol{b}) \cdot \boldsymbol{b} = 0$

(b) $\boldsymbol{b} \times \boldsymbol{a} = -\boldsymbol{a} \times \boldsymbol{b}$.

(c) $\boldsymbol{a} \times (t\boldsymbol{x} + s\boldsymbol{y}) = t(\boldsymbol{a} \times \boldsymbol{x}) + s(\boldsymbol{a} \times \boldsymbol{y})$

(d) $(\boldsymbol{a} \times \boldsymbol{b}) \cdot \boldsymbol{c} = \boldsymbol{a} \cdot (\boldsymbol{b} \times \boldsymbol{c})$

(e) $\boldsymbol{a} \times (\boldsymbol{b} \times \boldsymbol{c}) = (\boldsymbol{a} \cdot \boldsymbol{c})\boldsymbol{b} - (\boldsymbol{a} \cdot \boldsymbol{b})\boldsymbol{c}$

(f) $\|\boldsymbol{a} \times \boldsymbol{b}\| = \|\boldsymbol{a}\| \, \|\boldsymbol{b}\| \sin\theta$.

Chapter 2

Numerical Sequences

2.1 Limits of Sequences

Simply stated, a *sequence in a set* E is a function from $\mathbb{N}$ to E. It is more instructive, however, to think of a sequence as an infinite ordered list of members of E. The list may be written as

$$a_1, a_2, \ldots, a_n, \ldots$$

or abbreviated by the symbols $\{a_n\}_{n=1}^{\infty}$, or simply $\{a_n\}$, as we shall usually do. A sequence usually starts with the index 1, although this is not necessary, the index 0 being a common alternative. Sequences may be defined by a *closed formula*, such as $a_n = (-1)^n$, or *recursively*, such as the *Fibonacci sequence* defined by

$$a_0 = a_1 = 1 \quad \text{and} \quad a_{n+1} = a_n + a_{n-1}, \quad n \geq 1$$

(see Exercise 1.4.13). The set E in the definition of a sequence is arbitrary. However, in this chapter we consider only *numerical sequences*, that is, sequences contained in $\mathbb{R}$.

The following notion will occasionally be useful. A property P of a sequence $\{a_n\}$ is said to hold *eventually* if there exists an index N such that a_n has property P for all $n \geq N$. For example, by the Archimedean principle, the sequence $\{1/n\}$ is eventually less than .001. Or, consider the sequence defined by $a_n = n^2 + 100(-1)^n$; the reader may check that eventually $a_n < a_{n+1}$.

Convergence of a sequence to a real number a expresses the idea that eventually the terms of the sequence will be as close to a as desired. The following definition makes this precise.

2.1.1 Definition. A sequence $\{a_n\}$ in $\mathbb{R}$ is said to *converge to a real number* a, written

$$a_n \to a \quad \text{or} \quad \lim_n a_n = a \quad \text{or} \quad \lim_{n \to +\infty} a_n = a,$$

if for each $\varepsilon > 0$ there exists $N \in \mathbb{N}$ such that

$$|a_n - a| < \varepsilon \quad (\text{that is, } \ a - \varepsilon < a_n < a + \varepsilon) \quad \text{for all } n \geq N.$$

If no such real number a exists, then the sequence is said to *diverge*. ◊

DOI: 10.1201/9781003629382-2

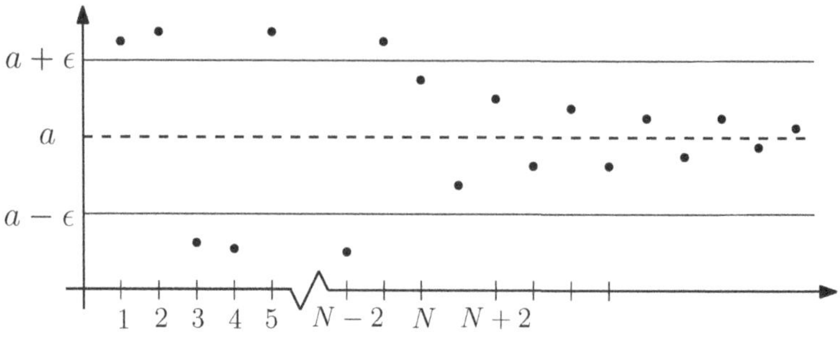

FIGURE 2.1: Convergence of a sequence to a.

It follows immediately from the definition that $a_n \to a$ iff the terms of sequence eventually lie in any open interval containing a. The definition also implies that $a_n \to a$ iff $|a_n - a| \to 0$.

Limits, if they exist, are unique. Indeed, if $a_n \to a$ and $a_n \to b$, then by the triangle inequality $|a-b| \le |a-a_n| + |b-a_n| \to 0$, hence $a = b$.

Examples. (a) The sequence $\{(-1)^n\}$ oscillates between -1 and 1 and so cannot converge. For a rigorous proof, suppose $(-1)^n \to a$ for some $a \in \mathbb{R}$. Choose N such that $a - 1 < (-1)^n < a + 1$ for all $n \ge N$. Taking $n \ge N$ to be even, we see that $1 < a + 1$ so $a > 0$. On the other hand taking $n \ge N$ to be odd, we have $a - 1 < -1$ or $a < 0$. Thus $\{(-1)^n\}$ cannot converge.

(b) To show that

$$\lim_n \frac{(-1)^n}{n} = 0,$$

let $\varepsilon > 0$ and choose an integer $N > 1/\varepsilon$ (Archimedean principle). Then $|(-1)^n/n - 0| = 1/n < \varepsilon$ for all $n \ge N$.

(c) To verify that

$$\lim_n \frac{2n+1}{3n+5} = \frac{2}{3},$$

note that

$$\left|\frac{2n+1}{3n+5} - \frac{2}{3}\right| = \frac{7}{3(3n+5)} < \frac{7}{n},$$

so any index $N > 7/\varepsilon$ satisfies the condition in 2.1.1. ♢

2.1.2 Definition. A sequence $\{a_n\}$ is said to be *bounded* (*above, below*) if the set of its terms is bounded (above, below). ♢

2.1.3 Proposition. *A convergent sequence in* $\mathbb{R}$ *is bounded.*

Proof. Let $a_n \to a \in \mathbb{R}$ and choose N such that $|a_n - a| < 1$ for all $n > N$. Since $|a_n| - |a| \le |a_n - a|$, we see that $|a_n| \le |a_n - a| + |a| < 1 + |a|$ for all $n > N$. Thus $|a_n| \le \max\{1 + |a|, |a_1|, \ldots, |a_N|\}$ for all $n \in \mathbb{N}$. □

2.1.4 Theorem. *Let $\{a_n\}$ and $\{b_n\}$ be sequences with $a_n \to a$ and $b_n \to b$. If $a_n \le b_n$ for infinitely many n, then $a \le b$.*

Proof. Suppose $b < a$. Then $b < (a+b)/2 < a$, hence we may choose indices N_1 and N_2 such that $b_n < (a+b)/2$ for all $n \ge N_1$ and $a_n > (a+b)/2$ for all $n \ge N_2$. But then $b_n < a_n$ for all $n \ge \max\{N_1, N_2\}$, contradicting the hypothesis. □

Note that as a consequence of the preceding theorem, a convergent sequence in a closed interval I must have its limit in I.

2.1.5 Theorem (Squeeze principle). *Let $\{a_n\}$, $\{b_n\}$, and $\{c_n\}$ be sequences in $\mathbb{R}$ such that $a_n \le b_n \le c_n$ for all n. If $\lim_n a_n = \lim_n c_n = x \in \mathbb{R}$, then $\lim_n b_n = x$.*

Proof. Given $\varepsilon > 0$, choose N_1, $N_2 \in \mathbb{N}$ such that $|a_n - x| < \varepsilon$ for all $n \ge N_1$ and $|c_n - x| < \varepsilon$ for all $n \ge N_2$. For $n \ge \max\{N_1, N_2\}$,

$$-\varepsilon < a_n - x \le b_n - x \le c_n - x < \varepsilon,$$

hence $|b_n - x| < \varepsilon$. □

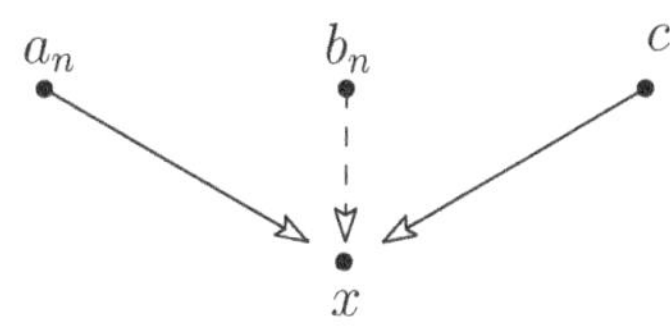

FIGURE 2.2: The squeeze principle.

2.1.6 Example. We show that $\lim_n nr^n = 0$ for any $r \in (0,1)$. Let $h = r^{-1} - 1$ and note that $h > 0$. By the binomial theorem,

$$r^{-n} = (1+h)^n = 1 + nh + \tfrac{1}{2}n(n-1)h^2 + \cdots > \tfrac{1}{2}n(n-1)h^2.$$

Thus

$$0 < nr^n < \frac{2}{(n-1)h^2}, \quad n > 1.$$

Since the term on the right tends to 0 as $n \to +\infty$, the squeeze principle implies that $nr^n \to 0$. (See Exercise 16 for an extension of this result.) ◊

For another illustration of the squeeze principle, we prove

2.1.7 Proposition. *For any real number x there exist sequences $\{a_n\}$ in $\mathbb{Q}$ and $\{b_n\}$ in $\mathbb{I}$ such that $\lim_n a_n = \lim_n b_n = x$.*

Proof. By 1.3.8 and Exercise 1.3.9, for each $n \in \mathbb{N}$ we may choose points $a_n \in (x - 1/n, x + 1/n) \cap \mathbb{Q}$ and $b_n \in (x - 1/n, x + 1/n) \cap \mathbb{I}$. The squeeze principle then implies that $a_n, b_n \to x$. □

2.1.8 Definition. (Infinite limits) A sequence $\{a_n\}$ in $\mathbb{R}$ is said to *diverge to* $+\infty$, written

$$a_n \to +\infty \quad \text{or} \quad \lim_n a_n = +\infty,$$

if, for each real number M, there exists an index N such that $a_n \geq M$ for all $n \geq N$. *Divergence to* $-\infty$ is defined analogously. The plus sign in $+\infty$ is frequently omitted. ◊

2.1.9 Example. If $r > 1$, then $r^n/n \to \infty$. This follows from 2.1.6: Given $M > 0$ there exists $N \in \mathbb{N}$ such that $n/r^n < 1/M$, and hence $r^n/n > M$, for all $n \geq N$. ◊

2.1.10 Example. If $r > 0$, then $a_n := r^n n! \to \infty$. Indeed, since

$$\frac{a_n}{a_{n-1}} = rn \to \infty,$$

there exists $N \in \mathbb{N}$ such that $a_n > 2a_{n-1}$, for all $n > N$. Iterating, we see that $a_n > 2^k a_{n-k} \geq k a_{n-k}$. Taking $k = n - N$ we have $a_n > (n - N)a_N$ for all $n > N$. Since $\lim_n (n - N)a_N = \infty$ (Archimedean principle), the assertion follows. ◊

For the next theorem, we shall need to observe the conventions regarding addition and multiplication in the extended real number system $\overline{\mathbb{R}}$ (1.3.12).

2.1.11 Theorem. *Let $\{a_n\}$ and $\{b_n\}$ be sequences in $\mathbb{R}$. The following limit properties hold in $\overline{\mathbb{R}}$ in the sense that if the expression on the right side of the equation exists in $\overline{\mathbb{R}}$, then the limit on the left side exists and equality holds.*

(a) $\lim_n (sa_n + tb_n) = s \lim_n a_n + t \lim_n b_n, \quad s, t \in \mathbb{R}$.

(b) $\lim_n a_n b_n = \lim_n a_n \lim_n b_n$.

(c) $\lim_n a_n/b_n = \lim_n a_n / \lim_n b_n$, *if* $\lim_n b_n \neq 0$.

(d) $\lim_n |a_n| = |\lim_n a_n|$.

(e) $\lim_n \sqrt{a_n} = \sqrt{\lim_n a_n}$ *if $a_n \geq 0$ for all n.*

Proof. Let $a_n \to a$, $b_n \to b$. We consider first the case $a, b \in \mathbb{R}$. Let $\varepsilon > 0$. For (a) choose N_1 and N_2 so that

$$|a_n - a| < \frac{\varepsilon}{2(|s|+1)} \text{ for all } n \geq N_1 \text{ and } |b_n - b| < \frac{\varepsilon}{2(|t|+1)} \text{ for all } n \geq N_2.$$

If $n \geq N := \max\{N_1, N_2\}$, then both of these inequalities hold, hence, by the triangle inequality,

$$|sa_n + tb_n - (sa + tb)| \leq |s|\,|a_n - a| + |t|\,|b_n - b| < \varepsilon/2 + \varepsilon/2 = \varepsilon.$$

For (b), choose $M \geq |a|$ so that $|b_n| \leq M$ for all n (2.1.3) and choose N so that $|a_n - a| < \varepsilon/2M$ and $|b_n - b| < \varepsilon/2M$ for all $n \geq N$. For such n,

$$\begin{aligned} |a_n b_n - ab| &= |(a_n - a)b_n + a(b_n - b)| \leq |a_n - a||b_n| + |a||b_n - b| \\ &\leq M|a_n - a| + M|b_n - b| < \varepsilon/2 + \varepsilon/2 = \varepsilon. \end{aligned}$$

For (c) it suffices to show that $1/b_n \to 1/b$. Choose N such that

$$|b_n - b| < \min\{|b|/2, \varepsilon b^2/2\} \quad \text{for all } n \geq N.$$

For such n, $|b_n| \geq |b| - |b_n - b| > |b|/2$, hence

$$\left| \frac{1}{b_n} - \frac{1}{b} \right| = \frac{|b_n - b|}{|bb_n|} \leq \frac{2|b_n - b|}{b^2} < \varepsilon.$$

Part (d) follows from the inequality $\big| |a_n| - |a| \big| \leq |a_n - a|$.
For (e), observe first that $a \geq 0$ (2.1.4). If $a = 0$, choose N such that $a_n < \varepsilon^2$ for all $n \geq N$. If $a > 0$, choose N such that $|a_n - a| < \varepsilon\sqrt{a}$ for all $n \geq N$. For such n,

$$|\sqrt{a_n} - \sqrt{a}| = \frac{|a_n - a|}{\sqrt{a_n} + \sqrt{a}} \leq \frac{|a_n - a|}{\sqrt{a}} < \varepsilon.$$

To illustrate the remaining cases $a = \pm\infty$ or $b = \pm\infty$, we prove part (b) for the case $-\infty < a < 0$ and $b_n \to \infty$. To show that $a_n b_n \to -\infty$, let $M < 0$ and choose N so that

$$a_n < a/2 \quad \text{and} \quad b_n > 2M/a \quad \text{for all } n \geq N.$$

For such n,

$$-a_n b_n > (-a/2)(2M/a) = -M,$$

hence $a_n b_n < M$. □

2.1.12 Example. To find

$$\lim_n \frac{\sqrt{4n^6 - 3n^2 + 5}}{2n^3 + 7n + 3},$$

divide the numerator and denominator of the general term a_n by n^3, the highest power of n occurring in the denominator, to obtain

$$a_n = \frac{\sqrt{4 - 3/n^4 + 5/n^6}}{2 + 7/n^2 + 3/n^3}.$$

The quotients in the numerator and denominator tend to 0, hence, by 2.1.11, $a_n \to \sqrt{4}/2 = 1$. ◊

Exercises

1. Let a, $b \in \mathbb{R}$. Find a closed formula for the nth term a_n of the sequences

   (a) $a, b, a, b, \ldots$ (b) $a, a, b, b, a, a, \ldots$ (c) $a, a, a, b, b, b, a, a, a \ldots$

   (d) $a, b, a, c, a, b, a, c \ldots$ (e) $1, 2, 3, 4, 1, 2, 3, 4, \ldots$

2. Find a recursive formula for the sequence $a, b, a, b, \ldots$

3. Use the ε, N definition of limit to prove that

   (a) $\lim_n \dfrac{4n-1}{2n+7} = 2$ (b) $\lim_n \dfrac{2n^2-n}{n^2+3} = 2$ (c) $\lim_n \dfrac{5\sqrt{n}+7}{3\sqrt{n}+2} = \dfrac{5}{3}$

   (d) $\lim_n \dfrac{n-1}{\sqrt{n}+1} = \infty$ (e) $\lim_n \left(2 + \dfrac{1}{n}\right)^3 = 8$ (f) $\lim_n \sqrt{\dfrac{n+2}{n+1}} = 1.$

4. Prove rigorously that the sequence $\{(-1)^n n/(n+1)\}$ has no limit.

5. Find $\lim_n \sin(n! r\pi)$ for $r \in \mathbb{Q}$.

6. Find $\lim_n \dfrac{1}{n}\left(n + \dfrac{1}{n}\right)^p$ for all $p \in \mathbb{R}$.

7. Let $\{a_n\}$ be contained in a finite set A. Prove that if $a_n \to a$, then there exists an index N such that $a_n = a$ for all $n \geq N$. In particular, $a \in A$.

8. Find $\lim_n b_n$ if

   (a) $a_n \to a$ and $3a_n + 2b_n \to c$

   (b) $a_n \to 2$ and $3a_n b_n + 5a_n^2 - 2b_n \to 1$.

9. Let $k \in \mathbb{N}$, $k > 1$, and a, $b > 0$. Evaluate $\lim_n a_n$ for the following a_n:

   (a) $\dfrac{2n+1}{(n^k+3n+1)^{1/k}}$ (b) $\left(\dfrac{an-1}{bn+1}\right)^{1/2}$

   (c) $\sqrt{n^2+kn} - n$ (d) $\sqrt{an + b\sqrt{n}} - \sqrt{an}$

   (e) $\dfrac{(n+k)!}{n!(n+k)^k}$ (f) $n^k\left(\sqrt{a^2+n^{-k}} - a\right)$

   (g) $n\left[(a-1/n)^k - a^k\right]$ (h) $n\left[1 - (1-a/n)^{1/k}\right]$

   (i) $(1-1/2)(1-1/3)\cdots(1-1/n)$ (j) $\displaystyle\sum_{j=1}^{n} (n^2+j)^{-1}$

   (k) $(1-1/2^2)(1-1/3^2)\cdots(1-1/n^2)$ (l) $\displaystyle\sum_{j=1}^{n} (n^k+j)^{-1/k}$.

10. Let $\{a_n\}$ be bounded and $b_n \to 0$. Prove that $a_n b_n \to 0$.

11. Let $a_n \to a \in \mathbb{R}$, $b_n \to b \in \mathbb{R}$, and $r > 0$ such that $|a_n - b_n| \le r$ for all n. Prove that $|a - b| \le r$.

12. Prove that if $na_n \to a \in \mathbb{R}$, then $\sqrt{n}\, a_n \to 0$. Show that the converse is false.

13. Let $a_n \ge 0$ for all n and $a_n \to a$. Prove that ${a_n}^{1/k} \to a^{1/k}$, $k \in \mathbb{N}$.

14. Let $r > 0$ and $k \in \mathbb{N}$. Prove in each case that $a_n \to 1$:

    (a) $a_n = r^{1/n}$ (b) $a_n = n^{1/n}$

    (c) $a_n = \left(r + n^k\right)^{1/n}$ (d) $a_n = \left[\sin(1/n)\right]^{1/n}$

15. Prove that $a_n \to a$ iff $a_n^+ \to a^+$ and $a_n^- \to a^-$. (See 1.2.7.)

16. Let $k \in \mathbb{N}$ and $r \in (-1, 1)$. Prove that $\lim_n n^k r^n = 0$.

17. Let $0 < r < 1$, $a_n > 0$, and $a_{n+1}/a_n < r$ for all n. Prove that $a_n \to 0$. Construct a sequence $\{a_n\}$ such that $a_n > 0$ and $a_{n+1}/a_n < 1$ for all n but $a_n \not\to 0$.

18. Suppose that $a_n \to a \in \overline{\mathbb{R}}$. Prove that

$$\lim_n (a_1 + \cdots + a_n)/n = a.$$

    Is the converse true?

19. Let $a_n \to a \in \mathbb{R}$ and suppose that $a_n \ge a$ for all n. Prove that

$$\lim_n \min\{a_1, \cdots, a_n\} = a.$$

    Does $\min\{a_1, \cdots, a_n\} \to a$ imply that $a_n \to a$?

20. Show that if $n^{-1} a_n \to 0$, then $n^{-1} \max\{a_1, \cdots, a_n\} \to 0$. Prove that the converse holds if $\{a_n\}$ is bounded below. Give an example to show that the converse is not generally true.

21. Let $0 < x_1 \le \cdots \le x_k$. Prove that

$$\lim_n (x_1^n + \cdots + x_k^n)^{1/n} = x_k.$$

22. Let $f(x)$ be any real-valued function on $\mathbb{R}$ such that $f(x) - x$ is bounded for all x (for example, $f(x) = \lfloor x \rfloor$). Use Exercise 1.4.4 to prove that

    (a) $(1/n^2) \sum_{j=1}^n f(jx) \to x/2$ (b) $(1/n^3) \sum_{j=1}^n f(j^2 x) \to x/3$.

23. Let a_0, $a_1 > 0$ and $a_n = \sqrt{a_{n-1} a_{n-2}}$, $n \ge 2$. Find $\lim_n a_n$.

24. Let $k \in \mathbb{N}$ and let $\{a_n\}$ be a sequence such that $a_{n+k} - a_n \to c \in \mathbb{R}$. Prove that $a_n/n \to c/k$. *Suggestion.* Consider first the case $k = 1$ to get the general idea.

2.2 Monotone Sequences

2.2.1 Definition. A sequence $\{a_n\}$ in $\mathbb{R}$ is said to be *increasing* (*strictly increasing*) if $a_n \le a_{n+1}$ ($a_n < a_{n+1}$) for all n. *Decreasing* and *strictly decreasing* sequences are defined analogously. A sequence that is either increasing or decreasing is called *monotone.* If $\{a_n\}$ is increasing (decreasing), we write $a_n \uparrow$ ($a_n \downarrow$). If $a_n \uparrow$ ($a_n \downarrow$) and $a_n \to a \in \overline{\mathbb{R}}$, we write $a_n \uparrow a$ ($a_n \downarrow a$). ◊

2.2.2 Theorem (Monotone sequence theorem). *If $\{a_n\}$ is increasing (decreasing), then $a_n \uparrow \sup_k a_k$ ($a_n \downarrow \inf_k a_k$). In particular, every bounded monotone sequence converges in $\mathbb{R}$.*

Proof. Assume $\{a_n\}$ is increasing and let $r < \sup_k a_k$. By the approximation property of suprema, $r < a_N \le \sup_k a_k$ for some N. Since $\{a_n\}$ is increasing, $r < a_n \le \sup_k a_k$ for all $n \ge N$. Therefore, $a_n \uparrow \sup_k a_k$. The proof for the decreasing case is similar. □

2.2.3 Example. Let $0 < a < x_1,\ y_1 < b := a + 1$ and define $\{x_n\}$ and $\{y_n\}$ recursively by

$$x_{n+1} = a + \sqrt{|x_n - a|} \quad \text{and} \quad y_{n+1} = b - \sqrt{|b - y_n|}.$$

By Exercise 1.4.1, $\{x_n\}$ is strictly increasing, $\{y_n\}$ is strictly decreasing, and $a < x_n,\ y_n < b$ for all n. By 2.2.2, $x_n \uparrow x$ and $y_n \downarrow y$ for some $x,\ y \in \mathbb{R}$. To find x, let $n \to \infty$ in the equation $x_{n+1} = a + \sqrt{x_n - a}$ to obtain $x = a + \sqrt{x - a}$. This has solutions $x = a$ and $x = b$. Since $\{x_n\}$ is increasing, $x = b$. Similarly, $y = a$. ◊

2.2.4 Example. We use the monotone sequence theorem to show that the sequence $\{(1 + 1/n)^n\}$ converges. By the binomial theorem (1.4.5) and the inequality $k! \ge 2^{k-1}$ (easily established by induction),

$$\begin{aligned}(1 + 1/n)^n &= \sum_{k=0}^{n} \binom{n}{k} 1/n^k \\ &= 2 + \sum_{k=2}^{n} (1 - 1/n)(1 - 2/n) \cdots (1 - (k-1)/n)/k! \\ &\le 2 + \sum_{k=2}^{n} 1/2^{k-1}.\end{aligned}$$

Since the sum in the last inequality is not bigger than 1, the sequence $\{(1 + 1/n)^n\}$ is bounded above by 3. Now let $m = n + 1$. Since $1/m < 1/n$,

$$1 - j/m \ge 1 - j/n \ge 0, \quad j = 1, \ldots, n - 1,$$

hence

$$\begin{aligned}(1+1/m)^m &\geq 2+\sum_{k=2}^{n}(1-1/m)(1-2/m)\cdots(1-(k-1)/m)/k! \\ &> 2+\sum_{k=2}^{n}(1-1/n)(1-2/n)\cdots(1-(k-1)/n)/k! \\ &= (1+1/n)^n .\end{aligned}$$

Thus $\{(1+1/n)^n\}$ is increasing. By 2.2.2, the sequence has a limit in $\mathbb{R}$. The limit is denoted by the letter e:

$$e := \lim_n (1+1/n)^n = 2.71828182845905\ldots$$

It will be seen later that e is the base of the natural logarithm. ◊

Exercises

1. Let $0 < a < 1 < b$. Prove that $a^{1/n} \uparrow 1$ and $b^{1/n} \downarrow 1$.
2. Let $a_n = a^n/n^k$ and $b_n = b^n/n^k$, where $0 < a < 1 < b$ and $k \in \mathbb{Z}^+$. Prove that $\{a_n\}$ is strictly decreasing and that $\{b_n\}$ is eventually strictly increasing.
3. Let $a_n = na/(1+n^2 b)$, a, $b > 0$. Prove: $a_n \downarrow 0$ (eventually) and $na_n \uparrow a/b$.
4. Let $x_n > 0$ and $x_n \uparrow x$. Prove that $(x_1^n + \cdots + x_n^n)^{1/n} \to x$.
5. Prove that for any nonempty set A of real numbers there exist sequences $\{a_n\}$ and $\{b_n\}$ in A such that $a_n \uparrow \sup A$ and $b_n \downarrow \inf A$.
6. Let $\{a_n\}$ be monotone and set $b_n := (a_1 + a_2 + \cdots + a_n)/n$. Prove that $\{b_n\}$ is monotone. (Compare with Exercise 2.1.18.)
7. Define $a_1 = 1$ and $a_n = 1 + (1 + a_{n-1})^{-1}$. Find $\lim_n a_n$ by first showing that $1 \leq a_n \leq 2$, $\{a_{2n}\}$ is decreasing, and $\{a_{2n+1}\}$ is increasing.
8. Let $r > 0$, $a_0 = \sqrt{r}$, and $a_n = \sqrt{r + a_{n-1}}$, $n \geq 1$. Find $\lim_n a_n$.
9. Let $r > 0$, $a_1 \geq \sqrt{r}$ and define
$$a_n = \tfrac{1}{2}(a_{n-1} + r/a_{n-1}), \quad n > 1.$$
Show that $a_n \geq a_{n+1} \geq \sqrt{r}$ and find $\lim_n a_n$.
10. Prove that $e = \lim_n (1 - 1/n)^{-n}$.
11. Let $0 < x_0 < y_0$ and define
$$x_{n+1} = \sqrt{x_n y_n} \quad \text{and} \quad y_{n+1} = (x_n + y_n)/2.$$
Prove that $0 < x_n < x_{n+1} < y_{n+1} < y_n$ and that $\lim_n x_n = \lim_n y_n$.

2.3 Subsequences and Cauchy Sequences

2.3.1 Definition. A *subsequence* of a sequence $\{a_n\}_{n=1}^{\infty}$ in $\mathbb{R}$ is a sequence $\{a_{n_k}\}_{k=1}^{\infty}$ where the indices satisfy $1 \le n_1 < n_2 < \cdots$. The limit in $\overline{\mathbb{R}}$ of a subsequence is called a *cluster point of* $\{a_n\}$. ◊

For example, in the following sequence the underlined terms define the beginning of a subsequence $\{a_{n_k}\}$ with $n_1 = 3$, $n_2 = 4$, $n_3 = 6$, etc.

$$a_1, a_2, \underline{a_3}, \underline{a_4}, a_5, \underline{a_6}, a_7, a_8, \underline{a_9}, \underline{a_{10}}, a_{12}, \underline{a_{13}}, a_{14}, \underline{a_{15}}, \ldots$$

Note that the indices n_k of a subsequence satisfy $n_k \ge k$.

Examples. **(a)** The sequence

$$\left\{\left[1 - (-1)^{\lfloor (n-1)/2 \rfloor}\right]\right\} = \{0, 0, 2, 2, 0, 0, 2, 2 \ldots\}$$

is a subsequence of

$$\{1 - (-1)^n\} = \{2, 0, 2, 0, \ldots\},$$

which has cluster points 0 and 2.

(b) The sequence $\{n \sin(n\pi/2)\}$ has cluster points 0 and $\pm\infty$.

(c) Let $\{r_1, r_2, \ldots\}$ be an arbitrary enumeration of the rational numbers (see Appendix A). Then every real number is a cluster point of $\{r_n\}$. Indeed, since every interval of the form $(x - 1/n, x + 1/n)$ contains infinitely many terms of the sequence, we may choose $n_1 \ge 1$ such that $|x - r_{n_1}| < 1$ and then choose $n_2 > n_1$ such that $|x - r_{n_2}| < 1/2$, etc. In this way we construct a subsequence $\{r_{n_2}\}$ inductively such that $|x - r_{n_k}| < 1/k$ for all k and so $r_{n_k} \to x$. ◊

Notation. It is occasionally convenient to use the following alternate method to describe a subsequence: If we set $b_k = a_{n_k}$ and change the index in $\{b_k\}_{k=1}^{\infty}$ to n, then $\{b_n\}$ may be used to denote the subsequence $\{a_{n_k}\}$. This provides a convenient way to denote a subsequence of a subsequence. In this regard, note that if $\{c_n\}$ is a subsequence of $\{b_n\}$ and $\{b_n\}$ is a subsequence of $\{a_n\}$, then $\{c_n\}$ is a subsequence of $\{a_n\}$.

The following proposition shows that a convergent sequence has a single cluster point.

2.3.2 Proposition. *If $\{a_n\}$ is a sequence in $\mathbb{R}$ and $a_n \to a \in \overline{\mathbb{R}}$, then $a_{n_k} \to a$ for any subsequence $\{a_{n_k}\}$ of $\{a_n\}$.*

Proof. We prove the proposition for the case $a \in \mathbb{R}$ and leave the other cases for the reader. Given $\varepsilon > 0$, choose N such that $|a_n - a| < \varepsilon$ for all $n \ge N$. Since $n_k \ge k$, $|a_{n_k} - a| < \varepsilon$ for all $k \ge N$. Therefore, $a_{n_k} \to a$. □

2.3.3 Example. We calculate $\lim_n(1+1/n^2)^{3n^2+5}$ by writing

$$\left(1+\frac{1}{n^2}\right)^{3n^2+5} = \left[\left(1+\frac{1}{n^2}\right)^{n^2}\right]^3 \left(1+\frac{1}{n^2}\right)^5.$$

The term in the square brackets is a subsequence of $(1+1/n)^n$ and hence converges to e (see 2.2.4). It follows that $(1+1/n^2)^{3n^2+5} \to e^3$. ◊

The following result will have important consequences in later chapters.

2.3.4 Theorem (Bolzano–Weierstrass Theorem). *Every bounded sequence in* $\mathbb{R}$ *has a convergent subsequence.*

Proof. The proof is based on the observation that if a union of two sets contains infinitely many terms of a sequence, then at least one of the sets must contain infinitely many of the terms of the sequence.

Let $\{a_n\}$ be a bounded sequence, say $c_0 \le a_n \le d_0$ for all n. Bisect the interval $I_0 := [c_0, d_0]$. By the preceding observation, one of the resulting subintervals, call it I_1, contains infinitely many terms of the sequence. Choose one such term, say a_{n_1}. Now bisect I_1. Again, one of the resulting subintervals, call it I_2, contains infinitely many terms of the sequence. Choose one such term a_{n_2} with $n_2 > n_1$. By repeating this procedure, we produce a subsequence $\{a_{n_k}\}_{k=1}^{\infty}$ of $\{a_n\}$ and a sequence of intervals $I_k = [c_k, d_k]$, $k = 0, 1, \ldots$, such that

$$c_0 \le c_{k-1} \le c_k \le a_{n_k} \le d_k \le d_{k-1} \le d_0, \text{ and } d_{k+1} - c_{k+1} = \tfrac{1}{2}(d_k - c_k).$$

Since $\{c_k\}$ and $\{d_k\}$ are monotone and bounded $c_k \to c$ and $d_k \to d$ for some c, $d \in \mathbb{R}$. Since $d_k - c_k = 2^{-k}(d_0 - c_0) \to 0$, $c = d$. Therefore, by the squeeze principle, $a_{n_k} \to c$. □

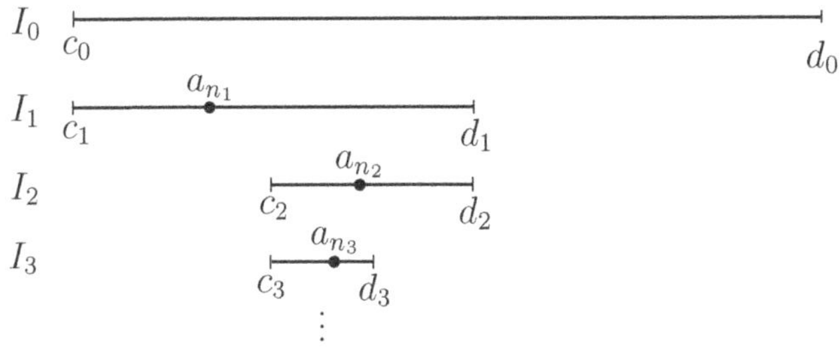

FIGURE 2.3: Interval halving process.

The Bolzano–Weierstrass theorem may be extended as follows:

2.3.5 Theorem. *Every sequence in* $\mathbb{R}$ *has a subsequence that converges in* $\overline{\mathbb{R}}$.

Proof. If $\{a_n\}$ is bounded, then the Bolzano–Weierstrass theorem applies. Suppose that $\{a_n\}$ is unbounded above. Then for each $k \in \mathbb{N}$ there exist infinitely many indices n such that $a_n > k$. We may then construct a subsequence $\{a_{n_k}\}$ with $a_{n_k} > k$ for all k and so $a_{n_k} \to \infty$. □

2.3.6 Corollary. *A sequence $\{a_n\}$ in $\mathbb{R}$ has a limit in $\overline{\mathbb{R}}$ iff it has exactly one cluster point in $\overline{\mathbb{R}}$.*

Proof. The necessity is 2.3.2. For the sufficiency, suppose that $\{a_n\}$ has exactly one cluster point $a \in \overline{\mathbb{R}}$. Consider first the case $a = +\infty$. We claim that $a_n \to +\infty$. If not, then there exists $M \in \mathbb{R}$ such that $a_n \le M$ for infinitely many n, hence there exists a subsequence $\{a_{n_k}\}$ of $\{a_n\}$ with $a_{n_k} \le M$ for all k. By 2.3.5, $\{a_{n_k}\}$ has a cluster point $b \in \overline{\mathbb{R}}$. But $b \le M < a$, so $\{a_n\}$ has more than one cluster point, a contradiction. Therefore, $a_n \to +\infty$, as claimed. The case $a = -\infty$ is treated similarly.

Now suppose $a \in \mathbb{R}$. Then $a_n \to a$. If not, then there exists $\varepsilon > 0$ such that $|a_n - a| \ge \varepsilon$ for infinitely many n. Choose a subsequence $\{a_{n_k}\}$ of $\{a_n\}$ with $|a_{n_k} - a| \ge \varepsilon$ for all k. If $\{a_{n_k}\}$ is unbounded, then it has a cluster point $+\infty$ or $-\infty$, giving an additional cluster point of $\{a_n\}$ in $\overline{\mathbb{R}}$. On the other hand, if $\{a_{n_k}\}$ is bounded, then $\{a_{n_k}\}$ has a cluster point b in $\mathbb{R}$. But then $|b - a| \ge \varepsilon$, so again $\{a_n\}$ has more than one cluster point. □

2.3.7 Definition. A sequence $\{a_n\}$ is said to be *Cauchy* if for each $\varepsilon > 0$ there exists an index N such that $|a_n - a_m| < \varepsilon$ for all $m, n \ge N$. We express this condition by writing $\lim_{m,n}(a_n - a_m) = 0$. ◊

The definition asserts that the terms of a Cauchy sequence get closer to one another. Thus the following result is not surprising.

2.3.8 Proposition. *Every convergent sequence is Cauchy.*

Proof. Let $a_n \to a$. Given $\varepsilon > 0$, choose N such that $|a_n - a| < \varepsilon/2$ for all $n \ge N$. Then for $n, m \ge N$,

$$|a_n - a_m| = |(a_n - a) + (a - a_m)| \le |a_n - a| + |a_m - a| < \varepsilon. \qquad \square$$

The converse of 2.3.8 is true, a fact of fundamental importance. For the proof we need the following lemma.

2.3.9 Lemma. *A Cauchy sequence is bounded.*

Proof. Let $\{a_n\}$ be a Cauchy sequence. Choose N such that $|a_n - a_m| < 1$ for all $m, n \ge N$. Then $|a_n| \le |a_n - a_N| + |a_N| < 1 + |a_N|$ for all $n \ge N$, hence

$$|a_n| \le \max\{1 + |a_N|, |a_1|, |a_2|, \ldots, |a_{N-1}|\} \text{ for all } n. \qquad \square$$

2.3.10 Theorem (Cauchy criterion). *Every Cauchy sequence in $\mathbb{R}$ converges.*

Proof. By 2.3.9 and the Bolzano–Weierstrass theorem, a Cauchy sequence $\{a_n\}$ has a convergent subsequence, say $a_{n_k} \to a \in \mathbb{R}$. We claim that $a_n \to a$. Let $\varepsilon > 0$ and choose N such that $|a_n - a_m| < \varepsilon$ for all $m, n \ge N$. In particular, $|a_n - a_{n_k}| < \varepsilon$ for $n, k \ge N$. Fixing $n \ge N$ and letting $k \to \infty$ in the last inequality yields $|a_n - a| \le \varepsilon$, verifying the claim. □

Exercises

1. Find all cluster points of $\{a_n\}$, where $a_n =$

   (a) $(-1)^n\left(\frac{2n+1}{4n+3}\right)\sin^2\left(\frac{n\pi}{3}\right)$ (b) $(-1)^n\left(\frac{2n+1}{n+5}\right)\cos^2\left(\frac{n\pi}{4}\right)$

   (c) $(-1)^{\lfloor n/3\rfloor}(1+1/n)^2+(-1)^{\lfloor n/4\rfloor}(2+1/n)^2+(-1)^{\lfloor n/5\rfloor}(3+1/n)^2$

   (d) $(-1)^n r_n + r_{2n}$, where r_k is the remainder on division of k by 3.

2. Construct a sequence with precisely the cluster points 1, 2, 3, $+\infty$.

3. Let $k \in \mathbb{N}$. Use the fact that $\lim_n(1+1/n)^n = e$ (2.2.4) to find $\lim_n a_n$ for the following a_n, where $k \in \mathbb{N}$.

   (a) $\left(1+\frac{1}{kn}\right)^n$ (b) $\left(1+\frac{1}{k+n}\right)^n$ (c) $\left(\frac{1}{k}+\frac{1}{n}\right)^n$

   (d) $\left(1+\frac{1}{2n+k}\right)^{kn}$ (e) $\left(1+\frac{1}{3n^3+5}\right)^{7n^3-4}$.

4. Let $\{a_n\}$ and $\{b_n\}$ be bounded sequences. Show that there exist convergent subsequences of $\{a_n\}$ and $\{b_n\}$ with the *same* indices.

5. Prove that a sequence contained in a finite set has a constant subsequence.

6. Let $-\infty < a_n < r \le +\infty$ with $a_n \to r$. Show that $\{a_n\}$ has a strictly increasing subsequence.

7. Show that every sequence of distinct real numbers has a strictly monotone subsequence.

8. Let $k \in \mathbb{N}$ and suppose that the series $\sum_{n=1}^{\infty}|a_{n+k}-a_n|$ converges (see Chapter 6). Prove that $\{a_n\}$ has a convergent subsequence.

9. Let a_0, a_1 be arbitrary and define $a_{n+1} = (a_n + a_{n-1})/2$, $n \ge 1$. Show directly that $\{a_n\}$ is a Cauchy sequence. (Its limit may be found from Exercise 1.4.14.)

10. Let $0 < p \le q$ and $a_n > 0$ for all n. Set $b_n = a_n^q/(1+a_n^p)$. Show that $a_n \to 0$ iff $b_n \to 0$. Is the assertion true if $0 < q < p$?

11. Let I be an open interval and let $\{a_n\}$ have the property that each open subinterval J of I contains a_n for infinitely many n. Prove that every point of I is a cluster point of $\{a_n\}$. Give an example of such a sequence.

12. Suppose that the cluster points of $\{a_n\}$ form a sequence $\{b_n\}$. Show that every cluster point b of $\{b_n\}$ is a cluster point of $\{a_n\}$. *Hint.* Choose a subsequence $\{b_{n_k}\}$ such that $|b_{n_k} - b| < 1/k$.

2.4 Limits Inferior and Superior

For an arbitrary sequence $\{a_n\}$ in $\mathbb{R}$, define

$$\underline{a}_n = \inf_{k\geq n} a_k \quad \text{and} \quad \overline{a}_n = \sup_{k\geq n} a_k, \quad n = 1, 2, \ldots.$$

Then $\{\underline{a}_n\}$ is increasing and $\{\overline{a}_n\}$ is decreasing, hence the limits

$$\liminf_n a_n := \lim_n \underline{a}_n \quad \text{and} \quad \limsup_n a_n := \lim_n \overline{a}_n$$

exist in $\overline{\mathbb{R}}$. These are called, respectively, the *limit inferior* and *limit superior* of the sequence $\{a_n\}$.

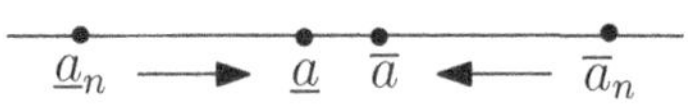

FIGURE 2.4: $\underline{a} = \liminf_n a_n$ and $\overline{a} = \limsup_n a_n$.

Clearly,

$$\underline{a}_n \leq a_n \leq \overline{a}_n \quad \text{and} \quad \liminf_n a_n \leq \limsup_n a_n.$$

Furthermore, if $\{a_n\}$ is unbounded below, then $\liminf_n a_n = -\infty$, and if $\{a_n\}$ is unbounded above, then $\limsup_n a_n = +\infty$.

Here are some examples:

(a) $\liminf_n \dfrac{(-1)^n n}{n+1} = -1,$ $\qquad \limsup_n \dfrac{(-1)^n n}{n+1} = 1,$

(b) $\liminf_n [(-1)^n + 1]n = 0,$ $\qquad \limsup_n [(-1)^n + 1]n = +\infty,$

(c) $\liminf_n \sin n = -1,$ $\qquad \limsup_n \sin n = 1.$

Example (c) follows from Example 8.3.10 (See Exercise 8.3.15)

The next proposition shows that lim sup and lim inf have properties similar to those of limits. Their usefulness derives from this fact together with the property that, in contrast to ordinary limits, the limits inferior and superior of a sequence always exist (in $\overline{\mathbb{R}}$).

2.4.1 Proposition. *For any sequences $\{a_n\}$ and $\{b_n\}$ in $\mathbb{R}$,*

(a) $\limsup_n(-a_n) = -\liminf_n a_n$.

(b) $\limsup_n(a_n + b_n) \leq \limsup_n a_n + \limsup_n b_n$ *if the right side is defined.*

(c) $\liminf_n(a_n + b_n) \geq \liminf_n a_n + \liminf_n b_n$ *if the right side is defined.*

(d) $\limsup_n ca_n = c \limsup_n a_n$, *if $c \geq 0$.*

(e) $\liminf_n ca_n = c \liminf_n a_n$, *if* $c \geq 0$.

(f) $\limsup_n (a_n b_n) \leq (\limsup_n a_n)(\limsup_n b_n)$ *if* $a_n, b_n \geq 0$ *for all* n.

(g) $\liminf_n (a_n b_n) \geq (\liminf_n a_n)(\liminf_n b_n)$ *if* $a_n, b_n \geq 0$ *for all* n.

(h) $\liminf_n a_n \leq \liminf_n b_n$, $\limsup_n a_n \leq \limsup_n b_n$ *if* $a_n \leq b_n$ *for all* n.

Proof. Part (a) follows from $\sup_{k \geq n}(-a_k) = -\inf_{k \geq n} a_k$ and part (h) is a direct consequence of the definitions. Part (b) follows by taking limits in the inequality

$$\sup_{k \geq n}(a_k + b_k) \leq \sup_{k \geq n} a_k + \sup_{k \geq n} b_k.$$

Part (f) follows similarly from

$$\sup_{k \geq n} a_k b_k \leq \sup_{k \geq n} a_k \sup_{k \geq n} b_k.$$

Part (d) is a consequence of

$$\sup_{k \geq n} ca_k = c \sup_{k \geq n} a_k, \quad c \geq 0.$$

Parts (c), (e), and (g) are proved in a similar manner. □

2.4.2 Theorem. *For any sequence* $\{a_n\}$ *in* $\mathbb{R}$, *the extended real numbers* $\underline{a} := \liminf_n a_n$ *and* $\overline{a} := \limsup_n a_n$ *are cluster points of* $\{a_n\}$. *All other cluster points of* $\{a_n\}$ *in* $\overline{\mathbb{R}}$ *(if any) lie between these.*

Proof. We leave the case $\overline{a} = -\infty$ to the reader. Assume then that $\overline{a} > -\infty$ and recall that $\overline{a}_n \downarrow \overline{a}$. Choose a strictly increasing sequence of real numbers r_n tending to $\overline{a}$. Since $r_1 < \overline{a}_1$, by the approximation property of suprema there exists an index n_1 such that $r_1 < a_{n_1} \leq \overline{a}_{n_1}$. Similarly, since $r_2 < \overline{a}_{n_1+1}$, there exists an index $n_2 > n_1$ such that $r_2 < a_{n_2} \leq \overline{a}_{n_2}$. In this way we may construct inductively a subsequence $\{a_{n_k}\}$ such that $r_k < a_{n_k} \leq \overline{a}_{n_k}$. By the squeeze principle, $a_{n_k} \to \overline{a}$. The limit infimum case is treated similarly.

Now let $\{a_{n_k}\}$ be *any* subsequence of $\{a_n\}$ with $a_{n_k} \to a \in \overline{\mathbb{R}}$. Then, for any m and $k \geq m$, $n_k \geq m$ and so $\underline{a}_m \leq a_{n_k} \leq \overline{a}_m$. Letting $k \to \infty$ yields $\underline{a}_m \leq a \leq \overline{a}_m$. Letting $m \to \infty$ we obtain $\underline{a} \leq a \leq \overline{a}$. □

Since $\lim_n a_n$ exists in $\overline{\mathbb{R}}$ iff $\{a_n\}$ has exactly one cluster point (2.3.6), the following result is immediate.

2.4.3 Corollary. *For any sequence* $\{a_n\}$ *in* $\mathbb{R}$, $\lim_n a_n$ *exists in* $\overline{\mathbb{R}}$ *iff* $\liminf_n a_n = \limsup_n a_n$. *In this case, all three limits are equal.*

Exercises

1. Find $\liminf_n a_n$ and $\limsup_n a_n$ if

   (a) $a_n = \dfrac{(-1)^n 5n + 7}{3n + 5}$

   (b) $a_n = n^{\sin(n\pi/2)} + (1/n)\cos(n)$

   (c) $a_n = (-1)^{\lfloor n/3 \rfloor}(1+1/n)^2 + (-1)^{\lfloor n/4 \rfloor}(2+1/n)^2 + (-1)^{\lfloor n/5 \rfloor}(3+1/n)^2$

   (d) $a_n = \dfrac{2nr_n + 1}{nr_{2n} + 1}$, r_k the remainder on division of $k \in \mathbb{N}$ by 3

   (e) $a_n = (-1)^{r_n} x_n + (-1)^{r_{n+1}} y_n + (-1)^{r_{n+2}} z_n$, where $x_n \to x$, $y_n \to y$, $z_n \to z$, $x < y < z$, and r_k as in (d)

   (f) $a_1 = 1$, $a_{2n} = ra_{2n-1}$, $a_{2n+1} = ar + a_{2n}$, $0 < r < 1$, $a > 0$

   (g) $a_n = 2^n + 2^{-n} + (-1)^n(2^n - 2^{-n})$

   (h) $a_n = \dfrac{3n\cos(n\pi/4) + 2}{2n\sin(n\pi/4) + 3}$.

2. Show by example that the inequalities (b), (c), (f), and (g) in 2.4.1 may be strict.

3. Let $a_n > 0$ for all n. Prove that

$$\limsup_n (1/a_n) = 1/\liminf_n a_n \quad \text{and} \quad \liminf_n (1/a_n) = 1/\limsup_n a_n.$$

4. Let $\{a_n\}$ be bounded and nonnegative and let $r \in \mathbb{Q}^+$. Prove that

$$\limsup_n a_n^r = \left(\limsup_n a_n\right)^r \quad \text{and} \quad \liminf_n a_n^r = \left(\liminf_n a_n\right)^r.$$

5. Show that for any subsequence $\{a_{n_k}\}$ of $\{a_n\}$,

$$\limsup_{k\to\infty} a_{n_k} \le \limsup_n a_n \quad \text{and} \quad \liminf_{k\to\infty} a_{n_k} \ge \liminf_n a_n.$$

6. Let $b_n \to b \in (0, +\infty)$. Prove that

$$\limsup_n (a_n + b_n) = b + \limsup_n a_n \quad \text{and} \quad \liminf_n (a_n + b_n) = b + \liminf_n a_n.$$

7. Let $a_n \ge 0$ for all n and $b_n \to b \in (0, +\infty)$. Prove that

$$\limsup_n a_n b_n = b \limsup_n a_n \quad \text{and} \quad \liminf_n a_n b_n = b \liminf_n a_n.$$

8. Prove that

$$\left|\limsup_n a_n\right| \le \limsup_n |a_n| \quad \text{and} \quad \left|\liminf_n a_n\right| \ge \liminf_n |a_n|.$$

   Show by examples that the inequalities may be strict.

9. Let $\{n_k\}$ be a sequence of positive integers that contains each positive integer exactly once, for example, $3, 2, 1, 6, 5, 4, 9, 8, 7, \ldots$. Show that

$$\limsup_k a_{n_k} = \limsup_n a_n \quad \text{and} \quad \liminf_k a_{n_k} = \liminf_n a_n.$$

In particular, if $a_n \to a$, then $a_{n_k} \to a$. (Note: $\{a_{n_k}\}_{k=1}^{\infty}$ is *not* necessarily a subsequence $\{a_n\}$.)

10. Let $a_n \to a > 0$ and $\liminf_n b_n > 0$. If $b_n^2 - a_n b_n - 6a_n^2 \to 0$, prove that $\limsup_{n\to\infty} b_n \le 3a$.

11. Prove that for any sequence $\{a_n\}$,

$$\liminf_n a_n \le \liminf_n \frac{1}{n}\sum_{j=1}^{n} a_j \le \limsup_n \frac{1}{n}\sum_{j=1}^{n} a_j \le \limsup_n a_n.$$

12. ($\Downarrow$ 7.4.2.) Let $a_n > 0$ for all n. Prove that

$$\liminf_n \frac{a_{n+1}}{a_n} \le \liminf_n a_n^{1/n} \le \limsup_n a_n^{1/n} \le \limsup_n \frac{a_{n+1}}{a_n}.$$

Use this to calculate $\lim_n n/(n!)^{1/n}$.

Chapter 3

Limits and Continuity on $\mathbb{R}$

3.1 Limit of a Function

The definition of limit of a function f, given in 3.1.3 below, is a precise formulation of the intuitive idea that as x gets closer to a number a, the function value $f(x)$ approaches some fixed number L. This notion is conveniently described in terms of certain subsets of $\mathbb{R}$ called *neighborhoods*.

3.1.1 Definition. Let $r > 0$. A *neighborhood of* $a \in \overline{\mathbb{R}}$ is an interval of the form

$$\mathcal{N}(a) = \mathcal{N}_r(a) := \begin{cases} (a-r, a+r) & \text{if } a \in \mathbb{R}, \\ (r, +\infty) & \text{if } a = +\infty, \\ (-\infty, -r) & \text{if } a = -\infty. \end{cases}$$

If $a \in \mathbb{R}$, the set $\mathcal{N}(a) \setminus \{a\} := (a-r, a) \cup (a, a+r)$ is called a *deleted neighborhood of* a. ◊

The reader should verify that the intersection of finitely many neighborhoods of a is again a neighborhood of a and that neighborhoods *separate points*, that is, if $a \neq b$ are extended real numbers, then there exist neighborhoods $\mathcal{N}(a)$ and $\mathcal{N}(b)$ such that $\mathcal{N}(a) \cap \mathcal{N}(b) = \emptyset$.

3.1.2 Definition. An *accumulation point* of a nonempty set E of real numbers is an extended real number a such that every neighborhood of a contains a point of E not equal to a. A member of E that is not an accumulation point of E is called an *isolated point* of E. ◊

For example, the set of accumulation points of $E := \big[\mathbb{Q} \cap (-1, 0)\big] \cup \mathbb{N}$ is $\big[-1, 0\big] \cup \{+\infty\}$, and the set of isolated points of E is $\mathbb{N}$.

The following definition of limit is sufficiently general to include the usual limits encountered in calculus, namely, one-sided limits, two-sided limits, limits at infinity, and infinite limits.

3.1.3 Definition. Let $E \subseteq \mathbb{R}$, let f be a real-valued function whose domain includes E, and let a, $L \in \overline{\mathbb{R}}$, where either $a \in E$ or a is an accumulation point of E (not necessarily in the domain of f). We write

$$L = \lim_{\substack{x \to a \\ x \in E}} f(x)$$

DOI: 10.1201/9781003629382-3

if for each neighborhood $\mathcal{N}(L)$ of L there is a neighborhood $\mathcal{N}(a)$ of a such that

$$x \in E \cap \mathcal{N}(a) \text{ implies } f(x) \in \mathcal{N}(L). \tag{3.1}$$

In this case we say that $f(x)$ *approaches* L *as* x *tends to* a *along* E. ◊

The restrictions on the extended real number a guarantee that $E \cap \mathcal{N}(a) \neq \emptyset$, hence condition (3.1) is not vacuously satisfied. Note that if $a \in E$ is not an accumulation point of E, then it must be an isolated point, in which case $\lim_{\{x \to a,\, x \in E\}} f(x)$ trivially exists and equals $f(a)$.

We single out the following important special cases, where $a \in \mathbb{R}$ and $s > 0$:

(a) *left-hand limit:* $\displaystyle\lim_{x \to a^-} f(x) := \lim_{\substack{x \to a \\ x \in E}} f(x), \quad E = (a - s, a).$

(b) *right-hand limit:* $\displaystyle\lim_{x \to a^+} f(x) := \lim_{\substack{x \to a \\ x \in E}} f(x), \quad E = (a, a + s).$

(c) *two-sided limit:* $\displaystyle\lim_{x \to a} f(x) := \lim_{\substack{x \to a \\ x \in E}} f(x), \quad E = (a - s, a + s) \setminus \{a\}.$

(d) *limit at* $+\infty$: $\displaystyle\lim_{x \to +\infty} f(x) := \lim_{\substack{x \to +\infty \\ x \in E}} f(x), \ E = (s, +\infty).$

(e) *limit at* $-\infty$: $\displaystyle\lim_{x \to -\infty} f(x) := \lim_{\substack{x \to -\infty \\ x \in E}} f(x), \ E = (-\infty, -s).$

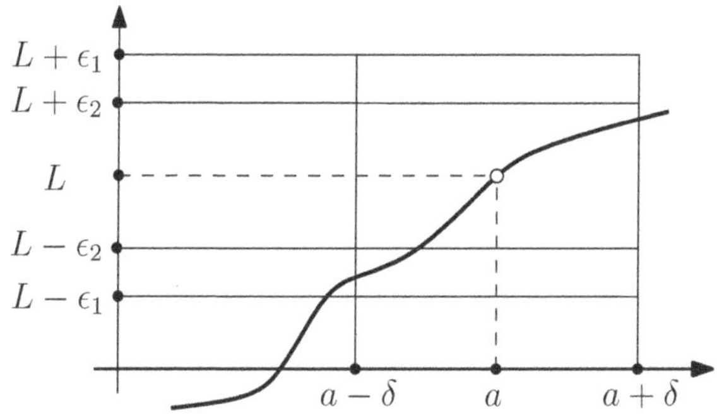

FIGURE 3.1: δ works for ε_1 but not for ε_2.

Applying the definition of limit to the cases (a)–(e) above produces the standard limit definitions encountered in beginning calculus. For example, if the limit L in (c) is finite, then, in the context of (c), 3.1.3 asserts that for each $\varepsilon > 0$ there exists a $\delta \in (0, s)$ such that

$$|f(x) - L| < \varepsilon \quad \text{for all } x \text{ with } 0 < |x - a| < \delta.$$

(See Figure 3.1.) For (e) and the case $L = +\infty$, 3.1.3 asserts that for each $M \in \mathbb{R}$, there exists an $r > s$ such that

$$f(x) > M \quad \text{for all } x \text{ with } x < -r.$$

The advantage of having a single definition of limit is that it provides a unified theory and allows for economy of thought and presentation.

As in the case of sequences, limits of functions are unique. Indeed, if $L_1 \neq L_2$ both satisfy criterion (3.1), then, given neighborhoods $\mathcal{N}(L_1)$ and $\mathcal{N}(L_2)$, there would exist a neighborhood $\mathcal{N}(a)$ such that

$$x \in E \cap \mathcal{N}(a) \Rightarrow f(x) \in \mathcal{N}(L_1) \cap \mathcal{N}(L_2).$$

However, $\mathcal{N}(L_1)$ and $\mathcal{N}(L_2)$ may be taken to be disjoint, and choosing any $x \in E \cap \mathcal{N}(a)$ then results a contradiction.

In any discussion of limits we shall tacitly assume that a and E satisfy the conditions of 3.1.3.

3.1.4 Example. Let $f(x) = (3x+2)/(2x-1)$. Then

(a) $\lim_{x\to\infty} f(x) = \lim_{x\to-\infty} f(x) = 3/2$.

(b) $\lim_{x\to a} f(x) = f(a)$, $(a \neq 1/2)$.

(c) $\lim_{x\to 1/2^+} f(x) = +\infty$.

(d) $\lim_{x\to 1/2^-} f(x) = -\infty$.

To verify (a,) let $\varepsilon > 0$ and note that the quantity

$$\left| f(x) - \frac{3}{2} \right| = \frac{7}{2|(2x-1)|}$$

will be less than ε if $|2x-1| > 7/2\varepsilon$. The latter inequality is satisfied if either $x > (1 + 7/2\varepsilon)/2$ or $x < (1 - 7/\varepsilon)/2$.

For (b), observe first that

$$|f(x) - f(a)| = \left| \frac{3x+2}{2x-1} - \frac{3a+2}{2a-1} \right| = \frac{7|x-a|}{|2x-1||2a-1|}.$$

By the triangle inequality,

$$|2x-1| = |2(x-a) + (2a-1)| =\geq |2a-1| - 2|a-x|.$$

Hence if $|a-x| < |2a-1|/4$, then $|2x-1| > |2a-1|/2$ and therefore

$$|f(x) - f(a)| < \frac{14|x-a|}{|2a-1|^2}.$$

It follows that $|f(x) - f(a)|$ will be less than ε if we require additionally that $|x-a| < \varepsilon|2a-1|^2/14$. Therefore, any $\delta < \min\{|2a-1|/4, \varepsilon|2a-1|^2/14\}$ will satisfy criterion (3.1).

To prove (c), note that if $0 < |x - 1/2| < 1/2$, then $x > 0$, hence

$$f(x) = \frac{1}{2}\,\frac{3x+2}{x-1/2} = \left[\frac{3x}{2} + 1\right] \frac{1}{x-1/2} > \frac{1}{x-1/2}.$$

Given $M > 2$, let $\delta = 1/M$. Then $|x - 1/2| < \delta \Rightarrow 0 < x - 1/2 < 1/M \Rightarrow f(x) > M$, proving (c). The proof of part (d) is similar. ◊

3.1.5 Theorem. *Let f be a function with domain D and let $E = E_1 \cup E_2 \subseteq D$. Suppose that one of the following holds:*

- *a is an accumulation point of both E_1 and E_2.*
- *a is an isolated point of both E_1 and E_2.*
- *a is an accumulation point of E_1 and an isolated point of E_2.*
- *a is an accumulation point of E_2 and an isolated point of E_1.*

Then $\lim_{\{x \to a,\, x \in E\}} f(x)$ exists in $\overline{\mathbb{R}}$ iff both limits $\lim_{\{x \to a,\, x \in E_1\}} f(x)$ and $\lim_{\{x \to a,\, x \in E_2\}} f(x)$ exist in $\overline{\mathbb{R}}$ and are equal. In this case all three limits are equal.

Proof. If a is an accumulation point of E_1 or E_2, then a is an accumulation point of E. If a is an isolated point of E_1 and E_2, then a is an isolated point of E. This shows that in each case, $\lim_{\{x \to a,\, x \in E\}} f(x)$ is at least *defined.*

Now suppose that $L := \lim_{\{x \to a,\, x \in E\}} f(x)$ exists. Then (3.1) holds for E, so it must hold for each of the subsets E_1 and E_2 as well. Therefore, $\lim_{\{x \to a,\, x \in E_1\}} f(x)$ and $\lim_{\{x \to a,\, x \in E_2\}} f(x)$ exist and equal L.

Conversely, suppose that the limits along E_1 and E_2 exist and equal $K \in \overline{\mathbb{R}}$. Then, given a neighborhood $\mathcal{N}(K)$, there exists a neighborhood $\mathcal{N}_j(a)$ of a such that $x \in E_j \cap \mathcal{N}_j(a)$ implies $f(x) \in \mathcal{N}(K)$ $(j = 1, 2)$. It follows that $\mathcal{N}(a) := \mathcal{N}_1(a) \cap \mathcal{N}_2(a)$ is a neighborhood of a and $x \in E \cap \mathcal{N}(a)$ implies $f(x) \in \mathcal{N}(K)$. This proves that $\lim_{\{x \to a,\, x \in E\}} f(x) = K$. □

3.1.6 Example. Take $E_1 = \mathbb{N}$ and $E_2 = (0, 2)$. Then 2 is an isolated point of E_1 and an accumulation point of E_2, and $\lim_{\{x \to 2,\, x \in E_1\}} f(x) = f(2)$. Therefore, by the theorem, $\lim_{\{x \to 2,\, x \in E\}} f(x)$ exists iff $\lim_{x \to 2^-} f(x) = f(2)$. ◊

3.1.7 Example. (Dirichlet function). Let

$$d(x) = \begin{cases} 1 & \text{if } x \in \mathbb{Q}, \\ 0 & \text{otherwise.} \end{cases}$$

Since $\lim_{\{x \to a,\, x \in \mathbb{Q}\}} d(x) = 1$ and $\lim_{\{x \to a,\, x \in \mathbb{I}\}} d(x) = 0$, $\lim_{x \to a} d(x)$ cannot exist. ◊

The following is an immediate consequence of 3.1.5.

3.1.8 Corollary. $\lim_{x \to a} f(x)$ *exists iff* $\lim_{x \to a^-} f(x)$ *and* $\lim_{x \to a^+} f(x)$ *exist and are equal. In this case all three limits are equal.*

The next result shows that function limits may be characterized in terms of limits of sequences.

3.1.9 Sequential Characterization of Limit. *Let f be a function whose domain includes E and let $a \in \mathbb{R}$ be an accumulation point of E. Then, $\lim_{\{x \to a,\, x \in E\}} f(x)$ exists in $\overline{\mathbb{R}}$ and equals L iff $f(a_n) \to L$ for all sequences $\{a_n\}$ in E with $a_n \to a$.*

Proof. Assume that $\lim_{\{x\to a,\, x\in E\}} f(x) = L$ and let $\{a_n\}$ be a sequence in E with $a_n \to a$. Given a neighborhood $\mathcal{N}(L)$, choose $\mathcal{N}(a)$ as in (3.1) and then choose N such that $a_n \in \mathcal{N}(a)$ for all $n \geq N$. For such n, $f(a_n) \in \mathcal{N}(L)$. Therefore, $f(a_n) \to L$.

Now suppose that $\lim_{\{x\to a,\, x\in E\}} f(x) \neq L$. Then there is a neighborhood of L such that (3.1) fails for each neighborhood $\mathcal{N}(a)$ of a. Consider the case $a, L \in \mathbb{R}$. Then $\mathcal{N}(L)$ is of the form $(L-r, L+r)$ for some $r > 0$. Taking $\mathcal{N}(a) = (a - 1/n, a + 1/n)$ we see that for each $n \in \mathbb{N}$, there exists $a_n \in E$ with $|a_n - a| < 1/n$ and $|f(a_n) - L| \geq r$. Thus $a_n \to a$ and $f(a_n) \not\to L$, so the sequential condition does not hold. A similar argument works if either a or L is infinite. □

3.1.10 Example. Let $f(x) = \sin(1/x)$, $x \neq 0$. Since $f\big(1/n\pi\big) = 0$ and $f\big(2/(4n+1)\pi\big) = 1$, $\lim_{x\to 0^+} f(x)$ does not exist. ◊

3.1.11 Cauchy Criterion for Functions. *Let a be an accumulation point of E. Then $\lim_{\{x\to a,\, x\in E\}} f(x)$ exists in $\mathbb{R}$ iff given $\varepsilon > 0$, there exists $\delta > 0$ such that $|f(x) - f(y)| < \varepsilon$ for all $x, y \in E$ with $|x - a| < \delta$ and $|y - a| < \delta$.*

Proof. If $\lim_{\{x\to a,\, x\in E\}} f(x)$ exists in $\mathbb{R}$, then an application of the triangle inequality shows that the ε, δ-condition of the theorem holds.

Conversely, assume that the ε, δ-condition holds and let $\{a_n\}$ be a sequence in E with $a_n \to a$. By the hypothesis, $\{f(a_n)\}$ is a Cauchy sequence and so converges to some real number L. Suppose $\{b_n\}$ is another sequence in E converging to a. It follows from the ε, δ-condition that $f(b_n) - f(a_n) \to 0$, hence $f(b_n) = f(b_n) - f(a_n) + f(a_n) \to L$. By 3.1.9, $\lim_{\{x\to a,\, x\in E\}} f(x) = L$. □

3.1.12 Theorem. *Let f be a function whose domain includes E and let $a \in \overline{\mathbb{R}}$ be an accumulation point of E. Then the following properties hold in the sense that if the expressions on the right exist in $\overline{\mathbb{R}}$, then the limits on the left exist and the equality holds.*

(a) $\lim\limits_{\substack{x\to a\\ x\in E}} [sf(x) + tg(x)] = s \lim\limits_{\substack{x\to a\\ x\in E}} f(x) + t \lim\limits_{\substack{x\to a\\ x\in E}} g(x), \quad s, t \in \mathbb{R}.$

(b) $\lim\limits_{\substack{x\to a\\ x\in E}} f(x)g(x) = \lim\limits_{\substack{x\to a\\ x\in E}} f(x) \lim\limits_{\substack{x\to a\\ x\in E}} g(x).$

(c) $\lim\limits_{\substack{x\to a\\ x\in E}} \dfrac{f(x)}{g(x)} = \dfrac{\lim_{\{x\to a,\, x\in E\}} f(x)}{\lim_{\{x\to a,\, x\in E\}} g(x)}$ *if* $\lim\limits_{\substack{x\to a\\ x\in E}} g(x) \neq 0.$

(d) $\lim\limits_{\substack{x\to a\\ x\in E}} |f(x)| = \Big| \lim\limits_{\substack{x\to a\\ x\in E}} f(x) \Big|.$

Proof. The assertions follow immediately from 2.1.11 and 3.1.9. However, it is instructive to formulate direct proofs. We do this for the finite version of part (c). Assume that the limits

$$L := \lim_{\substack{x\to a\\ x\in E}} f(x) \quad \text{and} \quad M := \lim_{\substack{x\to a\\ x\in E}} g(x) \neq 0$$

are finite and let $\varepsilon > 0$. Choose $\mathcal{N}_1(a)$ such that

$$|g(x) - M| < |M|/2 \ \text{ for all } \ x \in E \cap \mathcal{N}_1(a).$$

For such x, $|g(x)| = |M - (M - g(x))| \geq |M| - |M - g(x)| \geq |M|/2$, hence

$$\begin{aligned}
\left|\frac{f(x)}{g(x)} - \frac{L}{M}\right| &= \frac{|Mf(x) - Lg(x)|}{|Mg(x)|} \\
&= \frac{|M(f(x) - L) + L(M - g(x))|}{|M|\,|g(x)|} \\
&\leq \frac{|M|\,|f(x) - L| + |L|\,|M - g(x)|}{|M|^2/2} \\
&= \frac{2}{|M|}\,|f(x) - L| + \frac{2|L|}{M^2}\,|M - g(x)| \\
&\leq K\big(|f(x) - L| + |M - g(x)|\big), \quad K := 2/|M| + 2|L|/M^2.
\end{aligned}$$

Now choose $\mathcal{N}_2(a)$ so that $|f(x) - L| < \varepsilon/2K$ and $|M - g(x)| < \varepsilon/2K$ for all $x \in E \cap \mathcal{N}_2(a)$. Then $x \in E \cap \mathcal{N}_1(a) \cap \mathcal{N}_2(a) \Rightarrow |f(x)/g(x) - L/M| < \varepsilon$. □

3.1.13 Example. (Limits at infinity of rational functions). Let $f(x) = P(x)/Q(x)$, where

$$P(x) = a_0 + a_1 x + \cdots + a_n x^n \ \text{ and } \ Q(x) = b_0 + b_1 x + \cdots + b_m x^m, \ \ a_n, b_m \neq 0.$$

For any $a, c \in \mathbb{R}$, $\lim_{x\to c} a = a$ and $\lim_{x\to c} x = c$, hence, by 3.1.12, $\lim_{x\to c} f(x) = f(c)$, provided $Q(c) \neq 0$. To calculate limits at $+\infty$, write

$$f(x) = \frac{a_0 x^{-n} + a_1 x^{-n+1} + \cdots + a_{n-1}x^{-1} + a_n}{b_0 x^{-m} + b_1 x^{-m+1} + \cdots + b_{m-1}x^{-1} + b_m}\, x^{n-m}.$$

Since $\lim_{x\to+\infty} x^{-j} = 0$ for $j \in \mathbb{N}$, we see that

$$\lim_{x\to+\infty} f(x) = \begin{cases} 0 & \text{if } m > n, \\ a_n/b_n & \text{if } m = n, \text{ and} \\ \pm\infty & \text{if } m < n, \end{cases}$$

where the sign in the last case is that of a_n/b_m. ◊

3.1.14 Theorem. *Let f and g be functions domains including E and let $a \in \mathbb{R}$ be an accumulation point of E. If $f(x) \leq g(x)$ for all $x \in E$ and if $L := \lim_{\{x\to a,\, x\in E\}} f(x)$ and $M := \lim_{\{x\to a,\, x\in E\}} g(x)$ exist in $\overline{\mathbb{R}}$, then $L \leq M$.*

Proof. Assume, for a contradiction, that $M < L$. Choose any $K \in (M, L)$ and then choose neighborhoods $\mathcal{N}(L) \subseteq (K, +\infty)$ and $\mathcal{N}(M) \subseteq (-\infty, K)$ (see Figure 3.2). Then there exists a neighborhood $\mathcal{N}(a)$ such that $f(x) \in \mathcal{N}(L)$ and $g(x) \in \mathcal{N}(M)$ for all $x \in E \cap \mathcal{N}(a)$. But for any such x, $g(x) < f(x)$, contradicting the hypothesis. □

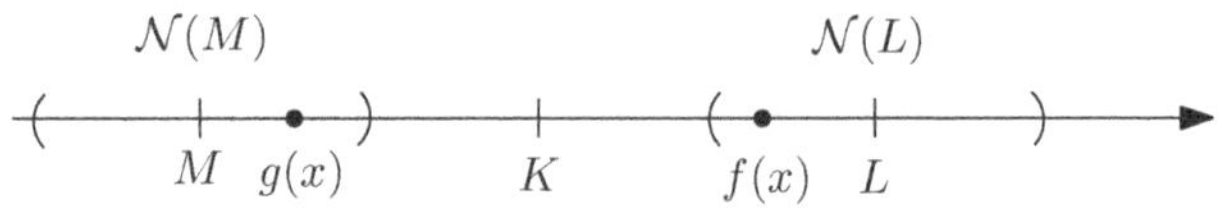

FIGURE 3.2: L can't be greater than M.

3.1.15 Theorem (Squeeze principle for functions). *Let f be a function whose domain contains E and let $a \in \mathbb{R}$ be an accumulation point of E. If $f(x) \le g(x) \le h(x)$ for all $x \in E$ and if the limits $\lim_{\{x \to a,\, x \in E\}} f(x)$ and $\lim_{\{x \to a,\, x \in E\}} h(x)$ exist in $\overline{\mathbb{R}}$ and are equal, then $\lim_{\{x \to a,\, x \in E\}} g(x)$ exists in $\overline{\mathbb{R}}$ and all three limits are equal.*

Proof. Let L denote the common limit. For the case $L \in \mathbb{R}$, given $\varepsilon > 0$ there exists a neighborhood $\mathcal{N}(a)$ of a such that

$$L - \varepsilon \le f(x) \le g(x) \le h(x) < L + \varepsilon \text{ for all } x \in E \cap \mathcal{N}(a).$$

The cases $L = \pm\infty$ are proved in a similar manner. □

3.1.16 Definitions. A function f is said to be *increasing on E* (*strictly increasing on E*) if $f(x) \le f(y)$ ($f(x) < f(y)$) for all $x, y \in E$ with $x < y$. The notions of *decreasing* and *strictly decreasing* are defined analogously. If f is either increasing or decreasing, then f is said to be *monotone*. Finally, f is *bounded on E* if there exists a real number M such that $|f(x)| \le M$ for all $x \in E$. ◊

3.1.17 Monotone Function Theorem. *Let $a, b, c \in \overline{\mathbb{R}}$ with $a < c < b$. If f is monotone on (a, b), then $\lim_{x \to a^+} f(x)$, $\lim_{x \to b^-} f(x)$ exist in $\overline{\mathbb{R}}$ and $\lim_{x \to c^-} f(x)$, $\lim_{x \to c^+} f(x)$ exist in $\mathbb{R}$.*

Proof. Assume that f is increasing. Let $s := \sup_{a<x<b} f(x)$. By the approximation property of suprema, for each $r < s$ there exists $x_r \in (a, b)$ such that $r < f(x_r) \le s$. Since $f(x)$ is increasing, $r < f(x) \le s$ for all $x \in (x_r, b)$. Therefore,

$$\lim_{x \to b^-} f(x) = \sup_{a<x<b} f(x).$$

Similarly,

$$\lim_{x \to a^+} f(x) = \inf_{a<x<b} f(x).$$

The assertion regarding limits at c is proved in the same way noting that, since f is bounded in a neighborhood of c, the limits are finite.

For an alternate proof one could use the monotone sequence theorem together with the sequential characterization of limit. □

Exercises

1. Show that all points of a finite set E are isolated.

2. Find all the accumulation points of the set $\{2/m - 3/n : m, n \in \mathbb{N}\}$.

3. Prove that if E has an accumulation point a, then there exists a sequence $\{a_n\}$ of distinct points in E such that $a_n \to a$.

4. Determine the limit and then use the definition to prove your result:

   (a) $\lim_{x\to 1}(3x^2 - 2x + 1)$ (b) $\lim_{x\to 1}\dfrac{x+3}{3x+1}$ (c) $\lim_{x\to 4}\dfrac{\sqrt{x}+1}{\sqrt{x}-1}$

   (d) $\lim_{x\to -\infty}(x^2 + x)$ (e) $\lim_{x\to -\infty}\sqrt{x^2+x} - x$ (f) $\lim_{x\to +\infty}\dfrac{\sqrt{x}+1}{\sqrt{x}-1}$.

5. Discuss the possible values of $\lim_{x\to -\infty} f(x)$ for the function f in 3.1.13.

6. Use the results of this section together with standard trig identities and the limits
$$\lim_{x\to 0}\cos x = \lim_{x\to 0}\frac{\sin x}{x} = 1$$
   to evaluate the following limits without using l'Hospital's rule.

   (a) $\lim_{x\to 0}\dfrac{\sin(2x)}{\sin(3x)}$ (b) $\lim_{x\to 1}\dfrac{\sin(x-1)}{x^3-1}$ (c) $\lim_{x\to 0^+}\dfrac{\sin x^2}{\sqrt{x}}$

   (d) $\lim_{x\to 0^+}\dfrac{\sin\sqrt{x}}{x^2}$ (e) $\lim_{x\to 0}\dfrac{1-\cos(3x)}{\sin^2 x}$ (f) $\lim_{x\to 0}\dfrac{3x+2\sin(5x)}{2x-3\sin(7x)}$

   (g) $\lim_{x\to +\infty}\dfrac{1-\sec(3/x)}{1-\sec(5/x)}$ (h) $\lim_{x\to 1}\dfrac{\tan(x^2-3x+2)}{\tan(x^2-4x+3)}$ (i) $\lim_{x\to +\infty}\dfrac{\sin(1/x)}{\sin(1/\sqrt{x})}$.

7. Evaluate the following limits without using l'Hospital's rule, where m, $n \in \mathbb{N}$, and a, b, c, $d > 0$.

   (a) $\lim_{x\to 3}\dfrac{1}{x-3}\left[1 + \dfrac{(x-15)}{(x^2+x)}\right]$ (b) $\lim_{x\to 0}\dfrac{1}{x}\left[\dfrac{1}{x+1} - \dfrac{1}{\sqrt{x+1}}\right]$

   (c) $\lim_{x\to a}\dfrac{x^n - a^n}{x^m - a^m}$ (d) $\lim_{x\to b}\dfrac{x^{1/n} - b^{1/n}}{x^{1/m} - b^{1/m}}$

   (e) $\lim_{x\to 2b^+}\dfrac{\sqrt{x-b} - \sqrt{b}}{x-2b}$ (f) $\lim_{x\to 0}\dfrac{\sqrt{b+x} - \sqrt{b-x}}{\sqrt{c+x} - \sqrt{c-x}}$

(g) $\lim_{x \to 0} \frac{\sqrt{(x+1)(x+2)} - \sqrt{2}}{x}$ (h) $\lim_{x \to 0} \frac{\sqrt{ax+b} - \sqrt{b}}{\sqrt{cx+d} - \sqrt{d}}$

(i) $\lim_{x \to n^-} \frac{x + \lfloor x \rfloor}{x - \lfloor x \rfloor}$ (j) $\lim_{x \to n^+} \frac{x + \lfloor x \rfloor}{x - \lfloor x \rfloor}$.

8. Let $a_j \in \mathbb{R}$ and $\varepsilon > 0$. Show that there exists $a \in \mathbb{R}$ such that $at^n + \sum_{j=1}^{n-1} a_j t^j > -\varepsilon$ for all $t \geq 0$.

9. Define
$$f(x) = \begin{cases} 4x^2 + 2x - 11 & \text{if } x \text{ is rational} \\ 3x^2 + x - 5 & \text{if } x \text{ is irrational.} \end{cases}$$
Show that $\lim_{x \to a} f(x)$ exists for precisely two values of a.

10. Let $c \in \mathbb{R}$. Show that $\lim_{x \to c} f(x)$ exists in $\overline{\mathbb{R}}$ iff for each strictly increasing sequence $\{a_n\}$ converging to c and each strictly decreasing sequence $\{b_n\}$ converging to c, $\lim_n f(a_n) = \lim_n f(b_n)$.

11. Evaluate the following limits without using l'Hospital's rule, where $a,\ b,\ c > 0,\ m,\ n \in \mathbb{N}$.

(a) $\lim_{x \to +\infty} \frac{\lfloor ax \rfloor}{x}$ (b) $\lim_{x \to +\infty} \sqrt{x} \left\lfloor \frac{b}{x} \right\rfloor x$

(c) $\lim_{x \to +\infty} \frac{\lfloor mx \rfloor + n}{\lfloor nx \rfloor + m}$ (d) $\lim_{x \to +\infty} \frac{\lfloor ax + \sin x \rfloor}{\lfloor bx + \cos x \rfloor}$

(e) $\lim_{x \to +\infty} \sqrt{ax} \left\lfloor \sqrt{bx+c} - \sqrt{bx} \right\rfloor$ (f) $\lim_{x \to +\infty} \frac{\sqrt{ax+b} - \sqrt{b}}{\sqrt{cx+d} - \sqrt{d}}$

(g) $\lim_{x \to +\infty} \left\lfloor \sqrt{ax + \sqrt{bx}} - \sqrt{ax + \sqrt{cx}} \right\rfloor$ (h) $\lim_{x \to +\infty} \left\lfloor x\sqrt{ax^2+b} - x^2\sqrt{a} \right\rfloor$.

12. (⇓ 3.4.5.) Let $f(x) = x^n$, $n \in \mathbb{N}$. Use Exercise 1.1.4 to prove that f is strictly increasing on $[0, +\infty)$, and if n is odd, then f is strictly increasing on $\mathbb{R}$.

13. Suppose that f is monotone on $(0, +\infty)$ and $a_n \to +\infty$. Prove that $\lim_{x \to \infty} f(x) = \lim_n f(a_n)$.

---

*3.2 Limits Inferior and Superior

Let f be a function whose domain includes $E \subseteq \mathbb{R}$, and let $a \in \overline{\mathbb{R}}$ be either a member of E or an accumulation point of E (not necessarily in the domain of f). For each neighborhood $\mathcal{N}_r(a)$, define

$$\underline{f}(r) = \inf\{f(x) : x \in E \cap \mathcal{N}_r(a)\} \text{ and } \overline{f}(r) = \sup\{f(x) : x \in E \cap \mathcal{N}_r(a)\}.$$

If $a \in \mathbb{R}$, then $\underline{f}(r)$ increases and $\overline{f}(r)$ decreases as $r \downarrow 0$. Similarly, if $a = \pm\infty$, then $\underline{f}(r)$ increases and $\overline{f}(r)$ decreases as $r \uparrow +\infty$. By the monotone function theorem, in the first case, $\underline{f}$ and $\overline{f}$ have limits as $r \to 0^+$, and in the second case, $\underline{f}$ and $\overline{f}$ have limits as $r \to +\infty$. We then define

$$\liminf_{\substack{x\to a\\ x\in E}} f(x) := \lim_{r\to 0^+} \underline{f}(r) \quad \text{and} \quad \limsup_{\substack{x\to a\\ x\in E}} f(x) := \lim_{r\to 0^+} \overline{f}(r) \quad \text{if } a \in \mathbb{R},$$

$$\liminf_{\substack{x\to a\\ x\in E}} f(x) := \lim_{r\to +\infty} \underline{f}(r) \quad \text{and} \quad \limsup_{\substack{x\to a\\ x\in E}} f(x) := \lim_{r\to +\infty} \overline{f}(r) \quad \text{if } a = \pm\infty.$$

The extended real numbers

$$\liminf_{\substack{x\to a\\ x\in E}} f(x) \quad \text{and} \quad \limsup_{\substack{x\to a\\ x\in E}} f(x)$$

are called, respectively, the *limit inferior* and *limit superior of f, as x tends to a along E*. Clearly,

$$\liminf_{\substack{x\to a\\ x\in E}} f(x) \le \limsup_{\substack{x\to a\\ x\in E}} f(x).$$

The above definitions include all the standard formulations of limit superior and limit inferior. For these we shall use notation analogous to that for limits. For example, taking $E = (0 + \infty)$, we have

$$\liminf_{x\to 0^+} \sin(1/x) = -1 \quad \text{and} \quad \limsup_{x\to +\infty} \sin x = 1.$$

Limits superior and inferior of a function have properties similar to ordinary limits but unlike the latter, always exist (in $\overline{\mathbb{R}}$).

The following theorem establishes a connection with limits superior and inferior of sequences.

3.2.1 Theorem. *Let $a \in \overline{\mathbb{R}}$ be an accumulation point of E. Then there exist sequences $\{x_n\}$ and $\{y_n\}$ in E tending to a such that*

$$\liminf_{\substack{x\to a\\ x\in E}} f(x) = \lim_n f(x_n) \quad \textit{and} \quad \limsup_{\substack{x\to a\\ x\in E}} f(x) = \lim_n f(y_n). \tag{3.2}$$

Moreover, if $\mathcal{A}$ denotes the set of all sequences $\{a_n\}$ in E tending to a, then

$$\liminf_{\substack{x\to a\\ x\in E}} f(x) = \inf_{\{a_n\}\in\mathcal{A}} \liminf_n f(a_n) \ \textit{and} \ \limsup_{\substack{x\to a\\ x\in E}} f(x) = \sup_{\{a_n\}\in\mathcal{A}} \limsup_n f(a_n).$$

Proof. We prove only the lim sup case. Let

$$s := \limsup_{\substack{x\to a\\ x\in E}} f(x) \quad\text{and}\quad t := \sup_{\{a_n\}\in\mathcal{A}} \limsup_{n\to\infty} f(a_n).$$

We assume that both a and s are finite; the proofs for the other cases are similar. Choose a sequence $r_n \to 0^+$ such that $\overline{f}(r_n) \to s$. For each n, use the approximation property of supremum to obtain a point $y_n \in E \cap \mathcal{N}_{r_n}(a)$ such that $\overline{f}(r_n) - 1/n < f(y_n) \le \overline{f}(r_n)$. Then $y_n \to a$ and $f(y_n) \to s$. This proves the limsup part of (3.2) and also shows that $s \le t$.

For the reverse inequality, let $\{a_n\}$ be in $\mathcal{A}$ and let $L := \limsup_{n\to\infty} f(a_n)$. Let $r > 0$ and choose $N \in \mathbb{N}$ such that $a_n \in \mathcal{N}_r(a)$ for all $n \ge N$. For such n, $f(a_n) \le \overline{f}(r)$, hence $L \le \overline{f}(r)$. Letting $r \to 0^+$ yields $L \le s$. This shows that $t \le s$ and hence that $t = s$, as required. □

3.2.2 Corollary. *Let $a \in \overline{\mathbb{R}}$ be an accumulation point of E. Then $\lim_{\{x\to a,\, x\in E\}} f(x)$ exists in $\overline{\mathbb{R}}$ iff*

$$\liminf_{\substack{x\to a\\ x\in E}} f(x) = \limsup_{\substack{x\to a\\ x\in E}} f(x), \tag{3.3}$$

in which case the three limits are equal.

Proof. Assume (3.3) holds let L denote the common value. By the theorem, for any sequence $\{a_n\}$ in E tending to a, $\limsup_n f(a_n) \le L \le \liminf_n f(a_n)$, hence $\lim_n f(a_n)$ exists and equals L. From the sequential characterization of limit, $\lim_{\{x\to a,\, x\in E\}} f(x)$ exists and equals L.

Conversely, assume $K := \lim_{\{x\to a,\, x\in E\}} f(x)$ exists in $\overline{\mathbb{R}}$ and let $a_n \in E$ with $a_n \to a$. Then $K = \lim_n f(a_n)$, hence $\limsup_n f(a_n) = \liminf_n f(a_n)$. By the theorem, (3.3) holds and the common value is K. □

3.2.3 Proposition. *Let f and g have domain containing $E \subseteq \mathbb{R}$ and let $a \in \overline{\mathbb{R}}$ be an accumulation point of E. Then*

(a) $\displaystyle\limsup_{\substack{x\to a\\ x\in E}} [-f(x)] = -\liminf_{\substack{x\to a\\ x\in E}} f(x)$

(b) $\displaystyle\limsup_{\substack{x\to a\\ x\in E}} [f(x) + g(x)] \le \limsup_{\substack{x\to a\\ x\in E}} f(x) + \limsup_{\substack{x\to a\\ x\in E}} g(x)$

(c) $\displaystyle\liminf_{\substack{x\to a\\ x\in E}} [f(x) + g(x]) \ge \liminf_{\substack{x\to a\\ x\in E}} f(x) + \liminf_{\substack{x\to a\\ x\in E}} g(x)$

(d) $\displaystyle\limsup_{\substack{x\to a\\ x\in E}} cf(x) = c\limsup_{\substack{x\to a\\ x\in E}} f(x), \quad \liminf_{\substack{x\to a\\ x\in E}} cf(x) = c\liminf_{\substack{x\to a\\ x\in E}} f(x)$ *if* $c \ge 0$

(e) $\displaystyle\limsup_{\substack{x\to a\\ x\in E}} f(x)g(x) \le \Big(\limsup_{\substack{x\to a\\ x\in E}} f(x)\Big)\Big(\limsup_{\substack{x\to a\\ x\in E}} g(x)\Big)$ *if* $f,\, g \ge 0$

(f) $\liminf\limits_{\substack{x\to a\\ x\in E}} f(x)g(x) \ge \left(\liminf\limits_{\substack{x\to a\\ x\in E}} f(x)\right)\left(\liminf\limits_{\substack{x\to a\\ x\in E}} g(x)\right)$ *if* $f,\, g \ge 0$

(g) $\liminf\limits_{\substack{x\to a\\ x\in E}} f(x) \le \liminf\limits_{\substack{x\to a\\ x\in E}} g(x)$ *and* $\limsup\limits_{\substack{x\to a\\ x\in E}} f(x) \le \limsup\limits_{\substack{x\to a\\ x\in E}} g(x)$ *if* $f \le g$.

Proof. In (b) and (c) it is assumed that the expressions on the right are defined in $\overline{\mathbb{R}}$. The assertions of the proposition may be proved directly or by using 2.4.1 together with 3.2.1. We illustrate the latter approach in proving (b). The proofs of the remaining parts are similar.

Let $\{a_n\}$ be an arbitrary sequence in E tending to a. By part (b) of 2.4.1 and by 3.2.1,

$$\begin{aligned}\limsup_n \left[f(a_n) + g(a_n)\right] &\le \limsup_n f(a_n) + \limsup_n g(a_n)\\ &\le \limsup_{\substack{x\to a\\ x\in E}} f(x) + \limsup_{\substack{x\to a\\ x\in E}} g(x).\end{aligned}$$

Taking the supremum over all such sequences $\{a_n\}$ yields (b). □

Exercises

1. Calculate $\liminf_{x\to+\infty} f(x)$ and $\limsup_{x\to+\infty} f(x)$ for each of the functions $f(x)$ below, where $r(k)$ is the remainder on division of $k \in \mathbb{N}$ by 3 and $d(x)$ is the Dirichlet function (3.1.7).

   (a) $\sin[xd(x)]$. (b) $\dfrac{(-1)^{\lfloor x\rfloor}3x+7}{2x+5(-1)^{\lfloor x\rfloor}}$ (c) $\dfrac{(-1)^{\lfloor x\rfloor}\lfloor 2x\rfloor+3}{\lfloor 3x\rfloor+2}$

   (d) $\dfrac{4r(\lfloor x\rfloor)x+5}{7r(\lfloor x\rfloor)x+1}$ (e) $\dfrac{e^x+e^{-x}}{(-1)^{\lfloor x\rfloor}(e^x-e^{-x})}$ (f) $\cos x\sin x$

   (g) $\dfrac{1}{2+\cos x}$ (h) $\sin x+\cos x$ (i) $\dfrac{3\sin x}{2+\sin x}$

   (j) $\cos\left(\dfrac{\pi}{3}\sin x\right)$ (k) $\dfrac{1}{1+\sin^2 x}$ (l) $(-1)^{\lfloor x\rfloor-\lfloor x^2\rfloor}$.

2. Prove the remaining parts of 3.2.3. Give examples to show that the inequalities in (b), (c), (e), and (f) may be strict.

3. Let $E = E_1 \cup E_2$, where a is an accumulation point of both E_1 and E_2. Prove:

$$\limsup_{\substack{x\to a\\ x\in E}} f(x) = \max\left\{\limsup_{\substack{x\to a\\ x\in E_1}} f(x),\ \limsup_{\substack{x\to a\\ x\in E_2}} f(x)\right\}$$

$$\liminf_{\substack{x\to a\\ x\in E}} f(x) = \min\left\{\liminf_{\substack{x\to a\\ x\in E_1}} f(x),\ \liminf_{\substack{x\to a\\ x\in E_2}} f(x)\right\}.$$

Conclude in particular that

$$\limsup_{x\to a} f(x) = \max\left\{ \limsup_{x\to a^-} f(x), \limsup_{x\to a^+} f(x)\right\}$$
$$\liminf_{x\to a} f(x) = \min\left\{ \liminf_{x\to a^-} f(x), \liminf_{x\to a^+} f(x)\right\}.$$

4. Let $f(x) > 0$ for all $x \in E$. Prove that

$$\limsup_{\substack{x\to a\\ x\in E}} \frac{1}{f(x)} = \frac{1}{\liminf_{\{x\to a,\, x\in E\}} f(x)}.$$

5. Prove that

$$\left|\limsup_{\substack{x\to a\\ x\in E}} f(x)\right| \le \limsup_{\substack{x\to a\\ x\in E}} |f(x)| \quad\text{and}\quad \left|\liminf_{\substack{x\to a\\ x\in E}} f(x)\right| \ge \liminf_{\substack{x\to a\\ x\in E}} |f(x)|.$$

Show by examples that the inequalities may be strict.

6. Let $f : [a, b) \to \mathbb{R}$ and $g(x) = \sup_{a\le t\le x} f(t)$, $a \le x < b$. Prove that $g(x_0) \le \lim_{x\to x_0^+} g(x)$ for every $x_0 \in [a, b)$.

3.3 Continuous Functions

3.3.1 Definition. A function $f : D \to \mathbb{R}$ is said to be *continuous at* $a \in D$ if $\lim_{\{x\to a,\, x\in D\}} f(x) = f(a)$, that is, for each $\varepsilon > 0$ there exists a $\delta > 0$ such that

$$|f(x) - f(a)| < \varepsilon \quad \text{for all } x \in D \text{ with } |x - a| < \delta.$$

If f is continuous at each point of a subset E of D, then f is said to be *continuous on* E. If f is continuous on D, then f is simply said to be *continuous*. A point in D at which f is not continuous is called a *discontinuity of* f. ◊

The definition of continuity implies that a function $f : D \to \mathbb{R}$ is always continuous at isolated points of D. For example, if D is a finite set or a set of integers, then every function $f : D \to \mathbb{R}$ is continuous.

Continuity of f on E is not the same as continuity of the restriction $f|_E$. For example, the function on $\mathbb{R}$ that is identically equal to one on $\mathbb{Z}$ and zero elsewhere is not continuous on $\mathbb{Z}$, yet its restriction to $\mathbb{Z}$ is continuous (as a function with domain $\mathbb{Z}$).

From the sequential characterization of limit we have

3.3.2 Sequential Characterization of Continuity. *A function f with domain D is continuous at $a \in D$ iff $f(a_n) \to f(a)$ for all sequences $\{a_n\}$ in D with $a_n \to a$.*

3.3.3 Example. Let $\{r_1, r_2, \ldots\}$ be an enumeration of the rationals in $(0,1)$. Define f on $(0,1)$ by $f(r_n) = 1/n$ and $f(x) = 0$ if x is irrational. We use the sequential characterization of continuity to show that f is continuous precisely at the irrational numbers in $(0,1)$.

Let $x \in (0,1)$ be rational, and choose a sequence $\{x_n\}$ of irrational numbers converging to x. Since $f(x_n) = 0$ for all n and $f(x) \neq 0$, $f(x_n) \not\to f(x)$. Therefore, f is not continuous at any rational.

Now let $x \in (0,1)$ be irrational and let $\{x_n\}$ be *any* sequence converging to x. If $f(x_n) \not\to f(x)$, then, since $f(x) = 0$, there exists an $N \in \mathbb{N}$ and a subsequence $\{y_n\}$ of $\{x_n\}$ such that $f(y_n) \geq 1/N$ for all n. Thus y_n is rational, say $y_n = r_m$. Since $1/N \leq f(r_m) = 1/m$, $y_n \in \{r_1, r_2, \ldots, r_N\}$. Since this holds for all n, $x \in \{r_1, r_2, \ldots, r_N\}$. But this contradicts the fact that x is irrational. (For a variation of this example, see Exercise 10.) ◊

The following is an immediate consequence of 3.1.12.

3.3.4 Theorem. *Let f and g be functions with domain D, let $\alpha, \beta \in \mathbb{R}$, and let $a \in D$. If f and g are continuous at a, then so are $\alpha f + \beta g$, fg, f/g (the last provided that $g(a) \neq 0$).*

3.3.5 Theorem. *Let $g : D \to \mathbb{R}$ and $f : E \to \mathbb{R}$ with $g(D) \subseteq E$. If g is continuous at $a \in D$ and f is continuous at $g(a)$, then $f \circ g$ is continuous at a.*

Proof. Let $b := g(a)$. Given $\varepsilon > 0$, choose $\eta > 0$ such that $|f(y) - f(b)| < \varepsilon$ for all $y \in E$ with $|y - b| < \eta$. Next, choose $\delta > 0$ such that $|g(x) - b| < \eta$ for all $x \in D$ with $|x - a| < \delta$. Then $|x - a| < \delta$ implies $|f(g(x)) - f(b)| < \varepsilon$.

A more succinct proof uses the sequential characterization of continuity: $a_n \to a$ in $D \Rightarrow g(a_n) \to g(a) \Rightarrow f\big(g(a_n)\big) \to f\big(g(a)\big)$. □

Constant functions and the function $f(x) = x$ are clearly continuous. It follows from 3.3.4 that polynomials and rational functions are continuous. Continuity of trigonometric, logarithmic, and exponential functions will follow from results in Chapter 4. As a consequence, power functions $x^\alpha := e^{\alpha \ln x}$ are continuous, being compositions of continuous functions. Of course, in each case the domain of the function must be carefully specified. The Dirichlet function (3.1.7) is an example of a function that is nowhere continuous.

3.3.6 Theorem. *A monotone function on an open interval I has at most countably many discontinuities.*

Proof. Assume without loss of generality that f is increasing on I. Let D denote the set of discontinuities of f on I. For each $t \in I$, let

$$a_t = \lim_{x \to t^-} f(x) \quad \text{and} \quad b_t = \lim_{x \to t^+} f(x)$$

and let $I_t = (a_t, b_t)$. Clearly, $I_t \neq \emptyset$ iff $t \in D$ (see Figure 3.3). Furthermore, by monotonicity, $s < t \Rightarrow b_s \leq a_t$. Therefore, the sets I_t are pairwise disjoint. For each $t \in D$, choose a rational number r_t in I_t. Since the correspondence $t \to r_t$ is one-to-one and the set of rationals is countable, D is countable. □

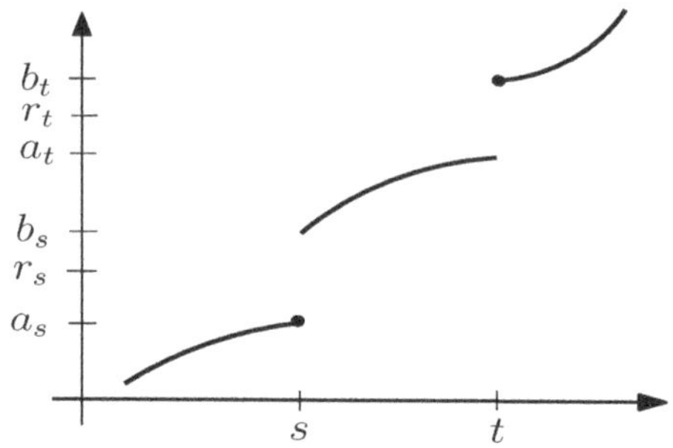

FIGURE 3.3: One-to-one correspondence between $t \in D$ and $r_t \in \mathbb{Q}$.

Exercises

1. Define
$$f(x) = \begin{cases} mx + 3 & \text{if } x < 2, \\ 3x^2 + 7 & \text{if } x > 2. \end{cases}$$
If f is continuous at $x = 2$, find the values of $f(2)$ and m.

2. Find all values of a for which the following function is continuous on $\mathbb{R}$.
$$f(x) = \begin{cases} 3x^2 + 5x - 7 & \text{if } x < a \\ 2x^2 + 2x + 3 & \text{if } x \geq a. \end{cases}$$

3. Let $f : (a, b) \to \mathbb{R}$ and $g : (b, c) \to \mathbb{R}$ be continuous and suppose that
$$\lim_{x \to b^-} f(x) = \lim_{x \to b^+} g(x).$$
Show that there exists a continuous function $h : (a, c) \to \mathbb{R}$ such that $h = f$ on (a, b) and $h = g$ on (b, c).

4. Let g be continuous on $\mathbb{R}$ and let $d(x)$ be the Dirichlet function. Show that $f(x) := g(x)d(x)$ is continuous precisely at the zeros of g.

5. Let f be defined on an open interval I and let $c \in I$. Show that f is continuous at c iff for each strictly increasing sequence $\{a_n\}$ converging to c and each strictly decreasing sequence $\{b_n\}$ converging to c, $f(a_n) \to f(c)$ and $f(b_n) \to f(c)$.

6. Let f be a continuous function on $[a, b]$ and let $\{a_n\}$ be a sequence in $[a, b]$. Prove:

   (a) $f\Big(\limsup_{n \to \infty} a_n\Big) \leq \limsup_{n \to \infty} f(a_n)$ (b) $f\Big(\liminf_{n \to \infty} a_n\Big) \geq \liminf_{n \to \infty} f(a_n)$

   Show that equality holds in each case if f is increasing. Give examples to show that the inequalities may be strict.

7. Let $f_1, \ldots, f_n$ be continuous at x_0. Prove that the functions

$$M_n(x) := \max_{1 \le j \le n} f_j(x) \quad \text{and} \quad m_n(x) := \min_{1 \le j \le n} f_j(x)$$

are continuous at x_0. Give examples to show that the corresponding result is not true for infinitely many functions, where max is replaced sup and min by inf.

8. Let $f : \mathbb{R} \to \mathbb{R}$ be continuous at zero and satisfy $f(x+y) = f(x) + f(y)$ for all $x, y \in \mathbb{R}$. Prove that $f(tx) = tf(x)$ for all $t, x \in \mathbb{R}$. Conclude that $f(x) = f(1)x$ for all $x \in \mathbb{R}$.

9. A function f is *right continuous at* a if $\lim_{x \to a^+} = f(a)$ and *left continuous at* a if $\lim_{x \to a^-} f(x) = f(a)$.

(a) Prove that f is continuous at a iff f is both right and left continuous at a.

(b) Prove that the greatest integer function $\lfloor x \rfloor$ is right continuous on $\mathbb{R}$ but not left continuous at any integer.

(c) Let $\{c_n\}$ be any sequence in $\mathbb{R}$. For $x \in \mathbb{R}$ define

$$f(x) = \sum_{n : c_n \le x} 2^{-n},$$

where the notation indicates that the sum, possibly infinite, is taken over all indices n for which $c_n \le x$. (If there are no such indices, the sum is defined to be 0.) Prove that f is right continuous everywhere. Prove also that f is left continuous at a iff a is not equal to any c_n. (Note that, because the series $\sum_{n=1}^{\infty} 2^{-n}$ converges, the order of summation is irrelevant (6.4.9). Thus $f(x)$ is well-defined.)

(d) Let f be increasing on an interval I. Define g on I by

$$g(x) = \lim_{t \to x^+} f(t) \left(= \inf_{t > x} f(t) \right).$$

Prove that g is increasing and right continuous on I and that g is continuous at a iff f is continuous at a.

10. Define $f : (0,1) \to \mathbb{R}$ by

$$f(x) = \begin{cases} 0 & \text{if } x \text{ is irrational} \\ 1/n & \text{if } x = m/n, \text{ reduced.} \end{cases}$$

Use the sequential characterization of continuity to show that f is continuous precisely at the irrational numbers in $(0,1)$.

11. Let $f : [0,1] \to \mathbb{R}$ have the property that the limit $g(x) := \lim_{t\to x} f(t)$ exists in $\mathbb{R}$ for all $x \in [0,1]$. Prove that

 (a) g is continuous.

 (b) f has at most countably many discontinuities.

 Hint. For (a), use the sequential criterion. For (b), use ideas similar to those used in the proof of 3.3.6.

3.4 Properties of Continuous Functions

3.4.1 Extreme Value Theorem. *If f is continuous on a closed bounded interval $[a,b]$, then f has a maximum and a minimum; that is, there exist $x_m, x_M \in [a,b]$ such that $f(x_m) \le f(x) \le f(x_M)$ for all $x \in [a,b]$.*

Proof. We prove the maximum case. First, we claim that f is bounded above. If not, then for each $n \in \mathbb{N}$ there exists $a_n \in [a,b]$ such that $f(a_n) > n$. By the Bolzano–Weierstrass theorem, $\{a_n\}$ has a convergent subsequence, say $a_{n_k} \to x_0$. But then, by continuity, $f(a_{n_k}) \to f(x_0)$, impossible because $f(a_{n_k}) > n_k \to \infty$. Thus f must be bounded above.

From the first paragraph, $M := \sup\{f(x) : x \in [a,b]\}$ is finite. Using the approximation property for suprema, we can construct sequence $x_n \in [a,b]$ such that $f(x_n) \to M$. By the Bolzano–Weierstrass theorem again, there exists a subsequence x_{n_k} converging to some $x_M \in [a,b]$. By continuity, $f(x_M) = M$ and so $f(x_M)$ is the maximum of f. □

The examples $f(x) = 1/x$ on $(0,1)$ and $f(x) = x$ on $[0,+\infty)$ show that the interval in the theorem must be both closed and bounded.

3.4.2 Definition. A function f is said to have the *intermediate value property* on an interval I if, for each a, $b \in I$ with $a < b$ and each y_0 between $f(a)$ and $f(b)$, there exists an $x_0 \in (a,b)$ such that $f(x_0) = y_0$. Equivalently, $f(I)$ is an interval whenever I is an interval. ◊

3.4.3 Intermediate Value Theorem. *A continuous function f on an interval I has the intermediate value property.*

Proof. Let a, $b \in I$ with $a < b$ and suppose that $f(a) < y_0 < f(b)$. The set $E := \{x \in [a,b] : f(x) < y_0\}$ contains a and is bounded, hence $x_0 := \sup E$ exists and lies in $[a,b]$. By continuity of f at a, E contains an interval $[a, a+\delta)$, hence $x_0 > a$. Since $f(x) < y_0$ for all $x \in E$, 3.1.14 and the continuity of f at x_0 imply that

$$f(x_0) = \lim_{\substack{x\to x_0\\ x\in E}} f(x) \le y_0.$$

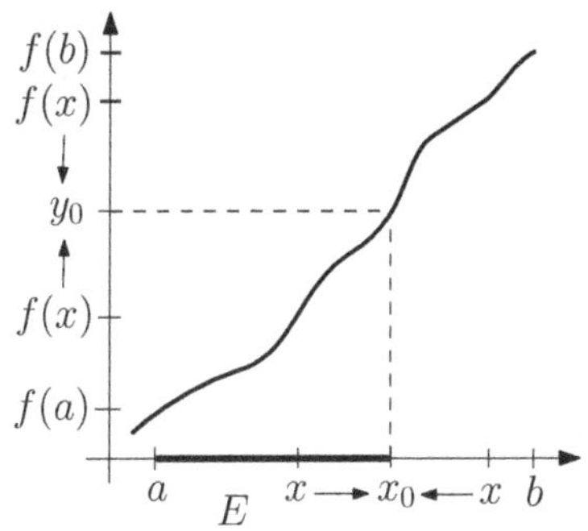

FIGURE 3.4: $y_0 = f(x_0)$.

In particular, $x_0 \neq b$. Similarly, since $f(x) \geq y_0$ for all $x \in (x_0, b)$,

$$f(x_0) = \lim_{x \to x_0^+} f(x) \geq y_0.$$

Therefore, $y_0 = f(x_0)$. Figure 3.4 illustrates the proof. □

Simple examples show that the continuity hypothesis is essential. Of course, there are many discontinuous functions that have the intermediate value property (see Exercise 5). Interestingly, all derivatives have the intermediate value property, whether or not they are continuous (Exercise 4.2.25). Thus a function without the intermediate value property cannot have an antiderivative.

Combining the extreme and intermediate value theorems we obtain

3.4.4 Corollary. *If f is continuous on $[a, b]$, then $f\big([a, b]\big) = [f(x_m), f(x_M)]$.*

3.4.5 Corollary (Existence of nth roots). *For each $b > 0$ and $n \in \mathbb{N}$, the equation $x^n = b$ has a unique positive solution.*

Proof. Let $f(x) = x^n$. Since $\lim_{x \to +\infty} x^n = +\infty$, we may choose $c > 0$ such that $f(c) > b > f(0) = 0$. By the intermediate value theorem, the equation $f(x) = b$ has a positive solution. By Exercise 3.1.12, x^n is strictly increasing on $(0 + \infty)$, hence the solution is unique. □

Here is an example that illustrates another use of the intermediate value theorem.

3.4.6 Example. The equation

$$f(x) := \frac{2\sqrt{x} + \sin(3x^2)}{(x-1)^3} + \frac{5x^2 + e^{2x+7}}{(x-2)^5} = 0$$

has a solution $x = x_0$ between 1 and 2. Indeed, since

$$\lim_{x \to 1+} f(x) = +\infty \quad \text{and} \quad \lim_{x \to 2^-} f(x) = -\infty,$$

there must exist $1 < a < b < 2$ such that $f(a) > 0 > f(b)$. By the intermediate value theorem, $f(x_0) = 0$ for some $x_0 \in (a, b)$. ♢

The example merely asserts the *existence* of solutions of $f(x) = 0$. Once this is established, solutions may be found, or at least approximated, using numerical methods. We describe one such method, the *interval halving process.* Suppose $f(a) < 0 < f(b)$ so that a zero of f lies in (a, b). Bisect the interval $[a, b]$ and compute the values of f at the endpoints of the resulting two intervals. If one of these values is zero, stop. If neither is zero, then for one of the intervals, denote it by $[a_1, b_1]$, the values of f at the endpoints have opposite signs. The intermediate value theorem then implies that a zero of f lies in (a_1, b_1), and we may approximate the zero by either a_1 or b_1. Continuing this process, we may (theoretically) approximate a zero of f to any desired degree of accuracy. The procedure is easily programmable.

Exercises

1. Find an example of a bounded function on $[0, 1]$ with a single discontinuity that has no maximum or minimum.

2. Let f be continuous and positive on $\mathbb{R}$ with $\lim_{x \to \pm\infty} f(x) = 0$. Prove that f has a maximum value on $\mathbb{R}$.

3. Let f be continuous on $\mathbb{R}$ with $\lim_{x \to \pm\infty} f(x) = +\infty$. Prove that f has a minimum value on $\mathbb{R}$.

4. A function f defined on an interval J and taking values in $\overline{\mathbb{R}}$ is said to be *upper (lower) semicontinuous at* $x_0 \in J$ if

$$f(x_0) \geq \limsup_{x \to x_0} f(x) \quad \Big(f(x_0) \leq \liminf_{x \to x_0} f(x)\Big),$$

where the limits are one-sided if x_0 is an endpoint of J. If f is upper (lower) semicontinuous at each point of J, then f is said to be *upper (lower) semicontinuous on J*

(a) Prove that f is upper semicontinuous at x_0 iff $-f$ is lower semicontinuous at x_0.

(b) Prove that f is continuous at x_0 iff it is both upper and lower semicontinuous at x_0.

(c) Show that, at any integer n, $\lfloor x \rfloor$ is upper semicontinuous but not lower semicontinuous.

(d) Let $f(x) = \sin(1/x)$, $x \neq 0$, and $f(0) = a$. Show that f is upper (lower) semicontinuous at 0 iff $a \geq 1$ ($a \leq -1$).

(e) Let f_i be defined on J and upper semicontinuous at x_0 for every i in some index set $\mathfrak{I}$. Define $f(x) = \inf_{i \in \mathfrak{I}} f_i(x)$, $x \in J$. Show that f is upper semicontinuous at x_0. Give an example to show that f may not be continuous at x_0 even if each f_i is continuous on J.

(f) (Semi-extreme value property) Prove: If f is upper (lower) semi-continuous at each point of $[a, b]$, then f is bounded above (below) on $[a, b]$ and there exists $x_0 \in [a, b]$ such that $f(x_0) \ge f(x)$ $(f(x_0) \le f(x))$ for all $x \in [a, b]$.

5. Give an example of a function on $[0, 1]$ with the intermediate value property that is discontinuous at precisely the points $1/n$, $n = 1, 2, \ldots$.

6. Prove that a polynomial P of odd degree maps $\mathbb{R}$ *onto* $\mathbb{R}$. In particular, P has a real zero.

7. Use the intermediate value theorem to show that each of the following equations has a solution in the indicated interval I.

   (a) $\ln x + x = e, \quad I = (1, e)$

   (b) $\sin x = ax, \quad I = (\pi/2, \pi), \ 0 < a < 2/\pi$

   (c) $\tan x = x, \quad I = (n\pi, (n + 1/2)\pi), \ n \in \mathbb{N}$

   (d) $e^x = 4.82 \sin x, \quad I = (0, \pi/2)$ and $I = (\pi/2, \pi)$

   (e) $\dfrac{x^4 + x^2 + 1}{x + 1} + \dfrac{x^3 + 1}{x} + \dfrac{e^{-x} + x}{x - 1} = 0, \quad I = (-1, 0)$ and $I = (0, 1)$

   (f) $\dfrac{e^{1-x^2} - x^2}{\sin x} = \dfrac{2x^2 - 4.8}{\cos x}, \quad I = (0, \pi/2)$.

8. Prove that the equation $e^x = x^n$ $(n \in \mathbb{N})$ has a solution in $(0, +\infty)$ iff $n \ge 3$. *Hint.* Find the minimum of e^x/x^n on $(0, +\infty)$.

9. Let $f : [a, b] \to [a, b]$ be continuous. Prove that there exists $x \in [a, b]$ such that $f(x) = x$.

10. Prove that if $n \in \mathbb{N}$ is odd, then *every* real number has a unique nth root.

11. Let f be continuous and nonzero on $\mathbb{R}$. Let a_0 be arbitrary and define $\{a_n\}$ recursively by $a_n = a_{n-1} + f(a_{n-1}), \quad n \ge 1$. Show that either $a_n \uparrow +\infty$ or $a_n \downarrow -\infty$.

3.5 Uniform Continuity

Recall that a function f is continuous on a set E if for each $y \in E$ and $\varepsilon > 0$ there exists a $\delta > 0$ such that $|f(x) - f(y)| < \varepsilon$ for all x in the domain of f with $|x - y| < \delta$. The number δ typically depends on both ε and y. Removing the dependence on y results in the notion of *uniform continuity*:

3.5.1 Definition. A function f is said to be *uniformly continuous* on a subset E of the domain of f if for each $\varepsilon > 0$ there exists $\delta > 0$ such that

$$|f(x) - f(y)| < \varepsilon \text{ for all } x, y \in E \text{ with } |x - y| < \delta. \qquad \Diamond$$

The following result is sometimes useful in determining whether or not a function is uniformly continuous.

3.5.2 Sequential Characterization of Uniform Continuity. *A function f is uniformly continuous on E iff $f(x_n) - f(y_n) \to 0$ for all sequences $\{x_n\}$ and $\{y_n\}$ in E with $x_n - y_n \to 0$.*

Proof. Let f be uniformly continuous on E and let $\{x_n\}$ and $\{y_n\}$ be sequences in E with $x_n - y_n \to 0$. Given $\varepsilon > 0$, choose $\delta > 0$ so that $|f(x) - f(y)| < \varepsilon$ for all $x, y \in E$ with $|x - y| < \delta$. Next, choose $N \in \mathbb{N}$ such that $|x_n - y_n| < \delta$ for all $n \geq N$. For such n, $|f(x_n) - f(y_n)| < \varepsilon$. Thus $f(x_n) - f(y_n) \to 0$.

Now assume that f is not uniformly continuous on E. Then there exists an $\varepsilon > 0$ and sequences $x_n, y_n \in E$ with $|x_n - y_n| < 1/n$ and $|f(x_n) - f(y_n)| \geq \varepsilon$. Since $x_n - y_n \to 0$ and $f(x_n) - f(y_n) \not\to 0$, f does not satisfy the sequential condition. $\square$

3.5.3 Example. The function $f(x) = 1/x$, $x > 0$, is uniformly continuous on intervals of the form $[r, +\infty)$, $r > 0$, as may be seen from the inequality

$$|f(x) - f(y)| = \frac{|x - y|}{xy} \leq \frac{|x - y|}{r^2}, \quad x, y \geq r.$$

However, f is not uniformly continuous on $(0, +\infty)$. Indeed, if $x_n = 1/2n$ and $y_n = 1/n$, then $x_n - y_n \to 0$ yet $f(x_n) - f(y_n) = n \to +\infty$. $\Diamond$

3.5.4 Theorem. *Let f, g be uniformly continuous on E and let $\alpha, \beta \in \mathbb{R}$. Then*

(a) *$\alpha f + \beta g$ is uniformly continuous on E.*

(b) *If f and g are bounded, then fg is uniformly continuous on E.*

(c) *If $g \neq 0$ and $1/g$ is bounded on E, then $1/g$ is uniformly continuous on E.*

Proof. Part (a) follows easily from the sequential characterization of uniform continuity. For (b), let $M > 0$ such that $|f(x)|, |g(x)| \leq M$ for all $x \in E$. Uniform continuity of fg then follows from the inequalities

$$\begin{aligned} |f(x)g(x) - f(y)g(y)| &\leq |f(x)g(x) - f(y)g(x)| + |f(y)g(x) - f(y)g(y| \\ &\leq M|f(x) - f(y)| + M|g(x) - g(y)|. \end{aligned}$$

For (c), choose $K > 0$ such that $1/|g(x)| < K$ for all $x \in E$. Uniform continuity of $1/g$ then follows from

$$\left| \frac{1}{g(x)} - \frac{1}{g(y)} \right| = \frac{|g(x) - g(y)|}{|g(x)g(y)|} \leq K^2 |g(x) - g(y)|, \quad x, y \in E. \qquad \square$$

The following theorem may be given a short proof based on the sequential criterion for uniform continuity. We leave the details to the reader.

3.5.5 Theorem. *Suppose that g is uniformly continuous on D, f is uniformly continuous on E, and $g(D) \subseteq E$. Then $f \circ g$ is uniformly continuous on D.*

The next theorem makes an important connection between continuity and uniform continuity on closed bounded intervals.

3.5.6 Theorem. *If f is continuous on a closed bounded interval $[a, b]$, then f is uniformly continuous there.*

Proof. We use the sequential characterization of uniform continuity. Let $\{x_n\}$ and $\{y_n\}$ be sequences in $[a, b]$ with $x_n - y_n \to 0$. Suppose, for a contradiction, that $f(x_n) - f(y_n) \not\to 0$. Then $|f(x_n) - f(y_n)| \geq \varepsilon$ for some $\varepsilon > 0$ and infinitely many n and hence for a subsequence of $\{n\}$. Changing notation if necessary, we may suppose that the inequality holds for all n. By the Bolzano–Weierstrass theorem, $\{x_n\}$ has a convergent subsequence, say $x_{n_k} \to x_0$. Since $x_{n_k} - y_{n_k} \to 0$, $y_{n_k} \to x_0$. But then, by continuity, $|f(x_{n_k}) - f(y_{n_k})| \to 0$, impossible. □

The connection between continuity and uniform continuity on open intervals is more complicated. For this we need some definitions.

3.5.7 Definition. A continuous function f on D is said to have a *continuous extension to a set* $D_1 \supseteq D$ if there exists a continuous function $f_1 : D_1 \to \mathbb{R}$ such that $f_1|_D = f$. In the special case $D_1 = D \cup \{a\}$, where $a \notin D$, $f(x)$ is said to have a *removable discontinuity at* $x = a$. ◊

3.5.8 Proposition. *Let f be defined and continuous on D and let a be an accumulation point of D, $a \notin D$. Then f has a removable discontinuity at $x = a$ iff $L := \lim_{\{x \to a,\, x \in D\}} f(x)$ exists in $\mathbb{R}$.*

Proof. The necessity is clear. For the sufficiency, simply set $f(a) = L$ to obtain a continuous extension of f to $D \cup \{a\}$. □

For example, the functions

$$x \sin \frac{1}{x}, \quad \frac{\sin x}{x}, \quad \text{and} \quad \frac{x}{\sqrt{|x|}}$$

defined for $x \neq 0$, all have limits at $x = 0$ and hence have unique continuous extensions to $\mathbb{R}$. On the other hand, since $\lim_{x \to 0^+} \sin(1/x)$ does not exist, the function $\sin(1/x)$ does *not* have a removable discontinuity at $x = 0$.

Here the main result regarding uniform continuity of functions on bounded open intervals.

3.5.9 Theorem. *Let f be continuous on the bounded interval (a, b). The following statements are equivalent:*

(a) $\lim_{x\to a^+} f(x)$ *and* $\lim_{x\to b^-} f(x)$ *exist in* $\mathbb{R}$.

(b) f *has a continuous extension to* $[a, b]$.

(c) f *is uniformly continuous on* (a, b).

Proof. (a) $\Rightarrow$ (b) is immediate from 3.5.8.

(b) $\Rightarrow$ (c): By 3.5.6, a continuous extension g of f to $[a, b]$ is uniformly continuous. Therefore, $f = g|_{(a,b)}$ is uniformly continuous.

(c) $\Rightarrow$ (a): Let $\{a_n\}$ be any sequence in (a, b) converging to a. Then $\{a_n\}$ is Cauchy and since f is uniformly continuous, $\{f(a_n)\}$ is Cauchy (Exercise 7). Therefore, $L := \lim_{n\to\infty} f(a_n)$ exists. We claim that $\lim_{x\to a^+} f(x)$ exists and equals L. To see this, let $\{a'_n\}$ be any sequence in (a, b) converging to a. Then $a_n - a'_n \to 0$, hence, by uniform continuity, $f(a_n) - f(a'_n) \to 0$, so $f(a'_n) \to L$. The assertion now follows from the sequential characterization of limit (3.1.9). A similar argument shows that $\lim_{x\to b^-} f(x)$ exists. □

For example, since $\sin(1/x)$ has no continuous extension to $[0, 1]$, it is not uniformly continuous on $(0, 1]$. On the other hand, for any $p > 0$, $\lim_{x\to 0^+} x^p \sin(1/x) = 0$, hence $x^p \sin(1/x)$ is uniformly continuous on $(0, 1]$. For another example, consider $f(x) = (1 - \cos x)/x$ on $\mathbb{R} \setminus \{0\}$. By l'Hospital's rule, proved in the next chapter, $\lim_{x\to 0} f(x) = \lim_{x\to 0} \sin x = 0$, hence f has a continuous extension to $\mathbb{R}$. Moreover, since $\lim_{x\to\pm\infty} f(x) = 0$, f is uniformly continuous on $\mathbb{R}$ (Exercise 5).

3.5.10 Corollary. *A bounded, continuous, monotone function f on a bounded interval (a, b) is uniformly continuous there.*

Proof. By 3.1.17, $\lim_{x\to a^+} f(x)$ and $\lim_{x\to b^-} f(x)$ exist in $\mathbb{R}$. □

The following result relies on the mean value theorem proved in the next chapter.

3.5.11 Theorem. *If f has a bounded derivative on an interval I, then f is uniformly continuous on I.*

Proof. Let M be a bound for $|f'|$ on I. By the mean value theorem, for any x, $y \in I$ there exists a z between x and y such that $f(x) - f(y) = f'(z)(x - y)$. Thus $|f(x) - f(y)| \le M|x - y|$, which implies uniform continuity. □

For example, $\sin^n x$ and $\cos^n x$ have bounded derivatives for every $n \in \mathbb{N}$, hence are uniformly continuous on $\mathbb{R}$. This also follows from periodicity (see Exercise 11). On the other hand, x^p is not uniformly continuous on $(0, +\infty)$ for $p > 1$. Indeed, if $x_n = n + n^{(1-p)/2}$ and $y_n = n$, then, by the mean value theorem, for each n there exists $z_n \in (y_n, x_n)$ such that

$$x_n^p - y_n^p = \frac{pz_n^{p-1}}{n^{(p-1)/2}} \ge \frac{py_n^{p-1}}{n^{(p-1)/2}} = pn^{(p-1)/2} \to +\infty.$$

Since $x_n - y_n \to 0$, 3.5.2 implies that x^p is not uniformly continuous.

Exercises

1. Find functions f and g with f continuous and g uniformly continuous such that neither $f \circ g$ nor $g \circ f$ is uniformly continuous.

2. Let $r > 0$. Show that the function $f(x) = (3x+2)/(2x-1)$ in 3.1.4 is uniformly continuous on D_r but not on its domain D, where

$$D_r := (-\infty, 1/2 - r] \cup [1/2 + r, +\infty) \text{ and } D = (-\infty, 1/2) \cup (1/2, +\infty).$$

3. Let $a, b > 0$. Give a careful ε, δ proof that each of the following functions is uniformly continuous on $\mathbb{R}$.

   (a) $\sqrt{ax^2+b}$ (b) $1/\sqrt{ax^2+b}$ (c) $|ax+b|$

4. Show that $\ln x$ is uniformly continuous on $(r, +\infty)$ for every $r > 0$ but is not uniformly continuous on $(0, 1)$.

5. Let f be continuous on $[0, \infty)$. Prove that if $\lim_{x \to +\infty} f(x)$ exists and is finite, then f is uniformly continuous on $[0, +\infty)$. Give an example of a bounded continuous function on $[0, +\infty)$ that is not uniformly continuous.

6. Prove that each of the following functions is uniformly continuous on the indicated interval, where $n \in \mathbb{N}$:

   (a) $\sin(1/x)$, $[r, +\infty)$, $r > 0$ (b) $x \sin(1/x)$, $[0, +\infty)$

   (c) $\arctan x$, $(-\infty, +\infty)$ (d) $x^n e^{-x}$, $[0, +\infty)$

   (e) $\cos\sqrt{x^2+1}$, $(-\infty, +\infty)$ (f) x^p, $0 < p \le 1$, $[0, +\infty)$

   (g) $(1+x^n)^{1/n}$, $[0, +\infty)$ (h) $(1+x^n)^{-1/n}$, $[0, +\infty)$.

7. Let f be uniformly continuous on E and let $\{a_n\}$ be a Cauchy sequence in E. Prove that $\{f(a_n)\}$ is Cauchy.

8. Suppose that $f(x)$ is uniformly continuous on $[0, +\infty)$. Prove that the function g is uniformly continuous on $\mathbb{R}$, where

$$g(x) := \begin{cases} f(x) & \text{if } x \ge 0, \\ f(-x) & \text{if } x < 0. \end{cases}$$

9. Let f be uniformly continuous on $\mathbb{R}$. Prove that $f(|x|)$, $|f(x)|$, and $|f(|x|)|$ are uniformly continuous on $\mathbb{R}$.

10. Let f be uniformly continuous on each of the intervals (a, b) and (c, d), where $a < b < c < d$. Prove that f is uniformly continuous on the set $(a, b) \cup (c, d)$. What if $b = c$?

11. Let $f : \mathbb{R} \to \mathbb{R}$ be periodic with period $p > 0$, that is, $f(x+p) = f(x)$ for all $x \in \mathbb{R}$. If f is continuous on $[0, p]$, prove that f is uniformly continuous and bounded on $\mathbb{R}$.

12. Let $f_1, \ldots, f_n$ be uniformly continuous on E. Prove that the functions

$$M(x) := \max_{1 \le j \le n} f_j(x) \quad \text{and} \quad m(x) := \min_{1 \le j \le n} f_j(x)$$

are uniformly continuous on E.

13. Find all values of $p > 0$ for which the function $f(x) = x^{-p} \sin x$, $x > 0$, has a continuous extension to $[0, +\infty)$. Prove that for all such p the extension is uniformly continuous.

14. Let $r > 0$. Prove that $f(x) := \sin(x^p)$ is uniformly continuous on $(r, +\infty)$ iff $p \le 1$.

15. Prove that a uniformly continuous function f on a bounded interval (a, b) is bounded. Give examples to show that the result is not true if (a, b) is unbounded or if f is merely continuous.

16. Give examples to show that parts (b) and (c) of 3.5.4 are not necessarily true if the hypotheses of boundedness are removed.

17. Let f be continuous on $[a, b]$. Prove that

$$g(x) := \sup_{a \le t \le x} f(t)$$

is continuous on $[a, b]$.

18. Let

$$f(x) = (1 - e^{1/x})^{-1}, \quad x \ne 0.$$

Is it possible to define $f(0)$ so that f is continuous on $\mathbb{R}$? What about for the function

$$g(x) = x(1 - e^{1/x})^{-1}, \quad x \ne 0?$$

Chapter 4

Differentiation on $\mathbb{R}$

The notion of rate of change of one quantity with respect to another is fundamental to many disciplines. It is expressed mathematically as the derivative of a function. In this chapter we establish the main properties of this important construct.

4.1 Definition of Derivative and Examples

4.1.1 Definition. A real-valued function f defined in a neighborhood of $a \in \mathbb{R}$ is said to be *differentiable at* a if the limit

$$f'(a) = Df(a) = \left.\frac{df}{dx}\right|_a := \lim_{x \to a} \frac{f(x) - f(a)}{x - a} = \lim_{h \to 0} \frac{f(a+h) - f(a)}{h}$$

exists in $\mathbb{R}$. The limit is then called the *derivative of* f *at* a. If f is differentiable at each member of a set E, then f is said to be *differentiable on* E and the function

$$f' = Df = \frac{df}{dx}$$

is called the *derivative of* f *on* E. If f' is continuous on E, then f is said to be *continuously differentiable on* E. ◊

It follows immediately from the definition that the derivative of a constant function is zero. Here are some nontrivial examples.

4.1.2 Example. We prove the following special cases of the power rule (the general power rule will be proved later): Let $n \in \mathbb{N}$ and $r = n$ or $1/n$. Then

$$Dx^r = rx^{r-1}.$$

The case $r = n$ is obtained by letting $h \to 0$ in the identity

$$\frac{(x+h)^n - x^n}{h} = \sum_{j=1}^{n} (x+h)^{n-j} x^{j-1}$$

(see Exercise 1.1.4). Each term in the sum tends to x^{n-1}, and since there are n terms, the formula follows.

DOI: 10.1201/9781003629382-4

For the case $r = 1/n$, we assume $x \neq 0$, and additionally, $x > 0$ if n is even. Consider the identity

$$\frac{(x+h)^{1/n} - x^{1/n}}{h} = \left[\sum_{j=1}^{n} (x+h)^{1-j/n} x^{(j-1)/n}\right]^{-1}$$

(see Exercise 1.3.15). As $h \to 0$, the term in square brackets tends to $nx^{1-1/n}$, which gives the desired formula. $\Diamond$

For the next example and indeed for the remainder of the book, we use the standard geometric definitions of cosine and sine as coordinates of points on the unit circle.[1] From these definitions one can derive the usual trigonometric identities, which we shall employ as needed.

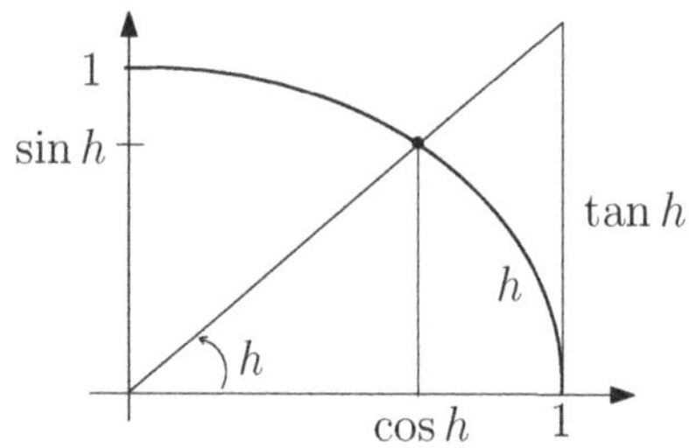

FIGURE 4.1: $\sin h < h < \tan h$.

4.1.3 Example. $D \sin x = \cos x$. To verify this we use the inequalities

$$\sin h < h < \tan h, \ 0 < h < \pi/2,$$

which may be derived with the help of Figure 4.1. It follows from these and the identity $\sin^2 h + \cos^2 h = 1$ that

$$\sqrt{1 - h^2} < \sqrt{1 - \sin^2 h} = \cos h < \frac{\sin h}{h} < 1, \quad 0 < h < \pi/2. \qquad (4.1)$$

Since $\sin(-h) = -\sin h$ and $\cos(-h) = \cos h$, (4.1) holds for $-\pi/2 < h < 0$ as well. By the squeeze principle,

$$\lim_{h \to 0} \cos h = \lim_{h \to 0} \frac{\sin h}{h} = 1.$$

From this and the calculation

$$\frac{\cos h - 1}{h} = \frac{\cos^2 h - 1}{h(\cos h + 1)} = -\left(\frac{\sin h}{h}\right)^2 \frac{h}{(\cos h + 1)}$$

[1]A more rigorous approach to the calculus of trigonometric functions may be based on the inverse sine function. We briefly describe this approach in Section 4.4.

we see that

$$\lim_{h\to 0}\frac{\cos h - 1}{h} = 0.$$

Therefore,

$$\begin{aligned}\frac{\sin(x+h)-\sin x}{h} &= \frac{\sin x\cos h + \cos x \sin h - \sin x}{h}\\ &= \frac{\cos h - 1}{h}\sin x + \frac{\sin h}{h}\cos x\\ &\to \cos x \quad\text{as}\quad h\to 0. \end{aligned}$$ ◊

It is sometimes necessary to consider *one-sided derivatives*, which are defined by using one-sided limits in 4.1.1. Specifically, the *left-hand* and *right-hand* derivatives are, respectively,

$$\begin{aligned} D_\ell f(a) = f'_\ell(a) &:= \lim_{x\to a^-}\frac{f(x)-f(a)}{x-a} \quad\text{and}\\ D_r f(a) = f'_r(a) &:= \lim_{x\to a^+}\frac{f(x)-f(a)}{x-a}. \end{aligned}$$

From the general theory of limits, a function is differentiable at a iff it has equal right-hand and left-hand derivatives at a. For example, at $x=0$ the function $f(x)=|x|$ has right-hand derivative 1 and left-hand derivative -1 and so is not differentiable there.

Although we shall have no need to do so, one may even consider the more general expressions

$$\liminf_{\substack{x\to a\\ x\in E}}\frac{f(x)-f(a)}{x-a} \quad\text{and}\quad \limsup_{\substack{x\to a\\ x\in E}}\frac{f(x)-f(a)}{x-a},$$

where a is an accumulation point of E. In particular, the *Dini derivates* are obtained by taking E to be intervals of the form (c,a) and (a,c).

The following proposition provides a useful characterization of differentiability. It asserts that for x near a, $f(x)$ is approximated by the linear function $y=f(a)+f'(a)(x-a)$, the equation of the tangent line at $(a, f(a))$.

4.1.4 Proposition. *Let f be defined in a neighborhood $\mathcal{N}(a)$ of a. Then f is differentiable at a iff there exists a function η on $\mathcal{N}(a)$, continuous at a, such that*

$$f(x) = f(a) + \eta(x)(x-a) \quad \textit{for all} \quad x\in\mathcal{N}(a).$$

In this case, $f'(a)=\eta(a)$.

Proof. If such a function η exists, then

$$\frac{f(x)-f(a)}{x-a} = \eta(x)\to\eta(a) \text{ as } x\to a,$$

hence $f'(a)$ exists and equals $\eta(a)$. Conversely, if f is differentiable at a, define

$$\eta(x) = \begin{cases} \dfrac{f(x) - f(a)}{x - a} & \text{if } x \in \mathcal{N}(a) \setminus \{a\}, \\ f'(a) & \text{if } x = a. \end{cases}$$

Then η has the required properties. □

4.1.5 Corollary. *If f is differentiable at a, then f is continuous there.*

Proof. Simply note that $f(x) = f(a) + \eta(x)(x - a) \to f(a)$ as $x \to a$. □

The example $|x|$ considered above shows that the converse of the corollary is false: $|x|$ is continuous at 0 but not differentiable there. It is a remarkable fact that there are continuous functions on $\mathbb{R}$ that are *nowhere* differentiable (see 8.9.7).

4.1.6 Theorem. *If $c \in \mathbb{R}$ and f and g are differentiable a, then so are $f + g$, cf, fg, and f/g, the last provided that $g(a) \neq 0$. Moreover, in this case,*

(a) $(f + g)'(a) = f'(a) + g'(a)$, (b) $(cf)'(a) = cf'(a)$,

(c) $(fg)'(a) = f(a)g'(a) + f'(a)g(a)$, (d) $\left(\dfrac{f}{g}\right)'(a) = \dfrac{g(a)f'(a) - f(a)g'(a)}{g^2(a)}$.

Proof. We prove only (d). Let $h = f/g$. Since g is continuous at a and $g(a) \neq 0$, h is defined in a neighborhood $\mathcal{N}(a)$ on which g is not 0. For $x \in \mathcal{N}(a) \setminus \{a\}$, a little algebra shows that

$$\frac{h(x) - h(a)}{x - a} = \frac{g(a)\dfrac{f(x) - f(a)}{x - a} - f(a)\dfrac{g(x) - g(a)}{x - a}}{g(x)g(a)}.$$

Letting $x \to a$, using the continuity of g at a, yields the formula in (d). □

The preceding theorem, together with 4.1.2 and 4.1.3, show that polynomials, rational functions, and trigonometric functions are differentiable (see Exercise 2). The following important result will yield additional examples.

4.1.7 Chain Rule. *Let g be differentiable at a and let f be differentiable at $g(a)$. Then $f \circ g$ is differentiable at a and $(f \circ g)'(a) = f'(g(a))g'(a)$.*

Proof. Set $b := g(a)$. By 4.1.4, there exists a function η, defined in a neighborhood $\mathcal{N}(b)$ of b and continuous at b with $\eta(b) = f'(b)$, such that

$$f(y) = f(b) + \eta(y)(y - b), \quad y \in \mathcal{N}(b). \tag{4.2}$$

Since g is continuous at a, we may choose a neighborhood $\mathcal{N}(a)$ of a such that $g(\mathcal{N}(a)) \subseteq \mathcal{N}(b)$. Then $f \circ g$ is defined on $\mathcal{N}(a)$, and by (4.2) with $y = g(x)$,

$$\frac{f(g(x)) - f(g(a))}{x - a} = \eta(g(x))\,\frac{g(x) - g(a)}{x - a}, \quad x \in \mathcal{N}(a) \setminus \{a\}.$$

Letting $x \to a$ produces the desired formula. □

The formula $(f \circ g)'(x) = f'(g(x))g'(x)$ is sometimes easier to apply when written in Leibniz notation:

$$\frac{dy}{dx} = \frac{dy}{du}\frac{du}{dx}, \quad \text{where } y = f(u) \text{ and } u = g(x).$$

4.1.8 Example. The power rule

$$Dx^r = rx^{r-1}, \quad r \in \mathbb{Q},$$

follows from 4.1.2 and the chain rule: Let $r = m/n$, $m, n \in \mathbb{N}$, and set $u = x^{1/n}$ and $y = u^m$. Then $y = x^r$ and

$$\frac{dy}{dx} = \frac{dy}{du}\frac{du}{dx} = mu^{m-1}\frac{1}{n}x^{(1/n)-1} = \frac{m}{n}x^{(m/n)-1} = rx^{r-1}.$$

The case $r < 0$ follows from the quotient rule. ◊

Higher-order derivatives of $y = f(x)$ are defined inductively by

$$f'' = D^2 f = \frac{d^2y}{dx^2} := \frac{d}{dx}\frac{dy}{dx},$$

$$\vdots$$

$$f^{(n)} = D^n f = \frac{d^n f}{dx^n} := \frac{d}{dx}\frac{d^{n-1}f}{dx^{n-1}}.$$

By convention, we set $f^{(0)} = D^0 f = f$.

Exercises

1. Use the limit definition to find the derivative of

   (a) $x^2 + x + 1$ (b) $\sqrt{2x+1}$ (c) $\dfrac{1}{x^2+1}$ (d) $\dfrac{1}{\sqrt{3x+2}}$

2. Use the techniques of 4.1.3 to find the derivative of $\cos x$. Use rules of differentiation to obtain the derivatives of $\tan x$, $\cot x$, $\sec x$, and $\csc x$.

3. Use rules of differentiation to find f' for each of the functions f:

   (a) $\sqrt[3]{5x+7}\,\sqrt[5]{3x+2}$ (b) $\left(\dfrac{2x+5}{7x+2}\right)^{2/3}$ (c) $\sin\left(\dfrac{x^2-1}{x^2+1}\right)$

   (d) $\dfrac{\sin^2 x - 1}{\sin^2 x + 1}$ (e) $\tan\left[\cos(1/x)\right]$ (f) $\sqrt{ax + \sqrt{bx+c}}$

4. Assuming that y is a differentiable function of x that satisfies the given equation, use the rules of differentiation to find $\dfrac{dy}{dx}$:

   (a) $x^3 + y^3 - xy = 1$ (b) $\sin(xy^2) + x^2 = 1$ (c) $\tan(x+y) + y^2 = x$.

5. Let $f(x) = x^n|x|$, $n \in \mathbb{N}$. Find $f^{(n-1)}$ and $f^{(n)}$.

6. Let $f(x) = x^m \lfloor x \rfloor$, $m \in \mathbb{N}$. Find $f'_\ell(n)$ and $f'_r(n)$, $n \in \mathbb{Z}$.

7. Find all values of a, b, such that f' exists on $\mathbb{R}$, where
$$f(x) = \begin{cases} ax^2 + bx + a/x & \text{if } x > 1, \\ x^3 & \text{if } x \le 1. \end{cases}$$

8. Find all values of a, b, and c such that f' is continuous on $(0, +\infty)$, where
$$f(x) = \begin{cases} ax^2 + bx & \text{if } x > 1, \\ c\sqrt{x} & \text{if } 0 < x \le 1. \end{cases}$$

9. Let
$$f(x) = \begin{cases} ax^2 + bx + c & \text{if } x > 1, \\ x^3 & \text{if } x \le 1. \end{cases}$$
Find all values of a, b, and c such that
   (a) f is continuous on $\mathbb{R}$
   (b) f is differentiable on $\mathbb{R}$
   (c) f' is continuous on $\mathbb{R}$
   (d) f'' exists on $\mathbb{R}$.

10. Find all values of c such that $f'(c)$ exists, where
$$f(x) = \begin{cases} ax - 4 & \text{if } x > c, \\ 9x^2 & \text{if } x \le c. \end{cases}$$
Is f' continuous at these values?

11. Let f be differentiable at a. Use the limit definition of derivative to calculate
   (a) $\displaystyle\lim_{h \to 0} \frac{f(a + 5\sin h) - f(a + 2\sin h)}{h}$ (b) $\displaystyle\lim_{h \to 0} \frac{f(a + h^2) - f(a - h)}{h}$.

12. Let g be differentiable on an open interval I and let $f(x) = g(x)d(x)$, where $d(x)$ is the Dirichlet function (3.1.7). Let a be a zero of g. Prove that $f'(a)$ exists iff a is a zero of g'.

13. Let f be differentiable at c and let $\{a_n\}$ and $\{b_n\}$ be sequences such that $a_n < c < b_n$ and a_n, $b_n \to c$. Prove that
$$\lim_{n \to \infty} \frac{f(b_n) - f(a_n)}{b_n - a_n} = f'(c).$$

14. Let f be differentiable and increasing on (a, b). Prove that $f'(x) \ge 0$ for all $x \in (a, b)$.

15. Let f be differentiable at a and nonnegative in a neighborhood of a with $f(a) = 0$. Prove that $f'(a) = 0$.

16. Prove Leibniz's rule: If f and g are n times differentiable, then

$$D^n(fg) = \sum_{k=0}^{n} \binom{n}{k} (D^k f)\,(D^{n-k} g).$$

17. Prove that if f has right-hand and left-hand derivatives at a (not necessarily equal), then f is continuous at a.

18. Assuming that f, g, and h have the necessary differentiability, find general formulas for

    (a) $D\big[f \circ (gh)\big]$ (b) $D\big[f \circ (g/h)\big]$ (c) $D^2\big[f \circ g\big]$ (d) $D\big[f \circ g \circ h\big]$.

19. Find a formula for the nth derivative of

    (a) $1/x$. (b) $1/\sqrt{x}$ (c) xe^x (d) xe^{-x}.

20. Find all values of $p \in \mathbb{R}$ for which the function

$$f(x) = \begin{cases} |x|^p \sin(1/x) & \text{if } x \neq 0, \\ 0 & \text{otherwise} \end{cases}$$

    is (a) continuous, (b) differentiable, (c) continuously differentiable on $\mathbb{R}$.

21. Define $f(0) = 0$ and $f(x) = x^m \sin x^n$, $x \neq 0$, where $m \in \mathbb{Z}$, $n \in \mathbb{N}$. For what values of m and n does $f'(0)$ exist? For which of these values is f' continuous on $\mathbb{R}$?

22. A function f defined on a symmetric neighborhood $(-a, a)$ of 0 is said to be *odd* if $f(-x) = -f(x)$ and *even* if $f(-x) = f(x)$.

    (a) Prove that any function $h : (-a, a) \to \mathbb{R}$ is the sum of an even function f and an odd function g.

    (b) Prove that if f is differentiable and odd (even), then f' is even (odd).

    (c) Is the converse true? That is, if f' is even (odd), is f odd (even)?

23. Let f_j, g_j, and h_j be differentiable, $j = 1, 2, 3$. Prove that

$$\begin{vmatrix} f_1 & f_2 \\ g_1 & g_2 \end{vmatrix}' = \begin{vmatrix} f_1' & f_2' \\ g_1 & g_2 \end{vmatrix} + \begin{vmatrix} f_1 & f_2 \\ g_1' & g_2' \end{vmatrix} \quad \text{and}$$

$$\begin{vmatrix} f_1 & f_2 & f_3 \\ g_1 & g_2 & g_3 \\ h_1 & h_2 & g_3 \end{vmatrix}' = \begin{vmatrix} f_1' & f_2' & f_3' \\ g_1 & g_2 & g_3 \\ h_1 & h_2 & h_3 \end{vmatrix} + \begin{vmatrix} f_1 & f_2 & f_3 \\ g_1' & g_2' & g_3' \\ h_1 & h_2 & h_3 \end{vmatrix} + \begin{vmatrix} f_1 & f_2 & f_3 \\ g_1 & g_2 & g_3 \\ h_1' & h_2' & h_3' \end{vmatrix}.$$

4.2 The Mean Value Theorem

The mean value theorem relates the average rate of change of a function to its instantaneous rate of change. It is one of the most useful theorems in analysis and will play a central role in the proof of the fundamental theorem of calculus in Chapter 5. The proof is based on the existence of local extrema.

4.2.1 Definition. A function f is said to have a *local maximum* (*local minimum*) at c if f is defined on an open interval I containing c and $f(x) \le f(c)$ ($f(x) \ge f(c)$) for all $x \in I$. In either case, f is said to have a *local extremum at* c. ◊

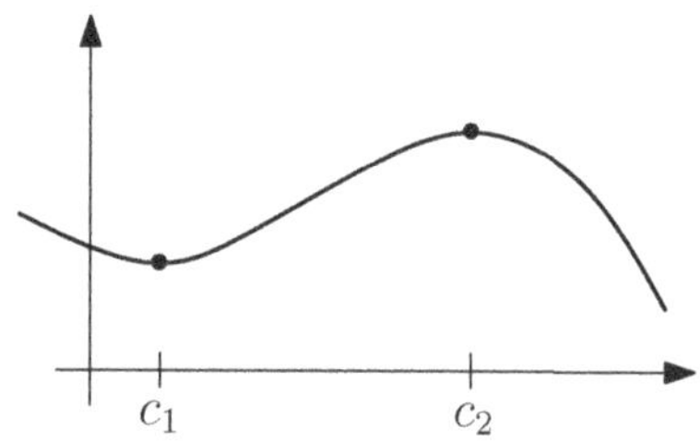

FIGURE 4.2: Local extrema of f.

4.2.2 Local Extremum Theorem. *If f has a local extremum at c and if f is differentiable at c, then $f'(c) = 0$.*

Proof. Suppose that f has a local maximum at c. Let I be an open interval containing c such that $f(x) \le f(c)$ for all $x \in I$. Then

$$\frac{f(x) - f(c)}{x - c} \begin{cases} \ge 0 & \text{if } x \in I \text{ and } x < c \\ \le 0 & \text{if } x \in I \text{ and } x > c. \end{cases}$$

It follows that the left-hand derivative of f at c is ≥ 0 and the right-hand derivative is ≤ 0. Since the derivative is assumed to exist, $f'(c) = 0$. The proof for a local minimum is similar. □

4.2.3 Rolle's Theorem. *Let f be continuous on $[a, b]$ and differentiable on (a, b). If $f(a) = f(b)$, then there exists a point $c \in (a, b)$ such that $f'(c) = 0$.*

Proof. By the extreme value theorem, there exist $x_m, x_M \in [a, b]$ such that $f(x_m) \le f(x) \le f(x_M)$ for all $x \in [a, b]$. If $f(x_m) = f(x_M)$, then f is a constant function and the assertion of the theorem holds trivially. If $f(x_m) \ne f(x_M)$, then either $x_m \in (a, b)$ or $x_M \in (a, b)$, and the conclusion follows from the local extremum theorem. □

The following result will be used to prove the mean value theorem. It is also the key ingredient in the proof of l'Hospital's rule in Section 4.5.

4.2.4 Cauchy Mean Value Theorem. *Let f and g be continuous on $[a, b]$ and differentiable on (a, b). Then there exists a point $c \in (a, b)$ such that*

$$[f(b) - f(a)]g'(c) = [g(b) - g(a)]f'(c).$$

Proof. The function

$$h(x) := [f(b) - f(a)]g(x) - [g(b) - g(a)]f(x)$$

is continuous on $[a, b]$, differentiable on (a, b), and satisfies $h(a) = h(b)$. By Rolle's theorem, $h'(c) = 0$ for some $c \in (a, b)$, which is the assertion of the theorem. □

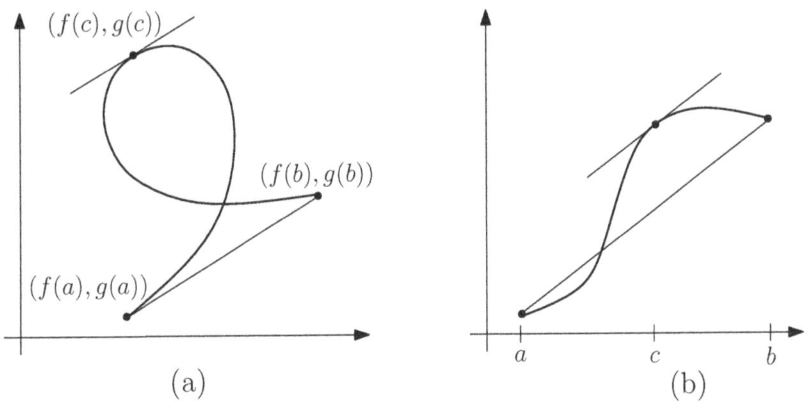

FIGURE 4.3: (a) Cauchy mean value theorem. (b) Mean value theorem.

If $f(a) \neq f(b)$ and $f'(x) \neq 0$ on (a, b), then the conclusion of 4.2.4 may be written

$$\frac{g(b) - g(a)}{f(b) - f(a)} = \frac{g'(c)}{f'(c)}.$$

For smooth functions f and g, this equation asserts that at some point $\big(f(c), g(c)\big)$ on the curve given parametrically by the equations $x = f(t)$ and $y = g(t)$, the line through the endpoints $(f(a), g(a))$ and $(f(b), g(b))$ is parallel to the line tangent to the curve at $\big(f(c), g(c)\big)$. See Figure 4.3(a).

Taking $g(x) = x$ in the Cauchy mean value theorem yields the standard mean value theorem (Figure 4.3(b)):

4.2.5 Mean Value Theorem. *If f is continuous on $[a, b]$ and differentiable on (a, b), then there exists $c \in (a, b)$ such that*

$$\frac{f(b) - f(a)}{b - a} = f'(c).$$

4.2.6 Corollary. *Let $f(x)$ and $g(x)$ be differentiable on an open interval I such that $f'(x) = g'(x)$ for all $x \in I$. Then there exists a constant k such that $f = g + k$ on I.*

Proof. Let a, $b \in I$. By the mean value theorem applied to $h := f - g$, there exists $c \in (a, b)$ such that $h(a) - h(b) = h'(c)(a - b)$. Since $h' = 0$, $h(a) = h(b)$. Since a and b were arbitrary, h must be constant. □

4.2.7 Corollary. *Let f be differentiable on an open interval I.*

(a) *If $f' \geq 0$ ($f' > 0$) on I, then f is increasing (strictly increasing) on I.*

(b) *If $f' \leq 0$ ($f' < 0$) on I, then f is decreasing (strictly decreasing) on I.*

Proof. We prove (a) for the strictly increasing case. Let a, $b \in I$, $a < b$. By the mean value theorem, $f(b) - f(a) = f'(c)(b - a)$ for some $c \in (a, b)$. Since $f'(c) > 0$, $f(b) > f(a)$. □

Exercises

1. Show that $\cos x = \sqrt{x} - 1$ has exactly one solution x in the interval $(0, \pi/2)$.

2. Find an interval I such that for each $c \in I$, $\sin x = x^2/2 + x + c$ has exactly one solution x in the interval $(0, \pi/2)$.

3. Show that $f(x) = x^4 - 4x^3 + 4x^2 + c$ has at most one zero in the interval $(1, 2)$. For what interval of values of the parameter c does f have *exactly* one zero in $(1, 2)$?

4. Let f have k derivatives and n distinct zeros on an interval I. Prove that $f^{(k)}$ has at least $n - k$ distinct zeros in I.

5. Let f have a continuous second derivative on $[-1, 3]$, $f(1) = 0$, and set $g(x) = x^2 f(x)$. Prove that g'' has at least one zero in $[-1, 2]$. *Hint.* Consider the function $g_n(x) := x(x + 1/n)f(x)$.

6. Let $P(x)$ be a polynomial of degree n and let $a \neq 0$. Prove that the equation $e^{ax} = P(x)$ has at most $n + 1$ solutions.

7. Let $P(x)$ be a polynomial of degree n and let $a \neq 0$. Prove that the equation $\sin(ax) = P(x)$ has at most $n + 1$ solutions in the interval $(-\pi/2|a|, \pi/2|a|)$.

8. Prove Bernoulli's inequality: $(1 + x)^r \geq 1 + rx$ for all $x \geq -1$ and all rational numbers $r \geq 1$. (Cf. Exercise 1.4.10.)

9. Let f and g be continuous on $[a, b]$ and differentiable on (a, b) such that $|f'| \leq |g'|$. If g' is never zero on (a, b), prove that

$$|f(x) - f(y)| \leq |g(x) - g(y)| \quad \text{for all } x,\ y \in [a, b].$$

10. Let f and g be differentiable on an open interval I and let $a, b \in I$ with $a < b$. Prove that if $f(a) = g(a)$ and $f' > g'$ on (a, b), then $f > g$ on (a, b). Use this to show that

    (a) $\ln x < x - 1$ on the interval $(1, +\infty)$

    (b) $\sin x < x$ on the interval $(0, \pi/2)$

    (c) $\cos x > 1 - x$ on the interval $(0, \pi/2)$

    (d) $\tan x > x$ on the interval $(0, \pi/2)$

    (e) $e^x > 1 + x + x^2/2! + \cdots + x^n/n!$ on the interval $(0, +\infty)$. (Use induction.)

11. Show that $\dfrac{\sin x}{x}$ is a decreasing function on $(0, \pi/2)$.

12. Show that on $(0, \pi/2)$

    (a) $x \sin x + \cos x > 1$

    (b) $x \sin x + p \cos x < p$, $p \geq 2$

    (c) $x^{-1}(1 - \cos x)$ is increasing

    (d) $x^{-2}(1 - \cos x)$ is decreasing.

13. Let $a, b, p > 0$, and for $x \geq 0$ define $f(x) = a^p + x^p - (a + x)^p$. Show that for $x > 0$,

$$f'(x) \begin{cases} > 0 & \text{if } 0 < p < 1, \\ < 0 & \text{if } p > 1. \end{cases}$$

    Conclude that

$$(a + b)^p \begin{cases} < a^p + b^p & \text{if } 0 < p < 1, \\ > a^p + b^p & \text{if } p > 1. \end{cases}$$

14. Let f and g have derivatives of order n on an open interval I and let $a \in I$. Suppose that

$$f^{(j)}(a) = g^{(j)}(a) = 0, \quad j = 0, \ldots, n - 1, \quad \text{and}$$

$$f^{(j)}(x) g^{(j)}(x) \neq 0 \quad \text{for } x > a \text{ and } j = 0, \ldots, n.$$

    Prove that for any $b \in I$ with $b > a$ there exists $c \in (a, b)$ such that

$$\frac{f(b)}{g(b)} = \frac{f^{(n)}(c)}{g^{(n)}(c)}.$$

15. Suppose that f has a local maximum at c. Prove that

$$\liminf_{x \to c^-} \frac{f(x) - f(c)}{x - c} \geq 0 \geq \limsup_{x \to c^+} \frac{f(x) - f(c)}{x - c}.$$

16. Let f and g be continuous on $[a, b]$, differentiable on (a, b) and let $f(a) = f(b) = 0$. Show that there exists $c \in (a, b)$ such that $f'(c) = g'(c)f(c)$.

17. Show that for any polynomial $P(x)$ there exist finitely many intervals with union $\mathbb{R}$ such that P is strictly monotone on each interval.

18. Suppose that f has the property

$$|f(x) - f(y)| \le c|x - y|^{1+\varepsilon} \text{ for all } x,\ y \in \mathbb{R},$$

where $c,\ \varepsilon > 0$. Prove that f is constant.

19. Let f have a bounded derivative on $\mathbb{R}$. Prove that for sufficiently large r, the function $g(x) := rx + f(x)$ is one-to-one and maps $\mathbb{R}$ onto $\mathbb{R}$.

20. Suppose $f > 0$ on $(1, +\infty)$ and $\lim_{x\to+\infty} xf'(x)/f(x) \in (1, +\infty)$. Prove that $x/f(x)$ is decreasing on $(b, +\infty)$ for some $b > 1$.

21. Let f be twice differentiable on $(0, a)$, $f'' \ge 0$, and $\lim_{x\to 0+} f(x) = 0$. Prove that $f(x)/x$ is increasing on $(0, a)$. Show that the conclusion is false if the hypothesis $f'' \ge 0$ is dropped.

22. Let $g(x) = x^2 \sin(1/x)$ if $x \ne 0$ and $g(0) = 0$. Set $f(x) = x + g(x)$. Show that $f'(0) > 0$ but f is not monotone on any neighborhood of 0.

23. Let $\lim_{x\to+\infty} f'(x) = 0$. Prove that if $g \ge c > 0$ on $(a, +\infty)$, then

$$\lim_{x\to+\infty} \left[f\left(x + g(x)\right) - f(x)\right] = 0.$$

24. Let f be differentiable on $\mathbb{R}$ with $\sup_{x\in\mathbb{R}} |f'(x)| < 1$. Prove that the sequence $\{x_n\}$ defined by $x_{n+1} = f(x_n)$ converges, where x_1 is arbitrary. Conclude that f has a unique *fixed point*; that is, there exists a unique $x \in \mathbb{R}$ such that $f(x) = x$.

25. Suppose f is differentiable on an open interval I. Show that f' has the intermediate value property. Conclude that if $f'(x) \ne 0$ on I, then f is strictly monotone on I. *Hint.* Apply the extreme value theorem to the function $g(x) = f(x) - y_0(x - a)$, $a \le x \le b$.

26. Let f be differentiable on $I := (1, +\infty)$. Prove that if f' has finitely many zeros in I, then $\lim_{x\to+\infty} f(x)$ exists in $\overline{\mathbb{R}}$.

27. Let f and g have continuous derivatives on an interval I with $g' \ne 0$ and let $a_j,\ b_j \in I$ with $a_j < b_j$, $j = 1, \ldots, n$. Prove that there exists $c \in I$ such that

$$\sum_{j=1}^{n} [f(b_j) - f(a_j)]g'(c) = \sum_{j=1}^{n} [g(b_j) - g(a_j)]f'(c).$$

28. A function f is said to be *uniformly differentiable* on an open interval I if, given $\varepsilon > 0$, there exists $\delta > 0$ such that

$$\left|\frac{f(x)-f(y)}{x-y} - f'(y)\right| < \varepsilon$$

for all x and y in I with $0 < |x-y| < \delta$. Prove that f is uniformly differentiable on I iff f' exists and is uniformly continuous on I.

29. Generalize the preceding exercise as follows: Let f and g be differentiable on an open interval I with $g' \neq 0$ on I. Prove that f'/g' is uniformly continuous on I iff, given $\varepsilon > 0$, there exists a $\delta > 0$ such that

$$\left|\frac{f(x)-f(y)}{g(x)-g(y)} - \frac{f'(y)}{g'(y)}\right| < \varepsilon,$$

for all x and y in I with $0 < |x-y| < \delta$.

30. Let f be differentiable on $[a, +\infty)$ and suppose that the zeros of f' form a strictly increasing sequence $a_n \uparrow +\infty$. Prove that if $L := \lim_n f(a_n)$ exists in $\mathbb{R}$, then $\lim_{x\to+\infty} f(x) = L$.

31. Prove that a function f is continuously differentiable on an open interval I iff there exists a continuous function φ on I^2 such that

$$f(x) - f(y) = \varphi(x,y)(x-y) \quad \text{for all } x,\ y \in I.$$

32. Let f be continuous on $(-r, r)$ and differentiable on $(-r, 0) \cup (0, r)$. If $\lim_{x\to 0} f'(x)$ exists, prove that $f'(0)$ exists and f' is continuous at 0.

---

*4.3 Convex Functions

4.3.1 Definition. A function f is said to be *convex on an interval* (a, b) if

$$f\big((1-t)u + tv\big) \le (1-t)f(u) + tf(v) \tag{4.3}$$

for all $a < u < v < b$ and all $t \in [0, 1]$. f is *concave* if $-f$ is convex. ◊

For example, $|x|$ is convex on $\mathbb{R}$, as is easily established using the triangle inequality.

To see the geometric significance of convexity, let $L_{uv} : [u, v] \to \mathbb{R}$ denote the function whose graph is the line segment from $(u, f(u))$ to $(v, f(v))$. Since a typical point on the line segment may be written

$$(1-t)\big(u, f(u)\big) + t\big(v, f(v)\big) = \big((1-t)u + tv, (1-t)f(u) + tf(v)\big), \quad t \in [0,1],$$

we see that

$$L_{uv}\big((1-t)u+tv\big) = (1-t)f(u)+tf(v),$$

which is the right side of (4.3). This shows that f is convex iff the line segment connecting any two points on the graph of f lies above the part of the graph between the two points (See Figure 4.4)

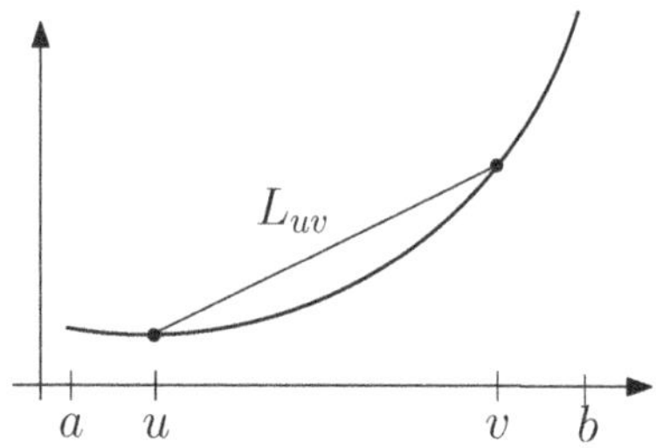

FIGURE 4.4: Convex function.

Now let x be in the open interval (u,v). Then for some $t \in (0,1)$,

$$x = (1-t)u+tv = t(v-u)+u = (1-t)(u-v)+v.$$

Solving for t and $1-t$ we have

$$t = (x-u)/(v-u) \quad \text{and} \quad 1-t = (v-x)/(v-u).$$

Therefore, we may write

$$L_{uv}(x) = f(u)\frac{v-x}{v-u} + f(v)\frac{x-u}{v-u}, \qquad a<u<x<v<b.$$

It follows that f is convex on (a,b) iff

$$f(x) \le f(u)\frac{v-x}{v-u} + f(v)\frac{x-u}{v-u} \quad \text{for all } a<u<x<v<b. \tag{4.4}$$

4.3.2 Theorem. *If $f:(a,b)\to\mathbb{R}$ has an increasing derivative, then f is convex. In particular, f is convex if $f''\ge 0$.*

Proof. Let $a<u<x<v<b$. By the mean value theorem applied to f on each of the intervals $[u,x]$ and $[x,v]$, there exist points $y\in\big(u,x\big)$ and $z\in\big(x,v\big)$ such that

$$\frac{f(x)-f(u)}{x-u} = f'(y) \le f'(z) = \frac{f(v)-f(x)}{v-x}.$$

Solving the inequality for $f(x)$ yields (4.4). □

For example, x^{2n} is convex on $\mathbb{R}$ for any $n\in\mathbb{N}$, $\ln(x)$ is concave on $(0,+\infty)$, and x^p is convex on $(0,+\infty)$ if $p\ge 1$ and concave if $p<1$.

There is a partial converse to 4.3.2. For this we need following lemma.

4.3.3 Lemma. *If f is convex and $a < u < x \le y < v < b$, then*

(a) $\dfrac{f(x) - f(u)}{x - u} \le \dfrac{f(y) - f(u)}{y - u} \le \dfrac{f(v) - f(y)}{v - y}$, *and*

(b) $\dfrac{f(v) - f(x)}{v - x} \le \dfrac{f(v) - f(y)}{v - y}$.

Proof. Referring to Figure 4.5, for (a) we have

$$\begin{aligned}
\frac{f(x) - f(u)}{x - u} &\le \frac{L_{uy}(x) - f(u)}{x - u} && \text{by convexity, since } u < x < y,\\
&= \frac{f(y) - f(u)}{y - u} && \text{by equality of slopes on } L_{uy},\\
&\le \frac{L_{uv}(y) - f(u)}{y - u} && \text{by convexity, since } u < y < v,\\
&= \frac{L_{uv}(v) - L_{uv}(y)}{v - y} && \text{by equality of slopes on } L_{uv},\\
&\le \frac{f(v) - f(y)}{v - y} && \text{by convexity, since } u < y < v.
\end{aligned}$$

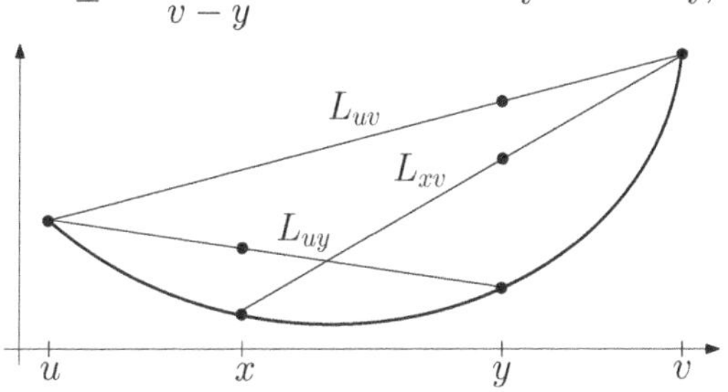

FIGURE 4.5: Convex function inequalities.

A similar calculation verifies (b):

$$\frac{f(v) - f(y)}{v - y} \ge \frac{L_{xv}(v) - L_{xv}(y)}{v - y} = \frac{L_{xv}(v) - L_{xv}(x)}{v - x} = \frac{f(v) - f(x)}{v - x}. \quad \square$$

4.3.4 Theorem. *If f is convex, then f_r' and f_ℓ' exist, are increasing, and $f_\ell'(x) \le f_r'(x)$.*

Proof. Let $a < u < x \le y < v < b$. By (a) of the lemma, the difference quotients $[f(x) - f(u)]/(x - u)$ decrease as $x \to u^+$, so $f_r'(u)$ exists in $\overline{\mathbb{R}}$ and

$$f_r'(u) \le \frac{f(v) - f(y)}{v - y} < +\infty.$$

Letting $v \to y^+$ shows that $f_r'(u) \le f_r'(y)$. Therefore, f_r' is increasing. Similarly, by (b) the difference quotients $[f(v) - f(y)]/(v - y)$ increase as $y \to v^-$, so $f'_\ell(v)$ exists in $\overline{\mathbb{R}}$ and

$$f'_\ell(v) \geq \frac{f(v) - f(x)}{v - x} > -\infty.$$

Taking $x = y$ in (a) of the lemma, we have

$$\frac{f(x) - f(u)}{x - u} \leq \frac{f(v) - f(x)}{v - x}.$$

Letting $u \uparrow x$ and $v \downarrow x$, we obtain $f'_\ell(x) \leq f'_r(x)$. In particular, $f'_\ell(x)$ and $f'_r(x)$ are finite. □

4.3.5 Corollary. *A convex function f is continuous.*

Proof. By the theorem, f has finite left-hand and right-hand derivatives and hence is left and right continuous. □

4.3.6 Theorem. *If a convex function f is differentiable at $x \in (u, v)$, then*

$$f'(x)(t - x) + f(x) \leq f(t) \quad \textit{for all } t \in (u, v).$$

That is, the tangent line at $(x, f(x))$ lies below the graph of f on (u, v).

Proof. Since the difference quotients $\left[f(t) - f(x)\right]/(t - x)$ decrease as $t \downarrow x$,

$$f'_r(x) \leq \frac{f(t) - f(x)}{t - x}, \quad t > x.$$

The same difference quotients increase as $t \uparrow x$, hence

$$f'_\ell(x) \geq \frac{f(t) - f(x)}{t - x}, \quad t < x.$$

Therefore, if $f'(x)$ exists, then $f'(x)(t - x) + f(x) \leq f(t)$ for all t. □

4.4 Inverse Functions

In this section we prove that, under suitable conditions, the inverse of a one-to-one continuous (differentiable) function is continuous (differentiable). For this we need the following two lemmas. The proof of the first is illustrated in Figures 4.6 and 4.7.

4.4.1 Lemma. *Let f be one-to-one on an interval I. If f has the intermediate value property, then f is strictly monotone and continuous on I.*

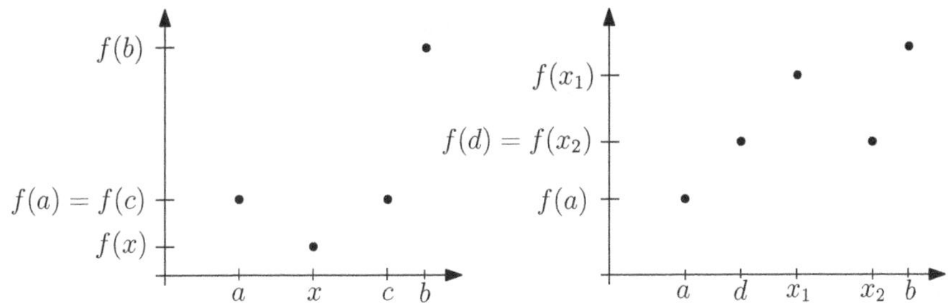

FIGURE 4.6: $f(x) < f(a)$ or $f(x_1) > f(x_2)$ violates one-to-one hypothesis.

Proof. Let a, b be arbitrary points in I with $a < b$. Assume, for definiteness, that $f(a) < f(b)$. We claim that $f(a) < f(x) < f(b)$ for all $a < x < b$. If, to the contrary, say $f(x) < f(a)$, then $f(a)$ lies between $f(x)$ and $f(b)$, hence, by the intermediate value property, there exists $c \in (x, b)$ such that $f(c) = f(a)$, contradicting that f is one-to-one.

Next we show that f is strictly increasing on $[a, b]$. Let $a < x_1 < x_2 < b$ and suppose that $f(x_2) < f(x_1)$. Then $f(x_2)$ lies between $f(a)$ and $f(x_1)$, hence there exists $d \in (a, x_1)$ such that $f(d) = f(x_2)$, again contradicting that f is one-to-one. Thus f is strictly increasing on $[a, b]$. It follows that f must be strictly increasing on any closed and bounded subinterval of I containing $[a, b]$. Since every pair of points in I lies in such a subinterval, f is strictly increasing on I.

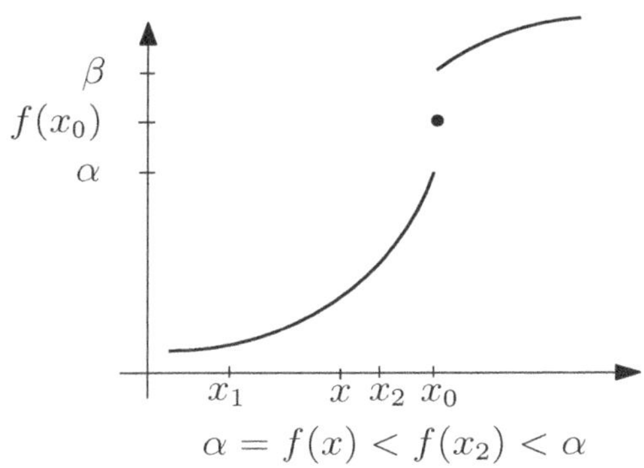

FIGURE 4.7: Intermediate value property implies continuity.

Now let $x_0 \in I$. To verify continuity of f at x_0, note that by monotonicity

$$\alpha := \lim_{x \to x_0^-} f(x) \le f(x_0) \le \beta := \lim_{x \to x_0^+} f(x).$$

(If x_0 is an endpoint, only one of these inequalities holds.) Continuity of f at x_0 will then follow if we show that $\alpha = f(x_0) = \beta$. Suppose, for example, that

$\alpha < f(x_0)$. Choose any $x_1 < x_0$ in I. Since $f(x_1) < \alpha < f(x_0)$, there exists some $x \in (x_1, x_0)$ such that $f(x) = \alpha$. But choosing $x_2 \in (x, x_0)$ then produces the contradiction $f(x) = \alpha < f(x_2) < \alpha$. □

4.4.2 Lemma. *If f is strictly increasing (decreasing) on an interval I, then f^{-1} is strictly increasing (decreasing) on $f(I)$.*

Proof. Assume that f is strictly increasing. If $y_1 = f(x_1) < y_2 = f(x_2)$, then $x_1 < x_2$ (that is, $f^{-1}(y_1) < f^{-1}(y_2)$), since otherwise $f(x_1) \geq f(x_2)$. Therefore, f^{-1} is strictly increasing on I. □

The next two theorems are the main results on inverse functions. They assert that the properties of continuity or differentiability of a one-to-one function are inherited by the inverse function.

4.4.3 Theorem. *Let f be continuous and one-to-one on an interval I. Then $J := f(I)$ is an interval and $f^{-1} : J \to I$ is continuous. Moreover, f and f^{-1} are strictly monotone.*

Proof. Since f is continuous, it has the intermediate value property, hence J is an interval. Moreover, by 4.4.1 and 4.4.2, f and f^{-1} are strictly monotone. Since $I = f^{-1}(J)$ is an interval, f^{-1} has the intermediate value property. The continuity of f^{-1} now follows from 4.4.1. □

4.4.4 Theorem. *Let I be an open interval and let $f : I \to \mathbb{R}$ be continuous and one-to-one on I. If f is differentiable at $a \in I$ and $f'(a) \neq 0$, then f^{-1} is differentiable at $f(a)$, and*

$$(f^{-1})'(f(a)) = \frac{1}{f'(a)}.$$

Proof. Let $y = f(x)$ and $b = f(a)$. For x near a,

$$\frac{f^{-1}(y) - f^{-1}(b)}{y - b} = \frac{x - a}{f(x) - f(a)} = \left[\frac{f(x) - f(a)}{x - a}\right]^{-1}.$$

Since f^{-1} is continuous, $x = f^{-1}(y) \to f^{-1}(b) = a$ as $y \to b$ and the conclusion follows. □

If f is differentiable and nonzero on I and $y = f^{-1}(x)$, then $x = f(y)$ and assertion of the theorem may be written pleasingly in Leibniz notation as

$$\frac{dy}{dx} = \frac{1}{\dfrac{dx}{dy}}.$$

From 4.4.3 we obtain the following result, which will be generalized in Chapter 9 to functions on open subsets of $\mathbb{R}^n$.

4.4.5 Inverse Function Theorem. *Let f be continuously differentiable on an open interval I. If $f'(a) \neq 0$, then there exist open intervals $I_a \subseteq I$ and $J_a = f(I_a)$ with $a \in I_a$ such that f is one-to-one on I_a and $f^{-1} : J_a \to I_a$ is continuously differentiable.*

Proof. Since f' is continuous and $f'(a) \neq 0$, there exists an open interval I_a containing a on which $f' \neq 0$. By the mean value theorem, f is one-to-one on I_a, hence, by 4.4.3, $J_a = f(I_a)$ is an interval, and, by 4.4.4, $f^{-1} : J_a \to I_a$ is continuously differentiable. □

4.4.6 Global Inverse Function Theorem. *Let f be continuously differentiable with f' nonzero on an open interval I. Then f is one-to-one on I, $J := f(I)$ is an open interval, and $f^{-1} : J \to I$ is continuously differentiable.*

Proof. That f is one-to-one follows from the mean value theorem. By 4.4.3, J is an interval and $f^{-1} : J \to I$ is continuous. Since continuous differentiability is a local property, 4.4.5 implies that f^{-1} is continuously differentiable. □

The following examples, as well as exercises below, establish the existence and basic properties of several well-known functions.

4.4.7 Example. Since $x = \sin y$ is strictly increasing and continuous on $[-\pi/2, \pi/2]$, the inverse function $y = \sin^{-1} x$ exists, is strictly increasing and continuous on $[-1, 1]$, and

$$\frac{dy}{dx} = \left(\frac{dx}{dy}\right)^{-1} = \frac{1}{\cos y} = \frac{1}{\sqrt{1-x^2}}, \quad -1 < x < 1.$$

Similarly, $x = \cos y$ is strictly decreasing and continuous on $[0, \pi]$, hence $y = \cos^{-1} x$ exists, is strictly decreasing and continuous on $[-1, 1]$, and

$$\frac{dy}{dx} = \left(\frac{dx}{dy}\right)^{-1} = \frac{-1}{\sin y} = \frac{-1}{\sqrt{1-x^2}}, \quad -1 < x < 1. \qquad \lozenge$$

An alternate approach to the preceding example is to *define* the inverse sine by

$$\sin^{-1} x = \int_0^x \frac{dt}{\sqrt{1-t^2}}, \quad -1 < x < 1$$

and then obtain the sine function as the inverse of $\sin^{-1}$. This allows the derivation of the standard properties of $\sin x$, and ultimately of the other trig functions, without relying on geometric arguments. The disadvantage of this approach is that verification of these properties is detailed and lengthy. Yet another approach is based on complex infinite series. For the latter, the reader may wish to consult [7].

The following example defines the natural logarithm $\ln x$ as an integral and obtains the exponential function as the inverse of $\ln x$. The material in the example relies on integration results from Chapter 5 but should be familiar to the reader.

4.4.8 Example. The *natural logarithm function* is defined by

$$\ln x := \int_1^x \frac{1}{t}\, dt, \quad x > 0.$$

One may show that all the familiar algebraic properties of the natural log follow from this definition (See Exercise 5) Since $\ln x$ is strictly increasing on $(0, +\infty)$, the inverse function

$$\exp x := \ln^{-1} x$$

exists and is strictly increasing. Since $\ln 2 > 0$,

$$\ln 2^n = n \ln 2 \to +\infty \quad \text{and} \quad \ln 2^{-n} = -n \ln 2 \to -\infty,$$

hence

$$\lim_{x\to+\infty} \ln x = +\infty \quad \text{and} \quad \lim_{x\to 0+} \ln x = -\infty.$$

It follows from these limits and the intermediate value theorem that the range of $\ln x$, which is the domain of $\exp x$, must be $\mathbb{R}$. Thus, by Exercise 4,

$$\lim_{x\to-\infty} \exp x = 0 \quad \text{and} \quad \lim_{x\to+\infty} \exp x = +\infty.$$

For the calculus properties of ln and exp, we apply the fundamental theorem of calculus (proved in Chapter 5) to obtain

$$\frac{d \ln y}{dy} = \frac{1}{y}$$

hence

$$\frac{d \exp x}{dx} = \left(\frac{d \ln y}{dy}\right)^{-1} = y = \exp x.$$

Moreover, since

$$1 = \left.\frac{d \ln y}{dy}\right|_{y=1} = \lim_{n\to+\infty} \frac{\ln(1 + 1/n) - \ln 1}{1/n} = \lim_{n\to+\infty} \ln(1 + 1/n)^n,$$

continuity of exp and 2.2.4 imply that

$$\exp 1 = \lim_{n\to+\infty} \exp\big(\ln(1 + 1/n)^n\big) = \lim_{n\to+\infty} (1 + 1/n)^n = e.$$

Additional properties of $\exp x$ may be found in the exercises, including the identity $\exp r = e^r$, $r \in \mathbb{Q}$. Because of this identity, we frequently write e^x for $\exp x$. Indeed, the function exp is the basis for rigorous definitions of the general exponential function a^x, $a > 0$, and the power function x^a, $x \geq 0$ (See Exercises 8 and 9) ◊

Exercises

1. Find f^{-1} and its domain for each of the following functions f with the given domain:

(a) $x^2 - 4x + 5, \ [2, +\infty)$ (b) $\dfrac{3x+2}{2x+3}, \ \mathbb{R} \setminus \{-3/2\}$

(c) $\dfrac{5e^{-x}+2}{3e^{-x}+7}, \ (-\infty, +\infty)$ (d) $\sin^2 x - 4\sin x + 3, \ [-\pi/2, \pi/2]$

(e) $e^x - 2e^{-x}, \ (-\infty, +\infty)$ (f) $\dfrac{2+\cos x}{3+\cos x}, \ (0, \pi)$.

2. Let $f(x) = ax + |x| + |x-1|$. Find all values of a for which f^{-1} exists on $\mathbb{R}$. Find f^{-1} for these values.

3. Give an example of a one-to-one continuous function on the union of two intervals that is (a) not monotone, (b) strictly monotone but with discontinuous inverse.

4. Let f be defined, continuous, and strictly increasing on (a, b), so the limits

$$c := \lim_{x \to a+} f(x) \quad \text{and} \quad d := \lim_{x \to b-} f(x)$$

exist in $\overline{\mathbb{R}}$. Show that the domain of f^{-1} is (c, d) and that

$$\lim_{x \to c+} f^{-1}(x) = a \quad \text{and} \quad \lim_{x \to d-} f^{-1}(x) = b.$$

5. Verify the following properties of $\ln x$.

(a) $\ln 1 = 0, \ \ln e = 1$ (b) $\ln(xy) = \ln x + \ln y$

(c) $\ln(x/y) = \ln x - \ln y$ (d) $\ln x^r = r \ln x, \ r \in \mathbb{Q}$.

6. Prove that $\exp(x+y) = \exp(x)\exp(y)$.

7. For $c, d \in \mathbb{R}$ with $c > 0$, define $c^d = \exp(d \ln c)$. Show that this definition agrees with the usual one if d is rational and verify the following properties, where $x, y \in \mathbb{R}$ and $a, b > 0$.

(a) $\ln a^x = x \ln a$ (b) $a^x a^y = a^{x+y}$ (c) $a^x / a^y = a^{x-y}$
(d) $(a^x)^y = a^{xy}$ (e) $(ab)^x = a^x b^x$ (f) $a^{\ln b} = b^{\ln a}$.

8. Let $a > 0$, $a \neq 1$, and define a^x as in Exercise 7. Find $\lim_{x \to -\infty} a^x$, $\lim_{x \to +\infty} a^x$, and $(a^x)'$.

9. Let $a \in \mathbb{R}$ and for $x > 0$ define x^a as in Exercise 7. Prove the power rule $(x^a)' = a x^{a-1}$.

10. Prove that $\tan x$ restricted to $(-\pi/2, \pi/2)$ has a differentiable inverse defined on $\mathbb{R}$. Find $\lim_{x \to -\infty} \tan^{-1} x$, $\lim_{x \to +\infty} \tan^{-1} x$, and $(\tan^{-1} x)'$.

11. Prove that $\sec x$ restricted to $[0, \pi/2) \cup [\pi, 3\pi/2)$ has a continuous inverse defined on $(-\infty, -1] \cup [1, +\infty)$. Show that $\sec^{-1} x$ is differentiable on $(-\infty, -1) \cup (1, +\infty)$ and compute its derivative. Also, find $\lim_{x \to -\infty} \sec^{-1} x$ and $\lim_{x \to +\infty} \sec^{-1} x$.

12. Verify the inequalities

    (a) $\dfrac{x-1}{x} < \ln x < x - 1,\ x > 1$ (b) $|\tan^{-1} x - \tan^{-1} y| \le |x - y|$

    (c) $\dfrac{y-x}{\sqrt{1-x^2}} < |\sin^{-1} y - \sin^{-1} x| < \dfrac{y-x}{\sqrt{1-y^2}}, \quad -1 < x < y < 1.$

13. Verify the following identities. (One way is to show that the right and left sides have the same derivative.)

    (a) $\tan\left(\sin^{-1} x\right) = \dfrac{x}{\sqrt{1-x^2}}, \quad -1 < x < 1$

    (b) $\sin^{-1} x + \cos^{-1} x = \pi/2, \quad -1 \le x \le 1$

    (c) $\cos^{-1}\left(\dfrac{x^2-1}{x^2+1}\right) + 2\tan^{-1} x = \pi, \quad x \ge 0$

    (d) $\cos^{-1} x = 2\sin^{-1}\sqrt{\dfrac{1-x}{2}}, \quad -1 \le x \le 1$

    (e) $\tan^{-1} x + \tan^{-1}(2/x) + \tan^{-1}(x + 2/x) = \pi, \quad x \ne 0.$

14. Suppose f satisfies $f(x+y) = f(x)f(y)$ for all $x, y \in \mathbb{R}$. Show that if $a := f'(0)$ exists, then $f(x) = f(0)e^{ax}$.

15. Suppose that $f : [0,1] \to [0,1]$ is continuous, one-to-one, onto, and $f = f^{-1}$. Prove that either $f(x) = x$ for all x or f is monotone decreasing.

16. Suppose f' is one-to-one on an open interval I. Show that f' is continuous and strictly monotone on I. (See Exercise 4.2.25.)

17. Let f be differentiable on an open interval I with $f' \ne 0$. Let $a, b \in I$ with $a < b$ and suppose that $f : [a,b] \to [a,b]$ is one-to-one and onto. Prove that there exists $c \in (a,b)$ such that

$$\frac{f(b) - f(a)}{f^{-1}(b) - f^{-1}(a)} = f'(c)f'\left(f^{-1}(c)\right).$$

18. Let f be twice differentiable and $f' \ne 0$ on an open interval I. Show that $(f^{-1})''(x)$ exists on $f(I)$ and find a formula.

4.5 L'Hospital's Rule

The rule for calculating the limit of a quotient of functions, namely,

$$\lim_{\substack{x\to a\\ x\in E}} \frac{f(x)}{g(x)} = \frac{\lim_{\{x\to a,\, x\in E\}} f(x)}{\lim_{\{x\to a,\, x\in E\}} g(x)}, \tag{4.5}$$

requires that the limits on the right are finite and the denominator is not 0. If the limits in the quotient are both zero or both $\pm\infty$, then the expression on the left in (4.5) is called an *indeterminate form of type* $\frac{0}{0}$ or $\frac{\pm\infty}{\pm\infty}$, respectively. There are other types of indeterminate forms, but all may be converted to one of these. The following theorem describes a method for evaluating these limits.

4.5.1 l'Hospital's Rule. *Let J be an open interval, finite or infinite, and let $a \in \overline{\mathbb{R}}$ be an accumulation point of J. Suppose that f and g are differentiable on $E := J \setminus \{a\}$ and that $g'(x) \neq 0$ for every $x \in E$. If the limits*

$$A := \lim_{\substack{x\to a\\ x\in E}} f(x), \quad B := \lim_{\substack{x\to a\\ x\in E}} g(x), \quad \text{and} \quad L := \lim_{\substack{x\to a\\ x\in E}} \frac{f'(x)}{g'(x)}$$

exist in $\overline{\mathbb{R}}$ and either $A = B = 0$ or $B = \pm\infty$, then

$$\lim_{\substack{x\to a\\ x\in E}} \frac{f(x)}{g(x)} = L.$$

Proof. By Rolle's theorem, the condition $g' \neq 0$ on E implies that g is zero at most at one point in E. By making E smaller if necessary, we may suppose that g in never zero on E.

There are a number of cases to consider, but the proofs of many of these are essentially the same. We prove the theorem for four fundamentally different cases and for one-sided limits, so $E = (a, c)$ or (c, a) for some c.

As a first step, we use the Cauchy mean value theorem to obtain, for every pair of distinct numbers x, $b \in E$, a number $\xi = \xi(x, b)$ between x and b such that

$$[f(x) - f(b)]g'(\xi) = [g(x) - g(b)]f'(\xi).$$

Now set

$$h(x) := \frac{f(x)}{g(x)} = \frac{f(b)}{g(x)} + \left[1 - \frac{g(b)}{g(x)}\right]\frac{f'(\xi)}{g'(\xi)}, \quad x,\, b \in E. \tag{4.6}$$

Case 1: $A = B = 0$, a and L are finite, and $E = (a, c)$. Extend f and g continuously to $[a, c)$ by defining $f(a) = g(a) = 0$. Taking $b = a$ and $x \in (a, c)$ in (4.6) we see that

$$h(x) = \frac{f'(\xi)}{g'(\xi)}, \qquad a < \xi < x.$$

Since $\xi \to a$ as $x \to a$, $\lim_{x\to a^+} h(x) = L$, as required.

Case 2: $A = B = 0$, $a = L = +\infty$, and $E = (c, +\infty)$. Let $M > 0$ and choose $x_0 \in E$ such that

$$\frac{f'(x)}{g'(x)} > 2M \quad \text{for } x > x_0.$$

Let $b > x > x_0$. For large b, $g(b)/g(x) < 1/2$, hence from (4.6)

$$h(x) \ge \frac{f(b)}{g(x)} + \frac{1}{2}(2M) = \frac{f(b)}{g(x)} + M.$$

Letting $b \to +\infty$ we see that $h(x) \ge M$. Therefore, $\lim_{x\to+\infty} h(x) = +\infty$.

Case 3: $B = +\infty$, a and L are finite and $E = (c, a)$. Given $\varepsilon > 0$, choose $b \in E$ such that

$$\left|\frac{f'(t)}{g'(t)} - L\right| < \varepsilon/2 \quad \text{for all } t \in (b, a).$$

Let $x \in (b, a)$. By (4.6),

$$h(x) - \frac{f'(\xi)}{g'(\xi)} = \frac{f(b)}{g(x)} - \frac{g(b)}{g(x)}\frac{f'(\xi)}{g'(\xi)}.$$

Since the right side tends to 0 as $x \to a$,

$$|h(x) - L| \le \left|h(x) - \frac{f'(\xi)}{g'(\xi)}\right| + \left|\frac{f'(\xi)}{g'(\xi)} - L\right| < \varepsilon/2 + \varepsilon/2 = \varepsilon$$

for all x near a. Therefore, $\lim_{x\to a^-} h(x) = L$.

Case 4: $B = +\infty$, $a = L = +\infty$, and $E = (c, +\infty)$. Given $M > 0$, choose $b > c$ such that

$$\frac{f'(t)}{g'(t)} > 3M \quad \text{for all } t > b.$$

For all sufficiently large x, $g(x) > 2g(b)$, hence by (4.6)

$$h(x) \ge \frac{f(b)}{g(x)} + \left[1 - \frac{g(b)}{g(x)}\right] 3M \ge \frac{f(b)}{g(x)} + \left[1 - \tfrac{1}{2}\right] 3M.$$

Since the quotients on the right side tend to zero, for all sufficiently large x we have

$$h(x) \ge -\tfrac{1}{2}M + \left[1 - \tfrac{1}{2}\right](3M) = M.$$

Therefore, $\lim_{x\to+\infty} h(x) = +\infty$. □

Here are some typical applications of l'Hospital's rule:

Examples. (a) The limit

$$L := \lim_{x\to 0} \frac{x - \tan x}{x^3}$$

is of the form $\frac{0}{0}$, hence

$$L = \lim_{x\to 0} \frac{1-\sec^2 x}{3x^2} = \lim_{x\to 0} \frac{2\sec^2 x \tan x}{-6x} = \lim_{x\to 0} \frac{\sec^4 x + 2\sec x \tan^2 x}{-3} = -\frac{1}{3}.$$

Note that each step except the last produces a limit of the form $\frac{0}{0}$, allowing another application of l'Hospital's rule. The validity of each step is therefore ultimately justified by the existence of the final limit.

(b) The limit

$$L := \lim_{x\to+\infty} \frac{\sin(1/x)}{e^{1/x}-1}$$

is of the form $\frac{0}{0}$; however, it is complicated to apply l'Hospital's rule directly. Making the substitution $y = 1/x$ produces a more tractable problem:

$$L = \lim_{y\to 0+} \frac{\sin y}{e^y - 1} = \lim_{y\to 0+} \frac{\cos y}{e^y} = 1.$$

(c) The limit

$$L := \lim_{x\to+\infty} x\sin(1/x)$$

is of the form $\infty \cdot 0$, but the preceding substitution produces the form $\frac{0}{0}$:

$$L = \lim_{x\to+\infty} \frac{\sin(1/x)}{1/x} = \lim_{y\to 0+} \frac{\sin y}{y} = 1.$$

Here, l'Hospital's rule was unnecessary, since we could use a known limit.[2]

(d) The limit

$$L := \lim_{x\to 1+} x^{1/(x^p-1)}, \quad p > 0,$$

is of the form 1^∞, so we take logarithms to obtain the form $\frac{0}{0}$:

$$\lim_{x\to 1+} \ln\left[x^{1/(x^p-1)}\right] = \lim_{x\to 1+} \frac{\ln x}{x^p - 1} = \lim_{x\to 1+} \frac{1/x}{px^{p-1}} = \frac{1}{p}.$$

Thus $L = e^{1/p}$.

(e) The technique used in **(d)** shows that

$$\lim_{x\to+\infty} \left(1 + \frac{t}{x}\right)^x = e^t,$$

since

$$\lim_{x\to+\infty} \ln\left(1 + \frac{t}{x}\right)^x = \lim_{y\to 0+} \frac{\ln(1+ty)}{y} = \lim_{y\to 0+} \frac{t}{1+ty} = t.$$

[2]In fact, strictly speaking, applying l'Hospital's rule here would constitute circular reasoning, since the limit, established geometrically, was crucial in our development of the derivative of the sine function. So, basically, we would be using the limit to prove the limit.

(f) The limit

$$L := \lim_{x \to \pi/2^+} \left[\frac{1}{x - \pi/2} + \sec x \right]$$

is of the form $\infty - \infty$. Combining fractions we obtain a limit of the form $\frac{0}{0}$ and so may apply l'Hospitals' rule:

$$\begin{aligned} L &= \lim_{x \to \pi/2^+} \frac{\cos x + x - \pi/2}{(x - \pi/2)\cos x} \\ &= \lim_{x \to \pi/2^+} \frac{1 - \sin x}{(\pi/2 - x)\sin x + \cos x} \\ &= \lim_{x \to \pi/2^+} \frac{-\cos x}{(\pi/2 - x)\cos x - 2\sin x} \\ &= 0. \end{aligned}$$

◊

Exercises

1. Evaluate the following limits, where $p,\ q > 0$:

   (a) $\displaystyle\lim_{x\to 0} \frac{e^{px} - e^{qx}}{\sin x}$ (b) $\displaystyle\lim_{x\to 1} \frac{e^{px} - e^{p}}{\tan(x-1)}$ (c) $\displaystyle\lim_{x\to +\infty} \frac{\ln(3x^2-1)}{\ln(5x^2-1)}$

   (d) $\displaystyle\lim_{x\to +\infty} x\left(1 - e^{1/x}\right)$ (e) $\displaystyle\lim_{x\to 0^+} \frac{\ln(\sin px)}{\ln(\sin qx)}$ (f) $\displaystyle\lim_{x\to 0} \frac{p\sin(px) - p^2 x}{x^3}$

   (g) $\displaystyle\lim_{x\to 0} \frac{x - \tan^{-1} x}{x - \sin^{-1} x}$ (h) $\displaystyle\lim_{x\to 0^+} \left(\frac{1}{x} - \frac{1}{\sin x}\right)$ (i) $\displaystyle\lim_{x\to 1^+} \ln(x-1)\ln x$

   (j) $\displaystyle\lim_{x\to 0^+} (\sin x)(\ln x)$ (k) $\displaystyle\lim_{x\to 0^+} x^{(x^p)}$ (l) $\displaystyle\lim_{x\to 0} (\cos x)^{1/x^2}$

   (m) $\displaystyle\lim_{x\to 0^+} \left[\frac{x}{\tan x} - \frac{1}{x}\right]$ (n) $\displaystyle\lim_{x\to 1^+} \left(\frac{x+1}{x-1}\right)^{x-1}$ (o) $\displaystyle\lim_{x\to 1} \frac{\sqrt{x}\ln x}{x-1}$

   (p) $\displaystyle\lim_{x\to 0} \frac{1 - \cos x^2}{x^2 + x^3 \sin x}$ (q) $\displaystyle\lim_{x\to 0} \frac{\sin x + \cos x - 1}{\ln(1+x)}$ (r) $\displaystyle\lim_{x\to +\infty} \left(\frac{x-1}{x+1}\right)^{x}$

   (s) $\displaystyle\lim_{x\to +\infty} x^{1/(\ln\ln x)^p}$ (t) $\displaystyle\lim_{x\to +\infty} \left(1 - \frac{1}{\sqrt{x}}\right)^{-x}$ (u) $\displaystyle\lim_{x\to 0^+} (\sin x)^{x}$

   (v) $\displaystyle\lim_{x\to 0^+} x^{\sin x}$ (w) $\displaystyle\lim_{x\to 0} \frac{x\cos x - \sin x}{x^2 \sin x}$ (x) $\displaystyle\lim_{x\to 0^+} \frac{(1+x)^{1/x} - e}{x}$.

2. For each function $f : (0, 1] \to \mathbb{R}$ below, define $f(0)$ so that f continuous on $[0, 1]$.

   (a) $\dfrac{1 - e^x}{x}$ (b) $\dfrac{\ln(1+x)}{x}$ (c) $\dfrac{\sin 5x}{\sin 3x}$

   (d) $\dfrac{x}{\tan x}$ (e) $x \ln x$ (f) $\dfrac{1 - \cos 2x}{1 - \cos 3x}$.

3. Find $\lim_n a_n$, where $a_n =$

   (a) $\sin^{1/n}(1/n)$ (b) $n - n^2 \ln(1 + 1/n)$ (c) $n\left[(1 + 1/n)^n - e\right]$.

4. Show that

$$(n + 1/n)^p - n^p \to \begin{cases} 0 & \text{if } p < 2, \\ 2 & \text{if } p = 2, \\ +\infty & \text{if } p > 2. \end{cases}$$

5. By considering the sequences $\{n\}$ and $\{n + 1/n\}$, use l'Hospital's rule to prove that e^x is not uniformly continuous on $[0, +\infty)$.

6. Let $f(x) = x^{1+1/x}$. Evaluate $\lim_n \left[f(n+1) - f(n)\right]$.

7. Let f be differentiable on (a, b) and suppose that $\lim_{x\to a^+} f(x)$ and $\lim_{x\to a^+} f'(x)$ exist in $\mathbb{R}$. Find a continuous extension of f to $[a, b)$ such that f' exists and is continuous at a.

8. Let g be differentiable on $(1, +\infty)$ and h differentiable on $(-\infty, 1]$ with

$$\lim_{x\to 1^+} g(x) = h(1) \quad \text{and} \quad \lim_{x\to 1^+} g'(x) = h'_\ell(1). \qquad (\dagger)$$

   Define

$$f(x) = \begin{cases} g(x) & \text{if } x > 1 \\ h(x) & \text{if } x \le 1. \end{cases}$$

   Show that f is differentiable at $x = 1$ and hence on $\mathbb{R}$. Conversely, suppose that $f'(1)$ exists. Do the limit equations in $(\dagger)$ hold?

9. Let f and g be differentiable on $(0, +\infty)$ with

$$\lim_{x\to+\infty} f(x) = \lim_{x\to+\infty} g(x) = +\infty \quad \text{and} \quad \lim_{x\to+\infty} \frac{f'(x)}{g'(x)} \in (0, +\infty).$$

   Evaluate

$$\lim_{x\to+\infty} \frac{\ln f(x)}{\ln g(x)}.$$

10. Let f be differentiable in a neighborhood of a and suppose that $f''(a)$ exists. For α, $\beta \in \mathbb{R}$ calculate

    (a) $$\lim_{h\to 0} \frac{\beta f(a + \alpha h) - \alpha f(a + \beta h) + (\alpha - \beta) f(a)}{h^2}$$

    (b) $$\lim_{h\to 0} \frac{f(a + \alpha h) + f(a + \beta h) - 2f(a)}{h^2} \quad \text{if } f'(a) = 0.$$

11. Suppose that f has n derivatives on $[a, +\infty)$ and that $\lim_{x\to+\infty} f^{(n)}(x)$ exists in $\overline{\mathbb{R}}$. Prove that $\lim_{x\to+\infty} f(x)/x^n$ exists in $\overline{\mathbb{R}}$.

12. Suppose that f has n derivatives on $(0,a)$ and $L := \lim_{x\to 0^+} x^{2n} f^{(n)}(x)$ exists in $\mathbb{R}$. Find $\lim_{x\to 0^+} x^n f(x)$ in terms of L.

13. Let f be differentiable on $(1,+\infty)$ and $\lim_{x\to+\infty} f(x) = 0$. Prove that if $\lim_{x\to+\infty} x^2 f'(x)$ exists in $\mathbb{R}$, then $\lim_{x\to+\infty} xf(x)$ also exists in $\mathbb{R}$. Is the converse true?

14. Suppose that, in a deleted neighborhood of 0, f is differentiable with $f' \neq 0$ and that $\lim_{x\to 0} f(x) = 0$. Prove that if $\lim_{x\to 0} f(x)/f'(x)$ exists, then it must equal 0.

15. Let $g(x)$ be differentiable on $(1,\infty)$ with g and g' nonzero there and let $f(x)$ be differentiable in a neighborhood of 0. Suppose that $\lim_{x\to+\infty} g(x) = 0$, $f(0) = 0$ and f' is continuous at 0. Find

$$\lim_{x\to+\infty} \frac{f(g(x))}{g(x)}.$$

Give nontrivial examples of functions f and g that satisfy these conditions.

16. Let f and g be differentiable on $(1,+\infty)$ with $g' \neq 0$ and suppose that $\lim_{x\to+\infty} f(x) = \lim_{x\to+\infty} g(x) = +\infty$ and that the limit $L := \lim_{x\to+\infty} f'(x)$ exists in $\overline{\mathbb{R}}$. Find

$$\lim_{x\to+\infty} \frac{f(g(x))}{g(x)}.$$

Give nontrivial examples that satisfy these conditions with L finite.

17. Let f be differentiable on $(1,+\infty)$ and suppose that $\lim_{x\to+\infty} f(x)$ and $\lim_{x\to+\infty} f'(x)$ exist in $\mathbb{R}$. Prove that the second limit must be zero. Does the assertion still hold if $\lim_{x\to+\infty} f(x)$ is infinite?

18. Let f be differentiable on $(1,+\infty)$ and suppose that $\lim_{x\to+\infty} f(x)$ and $\lim_{x\to+\infty} xf'(x)$ exist in $\mathbb{R}$. Prove that the second limit is zero. Does the assertion still hold if $\lim_{x\to+\infty} f(x)$ is infinite?

19. Let f be differentiable in a deleted neighborhood of 0 and suppose that $\lim_{x\to 0} f(x)$ and $\lim_{x\to 0} \left[f'(x)\tan x\right]$ exist in $\mathbb{R}$. Prove that the second limit must be 0. Does the assertion still hold if $\lim_{x\to 0} f(x)$ is infinite?

20. Let f be differentiable on $(0,b)$ and suppose that the limits $\lim_{x\to 0^+} f(x)$ and $\lim_{x\to 0^+} x^2 f'(x)$ exist in $\mathbb{R}$. Prove that one of these limits must be zero. Does the assertion still hold if $\lim_{x\to 0^+} f(x)$ is infinite?

21. Let f'' exist and be continuous on $(-1,1)$ and $f(0) = f'(0) = 0$. Prove that there exists a continuous function g on $(-1,1)$ such that $f(x) = x^2 g(x)$. Must g be differentiable at 0?

4.6 Taylor's Theorem on $\mathbb{R}$

Taylor's theorem may be viewed as a generalization of the mean value theorem. Its importance derives from its use in establishing various inequalities and from its fundamental connection with power series.

4.6.1 Taylor's Theorem. *Let f have $n+1$ derivatives in an open interval I. Then, for each x, $a \in I$ with $x \neq a$, there exists a number c between x and a such that*

$$f(x) = \sum_{k=0}^{n} \frac{f^{(k)}(a)}{k!}(x-a)^k + \frac{f^{(n+1)}(c)}{(n+1)!}(x-a)^{n+1}. \tag{4.7}$$

Proof. Assume for definiteness that $a < x$. Define a function g on $[a, x]$ by

$$g(t) = \sum_{k=0}^{n} \frac{f^{(k)}(t)}{k!}(x-t)^k + \alpha\frac{(x-t)^{n+1}}{(n+1)!} - f(x), \tag{4.8}$$

where α is chosen so that $g(a) = 0$. Since g is continuous on $[a, x]$, differentiable on (a, x) and $g(x) = 0$, there exists, by Rolle's theorem, $c \in (a, x)$ such that $g'(c) = 0$. From the calculations

$$\frac{d}{dt}\frac{f^{(k)}(t)}{k!}(x-t)^k = \begin{cases} \dfrac{f^{(k+1)}(t)}{k!}(x-t)^k - \dfrac{f^{(k)}(t)}{(k-1)!}(x-t)^{k-1} & \text{if } k \geq 1, \\ f'(t) & \text{if } k = 0, \end{cases}$$

we have

$$\begin{aligned} g'(t) &= \sum_{k=0}^{n} \frac{f^{(k+1)}(t)}{k!}(x-t)^k - \sum_{k=0}^{n-1} \frac{f^{(k+1)}(t)}{k!}(x-t)^k - \alpha\frac{(x-t)^n}{n!} \\ &= \frac{f^{(n+1)}(t)}{n!}(x-t)^n - \alpha\frac{(x-t)^n}{n!}. \end{aligned}$$

In particular,

$$0 = g'(c) = \frac{f^{(n+1)}(c)}{n!}(x-c)^n - \alpha\frac{(x-c)^n}{n!},$$

hence $\alpha = f^{(n+1)}(c)$. Thus, from (4.8),

$$0 = g(a) = \sum_{k=0}^{n} \frac{f^{(k)}(a)}{k!}(x-a)^k + \frac{f^{(n+1)}(c)}{(n+1)!}(x-a)^{n+1} - f(x),$$

which is (4.7). □

Equation (4.7) is frequently written

$$f(x) = T_n(x,a) + R_n(x,a), \text{ where}$$

$$T_n(x,a) = \sum_{k=0}^{n} \frac{f^{(k)}(a)}{k!}(x-a)^k \text{ and } R_n(x,a) = \frac{f^{(n+1)}(c)}{(n+1)!}(x-a)^{n+1}.$$

The expression $T_n(x,a)$ is called the *nth Taylor polynomial of f about a* and $R_n(x,a)$ is called the *remainder.* It may be shown that $T_n(x,a)$ is the unique polynomial of degree $\le n$ that best approximates f near a in the sense that

$$\lim_{x\to a} \frac{f(x) - T_n(x,a)}{(x-a)^n} = 0$$

(See Exercise 4)

The remainder term $R_n(x,a)$ has other forms, one of which is given in Exercise 3. Observe that if $R_n(x,a) \to 0$ as $n \to +\infty$, then $T_n(x,a) \to f(x)$, which implies that $f(x)$ is expressible as a power series about a. We exploit this idea in Section 7.4.

The following application of Taylor's theorem is a generalization of the second derivative test.

4.6.2 *n*th Derivative Test. *Let f have n continuous derivatives on an open interval I and let $a \in I$ with $f^{(j)}(a) = 0$, $1 \le j \le n-1$, and $f^n(a) \ne 0$.*

(a) *If n is even and $f^{(n)}(a) > 0$ ($f^{(n)}(a) < 0$), then f has a local minimum (local maximum) at a.*

(b) *If n is odd, then f has a neither a local minimum nor a local maximum at a.*

Proof. Assume $f^{(n)}(a) > 0$. By continuity, $f^{(n)}(x) > 0$ for all x in an open interval J containing a. Let $x \in J$, $x \ne a$. By Taylor's theorem, there exists a number c between a and x such that

$$f(x) = f(a) + f^{(n)}(c)\frac{(x-a)^n}{n!}.$$

Thus if n is even, then $f(x) > f(a)$, hence f has a local minimum at a. If n is odd, then $f(x) > f(a)$ if $x > a$ and $f(x) < f(a)$ if $x < a$, so f has a neither a local maximum nor a local minimum at a. A similar argument works for the case $f^{(n)}(a) < 0$. □

Note that the familiar second derivative test, obtained by taking $n = 2$ in the theorem, is inconclusive for the function $f(x) = x^4$ at $a = 0$. Here, one must take $n = 4$.

Exercises

1. Define $f(0) = 0$ and $f(x) = e^{-1/x^2}$ $x \neq 0$. Prove that $f^{(n)}$ exists on $\mathbb{R}$ and $f^{(n)}(0) = 0$ for all n. Conclude that every Taylor polynomial for f about 0 is identically 0.

2. Verify the following inequalities:

   (a) $\displaystyle\sum_{k=0}^{2n-1}(-1)^k x^k < \frac{1}{1+x} < \sum_{k=0}^{2n}(-1)^k x^k, \ x > 0$

   (b) $\displaystyle\sum_{k=0}^{2n-1}\frac{(-1)^k}{k!}x^k < e^{-x} < \sum_{k=0}^{2n}\frac{(-1)^k}{k!}x^k, \ x > 0$

   (c) $\displaystyle\sum_{k=1}^{2n}\frac{(-1)^{k+1}}{(2k-1)!}x^{2k-1} < \sin x < \sum_{k=1}^{2n+1}\frac{(-1)^{k+1}}{(2k-1)!}x^{2k-1}, \quad 0 < x < \pi$

   (d) $\displaystyle\sum_{k=0}^{2n-1}\frac{(-1)^k}{(2k)!}x^{2k} < \cos x < \sum_{k=0}^{2n}\frac{(-1)^k}{(2k)!}x^{2k}, \quad 0 < x < \pi$

   (e) $\displaystyle\sum_{k=1}^{n-1}\frac{(-1)^{k-1}}{k}x^k < \ln(1+x) < \sum_{k=1}^{n}\frac{(-1)^{k-1}}{k}x^k$ if n is odd,
   the reverse inequalities if n is even.

3. ($\Downarrow$ 5.6.3.) Show that if $f^{(n+1)}$ is continuous on I, then
$$R_n(x,a) = \frac{1}{n!}\int_a^x (x-t)^n f^{(n+1)}(t)\,dt.$$
   Hint. Integrate by parts n times.

4. Assuming the necessary differentiability of f, prove that a polynomial $P_n(x) = \sum_{k=0}^{n} a_k(x-a)^k$ satisfies
$$\lim_{x\to a}\frac{f(x) - P_n(x)}{(x-a)^n} = 0$$
   iff $P_n = T_n$, the nth Taylor polynomial of f about a.

5. Let $P(x) = \sum_{k=0}^{n} a_k(x-a)^k = \sum_{k=0}^{n} b_k(x-b)^k$. Show that
$$b_k = \sum_{j=0}^{n-k}\binom{j+k}{k}(b-a)^j a_{k+j}.$$

6. Let P be a polynomial of degree n. Prove that the polynomials $P(x \pm 1)$ may be written as linear combinations of $P^{(k)}(x)$, $k = 0, \dots, n$. Find simplified expressions for $P(x+1) \pm P(x-1)$.

7. Let f have n derivatives on $[0,1]$. Show that for each $y \neq f(1)$ there exists an extension g of f to $[0, +\infty)$ with n derivatives such that $g(b) = y$ for some $b > 1$.

*4.7 Newton's Method

A *simple zero* of a differentiable function f is a number z such that $f(z) = 0$ and $f'(z) \neq 0$. Newton's method is a rapidly converging recursion scheme for approximating such a zero. The idea is to choose x_1 near z and then define a sequence $\{x_n\}$ recursively by

$$x_{n+1} = x_n - \frac{f(x_n)}{f'(x_n)}, \quad n = 1, 2, \ldots, \tag{4.9}$$

as illustrated in Figure 4.8. Under suitable conditions, the sequence converges

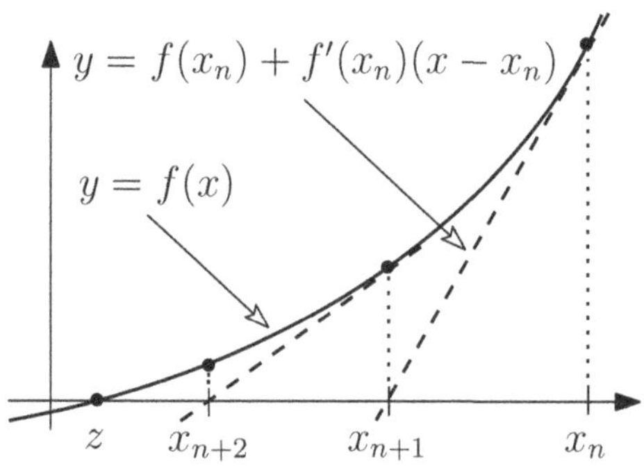

FIGURE 4.8: Newton's method.

to z and so may be used (theoretically) to approximate z to any desired degree of accuracy. Here is the precise formulation.

4.7.1 Newton's Method. *Let f'' be continuous on an open interval I and let z be a simple zero of f in I. If x_1 is chosen sufficiently near z, then the sequence $\{x_n\}$ lies in I and converges to z.*

Proof. Since $f'(z) \neq 0$, there exists a neighborhood I_z of z contained in I on which $|f'| \geq c > 0$. Suppose that $x_n \in I_z$. By Taylor's theorem, for each $x \in I$ there exists ξ between x and x_n such that

$$f(x) = f(x_n) + f'(x_n)(x - x_n) + \tfrac{1}{2} f''(\xi)(x - x_n)^2.$$

In particular,

$$0 = f(z) = f(x_n) + f'(x_n)(z - x_n) + \tfrac{1}{2} f''(\xi)(z - x_n)^2,$$

hence

$$-f(x_n) - f'(x_n)(z - x_n) = \tfrac{1}{2} f''(\xi)(z - x_n)^2.$$

Dividing by $f'(x_n)$ and using 4.9 we have

$$x_{n+1} - z = x_n - z - \frac{f(x_n)}{f'(x_n)} = \frac{f''(\xi)}{2f'(x_n)}(z - x_n)^2.$$

Thus if d is the maximum of $|f''|$ on I_z, then

$$|x_{n+1} - z| \le \left|\frac{f''(\xi)}{2f'(x_n)}\right| (z - x_n)^2 \le \alpha |x_n - z|^2, \quad \alpha := \frac{d}{2c}.$$

Iterating, we have

$$|x_{n+1} - z| \le \cdots \le \alpha^{2^{k+1}-1}|x_{n-k} - z|^{2^{k+1}} \le \cdots \le \alpha^{2^n-1}|x_1 - z|^{2^n}.$$

Thus if x_1 is sufficiently near z, and in particular if $\alpha|x_1 - z| < 1$, then $x_n \in I_z$ for all n and $x_n \to z$. □

4.7.2 Example. Let $f(x) = \sin x - x/3$. Since

$$f(3\pi/4) = 1/\sqrt{2} - \pi/4 < 0 < 1 - \pi/6 = f(\pi/2),$$

f has a zero in $[3\pi/4, \pi/2]$ by the intermediate value theorem. Taking $x_1 = 3\pi/4$ yields the zero 2.27886266, accurate to eight decimal places. Taking $x_1 = 1$ produces the symmetric zero -2.27886266, while $x_1 = \pi/4$ produces 0. ♢

If x_1 is not sufficiently near z, then the sequence $\{x_n\}$ may converge more slowly to z or may not converge at all (see Exercise 6).

4.7.3 Example. For an approximate solution of $e^x = 2 - x$ we apply Newton's method to $f(x) = e^x + x - 2$. By the intermediate value theorem, f has a zero in $(0, 1)$. The recursion formula for f is

$$x_{n+1} = x_n - (e^{x_n} + x_n - 2)(e^{x_n} + 1)^{-1}.$$

Table 4.1 gives the first few terms of the sequence $\{x_n\}$ and the corresponding

TABLE 4.1: Newton's method for $e^x + x - 2 = 0$.

x_1	x_2	x_3	x_4	x_5
1	.5378828	.4456167	.4428567	.4428544
$f(x_1)$	$f(x_2)$	$f(x_3)$	$f(x_4)$	$f(x_5)$
1.7182818	.2502604	.0070696	.0000059	.0000000
x_1	x_2	x_3	x_4	x_5
5	3.9866142	2.9686340	1.9701667	1.0961884
$f(x_1)$	$f(x_2)$	$f(x_3)$	$f(x_4)$	$f(x_5)$
151.4131591	55.8587993	20.4339472	7.1420387	2.0889256

values of f (accurate up to seven decimal places) for the initial values $x_1 = 1$ and $x_1 = 5$. The convergence is significantly slower for the larger value of x_1. The solution z, accurate to 10 decimal places, is .4428544010. ♢

Exercises

1. Find a zero, accurate to eight decimal places, of the given polynomial in the indicated interval.

   (a) $x^3 - x + 2,\ [-2,-1]$ (b) $x^3 + x + 1,\ [-1,0]$
   (c) $x^3 - 2x + 2,\ [-2,-1]$ (d) $x^5 - 2x + 3,\ [-2,-1]$
   (e) $x^7 - x - 1,\ [1,2]$ (f) $x^4 - 2x^3 + 5x^2 - 8x - 6,\ [2,3]$
   (g) $20x^4 - 20x^3 - 8x^2 + 4x - 1,\ [1,2]$
   (h) $20x^4 - 20x^3 - 4x + 1,\ [1,2]$.

2. Find a solution of the given equation in the indicated interval, correct to eight decimal places.

   (a) $\sin x = x^2,\ [.5,1]$ (b) $\sin x = x^3,\ [.5,1]$
   (c) $\ln x + x = 2,\ [1,2]$ (d) $2\cos x = e^x,\ [0,1]$
   (e) $\ln x = e^{-x},\ [1,2]$ (f) $\tan x + x = 1,\ [0,1]$.

3. Show that Newton's method applied to the function $x^{-1} - c$ produces the equation $x_{n+1} = 2x_n - cx_n^2$. Use this to convert $1/2.34567$ to a decimal, correct to eight decimal places. Check your answer with a calculator.

4. Use Newton's method to find $\sqrt{63}$ correct to eight decimal places. Check your answer with a calculator.

5. What happens when you apply Newton's method with $x_1 = 1$ to the polynomial in part (c) of Exercise 1?

6. Show that the sequence generated by Newton's method applied to $f(x) = x^{1/3}$ cannot converge for any value of $x_1 \neq 0$.

Chapter 5

Riemann Integration on $\mathbb{R}$

In this chapter we construct the Riemann integral of a bounded function using Darboux sums and then Riemann sums. Properties of the integral, as well as analytical methods of evaluation and numerical techniques for approximation, are examined. Improper integrals, which will have important uses in the next chapter, are also discussed.

5.1 The Riemann–Darboux Integral

Throughout this section f denotes an arbitrary bounded function defined on a closed and bounded interval $[a, b]$. The first step in the development of the Riemann–Darboux integral of f is partitioning the interval into finitely many subintervals and forming the so-called *upper* and *lower sums* of f. Under suitable conditions, the sums converge to the integral. Here are the details.

5.1.1 Definition. A *partition of* $[a, b]$ is a set

$$\mathcal{P} = \{x_0, x_1, \ldots, x_{n-1}, x_n\},$$

where

$$x_0 := a < x_1 < \cdots < x_{n-1} < x_n := b.$$

The points $x_1, \ldots, x_{n-1}$ are called the *interior points* of the partition. The *mesh of the partition* is

$$\|\mathcal{P}\| := \max_{1 \le j \le n} \Delta x_j, \quad \text{where} \quad \Delta x_j := x_j - x_{j-1}, \; 1 \le j \le n.$$

A *refinement* of $\mathcal{P}$ is a partition containing $\mathcal{P}$. The *common refinement of partitions* $\mathcal{P}$ *and* $\mathcal{Q}$ is the partition $\mathcal{P} \cup \mathcal{Q}$ (duplicates removed). ◊

5.1.2 Example. Let $p \in \mathbb{N}$. Then, for each $n \in \mathbb{N}$,

$$\mathcal{P}_n := \{j/p^n : j = 0, 1, \ldots, p^n\}$$

is a partition of $[0, 1]$, $\|\mathcal{P}_n\| = p^{-n}$, and $\mathcal{P}_{n+1}$ is a refinement of $\mathcal{P}_n$. ◊

DOI: 10.1201/9781003629382-5

5.1.3 Definition. The *lower* and *upper (Darboux) sums of f over a partition $\mathcal{P}$ of* $[a, b]$ are defined, respectively, by

$$\underline{S}(f, \mathcal{P}) := \sum_{j=1}^{n} m_j \Delta x_j \quad \text{and} \quad \overline{S}(f, \mathcal{P}) := \sum_{j=1}^{n} M_j \Delta x_j,$$

where

$$m_j = m_j(f) := \inf_{x_{j-1} \le x \le x_j} f(x) \quad \text{and} \quad M_j = M_j(f) := \sup_{x_{j-1} \le x \le x_j} f(x). \quad \Diamond$$

A geometric interpretation of the upper and lower sums for a positive continuous function f is given in Figure 5.1. The lower (upper) sum is the total area of the smaller (larger) rectangles.

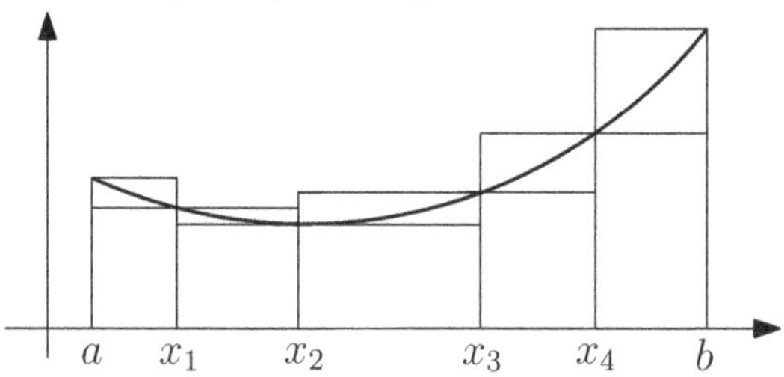

FIGURE 5.1: Upper and lower sums of f.

The following proposition asserts that refinements increase lower sums and decrease upper sums.

5.1.4 Proposition. *If $\mathcal{Q}$ is a refinement of $\mathcal{P}$, then*

$$\underline{S}(f, \mathcal{P}) \le \underline{S}(f, \mathcal{Q}) \le \overline{S}(f, \mathcal{Q}) \le \overline{S}(f, \mathcal{P}).$$

Proof. The middle inequality is clear. To prove the rightmost inequality, let $\mathcal{P} = \{x_0 = a < x_1 < \cdots < x_{n-1} < x_n = b\}$ and assume first that $\mathcal{Q} = \mathcal{P} \cup \{c\}$. Choose k so that $x_{k-1} < c < x_k$ and set

$$M_k' = \sup_{x_{k-1} \le x \le c} f(x) \quad \text{and} \quad M_k'' = \sup_{c \le x \le x_k} f(x).$$

Then M_k', $M_k'' \le M_k$, hence

$$\begin{aligned}
\overline{S}(f, \mathcal{Q}) &= \sum_{j=1}^{k-1} M_j \Delta x_j + \sum_{j=k+1}^{n} M_j \Delta x_j + M_k'(c - x_{k-1}) + M_k''(x_k - c) \\
&\le \sum_{j=1}^{k-1} M_j \Delta x_j + \sum_{j=k+1}^{n} M_j \Delta x_j + M_k(c - x_{k-1}) + M_k(x_k - c) \\
&= \overline{S}(f, \mathcal{P}),
\end{aligned}$$

verifying the inequality for the special case.

For the general case, observe that any refinement $\mathcal{Q}$ of $\mathcal{P}$ may be obtained by successively adding points to $\mathcal{P}$. By the special case, at each step, the upper sum is decreased so that ultimately one obtains the desired inequality. The proof for lower sums is similar. □

5.1.5 Corollary. *For any partitions* $\mathcal{P}$ *and* $\mathcal{Q}$ *of* $[a, b]$,

$$\underline{S}(f, \mathcal{Q}) \le \overline{S}(f, \mathcal{P}). \tag{5.1}$$

Proof. By 5.1.4, $\underline{S}(f, \mathcal{Q}) \le \underline{S}(f, \mathcal{P} \cup \mathcal{Q}) \le \overline{S}(f, \mathcal{P} \cup \mathcal{Q}) \le \overline{S}(f, \mathcal{P})$. □

5.1.6 Definition. The *lower* and *upper* (*Darboux*) *integrals of* f *on* $[a, b]$ are defined, respectively, by

$$\underline{\int_a^b} f = \underline{\int_a^b} f(x)\,dx := \sup_{\mathcal{P}} \underline{S}(f, \mathcal{P}) \quad \text{and} \quad \overline{\int_a^b} f = \overline{\int_a^b} f(x)\,dx := \inf_{\mathcal{P}} \overline{S}(f, \mathcal{P}),$$

where the supremum and infimum are taken over all partitions $\mathcal{P}$ of $[a, b]$. In each case, f is called the *integrand* and x the *integration variable.* ◊

5.1.7 Proposition. *For any partition* $\mathcal{P}$ *of* $[a, b]$,

$$\underline{S}(f, \mathcal{P}) \le \underline{\int_a^b} f \le \overline{\int_a^b} f \le \overline{S}(f, \mathcal{P}).$$

Proof. The left and right inequalities are immediate from the definition of lower and upper integrals. The middle inequality follows by taking the infimum over $\mathcal{P}$ in (5.1) and then the supremum over $\mathcal{Q}$ in the resulting inequality. □

5.1.8 Proposition. *The following statements are equivalent:*

(a) $\displaystyle \underline{\int_a^b} f = \overline{\int_a^b} f.$

(b) *For each* $\varepsilon > 0$ *there exists a partition* $\mathcal{P}_\varepsilon$ *of* $[a, b]$ *such that*

$$\overline{S}(f, \mathcal{P}_\varepsilon) - \underline{S}(f, \mathcal{P}_\varepsilon) \le \varepsilon.$$

Proof. (a) ⇒ (b): Given $\varepsilon > 0$, by the approximation property of suprema and infima there exist partitions $\mathcal{P}'$ and $\mathcal{P}''$ such that

$$\underline{\int_a^b} f - \varepsilon/2 < \underline{S}(f, \mathcal{P}') \quad \text{and} \quad \overline{S}(f, \mathcal{P}'') < \overline{\int_a^b} f + \varepsilon/2.$$

By 5.1.4, the inequalities still hold if $\mathcal{P}'$ and $\mathcal{P}''$ are each replaced by their common refinement $\mathcal{P}_\varepsilon := \mathcal{P}' \cup \mathcal{P}''$. Subtracting the resulting inequalities and applying (a) yields (b).

(b) ⇒ (a): If the inequality in (b) holds, then, by 5.1.7,

$$0 \le \overline{\int_a^b} f - \underline{\int_a^b} f \le \overline{S}(f, \mathcal{P}_\varepsilon) - \underline{S}(f, \mathcal{P}_\varepsilon) < \varepsilon.$$

Since ε is arbitrary, the integrals must be equal. □

5.1.9 Definition. The function f is said to be (*Darboux*) *integrable on* $[a, b]$ if one (hence both) of the conditions (a), (b) of 5.1.8 hold. In this case, the common value of the integrals in (a) is called the (*Riemann–Darboux*) *integral of* f *on* $[a, b]$ and is denoted by

$$\int_a^b f = \int_a^b f(x)\, dx.$$

The definition is extended in the following manner:

$$\int_b^a f = -\int_a^b f \quad \text{and} \quad \int_a^a f = 0.$$

The collection of all integrable functions on $[a, b]$ is denoted by $\mathcal{R}_a^b$. ◊

The following theorem guarantees a rich supply of integrable functions.

5.1.10 Theorem. *If f is continuous on $[a, b]$ except possibly at finitely many points, then $f \in \mathcal{R}_a^b$.*

Proof. Denote the points of discontinuity of f, if any, by $d_1 < \cdots < d_n$. For convenience, we assume that these lie in (a, b); only a minor modification of the proof is needed if $d_1 = a$ or $d_n = b$. Let $\varepsilon > 0$. For each j, remove an open interval of width r centered at d_j, the value of r to be determined. Since f is continuous on each of the resulting $n + 1$ closed intervals $I_0, \ldots, I_n$, it is uniformly continuous there. (If f is continuous on $[a, b]$, then $n = 0$ and $I_0 = [a, b]$.) Thus there exists a $\delta > 0$ such that for each j,

$$|f(x) - f(y)| < \varepsilon/2(b - a) \quad \text{for all } x,\ y \in I_j \text{ with } |x - y| < \delta.$$

The endpoints of the intervals I_j form a partition $\mathcal{P}$ of $[a, b]$. If necessary, refine $\mathcal{P}$ by inserting points (marked by $*$ in Figure 5.2) into these intervals so that the distance between consecutive points is less than δ. The subintervals of the resulting partition $\mathcal{Q}$ are of two types: those that contain some d_j, which we mark by α, and those that do not, which we mark by β. Thus, in the obvious notation,

$$\overline{S}(f, \mathcal{Q}) - \underline{S}(f, \mathcal{Q}) = \sum_\alpha (M_j - m_j)\Delta x_j + \sum_\beta (M_j - m_j)\Delta x_j.$$

In the first sum, $\Delta x_j = r$ and in the second, $M_j - m_j \le \varepsilon/2(b - a)$. Since the first sum has n terms (corresponding to the n discontinuities d_j),

$$\overline{S}(f, \mathcal{Q}) - \underline{S}(f, \mathcal{Q}) < 2Mnr + \varepsilon/2,$$

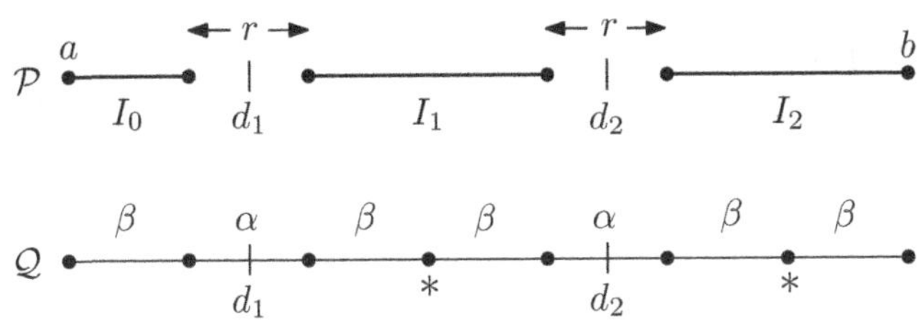

FIGURE 5.2: The partitions $\mathcal{P}$ and $\mathcal{Q}$.

where M is a bound for $|f|$ on $[a,b]$. Choosing $r < \varepsilon/4Mn$, we then have $\overline{S}(f,\mathcal{Q}) - \underline{S}(f,\mathcal{Q}) < \varepsilon$, which shows that f is integrable on $[a,b]$. □

The set of discontinuities of an integrable function can be infinite but cannot be too large. We make this precise in Section 5.8. In the meantime, we offer the following examples to illustrate the basic idea. In the first example, the function is discontinuous precisely on a countably infinite set, while in the second the function is discontinuous everywhere.

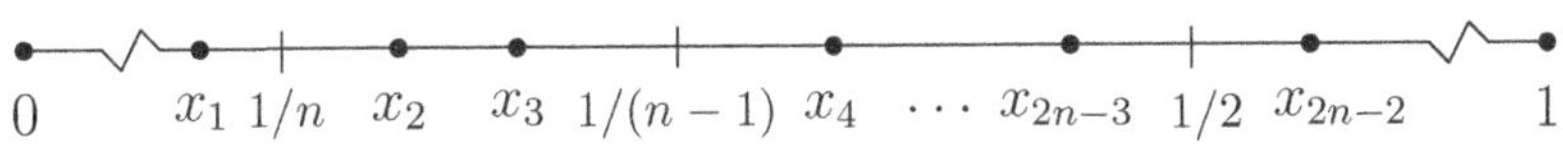

FIGURE 5.3: The partition $\mathcal{P}_n$ of Example 5.1.11.

5.1.11 Example. Let f be any bounded function on $[0,1]$ such that $f(x) = 0$ if $x \notin \{1/n : n = 2, 3 \ldots\}$. We claim that f is integrable and that $\int_0^1 f = 0$. The idea is to enclose the points of discontinuity of f in small intervals, as in the proof of 5.1.10. Fix n and let

$$\mathcal{P}_n = \{x_0 = 0, x_1, x_2, \ldots, x_{2n-2}, x_{2n-1} = 1\},$$

where

$$x_{2j-1} < 1/(n-j+1) < x_{2j} < x_{2j+1}, \quad j = 1, 2, \ldots, n-1, \quad \text{and}$$
$$\Delta x_{2j} = x_{2j} - x_{2j-1} < 1/n^2, \quad j = 1, 2, \ldots, n.$$

(See Figure 5.3.) Let $|f| \le M$ on $[0,1]$. Since $f = 0$ on $[x_{2j}, x_{2j+1}]$ and $m_j \ge -M$,

$$\begin{aligned}\underline{S}(f,\mathcal{P}_n) &= m_1 x_1 + m_2(x_2 - x_1) + \cdots + m_{2n-2}(x_{2n-2} - x_{2n-3})\\ &\ge -M\big[x_1 + (x_2 - x_1) + (x_4 - x_3) + \cdots + (x_{2n-2} - x_{2n-3})\big]\\ &\ge -M(1/n + (n-1)/n^2) = -M(2/n - 1/n^2).\end{aligned}$$

A similar calculation shows that $\overline{S}(f,\mathcal{P}_n) \le M(2/n - 1/n^2)$. Therefore,

$$\lim_n \underline{S}(f,\mathcal{P}_n) = \lim_n \overline{S}(f,\mathcal{P}_n) = 0,$$

hence f is integrable with zero integral. ◊

5.1.12 Example. The Dirichlet function $d(x)$ (3.1.7) is not integrable on any (nondegenerate) interval $[a,b]$. Indeed, every upper sum of $d(x)$ has the value $b-a$ and every lower sum has the value 0. ◊

A useful characterization of integrability may be given in terms of the limits of $\underline{S}(f,\mathcal{P})$ and $\overline{S}(f,\mathcal{P})$ as $\|\mathcal{P}\| \to 0$, defined as follows:

5.1.13 Definition. Let $L \in \mathbb{R}$. We write $L = \lim_{\|\mathcal{P}\|\to 0} \overline{S}(f,\mathcal{P})$ if, given $\varepsilon > 0$, there exists $\delta > 0$ such that $|\overline{S}(f,\mathcal{P}) - L| < \varepsilon$ for all partitions $\mathcal{P}$ with $\|\mathcal{P}\| < \delta$. The limit $\lim_{\|\mathcal{P}\|\to 0} \underline{S}(f,\mathcal{P})$ is defined analogously. ◊

For the promised characterization we need the following lemma

5.1.14 Lemma. *Let $\mathcal{P}' = \{x_0' = a < x_1' < \cdots x_n' < x_{n+1}' = b\}$ be a partition of $[a,b]$ and let $|f| \le M$ on $[a,b]$. Then*

$$\overline{S}(f,\mathcal{P}) \le \overline{S}(f,\mathcal{P}') + 3nM\|\mathcal{P}\|$$

for all partitions $\mathcal{P}$ of $[a,b]$ with $\|\mathcal{P}\| < \delta' := \min_j \Delta x_j'$.

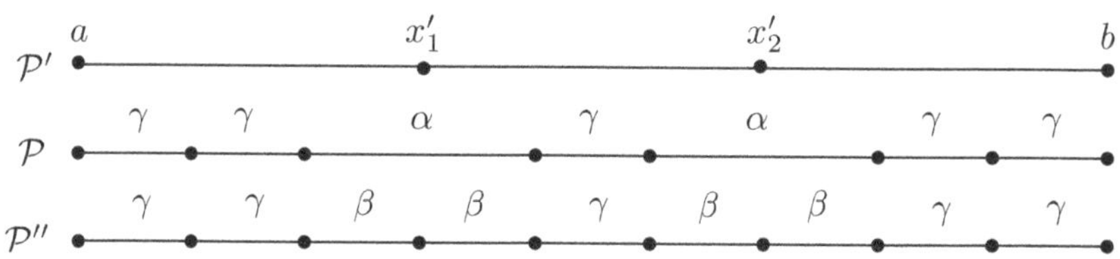

FIGURE 5.4: The partitions $\mathcal{P}'$, $\mathcal{P}$, and $\mathcal{P}''$.

Proof. Since $\|\mathcal{P}\| < \Delta x_j'$, no interval of $\mathcal{P}$ can contain more that one interior point of $\mathcal{P}'$. Mark the intervals of $\mathcal{P}$ that contain exactly one interior point of $\mathcal{P}'$ by α and mark those that contain no interior point of $\mathcal{P}'$ by γ. Consider the common refinement $\mathcal{P}'' = \mathcal{P} \cup \mathcal{P}'$ of $\mathcal{P}$ and $\mathcal{P}'$. Some of the intervals of $\mathcal{P}''$ were formed from an interval of $\mathcal{P}$ of type α; we mark those by β. The remaining intervals of $\mathcal{P}''$, intervals that were not formed from an interval of $\mathcal{P}$ of type α are precisely the intervals marked γ in $\mathcal{P}$. Thus the terms of $\overline{S}(f,\mathcal{P})$ and $\overline{S}(f,\mathcal{P}'')$ corresponding to intervals of type γ are identical, hence cancel under substraction of upper sums. Therefore, in the obvious notation,

$$\begin{aligned}
\overline{S}(f,\mathcal{P}) - \overline{S}(f,\mathcal{P}'') &= \sum_{\alpha} M_j(f)\Delta x_j - \sum_{\beta} M_j''(f)\Delta x_j'' \\
&\le M\Big[\sum_{\alpha} \Delta x_j + \sum_{\beta} \Delta x_j''\Big] \\
&\le M\big(n\|\mathcal{P}\| + 2n\|\mathcal{P}''\|\big),
\end{aligned}$$

the last inequality because there are at most n intervals of type α and at most $2n$ intervals of type β. Since $\mathcal{P}''$ is a refinement of $\mathcal{P}'$ and $\mathcal{P}$,

$$\overline{S}(f,\mathcal{P}) - \overline{S}(f,\mathcal{P}') \le \overline{S}(f,\mathcal{P}) - \overline{S}(f,\mathcal{P}'') \le 3nM\|\mathcal{P}\|. \qquad \square$$

5.1.15 Theorem. *For any bounded function f on $[a,b]$,*

$$\overline{\int_a^b} f = \lim_{\|\mathcal{P}\|\to 0} \overline{S}(f,\mathcal{P}) \quad \text{and} \quad \underline{\int_a^b} f = \lim_{\|\mathcal{P}\|\to 0} \underline{S}(f,\mathcal{P}). \tag{5.2}$$

Thus f is integrable on $[a,b]$ iff the limits in (5.2) are equal, in which case

$$\int_a^b f = \lim_{\|\mathcal{P}\|\to 0} \overline{S}(f,\mathcal{P}) = \lim_{\|\mathcal{P}\|\to 0} \underline{S}(f,\mathcal{P}). \tag{5.3}$$

Proof. By the approximation property for infima, given $\varepsilon > 0$ we may choose a partition $\mathcal{P}'$ such that

$$\overline{S}(f,\mathcal{P}') < \overline{\int_a^b} f + \varepsilon/2.$$

In the notation of 5.1.14, for any partition $\mathcal{P}$ with $\|\mathcal{P}\| < \delta'$,

$$\overline{S}(f,\mathcal{P}) \le \overline{S}(f,\mathcal{P}')| + 3nM\|\mathcal{P}\| < \overline{\int_a^b} f + \varepsilon/2 + 3nM\|\mathcal{P}\|.$$

Hence if $\|\mathcal{P}\| < \min\{\delta', \varepsilon/6nM\}$, then

$$\overline{\int_a^b} f \le \overline{S}(f,\mathcal{P}) < \overline{\int_a^b} f + \varepsilon.$$

Letting $\varepsilon \to 0$ we obtain the first limit in (5.2). The second limit follows by a similar argument or by using the first limit on $-f$ and Exercise 5.1.3. $\square$

Equation (5.3) represents the integral as a limit of upper and lower sums. It is also possible to represent the integral as a limit of intermediate sums, called *Riemann sums.*

5.1.16 Definition. Let $\mathcal{P} = \{x_0 = a < x_1 < \cdots < x_n = b\}$ be a partition of $[a,b]$ and let $\boldsymbol{\xi} = (\xi_1, \ldots, \xi_n)$, where $\xi_j \in [x_{j-1}, x_j]$. The sum

$$S(f,\mathcal{P},\boldsymbol{\xi}) := \sum_{j=1}^{n} f(\xi_j)\Delta x_j$$

is called the *Riemann sum of f determined by $\mathcal{P}$ and $\boldsymbol{\xi}$.* $\Diamond$

Figure 5.5 illustrates a Riemann sum for a positive continuous function f. In this case $S(f,\mathcal{P},\boldsymbol{\xi})$ is the total area of the rectangles with heights $f(\xi_j)$ and bases Δx_j.

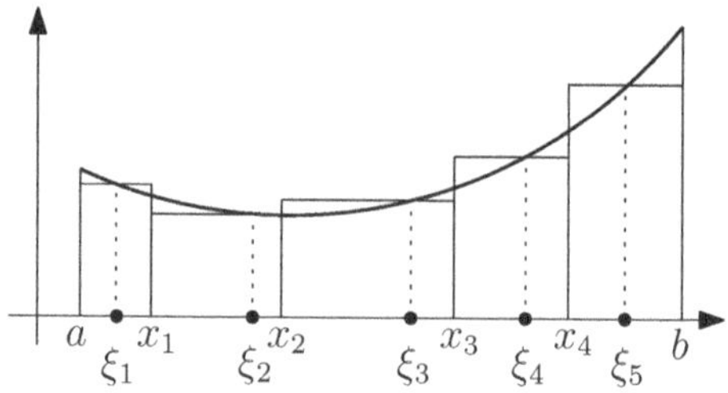

FIGURE 5.5: A Riemann sum.

5.1.17 Definition. Let $\mathcal{P} = \{x_0 = a < x_1 < \cdots < x_n = b\}$ be a partition of $[a, b]$ and let $\boldsymbol{\xi} = (\xi_1, \ldots, \xi_n)$, where $\xi_j \in [x_{j-1}, x_j]$. We write

$$L = \lim_{\|\mathcal{P}\| \to 0} S(f, \mathcal{P}, \boldsymbol{\xi})$$

if for each $\varepsilon > 0$ there exists $\delta > 0$ such that $|S(f, \mathcal{P}, \boldsymbol{\xi}) - L| < \varepsilon$ for all partitions $\mathcal{P}$ with $\|\mathcal{P}\| < \delta$ and all choices of $\boldsymbol{\xi}$. Similarly, we write

$$L = \lim_{\mathcal{P}} S(f, \mathcal{P}, \boldsymbol{\xi})$$

if for each $\varepsilon > 0$ there exists a partition $\mathcal{P}_\varepsilon$ such that $|S(f, \mathcal{P}, \boldsymbol{\xi}) - L| < \varepsilon$ for all refinements $\mathcal{P}$ of $\mathcal{P}_\varepsilon$ and all choices of $\boldsymbol{\xi}$. $\diamond$

We may now give Riemann's characterization of integrability.

5.1.18 Theorem. *The following statements are equivalent:*

(a) $f \in \mathcal{R}_a^b$.

(b) $\lim_{\|\mathcal{P}\| \to 0} S(f, \mathcal{P}, \boldsymbol{\xi})$ *exists in* $\mathbb{R}$.

(c) $\lim_{\mathcal{P}} S(f, \mathcal{P}, \boldsymbol{\xi})$ *exists in* $\mathbb{R}$.

If these conditions hold, then

$$\int_a^b f = \lim_{\|\mathcal{P}\| \to 0} S(f, \mathcal{P}, \boldsymbol{\xi}) = \lim_{\mathcal{P}} S(f, \mathcal{P}, \boldsymbol{\xi}).$$

Proof. (a) $\Rightarrow$ (b): Let $L = \int_a^b f$. For any partition $\mathcal{P}$ and any $\boldsymbol{\xi}$, we have

$$\underline{S}(f, \mathcal{P}) - L \le S(f, \mathcal{P}, \boldsymbol{\xi}) - L \le \overline{S}(f, \mathcal{P}) - L,$$

hence (b) follows from 5.1.15.

(b) $\Rightarrow$ (c): Let $L := \lim_{\|\mathcal{P}\| \to 0} S(f, \mathcal{P}, \boldsymbol{\xi})$. Given $\varepsilon > 0$, choose $\delta > 0$ such that

$$|S(f, \mathcal{P}, \boldsymbol{\xi}) - L| < \varepsilon \tag{5.4}$$

for all partitions $\mathcal{P}$ with $\|\mathcal{P}\| < \delta$ and all $\boldsymbol{\xi}$. Choose any partition $\mathcal{P}_\varepsilon$ with $\|\mathcal{P}_\varepsilon\| < \delta$. If $\mathcal{P}$ is any refinement of $\mathcal{P}_\varepsilon$, then $\|\mathcal{P}\| \le \|\mathcal{P}_\varepsilon\| < \delta$, hence (5.4) holds for $\mathcal{P}$.

(c) $\Rightarrow$ (a): Let $L := \lim_{\mathcal{P}} S(f, \mathcal{P}, \boldsymbol{\xi})$. Given $\varepsilon > 0$, choose a partition $\mathcal{P}_\varepsilon$ such that

$$|S(f, \mathcal{P}, \boldsymbol{\xi}) - L| < \varepsilon \tag{5.5}$$

for all refinements $\mathcal{P}$ of $\mathcal{P}_\varepsilon$ and all $\boldsymbol{\xi}$. For such a partition $\mathcal{P}$, by the approximation property of suprema there exists for each j a sequence $\{\xi_{j,k}\}_{k=1}^\infty$ in $[x_{j-1}, x_j]$ such that $\lim_k f(\xi_{j,k}) = M_j(f)$. It follows that

$$\lim_k S(f, \mathcal{P}, \boldsymbol{\xi}_k) = \overline{S}(f, \mathcal{P}), \quad \text{where } \boldsymbol{\xi}_k = (\xi_{1k}, \xi_{2k}, \ldots, \xi_{nk}).$$

From (5.5), $\overline{\int_a^b} f - L \le \overline{S}(f, \mathcal{P}) - L \le \varepsilon$. Since ε was arbitrary, $\overline{\int_a^b} f \le L$. Similarly, $\underline{\int_a^b} f \ge L$. Therefore $\underline{\int_a^b} f = \overline{\int_a^b} f$. $\square$

Exercises

1. Prove that if k is a constant, then $\underline{\int_a^b} k = \overline{\int_a^b} k = k(b-a)$.

2. Let $a \le c < d \le b$. Define f on $[a, b]$ by $f(x) = 1$ if $x \in [c, d]$ and $f(x) = 0$ otherwise. Show that $f \in \mathcal{R}_a^b$ and evaluate $\int_a^b f$.

3. ($\Downarrow$ 5.2.2.) Prove:

   (a) $\overline{S}(-f, \mathcal{P}) = -\underline{S}(f, \mathcal{P})$ and $\overline{\int_a^b}(-f) = -\underline{\int_a^b} f$

   (b) $f \in \mathcal{R}_a^b \Rightarrow -f \in \mathcal{R}_a^b$ and $\int_a^b(-f) = -\int_a^b f$

4. ($\Downarrow$ 5.9.8.) Prove that a monotone function is integrable.

5. Let $f \in \mathcal{R}_a^b$ and let $g : [a, b] \to \mathbb{R}$ be any function that differs from f at finitely many points in $[a, b]$. Prove that $g \in \mathcal{R}_a^b$ and that $\int_a^b f = \int_a^b g$. Does the same result hold if g differs from f at countably many points?

6. Let $f \in \mathcal{R}_a^b$. Prove:

   (a) If $\inf_{a \le x \le b} f(x) > 0$, then $1/f \in \mathcal{R}_a^b$.

   (b) If $f(x) \ge 0$ for all $x \in [a, b]$, then $\sqrt{f} \in \mathcal{R}_a^b$.

   (c) $\sin(f) \in \mathcal{R}_a^b$.

7. Let $F(\mathcal{P})$ be a real-valued function of partitions $\mathcal{P}$ on an interval $[a, b]$. Write $L = \lim_{\mathcal{P}} F(\mathcal{P})$ if given $\varepsilon > 0$ there exists a partition $\mathcal{P}_\varepsilon$ such that $|F(\mathcal{P}) - L| < \varepsilon$ for all partitions $\mathcal{P}$ refining $\mathcal{P}_\varepsilon$.

   (a) Show that the limit is linear, that is,

   $$\lim_{\mathcal{P}} \left[aF(\mathcal{P}) + bG(\mathcal{P})\right] = a \lim_{\mathcal{P}} F(\mathcal{P}) + b \lim_{\mathcal{P}} G(\mathcal{P}),$$

   provided the right side exists.

(b) Let f be a bounded function on $[a, b]$. With this definition, show that

$$\underline{\int_a^b} f = \lim_{\mathcal{P}} \underline{S}(f, \mathcal{P}) \quad \text{and} \quad \overline{\int_a^b} f = \lim_{\mathcal{P}} \overline{S}(f, \mathcal{P}).$$

8. Let $f \in \mathcal{R}_0^1$ and set $g(x) = x^q$, where $q > 0$. Prove that $f \circ g \in \mathcal{R}_0^1$.

5.2 Properties of the Integral

The following lemma will be useful in proving certain combinatorial properties of the integral.

5.2.1 Lemma. *Let $f : [a, b] \to \mathbb{R}$ be bounded. Then there exists a sequence of partitions $\{\mathcal{P}_n\}$ of $[a, b]$ such that*

$$\lim_{n\to\infty} \overline{S}(f, \mathcal{P}_n) = \overline{\int_a^b} f \quad \textit{and} \quad \lim_{n\to\infty} \underline{S}(f, \mathcal{P}_n) = \underline{\int_a^b} f.$$

Moreover, the limits still hold if each $\mathcal{P}_n$ is replaced by a refinement.

Proof. By the approximation property of infima and suprema, for each n there exist partitions $\mathcal{P}_n'$ and $\mathcal{P}_n''$ of $[a, b]$ such that

$$\underline{\int_a^b} f - 1/n < \underline{S}(f, \mathcal{P}_n') \le \underline{\int_a^b} f \quad \text{and} \quad \overline{\int_a^b} f \le \overline{S}(f, \mathcal{P}_n'') < \overline{\int_a^b} f + 1/n.$$

Since refinements decrease upper sums and increase lower sums, the inequalities still hold if $\mathcal{P}_n'$ and $\mathcal{P}_n''$ are replaced by their common refinement $\mathcal{P}_n$ or by any refinement of $\mathcal{P}_n$. Letting $n \to \infty$ yields the desired conclusions. □

5.2.2 Theorem. *If f, $g \in \mathcal{R}_a^b$ and α, $\beta \in \mathbb{R}$, then $\alpha f + \beta g \in \mathcal{R}_a^b$ and*

$$\int_a^b (\alpha f + \beta g) = \alpha \int_a^b f + \beta \int_a^b g.$$

Proof. By 5.2.1, we may choose a sequence of partitions $\mathcal{P}_n$ such that

$$\lim_n \overline{S}(f, \mathcal{P}_n) = \int_a^b f \quad \text{and} \quad \lim_n \overline{S}(g, \mathcal{P}_n) = \int_a^b g.$$

(There exists one such sequence for f, another for g; the sequence of common refinements then works for both functions.) Letting $n \to \infty$ in

$$\overline{\int_a^b} (f + g) \le \overline{S}(f + g, \mathcal{P}_n) \le \overline{S}(f, \mathcal{P}_n) + \overline{S}(g, \mathcal{P}_n)$$

yields

$$\overline{\int_a^b}(f+g) \le \int_a^b f + \int_a^b g.$$

Similarly,

$$\underline{\int_a^b}(f+g) \ge \int_a^b f + \int_a^b g.$$

It follows that $f+g$ is integrable and

$$\int_a^b (f+g) = \int_a^b f + \int_a^b g.$$

It remains to prove that αf is integrable and that $\int_a^b \alpha f = \alpha \int_a^b f$. If $\alpha > 0$, then

$$\overline{S}(\alpha f, \mathcal{P}) = \alpha \overline{S}(f, \mathcal{P}) \text{ and } \underline{S}(\alpha f, \mathcal{P}) = \alpha \underline{S}(f, \mathcal{P}).$$

Taking the infimum and supremum over $\mathcal{P}$ yields

$$\overline{\int_a^b} \alpha f = \alpha \int_a^b f = \underline{\int_a^b} \alpha f.$$

If $\alpha < 0$, then $-\alpha > 0$, hence

$$\int_a^b \alpha f = \int_a^b (-\alpha)(-f) = (-\alpha)\int_a^b (-f) = \alpha \int_a^b f,$$

the last equality by Exercise 5.1.3. □

5.2.3 Proposition. *If $f \in \mathcal{R}_a^b$ and $a \le c < d \le b$, then $f|_{[c,d]} \in \mathcal{R}_c^d$.*

Proof. Given $\varepsilon > 0$, let $\mathcal{P}$ be a partition of $[a,b]$ with $\overline{S}(f,\mathcal{P}) - \underline{S}(f,\mathcal{P}) < \varepsilon$. We may assume that c, $d \in \mathcal{P}$, otherwise replace $\mathcal{P}$ by the refinement $\mathcal{P} \cup \{c, d\}$. If $\mathcal{Q} = \mathcal{P} \cap [c,d]$, then clearly

$$\overline{S}(f|_{[c,d]}, \mathcal{Q}) - \underline{S}(f|_{[c,d]}, \mathcal{Q}) \le \overline{S}(f,\mathcal{P}) - \underline{S}(f,\mathcal{P}) < \varepsilon,$$

hence $f|_{[c,d]} \in \mathcal{R}_c^d$. □

The following theorem is a converse of 5.2.3.

5.2.4 Theorem. *Let $a < c < b$. If $f|_{[a,c]} \in \mathcal{R}_a^c$ and $f|_{[c,b]} \in \mathcal{R}_c^b$, then $f \in \mathcal{R}_a^b$ and*

$$\int_a^b f = \int_a^c f + \int_c^b f.$$

Proof. By 5.2.1, we may choose sequences of partitions $\mathcal{P}_n'$ of $[a,c]$ and $\mathcal{P}_n''$ of $[c,b]$ such that

$$\lim_n \overline{S}(f|_{[a,c]}, \mathcal{P}_n') = \int_a^c f \quad \text{and} \quad \lim_n \overline{S}(f|_{[c,b]}, \mathcal{P}_n'') = \int_c^b f.$$

Then $\mathcal{P}_n := \mathcal{P}_n' \cup \mathcal{P}_n''$ is a partition of $[a, b]$ and

$$\overline{\int_a^b} f \le \overline{S}(f, \mathcal{P}_n) = \overline{S}(f|_{[a,c]}, \mathcal{P}_n') + \overline{S}(f|_{[c,b]}, \mathcal{P}_n'').$$

Letting $n \to \infty$, we obtain

$$\overline{\int_a^b} f \le \int_a^c f + \int_c^b f.$$

Replacing f by $-f$ produces the reverse inequality for the lower integral of f, proving the theorem. □

5.2.5 Theorem. *If f, $g \in \mathcal{R}_a^b$ and $f \le g$ on $[a, b]$, then*

$$\int_a^b f \le \int_a^b g.$$

In particular, if $m \le f(x) \le M$ for all $x \in [a, b]$, then

$$m(b-a) \le \int_a^b f \le M(b-a).$$

Proof. Let $\mathcal{P}$ be a partition of $[a, b]$. By hypothesis, $M_j(f) \le M_j(g)$ for each j, hence $\overline{S}(f, \mathcal{P}) \le \overline{S}(g, \mathcal{P})$. Taking the infimum over $\mathcal{P}$ yields the first inequality. The remaining inequalities follow from the first and Exercise 5.1.1. □

5.2.6 Theorem. *If $f \in \mathcal{R}_a^b$, then $|f| \in \mathcal{R}_a^b$ and* $\left|\int_a^b f\right| \le \int_a^b |f|$.

Proof. By Exercise 1.3.5, for any partition $\mathcal{P}$ of $[a, b]$,

$$M_j(|f|) - m_j(|f|) \le M_j(f) - m_j(f).$$

Summing over j we have

$$\overline{S}(|f|, \mathcal{P}) - \underline{S}(|f|, \mathcal{P}) \le \overline{S}(f, \mathcal{P}) - \underline{S}(f, \mathcal{P}).$$

Since the right side can be made arbitrarily small, $|f| \in \mathcal{R}_a^b$. Applying 5.2.5 to the inequality $\pm f \le |f|$ yields

$$\pm \int_a^b f = \int_a^b (\pm f) \le \int_a^b |f|,$$

which gives the desired inequality. □

5.2.7 Theorem. *If f, $g \in \mathcal{R}_a^b$, then $fg \in \mathcal{R}_a^b$.*

Proof. Since $fg = \frac{1}{2}\big[(f+g)^2 - f^2 - g^2\big]$, it suffices to prove that the square of an integrable function f is integrable. To this end, let $\mathcal{P}$ be any partition of $[a, b]$ and let $|f| \le M$. Then

$$M_j(f^2) - m_j(f^2) = M_j^2(|f|) - m_j^2(|f|) \le 2M\big[M_j(|f|) - m_j(|f|)\big].$$

Multiplying by Δx_j and summing over j we obtain

$$\overline{S}(f^2, \mathcal{P}) - \underline{S}(f^2, \mathcal{P}) \le 2M\big[\overline{S}(|f|, \mathcal{P}) - \underline{S}(|f|, \mathcal{P})\big].$$

Since $|f| \in \mathcal{R}_a^b$, the right side of the last inequality may be made arbitrarily small. Therefore, $f^2 \in \mathcal{R}_a^b$. □

Exercises

1. Let $\{c_n\}$ be a convergent sequence in $[a, b]$ and let f be a bounded function on $[a, b]$ with $f(x) = 0$ for all $x \notin \{c_n\}$. Prove that $f \in \mathcal{R}_a^b$ and find $\int_a^b f$.

2. Define f on $[0, 1]$ by $f(0) = 0$ and

$$f(x) = 2^{-n} \quad \text{if } 2^{-n-1} < x \le 2^{-n}, \quad n \ge 0.$$

   Prove that $f \in \mathcal{R}_0^1$ and evaluate $\int_0^1 f$.

3. Prove or disprove: $|f| \in \mathcal{R}_a^b$ implies $f \in \mathcal{R}_a^b$.

4. A function s on $[a, b]$ is called a *step function* if there exists a partition of $[a, b]$ such that s is constant on the interior of each partition interval. Show that a step function is integrable. Prove that a bounded function f is integrable on $[a, b]$ iff for each $\varepsilon > 0$ there exist step functions s_ℓ and s_u such that $s_\ell \le f \le s_u$ and $\int_a^b (s_u - s_\ell) < \varepsilon$.

5. Prove that if $f_j \in \mathcal{R}_a^b$, $1 \le j \le n$, then $\max\{f_1, \ldots, f_n\} \in \mathcal{R}_a^b$ and $\min\{f_1, \ldots, f_n\} \in \mathcal{R}_a^b$.

6. Let f be continuous and $f(x) < M$ for all $x \in [a, b]$. Prove that $\int_a^b f < M(b-a)$. (Compare with 5.2.5.)

7. Let $f \in \mathcal{R}_a^b$ be nonnegative. Prove that if f is continuous at some point $x_0 \in [a, b]$ and $f(x_0) \ne 0$, then $\int_a^b f > 0$.

8. Let $f \in \mathcal{R}_a^b$ such that either

   (a) $\int_a^b fg = 0$ for every continuous function g, or

   (b) $\int_a^b fg = 0$ for every step function g.

   Prove that f is zero at each point of continuity of f.

9. Let $f \in \mathcal{R}_a^b$ and for $x, y \in [a, b]$ define $F(x, y) = \int_x^y f$. Prove that $F(x, y)$ is continuous in y for each x and continuous in x for each y.

10. Let f be bounded on $[a, b]$ and integrable $[c, b]$ for every $a < c < b$. Prove that the following statements are equivalent:

    (a) $\displaystyle\lim_{x \to a^+} \int_x^b f$ exists in $\mathbb{R}$.

    (b) $\displaystyle\lim_{n \to +\infty} \int_{a_n}^b f$ exists in $\mathbb{R}$ for some sequence $a_n \downarrow a$.

    (c) $f \in \mathcal{R}_a^b$.

    Conclude from Exercise 9 that if $f \in \mathcal{R}_a^b$, then the limit in (a) is $\int_a^b f$.

11. Let f be integrable on $[0, x]$ for all $x > 0$. Prove that

$$\liminf_{x \to +\infty} f(x) \le \liminf_{x \to +\infty} \frac{1}{x} \int_0^x f \le \limsup_{x \to +\infty} \frac{1}{x} \int_0^x f \le \limsup_{x \to +\infty} f(x).$$

    Conclude that if $L := \lim_{x \to +\infty} f(x)$ exists in $\mathbb{R}$, then

$$\lim_{x \to +\infty} \frac{1}{x} \int_0^x f(t)\, dt = L.$$

12. Let f be continuous on $[a, b]$ and let $M = \sup_{a \le x \le b} |f(x)|$. Prove:

    (a) For each $\varepsilon > 0$ there exists $\delta > 0$ such that

$$\delta(M - \varepsilon) \le \int_a^b |f(x)|\, dx \le M(b - a).$$

    (b) $\displaystyle M = \lim_{p \to +\infty} \left(\int_a^b |f|^p \right)^{1/p}$.

13. ($\Downarrow$ 5.7.19.) Let $f, g : [a, b] \to \mathbb{R}$ be continuous. Supply the details in the following outline of a proof of the *Cauchy–Schwarz inequality*.

$$\left(\int_a^b fg \right)^2 \le \left(\int_a^b f^2 \right) \left(\int_a^b g^2 \right).$$

    (a) The inequality holds if $\displaystyle\int_a^b g^2 = 0$.

    (b) For any real number t,

$$0 \le \int_a^b (f - tg)^2 = \int_a^b f^2 - 2t \int_a^b fg + t^2 \int_a^b g^2.$$

    (c) Let $\displaystyle t = \left(\int_a^b fg \right) \left(\int_a^b g^2 \right)^{-1}$ in (b).

5.3 Evaluation of the Integral

The theorems in this section describe standard methods for evaluating integrals. The first of these expresses the integral of a function f in terms of a *primitive* or *antiderivative*, that is, a function whose derivative is f. It also shows that the process of integration is the inverse of that of differentiation.

5.3.1 Fundamental Theorem of Calculus. *Let $f : [a, b] \to \mathbb{R}$ be continuous. Then the following hold:*

(a) *The function $G(x) := \int_a^x f(t)\,dt$, $x \in [a, b]$, is a primitive of f.*

(b) *For any primitive F of f, $\int_a^b f = F(x)\Big|_a^b := F(b) - F(a)$.*

(c) *If $f' \in \mathcal{R}_a^b$, then $\int_a^b f' = f(b) - f(a)$. In particular, $f(x) = f(a) + \int_a^x f'$.*

Proof. (a) For definiteness we assume that $a \le x < b$ and prove that

$$\lim_{h \to 0+} \frac{G(x+h) - G(x)}{h} = f(x). \tag{5.6}$$

By 5.2.4 and 5.2.6, if $h > 0$ and $x + h < b$, then

$$\left| \frac{G(x+h) - G(x)}{h} - f(x) \right| = \left| \frac{1}{h} \int_x^{x+h} [f(t) - f(x)]\,dt \right| \le \frac{1}{h} \int_x^{x+h} |f(t) - f(x)|\,dt.$$

By continuity of f at x, given $\varepsilon > 0$ we may choose $\delta > 0$ such that $|t - x| < \delta$ implies $|f(t) - f(x)| < \varepsilon$. Thus if $h < \delta$, then the term on the right in the above inequality is $\le \varepsilon$, proving (5.6).

(b) Let F be any primitive of f. Then $F' = f = G'$, hence $F = G + c$ for some constant c. Thus from (a),

$$\int_a^b f = G(b) - G(a) = F(b) - F(a).$$

(c) For any partition $\mathcal{P}$, by the mean value theorem

$$f(x_j) - f(x_{j-1}) = f'(\xi_j)\Delta x_j \quad \text{for some } \xi_j \in [x_{j-1}, x_j],\ j = 1, \ldots, n.$$

For this choice of ξ_j,

$$S(f', \mathcal{P}, \boldsymbol{\xi}) = \sum_{j=1}^{n} f'(\xi_j)\Delta x_j = \sum_{j=1}^{n} [f(x_j) - f(x_{j-1})] = f(b) - f(a).$$

Since we may choose $\mathcal{P}$ so that $S(f', \mathcal{P}, \boldsymbol{\xi})$ is arbitrarily near $\int_a^b f'$, (c) follows. $\square$

The general primitive of a continuous function f is denoted by $\int f$ and is called the *indefinite integral of* f. (In this context, $\int_a^b f$ is called a *definite integral.*) For example, one writes $\int 3x^2\, dx = x^3 + c$, where c is the so-called *constant of integration.* In general, since primitives of a function differ only by a constant, we write

$$\int f(x)\, dx = F(x) + c,$$

where F is any *particular* primitive of f.

5.3.2 Change of Variables Theorem. *Let* $\varphi : [a, b] \to \mathbb{R}$ *be continuously differentiable with* φ' *never zero and let* f *be integrable on* $[c, d] := \varphi([a, b])$. *Then* $(f \circ \varphi)|\varphi'| \in \mathcal{R}_a^b$ *and*

$$\int_a^b f(\varphi(x))|\varphi'(x)|\, dx = \int_c^d f(y)\, dy. \tag{5.7}$$

Proof. By the intermediate value theorem, we may assume that $\varphi'(x) > 0$ for all x, so φ is strictly increasing with $c = \varphi(a)$ and $d = \varphi(b)$.

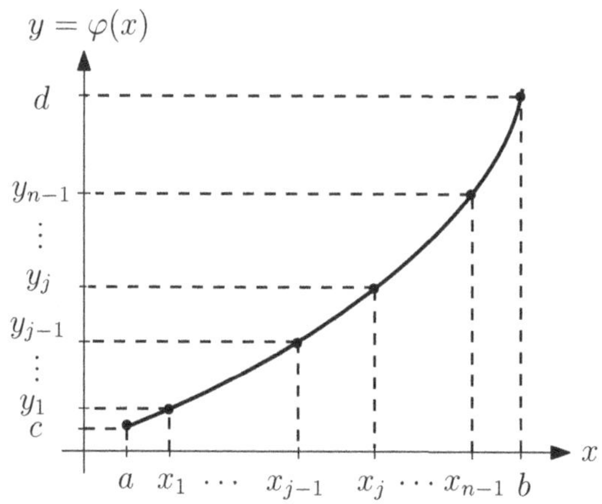

FIGURE 5.6: The partitions $\mathcal{P}^x$ and $\mathcal{P}^y$.

We show first that $f \circ \varphi \in \mathcal{R}_a^b$. For this we use the fact that φ induces a one-to-one correspondence between partitions $\mathcal{P}^x = \{x_0, \ldots, x_n\}$ of $[a, b]$ and partitions $\mathcal{P}^y = \{y_0, \ldots, y_n\}$ of $[c, d]$, where $y_j = \varphi(x_j)$ ($x_j = \varphi^{-1}(y_j)$) (see Figure 5.6). Since $\varphi([x_{j-1}, x_j]) = [y_{j-1}, y_j]$,

$$M_j^x(f \circ \varphi) = \sup_{x_{j-1} \le x \le x_j} f(\varphi(x)) = \sup_{y_{j-1} \le y \le y_j} f(y) = M_j^y(f). \tag{5.8}$$

Moreover, by the mean value theorem, there exists $z_j \in [y_{j-1}, y_j]$ such that

$$\Delta x_j = \varphi^{-1}(y_j) - \varphi^{-1}(y_{j-1}) = (\varphi^{-1})'(z_j)\Delta y_j \le C \Delta y_j, \tag{5.9}$$

where C is a bound for $(\varphi^{-1})'$ on $[c, d]$. From (5.8) and (5.9),

$$\overline{S}(f \circ \varphi, \mathcal{P}^x) = \sum_j M_j^x(f \circ \varphi)\Delta x_j = \sum_j M_j^y(f)(\varphi^{-1})'(z_j)\Delta y_j \le C\overline{S}(f, \mathcal{P}^y).$$

The same inequality evidently holds for $-f$, hence

$$-\underline{S}(f \circ \varphi, \mathcal{P}^x) \le -C\underline{S}(f, \mathcal{P}^y).$$

Adding these inequalities, we have

$$\overline{S}(f \circ \varphi, \mathcal{P}^x) - \underline{S}(f \circ \varphi, \mathcal{P}^x) \le C[\overline{S}(f, \mathcal{P}^y) - \underline{S}(f, \mathcal{P}^y)].$$

Since the right side may be made arbitrarily small, $f \circ \varphi \in \mathcal{R}_a^b$. Since φ' is continuous, $(f \circ \varphi)\varphi' \in \mathcal{R}_a^b$, proving the first part of the theorem.

To prove (5.7), we argue as in the first part of the proof, but now compare the Riemann sums $S((f \circ \varphi)\varphi', \mathcal{P}^x, \boldsymbol{\xi})$ and $S(f, \mathcal{P}^y, \boldsymbol{\zeta})$, where the intermediate points in each case are taken to be left endpoints:

$$\boldsymbol{\xi} := (x_0, \ldots, x_{n-1}), \quad \boldsymbol{\zeta} := (y_0, \ldots, y_{n-1}) = (\varphi(x_0), \ldots, \varphi(x_{n-1})).$$

Then

$$S((f \circ \varphi)\varphi', \mathcal{P}^x, \boldsymbol{\xi}) = \sum_{j=1}^n f(\zeta_j)\varphi'(x_j)\Delta x_j$$

and, by the mean value theorem,

$$S(f, \mathcal{P}^y, \boldsymbol{\zeta}) = \sum_{j=1}^n f(\zeta_j)\Delta\varphi(x_j) = \sum_{j=1}^n f(\zeta_j)\varphi'(t_j)\Delta x_j,$$

for some $t_j \in [x_{j-1}, x_j]$. Subtracting these equations and using the triangle inequality, we obtain

$$\begin{aligned}\left|S((f \circ \varphi)\varphi', \mathcal{P}_x, \boldsymbol{\xi}) - S(f, \mathcal{P}_y, \boldsymbol{\zeta})\right| &\le \sum_{j=1}^n |f(\zeta_j)|\,|\varphi'(x_j) - \varphi'(t_j)|\Delta x_j \\ &\le M\sum_{j=1}^n |\varphi'(x_j) - \varphi'(t_j)|\Delta x_j,\end{aligned}$$

where M is a bound for $|f|$ on $[c, d]$. By the uniform continuity of φ' on $[a, b]$, given $\varepsilon > 0$ there exists a $\delta > 0$ such that $|\varphi'(s) - \varphi'(t)| < \varepsilon/M(b - a)$ for all s, t with $|s - t| < \delta$. Hence if $\|\mathcal{P}^x\| < \delta$, then

$$\left|S((f \circ \varphi)\varphi', \mathcal{P}^x, \xi) - S(f, \mathcal{P}^y, \boldsymbol{\zeta})\right| < \varepsilon.$$

Letting $\|\mathcal{P}^x\| \to 0$ and noting that one then also has $\|\mathcal{P}^y\| \to 0$ (because $\Delta y_j = \varphi'(c_j)\Delta x_j \le B\|\mathcal{P}^x\|$, where B is a bound for $|\varphi'|$), we see that

$$\left|\int_a^b f(\varphi(x))\varphi'(x)\,dx - \int_a^b f(y)\,dy\right| \le \varepsilon.$$

Since ε was arbitrary, the two integrals are equal, completing the proof. □

Remark. Whether φ is increasing or decreasing, (5.7) may be written as

$$\int_a^b f(\varphi(x))\varphi'(x)\,dx = \int_{\varphi(a)}^{\varphi(b)} f(y)\,dy.$$

This formula has an easy proof if f is assumed to be continuous. Indeed, in this case f has a primitive F on $[c,d]$, hence, by the chain rule, $F\circ\varphi$ is a primitive for $(f\circ\varphi)\varphi'$. The desired formula now follows from the fundamental theorem of calculus:

$$\int_a^b f(\varphi(x))\varphi'(x)\,dx = F(\varphi(b)) - F(\varphi(a)) = \int_{\varphi(a)}^{\varphi(b)} f(y)\,dy.$$

Note that in this case it is not necessary to assume that $\varphi' \neq 0$. ◊

5.3.3 Integration by Parts Formula. *Let f and g be differentiable on $[a,b]$ with f', $g' \in \mathcal{R}_a^b$. Then*

$$\int_a^b f(x)g'(x)\,dx = f(x)g(x)\Big|_a^b - \int_a^b f'(x)g(x)\,dx. \tag{5.10}$$

Proof. Since $(fg)' = f'g + fg' \in \mathcal{R}_a^b$, 5.3.1(c) implies that

$$f(x)g(x)\Big|_a^b = \int_a^b (fg)' = \int_a^b f'g + \int_a^b fg'.$$ □

5.3.4 Example. We show that for $k > 1$,

$$\int_0^{\pi/2} \sin^k x\,dx = \begin{cases} \dfrac{(k-1)(k-3)\cdots 4\cdot 2}{k(k-2)\cdots 5\cdot 3}, & k \text{ odd},\\[2ex] \dfrac{\pi}{2}\,\dfrac{(k-1)(k-3)\cdots 5\cdot 3}{k(k-2)\cdots 4\cdot 2}, & k \text{ even}.\end{cases}$$

Let $I_k = \displaystyle\int_0^{\pi/2} \sin^k x\,dx$. Integrating by parts we have

$$I_k = \int_0^{\pi/2} \sin^{k-1} x \sin x\,dx = (k-1)\int_0^{\pi/2} \sin^{k-2} x\cos^2 x\,dx.$$

Since $\cos^2 x = 1 - \sin^2 x$, $I_k = (k-1)(I_{k-2} - I_k)$, hence

$$I_k = \frac{k-1}{k} I_{k-2}.$$

Iterating, we obtain

$$I_k = \frac{(k-1)(k-3)\cdots(k-2j+1)}{k(k-2)\cdots(k-2j+2)} I_{k-2j}.$$

If $k =$ is odd, take $j = (k-1)/2$ so

$$I_k = \frac{(k-1)(k-3)\cdots 4\cdot 2}{k(k-2)\cdots 3\cdot 1} I_1.$$

If k is even, take $j = (k-2)/2$ so

$$I_k = \frac{(k-1)(k-3)\cdots 5\cdot 3}{k(k-2)\cdots 6\cdot 4} I_2.$$

Since $I_1 = 1$ and $I_2 = \pi/4$, the desired formula follows. ◊

If f' and g' are continuous, then (5.10) has the following analog for indefinite integrals:

$$\int f(x)g'(x)\,dx = f(x)g(x) - \int f'(x)g(x)\,dx. \tag{5.11}$$

Setting $h = g'$ and using the symbols $\mathcal{D}$ for differentiation and $\mathcal{I}$ for integration, we may write (5.11) as

$$\mathcal{I}(fh) = f\cdot\mathcal{I}h - \mathcal{I}(\mathcal{D}f\cdot\mathcal{I}h).$$

By induction we obtain

$$\mathcal{I}(fh) = \sum_{k=1}^{n}(-1)^{(k-1)}\mathcal{D}^{k-1}f\cdot\mathcal{I}^k h + (-1)^n\mathcal{I}(\mathcal{D}^n f\cdot\mathcal{I}^n h). \tag{5.12}$$

The fundamental theorem of calculus may then be used to calculate $\int_a^b fh$.

Formula (5.12) may be expressed in tabular form as in Table 5.1. For each k, the entries in column k are multiplied and the resulting products are added. The exception is column $n+1$, where the product must be integrated before adding. The process terminates if and when $\mathcal{D}^n f = 0$.

TABLE 5.1: Table for evaluating $\int fh$ by parts.

k	1	2	3	$\cdots$	n	$n+1$
$(-1)^{k-1}$	$+1$	-1	$+1$	$\cdots$	$(-1)^{n-1}$	$(-1)^n$
$\mathcal{D}^{k-1}f$	f	$\mathcal{D}f$	$\mathcal{D}^2 f$	$\cdots$	$\mathcal{D}^{n-1}f$	$\mathcal{D}^n f$
$\mathcal{I}^k h$	$\mathcal{I}h$	$\mathcal{I}^2 h$	$\mathcal{I}^3 h$	$\cdots$	$\mathcal{I}^n h$	$\mathcal{I}^n h$

5.3.5 Example. Using Table 5.1 with $f(x) = (x+1)^3$ and $h(x) = e^{5x}$, we have

$$\int (x+1)^3 e^{5x}\,dx = e^{5x}\left[\frac{(x+1)^3}{5} - \frac{3(x+1)^2}{5^2} + \frac{6(x+1)}{5^3} - \frac{6}{5^4}\right] + c.$$

TABLE 5.2: Table for evaluating $\int (x+1)^3 e^{5x}\,dx$ by parts.

k	1	2	3	4	5
$(-1)^{k-1}$	$+1$	-1	$+1$	-1	$+1$
$\mathcal{D}^{k-1}f$	$(x+1)^3$	$3(x+1)^2$	$6(x+1)$	6	0
$\mathcal{I}^k h$	$e^{5x}/5$	$e^{5x}/5^2$	$e^{5x}/5^3$	$e^{5x}/5^4$	$e^{5x}/5^5$

◊

Exercises

1. (⇓ 13.6.4.) Let $f:\mathbb{R}\to\mathbb{R}$ be continuous and periodic with period $p>0$, that is, $f(x+p)=f(x)$ for all x. Prove that

$$\int_0^p f(x+y)\,dx = \int_0^p f(x)\,dx \quad \text{for all } y\in\mathbb{R}.$$

2. Let $f:(a,b)\to\mathbb{R}$ have a uniformly continuous derivative. Prove that the continuous extension of f' to $[a,b]$ (see 3.5.9) satisfies

$$\int_a^b f' = \lim_{\varepsilon\to 0+}\big[f(b-\varepsilon)-f(a+\varepsilon)\big].$$

3. Verify the following inequalities:

(a) $\displaystyle \frac{2}{\pi}(\sqrt{2}-1) \le \int_0^1 \frac{\sin x\,dx}{\sqrt{1+x^2}} \le \sqrt{2}-1$

(b) $\displaystyle \frac{1}{2^q(p+1)} \le \int_0^1 \frac{x^p\,dx}{(1+x^p)^q} \le \frac{2^{1-q}-1}{(p+1)(1-q)},\quad p,\ q>0,\ q\ne 1.$

4. Establish the formula

$$\int_0^1 (1-x)^m x^n\,dx = \frac{m!}{(n+1)(n+2)\cdots(n+m+1)}.$$

5. Let $n\in\mathbb{N}$. Evaluate

(a) $\displaystyle \int_0^1 \exp(x^{1/n})\,dx$ (b) $\displaystyle \int_1^e \ln^n x\,dx.$

6. Let $k\in\mathbb{N}$. Show that

$$\int_0^{\pi/2}\cos^k x\,dx = \begin{cases} \dfrac{(k-1)(k-3)\cdots 4\cdot 2}{k(k-2)\cdots 5\cdot 3}, & k \text{ odd},\\[2ex] \dfrac{\pi}{2}\,\dfrac{(k-1)(k-3)\cdots 5\cdot 3}{k(k-2)\cdots 4\cdot 2}, & k \text{ even}.\end{cases}$$

7. ($\Downarrow$ 13.4.2) Let $k \in \mathbb{N}$. Show that

$$\int_0^1 \frac{x^k}{\sqrt{1-x^2}}\,dx = \begin{cases} \dfrac{(k-1)(k-3)\cdots 4\cdot 2}{k(k-2)\cdots 3\cdot 1} & \text{if } k \text{ is odd} \\ \dfrac{(k-1)(k-3)\cdots 3\cdot 1}{k(k-2)\cdots 4\cdot 2}\dfrac{\pi}{2} & \text{if } k \text{ is even.} \end{cases}$$

   N.B. The integral is improper but converges by Exercise 5.7.7. For the even case, use Exercise 6.

8. Let f' be continuous and positive on $[a, b]$. Prove that

$$\int_a^b f(x)\,dx + \int_{f(a)}^{f(b)} f^{-1}(y)\,dy = bf(b) - af(a).$$

   Interpret geometrically for $f > 0$ and $a > 0$.

9. (Young's inequality). Let f be continuous and strictly increasing on $[0, a]$ with $f(0) = 0$. Prove that

$$\int_0^x f + \int_0^y f^{-1} = yf^{-1}(y) + \int_{f^{-1}(y)}^x f.$$

   Deduce that

$$\int_0^x f + \int_0^y f^{-1} \geq xy, \quad 0 \leq x \leq a, \quad 0 \leq y \leq f(a).$$

10. Use Young's inequality to verify the following:
    (a) $\sqrt{1-y^2} + y\sin^{-1} y \geq xy + \cos x, \ 0 \leq x \leq \pi/2, \ 0 \leq y \leq 1$
    (b) $x \ln x + e^y \geq xy + x, \ 1 \leq x \leq 2, \ 0 \leq y \leq \ln 2.$

11. Give an example of a discontinuous function that
    (a) has a primitive, (b) has no primitive.

12. Let f and g be continuously differentiable with $g > 0$. Prove that

$$\int \frac{f(x)g'(x)}{g^2(x)}\,dx = \int \frac{f'(x)}{g(x)}\,dx - \frac{f(x)}{g(x)}.$$

13. Let $f' \in \mathcal{R}_a^b$. Prove that

$$\lim_n \int_a^b f(x)\sin(nx)\,dx = 0.$$

14. Let f be continuous on $[0, +\infty)$ such that $\lim_{x\to+\infty} f(x)$ exists in $\overline{\mathbb{R}}$ and let $a > 0$. Find

$$\lim_{n\to+\infty} \int_0^a f(nx)\,dx.$$

15. Let h' be continuous and positive on $[a,b]$ and let g' be continuous on $[c,d] = [h(a), h(b)]$. Prove that
$$\int_a^b g\big(h(x)\big)\,dx = g(d)b - g(c)a - \int_c^d g'(t)h^{-1}(t)\,dt.$$

16. Let $f \in \mathcal{R}_{-a}^{a}$, $a > 0$. Show that
$$\int_{-a}^{a} f = \begin{cases} 0 & \text{if } f \text{ is an odd function,} \\ 2\int_0^a f & \text{if } f \text{ is an even function.} \end{cases}$$

17. Let $f : [a,b] \to \mathbb{R}$ be continuous and let u, v be differentiable functions with range contained in $[a,b]$. Prove that
$$\frac{d}{dx}\int_{u(x)}^{v(x)} f = f\big(v(x)\big)v'(x) - f\big(u(x)\big)u'(x).$$

18. Let functions a, b, c, $d : [0,1] \to [0,1]$ have continuous derivatives and let $f : [0,1] \to \mathbb{R}$ be continuous. Suppose that
$$\int_{a(x)}^{b(x)} f = \int_{c(x)}^{d(x)} f \quad \text{for all } x \in [0,1].$$
Prove that
$$\int_{b(0)}^{b(1)} f + \int_{c(0)}^{c(1)} f = \int_{a(0)}^{a(1)} f + \int_{d(0)}^{d(1)} f.$$

19. Let f be continuous and g differentiable with bounded derivative on $[a,b]$. Evaluate
$$\lim_{x\to a} \frac{g(x)}{x-a}\int_a^x f.$$

20. Let $p > 0$, $q > 1$, and m, $k \in \mathbb{N}$ with $m > k$. Evaluate $\lim\limits_{n\to+\infty} s_n$ if $s_n =$
$$\text{(a)} \sum_{k=1}^{n} \frac{k^p}{n^{p+1}} \qquad \text{(b)} \sum_{k=1}^{n} \frac{k^{q-1}}{n^q + k^q} \qquad \text{(c)} \sum_{k=1}^{n} \left(\frac{(mn)!}{n^{kn}[(m-k)n]!}\right)^{1/n}.$$

21. Let $|f'| \le M$ on $[a,b]$. For $n \in \mathbb{N}$ set $h = (b-a)/n$ and $x_k = a + kh$, $k = 0, 1, \ldots, n-1$. Prove that
$$\left|\int_a^b f - h\sum_{k=1}^{n} f(x_{k-1})\right| \le hM(b-a).$$

22. Let f be continuous on $[0,1]$. Prove that
$$\int_0^1 \int_0^x f(t)\,dt\,dx = \int_0^1 (1-x)f(x)\,dx.$$

23. Let f, $g : [0,1] \to \mathbb{R}$ be continuously differentiable, f monotone, and $g(x) > g(0) = g(1)$ on $(0,1)$. Prove that $\int_0^1 fg' = 0$ iff f is constant.

*5.4 Stirling's Formula

Stirling's formula gives an estimate for $n!$ when n is large. The proof relies on material from Section 4.3. We begin with the following lemma, which gives the fundamental inequality needed to establish the formula.

5.4.1 Lemma. *If f is concave and differentiable on (a, b), then*

$$\frac{f(u)+f(v)}{2} \le \frac{1}{v-u}\int_u^v f(t) \le f\left(\frac{u+v}{2}\right), \quad a < u < v < b.$$

Proof. By the concave versions of 4.3.6 and (4.4),

$$f(u)\frac{v-t}{v-u} + f(v)\frac{t-u}{v-u} \le f(t) \le f'(x)(t-x) + f(x)$$

for all $a < u < v < b$ and all $x,\ t \in [u, v]$. Integrating with respect to t,

$$[f(u)+f(v)]\frac{v-u}{2} \le \int_u^v f(t) \le f'(x)\,\frac{(v-x)^2-(x-u)^2}{2} + f(x)(v-u).$$

Taking $x = (u+v)/2$ and dividing by $v-u$ produces the desired inequalities. □

5.4.2 Stirling's Inequalities. *For all n,*

$$e^{7/8} \le \frac{e^n n!}{n^n\sqrt{n}} \le e, \tag{5.13}$$

where the middle term is decreasing in n.

Proof. Taking $f(x) = \ln x$, $u = k \in \mathbb{N}$, and $v = k+1$ in the lemma, we have

$$\tfrac{1}{2}\ln(k^2+k) \le \tfrac{1}{2}\left[\ln(k) + \ln(k+1)\right] \le \int_k^{k+1} \ln(t)\,dt \le \ln\left(k+\tfrac{1}{2}\right).$$

Rearranging,

$$0 \le \int_k^{k+1} \ln(t)\,dt - \tfrac{1}{2}\ln(k^2+k) \le \ln\left(k+\tfrac{1}{2}\right) - \tfrac{1}{2}\ln(k^2+k). \tag{5.14}$$

Now observe that

$$\sum_{k=1}^{n-1}\int_k^{k+1} \ln t\,dt = \int_1^n \ln t\,dt = n\ln n - n + 1,$$

$$\sum_{k=1}^{n-1}\ln(k^2+k) = \sum_{k=1}^{n-1}[\ln(k+1)+\ln k] = 2\sum_{k=2}^{n}\ln k - \ln n = 2\ln n! - \ln n, \text{ and}$$

$$\ln(k+\tfrac{1}{2}) - \tfrac{1}{2}\ln(k^2+k) = \tfrac{1}{2}\ln\left(1+\frac{1}{4(k^2+k)}\right) \le \frac{1}{8(k^2+k)},$$

where, for the last inequality, we used the fact that $\ln(1+x) < x$ for $x > 0$, which follows directly from the integral definition of $\ln(x+1)$. Summing in (5.14) and using the above inequalities, we obtain

$$0 \le \left(n + \tfrac{1}{2}\right)\ln n - n + 1 - \ln n! \le \sum_{k=1}^{n-1} \frac{1}{8(k^2+k)} = \frac{1}{8}\sum_{k=1}^{n-1}\left[\frac{1}{k} - \frac{1}{k+1}\right] \le \frac{1}{8}.$$

Note that the term $\left(n + \frac{1}{2}\right)\ln n - n + 1 - \ln n!$ is increasing in n since it was obtained as a sum of nonnegative terms in (5.14). Rearranging, we have

$$\frac{7}{8} \le -\left(n + \tfrac{1}{2}\right)\ln n + n + \ln n! \le 1,$$

where the middle term is decreasing in n. Exponentiating yields the desired inequalities. □

5.4.3 Stirling's Formula. $\lim_n \dfrac{e^n n!}{n^n\sqrt{n}} = \sqrt{2\pi}.$

Proof. By 5.4.2, the limit L in the formula exists in $\mathbb{R}$. Set $I_n = \int_0^{\pi/2} \sin^n x\,dx$. By 5.3.4,

$$I_{2n+1} = \frac{(2n)(2n-2)\cdots 4\cdot 2}{(2n+1)(2n-1)\cdots 5\cdot 3} \quad\text{and}\quad I_{2n} = \frac{\pi}{2}\,\frac{(2n-1)(2n-3)\cdots 5\cdot 3}{2n(2n-2)\cdots 4\cdot 2}.$$

For $x \in [0, \pi/2]$ and $n \ge m$, $\sin^n x \le \sin^m x$, hence

$$\frac{I_{2n+2}}{I_{2n}} \le \frac{I_{2n+1}}{I_{2n}} \le \frac{I_{2n}}{I_{2n}} = 1.$$

It follows that

$$\frac{2n+1}{2n+2}\,\frac{\pi}{2} \le \frac{2^2\cdot 4^2\cdot 6^2\cdots(2n-2)^2\cdot(2n)^2}{1^2\cdot 3^2\cdot 5^2\cdots(2n-1)^2(2n+1)} \le \frac{\pi}{2},$$

from which we obtain *Wallis's product*

$$\lim_n \frac{2^2\cdot 4^2\cdot 6^2\cdots(2n-2)^2\cdot(2n)^2}{1^2\cdot 3^2\cdot 5^2\cdots(2n-1)^2(2n+1)} = \frac{\pi}{2}.$$

Denote the general term in Wallis's product by α_n. Since

$$2\cdot 4\cdots(2n-2)\cdot(2n) = 2^n n! \;\text{ and }\; 3\cdot 5\cdots(2n-1) = \frac{(2n)!}{2^n n!},$$

we see that

$$\sqrt{\alpha_n} = \frac{2^{2n}(n!)^2}{(2n)!\sqrt{2n+1}}.$$

Now set $\beta_n = \dfrac{e^n n!}{n^n\sqrt{n}}$ and note that

$$\frac{\beta_n^2}{\beta_{2n}} = \frac{e^{2n}(n!)^2}{n^{2n+1}}\,\frac{(2n)^{2n}\sqrt{2n}}{e^{2n}(2n)!} = \frac{(n!)^2 2^{2n}\sqrt{2}}{(2n)!\sqrt{n}}.$$

Dividing by $\sqrt{\alpha_n}$ we have

$$\frac{\beta_n^2}{\beta_{2n}\sqrt{\alpha_n}} = \frac{(n!)^2 2^{2n}\sqrt{2}}{(2n)!\sqrt{n}}\,\frac{(2n)!\sqrt{2n+1}}{2^{2n}(n!)^2} = \sqrt{2}\,\sqrt{2+1/n} \to 2.$$

On the other hand, since $\sqrt{\alpha_n} \to \sqrt{\pi/2}$ and $\beta_n \to L$, we also have

$$\frac{\beta_n^2}{\beta_{2n}\sqrt{\alpha_n}} \to L\sqrt{\frac{2}{\pi}}.$$

Therefore, $L\sqrt{\frac{2}{\pi}} = 2$ and so $L = \sqrt{2\pi}$. □

5.5 Integral Mean Value Theorems

The following theorem asserts that the average value of a continuous function over an interval $[a, b]$ is actually assumed by the function at some intermediate point c.

5.5.1 First Mean Value Theorem for Integrals. *If f is continuous on $[a, b]$, then there exists $c \in (a, b)$ such that*

$$\frac{1}{b-a}\int_a^b f = f(c).$$

Proof. Apply the mean value theorem for derivatives and the fundamental theorem of calculus to the function $G(x) := \int_a^x f(t)\,dt$. □

The next theorem is a weighted average generalization of 5.5.1.

5.5.2 Weighted Mean Value Theorem for Integrals. *Let f be continuous on $[a, b]$ and let $g \in \mathcal{R}_a^b$. If g does not change sign in $[a, b]$, then there exists $c \in [a, b]$ such that*

$$\int_a^b fg = f(c)\int_a^b g. \tag{5.15}$$

Proof. We may assume that $g \geq 0$ on $[a, b]$, so $\int_a^b g \geq 0$. Suppose first that $\int_a^b g = 0$. If C is an upper bound for $|f|$ on $[a, b]$, then

$$\left|\int_a^b fg\right| \leq \int_a^b |f|g \leq C\int_a^b g = 0,$$

hence both sides of (5.15) are zero. Now assume that $\int_a^b g > 0$. Let $m = f(x_m)$ and $M = f(x_M)$ denote the minimum and maximum values of f on $[a, b]$. Since $mg \le fg \le Mg$,

$$m \int_a^b g \le \int_a^b fg \le M \int_a^b g,$$

hence

$$m \le \frac{\displaystyle\int_a^b fg}{\displaystyle\int_a^b g} \le M.$$

An application of the intermediate value theorem completes the proof. □

5.5.3 Second Mean Value Theorem for Integrals. *Let f be continuous and g differentiable and monotone on $[a, b]$ with $g' \in \mathcal{R}_a^b$. Then there exists $c \in [a, b]$ such that*

$$\int_a^b fg = g(a) \int_a^c f + g(b) \int_c^b f.$$

Proof. Let $F(x) = \int_a^x f$. Integrating by parts,

$$\int_a^b fg = \int_a^b F'g = F(b)g(b) - \int_a^b g'F.$$

Since g is monotone, the sign of g' does not change, hence, by 5.5.2, there exists $c \in [a, b]$ such that

$$\int_a^b g'F = F(c) \int_a^b g' = F(c)[g(b) - g(a)].$$

Therefore,

$$\int_a^b fg = F(b)g(b) - F(c)[g(b) - g(a)] = g(a)F(c) + g(b)[F(b) - F(c)],$$

which is the assertion of the theorem. □

Remarks. (a) Because derivatives have the intermediate value property (Exercise 4.2.25), the monotonicity requirement on g holds if $g' \ne 0$ on $[a, b]$.

(b) The second mean value theorem for integrals holds under the less restrictive hypotheses that f is integrable and g is monotone. A proof may be found in [3]. ◊

Exercises

1. Let $0 \le a < b$ and let f be continuous on $[\sqrt{a}, \sqrt{b}]$. Prove that there exists $c \in [a, b]$ such that
$$\frac{1}{2}\int_a^b f(\sqrt{x})\,dx = \sqrt{a}\int_{\sqrt{a}}^{\sqrt{c}} f(x)\,dx + \sqrt{b}\int_{\sqrt{c}}^{\sqrt{b}} f(x)\,dx.$$

2. Let $0 < a < b$ and let f be continuous on $[b^{-1}, a^{-1}]$. Prove that there exists $c \in [a, b]$ such that
$$\int_a^b f(1/x)\,dx = b^2\int_{1/b}^{1/c} f(x)\,dx + a^2\int_{1/c}^{1/a} f(x)\,dx.$$

3. Let f be continuous on $[0, 1]$. Prove that there exists $c \in [1/2, \sqrt{3}/2]$ such that
$$\int_{\pi/6}^{\pi/3} f(\sin x)\,dx = \frac{2}{\sqrt{3}}\int_{1/2}^{c} f(x)\,dx + 2\int_c^{\sqrt{3}/2} f(x)\,dx.$$

4. Let f be continuous on $[0, 1]$. Prove that there exists $c \in [0, 1]$ such that
$$\int_0^{\pi/4} f(\tan x)\,dx = \int_0^c f(x)\,dx + \frac{1}{2}\int_c^1 f(x)\,dx.$$

5. Let f and g be continuous on $[a, b]$. Show that there exists $c \in (a, b)$ such that
$$g(c)\int_a^b f = f(c)\int_a^b g.$$

6. Prove: If f is continuous, $g \in \mathcal{R}_a^b$, and m is lower bound for g, then there exist c, $d \in [a, b]$ such that
$$\int_a^b fg = f(c)\int_a^b g + m(b-a)[f(d) - f(c)].$$

7. Prove the following variant of the second mean value theorem for integrals: Let f, $g \in \mathcal{R}_a^b$ with $g \ge 0$. If $m \le f \le M$ on $[a, b]$, then there exists $c \in [a, b]$ such that
$$\int_a^b fg = m\int_a^c g + M\int_c^b g.$$
Hint. Consider $G(x) := m\int_a^x g + M\int_x^b g$.

8. Let g have a nonnegative integrable derivative on $[0,1]$ with $g(0)=0$ and $g(1)=1$. Show that there exists $c \in [0,1]$ such that

$$\int_0^1 x^n g(x)\,dx = \frac{1-c^{n+1}}{n+1}.$$

9. Let g have a nonnegative integrable derivative on $[0,\pi]$ with $g(0)=0$ and $g(\pi)=1$. Show that there exists $c \in [0,\pi]$ such that

$$\int_0^\pi g(x)\sin x\,dx = \cos c + 1.$$

10. Let g be twice differentiable on $[a,b]$ with $g''<0$ and $g'' \in \mathcal{R}_a^b$, and let f be continuous on $g([a,b])$. Show that if $g' \geq m > 0$ and $|f| \leq M$, then

$$\left|\int_a^b f' \circ g\right| \leq \frac{2M}{m}.$$

Hint. Use the second mean value theorem for integrals.

---

*5.6 Estimation of the Integral

It is frequently the case that an integral cannot be evaluated exactly. It may be possible, however, to approximate such an integral by a numerical method. Of course, Riemann sums can always be used to approximate an integral. However, unless the intermediate points of the subintervals are chosen judiciously, a Riemann sum usually offers only a coarse approximation of the integral. In this section we discuss three techniques, the *trapezoidal rule*, the *midpoint rule*, and *Simpson's rule*, that typically yield good numerical estimates of an integral.

The approximation techniques are given in order of increasing precision. For each of these, we use partitions of the form

$$x_k = a + kh_n, \quad k = 0, 1, \ldots, n, \quad \text{where } h_n := \frac{b-a}{n}. \tag{5.16}$$

The integral $\int_a^b f$ is then estimated by replacing f on the interval $[x_k, x_{k+1}]$ by a simpler function f_k. The approximation therefore takes the form

$$\int_a^b f(x)\,dx \approx \sum_{k=0}^{n-1} \int_{x_k}^{x_{k+1}} f_k(x)\,dx,$$

where the *error* in the approximation is the difference between the left and right sides. The main goal in the approximation schemes described below is to obtain, for a given class of functions, the sharpest upper bound for the magnitude of the error

The reader may wish to compare the error bounds in the three approximation techniques described below with the error bound for the approximation given by the Riemann sum

$$R_n = \frac{b-a}{n}\big[f(x_0) + f(x_1) + \cdots + f(x_{n-1})\big]. \tag{5.17}$$

By Exercise 5.3.21, for functions f with a bounded derivative one has in general only the *first-order* error bound

$$\left|\int_a^b f - R_n\right| \le h_n(b-a)\|f'\|_\infty,$$

implying that a reasonable estimate demands a large n. Here, for a bounded function g on $[a, b]$, we define

$$\|g\|_\infty := \sup\{|g(x)| \,:\, a \le x \le b\}.$$

Trapezoidal Rule

Define points

$$P_k := (x_k, f(x_k)) = (x_k, y_k), \quad k = 0, 1, \ldots, n, \tag{5.18}$$

where the x_k are given in (5.16). The trapezoidal rule uses the line segment from P_k to P_{k+1} to approximate f on $[x_k, x_{k+1}]$, $k = 0, 1, \ldots n-1$. Thus the approximating function f_k is given by

$$f_k(x) = y_k + m_k(x - x_k), \quad x_k \le x \le x_{k+1}, \quad m_k := \frac{y_{k+1} - y_k}{x_{k+1} - x_k}.$$

A simple calculation shows that

$$\int_{x_k}^{x_{k+1}} f_k = \frac{h_n}{2}(y_{k+1} + y_k),$$

The sum

$$T_n := \sum_{k=0}^{n-1} \int_{x_k}^{x_{k+1}} f_k = \frac{h_n}{2}(y_0 + 2y_1 + \cdots + 2y_{n-1} + y_n)$$

is then used to approximate $\int_a^b f$. If $f > 0$, T_n may be realized as the sum of areas of trapezoids (See Figure 5.7)

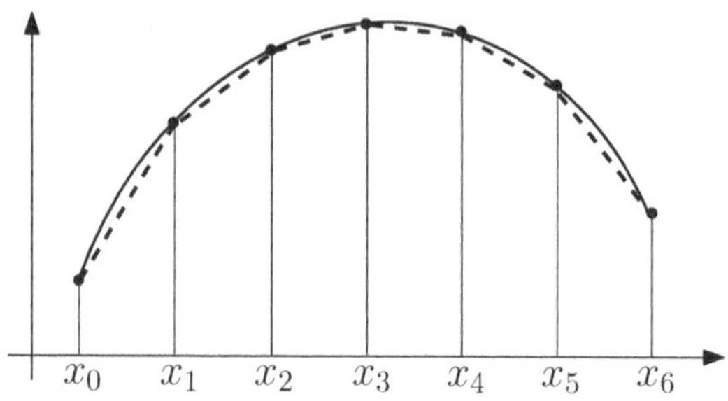

FIGURE 5.7: Trapezoidal rule approximation.

5.6.1 Trapezoidal Rule. *If $f \in \mathcal{R}_a^b$, then $\lim_n T_n = \int_a^b f$. Moreover, if f'' exists and is continuous on $[a, b]$, then the following error estimate holds:*

$$\left| \int_a^b f - T_n \right| \leq \frac{h_n^2}{12}(b-a)\|f''\|_\infty.$$

Proof. For the Riemann sum R_n in (5.17),

$$R_n - T_n = \frac{b-a}{2n}[f(x_0) - f(x_n)] = \frac{b-a}{2n}[f(a) - f(b)] \to 0,$$

hence $T_n = (T_n - R_n) + R_n \to \int_a^b f$.

To obtain the error estimate, consider the function

$$g_k(x) := \frac{f(x) - f_k(x)}{(x-x_k)(x-x_{k+1})} = \frac{f(x) - y_k - m_k(x-x_k)}{(x-x_k)(x-x_{k+1})},$$

which has singularities at x_k and x_{k+1}. Since both the numerator and the denominator vanish at these points, the singularities may be removed using l'Hospital's rule. Therefore, $g_k(x)$ has a continuous extension to $[x_k, x_{k+1}]$. Since $(x-x_k)(x-x_{k+1})$ does not change sign on $[x_k, x_{k+1}]$, by the weighted mean value theorem for integrals (5.5.2), there exists a point $z_k \in [x_k, x_{k+1}]$ such that

$$\begin{aligned}\int_{x_k}^{x_{k+1}} [f(x) - f_k(x)]\,dx &= \int_{x_k}^{x_{k+1}} g_k(x)(x-x_k)(x-x_{k+1})\,dx \\ &= g_k(z_k)\int_{x_k}^{x_{k+1}} (x-x_k)(x-x_{k+1})\,dx \\ &= -g_k(z_k)\frac{h^3}{6}.\end{aligned}$$

It follows that

$$\int_a^b f(t)\,dt - T_n = \sum_{k=0}^{n-1}\int_{x_k}^{x_{k+1}} [f(x) - f_k(x)]\,dx = -\frac{h_n^3}{6}\sum_{k=0}^{n-1} g_k(z_k). \qquad (5.19)$$

Now fix $x \in (x_k, x_{k+1})$ and define $\psi(z)$ on $[x_k, x_{k+1}]$ by

$$\psi(z) = f(z) - f_k(z) - g_k(x)(z - x_k)(z - x_{k+1}).$$

Since ψ has distinct zeros x, x_k, and x_{k+1}, Rolle's theorem applied twice shows that ψ'' has a zero $v_k \in (x_k, x_{k+1})$. It follows that $f''(v_k) = 2g_k(x)$. Since x was arbitrary,

$$|g_k(x)| \le \tfrac{1}{2}\|f''\|_\infty \quad \text{for all } x \in [x_k, x_{k+1}].$$

From this and (5.19) we see that

$$\left|\int_a^b f(t)\,dt - T_n\right| \le \frac{nh_n^3}{12}\|f''\|_\infty = \frac{h_n^2}{12}(b-a)\|f''\|_\infty,$$

which is the desired error bound. □

Midpoint Rule

Define

$$\overline{x}_k := \frac{x_k + x_{k+1}}{2} = a + \left(k + \tfrac{1}{2}h_n\right), \quad k = 0, 1, \ldots, n-1,$$

where the x_k are given in (5.16). The midpoint rule uses the constant function

$$f_k(x) = f\left(\overline{x}_k\right), \quad x_k \le x \le x_{k+1},$$

to approximate f on $[x_k, x_{k+1}]$. This amounts to approximating $\int_a^b f$ by Riemann sums M_n with intermediate points the midpoints of the intervals:

$$M_n = \frac{b-a}{n}\left[f(\overline{x}_0) + f(\overline{x}_1) + \cdots + f(\overline{x}_{n-1})\right].$$

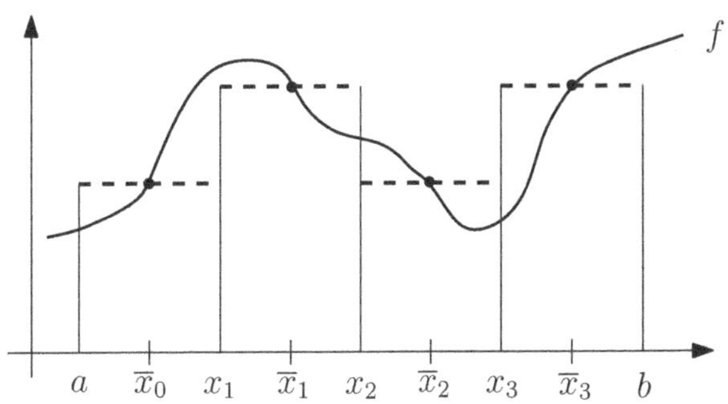

FIGURE 5.8: Midpoint rule approximation.

5.6.2 Midpoint Rule. *If f'' exists and is continuous on $[a,b]$, then the following error estimate holds:*

$$\left|\int_a^b f - M_n\right| \le \frac{h_n^2}{24}(b-a)\|f''\|_\infty.$$

Proof. The function

$$g_k(x) = \frac{f(x) - f(\overline{x}_k) - f'(\overline{x}_k)(x-\overline{x}_k)}{(x-\overline{x}_k)^2}$$

has a double singularity at $\overline{x}_k$, which may be removed by applying l'Hospital's rule twice and defining $g_k(\overline{x}_k)$ to be the resulting limit. Since

$$f(x) - f(\overline{x}_k) - f'(\overline{x}_k)(x-\overline{x}_k) = g_k(x)(x-\overline{x}_k)^2$$

and

$$\int_{x_k}^{x_{k+1}} (x-\overline{x}_k)\,dx = 0,$$

we see that

$$\int_{x_k}^{x_{k+1}} [f(x) - f(\overline{x}_k)]\,dx = \int_{x_k}^{x_{k+1}} g_k(x)(x-\overline{x}_k)^2\,dx.$$

Since $(x-\overline{x}_k)^2$ has constant sign on $[x_k, x_{k+1}]$, the weighted mean value theorem for integrals implies that the integral on the right equals

$$g_k(z_k)\int_{x_k}^{x_{k+1}} (x-\overline{x}_k)^2\,dx = g_k(z_k)\frac{h_n^3}{12}$$

for some point $z_k \in [x_k, x_{k+1}]$. Therefore,

$$\int_{x_k}^{x_{k+1}} [f(x) - f(\overline{x}_k)]\,dx = g_k(z_k)\frac{h_n^3}{12}. \tag{5.20}$$

Now fix $x \in [x_k, \overline{x}_k) \cup (\overline{x}_k, x_{k+1}]$. By Taylor's theorem, there exists a point $\xi_k \in [x_k, x_k]$ such that

$$f(x) = f(\overline{x}_k) + f'(\overline{x}_k)(x-\overline{x}_k) + \frac{f''(\xi_k)}{2}(x-\overline{x}_k)^2.$$

Solving for $f''(\xi_k)$ we see that $f''(\xi_k) = 2g_k(x)$. Therefore, $|g_k(x)| \le \|f''\|_\infty/2$ for all $x \in [x_{k-1}, x_{k+1}]$, hence from (5.20),

$$-\frac{h_n^3}{24}\|f''\|_\infty \le \int_{x_k}^{x_{k+1}} [f(x) - f(\overline{x}_k)]\,dx \le \frac{h_n^3}{24}\|f''\|_\infty.$$

Summing, we obtain

$$-\frac{nh_n^3}{24}\|f''\|_\infty \le \int_a^b f(x)\,dx - M_n \le \frac{nh_n^3}{24}\|f''\|_\infty,$$

which is the assertion of the theorem. □

Note that the estimates in both the trapezoidal rule and the midpoint rule are exact for all linear functions f, since then $f'' = 0$. Thus, it is impossible to achieve better general error bounds.

Simpson's Rule

Simpson's rule assumes $n = 2m$ in (5.16) and uses a parabola through each triple of points

$$(P_{k-1}, P_k, P_{k+1}),\ k = 2j+1,\ j = 0, \ldots, m-1,\ P_k := (x_k, f(x_k)) = (x_k, y_k),$$

to approximate f. To obtain the rule, observe that any polynomial $p(x)$ of

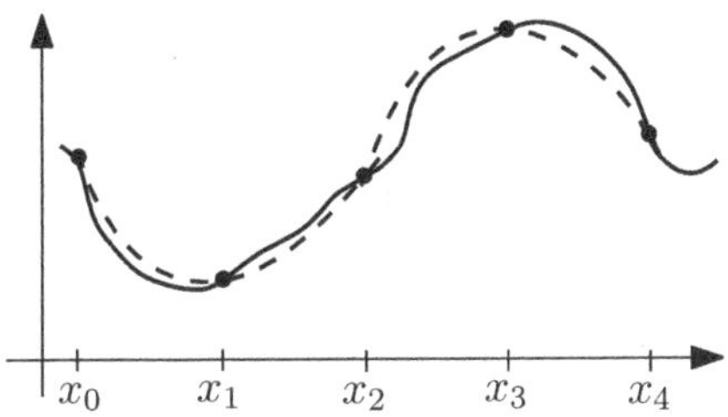

FIGURE 5.9: Simpson's rule approximation.

degree ≤ 2 may be written in the form

$$p(x) = b_k(x - x_{k-1})(x - x_k) + c_k(x - x_{k-1}) + d_k, \tag{5.21}$$

where

$$\begin{aligned} b_k &= \frac{p(x_{k+1}) - 2p(x_k) + p(x_{k-1})}{2h^2}, \\ c_k &= \frac{p(x_k) - p(x_{k-1})}{h}, \quad \text{and} \\ d_k &= p(x_{k-1}). \end{aligned}$$

It follows that the unique polynomial p_k of degree ≤ 2 that passes through the points P_{k-1}, P_k, and P_{k+1} is obtained by choosing

$$\begin{aligned} b_k = b_k(f) &:= \frac{f(x_{k+1}) - 2f(x_k) + f(x_{k-1})}{2h^2}, \\ c_k = c_k(f) &:= \frac{f(x_k) - f(x_{k-1})}{h}, \quad \text{and} \\ d_k = d_k(f) &:= f(x_{k-1}). \end{aligned} \tag{5.22}$$

With this choice, one readily calculates

$$S_{n,k} := \int_{x_{k-1}}^{x_{k+1}} p_k(x)\, dx = \frac{h_n}{3}[y_{k-1} + y_{k+1} + 4y_k],\ k = 2j+1,\ j = 0, \cdots, m-1,$$

which is taken as an approximation of $\int_{x_{k-1}}^{x_{k+1}} f$. Note that the approximation is exact for all polynomials f of degree ≤ 2, since such a polynomial may be written in the form (5.21). Summing this result, we see that the integral of the approximating function on $[a, b]$ is

$$S_n := \frac{b-a}{3n}\big(y_0 + 4y_1 + 2y_2 + 4y_3 + 2y_4 + \cdots + 2y_{n-2} + 4y_{n-1} + y_n\big).$$

5.6.3 Simpson's Rule. *If $f \in \mathcal{R}_a^b$, then $\lim_n S_n = \int_a^b f$. Moreover, if $f^{(4)}$ exists and is continuous on $[a, b]$, then the following error estimate holds:*

$$\Big|\int_a^b f - S_n\Big| \le \frac{h_n^4(b-a)\|f^{(4)}\|_\infty}{180}.$$

Proof. Set

$$R_n' := \big(y_0 + y_2 + \cdots + y_{n-2}\big)(2h_n) \quad\text{and}\quad R_n'' := \big(y_1 + y_3 + \cdots + y_{n-1}\big)(2h_n).$$

These are Riemann sums for f on $[a, b]$ and

$$6S_n = 2R_n' + 4R_n'' + (b-a)(2h_n).$$

It follows that $S_n \to \int_a^b f$.

To obtain the error estimate, let $f^{(4)}$ be continuous on $[a, b]$ and denote the errors by

$$E_{n,k} = \int_{x_{k-1}}^{x_{k+1}} f(x)\,dx - S_{n,k} \quad\text{and}\quad E_n = \sum_{j=0}^{m-1} E_{n,2j+1} = \int_a^b f(x)\,dx - S_n.$$

We show that there exists a point $\xi_k \in [x_k, x_{k+1}]$ such that

$$E_{n,k} = -\frac{h_n^5 f^{(4)}(\xi_k)}{90}. \tag{5.23}$$

It will follow that

$$|E_n| \le \frac{m h_n^5\|f^{(4)}\|_\infty}{90} = \frac{h_n^4(b-a)\|f^{(4)}\|_\infty}{180},$$

proving the theorem.

To verify (5.23), fix k and choose a point in $x_k^* \in (x_{k-1}, x_k) \cup (x_k, x_{k+1})$. For any function g, define a function Lg on $[x_{k-1}, x_{k+1}]$ by

$$\begin{aligned}(Lg)(x) = a_k(g)(x - x_{k-1})(x - x_k)(x - x_{k+1}) + b_k(g)(x - x_{k-1})(x - x_k)\\ + c_k(g)(x - x_{k-1}) + d_k(g),\end{aligned}$$

where $b_k(g)$, $c_k(g)$, and $d_k(g)$ are defined as in (5.22) and $a_k(g)$ is chosen so that $(Lg)(x_k^*) = g(x_k^*)$. Then Lg is the unique polynomial of degree ≤ 3 passing through the four points

$$(x_{k-1}, g(x_{k-1})), \quad (x_k, g(x_k)), \quad (x_{k+1}, g(x_{k+1})), \quad \text{and} \quad (x_k^*, g(x_k^*)).$$

Note that the coefficients in the definition of L are linear functions of g, hence L itself is a linear function. Furthermore, $Lg = g$ for all polynomials of degree ≤ 3. Since

$$(Lf)(x) = a_k(f)(x - x_{k-1})(x - x_k)(x - x_{k+1}) + p_k(x)$$

and

$$\int_{x_{k-1}}^{x_{k+1}} (x - x_{k-1})(x - x_k)(x - x_{k+1})\, dx = 0,$$

we see that

$$\int_{x_{k-1}}^{x_{k+1}} Lf = \int_{x_{k-1}}^{x_{k+1}} p_k = S_{n,k}.$$

By Taylor's formula with integral remainder (Exercise 4.6.3), there exists a polynomial $T_3(x)$ of degree ≤ 3 such that

$$f(x) = T_3(x) + R_3(x), \quad \text{where} \quad R_3(x) := \frac{1}{3!}\int_{x_{k-1}}^{x} (x-t)^3 f^{(4)}(t)\, dt.$$

The remainder may be written

$$R_3(x) = \frac{1}{3!}\int_{x_{k-1}}^{x_{k+1}} q_t(x) f^{(4)}(t)\, dt \quad \text{where} \quad q_t(x) := \begin{cases} (x-t)^3 & \text{if } t \le x \\ 0 & \text{if } t > x. \end{cases}$$

Since

$$Lf = LT_3 + LR_3 = T_3 + LR_3 = f - R_3 + LR_3,$$

we see that

$$E_{n,k} = \int_{x_{k-1}}^{x_{k+1}} (f - Lf) = \int_{x_{k-1}}^{x_{k+1}} (R_3 - LR_3).$$

In the remaining calculations, for ease of notation we assume that $[x_{k-1}, x_{k+1}] = [-h, h]$. By Fubini's theorem for continuous functions,

$$\begin{aligned} \int_{-h}^{h} R_3(x)\, dx &= \frac{1}{3!}\int_{-h}^{h} f^{(4)}(t) \int_{-h}^{h} q_t(x)\, dx\, dt \\ &= \frac{1}{4!}\int_{-h}^{h} f^{(4)}(t)(h-t)^4\, dt. \end{aligned} \tag{5.24}$$

Also, because L is linear,[1]

$$(LR_3)(x) = \frac{1}{3!}\int_{-h}^{h} f^{(4)}(t)(Lq_t)(x)\, dt.$$

[1]This may be proved using the dominated convergence theorem (See Exercise 14).

Therefore, by Fubini's theorem,

$$\int_{-h}^{h} (LR_3)(x)\,dx = \frac{1}{3!}\int_{-h}^{h} f^{(4)}(t)\int_{-h}^{h} (Lq_t)(x)\,dx\,dt. \tag{5.25}$$

Now, by definition of L,

$$(Lq_t)(x) = a_t(x+h)x(x-h) + b_t(x+h)x + c_t(x+h) + d_t,$$

where

$$b_t = \frac{q_t(h) - 2q_t(0) + q_t(-h)}{2h^2}, \quad c_t = \frac{q_t(0) - q_t(-h)}{h}, \quad \text{and} \quad d_t = q_t(-h).$$

Since $q_t(-h) = 0$ and $q_t(h) = (h-t)^3$, $t \in [-h,h]$,

$$\int_{-h}^{h} (Lq_t)(x)\,dx = \tfrac{2}{3}h^3 b_t + 2h^2 c_t = \tfrac{1}{3}h[(h-t)^3 + 4q_t(0)]. \tag{5.26}$$

From (5.24), (5.25), and (5.26),

$$\int_{-h}^{h} (f - Lf) = \int_{-h}^{h} (R_3 - LR_3) = \frac{1}{72}\int_{-h}^{h} f^{(4)}(t)\alpha(t)\,dt, \tag{5.27}$$

where

$$\alpha(t) := 3(h-t)^4 - 4h[(h-t)^3 + 4q_t(0)].$$

Recalling the definition of $q_t(0)$, we see that

$$\alpha(t) = \begin{cases} (t-h)^3(3t+h) + 16ht^3 & \text{if } -h \le t \le 0, \\ (t-h)^3(3t+h) & \text{if } 0 \le t \le h. \end{cases} \tag{5.28}$$

Thus if $t \ge 0$,

$$\alpha(-t) = (t+h)^3(3t-h) - 16ht^3 \quad \text{and} \quad \alpha(t) = (t-h)^3(3t+h). \tag{5.29}$$

The cubic polynomials in (5.29) are easily seen to be equal at the conveniently chosen points $t = 0, \pm h, 2h$ and therefore must be identical. Thus α is an even function of t, so (5.28) may be rewritten

$$\alpha(t) = \begin{cases} (t+h)^3(3t-h) & \text{if } -h \le t \le 0, \\ (t-h)^3(3t+h) & \text{if } 0 \le t \le h. \end{cases}$$

Taking derivatives, we see that α is decreasing on $[-h,0]$ and increasing on $[0,h]$. Since $\alpha(-h) = \alpha(h) = 0$, it follows that $\alpha \le 0$ on $[-h,h]$. By (5.27) and the weighted mean value theorem for integrals, for some point $\xi \in [-h,h]$ we have

$$\int_{-h}^{h} (f - Lf) = \frac{f^{(4)}(\xi)}{72}\int_{-h}^{h} \alpha(t)\,dt = \frac{f^{(4)}(\xi)}{36}\int_{0}^{h} (t-h)^3(3t+h)\,dt = -\frac{h^5 f^{(4)}(\xi)}{90}.$$

The same result holds for $E_{n,k}$, with the point ξ depending on k. This completes the proof of the theorem. □

Comparison of the Approximations

Table 5.3 below gives the errors $\int_1^2 x^{-1}\,dx - A_n$, rounded to 10 decimal places, where A_n is the approximation. The *left point rule* simply refers to approximation by the Riemann sum R_n. The exact value of the integral, up to 10 decimal places, is $\ln 2 = .6931471805\ldots$

TABLE 5.3: A comparison of the methods.

Method	$n=4$	$n=8$
Left Point Rule	.1836233710	.0927753302
Trapezoidal Rule	−.0038766290	−.0009746698
Midpoint Rule	.0019272893	.0004866265
Simpson's Rule	−.0001067877	−.00000735011

5.7 Improper Integrals

In this section, the Riemann integral is extended in two ways: First, the integrand is allowed to be unbounded, and second, the integration interval can be infinite.

5.7.1 Definition. A function f is said to be *locally integrable on an interval* I if $f \in \mathcal{R}_c^d$ for every interval $[c,d]$ contained in I. ◊

For example, a continuous function is locally integrable on any interval I.

5.7.2 Definition. Each expression in (a)–(c) below is called an *improper integral.* The integral is said to *converge* if the limit exists in $\mathbb{R}$ and to *diverge* otherwise. In the former case, f is said to be *improperly integrable on I*.

(a) $\displaystyle\int_a^b f := \lim_{t\to b^-}\int_a^t f,$ where f is locally integrable on $[a,b)$.

(b) $\displaystyle\int_a^b f := \lim_{t\to a^+}\int_t^b f,$ where f is locally integrable on $(a,b]$.

(c) $\displaystyle\int_a^b f := \int_a^c f + \int_c^b f,$ where f is locally integrable on $(a,c)\cup(c,b)$. ◊

Note that the limits of integration in these definitions, where appropriate, may be infinite.

It is easy to see that if f is locally integrable on $(a, b]$, then $\int_a^b f$ converges iff $\int_a^c f$ converges for some (every) $c \in (a, b)$. In this case,

$$\int_a^b f = \int_a^c f + \int_c^b f.$$

The first integral on the right is improper, while the second is a Riemann integral. Moreover, if f is also bounded and $a,\ b \in \mathbb{R}$, then, by Exercise 5.2.10, the improper integral $\int_a^b f$ is simply the Riemann integral. Analogous remarks apply to the other cases.

5.7.3 Examples. (a) Let $p \in \mathbb{R}$. For $0 < s < t$,

$$\int_s^t \frac{dx}{x^p} = \begin{cases} (1-p)^{-1}\left(t^{1-p} - s^{1-p}\right) & \text{if } p \neq 1 \\ \ln t - \ln s & \text{if } p = 1. \end{cases}$$

It follows that

$$\int_1^\infty \frac{dx}{x^p} \text{ converges iff } p > 1 \quad \text{and} \quad \int_0^1 \frac{dx}{x^p} \text{ converges iff } p < 1.$$

(b) Let $r > 0$, $r \neq 1$. For $t > 1$, $\displaystyle\int_1^t r^x\, dx = \frac{r^t - r}{\ln r}$, hence

$$\int_1^\infty r^x\, dx \text{ converges iff } r < 1.$$

(c) Since $\displaystyle\int_s^t (1 + x^2)^{-1}\, dx = \tan^{-1} t - \tan^{-1} s$,

$$\int_0^\infty \frac{dx}{1+x^2} = \int_{-\infty}^0 \frac{dx}{1+x^2} = \frac{\pi}{2}, \text{ hence } \int_{-\infty}^\infty \frac{dx}{1+x^2} = \pi.$$

(d) $$\int_{-1}^1 \frac{dx}{\sqrt{|x|}} = \int_{-1}^0 \frac{dx}{\sqrt{-x}} + \int_0^1 \frac{dx}{\sqrt{x}} = 2 \lim_{t \to 0+} \int_t^1 \frac{dx}{\sqrt{x}} = 4. \qquad \lozenge$$

For ease of exposition, in the remainder of the section we consider only integrals that are improper at the upper limit. Analogous discussions hold for the other types of improper integrals.

The proof of the following theorem is left to the reader.

5.7.4 Theorem. *Let f and g be locally integrable on $[a, b)$ and let $\alpha,\ \beta \in \mathbb{R}$. If the improper integrals $\int_a^b f$ and $\int_a^b g$ converge, then so does $\int_a^b (\alpha f + \beta g)$ and*

$$\int_a^b (\alpha f + \beta g) = \alpha \int_a^b f + \beta \int_a^b g.$$

In contrast to the Riemann integral, the product of improperly integrable functions may not be improperly integrable. For example, $f(x) := 1/\sqrt{1-x}$ is improperly integrable on the interval $[0,1)$ but f^2 is not. The following example illustrates the same phenomenon, but on an unbounded interval. It is the first of several examples in this section that use the fact, proved in Chapter 6, that a series of the form $\sum_{n=1}^{\infty} 1/n^p$ converges iff $p > 1$.

5.7.5 Example. Define f on $[1, +\infty)$ by $f(x) = n$ if $n \le x < n + 1/n^{5/2}$, $n = 1, 2, \ldots$, and $f(x) = 0$ otherwise. Then

$$\int_1^{n+1} f = \sum_{k=1}^{n} \frac{1}{k^{3/2}} \quad \text{and} \quad \int_1^{n+1} f^2 = \sum_{k=1}^{n} \frac{1}{k^{1/2}},$$

hence $\int_1^{\infty} f$ converges, whereas $\int_1^{\infty} f^2$ diverges. ◊

We now have examples on both bounded and unbounded intervals of nonnegative improperly integrable functions whose squares are not improperly integrable. Conversely, there exist locally integrable nonnegative functions on unbounded intervals, for example, $f(x) = 1/x$ on $[1, +\infty)$, such that f^2 is improperly integrable but f is not. However, for bounded intervals this is not possible: If f^2 is improperly integrable on a bounded interval, then so is $|f|$. (Exercise 25.)

The remainder of this section describes various convergence tests for improper integrals. Many of these are analogs of convergence tests for infinite series, discussed in Chapter 6.

5.7.6 Comparison Test for Integrals. *Let f and g be locally integrable on $[a, b)$ such that $0 \le f \le g$. If $\int_a^b g$ converges, then so does $\int_a^b f$.*

Proof. Let $F(x) = \int_a^x f$ and $G(x) = \int_a^x g$, $a \le x < b$. Since f and g are nonnegative, F and G are increasing, hence, by the monotone function theorem (3.1.17),

$$\int_a^b f = \lim_{x \to b^-} F(x) \quad \text{and} \quad \int_a^b g = \lim_{x \to b^-} G(x)$$

exist in $\overline{\mathbb{R}}$. Since $F \le G$, the conclusion follows. □

5.7.7 Example. Let $f(x) = \dfrac{1 + \sin x}{\sqrt{x}(x+1)^2}$, $x > 0$. By definition,

$$\int_0^{\infty} f = \int_0^1 f + \int_1^{\infty} f,$$

provided the integrals on the right converge. That this is indeed the case follows from 5.7.3(a), 5.7.6, and the inequalities $f(x) \le 2/\sqrt{x}$ on $(0, 1]$ and $f(x) \le 2/(x+1)^2$ on $[1, +\infty)$. ◊

5.7.8 Example. The *gamma function* is defined by

$$\Gamma(x) = \int_0^\infty t^{x-1}e^{-t}\,dt, \quad x > 0.$$

To see that the integral converges for all $x > 0$, note that $t^{x-1}e^{-t} \le t^{x-1}$ for $t \in (0,1]$, hence $\int_0^1 t^{x-1}e^{-t}\,dt$ converges by comparison with $\int_0^1 t^{x-1}\,dt$ (see 5.7.3(a)). Furthermore, by l'Hospital's rule applied sufficiently many times,

$$\lim_{t\to+\infty} t^{x+1}e^{-t} = 0$$

so there exists $t_0 > 1$ such that $t^{x+1}e^{-t} \le 1$, or $t^{x-1}e^{-t} \le t^{-2}$, for all $t \ge t_0$. Therefore, $\int_1^\infty t^{x-1}e^{-t}\,dt$ converges by comparison with $\int_1^\infty t^{-2}\,dt$.

The gamma function has the following recursive property:

$$\Gamma(x+1) = x\Gamma(x).$$

To see this, integrate $\Gamma(x+1)$ by parts to obtain

$$\int_a^b t^x e^{-t}\,dt = t^x e^{-t}\Big|_{t=b}^{t=a} + x\int_a^b t^{x-1}e^{-t}\,dt,$$

and then let $a \to 0$ and $b \to +\infty$. In particular, for $n \in \mathbb{N}$

$$\Gamma(n+1) = n\Gamma(n) = n(n-1)\Gamma(n-1) = \cdots = n(n-1)\cdots 1\cdot\Gamma(1).$$

Since

$$\Gamma(1) = \int_0^\infty e^{-t}\,dt = 1,$$

we see that $\Gamma(n+1) = n!$. Thus $\Gamma(x)$ is a continuous (indeed, differentiable) extension of the factorial function on $\mathbb{N}$. ◊

5.7.9 Limit Comparison Test for Integrals. *Let f and g be locally integrable on $[a,b)$ with $f \ge 0$ and $g > 0$. If $L := \lim_{x\to b-} f(x)/g(x)$ exists and $0 < L < +\infty$, then $\int_a^b g$ converges iff $\int_a^b f$ converges.*

Proof. Since $f, g \ge 0$, $\int_a^b f$ and $\int_a^b g$ exist in $\overline{\mathbb{R}}$. Choose $c \in (a,b)$ such that $L/2 < f(x)/g(x) < 2L$ for all $x \in [c,b)$. For such x, $g(x) < 2f(x)/L$ and $f(x) < 2Lg(x)$. The assertion then follows from the inequalities

$$\int_c^b g \le \frac{2}{L}\int_c^b f \quad\text{and}\quad \int_c^b f \le 2L\int_c^b g. \qquad \square$$

5.7.10 Example. Let $f(x) = \dfrac{\sqrt{2x^5 - x^2 + 1}}{x^4 + 3x + 5}$, $x \ge 1$. For $g(x) = x^{-3/2}$,

$$\lim_{x\to+\infty}\frac{f(x)}{g(x)} = \lim_{x\to+\infty}\frac{\sqrt{2x^8 - x^5 + x^3}}{x^4 + 3x + 5} = \sqrt{2}.$$

Since $\int_1^\infty g$ converges, so does $\int_1^\infty f$. ◊

5.7.11 Root Test for Integrals. *Let f be locally integrable and nonnegative on $[a, \infty)$ and suppose that $L := \lim_{x\to\infty}[f(x)]^{1/x}$ exists in $\overline{\mathbb{R}}$. Then, $\int_a^\infty f$ converges if $L < 1$ and diverges if $L > 1$.*

Proof. Suppose $L < 1$. Choose $r \in (L, 1)$ and $x_0 > a$ such that $[f(x)]^{1/x} < r$ for all $x \geq x_0$. For such x, $f(x) < r^x$, hence, by the comparison theorem and 5.7.3(b), $\int_{x_0}^\infty f$ converges. Therefore, $\int_a^\infty f$ converges. A similar argument shows that $\int_a^\infty f$ diverges if $L > 1$. □

5.7.12 Example. For $p \in \mathbb{R}$ and $x \geq 1$, let

$$f(x) = \left(\frac{2x + \cos x}{3x + \sin x}\right)^{px}.$$

Since $\lim_{x\to+\infty} [f(x)]^{1/x} = (2/3)^p$, $\int_1^\infty f$ converges iff $p > 0$. ◊

There are examples of convergent integrals and divergent integrals with $L = 1$, so the root test in inconclusive in this case (see Exercise 3).

5.7.13 Definition. Let f be locally integrable on $[a, b)$. The improper integral $\int_a^b f$ is said to *converge absolutely* if $\int_a^b |f|$ converges. In this case f is said to be *improperly absolutely integrable on* $[a, b)$. If $\int_a^b f$ converges but not absolutely, then the integral is said to *converge conditionally.* ◊

5.7.14 Proposition. *If f is improperly absolutely integrable on $[a, b)$, then $\int_a^b f$ converges and*

$$\left|\int_a^b f\right| \leq \int_a^b |f|.$$

Proof. Set $g(x) := |f(x)| + f(x)$, so $0 \leq g \leq 2|f|$ on $[a, b]$. By the comparison test, $\int_a^b g$ converges. Since $f = g - |f|$ is the difference of two improperly integrable functions, f is improperly integrable. The inequality follows on letting $t \to +\infty$ in the inequality $|\int_a^t f| \leq \int_a^t |f|$. □

5.7.15 Example. For $p > 0$, define

$$f(x) = \frac{(-1)^{n+1}}{n^p}, \quad n \leq x < n+1, \quad n = 1, 2, \ldots.$$

Then

$$\int_1^{n+1} |f| = \sum_{k=1}^n \frac{1}{k^p} \quad \text{and} \quad \int_1^{n+1} f = \sum_{k=1}^n \frac{(-1)^{k+1}}{k^p}.$$

The first sum has a finite limit iff $p > 1$, while the second sum has a finite limit iff $p > 0$ (see Chapter 6). Therefore, $\int_1^\infty f$ converges absolutely iff $p > 1$ and conditionally iff $0 < p \leq 1$. ◊

The following theorem is sometimes useful in establishing conditional convergence of improper integrals.

5.7.16 Dirichlet's Test for Integrals. *Let f be continuous and g' improperly absolutely integrable on $[a,b)$. If the function $F(t) := \int_a^t f$ is bounded on $[a,b)$ and $\lim_{x\to b^-} g(x) = 0$, then $\int_a^b fg$ converges.*

Proof. Let M be a bound for $|F|$ on $[a,b)$. Then $|Fg'| \le M|g'|$, hence, by the comparison test, Fg' is absolutely integrable on $[a,b)$. Integrating by parts yields

$$\int_a^t fg = \int_a^t F'g = F(t)g(t) - \int_a^t Fg'.$$

Since $\int_a^b Fg'$ converges and $\lim_{t\to b^-} F(t)g(t) = 0$, $\int_a^b fg$ converges. □

5.7.17 Corollary. *Let f be continuous and g' locally integrable on $[a,b)$ with $\lim_{x\to b^-} g(x) = 0$. If the function $F(t) := \int_a^t f$ is bounded on $[a,b)$ and if g' has constant sign, then $\int_a^b fg$ converges.*

Proof. By the fundamental theorem of calculus, $\int_a^t g' = g(t) - g(a)$, hence g' is absolutely integrable on $[a,b)$ and Dirichlet's test applies. □

5.7.18 Example. Let $h(x) = x^{-p}\sin x$, $x \ge 1$, where $p > 0$. Taking $f(x) = \sin x$ and $g(x) = x^{-p}$ in 5.7.17 shows that $\int_1^\infty h$ converges. Since $|h(x)| \le 1/x^p$, h is improperly absolutely integrable on $[1,+\infty)$ if $p > 1$. If $0 < p \le 1$, then the sums on the extreme right in the inequality

$$\int_\pi^{n\pi} |h| = \sum_{k=2}^n \int_{(k-1)\pi}^{k\pi} |h| > \sum_{k=2}^n (k\pi)^{-p} \int_{(k-1)\pi}^{k\pi} |\sin x|\,dx = 2\pi^{-p}\sum_{k=2}^n k^{-p}$$

are unbounded (see Example 6.2.5), hence h is not improperly absolutely integrable in this case. ◊

5.7.19 Cauchy–Schwarz Inequality for Improper Integrals. *Let f and g be continuous with f^2 and g^2 improperly integrable on $[a,b)$. Then fg is improperly absolutely integrable on $[a,b)$ and*

$$\left(\int_a^b |fg|\right)^2 \le \int_a^b f^2 \cdot \int_a^b g^2.$$

Proof. By Exercise 5.2.13, for all $t \in [a,b)$

$$\left(\int_a^t |fg|\right)^2 \le \int_a^t f^2 \cdot \int_a^t g^2.$$

Now let $t \to b$. □

Exercises

1. Determine all values of p, $q > 0$ for which $\displaystyle\int_0^1 \frac{dx}{(\sin x)^p(1-x)^q}$ converges.

2. Let $p > 0$. Show that $\displaystyle\int_1^\infty x^{-px}\,dx$ converges and $\displaystyle\int_1^\infty x^{-p/x}\,dx$ diverges.

   Show that the same behavior holds for $\displaystyle\int_0^1 x^{-px}\,dx$ and $\displaystyle\int_0^1 x^{-p/x}\,dx$.

3. Find examples for which $\lim_{x\to+\infty}[f(x)]^{1/x} = 1$ and

   (a) $\displaystyle\int_1^\infty f$ converges (b) $\displaystyle\int_1^\infty f$ diverges.

4. Let f and g be positive and continuous on $[1, +\infty)$. Prove that if $L := \lim_{x\to+\infty} \dfrac{f(x)}{g(x)}$ exists in $\mathbb{R}$ and $\int_1^\infty f$ and $\int_1^\infty g$ converge, then $\displaystyle\lim_{x\to+\infty} \frac{\int_x^\infty f}{\int_x^\infty g} = L.$

5. Determine if the integrals converge or diverge:

   (a) $\displaystyle\int_0^1 \frac{\sin x - x}{x^3}\,dx$ (b) $\displaystyle\int_0^1 \frac{\sin x}{x^2}\,dx$ (c) $\displaystyle\int_0^1 \frac{\sqrt{\sin x}}{x}\,dx$

   (d) $\displaystyle\int_2^\infty \sin^2\frac{1}{\ln x}\,dx$ (e) $\displaystyle\int_2^\infty \frac{(\ln x)(\sin x)}{x}\,dx$ (f) $\displaystyle\int_1^\infty \frac{(\sin x)(\cos x^{-1})}{x}\,dx.$

6. Prove that $\displaystyle\int_0^{\pi/2} \cos(\sec^p x)\,dx$ converges for all $p > 0$.

7. Show that $\displaystyle\int_0^1 \frac{x^p}{\sqrt{1-x^2}}\,dx$ converges iff $p > -1$.

8. Show that $\displaystyle\int_0^1 \frac{\sin^p x}{x^q}\,dx$ converges iff $p < 1 + q$.

9. Find all values of p for which the integral converges:

   (a) $\displaystyle\int_1^\infty x^p e^{-x}\,dx$ (b) $\displaystyle\int_0^1 x^p e^{-x}\,dx$ (c) $\displaystyle\int_0^1 \sin x^p\,dx$

   (d) $\displaystyle\int_0^1 x^p \sin x^p\,dx$ (e) $\displaystyle\int_0^1 x^p \ln x\,dx$ (f) $\displaystyle\int_1^\infty x^p \ln x\,dx$

   (g) $\displaystyle\int_0^{\pi/2} \sin^p x\,dx$ (h) $\displaystyle\int_0^{\pi/2} x\sin^p x\,dx$ (i) $\displaystyle\int_0^{\pi/2} (1-\sin x)^p\,dx$

   (j) $\displaystyle\int_0^{\pi/2} \tan^p x\,dx$ (k) $\displaystyle\int_0^{\pi/2} x^p \cos x\,dx$ (l) $\displaystyle\int_0^{\pi/2} x^p \sin x\,dx.$

10. Find all values of $p > 0$ for which $\int_0^1 x^{-p}\sin e^x\,dx$ converges absolutely.

11. Prove that $\int_1^\infty \frac{x\sin x}{1+x^2}\,dx$ converges conditionally.

12. Prove that $\int_1^\infty x^p \sin e^x\,dx$ converges for all p. For what values of p does the integral converge conditionally? (See 5.7.18)

13. Find all values of $p,\ q > 0$ for which the integral converges:

(a) $\int_0^1 \frac{dx}{(1-x^p)^q}$ (b) $\int_0^1 \frac{x^p}{(1-x^{2p})^q}\,dx$ (c) $\int_1^\infty \frac{dx}{\sqrt{x^p+x^q}}$

(d) $\int_0^1 \frac{dx}{\sqrt{x^p+x^q}}$ (e) $\int_0^{\pi/2} \frac{\sin^p x}{\cos^q x}\,dx$ (f) $\int_0^{\pi/2} \frac{1}{\sin^p x^q}\,dx.$

14. Prove by induction that $\int_0^\infty x^n e^{-x}\,dx = n!.$

15. Given that $\int_{-\infty}^\infty e^{-x^2/2}\,dx = \sqrt{2\pi}$ (to be established in 11.5.3) show that,

$$\frac{1}{\sqrt{2\pi}}\int_{-\infty}^\infty x^{2n}e^{-x^2/2}\,dx = (2n-1)(2n-3)\cdots 3\cdot 1 = \frac{(2n)!}{n!2^n}.$$

16. Given that $\int_0^\infty e^{-s^2}\,ds = \sqrt{\pi}/2$, show that

$$\Gamma\left(\frac{1}{2}\right) = \sqrt{\pi},\quad \Gamma\left(\frac{3}{2}\right) = \frac{\sqrt{\pi}}{2},\quad \text{and}\quad \Gamma\left(\frac{5}{2}\right) = \frac{3\sqrt{\pi}}{4}.$$

17. The formula $\Gamma(x) = x^{-1}\Gamma(x+1)$ may be used to extend the gamma function to non-integer values $x < 0$. Use this to find $\Gamma\left(-\frac{1}{2}\right)$ and $\Gamma\left(-\frac{3}{2}\right)$.

18. Prove that if f is absolutely integrable on $[1,\infty)$, then

$$\lim_{n\to\infty}\int_1^\infty f(x^n)dx = 0.$$

19. (Log test for integrals). Let f be locally integrable and positive on $[0,+\infty)$ such that

$$L := \lim_{x\to\infty}\frac{-\ln f(x)}{\ln x}$$

exists in $\overline{\mathbb{R}}$. Prove that $\int_0^\infty f$ converges if $L > 1$ and diverges if $L < 1$.

20. Use Exercise 19 to determine the convergence behavior of

 (a) $\displaystyle\int_1^\infty (\ln x)^{-\ln x}\,dx.$ (b) $\displaystyle\int_1^\infty (\ln x)^{-\sqrt{x}}\,dx.$

 What does the root test reveal?

21. Prove that $L(a) := \displaystyle\lim_{t\to+\infty}\int_{1/t}^{t} \frac{\sin ax}{x}\,dx$ converges for all $a \in \mathbb{R}$ and that $L(a) = L(1)$ for all $a > 0$.

22. Let f be differentiable and nonzero on $[1, +\infty)$. If $\displaystyle\lim_{x\to+\infty} xf'(x)/f(x)$ exists in $\overline{\mathbb{R}}$ and is less than -1, prove that $\int_1^\infty f$ converges.

23. Prove that if $\int_0^\infty f(x)\,dx$ converges, then $\lim_n \int_0^1 f(nx)\,dx = 0$.

24. Prove that if $f \ge 0$ and $\int_1^\infty f$ converges, then $\int_1^\infty \sqrt{f(x)}/x\,dx$ converges.

25. Prove that if $[a, b)$ is finite and f^2 is improperly integrable on $[a, b)$, then $|f|$ is improperly integrable on $[a, b)$.

26. Let f be continuous and g locally integrable and positive on $[a, b)$. Suppose that the function $G(x) := \int_a^x g$ is bounded on $[a, b)$ and that $\lim_{x\to b^-} f(x) = 0$. Prove that $\int_a^b fg$ converges.

27. Let f be continuous on $[a, b)$ such that $\int_a^b f$ converges. Let g be bounded on $[a, b)$ such that g' is locally integrable and has constant sign on $[a, b)$. Prove that $\int_a^b fg$ converges.

28. Let f be improperly integrable on $(-\infty, +\infty)$ and $c \in \mathbb{R}$. Prove that

$$\int_{-\infty}^{\infty} f(x+c)\,dx = \int_{-\infty}^{+\infty} f(x)\,dx.$$

---

*5.8 A Deeper Look at Riemann Integrability

In this section we characterize Riemann integrability of a function in terms of the size of its set of discontinuities.

5.8.1 Definition. A set A of real numbers is said to have (*Lebesgue*) *measure zero* if for each $\varepsilon > 0$ there exists a finite or infinite sequence of intervals I_n with *total length* $\sum_n |I_n| < \varepsilon$ such that the sequence *covers* A, that is, every member of A is contained in some I_n. ◊

Any countable set has measure zero. Indeed, if $A = \{a_1, a_2, \ldots\}$ and $\varepsilon > 0$, then the intervals $I_n = (a_n - \varepsilon/2^{n+2}, a_n + \varepsilon/2^{n+2})$ obviously cover A and have total length $< \varepsilon$. In particular, the set of rational numbers has measure zero. An uncountable set of measure zero is constructed in Example 10.3.4.

The following result will be proved in Chapter 11.

5.8.2 Theorem. *Let f be bounded on $[a, b]$. Then $f \in \mathcal{R}_a^b$ iff its set of discontinuities has measure zero.*

Examples 5.1.11 and 5.1.12 are relevant here. The function in the first example, shown to be integrable, has a countable set of discontinuities. The function in the second example, shown not to be integrable, has $[0, 1]$ as its set of discontinuities, certainly *not* a set of measure zero.

Theorem 5.8.2 allows simple proofs of many of the properties discussed in this chapter. For example, if f and g are integrable with sets of discontinuity A and B, respectively, then $f + g$ and fg have sets of discontinuity contained in $A \cup B$, a set of measure zero (Exercise 2), and hence are integrable.

Exercises

1. Show that if B has measure zero and $A \subseteq B$, then A has measure zero.

2. Prove: If A_n has measure zero for every $n \in \mathbb{N}$, then so does $A_1 \cup A_2 \cup \cdots$.

3. Let A have measure zero. Prove that $A + \mathbb{Q}$ has measure zero.

4. Let $f : [a, b] \to [c, d]$ be integrable and $g : [c, d] \to \mathbb{R}$ continuous. Prove that $g \circ f$ is integrable.

5. A set A of real numbers has (*Jordan*) *content zero* if for each $\varepsilon > 0$, there exist finitely many intervals of total length $< \varepsilon$ that cover A. Show that

   (a) a convergent sequence has content zero.

   (b) $[0, 1] \cap \mathbb{Q}$ has measure zero but not content zero.

6. Prove that the function f in Exercise 3.3.10 is integrable on $[a, b]$ and find its integral.

---

*5.9 Functions of Bounded Variation

5.9.1 Definition. Let $\mathcal{P} = \{a = x_0 < x_1 < \cdots < x_n = b\}$ be a partition of $[a, b]$. For $f : [a, b] \to \mathbb{R}$, define

$$V_{\mathcal{P}}(f) = \sum_{j=1}^{n} |f(x_j) - f(x_{j-1})|.$$

The *total variation of f on $[a,b]$* is the extended real number

$$V_a^b(f) := \sup_{\mathcal{P}} V_{\mathcal{P}}(f).$$

The function f is said to have *bounded variation on* $[a,b]$ if $V_a^b(f) < +\infty$. The set of all functions with bounded variation on $[a,b]$ is denoted by $\mathcal{BV}_a^b$. $\lozenge$

Here is a list of elementary properties of functions of bounded variation:

5.9.2 Proposition. *Let $f : [a,b] \to \mathbb{R}$.*

(a) *If $f \in \mathcal{BV}_a^b$, then f is bounded.*

(b) *If f has a bounded derivative on $[a,b]$, then $f \in \mathcal{BV}_a^b$.*

(c) *If f is monotone on $[a,b]$, then $V_a^b(f) = |f(b) - f(a)|$.*

(d) *If $g \in \mathcal{R}_a^b$ and $f(x) = \int_a^x g(t)\,dt$, then $V_a^b(f) \le (b-a)\sup_{[a,b]}|g|$.*

(e) *If $\mathcal{P}$ is a partition of $[a,b]$ and $\mathcal{Q}$ is a refinement of $\mathcal{P}$, then $V_{\mathcal{P}}(f) \le V_{\mathcal{Q}}(f)$.*

(f) *If $f, g \in \mathcal{BV}_a^b$ and $c \in \mathbb{R}$, then $f+g, cf, fg \in \mathcal{BV}_a^b$.*

Proof. (a) Let $a < x < b$ and $\mathcal{P} = \{a, x, b\}$. Then

$$\begin{aligned} 2|f(x)| &\le |f(x) - f(a)| + |f(x) - f(b)| + |f(a)| + |f(b)| \\ &= V_{\mathcal{P}}(f) + |f(a)| + |f(b)| \\ &\le V_a^b(f) + |f(a)| + |f(b)|. \end{aligned}$$

(b) Let $|f'| \le C$ on $[a,b]$. By the mean value theorem, given a partition $\mathcal{P}$, there exists for each j a point $t_j \in (x_{j-1}, x_j)$ such that

$$V_{\mathcal{P}}(f) = \sum_{\mathcal{P}} |f(x_j) - f(x_{j-1})| = \sum_{\mathcal{P}} |f'(t_j)|(x_j - x_{j-1}) \le C(b-a).$$

Therefore, $V_a^b(f) \le C(b-a)$.

(c) If f is increasing, then

$$\sum_{\mathcal{P}} |f(x_j) - f(x_{j-1})| = \sum_{\mathcal{P}} \big(f(x_j) - f(x_{j-1})\big) = f(b) - f(a).$$

(d) Let $M := \sup_{a \le t \le b} |g(t)|$. Then, for any partition $\mathcal{P}$,

$$V_{\mathcal{P}}(f) \le \sum_{\mathcal{P}} \int_{x_{j-1}}^{x_j} |g(t)|\,dt \le M \sum_{\mathcal{P}} (x_j - x_{j-1}) = M(b-a).$$

(e) Let $\mathcal{P} = \{a = x_0 < x_1 < \cdots < x_n = b\}$ and $\mathcal{P}' = \mathcal{P} \cup \{c\}$, where $c \in [x_{i-1}, x_i]$. Then

$$\begin{aligned} V_{\mathcal{P}}(f) &= \sum_{j \neq i} |f(x_j) - f(x_{j-1})| + |f(x_i) - f(x_{i-1})| \\ &\leq \sum_{j \neq i} |f(x_j) - f(x_{j-1})| + |f(x_i) - f(c)| + |f(c) - f(x_{i-1})| \\ &= V_{\mathcal{P}'}(f). \end{aligned}$$

Adding points successively yields (e).

(f) Let $|f|$, $|g| \leq M$ on $[a, b]$. The inequality

$$|(fg)(x_j) - (fg)(x_{j-1})| \leq M|g(x_j) - g(x_{j-1})| + M|f(x_j) - f(x_{j-1})|$$

shows that $fg \in \mathcal{BV}_a^b$. The proofs of the other parts of (f) are similar. □

5.9.3 Example. For $\alpha > 0$, define a continuous function f_α on $[0, 1]$ by

$$f_\alpha(x) := \begin{cases} x^\alpha \sin(1/x) & \text{if } 0 < x \leq 1, \\ 0 & \text{if } x = 0. \end{cases}$$

We show that if $\alpha \leq 1$, then f_α does not have bounded variation on $[0, 1]$. To see this define

$$a_k := \frac{1}{2k\pi + \pi/2} = \frac{2}{(4k+1)\pi} \quad \text{and} \quad b_k := \frac{1}{2k\pi}$$

and note that

$$f_\alpha(b_k) = 0 \quad \text{and} \quad f_\alpha(a_k) = a_k^\alpha = \frac{c}{(4k+1)^\alpha}, \quad \text{where } c := \frac{2^\alpha}{\pi^\alpha}.$$

Since $b_{k+1} < a_k < b_k$, for sufficiently small $\varepsilon > 0$ we may form the partition

$$\mathcal{P}_\varepsilon = \{\varepsilon < a_p < b_p < a_{p-1} < \cdots < a_k < b_k < \cdots < b_{q+1} < a_q < b_q < 1\}$$

of $[\varepsilon, 1]$, where p and q are, respectively, the largest and smallest integers satisfying $\varepsilon < a_p < b_q < 1$, equivalently,

$$\frac{1}{2\pi} < q < p < \frac{2 - \pi\varepsilon}{4\pi\varepsilon}.$$

From $f_\alpha(a_k) - f_\alpha(b_k) = \dfrac{c}{(4k+1)^\alpha}$,

$$V_0^1(f_\alpha) \geq V_\varepsilon^1(f_\alpha) \geq c \sum_{k=q}^{p} \frac{1}{(4k+1)^\alpha}.$$

Since ε may be chosen arbitrarily small, the upper limit p of the sum on the right may be made arbitrarily large. Since the series $\sum_{k=1}^\infty (4k+1)^{-\alpha}$ diverges for $\alpha \leq 1$, $V_0^1(f_\alpha) = +\infty$. ◊

5.9.4 Theorem. *If* $f' \in \mathcal{R}_a^b$*, then* $f \in \mathcal{BV}_a^b$ *and*

$$V_a^b(f) = \int_a^b |f'(x)|\,dx. \tag{5.30}$$

Proof. Let $\mathcal{P}$ be a partition of $[a,b]$. By the mean value theorem, there exists $\xi_j \in [x_{j-1}, x_j]$ such that

$$V_{\mathcal{P}}(f) = \sum_{\mathcal{P}} |f'(\xi_j)|\,|x_j - x_{j-1}| = S(|f'|, \mathcal{P}, \boldsymbol{\xi}),$$

By 5.1.18,

$$\int_a^b |f'| = \lim_{\mathcal{P}} S(|f'|, \mathcal{P}, \boldsymbol{\xi}) = \lim_{\mathcal{P}} V_{\mathcal{P}}(f). \tag{5.31}$$

On the other hand, given $r < V_a^b(f)$, we may choose a partition $\mathcal{P}_r$ of $[a,b]$ such that $r < V_{\mathcal{P}_r}(f) \le V_a^b(f)$. By 5.9.2(e), $r < V_{\mathcal{P}}(f) \le V_a^b(f)$ for all refinements $\mathcal{P}$ of $\mathcal{P}_r$. Thus, by (5.31), $r \le \int_a^b |f'| \le V_a^b(f)$. Since r was arbitrary, (5.30) follows. □

5.9.5 Corollary. *If f is continuous at a and f' is locally integrable on $(a,b]$, then* (5.30) *holds, where the integral is improper. Thus $f \in \mathcal{BV}_a^b$ iff $\int_a^b |f'|$ converges.*

Proof. By the theorem and the definition of improper integral, it suffices to show that

$$V_a^b(f) = \lim_{t \to a^+} V_t^b(f) \ \left(= \sup_{a<t\le b} V_t^b(f)\right).$$

Clearly, we may assume that $V_a^b(f) > 0$. Let $0 < s < r < V_a^b(f)$ and choose a partition $\mathcal{P}_r = \{x_0 = a < x_1 < \cdots < x_n = b\}$ such that $r < V_{\mathcal{P}_r}(f) \le V_a^b(f)$. Next, choose $t \in (a, x_1)$ so that $|f(t) - f(a)| < r - s$. For such t, let $\mathcal{P}_t = \mathcal{P}_r \cup \{t\}$. Then

$$\begin{aligned} V_t^b(f) &\ge |f(x_1) - f(t)| + \sum_{j=1}^{n-1} |f(x_{j+1}) - f(x_j)| \\ &= V_{\mathcal{P}_t}(f) - |f(t) - f(a)| \\ &> V_{\mathcal{P}_r}(f) - (r - s) > s. \end{aligned}$$

It follows that $\lim_{t\to a^+} V_t^b(f) \ge s$. Since s was arbitrary, the assertion follows. □

5.9.6 Example. We use 5.9.5 to show that the function f_α in 5.9.3 has bounded variation on $[0,1]$ if $\alpha > 1$. We have

$$|f_\alpha'(x)| = |\alpha x^{\alpha-1}\sin(1/x) - x^{\alpha-2}\cos(1/x)| \le \alpha x^{\alpha-1} + x^{\alpha-2}.$$

If $\alpha > 1$, the integral $\int_0^1 x^{\alpha-2}\,dx$ converges, hence $\int_0^1 |f_\alpha|$ converges. ◊

5.9.7 Theorem. *If $f \in \mathcal{BV}_a^b$, then there exist monotone increasing functions g and h on $[a,b]$ such that $f = g - h$.*

Proof. For $x \in [a,b]$, define $g(x) := V_a^x(f)$ and $h(x) := g(x) - f(x)$. Clearly, g is increasing. To see that h is increasing, let $x < y$, let $\mathcal{P}_x$ be an arbitrary partition of $[a,x]$, and let $\mathcal{P}_y = \mathcal{P}_x \cup \{y\}$. Then

$$V_{\mathcal{P}_x}(f) + f(y) - f(x) = V_{\mathcal{P}_y}(f) \le g(y).$$

Taking suprema over all partitions $\mathcal{P}_x$ yields $g(x) + f(y) - f(x) \le g(y)$, that is, $h(x) \le h(y)$. □

From Exercise 5.1.4 we have

5.9.8 Corollary. $\mathcal{BV}_a^b \subseteq \mathcal{R}_a^b$.

---

*5.10 The Riemann–Stieltjes Integral

In this section we describe the main features of the Riemann–Stieltjes integral, an important generalization of the Riemann integral. These integrals have many of the properties of Riemann integrals; however, as we shall see, there are some striking differences.

Definition and General Properties

5.10.1 Definition. Let f and w be bounded, real-valued functions on an interval $[a,b]$. If $\mathcal{P} = \{x_0 = a < x_1 < \cdots < x_n = b\}$ and $\xi_j \in [x_{j-1}, x_j]$, then

$$S_w(f, \mathcal{P}, \boldsymbol{\xi}) := \sum_{j=1}^{n} f(\xi_j)\Delta w_j, \quad \Delta w_j := w(x_j) - w(x_{j-1}), \quad \boldsymbol{\xi} := (\xi_1, \ldots, \xi_n),$$

is called a *Riemann–Stieltjes sum of f with respect to w*. The function f is said to be *Riemann–Stieltjes integrable with respect to w* if for some $I \in \mathbb{R}$ and each $\varepsilon > 0$, there exists a partition $\mathcal{P}_\varepsilon$ such that

$$|S_w(f, \mathcal{P}, \boldsymbol{\xi}) - I| < \varepsilon \quad \text{for all refinements } \mathcal{P} \text{ of } \mathcal{P}_\varepsilon \text{ and all choices of } \boldsymbol{\xi}.$$

In this case, I is called the *Riemann–Stieltjes integral with respect to w* and is denoted by

$$\int_a^b f\,dw = \int_a^b f(x)\,dw(x) = \lim_{\mathcal{P}} S_w(f, \mathcal{P}, \boldsymbol{\xi}). \tag{5.32}$$

The function f is called the *integrand* and w the *integrator*. The collection of all functions that are Riemann–Stieltjes integrable with respect to w is denoted by $\mathcal{R}_a^b(w)$. ◊

From 5.1.18 we see that by taking $w(x) = x$ the Riemann–Stieltjes integral reduces to the Riemann integral.

It is clear that constant functions are Riemann–Stieltjes integrable. The following example shows that, in contrast to the Riemann integral, if f has a simple discontinuity, then $\int_a^b f\,dw$ may not exist.

5.10.2 Example. Let $f : [0,1] \to \mathbb{R}$ and define

$$w(x) := \begin{cases} 0 & \text{if } 0 \le x < 1, \\ 1 & \text{if } x = 1. \end{cases}$$

We show that $f \in \mathcal{R}_0^1(w)$ iff f is continuous at 1. To this end, let

$$\mathcal{P} = \{x_0 = 0 < x_1 < \cdots < x_n = 1\}$$

be any partition of $[0,1]$. Then

$$S_w(f, \mathcal{P}, \boldsymbol{\xi}) = f(\xi_n)[w(1) - w(x_{n-1})] = f(\xi_n).$$

Thus if $f \in \mathcal{R}_0^1(w)$ and $\boldsymbol{\xi}$ is chosen so that first $\xi_n = 1$ and second $\xi_n < 1$, we see that f is continuous at 1 and $\int_0^1 f\,dw = f(1)$.

Conversely, if f is continuous at 1, then for small $\|\mathcal{P}\|$ the quantities $f(\xi_n)$ approximate $f(1)$, so $f \in \mathcal{R}_0^1(w)$. ◊

5.10.3 Theorem. *If $f, g \in \mathcal{R}_a^b(w)$ and $\alpha, \beta \in \mathbb{R}$, then $\alpha f + \beta g \in \mathcal{R}_a^b(w)$ and*

$$\int_a^b (\alpha f + \beta g)\,dw = \alpha \int_a^b f\,dw + \beta \int_a^b g\,dw.$$

Proof. This follows from the identity

$$S_w(\alpha f + \beta g, \mathcal{P}, \boldsymbol{\xi}) = \alpha S_w(f, \mathcal{P}, \boldsymbol{\xi}) + \beta S_w(g, \mathcal{P}, \boldsymbol{\xi})$$

and the linearity of the limit in (5.32), as is readily established by a standard argument. □

5.10.4 Theorem. *Let $w := \alpha u + \beta v$, where $\alpha, \beta \in \mathbb{R}$ and $u, v : [a,b] \to \mathbb{R}$ are bounded. If $f \in \mathcal{R}_a^b(u) \cap \mathcal{R}_a^b(v)$, then $f \in \mathcal{R}_a^b(w)$ and*

$$\int_a^b f\,dw = \alpha \int_a^b f\,du + \beta \int_a^b f\,dv.$$

Proof. This follows from

$$S_w(f, \mathcal{P}, \boldsymbol{\xi}) = \alpha S_u(f, \mathcal{P}, \boldsymbol{\xi}) + \beta S_v(f, \mathcal{P}, \boldsymbol{\xi})$$

and the linearity of the limit in (5.32). □

5.10.5 Theorem. *Let $a < c < b$. If $f|_{[a,c]} \in \mathcal{R}_a^c(w)$ and $f|_{[c,b]} \in \mathcal{R}_c^b(w)$, then $f \in \mathcal{R}_a^b(w)$ and*

$$\int_a^b f\,dw = \int_a^c f\,dw + \int_c^b f\,dw.$$

Proof. Given $\varepsilon > 0$, choose partitions $\mathcal{P}'_\varepsilon$ of $[a,c]$ and $\mathcal{P}''_\varepsilon$ of $[c,b]$ such that the following hold:

$$\left|\int_a^c f\,dw - S_w(f,\mathcal{P}',\boldsymbol{\xi}')\right| < \varepsilon/2 \quad \text{for all refinements } \mathcal{P}' \text{ of } \mathcal{P}'_\varepsilon \text{ and all } \boldsymbol{\xi}',$$

$$\left|\int_c^b f\,dw - S_w(f,\mathcal{P}'',\boldsymbol{\xi}'')\right| < \varepsilon/2 \quad \text{for all refinements } \mathcal{P}'' \text{ of } \mathcal{P}''_\varepsilon \text{ and all } \boldsymbol{\xi}''.$$

Then $\mathcal{P}_\varepsilon := \mathcal{P}'_\varepsilon \cup \mathcal{P}''_\varepsilon$ is a partition of $[a,b]$ containing c. Moreover, if $\mathcal{P}$ is a refinement of $\mathcal{P}_\varepsilon$, then $\mathcal{P}' := \mathcal{P} \cap [a,c]$ and $\mathcal{P}'' = \mathcal{P} \cap [c,b]$ are refinements of $\mathcal{P}'_\varepsilon$ and $\mathcal{P}''_\varepsilon$, respectively. From

$$S_w(f,\mathcal{P},\boldsymbol{\xi}) = S_w(f,\mathcal{P}',\boldsymbol{\xi}') + S_w(f,\mathcal{P}'',\boldsymbol{\xi}'')$$

and the above inequalities we see that

$$\left|\int_a^c f\,dw + \int_c^b f\,dw - S_w(f,\mathcal{P},\boldsymbol{\xi})\right| < \varepsilon/2 + \varepsilon/2 = \varepsilon.$$

This establishes the existence of $\int_a^b f\,dw$ as well as the desired equality. □

5.10.6 Example. Consider the floor function integrator $w(x) = \lfloor x \rfloor$. A slight modification of the argument in 5.10.2 shows that $\int_0^n f(x)\,d\lfloor x \rfloor$ exists iff f is left continuous at the integers $1, 2, \ldots, n$, in which case $\int_{k-1}^k f(x)\,d\lfloor x \rfloor = f(k)$. For such a function, 5.10.5 implies that

$$\int_0^n f(x)\,d\lfloor x \rfloor = \sum_{k=1}^n \int_{k-1}^k f(x)\,d\lfloor x \rfloor = \sum_1^n f(k). \qquad \lozenge$$

The preceding example suggests that improper Riemann–Stieltjes integration could be used to provide a unified theory that includes both improper Riemann integrals and infinite series. This is indeed possible; however, it turns out that Lebesgue integration is a more efficient approach. Lebesgue theory on $\mathbb{R}^n$ is developed in Chapter 11.

The following theorem reveals a remarkable symmetry between integrand and integrator.

5.10.7 Integration by Parts Formula. *If $f \in \mathcal{R}_a^b(w)$, then $w \in \mathcal{R}_a^b(f)$ and*

$$\int_a^b f\,dw + \int_a^b w\,df = f(b)w(b) - f(a)w(a).$$

Proof. For any partition $\mathcal{P}\{x_0 = a, x_1, \ldots, x_{n-1}, x_n = b\}$,

$$f(b)w(b) - f(a)w(a) = \sum_{j=1}^{n} f(x_j)w(x_j) - \sum_{j=1}^{n} f(x_{j-1})w(x_{j-1}) \quad \text{and}$$

$$S_f(w, \mathcal{P}, \boldsymbol{\xi}) = \sum_{j=1}^{n} w(\xi_j)f(x_j) - \sum_{j=1}^{n} w(\xi_j)f(x_{j-1}).$$

Subtracting we obtain

$$\begin{aligned} &f(b)w(b) - f(a)w(a) - S_f(w, \mathcal{P}, \boldsymbol{\xi}) \\ &\qquad = \sum_{j=1}^{n} f(x_{j-1})[w(\xi_j) - w(x_{j-1})] + \sum_{j=1}^{n} f(x_j)[w(x_j) - w(\xi_j)] \\ &\qquad = S_w(f, \mathcal{Q}, \boldsymbol{\zeta}), \end{aligned}$$

where $\boldsymbol{\zeta} = (a, x_1, x_1, x_2, x_2, \ldots, x_{n-1}, x_{n-1}, b)$ and $\mathcal{Q}$ is the refinement of $\mathcal{P}$ obtained by adding the coordinates of $\boldsymbol{\xi}$ to $\mathcal{P}$ (see Figure 5.10). Therefore,

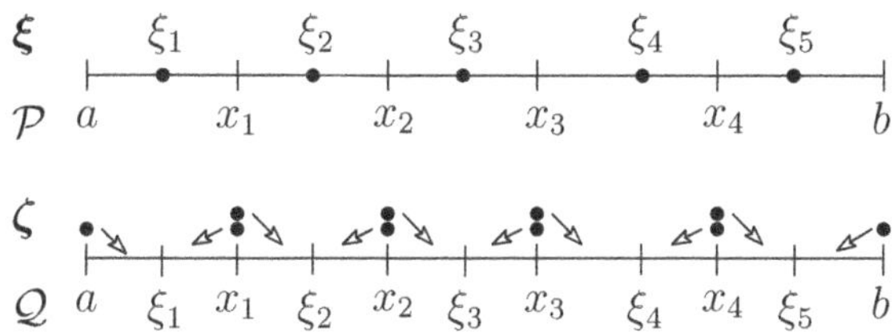

FIGURE 5.10: The partition $\mathcal{Q}$.

$$\left| f(b)w(b) - f(a)w(a) - \int_a^b f\,dw - S_f(w, \mathcal{P}, \boldsymbol{\xi}) \right| = \left| S_w(f, \mathcal{Q}, \boldsymbol{\zeta}) - \int_a^b f\,dw \right|.$$

Since $f \in \mathcal{R}_a^b(w)$, the right side may be made arbitrarily small, Therefore, $\int_a^b w\,df$ exists and equals $f(b)w(b) - f(a)w(a) - \int_a^b f\,dw$. □

The next result shows that under certain general conditions the Riemann–Stieltjes integral reduces to a Riemann integral.

5.10.8 Theorem. *Let $f \in \mathcal{R}_a^b(w)$. If w is continuously differentiable, then $fw' \in \mathcal{R}_a^b$ and*

$$\int_a^b f\,dw = \int_a^b f(x)w'(x)\,dx.$$

Proof. For any partition $\mathcal{P}$ of $[a, b]$ and any $\boldsymbol{\xi}$,

$$S_w(f, \mathcal{P}, \boldsymbol{\xi}) - S(fw', \mathcal{P}, \boldsymbol{\xi}) = \sum_{j=1}^{n} f(\xi_j)\Delta w_j - \sum_{j=1}^{n} f(\xi_j)w'(\xi_j)\Delta x_j.$$

By the mean value theorem, for each j there exists $t_j \in (x_{j-1}, x_j)$ such that

$$\Delta w_j = w(x_j) - w(x_{j-1}) = w'(t_j)\Delta x_j.$$

Therefore,

$$S_w(f, \mathcal{P}, \boldsymbol{\xi}) - S(fw', \mathcal{P}, \boldsymbol{\xi}) = \sum_{j=1}^{n} f(\xi_j)\big[w'(t_j) - w'(\xi_j)\big]\Delta x_j. \tag{5.33}$$

Let $|f| \le M$ on $[a, b]$. By uniform continuity of w', given $\varepsilon > 0$, there exists a $\delta > 0$ such that

$$|w'(x) - w'(y)| < \frac{\varepsilon}{2M(b-a)} \quad \text{whenever } |x - y| < \delta. \tag{5.34}$$

Let $\mathcal{P}'_\varepsilon$ be a partition of $[a, b]$ with $\|\mathcal{P}'_\varepsilon\| < \delta$. From (5.33) and (5.34),

$$|S_w(f, \mathcal{P}, \boldsymbol{\xi}) - S(fw', \mathcal{P}, \boldsymbol{\xi})| \le \frac{\varepsilon}{2(b-a)} \sum_{j=1}^{n} \Delta x_j = \frac{\varepsilon}{2} \tag{5.35}$$

for all refinements $\mathcal{P}$ of $\mathcal{P}'_\varepsilon$ and all $\boldsymbol{\xi}$. Next, choose a partition $\mathcal{P}''_\varepsilon$ such that

$$\left|\int_a^b f\,dw - S_w(f, \mathcal{P}, \boldsymbol{\xi})\right| < \varepsilon/2 \ \text{ for all } \boldsymbol{\xi} \text{ and all refinements } \mathcal{P} \text{ of } \mathcal{P}''_\varepsilon. \tag{5.36}$$

If $\mathcal{P}$ is a refinement of $\mathcal{P}'_\varepsilon \cup \mathcal{P}''_\varepsilon$, then both (5.35) and (5.36) hold, hence, by the triangle inequality,

$$\left|\int_a^b f\,dw - S(fw', \mathcal{P}, \boldsymbol{\xi})\right| < \varepsilon.$$

This shows that $fw' \in \mathcal{R}_a^b$ and establishes the desired equality. □

Monotone Increasing Integrators

If $w : [a, b] \to \mathbb{R}$ is monotone increasing, then the Riemann–Stieltjes integral may be characterized, as in the Darboux theory, by upper and lower sums. This fact will lead to an important existence theorem for integrators with bounded variation and continuous integrands.

Let $f : [a, b] \to \mathbb{R}$ be bounded and let $\mathcal{P}$ be a partition of $[a, b]$. Define the *upper and lower Darboux–Stieltjes sums of f with respect to w* by

$$\overline{S}_w(f, \mathcal{P}) = \sum_{j=1}^{n} M_j \Delta w_j \quad \text{and} \quad \underline{S}_w(f, \mathcal{P}) = \sum_{j=1}^{n} m_j \Delta w_j,$$

where

$$M_j = M_j(f) := \sup_{x_{j-1} \le x \le x_j} f(x) \quad \text{and} \quad m_j = m_j(f) := \inf_{x_{j-1} \le x \le x_j} f(x).$$

The *upper* and *lower Darboux–Stieltjes integrals of f with respect to w* are defined, respectively, by

$$\overline{\int_a^b} f\,dw := \inf_{\mathcal{P}} \overline{S}_w(f,\mathcal{P}) \quad \text{and} \quad \underline{\int_a^b} f\,dw := \sup_{\mathcal{P}} \underline{S}_w(f,\mathcal{P}).$$

As in the Darboux theory, if $\mathcal{Q}$ is a refinement of $\mathcal{P}$ then, because w is increasing,

$$\underline{S}_w(f,\mathcal{P}) \le \underline{S}_w(f,\mathcal{Q}) \le \underline{\int_a^b} f\,dw \le \overline{\int_a^b} f\,dw \le \overline{S}_w(f,\mathcal{Q}) \le \overline{S}_w(f,\mathcal{P}).$$

Here is the analog of 5.1.8 for Riemann–Stieltjes integrals.

5.10.9 Theorem. *The following statements are equivalent:*

(a) $f \in \mathcal{R}_a^b(w)$.

(b) *For each* $\varepsilon > 0$, *there exists a partition* $\mathcal{P}_\varepsilon$ *such that*

$$\overline{S}_w(f,\mathcal{P}) - \underline{S}_w(f,\mathcal{P}) < \varepsilon.$$

(c) $\displaystyle\underline{\int_a^b} f\,dw = \overline{\int_a^b} f\,dw.$

If these conditions hold, then $\displaystyle\int_a^b f\,dw = \underline{\int_a^b} f\,dw = \overline{\int_a^b} f\,dw.$

Proof. That (b) and (c) are equivalent is proved exactly as in 5.1.8. Assume that (a) holds. Given $\varepsilon > 0$, choose a partition $\mathcal{P}_\varepsilon$ such that

$$\left|\int_a^b f\,dw - S_w(f,\mathcal{P},\boldsymbol{\xi})\right| < \varepsilon/3 \ \text{ for all refinements } \mathcal{P} \text{ of } \mathcal{P}_\varepsilon \text{ and all } \boldsymbol{\xi}. \quad (5.37)$$

For such a partition $\mathcal{P}$ and for each j, there exists a sequence $\{\xi_{j,k}\}_{k=1}^\infty$ in $[x_{j-1}, x_j]$ such that $\lim_k f(\xi_{j,k}) = M_j(f)$. It follows that

$$\lim_k S_w(f,\mathcal{P},\boldsymbol{\xi}_k) = \overline{S}_w(f,\mathcal{P}), \quad \text{where } \ \boldsymbol{\xi}_k = (\xi_{1,k},\ldots,\xi_{n,k}).$$

From (5.37),

$$\left|\int_a^b f\,dw - \overline{S}_w(f,\mathcal{P})\right| \le \varepsilon/3.$$

Similarly,

$$\left|\int_a^b f\,dw - \underline{S}_w(f,\mathcal{P})\right| \le \varepsilon/3.$$

Part (b) now follows from the triangle inequality.

Now assume that (c) holds. Let I denote the common value of the integrals in (c). Given $\varepsilon > 0$, choose partitions $\mathcal{P}'_\varepsilon$ and $\mathcal{P}''_\varepsilon$ such that

$$I - \varepsilon < \underline{S}_w(f, \mathcal{P}'_\varepsilon) \text{ and } \overline{S}_w(f, \mathcal{P}''_\varepsilon) < I + \varepsilon.$$

The inequalities still hold if $\mathcal{P}'_\varepsilon$ and $\mathcal{P}''_\varepsilon$ are replaced by any refinement $\mathcal{P}$ of $\mathcal{P}_\varepsilon := \mathcal{P}'_\varepsilon \cup \mathcal{P}''_\varepsilon$. Thus

$$-\varepsilon < \underline{S}_w(f, \mathcal{P}) - I \le S_w(f, \mathcal{P}, \boldsymbol{\xi}) - I \le \overline{S}_w(f, \mathcal{P}) - I < \varepsilon.$$

This shows that $f \in \mathcal{R}_a^b(w)$ and $\int_a^b f\,dw = I$. □

Integrators of Bounded Variation

Recall that a function of bounded variation may be expressed as the difference of two monotone increasing functions (5.9.7). This, together with 5.10.9, allows for a simple proof of the following existence theorem.

5.10.10 Theorem. *If $f : [a, b] \to \mathbb{R}$ is continuous and $w : [a, b] \to \mathbb{R}$ has bounded variation, then $f \in \mathcal{R}_a^b(w)$.*

Proof. By the remark preceding the theorem and by 5.10.4, we may assume that w is increasing. By uniform continuity of f, given $\varepsilon > 0$, there exists a $\delta > 0$ such that

$$|f(x) - f(y)| < \frac{\varepsilon}{w(b) - w(a) + 1} \quad \text{for all } x,\, y \text{ with } |x - y| < \delta.$$

Let $\mathcal{P}_\varepsilon$ be a partition with $\|\mathcal{P}_\varepsilon\| < \delta$. For any refinement $\mathcal{P}$ of $\mathcal{P}_\varepsilon$, $\|\mathcal{P}\| < \delta$, hence

$$M_j(f) - m_j(f) \le \frac{\varepsilon}{w(b) - w(a) + 1}.$$

Therefore,

$$\overline{S}_w(f, \mathcal{P}) - \underline{S}_w(f, \mathcal{P}) = \sum_{j=1}^{n} \big[M_j(f) - m_j(f)\big]\Delta w_j \le \varepsilon,$$

which shows that $f \in \mathcal{R}_a^b(w)$. □

The conclusion of the theorem does not necessarily hold if w fails to have bounded variation, even if w is continuous:

5.10.11 Example. Let $f = w = f_{1/2}$, where f_α is defined as in Example 5.9.3. We show that $\int_0^1 f\,dw$ does not exist. Referring to that example, let $\mathcal{P}_\varepsilon$ be the partition

$$\varepsilon < a_p < b_p < a_{p-1} < \cdots < b_{k+1} < a_k < b_k < \cdots < b_{q+1} < a_q < b_q < 1,$$

of $[\varepsilon, 1]$, and let $\boldsymbol{\xi}$ consist of left endpoints of $\mathcal{P}_\varepsilon$. Then

$$S_w(f, \mathcal{P}, \boldsymbol{\xi}) = f(\varepsilon)\big[w(a_q) - w(\varepsilon)\big] + f(b_q)\big[w(1) - w(b_q)\big]$$
$$+ \sum_{k=q}^{p} f(a_k)\big[w(b_k) - w(a_k)\big] + \sum_{k=q}^{p-1} f(b_{k+1})\big[w(a_k) - w(b_{k+1})\big].$$

Since $f(b_k) = 0$ and $f(a_k) = w(a_k) = \sqrt{a_k}$,

$$S_w(f, \mathcal{P}_\varepsilon, \boldsymbol{\xi}) = f(\varepsilon)\big[\sqrt{a_q} - w(\varepsilon)\big] - \sum_{k=q}^{p} a_k.$$

Since the sums diverge as $\varepsilon \to 0$, $\lim_{\varepsilon\to 0} S_w(f, \mathcal{P}_\varepsilon, \boldsymbol{\xi}) = -\infty$. ◇

Chapter 6

Numerical Infinite Series

An *infinite series* is the limit of a sequence of expanding finite sums. The terms of these sums may be real numbers or functions. In this chapter we examine the convergence behavior of series of the former type; series whose terms are functions are treated in the next chapter. In the first section, we give examples of series that may be summed, that is, for which an explicit numerical value may be calculated. The remaining sections describe various tests for convergence of general series. Additional methods of summing series may be found in Section 7.4.

6.1 Definition and Examples

6.1.1 Definition. Let $\{a_n\}$ be a sequence of real numbers. The various symbols

$$\sum a_n = \sum_n a_n = \sum_{n=1}^{\infty} a_n = a_1 + a_2 + \cdots + a_n + \cdots$$

represent what is called an *infinite series with* n*th term* a_n or, simply, a *series*. The nth *partial sum* of the series is defined as

$$s_n = \sum_{k=1}^{n} a_k.$$

The series is said to *converge* if the sequence of partial sums converges. In this case we write

$$\sum a_n = \lim_n s_n$$

and call $\sum a_n$ the *sum of the series*. If the sequence $\{s_n\}$ diverges, then the series is said to *diverge*. ◊

6.1.2 Remark. A series may begin with an index other than 1, 0 being a common alternative. In this regard, note that, because

$$s_n = s_{m-1} + \sum_{k=m}^{n} a_k, \quad n \geq m > 1,$$

DOI: 10.1201/9781003629382-6

the series $s := \sum_{n=1}^{\infty} a_n$ converges iff $\sum_{n=m}^{\infty} a_n$ converges. Moreover, the series converges iff the "tail end" tends to zero:

$$\lim_{m\to+\infty} \sum_{n=m}^{\infty} a_n = \lim_{m\to+\infty} (s - s_{m-1}) = s - s = 0. \qquad \Diamond$$

The following is one of the most important examples of an infinite series.

6.1.3 Example. Using the definition $e := \lim_{n\to\infty} (1 + 1/n)^n$ (see 2.2.4), we show that

$$e = \sum_{n=0}^{\infty} \frac{1}{n!} = 1 + 1 + 1/2 + 1/6 + 1/12 + \cdots$$

First, since the partial sums $s_n := \sum_{k=0}^{n} 1/k!$ increase, the limit $s := \lim_n s_n$ exists in $\overline{\mathbb{R}}$. From the calculations in 2.2.4,

$$(1 + 1/n)^n = 2 + \sum_{k=2}^{n} \frac{1}{k!}(1 - 1/n)(1 - 2/n) \cdots (1 - (k-1)/n) \le s_n.$$

Letting $n \to \infty$, we obtain $e \le s$. On the other hand, if $n > m$ then

$$(1 + 1/n)^n > 2 + \sum_{k=2}^{m} \frac{1}{k!}(1 - 1/n)(1 - 2/n) \cdots (1 - (k-1)/n),$$

and letting $n \to \infty$ we see that $e \ge s_m$ for all m. Finally, letting $m \to \infty$ yields $e \ge s$. $\Diamond$

6.1.4 Example. The *geometric series* $\sum_{n=0}^{\infty} ar^n$, where $a, r \in \mathbb{R}$ and $a \ne 0$, converges iff $|r| < 1$, in which case

$$\sum_{n=0}^{\infty} ar^n = \frac{a}{1-r}.$$

This follows from the calculation

$$s_n = \sum_{k=0}^{n} ar^k = a\,\frac{1 - r^{n+1}}{1 - r}, \quad r \ne 1$$

and the fact that $r^n \to 0$ for $|r| < 1$. $\Diamond$

6.1.5 Example. For $m \in \mathbb{N}$,

$$\sum_{n=1}^{\infty} \frac{1}{n(m+n)} = \frac{1}{m} \sum_{k=1}^{m} \frac{1}{k}. \qquad (6.1)$$

To see this, we use partial fractions: For $n > m$

$$ms_n = \sum_{k=1}^{n} \frac{m}{k(m+k)} = \sum_{k=1}^{n} \left[\frac{1}{k} - \frac{1}{(k+m)}\right] = \sum_{k=1}^{m} \frac{1}{k} - \sum_{k=n+1}^{n+m} \frac{1}{k}. \tag{6.2}$$

Since

$$\sum_{k=n+1}^{n+m} \frac{1}{k} \le \sum_{k=n+1}^{n+m} \frac{1}{n+1} = \frac{n+m-n}{n+1} = \frac{m}{n+1},$$

the last summation in (6.2) tends to zero as $n \to \infty$. Division by m produces (6.1).

The series in (6.1) is an example of a *telescopic series*, the name referring to the cancellations occurring in (6.2). ◊

6.1.6 Theorem. *Let $\{a_n\}$ and $\{b_n\}$ be sequences and let $\alpha, \beta \in \mathbb{R}$. If $\sum a_n$ and $\sum b_n$ converge, then $\sum(\alpha a_n + \beta b_n)$ converges and*

$$\sum(\alpha a_n + \beta b_n) = \alpha \sum a_n + \beta \sum b_n. \tag{6.3}$$

Proof. Let $s_n = \sum_{k=1}^n a_k$ and $t_n = \sum_{k=1}^n b_k$. Then $\alpha s_n + \beta t_n$ is the nth partial sum of the series $\sum(\alpha a_n + \beta b_n)$ and

$$\lim_{n\to\infty} (\alpha s_n + \beta t_n) = \alpha \lim_{n\to\infty} s_n + \beta \lim_{n\to\infty} t_n,$$

which is (6.3). □

6.1.7 Example. By 6.1.6 and 6.1.4,

$$\begin{aligned}\sum_{n=1}^{\infty} \frac{2 \cdot 3^{n+1} + 3 \cdot 2^{n-1}}{6^n} &= 6 \sum_{n=1}^{\infty} \frac{1}{2^n} + \frac{3}{2} \sum_{n=1}^{\infty} \frac{1}{3^n} \\ &= 6 \frac{1/2}{1 - 1/2} + \frac{3}{2} \frac{1/3}{1 - 1/3} \\ &= 6.75.\end{aligned}$$

◊

The following result is a test for divergence. It implies that a series whose nth term does not tend to zero must diverge. For example, the series $\sum \sin n$ and $\sum 2^{-1/n}$ diverge.

6.1.8 Proposition. *If $\sum a_n$ converges, then $a_n \to 0$.*

Proof. $a_n = s_n - s_{n-1} \to s - s = 0$. □

The converse of 6.1.8 is false:

6.1.9 Example. The *harmonic series* $\sum_{n=1}^{\infty} 1/n$ diverges. Indeed, if s_n is the nth partial sum of the series, then for all n

$$s_{2^n} - s_{2^{n-1}} = \frac{1}{2^{n-1}+1} + \cdots + \frac{1}{2^{n-1}+2^{n-1}} > \frac{2^{n-1}}{2^{n-1}+2^{n-1}} = \frac{1}{2},$$

hence $\{s_n\}$ is not a Cauchy sequence and so cannot converge. It is of interest to note, however, that the related the sequence $t_n := s_n - \ln n$ *does* converge. To see this, observe first that

$$\ln n = \int_1^n \frac{dx}{x} = \sum_{k=1}^{n-1} \int_k^{k+1} \frac{dx}{x} < \sum_{k=1}^{n} \int_k^{k+1} \frac{dx}{k} = \sum_{k=1}^{n} \frac{1}{k} = s_n,$$

so $t_n > 0$, showing that the sequence $\{t_n\}$ is bounded below. Furthermore,

$$\ln(n+1) - \ln n = \int_n^{n+1} \frac{dx}{x} > \int_n^{n+1} \frac{dx}{n+1} = \frac{1}{n+1},$$

hence

$$t_n - t_{n+1} = \ln(n+1) - \ln n + s_n - s_{n+1} = \ln(n+1) - \ln n - \frac{1}{n+1} > 0.$$

showing $\{t_n\}$ is also decreasing. By the monotone sequence theorem 2.2.2, $\{t_n\}$ converges. Its limit is denoted by γ and is known as *Euler's constant*:

$$\gamma := \lim_{n\to\infty} \left[\sum_{k=1}^{n} \frac{1}{k} - \ln n\right].$$

The value of γ to eleven decimal places is $.57721566490\ldots$. As of this writing, it is not known whether γ is rational or irrational. Note that since $s_n = t_n + \ln n$ and $\{\ln n\}$ diverges, the convergence of $\{t_n\}$ provides another proof that the harmonic series diverges. ◊

Exercises

1. Let $m \in \mathbb{N}$. Sum the series $\sum_{n=1}^{\infty} a_n$, where $a_n =$

   (a) $\dfrac{m^{2n+1}}{(m+1)^{2n-1}}$ (b) $\dfrac{(-1)^{n+1} m^{3n+1}}{(m+1)^{3n-1}}$

   (c) $\ln\left[\dfrac{2^{n+1}+2}{2^{n+1}+1}\right]$ (d) $\dfrac{1}{\sqrt{n(n+1)}(\sqrt{n+1}+\sqrt{n})}$

   (e) $\dfrac{(-1)^n}{n(n+m)}$, m even (f) $\dfrac{12}{(n+1)(n+2)(n+3)}$

   (g) $\dfrac{1}{(n+1)(n+3)(n+5)}$ (h) $\dfrac{(-1)^n}{(n+1)(n+3)(n+5)}$

(i) $\ln \dfrac{m\sqrt{n}+\sqrt{n(n+1)}}{m\sqrt{n+1}+\sqrt{n(n+1)}}$ (j) $\ln \dfrac{n^2+4n+4}{n^2+4n+3}$

(k) $\dfrac{1}{(n+m)\sqrt{n}+\sqrt{n}\sqrt{n+m}}$ (l) $\dfrac{18}{n(n+1)(n+2)(n+3)}$

(m) $\dfrac{(-1)^n(n+m+1)}{(2n+1)(2n+4m+3)}$ (n) $\dfrac{(-1)^n(2n+2m+1)}{n(n+2m+1)}$.

2. Let $0 < r < 1$ and $m \in \mathbb{N}$. Sum the series $\sum_{n=0}^{\infty} a_n$ if $a_n =$
   (a) $r^n \cos\left[(n\pi)/2\right]$ (b) $(-1)^{\lfloor n/3 \rfloor} r^n$ (c) $(-1)^{\lfloor n/m \rfloor} r^n$.

3. Given that $e = \sum_{n=0}^{\infty} 1/n!$ and $e^{-1} = \sum_{n=0}^{\infty} (-1)^n 1/n!$, find the value of the series $\sum_{n=0}^{\infty} a_n$ if $a_n =$

   (a) $\dfrac{(2n+3)^3}{n!}$ (b) $\dfrac{1}{(2n)!}$ (c) $\dfrac{1}{(2n+1)!}$ (d) $\dfrac{n}{(2n+1)!}$ (e) $\dfrac{n}{(2n)!}$.

4. Let $p > 0$ and $s_n = \sum_{k=1}^{n} 1/k$. Prove that $s_n/n^p \to 0$, $s_n/\ln n \to 1$, and $s_n/\ln\ln n \to +\infty$.

5. Let γ denote Euler's constant (6.1.9). Prove that

   (a) $\displaystyle\sum_{k=1}^{n} \frac{1}{2k-1} - \ln\sqrt{n} \to \ln 2 + \frac{\gamma}{2}$

   (b) $\displaystyle\sum_{k=1}^{n} \frac{4k}{(2k-1)(2k+1)} - \ln n \to \ln 4 + \gamma - 1$

   (c) $\displaystyle\sum_{n=1}^{\infty} \frac{1}{n(2n-1)(2n+1)} = \ln 4 - 1.$

6. Prove that $\sum a_n$ converges iff for each $\varepsilon > 0$ there exists an index N such that

$$\left| \sum_{k=n}^{n+p} a_k \right| < \varepsilon \quad \text{for all } n \geq N \text{ and } p \geq 1.$$

7. Suppose a_n tends monotonically to 0 and $s := \sum_{n=1}^{\infty} a_n$ converges.
   (a) Prove that $na_n \to 0$.
   (b) Let $p \in \mathbb{N}$. Show that $t := \sum_{n=1}^{\infty} n(a_n - a_{n+p})$ converges and express t in terms of s.

   Suggestion. For (b), consider first the case $p = 1$.

8. Let $\sum a_n$ and $\sum b_n$ be convergent series with $b_n > 0$ for all n. Suppose that $L := \lim_n (a_n/b_n)$ exists in $\mathbb{R}$. Prove that

$$\lim_n \frac{\sum_{k=n}^{\infty} a_k}{\sum_{k=n}^{\infty} b_k} = L.$$

Use this to calculate $\lim_n \left[\sum_{k=n}^{\infty} \sin(3/k^2)\right]\left[\sum_{k=n}^{\infty} 1/k^2\right]^{-1}$.

9. For a sequence $\{c_n\}$, define $\Delta c_n = c_{n+1} - c_n$. Prove the following discrete analog of l'Hospital's rule: Let $\{a_n\}$ and $\{b_n\}$ be sequences with $\{b_n\}$ strictly monotone. Suppose that either (a) $a_n \to 0$ and $b_n \to 0$, or (b) $b_n \to \pm\infty$. Then
$$\lim_n \frac{a_n}{b_n} = \lim_n \frac{\Delta a_n}{\Delta b_n},$$
provided that the limit on the right exists in $\mathbb{R}$.

10. Let $\{a_n\}$ and $\{b_n\}$ be sequences with $b_n > 0$ for all n and $\sum_n b_n = +\infty$. Set $A_n = \sum_{k=1}^{n} a_k$ and $B_n = \sum_{k=1}^{n} b_k$. Use Exercise 9 to prove that
$$\lim_n \frac{A_n}{B_n} = \lim_n \frac{a_n}{b_n},$$
provided that the limit on the right exists in $\mathbb{R}$. Use this to calculate the limits of
$$\text{(a)}\ \frac{\sum_{k=1}^{n} \sin(1/k)}{\sum_{k=1}^{n} 1/k} \qquad \text{(b)}\ \frac{\sum_{k=1}^{n} \ln k}{\sum_{k=1}^{n} k^p} \qquad \text{(c)}\ \frac{\sum_{k=1}^{n} r^k}{\sum_{k=1}^{n} k^p}.$$
where $r, p > 0$.

11. Let $\{b_n\}_{n=1}^{\infty}$ be a sequence obtained by rearranging finitely many terms of a sequence $\{a_n\}_{n=1}^{\infty}$. Show that $\sum b_n$ converges in $\overline{\mathbb{R}}$ iff $\sum a_n$ converges in $\overline{\mathbb{R}}$, in which case the series are equal.

12. Let $\{b_k\}$ be a sequence obtained from a sequence $\{a_n\}$ by *grouping*, that is,
$$b_k = a_{n_{k-1}+1} + a_{n_{k-1}+2} + \cdots + a_{n_k}, \quad k = 1, 2, \ldots,$$
where $\{n_k\}_k$ is a strictly increasing sequence of nonnegative integers and $n_0 = 0$. Show that if $\sum_n a_n$ converges in $\overline{\mathbb{R}}$, then so does $\sum_k b_k$ and the series are equal. Show that the converse is true if $a_n \geq 0$ for all sufficiently large n. What if the terms a_n change sign infinitely often?

13. Let $\{a_n\}$ be decreasing and nonnegative. Prove that $\sum_{n=1}^{\infty} a_n$ converges iff $\sum_{k=0}^{\infty} 2^k a_{2^k}$ converges. *Hint.* Set
$$s_n = \sum_{j=1}^{n} a_j \quad \text{and} \quad t_k = \sum_{j=0}^{k} 2^j a_{2^j}.$$
Show that $s_n \leq t_k$ if $n \leq 2^{k+1} - 1$ and $s_n \geq t_k/2$ if $n \geq 2^k$.

14. (Decimal representation of real numbers). Prove that every real number $x \geq 0$ has a *decimal representation*

$$x = b_N b_{N-1} \cdots b_0.a_1 a_2 \cdots := \sum_{n=0}^{N} b_n 10^n + \sum_{n=1}^{\infty} a_n 10^{-n},$$

where the *digits* b_n, a_n are integers from 0 to 9.

Hint. By Exercise 1.4.15, it may be assumed that $x \in [0, 1)$. Prove by induction that for each n there exist $a_j \in \{0, 1, \ldots 9\}$ and $x_n \in [0, 10^{-n})$ such that

$$x = x_n + \sum_{j=1}^{n} a_j 10^{-j} = x_n + (.a_1 a_2 \cdots a_n).$$

15. Call a decimal representation $b_N b_{N-1} \cdots b_0.a_1 a_2 \cdots$ *standard* if no index n exists such that $a_k = 9$ for all $k \geq n$. Prove that every real number has a *unique* standard decimal representation.

16. A real number $x \geq 0$ is a *repeating decimal* if it has decimal representation of the form

$$x = b_N b_{N-1} \cdots b_0.a_1 a_2 \cdots a_m \overline{a_{m+1} a_{m+2} \cdots a_{m+k}},$$

where the upper bar indicates that the block repeats forever. (For example, $61/495 = .12323 \cdots = .1\overline{23}$.) Prove that every repeating decimal is rational.

17. Prove the converse of Exercise 16, that is, every rational number p/q is a repeating decimal. Conclude that if $f : \mathbb{N} \mapsto \mathbb{N}$ is strictly increasing, then $\sum 10^{-f(n)}$ is irrational.

Hint. By the division algorithm, you may assume that $1 \leq p < q$. Begin by showing that if $p/q = .a_1 a_2 \cdots$, then for each n

$$\frac{p}{q} = .a_1 a_2 \cdots a_n + \frac{r_n}{10^n q}, \quad \text{where } r_n \in \{0, 1, \ldots, q-1\},$$

and use this to show that $q a_n = 10 r_{n-1} - r_n$, where $r_0 := p$.

18. Let $s_n = \sum_{k=1}^{n} a_n$, where $a_n \to 0$. Suppose there exists a positive integer q such that $s_{nq} \to s \in \mathbb{R}$. Prove that $\sum a_n$ converges to s.

6.2 Series with Nonnegative Terms

There are a variety of tests for convergence of series with nonnegative terms. The most basic of these is the following theorem.

6.2.1 Theorem. *If $a_n \geq 0$ for all n, then the series $\sum a_n$ converges in $\mathbb{R}$ iff its partial sums are bounded.*

Proof. Since the terms of the series are nonnegative, the sequence of partial sums is increasing. The assertion therefore follows from the monotone sequence theorem (2.2.2). □

6.2.2 Remark. By 6.1.2, the theorem is still valid if the inequality $a_n \geq 0$ holds only eventually, that is, for all $n \geq$ some m. Many of the results in this chapter have similar extensions. Rather than make these explicit, we leave the straightforward formulations to the reader. ◊

6.2.3 Example. Let $a_n, b_n \geq 0$ for all n, and suppose that $\sum a_n$ and $\sum b_n$ converge. By the Cauchy–Schwarz inequality (1.5.3(e)),

$$\sum_{k=1}^{n} \sqrt{a_k b_k} \leq \left(\sum_{k=1}^{n} a_k\right)^{1/2} \left(\sum_{k=1}^{n} b_k\right)^{1/2}.$$

Since the sums on the right are bounded, so are the sums on the left. Therefore, $\sum \sqrt{a_n b_n}$ converges. ◊

The following test relates the convergence of a series to that of an improper integral.

6.2.4 Integral Test. *Let f be decreasing, positive, and locally integrable on the interval $[1, \infty)$. Then the series $\sum_{n=1}^{\infty} f(n)$ converges iff the improper integral $\int_1^{\infty} f$ converges. Moreover, for every $n \in \mathbb{N}$*

$$0 \leq s - s_n \leq \int_n^{\infty} f(x)\,dx. \tag{6.4}$$

Proof. For each $n \in \mathbb{N}$ let

$$s_n = \sum_{k=1}^{n} f(k) \quad \text{and} \quad t_n = \int_1^n f.$$

For each $k \in \mathbb{N}$ and $x \in [k, k+1]$, $f(k+1) \leq f(x) \leq f(k)$, hence

$$f(k+1) \leq \int_k^{k+1} f \leq f(k)$$

and so

$$s_n - f(1) = \sum_{k=2}^{n} f(k) = \sum_{k=1}^{n-1} f(k+1) \leq \sum_{k=1}^{n-1} \int_k^{k+1} f = t_n \leq \sum_{k=1}^{n-1} f(k) = s_{n-1}.$$

Therefore, $\{s_n\}$ is bounded iff $\{t_n\}$ is bounded. The first assertion of the theorem now follows from 6.2.1.

To verify (6.4) we apply a similar argument: for $m > n$

$$0 \le s_m - s_n = \sum_{k=n-1}^{m} f(k) = \sum_{k=n}^{m-1} f(k+1) \le \sum_{k=n}^{m-1} \int_k^{k+1} f = \int_n^m f.$$

Letting $m \to +\infty$ yields the desired inequality. □

Inequality (6.4) allows one to estimate the error made in approximating s by a partial sum s_n.

6.2.5 Example. (p-series). By 5.7.3(a), $\int_1^\infty 1/x^p\,dx$ converges iff $p > 1$. Therefore, the same is true of the series $s := \sum_{n=1}^\infty 1/n^p$. Furthermore, if $p > 1$, then

$$0 \le s - s_n \le \int_n^\infty x^{-p}\,dx = \frac{1}{(p-1)n^{p-1}}.$$

Thus if the partial sum s_n is to agree with s in, say, the first 10 decimal places, then n should be chosen so that $(p-1)n^{p-1} > 10^{10}$. ◊

6.2.6 Comparison Test. *Let $0 \le a_n \le b_n$ for all n. If $\sum b_n$ converges, then so does $\sum a_n$.*

Proof. The partial sums of $\sum b_n$ are bounded and dominate those of $\sum a_n$, hence assertion follows from 6.2.1. □

6.2.7 Limit Comparison Test. *Let a_n, $b_n > 0$ for all n.*

(a) *If $\overline{r} := \limsup(a_n/b_n) < +\infty$ and $\sum b_n$ converges, then $\sum a_n$ converges.*

(b) *If $\underline{r} := \liminf(a_n/b_n) > 0$ and $\sum a_n$ converges, then $\sum b_n$ converges.*

(c) *If $r := \lim(a_n/b_n)$ exists and $r \in (0, +\infty)$, then $\sum b_n$ converges iff $\sum a_n$ converges.*

Proof. For (a), let $r \in (\overline{r}, +\infty)$ and choose N so that $\sup_{n\ge N} a_n/b_n < r$. Then $a_n < b_n r$ for every $n \ge N$, hence the conclusion follows from the comparison test and 6.2.2. Part (b) follows similarly by choosing $r \in (0, \underline{r})$ and then N so that $\inf_{n\ge N} a_n/b_n > r$. Part (c) follows from (a) and (b). □

6.2.8 Examples. (a) The series

$$\sum_n \frac{2^n + n^3}{3^n + n^2}$$

converges by comparison with the convergent series $\sum_n (2/3)^n$, since

$$\frac{2^n + n^3}{3^n + n^2}(2/3)^{-n} = \frac{1 + n^3/2^n}{1 + n^2/3^n} \to 1.$$

(b) The series

$$\sum_n \frac{\sqrt{cn+d}-\sqrt{cn}}{n+1}, \quad c,\, d > 0,$$

converges by comparison with the convergent series $\sum_n n^{-3/2}$, since

$$\frac{\sqrt{cn+d}-\sqrt{cn}}{n+1} n^{3/2} = \frac{dn^{3/2}}{(n+1)(\sqrt{cn+d}+\sqrt{cn})} \to \frac{d}{2\sqrt{c}}. \qquad \diamond$$

6.2.9 Ratio Test. *Let $a_n > 0$ for all n.*

(a) *If* $\overline{r} := \limsup_n \dfrac{a_{n+1}}{a_n} < 1$, *then* $\sum a_n$ *converges.*

(b) *If* $\underline{r} := \liminf_n \dfrac{a_{n+1}}{a_n} > 1$, *then* $\sum a_n$ *diverges.*

Proof. (a) Let $r \in (\overline{r}, 1)$ and choose N so that $\sup_{n \ge N} a_{n+1}/a_n < r$. For $n > N$

$$a_n < a_{n-1}r < a_{n-2}r^2 < \cdots < a_N r^{n-N},$$

so $\sum a_n$ converges by comparison with the convergent series $\sum r^n$.

(b) If $\underline{r} > 1$ there exists N such that $\inf_{n \ge N} a_{n+1}/a_n > 1$. Therefore,

$$a_n > a_{n-1} > a_{n-2} > \cdots > a_N > 0, \quad n > N,$$

so a_n cannot converge to zero. Therefore, $\sum a_n$ diverges. $\square$

6.2.10 Examples. (a) Let a_n denote the general term of the series

$$\sum_{n=1}^{\infty} \frac{8 \cdot 14 \cdot 20 \cdots (6n+2)}{6 \cdot 11 \cdot 16 \cdots (5n+1)} c^n,$$

where $c > 0$. Then

$$\frac{a_{n+1}}{a_n} = \frac{6n+8}{5n+6} c \to \frac{6}{5} c,$$

hence the series converges if $c < 5/6$ and diverges if $c > 5/6$. If $c = 5/6$, then

$$a_n = \frac{8 \cdot 14 \cdot 20 \cdots (6n+2) 5^n}{6 \cdot 11 \cdot 16 \cdots (5n+1) 6^n} = \frac{(1+1/3)(2+1/3) \cdots (n+1/3)}{(1+1/5)(2+1/5) \cdots (n+1/5)} > 1,$$

so the series diverges in this case as well.

(b) For the series

$$\sum_{n=1}^{\infty} (n!)^p r^{n^2}, \quad r > 0,\ p \in \mathbb{R}$$

the ratios are

$$\frac{a_{n+1}}{a_n} = (n+1)^p r^{2n+1},$$

hence the series converges iff $r < 1$.

(c) For the series

$$\sum_{n=2}^{\infty} \frac{2^n \ln^2 n}{n!},$$

$$\frac{a_{n+1}}{a_n} = \frac{2\ln^2(n+1)}{(n+1)\ln^2 n} \le \frac{2\ln^2(n+1)}{(n+1)} \to 0,$$

hence the series converges. ◊

6.2.11 Root Test. *Let $a_n \ge 0$ for all n and set $\rho := \limsup_n a_n^{1/n}$.*

(a) *If $\rho < 1$, then $\sum a_n$ converges.*

(b) *If $\rho > 1$, then $\sum a_n$ diverges.*

Proof. (a) Let $r \in (\rho, 1)$ and choose N such that $\sup_{n \ge N} a_n^{1/n} < r$. Then $a_n < r^n$ for all $n \ge N$, hence $\sum a_n$ converges by the comparison test,

(b) By 2.4.2, there exists a subsequence $a_{n_k}^{1/n_k} \to \rho$. Then, for all sufficiently large k, $a_{n_k} > 1$, hence the series diverges. □

6.2.12 Example. For the series

$$\sum_{n=1}^{\infty} \left(a + (-1)^n b\right)^n, \quad \text{where } a > b > 0,$$

$\limsup_n a_n^{1/n} = a + b$, hence the series converges if $a + b < 1$ and diverges if $a + b > 1$. If $a + b = 1$, $a_n \not\to 0$ so the series diverges in this case as well. ◊

6.2.13 Remark. No conclusions regarding the convergence of the series $\sum a_n$ in 6.2.9 and 6.2.11 can be inferred from the relations $\overline{r} \ge 1$, $\underline{r} \le 1$, or $\rho = 1$. Indeed, the series $\sum 1/n^2$ and $\sum 1/n$ both satisfy $\overline{r} = \underline{r} = \rho = 1$, yet the first series converges while the second diverges. ◊

In Section 6.3, we consider more refined tests that can detect convergence or divergence in cases where the ratio or root test fails. Here's an example:

6.2.14 Example. Let a_n denote the nth term of the series

$$\sum_{n=1}^{\infty} \left(\sqrt{an + b\sqrt{n}} - \sqrt{an}\right)^n,$$

where a, $b > 0$. Then

$$a_n^{1/n} = \sqrt{an + b\sqrt{n}} - \sqrt{an} = \frac{b\sqrt{n}}{\sqrt{an + b\sqrt{n}} + \sqrt{an}} \to \frac{b}{2\sqrt{a}},$$

hence the series $\sum a_n$ converges if $b^2 < 4a$ and diverges if $b^2 > 4a$. If $b^2 = 4a$, the root test fails but the log test (6.3.4) shows that the series converges in this case. (Exercise 6.3.11.) ◊

6.2.15 Remark. By Exercise 2.4.12, if $a_n > 0$ for all n, then

$$\liminf_n \frac{a_{n+1}}{a_n} \le \liminf_n a_n^{1/n} \le \limsup_n a_n^{1/n} \le \limsup_n \frac{a_{n+1}}{a_n}.$$

This shows that if the ratio test determines convergence or divergence conclusively, then so does the root test. It also suggests that the root test may be effective when the ratio test fails. ◊

6.2.16 Example. Let $a_n = s^n \delta_{n-1} + t^n \delta_n$, where $\delta_n = \frac{1}{2}[1 + (-1)^n]$ and $0 < s < t < 1$. Then

$$a_n = \begin{cases} s^n & \text{if } n \text{ is odd,} \\ t^n & \text{if } n \text{ is even,} \end{cases}$$

so the ratios a_{n+1}/a_n are s^{n+1}/t^n or t^{n+1}/s^n, and the roots $a_n^{1/n}$ are s or t, depending on the parity of n. Therefore, $\underline{r} = 0$, $\overline{r} = +\infty$ and $\rho = t$, which shows that the root test detects the convergence of $\sum a_n$ while the ratio test does not. ◊

Exercises

1. Determine whether the series $\sum a_n$ converges or diverges, where $a_n =$

(a) $\dfrac{n!}{3 \cdot 5 \cdots (2n+1)}$ (b) $\dfrac{3 \cdot 5 \cdots (2n+1)}{(2n+1)!}$ (c) $\dfrac{3 \cdot 6 \cdots (3n)}{3 \cdot 5 \cdots (2n+1)}$

(d) $\dfrac{4^n n!}{5 \cdot 8 \cdots (3n+2)}$ (e) $\dfrac{2 \cdot 4 \cdots (2n)}{4 \cdot 7 \cdots (3n+1)}$ (f) $\dfrac{4 \cdot 7 \cdots (3n+1)}{5 \cdot 9 \cdots (4n+1)}$.

2. Determine whether the series $\sum a_n$ converges or diverges, where $a_n =$

(a) $\dfrac{n^3}{2^n}$ (b) $(1 + r/n)^{n^2}$, $r > 0$ (c) $\dfrac{\ln n}{n^{1.1}}$

(d) $\dfrac{n!}{n^n}$ (e) $(n^{1/n} - r)^n$ (f) $\dfrac{1}{n^{1+1/n}}$

(g) $\dfrac{1}{n(\ln n)(\ln \ln n)^p}$ (h) $\dfrac{2^n}{n!}$ (i) $\sin^2(1/n)$

(j) $\sin^2(1/\sqrt{n})$ (k) $\dfrac{r^n}{1 - r^n}$, $r \ne \pm 1$ (l) $\dfrac{1}{(1 - r^n)^2}$, $r \ne \pm 1$

(m) $\dfrac{n + \sin n}{n^3 + \sin n}$ (n) $\dfrac{n + \ln n}{n^r \ln n}$ (o) $\dfrac{1}{n^r \ln n}$

(p) $\dfrac{3^n n!}{n^n}$ (q) $\dfrac{n!}{(1.1)^{n^3}}$. (r) $\left(\dfrac{1 + an}{1 + bn}\right)^n$, $a, b > 0$

(s) $\dfrac{1}{2^{\ln n}}$ (t) $\dfrac{1}{3^{\ln n}}$ (u) $r^{\sin n}$, $r > 0$

(v) $\dfrac{3^n + 4^n}{8^n - 6^n}$ (w) $(1 - r/n)^{n^3}$, $r > 0$ (x) $1/r^{\ln n}$, $r > 0$.

3. Let $a_n > 0$ for all n and suppose that $\sum a_n$ diverges. Prove that $\sum a_n b_n$ diverges for all sequences $\{b_n\}$ with $\liminf_n b_n > 0$.

4. Let $b_n \to p > 0$. Prove that $\sum_{n=1}^{\infty} n^{-b_n}$ converges if $p > 1$ and diverges if $p < 1$. Give an example of a sequence $\{b_n\}$ with $b_n > 1$ for all n and $b_n \downarrow 1$ such that $\sum n^{-b_n}$ diverges.

5. Let $a_n > 0$ for all n. Prove that $\sum a_n$ converges iff $\sum n^{1/n} a_n$ converges.

6. Find all values of a, b, p, $q > 0$ for which $\sum_{n=1}^{\infty} a_n$ converges if $a_n =$

(a) $\dfrac{1}{\ln^p n}$ (b) $\dfrac{\ln^p n}{n^q}$ (c) $\dfrac{1}{n^q \ln^p n}$

(d) $(n^{qp}+1)^{1/q} - n^p$ (e) $\dfrac{(n+1)^p - n^p}{n^q}$ (f) $p^{n/2} n! \Big(\prod_{j=1}^{n} pj + 1\Big)^{-1}$

(g) $\dfrac{a+n^p}{b+n^q}$ (h) $\dfrac{1+an^p}{1+bn^q}$ (i) $\left(\dfrac{1+an^p}{1+bn^q}\right)^n$.

7. Let $\{a_n\}$ be positive and decreasing. Prove that $\sum a_n$ converges iff $\sum a_{2n}$ converges.

8. Let $a_n > 0$ for all n. Prove or disprove: If $\sum_{n=1}^{\infty} a_n$ converges, then $\sum_{n=1}^{\infty} b_n$ converges, where $b_n =$

(a) a_n^2 (b) $\sqrt{a_n}$. (c) $\sum_{n \le j \le n+m} a_j$ (d) $\min_{n \le j \le 2n} a_j$

(e) $\max_{n \le j \le 2n} a_j$ (f) $\sum_{n \le j \le 2n} a_j$ (g) $\prod_{n \le j \le 2n} a_j$ (h) $\prod_{1 \le j \le n} a_j$

(i) $\sum_{j=1}^{n} a_n a_j$ (j) $\dfrac{1}{a_n} \prod_{n < j \le n+m} a_j$ (k) $\dfrac{1}{a_n} \sum_{n < j \le n+m} a_j$ (l) $\left(\dfrac{a_n}{n}\right)^{3/4}$.

9. Let $a_n > 0$. Prove that if $\sum a_n$ converges, then $\sum (1 - \cos a_n)$ converges.

10. Let $r > 0$ and $p > 3/r$. Prove that $\sum \left[b_n/(rp-3)\right]^n$ converges for all sequences $\{b_n\}$ with $b_n \to r$.

11. Let $a_n, b_n > 0$ and $a_{n+1}/a_n \le b_{n+1}/b_n$ for all n. Prove that if $\sum b_n$ converges, then so does $\sum a_n$.

12. Let $0 < a_n < r$, where r is sufficiently small. Show that $\sum a_n$ converges iff $\sum f(a_n)$ converges, where $f(x) =$

(a) $\sin x$ (b) $\tan x$ (c) $\sin^{-1} x$ (d) $\tan^{-1} x$

(e) $\dfrac{x}{1+ax}$ (f) $\ln(1+x)$ (g) $e^x - 1$ (h) $x^3 + x^2 + x$.

13. Let $\{p_n\}$ be a sequence in $\mathbb{Z}^+$ and $\{a_n\}$ a sequence of positive reals.

    (a) Prove that if $\sum_n a_n$ converges, then $\sum_n \sqrt{a_n a_{n+p_n}}$ converges, provided that either $\{p_n\}$ is bounded or a_n is decreasing.

    (b) Suppose $\{p_n\}$ is bounded and $a_n \downarrow 0$. Prove that if $\sum_n \sqrt{a_n a_{n+p_n}}$ converges, then $\sum_n a_n$ converges.

    Does (b) hold if $\{a_n\}$ is not monotone or $\{p_n\}$ is not bounded?

14. Let g be positive and differentiable on $[1, \infty)$ such that $\lim_{x\to\infty} g(x) = 0$, and let f be differentiable in a neighborhood of 0 such that $f(0) = 0$, $f(x) > 0$ for $x > 0$, f' is continuous at 0, and $f'(0) > 0$. Prove that $\sum_{n=1}^{\infty} f(g(n))$ converges iff $\sum_{n=1}^{\infty} g(n)$ converges.

15. Let $f : \mathbb{R} \to [0, +\infty)$ be twice differentiable and $p > 0$. Prove:

    (a) If $p \le 1$ and $\sum f(1/n^p)$ converges, then $f(0) = f'(0) = 0$.

    (b) If $p \ge 1$ and $f(0) = f'(0) = 0$, then $\sum f(1/n^p)$ converges.

16. Let $a_n \ge 0$ for all n and suppose that $\sum a_n$ converges. Prove that if $\alpha > 1/2$, then $\sum n^{-\alpha}\sqrt{a_n}$ converges. Give an example which shows that the assertion is false if $\alpha = 1/2$.

17. This exercise shows that e is irrational. Assume, for a contradiction, that $e = m/n$, $m, n \in \mathbb{N}$. Let $s_n = \sum_{k=0}^{n} 1/k!$. Using the series representation $e = \sum_{k=0}^{\infty} 1/k!$, show that

    (a) $n!(e - s_n) \in \mathbb{N}$

    (b) $n!(e - s_n) < \sum_{k=1}^{\infty}(n+1)^{-k} = 1/n$.

    Conclude that e must be irrational.

18. Let $s_n = \sum_{k=1}^{n} k^{-p}$, $0 < p < 1$. Show that $\{s_n - (1-p)^{-1}n^{1-p}\}$ converges. Conclude that

$$\lim_{n\to+\infty} \frac{1}{n^q}\sum_{k=1}^{n}\frac{1}{k^p} = \begin{cases} 0 & \text{if } p+q > 1, \\ \dfrac{1}{1-p} & \text{if } p+q = 1, \\ +\infty & \text{if } p+q < 1. \end{cases}$$

19. Let $s_n = \sum_{k=1}^{n} a_k$ and suppose that $\sum a_n^2 n^{-p} < +\infty$, where $a_n, p > 0$. Prove that $\lim_n s_n n^{-q} = 0$ for all $q > (p+1)/2$.

20. Let $\{a_n\}$, $\{b_n\}$, and $\{c_n\}$ be positive sequences such that $\sum c_n$ diverges, $b_n \to b \in (0, +\infty]$, and $a_n/a_{n+1} = 1 + b_n c_n$. Prove that $a_n \to 0$.

    Hint. Let $r \in (0, b)$ and choose m so that $b_n > r$ for all $n \ge m$. Then $a_{m+k}/a_{m+k+1} > 1 + r c_{m+k}$ for all $k \ge 0$.

*6.3 More Refined Convergence Tests

The tests in this section are frequently useful when the root and ratio tests fail. The first is a generalization of the ratio test.

6.3.1 Kummer's Test. *Let* a_n, $b_n > 0$ *for all* n *and set*

$$c_n := \frac{a_n}{a_{n+1}} b_n - b_{n+1}.$$

(a) *If* $\underline{c} := \liminf_n c_n > 0$, *then* $\sum_n a_n$ *converges.*

(b) *If* $\overline{c} := \limsup_n c_n < 0$ *and* $\sum_n b_n^{-1}$ *diverges, then* $\sum_n a_n$ *diverges.*

Proof. (a) Set $s_n = \sum_{k=1}^n a_k$ and let $r \in (0, \underline{c})$. Choose N so that $c_n \geq r$ for all $n \geq N$. Since $a_n b_n - a_{n+1} b_{n+1} = c_n a_{n+1}$, for all $m > N$ we have

$$a_N b_N \geq a_N b_N - a_m b_m = \sum_{n=N}^{m-1} (a_n b_n - a_{n+1} b_{n+1}) \geq r \sum_{n=N}^{m-1} a_{n+1} = r(s_m - s_N),$$

hence $s_m \leq s_N + a_N b_N / r$. The partial sums of $\sum a_n$ are therefore bounded, so the series converges.

(b) If $\overline{c} < 0$, then for some N

$$a_k b_k - a_{k+1} b_{k+1} < 0 \quad \text{for all } k \geq N.$$

Summing we have

$$a_N b_N - a_n b_n = \sum_{k=N}^{n-1} (a_k b_k - a_{k+1} b_{k+1}) < 0$$

so $a_n > (a_N b_N)/b_n$, for all $n > N$. Since $\sum 1/b_n$ diverges, $\sum a_n$ diverges by the comparison test. □

A simple but important consequence of Kummer's test is

6.3.2 Raabe's Test. *Let* $a_n > 0$ *for all* n *and set*

$$d_n := n\left(\frac{a_n}{a_{n+1}} - 1\right).$$

(a) *If* $\underline{d} := \liminf_n d_n > 1$, *then* $\sum a_n$ *converges.*

(b) *If* $\overline{d} := \limsup_n d_n < 1$, *then* $\sum a_n$ *diverges.*

Proof. Taking $b_n = n$ in Kummer's test we have

$$c_n = \frac{a_n}{a_{n+1}} n - (n+1) = d_n - 1.$$

Therefore, $\underline{c} = \underline{d} - 1$ and $\overline{c} = \overline{d} - 1$, and the assertions follow. □

6.3.3 Example. We use Raabe's test to show that the series

$$\sum_n \frac{1}{(n+m)!} \prod_{k=1}^{n} (k+a), \quad \text{where } a > 0 \text{ and } m \in \mathbb{N},$$

converges iff $m > 1 + a$. Indeed, since

$$n\left(\frac{a_n}{a_{n+1}} - 1\right) = n\left(\frac{n+m+1}{n+1+a} - 1\right) = \frac{n(m-a)}{n+1+a} \to m - a,$$

the series converges if $m - a > 1$ and diverges if $m - a < 1$. If $m - a = 1$, then the general term reduces to

$$\frac{1}{(n+m)!} \prod_{k=1}^{n} (m-1+k) = \frac{m(m+1)\cdots(m+n-1)}{(n+m)!} = \frac{1}{(m+n)(m-1)!},$$

hence the series diverges in this case as well. Note that the ratio test is inconclusive in this example since $a_{n+1}/a_n \to 1$. ◊

The following test is sometimes useful when the root test fails.

6.3.4 Log Test. *Let $a_n > 0$ for all n and set $c_n := \ln(a_n^{-1})/\ln n$.*

(a) *If $\underline{c} := \liminf_n c_n > 1$, then $\sum a_n$ converges.*

(b) *If $\overline{c} := \limsup_n c_n < 1$, then $\sum a_n$ diverges.*

Proof. (a) Let $p \in (1, \underline{c})$. Then there exists N such that $c_n > p$ for all $n \geq N$. For such n, $\ln(a_n^{-1}) > \ln n^p$, hence $a_n < 1/n^p$. Since $p > 1$, $\sum a_n$ converges by the comparison test. The proof of (b) is similar. □

6.3.5 Example. Let a_n denote the general term of the series

$$\sum_{n=1}^{\infty} \left(\frac{a+n^p}{b+n^q}\right)^n, \quad a, b, p, q > 0.$$

The root test shows that the series converges if $p < q$ and diverges if $p > q$. If $p = q$, the test is inconclusive, so we consider cases: If $a \geq b$, then $a_n \geq 1$ and the series diverges. If $a < b$, we use the log test: By l'Hospital's rule, the sequence

$$c_n = \frac{-\ln a_n}{\ln n} = \frac{\ln(b+n^p) - \ln(a+n^p)}{(\ln n)/n}$$

has the same limit as

$$\frac{\dfrac{pn^{p-1}}{b+n^p} - \dfrac{pn^{p-1}}{a+n^p}}{\dfrac{1-\ln n}{n^2}} = \frac{pn^{p+1}}{1-\ln n} \frac{(a+n^p)-(b+n^p)}{(a+n^p)(b+n^p)}$$

$$= \frac{p(a-b)}{b/n^p + 1} \frac{n}{(1-\ln n)(a+n^p)}.$$

The first quotient in the last expression tends to $p(a-b) < 0$. By l'Hospital's rule, the second quotient has the same limit as

$$\frac{1}{(1-\ln n)(pn^{p-1}) - (a+n^p)/n} = \frac{-n^{1-p}}{p(\ln n - 1) + (a/n^p + 1)},$$

which converges to 0 if $p \geq 1$ and to $-\infty$ if $p < 1$. Thus if $p = q$ and $a < b$, then

$$\lim_n c_n = \begin{cases} 0 & \text{if } p \geq 1 \\ +\infty & \text{if } p < 1, \end{cases}$$

hence $\sum a_n$ converges iff $p < 1$. ◊

Exercises

1. Show that the ratio test is a consequence of Kummer's test.

2. Show that Raabe's test detects the convergence properties of the p-series $\sum 1/n^p$ for $p \neq 1$, whereas the ratio and root tests do not.

3. Use Raabe's test to determine the convergence of $\sum a_n$ if $a_n =$

   (a) $\displaystyle\prod_{k=1}^{n} \frac{3k-1}{3k+1}$ (b) $\displaystyle\frac{1}{2n}\prod_{k=1}^{n}\left(\frac{2k-1}{2k}\right)$ (c) $\displaystyle\frac{1}{3^n(n+1)!}\prod_{k=1}^{n}(3k+1)$.

   Show in each case that the ratio test is inconclusive.

4. Let $a, b > 0$ and $m \in \mathbb{N}$. Use Raabe's test to show that the following series converges iff $b - a > m$:

$$\sum_n \left(\prod_{k=1}^{n} mk + a\right)\left(\prod_{k=1}^{n} mk + b\right)^{-1}.$$

5. Find all values of $p > 0$ for which the series converge:

   (a) $\displaystyle\sum_n \frac{p^n n!}{n^n}$ (b) $\displaystyle\sum_n \frac{p^n n!}{(p+1)(2p+1)\cdots(np+1)}$.

   What does the ratio test reveal?

6. Show that the series

$$\sum_{n=1}^{\infty} \frac{1 \cdot 3 \cdots (2n-1)}{(2+p)\cdot(4+p)\cdots(2n+p)}$$

   converges iff $p > 1$.

7. Let $p \in \mathbb{N}$. Use Raabe's test to show that the series

$$\sum \frac{(pn)!}{p^{pn}(n!)^p}$$

converges if $p > 3$ and diverges if $p < 3$. What does the ratio test tell us for these values of p?

8. Let a, b, $c > 0$ and $m \in \mathbb{Z}^+$. Use Raabe's test to show that the series

$$\sum_{n=1}^{\infty} \frac{1}{n^m} \prod_{k=1}^{n} \left(\frac{ak+b}{ak+c} \right)$$

converges iff $c > (m+1)a + b$.

9. Let $b > 0$ and $m \in \mathbb{N}$. Use Raabe's test to show that

$$\sum_{n=1}^{\infty} \left(\prod_{k=1}^{n} \frac{kb}{kb+1} \right)^m$$

converges if $m > b$ and diverges if $m < b$. What does the ratio test reveal? What happens if $m = b = 1$?

10. Let $r > 0$. Use the log test to determine the convergence behavior of $\sum a_n$ if $a_n =$

(a) $r^{\ln \ln n}$ (b) $\dfrac{1}{n^{r \ln n}}$ (c) $\dfrac{1}{n^{r \ln \ln n}}$ (d) $\dfrac{1}{(\ln n)^{r^n}}$.

11. Let a_n be as in 6.2.14. Use the log test to verify that $\sum a_n$ converges if $b^2 = 4a$.

12. Let p, $r > 0$. Use the log test to verify that $\sum r^{(n^p)}$ converges iff $r < 1$.

13. Let $b_n \to b > 0$. Use the log test to verify that $\sum b_n^{-\ln n}$ converges if $b > e$ and diverges if $b < e$.

14. Use the log test to show that the series $\sum (1 - 1/n)^{(n^p)}$ converges iff $p > 1$.

15. Let $p > 0$ and $a \neq 0$. Use the log test to verify that $\sum (1 - a/n^p)^n$ diverges if $p \geq 1$, converges if $0 < p < 1$ and $a > 0$, and diverges if $0 < p < 1$ and $a < 0$. What does the root test reveal?

16. Let a, b, p, $q > 0$. Determine the convergence behavior of $\sum a_n$ if $a_n =$

(a) $\left(\dfrac{a+n^p}{b+n^q} \right)^{\ln n}$ (b) $\left(\dfrac{a+n^p}{b+n^q} \right)^{\ln \ln n}$ (c) $\left(\dfrac{1+an^p}{1+bn^q} \right)^{\ln \ln n}$.

17. Show that $\sum (\ln n)^{b_n}$ diverges if $\{b_n\}$ is bounded. What happens in the unbounded special cases (a) $b_n = -\ln n$ and (b) $b_n = -n^p$, $p > 0$? What does the root test reveal in (b)?

18. (Loglog test) Let $a_n > 0$ for all n and set

$$c_n = -\frac{\ln(na_n)}{\ln\ln n}, \quad \underline{c} := \liminf_n c_n, \quad \text{and} \quad \overline{c} := \limsup_n c_n.$$

Prove that $\sum a_n$ converges if $\underline{c} > 1$ and diverges if $\overline{c} < 1$. Use the test to determine the convergence behavior of

$$\sum_{n=2}^{\infty} \left(\frac{1+an}{1+bn}\right)^{\ln\ln n}, \qquad a, b > 0.$$

19. Let $a, b > 0$. Use the log test to show that

$$\sum_n \left(\frac{1+an^p}{1+bn^q}\right)^{\ln n}$$

diverges if $p > q$; converges if $p < q$; and if $p = q$, then converges if $b/a > e$ and diverges if $b/a < e$. Use the log log test to show that the series also diverges if $p = q$ and $b/a = e$.

20. Use Kummer's test to prove Gauss's test: Let $a_n > 0$ for all n and let $\{\alpha_n\}$ be a bounded sequence such that

$$\frac{a_n}{a_{n+1}} = 1 + \frac{r}{n} - \frac{\alpha_n}{n^s},$$

where $r, s \in \mathbb{R}$, $s > 1$. Then $\sum a_n$ converges iff $r > 1$.

21. Use Kummer's test to prove Bertrand's test: Let $a_n > 0$ for all n and let $\{\beta_n\}$ be a sequence such that

$$\frac{a_n}{a_{n+1}} = 1 + \frac{1}{n} - \frac{\beta_n}{n\ln n}.$$

Then $\sum a_n$ converges if $\liminf_n \beta_n > 1$ and diverges if $\limsup_n \beta_n < 1$.

6.4 Absolute and Conditional Convergence

The convergence tests in Sections 6.2 and 6.3 apply only to series with nonnegative terms. In this section we consider tests applicable to general series.

6.4.1 Definition. A series $\sum a_n$ is said to *converge absolutely* if the series $\sum |a_n|$ of absolute values converges. A convergent series that does not converge absolutely is said to *converge conditionally*. ◇

For the main result connecting convergence, absolute convergence, and conditional convergence, refer to 1.2.7 for definitions and properties of the positive and negative parts of a real number.

6.4.2 Theorem. (a) *If $\sum a_n$ converges absolutely, then the series*

$$\sum a_n, \quad \sum a_n^+, \quad \text{and} \quad \sum a_n^-$$

converge and

$$\sum a_n = \sum a_n^+ - \sum a_n^-, \quad \sum |a_n| = \sum a_n^+ + \sum a_n^-.$$

(b) *If $\sum a_n$ converges conditionally, then $\sum a_n^+$ and $\sum a_n^-$ diverge.*

Proof. (a) If $\sum |a_n|$ converges, then the inequalities

$$0 \le a_n^{\pm} = \tfrac{1}{2}\big(|a_n| \pm a_n\big) \le |a_n|$$

and the comparison test show that $\sum a_n^+$ and $\sum a_n^-$ converge. The remaining assertions in (a) follow from the identities $a_n = a_n^+ - a_n^-$ and $|a_n| = a_n^+ + a_n^-$.

(b) If $\sum a_n$ and $\sum a_n^-$ converge, then from

$$\sum |a_n| = \sum a_n + 2\sum a_n^-$$

we see that $\sum |a_n|$ converges. A similar identity shows that the same conclusion holds if $\sum a_n$ and $\sum a_n^+$ converge. Hence if $\sum a_n$ converges conditionally, then neither $\sum a_n^+$ nor $\sum a_n^-$ can converge. □

All series of the form $\sum_{n=1}^{\infty}(-1)^{n+1}/n^p$, $0 < p \le 1$ converge conditionally. This follows from the alternating series test given below. The following example is somewhat more interesting.

6.4.3 Example. We show that the series

$$s := \sum_{n=2}^{\infty} \big[(-1)^n n^p - 1\big]^{-1}$$

converges conditionally iff $1/2 < p \le 1$ and absolutely iff $p > 1$.

To see this, note first that if $p < 0$, then the nth term of the series does not tend to zero, and if $p = 0$ the series is undefined. So assume $p > 0$. Since

$$n^p - 1 \le \big|(-1)^{n+1} n^p - 1\big| \le n^p + 1,$$

we have

$$\frac{1}{n^p + 1} \le \frac{1}{|(-1)^{n+1} n^p - 1|} \le \frac{1}{n^p - 1},$$

so the series s converges absolutely iff $p > 1$.

Now assume that $0 < p \le 1$ and let s_n denote the nth partial sum of the series. Since the nth term of the series tends to 0, $s_{2n+1} - s_{2n} \to 0$, hence the series converges iff the partial sum

$$s_{2n+1} = \sum_{k=2}^{2n+1} \frac{1}{(-1)^k k^p - 1}$$

converges. Break it up into even and odd parts:

$$s_{2n+1} = \sum_{k=1}^{n} \left\{ \frac{1}{(2k)^p - 1} - \frac{1}{(2k+1)^p + 1} \right\} = \sum_{k=1}^{n} (\alpha_k + \beta_k), \tag{6.5}$$

where

$$\alpha_k := \frac{(2k+1)^p - (2k)^p}{\big[(2k)^p - 1\big]\big[(2k+1)^p + 1\big]}$$

and

$$\beta_k := \frac{2}{\big[(2k)^p - 1\big]\big[(2k+1)^p + 1\big]}.$$

Since

$$\begin{aligned}\frac{\beta_k}{k^{-2p}} &= \frac{1}{k^{-2p}} \frac{2}{\big[(2k)^p - 1\big]\big[(2k+1)^p + 1\big]} \\ &= \frac{2}{[2^p - k^{-p}][(2 + 1/k)^p + k^{-p}]} \to \frac{1}{2^{2p-1}} \in (0, +\infty),\end{aligned}$$

the limit comparison test shows that $\sum_{k=1}^{n} \beta_k$ converges iff $p > 1/2$.

By the mean value theorem applied to x^p on the interval $[2k, 2k+1]$, there exists some $x_k \in (2k, 2k+1)$ for which

$$\alpha_k = \frac{p x_k^{p-1}}{\big[(2k)^p - 1\big]\big[(2k+1)^p + 1\big]} = \frac{p}{(x_k^{1-p})\big[(2k)^p - 1\big]\big[(2k+1)^p + 1\big]}.$$

Since $0 < p \le 1$,

$$\alpha_k \le \frac{p}{(2k)^{1-p}\big[(2k)^p - 1\big]\big[(2k)^p + 1\big]} = \frac{1}{(2k)^{1-p}\big[(2k)^{2p} - 1\big]} \le \frac{1}{k^{p+1}},$$

the last inequality for all sufficiently large k. Therefore, $\sum_{k=1}^{n} \alpha_k$ converges by comparison with $\sum_k 1/k^{p+1}$. Thus the partial sum (6.5) has a finite limit iff $p > 1/2$, and so the series s converges iff $p > 1/2$. ◊

The tests of Sections 6.2 and 6.3 for positive-term series may be used in conjunction with 6.4.2 to test series with terms of mixed sign. For example, the inequality $\left|n^{-2} \sin n\right| \le n^{-2}$, together with the comparison test, shows that the series $\sum n^{-2} \sin n$ converges absolutely and hence converges.

The following test, useful in its own right, will give an easy proof of the alternating series test.

6.4.4 Dirichlet's Test. *Let $\{a_n\}$ and $\{b_n\}$ be sequences such that the following conditions hold:*

(a) *The partial sums of $\sum a_n$ are bounded.*

(b) $\lim_n b_n = 0$.

(c) *The series $\sum |b_{n+1} - b_n|$ converges, which is the case, for example, if $\{b_n\}$ is monotone.*

Then $\sum a_n b_n$ converges.

Proof. Let

$$s_n := \sum_{k=1}^{n} a_k \quad \text{and} \quad t_n := \sum_{k=1}^{n} a_k b_k.$$

Then $a_k = s_k - s_{k-1}$ for $k > 11$. Using Abel's "summation by parts" method, we have for $m > n+1$

$$\begin{aligned}
t_m - t_n &= \sum_{k=n+1}^{m} a_k b_k \\
&= \sum_{k=n+1}^{m} (s_k - s_{k-1}) b_k \\
&= \sum_{k=n+1}^{m} s_k b_k - \sum_{k=n+1}^{m} s_{k-1} b_k \\
&= \sum_{k=n+1}^{m} s_k b_k - \sum_{k=n}^{m-1} s_k b_{k+1} \\
&= \sum_{k=n+1}^{m-1} s_k (b_k - b_{k+1}) + s_m b_m - s_n b_{n+1}.
\end{aligned}$$

By hypothesis, there exists $M > 0$ such that $|s_n| \le M$ for every n. Then

$$|t_m - t_n| = \left| \sum_{k=n+1}^{m} a_k b_k \right| \le M \sum_{k=n+1}^{m} |b_k - b_{k+1}| + M(|b_m| + |b_{n+1}|).$$

Since the right side of the inequality tends to 0 as $m, n \to \infty$, $\{t_n\}$ is a Cauchy sequence and hence converges. [1]

If $\{b_n\}$ is monotone, say decreasing, then

$$\sum_{k=1}^{n} |b_{k+1} - b_k| = \sum_{k=1}^{n} (b_k - b_{k+1}) = b_1 - b_{n+1},$$

which converges. □

[1]The case $m = n+1$ is easily treated separately.

6.4.5 Example. We apply Dirichlet's test to the series $\sum_{n=1}^{\infty} b_n \sin(n\theta)$, where $\{b_n\}$ is monotone and $b_n \to 0$. To establish the boundedness of the sequence of partial sums $s_n := \sum_{k=1}^{n} \sin(k\theta)$, we use the identity

$$2\sin(\theta/2)\sin(k\theta) = \cos\big[(k-1/2)\theta\big] - \cos\big[(k+1/2)\theta\big].$$

Summing,

$$2\sin(\theta/2)\sum_{k=1}^{n}\sin(k\theta) = \cos(\theta/2) - \cos\big((n+1)\theta/2\big).$$

Thus if θ is not a multiple of 2π, then $|s_n| \le |\sin(\theta/2)|^{-1}$. By 6.4.4, $\sum_{n=1}^{\infty} b_n \sin(n\theta)$ converges for all θ. Note that if, for example, $\theta = \pi/2$ and $b_n = 1/n$, then the convergence is conditional. ◊

6.4.6 Alternating Series Test. *If $b_n \downarrow 0$, then $\sum_{n=1}^{\infty}(-1)^{n+1}b_n$ converges.*

Proof. The partial sums of $\sum_{n=1}^{\infty}(-1)^{n+1}$ are clearly bounded. Therefore, the assertion follows from 6.4.4. □

6.4.7 Example. (Alternating Harmonic Series). By 6.4.6, the series $\sum_{n=1}^{\infty}(-1)^{n+1}n^{-1}$ converges. We show that its value is $\ln 2$. Let

$$s_n = \sum_{k=1}^{n}\frac{(-1)^{k+1}}{k} \quad\text{and}\quad t_n = \sum_{k=1}^{n}\frac{1}{k} - \ln n.$$

By 6.1.9, the sequence $\{t_n\}$ converges. Also, by Exercise 1.4.3,

$$s_{2n} = \sum_{k=1}^{2n}\frac{(-1)^{k+1}}{k} = \sum_{k=n+1}^{2n}\frac{1}{k} = t_{2n} - t_n + \ln 2.$$

It follows that $s_{2n} \to \ln 2$. Since $s_{2n+1} - s_{2n} \to 0$, $s_n \to \ln 2$. ◊

The contrast between absolutely convergent and conditionally convergent series is strikingly demonstrated in the context of *rearrangements*.

6.4.8 Definition. A *rearrangement* of a series $\sum_{n=1}^{\infty} a_n$ is a series $\sum_{k=1}^{\infty} a_{m_k}$, where $\{m_k\}$ is a sequence of positive integers that contains every positive integer exactly once.[2] ◊

6.4.9 Theorem. *If $\sum_{n=1}^{\infty} a_n$ converges absolutely to s, then any rearrangement $\sum_{k=1}^{\infty} a_{m_k}$ converges absolutely to s.*

Proof. Assume first that $a_n \ge 0$ for all n. Let

$$t_n = \sum_{k=1}^{n} a_{m_k} \quad\text{and}\quad s_n = \sum_{k=1}^{n} a_k.$$

[2] In other words, $k \mapsto m_k$ is a one-to-one mapping of $\mathbb{N}$ onto itself.

For each N, choose K so large that the terms a_k, $1 \le k \le N$, are included among the terms a_{m_k}, $1 \le k \le K$. Then $s_N \le t_K \le \sum_{k=1}^{\infty} a_{m_k}$. Letting $N \to \infty$ shows that $\sum_n a_n \le \sum_k a_{m_k}$. Since $\sum_n a_n$ is a rearrangement of $\sum_k a_{m_k}$, the reverse inequality holds as well. The general case follows by considering $\sum a_n^+$ and $\sum a_n^-$ and using 6.4.2. □

6.4.10 Example. Consider the alternating series

$$s := 1 - \frac{1}{2^p} + \frac{1}{3^p} - \frac{1}{4^p} + \frac{1}{5^p} - \frac{1}{6^p} + \frac{1}{7^p} - \frac{1}{8^p} + \frac{1}{9^p} - \frac{1}{10^p} + \cdots,$$

and the rearrangement

$$t := 1 - \frac{1}{2^p} - \frac{1}{4^p} + \frac{1}{3^p} - \frac{1}{6^p} - \frac{1}{8^p} + \frac{1}{5^p} - \frac{1}{10^p} - \frac{1}{12^p} + \cdots,$$

If $p > 1$, then both series converge absolutely and $t = s$. If $p = 1$, then the two series converge to different values. Indeed, if s_n and t_n denote the nth partial sums of s and t, respectively, then

$$t_{3n} = \sum_{k=1}^{n} \left(\frac{1}{2k-1} - \frac{1}{4k-2} - \frac{1}{4k} \right) = \frac{1}{2} \sum_{k=1}^{n} \left(\frac{1}{2k-1} - \frac{1}{2k} \right) = \frac{s_{2n}}{2} \to \frac{s}{2}.$$

Since

$$t_{3n+1} = t_{3n} + \frac{1}{2n+1} \quad \text{and} \quad t_{3n+2} = t_{3n+1} - \frac{1}{4n+2},$$

we see that $t_n \to s/2$. ◊

The phenomenon illustrated in the last example holds generally, as shown by the following remarkable result of Riemann.

6.4.11 Theorem. *If $s := \sum_{n=1}^{\infty} a_n$ converges conditionally, then, for any real number x, some rearrangement of s converges to x.*

Proof. We may assume that $x \ge 0$. For $n \in \mathbb{N}$ let

$$s_n := \sum_{j=1}^{n} a_j, \quad s_n^+ := \sum_{j=1}^{n} a_j^+, \quad s_n^- := \sum_{j=1}^{n} a_j^-, \quad \text{and} \quad s_0^+ := 0.$$

Since $s_n^+ \to +\infty$ (6.4.2), there exists a smallest integer m_1 such that $s_{m_1}^+ > x$. Since $x \ge 0$, $m_1 \ne 0$. Because $s_n^- \to +\infty$, there exists a smallest positive integer n_1 such that $s_{m_1}^+ - s_{n_1}^- < x$ and then a smallest integer m_2 such that $s_{m_2}^+ - s_{n_1}^- > x$. Obviously, $m_2 > m_1$. Continuing in this manner, we obtain strictly increasing sequences $\{m_k\}$ and $\{n_k\}$ with the following properties:

- m_k is the smallest integer such that

$$t_k := s_{m_k}^+ - s_{n_{k-1}}^- = (a_1^+ + \cdots + a_{m_k}^+) - (a_1^- + \cdots + a_{n_{k-1}}^-) > x,$$

- n_k the smallest integer such that

$$r_k := s^+_{m_k} - s^-_{n_k} = (a^+_1 + \cdots + a^+_{m_k}) - (a^-_1 + \cdots + a^-_{n_k}) < x.$$

Now consider the series

$$s' := a^+_1 + \cdots + a^+_{m_1} - a^-_1 - \cdots - a^-_{n_1} + a^+_{m_1+1} + \cdots + a^+_{m_2} - a^-_{n_1+1} - \cdots.$$

The terms of s' are either a_j or 0, and s' contains each term of the series s exactly once. Thus s' is a rearrangement of s. We show that $s' = x$.

By the minimality properties of the sequences $\{m_k\}$ and $\{n_k\}$,

$$t_k - a^+_{m_k} \le x < t_k \quad \text{and} \quad r_k < x \le r_k + a^-_{n_k},$$

hence

$$x - a^+_{n_k} \le r_k < x < t_k \le x + a^+_{m_k}.$$

Since $a_n \to 0$,

$$\lim_k r_k = \lim_k t_k = x. \tag{6.6}$$

Now let s'_k denote the kth partial sum of the series s' and consider the partial sums

$$\begin{aligned} r_1 &= (a^+_1 + \cdots + a^+_{m_1}) - (a^-_1 + \cdots + a^-_{n_1}), \\ t_2 &= (a^+_1 + \cdots + a^+_{m_2}) - (a^-_1 + \cdots + a^-_{n_1}), \\ r_2 &= (a^+_1 + \cdots + a^+_{m_2}) - (a^-_1 + \cdots + a^-_{n_2}). \end{aligned}$$

If $m_1 + n_1 \le k \le m_2 + n_1$, then s'_k includes the terms of r_1, additional terms from $a^+_{m_1+1} + \cdots + a^+_{m_2}$, and no others, hence $r_1 \le s'_k \le t_2$. Similarly, if $m_2 + n_1 \le k \le m_2 + n_2$, then s'_k includes the terms of t_2, additional terms from $-a^-_{n_1+1} - \cdots - a^-_{n_2}$, and no others, so $r_2 \le s'_k \le t_2$. In general, for $j \ge 1$,

$$\begin{aligned} m_j + n_j \le k \le m_{j+1} + n_j &\Rightarrow r_j \le s'_k \le t_{j+1} \quad \text{and} \\ m_{j+1} + n_j \le k \le m_{j+1} + n_{j+1} &\Rightarrow r_{j+1} \le s'_k \le t_{j+1}. \end{aligned}$$

From (6.6), $s'_k \to x$. □

Exercises

1. Suppose that $\sum a_n$ converges absolutely. Prove that $\sum a_n b_n$ converges absolutely for all sequences $\{b_n\}$ with $\limsup_{n\to\infty} |b_n| < +\infty$.

2. Suppose the ratio test shows that $\sum a_n$ does not converge absolutely. Can $\sum a_n$ still converge conditionally?

3. For an alternating series $s = \sum_{n=1}^{\infty}(-1)^{n+1}b_n$ ($b_n \downarrow 0$), prove the inequality $|s - s_n| \le b_n$. This result is useful in estimating the error made by using s_n to approximate s. For example, use the estimate to determine how large n should be so that the partial sum s_n agrees with $s = \sum_{n=1}^{\infty}(-1)^{n+1}/n^4$ in nine decimal places.

4. Let $p > 0$. Determine whether the series $\sum a_n$ converges absolutely, conditionally, or diverges, where $a_n =$

(a) $\dfrac{(-1)^n}{n^{1/n}}$

(b) $(-1)^n(n^{1/n} - 1)$

(c) $(-1)^n \sin(1/n^p)$

(d) $(-1)^n \sin^{-1}(1/n^p)$

(e) $(-1)^n \tan(1/n^p)$

(f) $(-1)^n \tan^{-1}(1/n^p)$

(g) $\dfrac{\sin[(2n+1)\pi/2]}{\ln n}$

(h) $\dfrac{(-2)^n}{n!}$

(i) $(-1)^n \dfrac{\sqrt{n+1} - \sqrt{n}}{n^p}$

(j) $\dfrac{(-1)^n}{n \ln^p(n+1)}$

(k) $(-1)^n \dfrac{3n}{\sqrt{n^3+2}}$

(l) $(-1)^n p^n \dfrac{(n!)^2}{(2n)!}$

(m) $\dfrac{(-1)^n}{p^n + (-1)^n}$, $(p \neq 1)$

(n) $\dfrac{(-1)^n}{n^p + (-1)^n}$, $(n \geq 2)$

(o) $\dfrac{(-1)^n n!}{3 \cdot 5 \cdots (2n+1)}$

(p) $\dfrac{(-1)^n 3 \cdot 6 \cdots (3n)}{3 \cdot 5 \cdots (2n+1)}$

(q) $\dfrac{(-1)^n e^n n!}{5 \cdot 8 \cdots (3n+2)}$

(r) $(-1)^{n+1} n^{[(-1)^n - 3]/2}$.

5. Suppose that $\{b_n\}$ is monotone and $b_n \to 0$. Use the identity

$$2\sin(\theta/2)\cos(n\theta) = \sin\big[(n+1/2)\theta\big] - \sin\big[(n-1/2)\theta\big]$$

to verify that the series $\sum_{n=1}^{\infty} b_n \cos n\theta$ converges if $\theta/(2\pi) \notin \mathbb{Z}$.

6. Let $b_n \downarrow 0$ and $m \in \mathbb{N}$. Show that $\sum_{n=0}^{\infty} (-1)^{\lfloor n/m \rfloor} b_n$ converges.

7. Let $m \in \mathbb{N}$. Show that

$$\sum_{n=1}^{\infty} \frac{(-1)^{n+1} m}{n(n+m)} = \sum_{n=1}^{m} \frac{(-1)^{n+m+1}}{n} + \delta_m \ln 2,$$

where $\delta_m = 0$ or 2 according as m is even or odd.

8. (Abel) Prove that if $\sum_{n=1}^{\infty} a_n$ converges and $\{b_n\}$ is bounded and monotone, then $\sum_{n=1}^{\infty} a_n b_n$ converges.

9. Prove that $\displaystyle\sum_{n=2}^{\infty} \frac{(n-1/2)\sin(n\theta)}{n^p + (-1)^n}$ converges for all real θ iff $p > 1$.

10. Let $p > 1$. Express each of the series $\displaystyle\sum_{n=1}^{\infty} \frac{1}{(2n-1)^p}$ and $\displaystyle\sum_{n=1}^{\infty} \frac{(-1)^n}{n^p}$ in terms of $\displaystyle\sum_{n=1}^{\infty} \frac{1}{n^p}$.

11. Prove that if $\sum na_n$ converges, then $\sum a_n$ converges and, moreover, $\sum |a_n|^p$ converges for every $p > 1$. What if $p = 1$?

12. Prove that if $\sum n^{-p}a_n$ converges, then $\sum n^{-q}a_n$ converges for all $q > p$.

13. Show that the following series converges

$$s := 1 + \frac{1}{2} + \frac{1}{3} - \frac{1}{4} - \frac{1}{5} - \frac{1}{6} + \frac{1}{7} + \frac{1}{8} + \frac{1}{9} - \cdots,$$

where sums of length three alternate signs. Generalize your result to alternating sums of length $p > 1$. Show that by contrast, the following series diverges,

$$t := 1 + \frac{1}{2} + \frac{1}{3} - \frac{1}{4} - \frac{1}{5} + \frac{1}{6} + \frac{1}{7} + \frac{1}{8} - \frac{1}{9} - \cdots.$$

where sums of lengths $p = 3$ and $q = 2$ alternate signs.

---

*6.5 Double Sequences and Series

A *double sequence* is a doubly indexed infinite array

$$\{a_{m,n}\} = \{a_{m,n}\}_{m,n=1}^{\infty}$$

of real numbers $a_{m,n}$.[3] Associated with each double sequence are the so-called *iterated limits*

$$\lim_m \lim_n a_{m,n} \quad \text{and} \quad \lim_n \lim_m a_{m,n}.$$

For the first iterated limit to exist, each inner limit $b_m := \lim_n a_{m,n}$, as well as the outer limit $\lim_m b_m$, must exist. Similar remarks apply to the second iterated limit. The following scheme illustrates the case when the iterated limits exist and equal L.

$$\begin{array}{ccccc}
a_{1,1} & a_{1,2} & \cdots & a_{1,n} & \to b_1 \\
a_{2,1} & a_{2,2} & \cdots & a_{2,n} & \to b_2 \\
\vdots & \vdots & \cdots & \vdots & \vdots \\
a_{m,1} & a_{m,2} & \cdots & a_{m,n} & \to b_m \\
\downarrow & \downarrow & \cdots & \downarrow & \downarrow \\
c_1 & c_2 & \cdots & c_n & \to L
\end{array}$$

In addition to iterated limits, a double sequence gives rise to a third type of limit, frequently called a *double limit* to distinguish it from iterated limits.

[3] More precisely, a double sequence is a function $(m, n) \mapsto a_{m,n}$ from $\mathbb{N} \times \mathbb{N}$ to $\mathbb{R}$.

6.5.1 Definition. Let $L \in \mathbb{R}$. We write

$$L = \lim_{m,n} a_{m,n},$$

($m \to +\infty$ and $n \to +\infty$ together) and say that $a_{m,n}$ converges to L or has limit L if for each $\varepsilon > 0$ there exists $N \in \mathbb{N}$ such that $|a_{m,n} - L| < \varepsilon$ for all $n, m \geq N$. We also write

$$\lim_{m,n} a_{m,n} = +\infty \ (-\infty)$$

if for any $r \in \mathbb{R}$ there exists $N \in \mathbb{N}$ such that $a_{m,n} > r$ ($< r$) for all $n, m \geq N$. $\Diamond$

Double limits have properties similar to limits of single sequences. For example, double limit analogs of 2.1.3, 2.1.4, 2.1.5, and 2.1.11, are readily formulated and proved.

It is easy to find examples of iterated limits that exist but are unequal; $a_{m,n} = (1 - 1/n)^m$ is one such. When this happens, the double limit cannot exist, as shown in 6.5.2 below. However, even if the iterated limits are equal, the double limit may fail to exist. This is the case for the sequence defined by

$$a_{m,n} = \begin{cases} 1 & \text{if } m = n, \text{ and} \\ 0 & \text{otherwise,} \end{cases}$$

which has zero iterated limits. Finally, the example

$$a_{m,n} = (-1)^{m+n}(1/m + 1/n)$$

shows that a double limit may exist even if both iterated limits fail to exist.

The following theorem gives the basic connection between double limits and iterated limits.

6.5.2 Iterated Limit Theorem. *Let $\{a_{m,n}\}$ be a double sequence such that $\lim_n a_{m,n}$ exists for each m and $\lim_m a_{m,n}$ exists for each n. If the double limit $\lim_{m,n} a_{m,n}$ exists, then the iterated limits $\lim_m \lim_n a_{m,n}$ and $\lim_n \lim_m a_{m,n}$ exist and equal the double limit.*

Proof. Let $L := \lim_{m,n} a_{m,n}$, $b_m := \lim_n a_{m,n}$, and $c_n := \lim_m a_{m,n}$. Given $\varepsilon > 0$, choose $N \in \mathbb{N}$ such that

$$|a_{m,n} - L| < \varepsilon \quad \text{for all } m,\ n \geq N.$$

Letting $n \to +\infty$ yields $|b_m - L| \leq \varepsilon$ for all $m \geq N$. Therefore, $b_m \to L$. Similarly, $c_n \to L$. $\square$

For series we have the following analogous definitions:

6.5.3 Definition. Given a double sequence $\{a_{m,n}\}$, form the *partial sums*

$$s_{m,n} = \sum_{j=1}^{m}\sum_{k=1}^{n} a_{j,k}, \quad m,\, n \in \mathbb{N}.$$

The *double infinite series*

$$\sum a_{m,n} = \sum_{m,n} a_{m,n} = \sum_{m,n=1}^{\infty} a_{m,n}$$

is said to *converge to* $s \in \mathbb{R}$ if $\{s_{m,n}\}$ converges to s in the sense of 6.5.1. The series *converges absolutely* if $\sum |a_{m,n}|$ converges, and *converges conditionally* if $\sum a_{m,n}$ converges but not absolutely. ◇

As in the case of single series, an absolutely convergent double series converges (Exercise 7). Moreover, a double series with nonnegative terms converges absolutely iff the partial sums $\sum_{j=1}^{m}\sum_{k=1}^{n} a_{j,k}$ are bounded (Exercise 5).

The iterated limits

$$\lim_m \lim_n s_{m,n} = \lim_m \lim_n \sum_{j=1}^{m}\sum_{k=1}^{n} a_{j,k} = \sum_{j=1}^{\infty}\sum_{k=1}^{\infty} a_{j,k}$$

and

$$\lim_n \lim_m s_{m,n} = \lim_n \lim_m \sum_{k=1}^{n}\sum_{j=1}^{m} a_{j,k} = \sum_{k=1}^{\infty}\sum_{j=1}^{\infty} a_{j,k}$$

are called *iterated series*. The following result, a special case of the *Fubini–Tonelli theorem*, establishes a connection between double and iterated series.

6.5.4 Fubini–Tonelli Theorem for Series. *A double series* $\sum a_{m,n}$ *is absolutely convergent iff one (hence both) of the following conditions hold:*

$$\sum_{m=1}^{\infty}\sum_{n=1}^{\infty} |a_{m,n}| < +\infty \quad \textit{and} \quad \sum_{n=1}^{\infty}\sum_{m=1}^{\infty} |a_{m,n}| < +\infty. \tag{6.7}$$

In this case,

$$\sum_{m,n} a_{m,n} = \sum_{m=1}^{\infty}\sum_{n=1}^{\infty} a_{m,n} = \sum_{n=1}^{\infty}\sum_{m=1}^{\infty} a_{m,n}. \tag{6.8}$$

Proof. Set $s_{m,n} = \sum_{j=1}^{m}\sum_{k=1}^{n} a_{j,k}$ and $t_{m,n} = \sum_{j=1}^{m}\sum_{k=1}^{n} |a_{j,k}|$. The first assertion of the theorem is clear, since each condition in (6.7) implies that $T := \sup_{m,n} t_{m,n} < +\infty$, and conversely.

Now suppose that $\sum a_{m,n}$ is absolutely convergent. Let $s := \lim_{m,n} s_{m,n}$. For each j,

$$\sum_{k=1}^{n} |a_{j,k}| \le t_{j,n} \le T \quad \text{for all } n,$$

hence $\sum_{k=1}^{\infty} a_{j,k}$ converges. Set

$$r_m := \sum_{j=1}^{m}\sum_{k=1}^{\infty} a_{j,k}.$$

Given $\varepsilon > 0$, choose N such that

$$\left|\sum_{j=1}^{m}\sum_{k=1}^{n} a_{j,k} - s\right| < \varepsilon \text{ for all } m,\, n \geq N.$$

Fixing $m \geq N$ and letting $n \to +\infty$ in this inequality yields $|r_m - s| \leq \varepsilon$. This shows that $r_m \to s$, which is the first equality in (6.8). The proof of the second equality is similar. □

Exercises

1. Let $\alpha : \mathbb{N} \to \mathbb{N}$ be strictly increasing. Show that if $L := \lim_{m,n} a_{m,n}$ exists in $\mathbb{R}$, then $\lim_{m,n} a_{\alpha(n),n}$ exists and equals L.

2. A double sequence $\{a_{m,n}\}$ is said to be *Cauchy* if, given $\varepsilon > 0$, there exists $N \in \mathbb{N}$ such that $|a_{m,n} - a_{m+p,n+q}| < \varepsilon$ for all $m,\, n \geq N$ and all $p,\, q \geq 0$. Prove that $\{a_{m,n}\}$ converges iff it is Cauchy. *Hint.* Show that $\{a_{n,n}\}$ converges.

3. Determine the convergence behavior, double and iterated, of the following sequences, where $a,\, b > 0$:

   (a) $\sin(m/n)$ (b) $\dfrac{\ln(mn)}{n}$ (c) $\dfrac{(-1)^m m}{m+n}$

   (d) $\dfrac{m-n}{m+n}$ (e) $\dfrac{mn}{(m+n)^2}$ (f) $\dfrac{mn}{m^2+n^2}$

   (g) $\dfrac{1}{m^{1/n}}$ (h) $\dfrac{n}{m+n^2}$ (i) $\dfrac{n^3 m}{m^4+n^4}$

   (j) $\dfrac{n+nm\sin(1/n)}{am+bn}$ (k) $\dfrac{m^2 n}{an^2+bm^4}$ (l) $\dfrac{n^2\sin(1/n)}{m+n}$.

4. Show that if $\sum a_{m,n}$ converges, then $\lim_{m,n} a_{m,n} = 0$.

5. Let $a_{m,n} \geq 0$ for all $m,\, n \in \mathbb{N}$. Prove that $\sum_{m,n} a_{m,n}$ converges iff $s := \sup_{m,n} s_{m,n} < +\infty$, in which case the series sums to s.

6. State and prove a comparison test for double series with nonnegative terms.

7. Prove that an absolutely convergent double series converges.

8. For sequences $\{a_n\}$ and $\{b_n\}$, set $c_{m,n} = a_n b_m$. Prove that $c := \sum_{m,n} c_{m,n}$ converges absolutely iff $a := \sum_n a_n$ and $b := \sum_n b_n$ converge absolutely, in which case $c = ab$. Conclude that $\sum_{m,n} m^{-q}n^{-p}$ converges iff p, $q > 1$.

9. Given a double sequence $\{a_{m,n}\}$ with $a_{m,n} \geq 0$, let $\{b_n\}$ be the sequence obtained by summing $a_{m,n}$ along the diagonals $j + k = n + 1$, that is, $b_n := \sum_{j=1}^{n} a_{j,n+1-j}$. Prove that $\sum a_{m,n}$ converges iff $\sum_n b_n$ converges, in which case the two series are equal.

10. Use Exercise 9 to show that the double series

$$\text{(a)} \sum_{m,n} \frac{1}{(m+n)^p}, \quad \text{(b)} \sum_{m,n} \frac{1}{(m^2+n^2)^{p/2}}, \quad \text{and} \quad \text{(c)} \sum_{m,n} \frac{1}{m^p + n^p}$$

converge iff $p > 2$. Show that for $p > 2$,

$$\sum_{m,n=1}^{\infty} \frac{1}{(m+n)^p} = \sum_{n=2}^{\infty} \frac{1}{n^{p-1}} - \sum_{n=2}^{\infty} \frac{1}{n^p}.$$

11. Prove that $\sum_{m,n} r^{mn}$ converges iff $|r| < 1$, in which case the iterated series $\sum_{m=1}^{\infty} \sum_{n=1}^{\infty} r^{mn}$ converges.

12. Prove the root test for double series with nonnegative terms: Suppose that $L := \lim_{m,n} a_{m,n}{}^{1/mn}$ exists. Then $\sum_{m,n} a_{m,n}$ converges if $L < 1$ and diverges if $L > 1$.

13. Let $a_{m,n} = (-1)^m n^{-m-2}$. Prove that

$$\sum_{m \geq 0, n \geq 2} |a_{m,n}| = 1 \quad \text{and} \quad \sum_{m \geq 0, n \geq 2} a_{m,n} = 1/2.$$

Chapter 7

Sequences and Series of Functions

7.1 Convergence of Sequences of Functions

Unlike numerical sequences, sequences of functions have several modes of convergence. In this chapter we consider the two most common types: pointwise and uniform. Other types of convergence will be examined in Chapter 11.

7.1.1 Definition. Let S be a nonempty set. A sequence of real-valued functions f_n on S is said to *converge pointwise on S to a function* $f : S \to \overline{\mathbb{R}}$ if $f_n(x) \to f(x)$ for each $x \in S$. We then write $f = \lim_n f_n$ or $f_n \to f$ (on S). ◊

The following theorem is an immediate consequence of 2.1.11 and 3.1.9.

7.1.2 Theorem. *Let $f_n \to f$ and $g_n \to g$ pointwise on S and let h be continuous such that $h \circ f_n$ and $h \circ f$ are defined. Then, for $\alpha, \beta \in \mathbb{R}$,*

$$\alpha f_n + \beta g_n \to \alpha f + \beta g, \ f_n g_n \to fg, \ \frac{f_n}{g_n} \to \frac{f}{g} \text{ (if } g \neq 0) \text{ and } h \circ f_n \to h \circ f$$

pointwise on S.

The definition of pointwise convergence may be phrased as follows: For each $x \in S$ and $\varepsilon > 0$, there exists an index N such that $|f_n(x) - f(x)| < \varepsilon$ for all $n \geq N$. Here, the index N usually depends on both ε and x. Removing

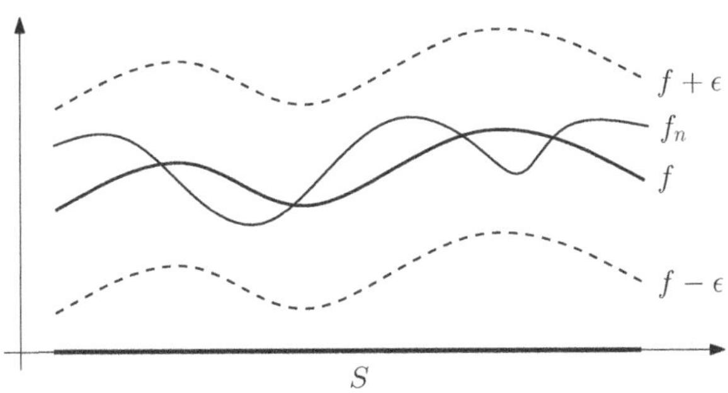

FIGURE 7.1: Uniform convergence of f_n to f.

the dependence on x results in the stronger property of *uniform convergence:*

DOI: 10.1201/9781003629382-7

7.1.3 Definition. A sequence of functions $f_n : S \to \mathbb{R}$ is said to *converge uniformly on* S *to a function* $f : S \to \mathbb{R}$ if for each $\varepsilon > 0$ there exists $N \in \mathbb{N}$ such that $|f_n(x) - f(x)| < \varepsilon$ for all $n \geq N$ and all $x \in S$ (See Figure 7.1) ◊

Clearly, uniform convergence implies pointwise convergence. The examples below show that the converse is not generally true. For these examples and for the exercises at the end of the section, the following propositions are useful.

7.1.4 Proposition. *Let* $f_n, f : S \to \mathbb{R}$. *Suppose that there exists a sequence* $\{a_n\}$ *of positive real numbers such that* $a_n \to 0$ *and* $|f_n(x) - f(x)| \leq a_n$ *for all* $x \in S$ *and all* n. *Then* f_n *converges uniformly to* f *on* S.

Proof. One need only choose N so that $a_n < \varepsilon$ for all $n \geq N$. □

7.1.5 Proposition. *Let* f_n, $f : S \to \mathbb{R}$. *Then* f_n *converges uniformly to* f *on* S *iff*

$$\lim_n \left[f_n(b_n) - f(b_n)\right] = 0$$

for any sequence $\{b_n\}$ *in* S.

Proof. If f_n converges uniformly to f on S, choose N so that $|f_n(x) - f(x)| < \varepsilon$ for all $n \geq N$ and all $x \in S$. For such n, $|f_n(b_n) - f(b_n)| < \varepsilon$.

Conversely, suppose f_n does not converge uniformly to f on S. Then there exists an $\varepsilon > 0$ and points $b_n \in S$ such that $|f_n(b_n) - f(b_n)| \geq \varepsilon$ for infinitely many n. Thus the sequential condition fails. □

7.1.6 Examples. **(a)** The sequence $\{x^n\}$ converges pointwise to zero on $(-1, 1)$ but not uniformly there. (Take $b_n = 1/2^{1/n}$ in 7.1.5.) The convergence is uniform on intervals $[-r, r]$, $0 < r < 1$, since on such an interval $|x^n| \leq r^n$ and $r^n \to 0$.

(b) The sequence $\{n/x^n\}$ converges pointwise to zero on $(1, +\infty)$ but the convergence is not uniform there, as can be seen by taking $b_n = 2^{1/n}$ in 7.1.5. The convergence is uniform for $x \in [r, +\infty)$, $r > 1$, since then $|n/x^n| \leq n/r^n \to 0$.

(c) The sequence $\{x^n e^{-nx}\}$ converges uniformly to zero on $[0, +\infty)$ since $x^n e^{-nx} \leq e^{-n}$ for $x \geq 0$.

(d) The sequence $\{n^{-1} \sin nx\}$ converges uniformly to zero on $\mathbb{R}$ since $|n^{-1} \sin nx| \leq 1/n$ for all x.

(e) The sequence $\{\sin(x/n)\}$ converges pointwise to zero on $\mathbb{R}$, but the convergence is not uniform, as can be seen, for example, by taking $b_n = \pi n/2$ in 7.1.5. The convergence is uniform on bounded intervals $[a, b]$ since on this interval $|\sin(x/n)| \leq n^{-1}|x| \leq n^{-1} \max\{|a|, |b|\}$. ◊

There is an analog of 7.1.2 for uniform convergence; however, it is more restrictive and requires the notion of *uniform boundedness*.

7.1.7 Definition. A sequence of functions f_n is said to be *uniformly bounded on S with uniform bound M* if $|f_n(x)| \leq M$ for all $x \in S$ and all n. ◊

7.1.8 Proposition. *Let $f_n \to f$ pointwise on a set S.*

(a) *If $\{f_n\}$ is uniformly bounded on S, then f is bounded on S.*

(b) *If each f_n is bounded on S and $f_n \to f$ uniformly on S, then $\{f_n\}$ is uniformly bounded on S, hence f is bounded.*

(c) *If $f_n \to f$ uniformly on S and f is bounded, then $\{f_n\}_{n=N}^{\infty}$ is uniformly bounded for some N.*

Proof. (a) This follows by letting $n \to +\infty$ in the inequality $|f_n(x)| \leq M$.

(b) Choose N such that

$$|f_n(x) - f(x)| \leq 1 \quad \text{for all } n \geq N \text{ and } x \in S.$$

For such n and for all $x \in S$,

$$|f_n(x)| \leq |f_n(x) - f(x)| + |f(x) - f_N(x)| + |f_N(x)| \leq 2 + M_N,$$

where M_N is a bound for f_N on S. Since the functions $f_1, \ldots, f_{N-1}$ are bounded, $\{f_n\}_{n=1}^{\infty}$ is uniformly bounded.

(c) Let $|f(x)| \leq M$ for all x. Choose N such that $|f_n(x) - f(x)| \leq 1$ for all $n \geq N$ and $x \in S$. For such n, $|f_n(x)| \leq 1 + M$ for all $x \in S$. □

The sequence $\{f_n\}$ on $(0, 1)$ defined by

$$f_n(x) = \begin{cases} n & \text{if } 0 < x < 1/n, \\ 0 & \text{if otherwise} \end{cases}$$

shows that the first assertion in (b) may be false if the convergence is merely pointwise.

7.1.9 Theorem. *Let $f_n \to f$ and $g_n \to g$ uniformly on S and let h be uniformly continuous such that $h \circ f$ and $h \circ f_n$ are defined. Then*

(a) *$\alpha f_n + \beta g_n \to \alpha f + \beta g$ uniformly on S, α, $\beta \in \mathbb{R}$.*

(b) *$h \circ f_n \to h \circ f$ uniformly S.*

(c) *$f_n g_n \to fg$ uniformly on S if $\{f_n\}$ and $\{g_n\}$ are uniformly bounded on S.*

(d) *$\dfrac{1}{g_n} \to \dfrac{1}{g}$ uniformly on S if $\left\{\dfrac{1}{g_n}\right\}$ is uniformly bounded on S.*

Proof. The proof of (a) is left to the reader. To prove (b), choose $\delta > 0$ such that $|h(u) - h(v)| < \varepsilon$ for all u, v with $|u - v| < \delta$ and choose N such that $|f_n(x) - f(x)| < \delta$ for all $x \in S$ and $n \geq N$. For such n, $|h \circ f_n(x) - h \circ f(x)| < \varepsilon$.

For (c), let $M > 0$ be a common uniform bound for the sequences $\{|f_n|\}$ and $\{|g_n|\}$. Given $\varepsilon > 0$, choose N such that

$$|f_n(x) - f(x)| < \varepsilon/2M \quad \text{and} \quad |g_n(x) - g(x)| < \varepsilon/2M$$

for all $x \in S$ and $n \geq N$. For such n and x,

$$\begin{aligned}
|f_n(x)g_n(x) - f(x)g(x)| &\leq |f_n(x)g_n(x) - f(x)g_n(x)| + |f(x)g_n(x) - f(x)g(x)| \\
&= |g_n(x)|\,|f_n(x) - f(x)| + |f(x)|\,|g_n(x) - g(x)| \\
&\leq M|f_n(x) - f(x)| + M|g_n(x) - g(x)| < \varepsilon.
\end{aligned}$$

For (d), let $1/|g_n(x)| \leq M$ for all n and x. Then the same inequality holds for g, and

$$\left|\frac{1}{g_n(x)} - \frac{1}{g(x)}\right| = \frac{|g_n(x) - g(x)|}{|g_n(x)g(x)|} \leq \frac{1}{M^2}|g_n(x) - g(x)|,$$

from which the assertion follows. □

The hypothesis of uniform boundedness in parts (c) and (d) of the theorem cannot be relaxed (See Exercises 6 and 7)

There are versions of the Cauchy criterion for pointwise and uniform convergence of sequences of functions. For the pointwise version, consider a sequence of functions f_n on S such that $\lim_{m,n} |f_n(x) - f_m(x)| = 0$ for each $x \in S$. Then $\{f_n(x)\}_{n=1}^{\infty}$ is a Cauchy sequence of real numbers and hence converges to a unique real number $f(x)$. Thus $f_n \to f$ on S. Here is the analogous result for uniform convergence:

7.1.10 Uniform Cauchy Criterion. *A sequence of functions f_n converges uniformly on a set S iff for each $\varepsilon > 0$, there exists an index N such that*

$$|f_n(x) - f_m(x)| < \varepsilon \text{ for all } x \in S \text{ and all } m,\, n \geq N. \tag{7.1}$$

Proof. If $f_n \to f$ uniformly on S, then, given $\varepsilon > 0$, there exists an index N such that $|f_n(x) - f(x)| < \varepsilon/2$ for all $x \in S$ and all $n \geq N$. An application of the triangle inequality yields (7.1).

Conversely, assume that the condition holds. Then, in particular, $\lim_{m,n} |f_n(x) - f_m(x)| = 0$ for every $x \in S$, hence, by the observation preceding the theorem, there exists a function f such that $f_n \to f$ pointwise on S. We claim that the convergence is in fact uniform. To see this, let $\varepsilon > 0$ and choose N as in (7.1). Letting $m \to +\infty$ in that inequality then yields $|f_n(x) - f(x)| \leq \varepsilon$ for all $x \in S$ and all $n \geq N$, verifying the claim. □

7.1.11 Definition. Let S be an arbitrary set and let $f_n : S \to \mathbb{R}$. If the sequence $\{f_n(x)\}$ is increasing (decreasing) for each $x \in S$ and $f_n \to f$ on S, we write $f_n \uparrow f$ ($f_n \downarrow f$). In either case we say that $\{f_n\}$ is *monotone*. ◊

The following theorem gives general conditions under which pointwise convergence implies uniform convergence.

7.1.12 Dini's Theorem. *Let f and f_n be continuous on $[a, b]$ for each n and suppose that either $f_n \downarrow f$ or $f_n \uparrow f$ on $[a, b]$. Then $f_n \to f$ uniformly.*

Proof. We may assume that $f_n \downarrow f$. Let $g_n = f_n - f$, so $g_n \downarrow 0$. Suppose the assertion of the theorem is false. Then there exists an $\varepsilon > 0$, a subsequence $\{h_n\}$ of $\{g_n\}$, and a sequence $\{x_n\}$ in $[a, b]$ such that $h_n(x_n) \geq \varepsilon$ for all n. (Why?) By the Bolzano–Weierstrass theorem, there exists a subsequence $\{x_{n_k}\}$ converging to some $x \in [a, b]$. Since $h_n \downarrow$, for any fixed n and all sufficiently large k, $h_n(x_{n_k}) \geq h_{n_k}(x_{n_k})$, hence $h_n(x_{n_k}) \geq \varepsilon$. Letting $k \to +\infty$ in the last inequality yields $h_n(x) \geq \varepsilon$ for all n, contradicting that $h_n(x) \to 0$. □

The examples x^n on $[0, 1)$ and x^{-n} on $[2, +\infty)$ show that Dini's theorem is false if the interval is not closed and bounded. The decreasing sequence defined by

$$f_n(x) = \begin{cases} 1 & \text{if } 0 \leq x \leq 1, \\ 1 + n(1 - x) & \text{if } 1 \leq x \leq 1 + 1/n, \\ 0 & \text{if } 1 + 1/n \leq x \leq 2 \end{cases} \tag{7.2}$$

shows that continuity of the limit function in Dini's theorem is essential.

Exercises

1. Find the largest subset of $\mathbb{R}$ on which the given sequence converges pointwise, and determine the intervals on which the convergence is uniform.

   (a) $x^n(1-x)^n$ (b) $n^p x^n (1-x)$ (c) $e^{x/n}$

   (d) $\dfrac{nx^2}{e^{nx^2}}$ (e) $\dfrac{1}{1 + x^{2n}(1-x)^2}$ (f) $n^{1/2} \sin\left(\dfrac{x}{n^{2/3}}\right)$

   (g) $\dfrac{\sqrt{n}\, x^2}{1 + nx^2}$ (h) $\dfrac{nx^2}{1 + nx^2}$ (i) $\left(\dfrac{x}{2+x}\right)^n$

   (j) $\dfrac{x^{2n}}{2 + x^{2n}}$ (k) $\dfrac{1}{1 + |x|^n}$ (l) $\dfrac{n \sin x^2}{1 + nx^2}$.

2. Describe the convergence behavior of the following sequences on $[0, 1]$:

   (a) $\dfrac{1}{nx+1}$ (b) $\dfrac{x}{nx+1}$ (c) $\dfrac{nx}{n^2 x + 1}$ (d) $\dfrac{nx}{n^2 x^2 + 1}$.

3. Describe the convergence behavior of the sequences on $(0, 1)$:

   (a) $\{x^{1/n}\}$ (b) $\{x^{1+1/n}\}$ (c) $\{x^{-1/n}\}$ (d) $\{x^{1-1/n}\}$.

4. Show directly that the sequence defined in (7.2) does not converge uniformly.

5. Let p, $q > 0$. Prove that the sequence of functions $\dfrac{x^p}{n + x^q}$ converges uniformly to zero on $[0, +\infty)$ iff $p < q$.

6. Give an example of sequences $\{f_n\}$, $\{g_n\}$ and functions f, g such that $f_n \to f$ and $g_n \to g$ uniformly, and $f_n g_n \to fg$ pointwise but not uniformly.

7. Give an example of a sequence $\{g_n\}$ and a function g such that $g_n \to g$ uniformly and $1/g_n \to 1/g$ pointwise but not uniformly.

8. Let $-\infty < a < b \le +\infty$. Suppose that $f_n \to f$ uniformly on $[a, r]$ for every $r \in (a, b)$. Prove that $f_n \to f$ uniformly on $[a, b)$ iff for each sequence $\{b_n\}$ with $b_n \uparrow b$, $f_n(b_n) - f(b_n) \to 0$.

9. Let f_n be bounded for each n and let $f_n \to f$ uniformly on a set S. Prove that $\sup_S f_n \to \sup_S f$ and $\inf_S f_n \to \inf_S f$.

10. Let f be uniformly continuous on $\mathbb{R}$ and $a_n \to a$. Set $f_n(x) = f(x + a_n)$. Show that $\{f_n\}$ converges uniformly on $\mathbb{R}$.

11. Let f_n be continuous on $[a, b]$ for each n and let f_n converge uniformly on $(a, b) \cap \mathbb{Q}$. Prove that f_n converges uniformly on $[a, b]$.

12. Prove: If $f_n \to f$ uniformly on each of the sets $S_1, \ldots, S_m$, then $f_n \to f$ uniformly on $S_1 \cup \cdots \cup S_m$. Show that the corresponding statement for a union of infinitely many sets is false.

13. For $x \in [0, 1]$ define

$$f_n(x) = \begin{cases} 1 & \text{if } x \in \mathbb{Q} \text{ and } x = k/m \text{ in reduced form with } m \le n, \\ 0 & \text{otherwise.} \end{cases}$$

Show that $\{f_n\}$ converges pointwise but not uniformly to the Dirichlet function.

14. Let $p \in \mathbb{N}$. For $x \in [0, 1]$ define

$$g_n(x) = \begin{cases} (m + 1/n)^p & \text{if } x \in \mathbb{Q},\ x = k/m \text{ in reduced form} \\ 0 & \text{if } x \text{ is irrational.} \end{cases}$$

Show that g_n converges uniformly on $[0, 1]$ iff $p = 1$.

15. Let $\{f_n\}$ be uniformly bounded, let f, g be bounded on $[0, 1]$, and suppose that $f_n \to f$ pointwise (uniformly) on $[r, 1]$ for each $0 < r < 1$. If g is continuous at 0 and $g(0) = 0$, prove that $f_n g \to fg$ pointwise (uniformly) on $[0, 1]$.

16. Let $\{f_n\}$ be uniformly bounded and $f_n \to f$ uniformly on S.
    (a) Prove that $(f_1 + f_2 + \cdots + f_n)/n \to f$ uniformly on S.
    (b) Suppose for some $r > 0$ that $f_n(x) \geq r$ for all n and all $x \in S$. Prove that $(f_1 f_2 \cdots f_n)^{1/n} \to f$ uniformly on S.

17. Let f_0 be a bounded function on a set S and $0 < r < 1$. Define a sequence $\{f_n\}$ recursively by

$$f_n(x) = \sin\big(r f_{n-1}(x)\big), \quad x \in S, \quad n \geq 1.$$

Prove that $\{f_n\}$ converges uniformly on S. Show that a similar result holds if S is an interval and $\sin x$ is replaced by any function g such that $\sup_x |g'(x)| < 1/r$, where r is any positive number.

18. Let g and h be positive and continuous on $[a, b]$ and define

$$f_n(x) := \frac{ng(x)}{1 + n^2 h(x)}.$$

Prove that the following convergence is uniform on $[a, b]$:

(a) $n \sin f_n \to \dfrac{g}{h}$ (b) $n\big(1 - \cos f_n\big) \to 0$ (c) $n^2\big(1 - \cos f_n\big) \to \dfrac{g^2}{2h^2}$.

7.2 Properties of the Limit Function

The theorems in this section give conditions under which the properties of continuity, integrability, or differentiability of functions in a sequence are passed along to the limit function. We shall see that pointwise convergence is generally insufficient for this—the stronger property of uniform convergence is needed.

The following theorem asserts that under suitable conditions two limit processes may be interchanged. It is one of several such results to be found in the book.

7.2.1 Interchange of Limits. *Let $f_n \to f$ uniformly on a subset E of $\mathbb{R}$ and let a be an accumulation point of E such that $L_n := \lim_{\{x \to a,\, x \in E\}} f_n(x)$ exists in $\mathbb{R}$ for each n. Then $L := \lim_n L_n$ exists in $\mathbb{R}$ and $\lim_{\{x \to a,\, x \in E\}} f(x) = L$. In other words, the equality*

$$\lim_n \lim_{\substack{x \to a \\ x \in E}} f_n(x) = \lim_{\substack{x \to a \\ x \in E}} \lim_n f_n(x)$$

holds provided that each inner limit exists in $\mathbb{R}$ and the convergence in the inner limit on the right is uniform.

Proof. Given ε, for each n choose $\delta_n > 0$ such that

$$|f_n(x) - L_n| < \varepsilon/3 \quad \text{for all } x \in E \text{ with } |x - a| < \delta_n.$$

Next, choose $N \in \mathbb{N}$ such that

$$|f_n(x) - f(x)| < \varepsilon/6 \quad \text{for all } x \in E \text{ and all } n \geq N.$$

For n, $m \geq N$, choose $x \in E$ such that $|x - a| < \min\{\delta_n, \delta_m\}$. Then

$$|L_n - L_m| \leq |L_n - f_n(x)| + |f_n(x) - f_m(x)| + |L_m - f_m(x)| < \varepsilon.$$

This shows that $\{L_n\}$ is a Cauchy sequence and hence converges to some $L \in \mathbb{R}$. Let $n \geq N$ be sufficiently large so that $|L_n - L| < \varepsilon/6$. If $x \in E$ and $|x - a| < \delta_n$, then

$$|f(x) - L| \leq |f(x) - f_n(x)| + |f_n(x) - L_n| + |L_n - L| < \varepsilon/6 + \varepsilon/3 + \varepsilon/6 < \varepsilon.$$

Therefore, $\lim_{\{x \to a,\, x \in E\}} f(x) = L$. □

7.2.2 Corollary. *If $f_n \to f$ uniformly on an interval I and if each f_n is continuous at some $a \in I$, then f is continuous at a.*

Proof. Take $L_n = f_n(a)$ in the theorem. □

The corollary is false if the convergence is only pointwise. For example, the sequence of continuous functions x^n converges pointwise on $[0, 1]$ to a function that is discontinuous at $x = 1$.

7.2.3 Theorem. *If $f_n \to f$ uniformly on $[a, b]$ and $f_n \in \mathcal{R}_a^b$ for all n, then $f \in \mathcal{R}_a^b$ and*

$$\lim_n \int_a^b f_n(t)\, dt = \int_a^b f(t)\, dt. \tag{7.3}$$

Proof. By 7.1.8, f is bounded. By uniform convergence, given $\varepsilon > 0$, there exists an N such that

$$f_n(x) - \frac{\varepsilon}{4(b-a)} < f(x) < f_n(x) + \frac{\varepsilon}{4(b-a)}$$

for all $x \in [a, b]$ and $n \geq N$. It follows that for fixed $n \geq N$ and any partition $\mathcal{P}$,

$$\underline{S}(f_n, \mathcal{P}) - \frac{\varepsilon}{4} \leq \underline{S}(f, \mathcal{P}) \leq \overline{S}(f, \mathcal{P}) \leq \overline{S}(f_n, \mathcal{P}) + \frac{\varepsilon}{4},$$

hence

$$\overline{S}(f, \mathcal{P}) - \underline{S}(f, \mathcal{P}) \leq \overline{S}(f_n, \mathcal{P}) - \underline{S}(f_n, \mathcal{P}) + \frac{\varepsilon}{2}.$$

Since f_n is integrable, $\mathcal{P}$ may be chosen so that the right side of this inequality is less than ε. Therefore, f is integrable.

Since $|f_n(t) - f(t)| < \varepsilon/4(b-a)$ for $n \geq N$ and all t,

$$\left| \int_a^b f_n(t)\, dt - \int_a^b f(t)\, dt \right| \leq \int_a^b |f_n(t) - f(t)|\, dt \leq \frac{\varepsilon}{4},$$

which shows that $\int_a^b f_n \to \int_a^b f$. □

The following examples show that the hypothesis of uniform convergence in 7.2.3 cannot be relaxed.

7.2.4 Example. Define $f_n : [0, \pi] \mapsto \mathbb{R}$ by

$$f_n(x) = \begin{cases} n\sin(nx) & \text{if } 0 \le x \le \pi/n, \\ 0 & \text{if } \pi/n \le x \le \pi. \end{cases}$$

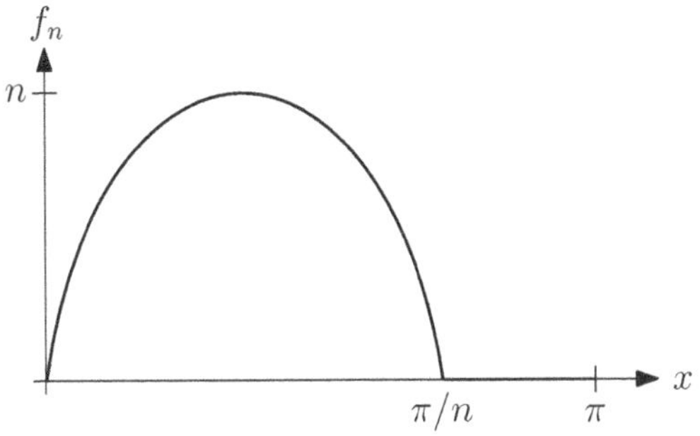

FIGURE 7.2: Pointwise convergence insufficient.

Each f_n is continuous and $\{f_n\}$ converges pointwise on $[0, \pi]$ to the zero function, yet $\int_0^\pi f_n = 2$ for all n. ♢

7.2.5 Example. Let $r_1, r_2, \ldots$ be an enumeration of the rationals in $[0, 1]$ and let

$$f_n(x) = \begin{cases} 1 & \text{if } x \in \{r_1, \ldots, r_n\}, \\ 0 & \text{otherwise.} \end{cases}$$

Then f_n is integrable with zero integral and f_n converges pointwise to the Dirichlet function, which is not Riemann integrable. ♢

In the two examples above, either the sequence was not uniformly bounded or the limit function was not integrable. It will follow from results in Chapter 11 that if $\{f_n\}$ is uniformly bounded, f_n, $f \in \mathcal{R}_a^b$, and $f_n \to f$ merely pointwise on $[a, b]$, then (7.3) holds.

Here is the derivative analog of the preceding theorems. The proof is somewhat more involved.

7.2.6 Theorem. *Let f_n be differentiable on (a, b) for each n and let $\{f_n'\}$ converge uniformly on (a, b). If $\{f_n(x_0)\}$ converges for some $x_0 \in (a, b)$, then $\{f_n\}$ converges uniformly to a differentiable function f on (a, b) and $f_n' \to f'$ on (a, b).*

Proof. Given $\varepsilon > 0$, choose N such that, for all m, $n \ge N$ and $x \in (a, b)$,

$$|f_n(x_0) - f_m(x_0)| < \frac{\varepsilon}{2} \text{ and } |f_n'(x) - f_m'(x)| < \frac{\varepsilon}{2(b-a)}.$$

Fix $m,\, n \geq N$. By the mean value theorem applied to $f_n - f_m$, for each pair $x, y \in (a,b)$, there exists $\xi_{m,n} \in (a,b)$ such that

$$\begin{aligned} \left|(f_n(x) - f_m(x)) - (f_n(y) - f_m(y))\right| &= |f_n'(\xi_{m,n}) - f_m'(\xi_{m,n})||x-y| \\ &\leq \frac{\varepsilon|x-y|}{2(b-a)} \leq \frac{\varepsilon}{2}. \end{aligned} \tag{7.4}$$

In particular, for all $x \in (a,b)$,

$$\begin{aligned} &|f_n(x) - f_m(x)| \\ &\qquad \leq \left|(f_n(x) - f_m(x)) - (f_n(x_0) - f_m(x_0))\right| + |f_n(x_0) - f_m(x_0)| \\ &\qquad < \varepsilon/2 + \varepsilon/2 = \varepsilon. \end{aligned}$$

By the uniform Cauchy criterion, $\{f_n\}$ converges uniformly on (a,b) to some function f. Also, from (7.4), for fixed y and for all $x \neq y$,

$$\left|\frac{f_n(x) - f_n(y)}{x-y} - \frac{f_m(x) - f_m(y)}{x-y}\right| \leq \frac{\varepsilon}{2(b-a)}.$$

Therefore, the sequence of functions $[f_n(x) - f_n(y)]/(x-y)$ converges uniformly in x on the set $E_y := (a,y) \cup (y,b)$. Since f_n converges to f,

$$\frac{f_n(x) - f_n(y)}{x-y} \to \frac{f(x) - f(y)}{x-y} \quad \text{uniformly in } x \text{ on } E_y.$$

By 7.2.1 with $E = E_y$,

$$\lim_n f_n'(y). = \lim_n \lim_{x \to y} \frac{f_n(x) - f_n(y)}{x-y} = \lim_{x\to y} \frac{f(x) - f(y)}{x-y} = f'(y). \qquad \square$$

The sequence given by $f_n(x) = x^n/n$, $0 < x < 1$ shows that uniform convergence of a sequence of functions does not guarantee that the derivatives converge uniformly.

Exercises

1. Prove: If $f_n \to f$ uniformly on an interval I and each f_n is continuous at $a \in I$, then, for any sequence $\{a_n\}$ in I with $a_n \to a$, $\lim_n f_n(a_n) = f(a)$.

2. Show that if $f_n \to f$ uniformly on a subset E of $\mathbb{R}$ and each f_n is uniformly continuous on E, then f is uniformly continuous on E.

3. Prove that $(1 + x/n)^n \to e^x$ uniformly on any bounded interval of $\mathbb{R}$. Conclude that $\int_0^1 (1 + x/n)^n \, dx \to e - 1$.

4. Show that $n^2 x e^{-nx} \to 0$ for all $x \geq 0$, yet $\int_0^1 n^2 x e^{-nx} \, dx \not\to 0$. Why does this not contradict 7.2.3?

5. Evaluate $\lim_n \int_0^1 f_n$ if $f_n(x) =$

(a) $\dfrac{1}{\cos(x/n)}$ (b) $\dfrac{x}{n\sin(x/n)}$ (c) $\dfrac{n(e^{x/n}-1)}{x}$

(d) $\dfrac{\sqrt{n}\left(e^{-x/n}-1\right)}{x}$ (e) $\arctan\left(\dfrac{ax(x+1)n+1}{nx+1}\right)$, $a > 0$.

6. Prove that $f_n(x) := \sqrt{n}/(1+n^2x^2)$ converges to 0 pointwise on $(0,+\infty)$, uniformly on $[r,+\infty)$ for every $r > 0$, but not uniformly on $(0,1)$. Show that, nonetheless, $\int_0^1 f_n \to 0$.

7. Let $\{a_n\}$ be a positive, strictly increasing sequence. Prove that

$$\lim_n \int_0^1 \frac{a_n x}{1+a_n x}\,dx = \int_0^1 \lim_n \frac{a_n x}{1+a_n x}\,dx.$$

8. Let f and f' be positive and continuous on $[a,b]$. Define

$$f_n(x) := \frac{nf'(x)}{1+n^2f(x)} \quad \text{and} \quad g_n(x) := \frac{2n\sqrt{f'(x)}}{1+n^2f(x)}.$$

Use Exercise 7.1.18 to find

(a) $\lim_n \int_a^b n\sin f_n$ (b) $\lim_n \int_a^b n(1-\cos f_n)$ (c) $\lim_n \int_a^b n(1-\cos g_n)$.

9. Show that if $f_n \to f$ uniformly on $[a,b]$ and f_n is integrable for each n then

$$\int_a^x f_n(t)\,dt \to \int_a^x f(t)\,dt$$

uniformly in x on $[a,b]$.

10. Suppose that f_n is improperly integrable on $[a,c)$, $f_n \to f$ uniformly on $[a,t]$ for all $t \in [a,c)$, and $|f_n| \le g$ on $[a,c)$ for all n, where g is improperly integrable on $[a,c)$. Prove that f is improperly integrable on $[a,c)$ and

$$\lim_n \int_a^c f_n = \int_a^c f.$$

11. Prove that if f is continuous on $[0,1]$, then

$$\lim_n \int_0^1 f(x^n)\,dx = f(0).$$

12. For each n, let f_n be continuous on $[a,+\infty)$, $a > 0$, and suppose that $c_n := \lim_{x\to+\infty} f_n(x)$ exists in $\mathbb{R}$. Prove that if $f_n \to f$ uniformly on

$[a, +\infty)$, then $\lim_n c_n$ and $\lim_{x\to+\infty} f(x)$ exist and are equal. Show also that

$$\lim_n \int_0^{1/a} f_n(x)\,dx = \int_0^{1/a} f(x)\,dx.$$

Hint. Let $g_n(x) = f_n(1/x)$, $0 < x \le 1/a$ and apply 7.2.1.

13. Let f_n be as in 7.2.4 and define

$$g_n(x) = \int_0^x f_n(t)\,dt, \quad h_n(x) = xg_n(x), \quad 0 \le x \le \pi.$$

Show that

(a) $\{g_n\}$ converges pointwise and monotonically on $[0, \pi]$ but not uniformly.

(b) $\{h_n\}$ converges uniformly on $[0, \pi]$.

(c) $\{h_n'\}$ does not converge uniformly on $[0, \pi]$.

7.3 Convergence of Series of Functions

As in the case of numerical series, the material on sequences of functions readily extends to series of functions.

7.3.1 Definition. Let $\{f_n\}$ be a sequence of real-valued functions on a set S. For each $x \in S$ and $n \in \mathbb{N}$, form the nth *partial sums*

$$s_n(x) = \sum_{k=1}^{n} f_n(x) \text{ and } t_n(x) = \sum_{k=1}^{n} |f_n(x)|.$$

The infinite series of functions $\sum_n f_n = \sum_{n=1}^{\infty} f_n$ is said to *converge*

- *pointwise on S* if $\sum_n f_n(x)$ converges for each $x \in S$;
- *absolutely pointwise on S* if $\sum_n f_n(x)$ converges absolutely for every $x \in S$;
- *uniformly on S* if $\{s_n\}$ converges uniformly on S;
- *absolutely uniformly on S* if $\{t_n\}$ converges uniformly on S. ◊

The methods of Chapter 6 may be applied at each x to test pointwise convergence of a series of functions. For uniform convergence, additional tests are required.

The following result is an immediate consequence of 7.1.9.

7.3.2 Theorem. *Let $\sum_n f_n$ and $\sum_n g_n$ converge uniformly on a set S and let α, $\beta \in \mathbb{R}$. Then $\sum_n(\alpha f_n + \beta g_n)$ converges uniformly on S and*

$$\sum_n(\alpha f_n + \beta g_n) = \alpha \sum_n f_n + \beta \sum_n g_n.$$

The next theorem is a useful for testing nonuniform convergence of a series. The proof is immediate from the identity $f_n = s_n - s_{n-1}$.

7.3.3 Theorem. *If $\sum_n f_n$ converges uniformly on a set S, then $f_n \to 0$ uniformly on S.*

For example, the geometric series

$$\sum_{n=0}^{\infty} x^n = \frac{1}{1-x}, \quad |x| < 1, \tag{7.5}$$

converges pointwise but not uniformly on $(-1, 1)$, since x^n does not tend to zero uniformly on $(-1, 1)$. We show below that the series converges uniformly on all closed subintervals of $(-1, 1)$.

The comparison test for uniform convergence of a series of functions takes the following form:

7.3.4 Uniform Comparison Test. *If $|f_n(x)| \le g_n(x)$ for all n and all $x \in S$ and if $\sum_n g_n$ converges uniformly on S, then $\sum_n f_n$ converges absolutely uniformly on S.*

Proof. Since $\sum_{k=m}^n |f_n(x)| \le \sum_{k=m}^n g_n(x)$, the assertion follows from the uniform Cauchy criterion. □

7.3.5 Corollary. *If $\sum_n f_n$ converges absolutely uniformly on a set S, then $\sum_n f_n$ converges uniformly on S.*

Proof. $0 \le f_n + |f_n| \le 2|f_n|$, hence, by 7.3.4, $\sum_n(f_n + |f_n|)$ converges uniformly on S and therefore so must

$$\sum_n f_n = \sum_n (f_n + |f_n|) - \sum_n |f_n|.$$ □

7.3.6 Weierstrass *M*-test. *If there exist positive constants M_n such that $\sum_n M_n < +\infty$ and $|f_n| \le M_n$ on S for all n, then $\sum_n f_n$ converges absolutely uniformly on S.*

Proof. Take g_n to be the constant function M_n in 7.3.4. □

For example, taking $M_n = r^n$, we see that the geometric series (7.5) converges uniformly in every interval $[-r, r]$, $0 < r < 1$.

The next results are uniform convergence analogs of Dirichlet's and Abel's tests for numerical series.

7.3.7 Theorem. *If $\sum_n f_n$ converges uniformly on a set S and if there exists a constant M such that*

$$|g_1(x)| + \sum_{n=1}^{\infty} |g_{n+1}(x) - g_n(x)| \le M \quad \textit{for all } x \in S,$$

then $\sum_n f_n g_n$ converges uniformly on S.

Proof. Let

$$s_k = \sum_{j=1}^{k} f_j - \sum_{n=1}^{\infty} f_n \quad \text{and} \quad t_n = \sum_{k=1}^{n} f_k g_k.$$

Then $s_k - s_{k-1} = f_k$, hence by **??** with $a_k = f_k(x)$ and $b_k = g_k(x)$, for $m > n > N$ and $x \in S$,

$$\begin{aligned} t_m(x) - t_{n-1}(x) &= \sum_{k=n}^{m} f_k(x) g_k(x) \\ &= \sum_{k=n}^{m-1} s_k(x)(g_k(x) - g_{k+1}(x)) + s_m(x) g_m(x) - s_{n-1}(x) g_n(x). \end{aligned}$$

Since for each $n > 1$

$$g_n = \sum_{k=1}^{n-1} (g_{k+1} - g_k) + g_1,$$

by hypothesis $|g_n| \le M$ on S. Given $\varepsilon > 0$, choose N so that $|s_n(x)| < \varepsilon$ for all $n, m \ge N$ and $x \in S$. For such m and n,

$$\begin{aligned} &|t_m(x) - t_{n-1}(x)| \\ &\le \sum_{k=n}^{m} |s_k(x)|\,|g_k(x) - g_{k+1}(x)| + |s_m(x)|\,|g_m(x)| + |s_{n-1}(x)|\,|g_n(x)| \qquad (7.6) \\ &\le M\varepsilon + M\varepsilon + M\varepsilon = 3M\varepsilon. \end{aligned}$$

Therefore, $\{t_n\}$ is uniformly Cauchy on S and hence converges uniformly. □

7.3.8 Theorem. *If, on a set S, the partial sums of $\sum_n f_n$ are uniformly bounded, $\sum_n |g_{n+1} - g_n|$ converges uniformly, and $g_n \to 0$ uniformly, then $\sum_n f_n g_n$ converges uniformly on S.*

Proof. Let t_n be as in the proof of 7.3.7, $s_n := \sum_{k=1}^{n} f_k$, and let M be a uniform bound for $\{s_n\}$ on S. Given $\varepsilon > 0$, choose N such that

$$|g_n(x)| < \varepsilon \;\text{ and }\; \sum_{k=n}^{m} |g_k(x) - g_{k+1}(x)| < \varepsilon, \quad m > n > N,\ x \in S. \qquad (7.7)$$

Since $s_n - s_{n-1} = f_n$, (7.6) holds in the current setting, hence (7.7) implies that

$$|t_m(x) - t_{n-1}(x)| \le 3M\varepsilon, \quad m > n > N,\ x \in S.$$

Therefore, $\{t_n\}$ converges uniformly on S. □

7.3.9 Corollary. *If the partial sums of $\sum_n f_n$ are uniformly bounded and if $g_n \downarrow 0$ or $g_n \uparrow 0$ uniformly on S, then $\sum_n f_n g_n$ converges uniformly on S.*

Proof. Assume that $\{g_n\}$ is decreasing. Then

$$\sum_{k=1}^{n} |g_{k+1} - g_k| = \sum_{k=1}^{n} (g_k - g_{k+1}) = g_1 - g_{n+1},$$

hence $\sum_{n=1}^{\infty} |g_{n+1} - g_n|$ converges uniformly. □

7.3.10 Example. Let g_n be continuous and $g_n \downarrow 0$ or $g_n \uparrow 0$ on $\mathbb{R}$. We apply the preceding corollary to the series

$$s(x) := \sum_n g_n(x) \sin nx$$

on closed bounded intervals I not containing any integer multiple of 2π.

By Dini's theorem, $g_n \to 0$ uniformly on I. Also, by Example 6.4.5, $s(x)$ converges pointwise on $\mathbb{R}$. Moreover, if x is not a multiple of 2π, then, by the same example,

$$\left|\sum_{k=1}^{n} \sin(kx)\right| \le \left|\frac{1}{\sin(x/2)}\right|.$$

Since $\inf_I |\sin(x/2)| > 0$, the sums $\sum_{k=1}^{n} \sin(kx)$ are uniformly bounded on I. By 7.3.9, $s(x)$ converges uniformly on I.

By 7.3.8, the same result holds if, instead of monotonicity of the sequence $\{g_n\}$, we require that $\sum_n |g_{n+1} - g_n|$ converges and $g_n \to 0$, both uniformly on I. Analogous results hold for series of the form $\sum_n g_n(x) \cos nx$. ◊

7.3.11 Uniform Alternating Series Test. *If $g_n \downarrow 0$ or $g_n \uparrow 0$ uniformly on a set S, then $\sum_{n=1}^{\infty} (-1)^{n+1} g_n$ converges uniformly on S.*

Proof. Take $f_n = (-1)^{n+1}$ in 7.3.9. □

7.3.12 Example. Let f be continuous on $\mathbb{R}$ and monotone in some neighborhood $\mathcal{N}$ of 0 with $f(0) = 0$. If $a_n \downarrow 0$, then the series

$$\sum_{n=1}^{\infty} (-1)^{n+1} f(a_n x)$$

converges uniformly on any closed, bounded interval I.

We verify this for the case $I \subseteq [0, +\infty)$ and f increasing. Choose N so that $a_n x \in \mathcal{N}$ for all $n \ge N$ and $x \in I$. Then $f(a_n x) \downarrow 0$ on I, hence, by Dini's theorem and 7.3.11, $\sum_{n=1}^{\infty} (-1)^n f(a_n x)$ converges uniformly on I.

For example, taking $a_n = 1/n$ we see that the three series

$$\sum_{n=1}^{\infty} (-1)^{n+1} \sin(x/n), \quad \sum_{n=1}^{\infty} (-1)^{n+1} n^{-1} x e^{x/n}, \quad \text{and} \quad \sum_{n=1}^{\infty} (-1)^{n+1} [1 - e^{-n^{-2} x^2}]$$

converge uniformly on closed bounded intervals. ◊

The following theorem is an immediate consequence of 7.2.2, 7.2.3, and 7.2.6 applied to the sequence of partial sums of the series.

7.3.13 Theorem. *Let* $f_n : [a, b] \to \mathbb{R}$ *and* $s := \sum_n f_n$.

(a) *If* s *converges uniformly on* $[a, b]$ *and each* f_n *is continuous, then* s *is continuous.*

(b) *If* s *converges uniformly on* $[a, b]$ *and* $f_n \in \mathcal{R}_a^b$ *for all* n, *then* $s \in \mathcal{R}_a^b$ *and* $\int_a^b s = \sum_n \int_a^b f_n$.

(c) *Let* f_n *be differentiable on* (a, b) *and suppose that the* derived series $\sum_n f_n'$ *converges uniformly on* (a, b) *and that* $\sum_n f_n(x_0)$ *converges for some* $x_0 \in (a, b)$. *Then* s *converges uniformly on* (a, b) *and* $s' = \sum_n f_n'$.

7.3.14 Example. (a) By 7.3.10, $\sum_n n^{-1} \sin(nx)$ converges uniformly on intervals $[a, b] \subseteq (0, 2\pi)$, hence

$$\int_a^b \sum_n n^{-1} \sin(nx)\, dx = \sum_n \int_a^b n^{-1} \sin(nx)\, dx = \sum_n \frac{\cos(na) - \cos(nb)}{n^2}.$$

On the other hand, the derived series $\sum_n \cos(nx)$ does not converge.

(b) Both $s(x) := \sum_n n^{-1} \sin(x/n)$ and its derived series $\sum_n n^{-2} \cos(x/n)$ converge uniformly on $\mathbb{R}$, hence the latter equals $s'(x)$. ◊

A *closed form* for a series $s := \sum_n f_n$ on a subset E of $\mathbb{R}$ is a "standard function" that equals s on E. Closed forms are typically combinations of rational, power, exponential, logarithmic, trigonometric, or inverse trigonometric functions.

7.3.15 Example. Since $1/(1-x)$ is a closed form for the geometric series (7.5) on $(-1, 1)$, the function

$$\frac{1}{1 - \dfrac{1}{2 + \sin x}} = \frac{2 + \sin x}{1 + \sin x}$$

is a closed form for the series $\displaystyle\sum_{n=0}^{\infty} \left(\frac{1}{2 + \sin x} \right)^n$ on intervals I not containing $(4n-1)\pi/2$ or $-(4n+1)\pi/2$, $n = 0, 1, 2, \ldots$. By the Weierstrass M-test, the series converges absolutely uniformly on closed subintervals of I, since on such subintervals $0 < 1/(2 + \sin x) < 1/(1 + \varepsilon)$ for some $\varepsilon > 0$. ◊

Exercises

1. For the functions f_n below, determine all subintervals of $[0, +\infty)$ on which $\sum_{n=0}^{\infty} f_n(x)$ converges pointwise or uniformly, where $p \in \mathbb{N}$.

   (a) $\dfrac{1}{1+x^n}$ (b) $\dfrac{x^n}{1+x^n}$ (c) $\dfrac{x}{n^2+x}$

   (d) $\dfrac{x}{1+n^2x}$ (e) $\left(\dfrac{x}{x-2}\right)^n$ (f) $\dfrac{\sin(nx)}{1+n^2x^2}$

   (g) $n^p e^{-nx}$ (h) n^{-x} (i) $\sin(x/n^p)$

   (j) $x^n(1-x)^n$ (k) $\left(\dfrac{1-x}{1+x}\right)^n$ (l) $x^n e^{-nx}$.

2. Find the largest intervals of pointwise convergence and uniform convergence and a closed form for the series $\sum_{n=0}^{\infty} f_n(x)$, where $f_n(x) =$

   (a) $\cos^n(\pi x/2)$, $x \in [0,1]$ (b) $\ln^n(1/x)$ (c) $\dfrac{(-1)^n}{e^{nx}}$ (d) $(x^2 \ln x)^n$.

3. Prove that if $\sum_n f_n$ converges absolutely uniformly on a set S, then $\sum_n f_n^+$ and $\sum_n f_n^-$ converge uniformly on set S, where, for each $x \in S$, $f_n^+(x)$ and $f_n^-(x)$ are, respectively, the positive and negative parts of $f_n(x)$.

4. Suppose that the numerical series $\sum_n a_n$ converges absolutely. Let

   $$s(t) = \sum_n a_n \sin\left[(2n+1)t\right] \quad \text{and} \quad c(t) = \sum_n a_n \cos(nt).$$

   Find series expansions for

   $$\int_x^{\pi/2} s(t)\,dt \quad \text{and} \quad \int_0^x c(t)\,dt.$$

5. Let $p > 0$ and $s(x) = \sum_{n=1}^{\infty} \sin(x/n^p)$. Prove:

   (a) If $p \le 1$, then $s(x)$ diverges for all $x \neq 0$.

   (b) If $p > 1$, then $s(x)$ converges absolutely uniformly on bounded intervals, (hence pointwise on $\mathbb{R}$) but not uniformly on $\mathbb{R}$.

6. Let $p > 0$ and $s(x) = \sum_{n=1}^{\infty} [1 - \cos(x/n^p)]$. Prove:

   (a) If $p \le 1/2$, then $s(x)$ diverges for all $x \neq 0$.

   (b) If $p > 1/2$, then $s(x)$ converges absolutely uniformly on bounded intervals, (hence pointwise on $\mathbb{R}$) but not uniformly on $\mathbb{R}$.

7. Let $f(x)$ be bounded on $[0,1]$ and

$$t(x) := \sum_{n=0}^{\infty} x^n f(x), \quad x \in [0,1].$$

(a) Prove that $t(x)$ converges pointwise on $[0,1)$ and uniformly on $[0,r]$ for $0 < r < 1$.

(b) Prove that if $f(1) \neq 0$, then the convergence of $t(x)$ is not uniform on $[0,1)$.

(c) Suppose that $L := \lim_{x\to 1^-}(1-x)^{-1}f(x)$ exists. Prove that the convergence of $t(x)$ is uniform on $[0,1)$ iff $L = 0$.

(d) Let $m \in \mathbb{N}$. Determine whether the convergence of $t(x)$ is uniform on $[0,1)$ for $f(x) =$

(i) $(1-x)^m$. (ii) $1 - x^m$. (iii) $1 - \sin(\pi x/2)$. (iv) $\cos(\pi x/2)$.

8. (Uniform limit comparison test). Let $f_n \geq 0$ and $g_n > 0$ on a set S and let $f_n/g_n \to h$ uniformly on S, where $h : S \to \mathbb{R}$ satisfies

$$0 < \inf_S h \leq \sup_S h < +\infty.$$

Prove that $\sum_n f_n$ converges uniformly on S iff $\sum_n g_n$ converges uniformly on S.

9. Suppose that f' exists, is bounded on $I := (-r,r)$, and $f(0) = 0$. Prove that the series

$$s(x) := \sum_{n=0}^{\infty} \frac{1}{n} f\left(\frac{x}{n+1}\right)$$

converges uniformly on I and that $s'(0) = f'(0)$.

10. Suppose that $|f(x)| \leq |x|$ on $I = (-r,r)$, $r > 0$. If f is differentiable on I and f' is continuous at 0, show that the series $s(x)$ in Exercise 9 converges uniformly on I and that $|s'(0)| \leq 1$.

11. Let $f_n(x)$ be continuous and nonnegative on $[a,b]$. Prove that if $\sum_n f_n$ converges pointwise on $[a,b]$ to a continuous function, then the convergence is uniform.

12. Let $\{a_n\}$ be a sequence such that $\sum_{n=1}^{\infty} a_n^{-1}$ converges absolutely. Prove that $\sum_{n=1}^{\infty} |x-a_n|^{-1}$ converges uniformly on closed and bounded intervals not containing any a_n.

13. Suppose that f_n is monotone on $[a,b]$ for each n. Prove that if $\sum_n f_n(a)$ and $\sum_n f_n(b)$ converge absolutely, then $\sum_n f_n \in \mathcal{R}_a^b$ and

$$\int_a^b \sum_n f_n = \sum_n \int_a^b f_n.$$

14. Let $\sum_n f_n$ converge uniformly on S and let $\{g_n\}$ be a uniformly bounded sequence of functions on a set S such that either $\{g_n\}$ is monotone increasing or monotone decreasing on S. Prove that $\sum_n f_n g_n$ converges uniformly on S.

15. Suppose that the partial sums of $\sum_n f_n$ are uniformly bounded on $I = [a, b]$, g_n is continuous for each n, and $g_n \downarrow 0$ or $g_n \uparrow 0$ on I. Prove that $\sum_n f_n g_n$ converges uniformly on I.

16. Suppose that $\sum_n f_n$ converges uniformly on $I = [a, b]$, g_n is continuous for each n, and $g_n \downarrow g$ or $g_n \uparrow g$ on I, where g is continuous. Prove that $\sum_n f_n g_n$ converges uniformly on I.

17. Suppose that g_n is continuous on $I = [a, b]$ for each n, $\{g_n\}$ is monotone, and $s(x) := \sum_n (-1)^n g_n(x)$ converges for each $x \in I$. Prove that $s(x)$ is continuous on I.

18. Let g be continuous and nonnegative on $\mathbb{R}$. Prove that the series

$$s(x) := \sum_{n=1}^{\infty} (-1)^n \frac{g(x) + n}{n^2}$$

converges uniformly on bounded intervals, hence pointwise on $\mathbb{R}$, but does not converge absolutely for any x.

19. Let g_n be continuous and $g_n \downarrow 0$ on $\mathbb{R}$. Show that if $[a, b]$ contains no odd multiple of π, then $\sum_n (-1)^n g_n(x) \cos nx$ converges uniformly on $[a, b]$.

7.4 Power Series

A *power series in x about a* is an infinite series of the form

$$s(x) = \sum_{n=0}^{\infty} c_n (x - a)^n, \qquad a,\ c_n \in \mathbb{R}.$$

In this section we examine the properties of these important series. The first step is to determine the convergence set.

7.4.1 Radius of Convergence Theorem. *For the above power series define the extended real number $R \in [0, +\infty]$ by*

$$R = \rho^{-1}, \quad \textit{where} \quad \rho := \limsup_n |c_n|^{1/n}.$$

Then

(a) $s(x)$ *converges absolutely pointwise for* $|x-a| < R$.

(b) $s(x)$ *converges absolutely uniformly for* $|x-a| \le r < R$.

(c) $s(x)$ *diverges for* $|x-a| > R$.

Proof. For the case $R = 0$ ($\rho = +\infty$), the theorem asserts that $s(x)$ diverges for all $x \neq a$. This is immediate from the root test. A similar application of the root test proves (c): If $|x-a| > R$, then

$$\limsup_n |c_n(x-a)^n|^{1/n} = \rho|x-a| > 1.$$

Clearly, (a) follows from (b). To prove (b), assume $R > 0$ ($\rho < +\infty$) and let $0 < r < R$. Choose $s \in (r, R)$. Then $\rho < 1/s$ so there exists an index N such that $|c_n|^{1/n} < 1/s$ for all $n \ge N$. For such n and for all x with $|x-a| \le r$, $|c_n(x-a)^n| \le (r/s)^n$. Since $r/s < 1$, the series converges uniformly on $[a-r, a+r]$ by the Weierstrass M-test. □

The number $R = \rho^{-1}$ is called the *radius of convergence* of the series. The set of all x for which the series $\sum_{n=0}^{\infty} c_n(x-a)^n$ converges is called the *interval of convergence.* By 7.4.1, this interval must be one of the following:

$$\{a\}, \quad (a-R, a+R), \quad (a-R, a+R], \quad [a-R, a+R), \quad \text{or} \quad [a-R, a+R].$$

Unless the interval is infinite, the theorem gives no information as to precisely which of these is the interval of convergence. However, frequently the methods of Chapter 6 may be applied to give convergence behavior at the endpoints.

Here is another characterization of R.

7.4.2 Theorem. *If* $c_n \neq 0$ *for all sufficiently large* n, *then*

$$R = \lim_n \frac{|c_n|}{|c_{n+1}|},$$

provided the limit exists in $\overline{\mathbb{R}}$.

Proof. Let L denote the limit and set $a_n = |c_n|$. Choose N so that $a_n > 0$ for all $n \ge N$. The assertion then follows from the inequalities

$$\frac{1}{L} = \liminf_n \frac{a_{n+1}}{a_n} \le \liminf_n a_n^{1/n} \le \rho = \limsup_n a_n^{1/n} \le \limsup_n \frac{a_{n+1}}{a_n} = \frac{1}{L},$$

(Exercise 2.4.12). □

Here are some typical examples using 7.4.2, where I denotes the convergence interval.

Examples.

(a) $\sum_{n=1}^{\infty} n^n x^n$, $I = \{0\}$.

(b) $\sum_{n=1}^{\infty} \dfrac{x^n}{n!}$, $I = (-\infty, +\infty)$.

(c) $\sum_{n=1}^{\infty} \dfrac{x^n}{\sqrt{n}}$, $I = [-1, 1)$, conditional convergence at -1.

(d) $\sum_{n=1}^{\infty} \dfrac{x^n}{n^2}$, $I = [-1, 1]$, absolute convergence at ± 1. ◊

The following example is somewhat more interesting.

7.4.3 Example. The *Fibonacci sequence* $\{c_n\}$ is defined by

$$c_0 = c_1 = 1, \quad c_n = c_{n-1} + c_{n-2}, \quad n \geq 2.$$

The *Fibonacci power series* is the series $\sum_{n=0}^{\infty} c_n x^n$. We use 7.4.2 to show that the radius of convergence of the series is $(\sqrt{5} - 1)/2$.

Set $r_n = c_{n+1}/c_n$. Note that the first few terms of the sequence $\{r_n\}$ are 1, 2, 3/2, 5/3, 8/5 and that

$$r_n = \frac{c_n + c_{n-1}}{c_n} = 1 + \frac{1}{r_{n-1}}, \quad n \geq 2. \tag{7.8}$$

An induction argument then shows that

$$3/2 \leq r_n \leq 5/3, \quad n \geq 2. \tag{7.9}$$

Now, from (7.8),

$$r_n - r_m = \frac{1}{r_{n-1}} - \frac{1}{r_{m-1}} = \frac{r_{m-1} - r_{n-1}}{r_{m-1} r_{n-1}}, \tag{7.10}$$

and in particular,

$$r_{2k+1} - r_{2k-1} = \frac{r_{2k-2} - r_{2k}}{r_{2k} r_{2k-2}} \quad \text{and} \quad r_{2k} - r_{2k-2} = \frac{r_{2k-3} - r_{2k-1}}{r_{2k-1} r_{2k-3}}.$$

Therefore,

$$r_{2k+1} - r_{2k-1} = \frac{r_{2k-1} - r_{2k-3}}{r_{2k} r_{2k-2} r_{2k-1} r_{2k-3}}.$$

Iterating, we obtain $r_{2k+1} - r_{2k-1} = (r_3 - r_1)/a_k$ for some $a_k > 0$, hence $\{r_{2k+1}\}$ is increasing. A similar argument shows that $\{r_{2k}\}$ is decreasing. The sequences therefore converge, say, $r_{2k+1} \to L$ and $r_{2k} \to M$. From (7.10),

$$|r_n - r_{n-1}| = \frac{|r_{n-1} - r_{n-2}|}{r_{n-1} r_{n-2}} = \ldots = \frac{|r_2 - r_1|}{b_n},$$

where b_n is a product of 2^{n-2} terms, each of which is an r_k. From (7.9), $b_n \to +\infty$, hence $r_n - r_{n-1} \to 0$. Therefore,

$$\frac{c_{n+1}}{c_n} = r_n \to L = M = 1/R,$$

where R is the radius of convergence of the series. Taking limits in (7.8) shows that $1/R = 1 + R$, which has positive solution $R = (\sqrt{5} - 1)/2$. ◊

Since a power series converges uniformly on closed bounded subintervals of $(a - R, a + R)$, 7.3.13 implies that the series is continuous on the entire interval. The following theorem asserts continuity at the endpoints as well.

7.4.4 Abel's Continuity Theorem. *Let $s(x) := \sum_{n=0}^{\infty} c_n (x - a)^n$ have radius of convergence R with $0 < R < +\infty$. If $s(x)$ converges at $x = a + R$, then $s(x)$ converges uniformly on $[b, a + R]$ for any $b \in (a - R, a + R)$. In particular, s is continuous on $(a - R, a + R]$.*

Proof. The transformation $x = Ry + a$ produces a power series in $y = (x - a)/R$ that converges on $(-1, 1]$. Hence we may assume in the original series that $a = 0$ and $s(x)$ converges on $(-1, 1]$. It suffices then to show that $s(x)$ converges uniformly on $[0, 1]$.

Set $s_k(x) = \sum_{j=0}^{k} c_j x^j$, $0 \le x \le 1$, and for $n > 1$, define

$$C_{k,n} = \sum_{j=n}^{k} c_j, \quad k \ge n,$$

so $c_k = C_{k,n} - C_{k-1,n}$ $(k > n)$ and $C_{n,n} = c_n$. For $m > n + 1$

$$\begin{aligned} s_m(x) - s_n(x) &= \sum_{k=n+1}^{m} c_k x^k \\ &= \sum_{k=n+1}^{m} C_{k,n} x^k - \sum_{k=n+1}^{m} C_{k-1,n} x^k \\ &= \sum_{k=n+1}^{m} C_{k,n} x^k - \sum_{k=n}^{m-1} C_{k,n} x^{k+1} \\ &= \sum_{k=n+1}^{m-1} C_{k,n}(x^k - x^{k+1}) + C_{m,n} x^m - C_{n,n} x^{n+1} \end{aligned}$$

Since $\sum_n c_n$ converges, given $\varepsilon > 0$ we may choose N such that $|C_{m,n}| < \varepsilon/3$ for all $m \ge n \ge N$. Noting that $0 \le x^{k+1} \le x^k \le 1$, we have for $m > n+1 \ge N$

$$\begin{aligned} |s_m(x) - s_n(x)| &\le \sum_{k=n+1}^{m-1} |C_{k,n}|(x^k - x^{k+1}) + |C_{m,n}| + |C_{n,n}| \\ &\le \frac{\varepsilon}{3} \sum_{k=n+1}^{m-1} (x^k - x^{k+1}) + \frac{2\varepsilon}{3}. \end{aligned}$$

Since the second summation in the above expression collapses to a difference of two powers of x, $|s_m(x) - s_n(x)| \le \varepsilon$. This shows that $\{s_n\}$ is uniformly Cauchy on $[0, 1]$ and so converges uniformly. □

The next result shows that a power series may be differentiated or integrated term by term over the interior of the interval of convergence.

7.4.5 Theorem. *Let* $s(x) := \sum_{n=0}^{\infty} c_n(x-a)^n$ *have radius of convergence* $R > 0$. *Then the* derived series *and the* integrated series

$$D(x) := \sum_{n=1}^{\infty} nc_n(x-a)^{n-1} \quad \textit{and} \quad I(x) := \sum_{n=0}^{\infty} \frac{c_n}{n+1}(x-a)^{n+1}$$

have radius of convergence R. *Moreover,* $s(x)$ *is differentiable on the interval* $(a-R, a+R)$, *and for* $x \in (a-R, a+R)$

$$s'(x) = D(x) \quad \textit{and} \quad \int_a^x s(t)\, dt = I(x).$$

Proof. Since $\lim_n n^{1/n} = \lim_n 1/(n+1)^{1/n} = 1$,

$$\limsup_n |nc_n|^{1/n} = \limsup_n |c_n/(n+1)|^{1/n} = \limsup_n |c_n|^{1/n}.$$

Therefore, the series $s(x)$, $D(x)$, and $I(x)$ have the same radius of convergence. Since the differentiation and integration takes place on closed subintervals where the convergence of each of the three series is uniform, the remaining assertions follow from 7.3.13. □

Representation of Functions by Power Series

A power series $s(x) = \sum_{n=0}^{\infty} c_n(x-a)^n$ is said to *represent a function* f *on an interval* I if $f = s$ on I. The largest interval for which the representation is valid is called the *representation interval.* Note that the representation interval may be smaller than the convergence interval. (See the examples below.)

Power series representations are unique. Indeed, if f is represented by $\sum_{n=0}^{\infty} c_n(x-a)^n$ on $I_a := (a-r, a+r)$, $r > 0$, then, by 7.4.5, f has derivatives of all orders on I_a, and repeated differentiation of the identity

$$f(x) = \sum_{n=0}^{\infty} c_n(x-a)^n, \quad x \in I_a$$

shows that $f^{(n)}(a) = c_n n!$. Therefore, if f has a power series representation about a, then

$$f(x) = \sum_{n=0}^{\infty} \frac{f^{(n)}(a)}{n!}(x-a)^n, \quad x \in I_a.$$

The last series is called the *Taylor series expansion of f about a*. For $a = 0$ it is called a *Maclaurin series.*

The following examples show how various power series representations may be obtained from the geometric series representation of $(1-x)^{-1}$ in (7.5).

7.4.6 Examples. **(a)** Differentiating (7.5) term by term and multiplying the result by x yields the representation

$$\frac{x}{(1-x)^2} = \sum_{n=1}^{\infty} nx^n, \quad |x| < 1. \tag{7.11}$$

(b) Replacing x in (7.5) by $-t$ and integrating produces

$$\ln(x+1) = \int_0^x \frac{1}{1+t}\, dt = \sum_{n=1}^{\infty} \frac{(-1)^{n+1}}{n} x^n, \quad |x| < 1. \tag{7.12}$$

Since the series converges at $x = 1$, Abel's continuity theorem shows that

$$\ln 2 = \sum_{n=1}^{\infty} \frac{(-1)^{n+1}}{n},$$

a result obtained in 6.4.7 by another method.

(c) Replacing x in (7.5) by $-t^2$ and integrating produces

$$\arctan x = \int_0^x \frac{1}{1+t^2}\, dt = \sum_{n=0}^{\infty} (-1)^n \frac{x^{2n+1}}{2n+1}, \quad |x| < 1. \tag{7.13}$$

(d) For an example with $a \neq 0$, consider

$$\frac{3}{5-2x} = \frac{3}{3-2(x-1)} = \frac{1}{1-2(x-1)/3} = \sum_{n=0}^{\infty} \frac{2^n}{3^n}(x-1)^n, \quad |x-1| < \frac{3}{2}. \ \Diamond$$

The next example and the theorem thereafter show that differentiation can be a powerful tool for finding a closed form of a power series.

7.4.7 Example. We show that

$$e^x = \sum_{n=0}^{\infty} \frac{x^n}{n!}, \qquad -\infty < x < +\infty. \tag{7.14}$$

Let $s(x)$ denote the series. By 7.4.2, the radius of convergence of s is

$$\lim_n \frac{(n+1)!}{n!} = \lim_n (n+1) = +\infty,$$

so $s(x)$ converges for all x. Differentiating the series term by term yields $s'(x) = s(x)$. Now set $g(x) = e^{-x}s(x)$. Then $g'(x) = e^{-x}[s'(x) - s(x)] = 0$, hence g is constant. Since $g(0) = 1$, $s(x) = e^x$. $\Diamond$

The following result is an extension of the binomial theorem. The coefficient $\binom{a}{n}$ in (7.15) is called a *generalized binomial coefficient.*

7.4.8 Binomial Series. *For any $a \in \mathbb{R}$ and $|x| < 1$,*

$$(1+x)^a = \sum_{n=0}^{\infty} \binom{a}{n} x^n, \quad \binom{a}{n} := \frac{a(a-1)\cdots(a-n+1)}{n!}, \quad \binom{a}{0} := 1. \quad (7.15)$$

Proof. Let $s(a,x)$ denote the series in (7.15). A simple calculation shows that

$$\left| \binom{a}{n} \binom{a}{n+1}^{-1} \right| = \frac{n+1}{|a-n|} \to 1.$$

Therefore, by 7.4.2, $s(a,x)$ converges for $|x| < 1$. For such x,

$$\begin{aligned}(1+x)s(a-1,x) &= \sum_{n=0}^{\infty} \binom{a-1}{n} x^n + \sum_{n=0}^{\infty} \binom{a-1}{n} x^{n+1} \\ &= 1 + \sum_{n=0}^{\infty} \left[\binom{a-1}{n+1} + \binom{a-1}{n} \right] x^{n+1} \\ &= 1 + \sum_{n=0}^{\infty} \binom{a}{n+1} x^{n+1} \\ &= s(a,x), \end{aligned} \quad (7.16)$$

where, for the third equality, we used the identity (Exercise 6)

$$\binom{a-1}{n} + \binom{a-1}{n+1} = \binom{a}{n+1}, \quad n \in \mathbb{Z}^+. \quad (7.17)$$

Now differentiate the series $s(a,x)$ term by term to obtain

$$\begin{aligned} s'(a,x) &= \sum_{n=1}^{\infty} \binom{a}{n} n x^{n-1} = \sum_{n=0}^{\infty} \binom{a}{n+1} (n+1) x^n = a \sum_{n=0}^{\infty} \binom{a-1}{n} x^n \\ &= a s(a-1,x). \end{aligned} \quad (7.18)$$

Set $g(x) = (1+x)^{-a} s(a,x)$, $|x| < 1$. By (7.18) and (7.16),

$$\begin{aligned} g'(x) &= -a(1+x)^{-a-1} s(a,x) + a(1+x)^{-a} s(a-1,x) \\ &= a(1+x)^{-a-1} \big[- s(a,x) + (1+x)s(a-1,x) \big] \\ &= 0. \end{aligned}$$

Therefore, $g(x) = g(0) = 1$, hence $s(a,x) = (1+x)^a$, as claimed. □

7.4.9 Example. Replacing x in (7.15) by $-x$, we have

$$\frac{1}{\sqrt{1-x}} = \sum_{n=0}^{\infty} \binom{-1/2}{n} (-1)^n x^n, \quad |x| < 1.$$

Since

$$\begin{aligned}\binom{-1/2}{n} &= \frac{1}{n!}\left(-\frac{1}{2}\right)\left(-\frac{3}{2}\right)\cdots\left(-\frac{2n-1}{2}\right)\\ &= \frac{(-1)^n 1\cdot 3\cdot 5\cdots(2n-1)}{n!\,2^n}\\ &= \frac{(-1)^n 1\cdot 2\cdot 3\cdot 4\cdots(2n-1)\cdot 2n}{n!\,2^n 2\cdot 4\cdots 2n}\\ &= \frac{(-1)^n (2n)!}{(n!)^2 4^n},\end{aligned}$$

we see that

$$\frac{1}{\sqrt{1-x}} = \sum_{n=0}^{\infty} \frac{(2n)!}{(n!)^2 4^n} x^n, \quad |x| < 1. \tag{7.19}$$

Replacing x by t^2 and integrating term by term from 0 to x yields the Maclaurin series for $\arcsin x$:

$$\arcsin x = \sum_{n=0}^{\infty} \frac{(2n)!}{(n!)^2(2n+1)4^n} x^{2n+1}, \quad |x| < 1. \tag{7.20}$$

7.4.10 Remark. If $a > 0$ and is not an integer, then the binomial series converges absolutely uniformly on $[-1, 1]$. Indeed, if $a_n = |\binom{a}{n}|$, then

$$\frac{a_n}{a_{n+1}} = \frac{|a(a-1)\cdots(a-n+1)|}{n!}\,\frac{(n+1)!}{|a(a-1)\cdots(a-n)|} = \frac{n+1}{|a-n|},$$

hence, for sufficiently large n,

$$n\left(\frac{a_n}{a_{n+1}} - 1\right) = n\left(\frac{n+1}{n-a} - 1\right) = \frac{n(1+a)}{n-a} \to 1 + a > 1.$$

By Raabe's test (6.3.2), the series converges absolutely at $x = \pm 1$, hence, by Abel's continuity theorem (7.4.4), the series converges absolutely uniformly on the interval $[-1, 1]$. ◊

Multiplication of Power Series

7.4.11 Definition. The *Cauchy product* of the power series $\sum_{n=0}^{\infty} a_n x^n$ and $\sum_{n=0}^{\infty} b_n x^n$ is the power series $\sum_{n=0}^{\infty} c_n x^n$, where

$$c_n = \sum_{k=0}^{n} a_k b_{n-k}.$$ ◊

Note that $\sum_n c_n x^n$ is precisely the series one obtains by formally carrying out the multiplication

$$(a_0 + a_1 x + a_2 x^2 + \cdots)(b_0 + b_1 x + b_2 x^2 + \cdots)$$

and collecting like powers. We show below that if the power series $\sum_{n=0}^{\infty} a_n x^n$ and $\sum_{n=0}^{\infty} b_n x^n$ converge for $|x| < R$, then so does the Cauchy product. For this we need the following result due to Mertens.

7.4.12 Lemma. *If the numerical series $A := \sum_{n=0}^{\infty} \alpha_n$ and $B := \sum_{n=0}^{\infty} \beta_n$ both converge, and if at least one of the series converges absolutely, then the Cauchy product*

$$C := \sum_{n=0}^{\infty} \gamma_n, \quad \gamma_n = \sum_{k=0}^{n} \alpha_k \beta_{n-k},$$

converges and $C = AB$.

Proof. Assume that $\sum_{n=0}^{\infty} \alpha_n$ converges absolutely. Let

$$A_n = \sum_{k=0}^{n} \alpha_k, \quad B_n = \sum_{k=0}^{n} \beta_k, \quad C_n = \sum_{k=0}^{n} \gamma_k, \quad \text{and} \quad A' = \sum_{n=0}^{\infty} |\alpha_n|.$$

Then

$$\begin{aligned} C_n &= \alpha_0\beta_0 + (\alpha_0\beta_1 + \alpha_1\beta_0) + \cdots + (\alpha_0\beta_n + \alpha_1\beta_{n-1} + \cdots + \alpha_n\beta_0) \\ &= \alpha_0 B_n + \alpha_1 B_{n-1} + \cdots + \alpha_n B_0 \\ &= \alpha_0(B_n - B + B) + \alpha_1(B_{n-1} - B + B) + \cdots + \alpha_n(B_0 - B + B) \\ &= \alpha_0(B_n - B) + \alpha_1(B_{n-1} - B) + \cdots + \alpha_n(B_0 - B) + A_n B. \end{aligned}$$

Thus to show that $C_n \to AB$ it suffices to verify that

$$\alpha_0(B_n - B) + \alpha_1(B_{n-1} - B) + \cdots + \alpha_n(B_0 - B) \to 0.$$

Given $\varepsilon > 0$, choose N such that

$$|B_n - B| < \varepsilon/(2A') \quad \text{for all } n > N.$$

Since $\alpha_n \to 0$, we may choose $N' > N$ so that for all $n > N'$

$$|\alpha_n(B_0 - B) + \alpha_{n-1}(B_1 - B) + \cdots + \alpha_{n-N}(B_N - B)| < \varepsilon/2.$$

For such n,

$$\begin{aligned} &|\alpha_0(B_n - B) + \alpha_1(B_{n-1} - B) + \cdots + \alpha_n(B_0 - B)| \\ &\quad \le |\alpha_n(B_0 - B) + \alpha_{n-1}(B_1 - B) + \cdots + \alpha_{n-N}(B_N - B)| \\ &\quad + |\alpha_{n-N-1}|\,|B_{N+1} - B| + |\alpha_{n-N-2}|\,|B_{N+2} - B| + \cdots + |\alpha_0|\,|B_n - B| \\ &\quad < \varepsilon/2 + \varepsilon/2 = \varepsilon. \end{aligned}$$

□

7.4.13 Cauchy Product Theorem. *For each x, let*

$$C(x) = \sum_{n=0}^{\infty} c_n x^n, \quad c_n := \sum_{k=0}^{n} a_k b_{n-k}$$

be the Cauchy product of series $A(x) = \sum_{n=0}^{\infty} a_n x^n$ *and* $B(x) = \sum_{n=0}^{\infty} b_n x^n$. *If* $A(x)$ *and* $B(x)$ *have radii of convergence* R_a *and* R_b*, respectively, then* $C(x)$ *has radius of convergence* $R_c \geq \min\{R_a, R_b\}$ *and*

$$C(x) = A(x)B(x), \quad |x| < \min\{R_a, R_b\}. \tag{7.21}$$

Moreover, if, say $R_b < R_a$ *and* $B(R_b)$ *converges, then* $C(R_b)$ *converges and* $C(R_b) = A(R_b)B(R_b)$.

Proof. Assume that $R_b \leq R_a$ and let $|x| < R_b$. By 7.4.12 applied to $\alpha_n = a_n x^n$ and $\beta_n = b_n x^n$, the series $C(x)$ converges, hence $R_c \geq |x|$ and 7.21 holds. Since $|x|$ was arbitrary, $R_c \geq R_b = \min\{R_a, R_b\}$. The last assertion of the theorem follows from 7.4.4 by letting $x \uparrow R_b$ in 7.21. □

7.4.14 Example. By (7.5) and (7.14), for $|x| < 1$ we have

$$\frac{e^x}{1+x} = \sum_{n=0}^{\infty} c_n x^n, \quad \text{where } c_n = \sum_{k=0}^{n} \frac{(-1)^{n-k}}{k!} = (-1)^n \sum_{k=0}^{n} \frac{(-1)^k}{k!}. \qquad \Diamond$$

Remark. If $R_a = R_b$ and both $A(R_a)$ and $B(R_b)$ in 7.4.13 converge, it does not necessarily follow that $C(R_a)$ converges. Consider, for example, $A(x) = B(x) = \sum_{n=1}^{\infty} (-1)^n x^n / \sqrt{n}$, which has radius of convergence 1 and converges conditionally at $x = 1$. The Cauchy product at $x = 1$ is $\sum_{n=1}^{\infty} c_n$, where

$$c_n = (-1)^n \sum_{k=1}^{n-1} \frac{1}{\sqrt{k(n-k)}}.$$

However, for odd n,

$$|c_n| = \sum_{k=1}^{n-1} \frac{1}{\sqrt{k(n-k)}} \geq \sum_{k=1}^{(n-1)/2} \frac{1}{\sqrt{k(n-k)}} \geq \sum_{k=1}^{(n-1)/2} \frac{1}{\sqrt{(n-1)^2/2}} = \frac{\sqrt{2}}{2},$$

hence $\sum_n c_n$ diverges. ◊

Analytic Functions

7.4.15 Definition. A function f is said to be (*real*) *analytic at a point* a if, for some $r > 0$, f has derivatives of all orders on $(a - r, a + r)$ and is represented there by its Taylor series at a, that is,

$$f(x) = \sum_{n=0}^{\infty} \frac{f^{(n)}(a)}{n!}(x-a)^n, \quad |x - a| < r.$$

If f is analytic at each point of a set E, then f is said to be *analytic on* E. ◊

A function that has derivatives of all orders on an interval may not be analytic there. This is the case for the function in Exercise 29 below. The following theorem gives a necessary and sufficient condition for analyticity at a point.

7.4.16 Taylor Series Representation. *Let f have derivatives of all orders on an open interval I containing a. Then f is analytic at a iff there exist positive constants M and r such that*

$$|f^{(k)}(x)| \le k!M^k \quad \text{for all } k \in \mathbb{N} \text{ and } x \in (a-r, a+r). \tag{7.22}$$

Proof. Assume condition (7.22) holds. To prove that f is analytic at a, we use Taylor's theorem (Section 4.6), which asserts that for each $n \in \mathbb{N}$ and $x \in (a-r, a+r)$ there exists a number $c = c(n,x)$ between x and a such that $f(x) = T_n(x) + R_n(x)$, where

$$T_n(x) := \sum_{k=0}^{n-1} \frac{f^{(k)}(a)}{k!}(x-a)^k, \quad \text{and} \quad R_n(x) := \frac{f^{(n)}(c)}{n!}(x-a)^n.$$

Let $r \in (0, 1/M)$ and $|x-a| < r$. By hypothesis,

$$|R_n(x)| \le M^n |x-a|^n \le (Mr)^n.$$

Since $Mr < 1$, $R_n(x) \to 0$, hence $T_n(x) \to f(x)$. Therefore, f is analytic at a.

Conversely, if f be analytic at a, then there exist constants $r_1 \in (0,1)$ and c_n such that

$$f(x) = \sum_{n=0}^{\infty} c_n(x-a)^n, \quad |x-a| \le r_1. \tag{7.23}$$

In particular, $c_n r_1^n \to 0$. Choose $M_1 > 1$ so that $|c_n r_1^n| < M_1$ for all n. Termwise differentiation of (7.23) yields

$$f^{(k)}(x) = \sum_{n=k}^{\infty} n(n-1)\cdots(n-k+1)c_n(x-a)^{n-k},$$

hence for $|x-a| \le r_1/2$,

$$\begin{aligned} |f^{(k)}(x)| &\le \sum_{n=k}^{\infty} n(n-1)\cdots(n-k+1)|c_n|(r_1/2)^{n-k} \\ &\le M_1 r_1^{-k} \sum_{n=k}^{\infty} n(n-1)\cdots(n-k+1)(1/2)^{n-k}. \end{aligned}$$

The last series is the kth derivative of the geometric series for $(1-x)^{-1}$ evaluated at $1/2$ and therefore equals

$$\left.\frac{d^k}{dx^k}\right|_{x=1/2}(1-x)^{-1} = k!\,(1-1/2)^{-k-1} = k!\,2^{k+1}.$$

Thus

$$|f^{(k)}(x)| \le M_1 r_1^{-k} k!\,2^{k+1}, \qquad |x-a| \le r_1/2.$$

Finally, to obtain (7.22) take $r = r_1/2$ and choose $M > 4M_1/r_1$, implying that $M^k > M_1 r_1^{-k} 2^{k+1}$ for all k. □

7.4.17 Example. Let $f(x) = \sin x$. Then $f^{(2k)}(0) = 0$ and $f^{(2k+1)}(0) = (-1)^k$. Since the derivatives of f are bounded, (7.22) holds for all x. Therefore,

$$\sin x = \sum_{n=0}^{\infty} \frac{(-1)^n}{(2n+1)!} x^{2n+1} = x - \frac{x^3}{3!} + \frac{x^5}{5!} - \dots, \qquad \infty < x < +\infty.$$

Similarly,

$$\cos x = \sum_{n=0}^{\infty} \frac{(-1)^n}{(2n)!} x^{2n} = 1 - \frac{x^2}{2!} + \frac{x^4}{4!} - \dots, \qquad \infty < x < +\infty. \qquad \Diamond$$

It is clear from 7.3.2 and 7.4.13 that sums and products of functions analytic at a are analytic at a. In Exercise 33 the reader is asked to show that the reciprocal of a nonzero analytic function is analytic. It follows that the ratio of two analytic functions, if defined, is analytic.

The next result extends the property of analyticity to nearby points.

7.4.18 Theorem. *If the series $f(x) = \sum_{n=0}^{\infty} a_n(x-a)^n$ converges on $I := (a-r, a+r)$, then f is analytic on I.*

Proof. By considering $g(x) = f(x+a)$, we may suppose that $a = 0$. Let $|b| < r$, $0 < s < r - |b|$, and $|x - b| < s$. We show that f has a power series expansion about b on the interval $|x - b| < s$. Since b is arbitrary, it will follow that f is analytic on I.

By the binomial theorem applied to $\big[(x-b)+b\big]^n$,

$$f(x) = \sum_{n=0}^{\infty}\sum_{k=0}^{n} a_n \binom{n}{k} (x-b)^k b^{n-k} = \sum_{n=0}^{\infty}\sum_{k=0}^{\infty} a_n d_{k,n} (x-b)^k b^{n-k}, \tag{7.24}$$

where $d_{k,n} = \binom{n}{k}$ for $k = 0, 1, \cdots, n$ and $d_{k,n} = 0$ for $k > n$. Now,

$$\begin{aligned} \sum_{n=0}^{\infty}\sum_{k=0}^{\infty} |a_n (x-b)^k b^{n-k} d_{k,n}| &= \sum_{n=0}^{\infty} |a_n| \sum_{k=0}^{n} \binom{n}{k} |x-b|^k |b^{n-k}| \\ &= \sum_{n=0}^{\infty} |a_n| (|x-b| + |b|)^n. \end{aligned}$$

If $|x - b| < s$ then $|x - b| + |b| < s + |b| < r$ and the last series converges. Therefore, (7.24) converges uniformly for $|x - b| < s$. By 6.5.4, the order of summation may be interchanged, so

$$\begin{aligned} f(x) &= \sum_{k=0}^{\infty}\sum_{n=0}^{\infty} a_n d_{k,n} (x-b)^k b^{n-k} \\ &= \sum_{k=0}^{\infty}\sum_{n=k}^{\infty} a_n \binom{n}{k} (x-b)^k b^{n-k} \\ &= \sum_{k=0}^{\infty} b_k (x-b)^k, \quad b_k := \sum_{n=k}^{\infty} a_n \binom{n}{k} b^{n-k}. \end{aligned}$$

This shows that f has a power series expansion about b on $(b-s, b+s)$. □

7.4.19 Theorem. *Let f be analytic on an open interval I and let $f=0$ on a subinterval (a,b) of I. Then $f=0$ on I.*

Proof. Let $c \in I$, $c > b$, and define

$$A = \{t \in (a,c) \mid f^{(n)} = 0 \text{ on } (a,t] \text{ for all } n \geq 0\}.$$

Then $A \neq \emptyset$ and $t_0 := \sup A \leq c$. We claim that $t_0 = c$. If this is not the case, then, since f is analytic at t_0, f has a Taylor series representation about t_0 on $J := (t_0 - r, t_0 + r)$ for some $r > 0$. By continuity and the approximation property of suprema, $f^{(n)}(t_0) = 0$ for each n. It follows that f is identically zero on J, contradicting the definition of t_0. Therefore, $t_0 = c$ as claimed and so $f = 0$ on (a,c). Since c was arbitrary, $f(x) = 0$ for all $x \in I$ with $x \geq a$. Similarly, $f(x) = 0$ for all $x \in I$ with $x \leq b$. □

The proof of the following corollary is left to the reader.

7.4.20 Corollary. *Let f and g be analytic on the open intervals I and J, respectively. If $I \cap J \neq \emptyset$ and $f = g$ on an open subinterval of $I \cap J$, then there exists an analytic function h on $I \cup J$ such that $h|_I = f$ and $h|_J = g$.*

The result is known as *analytic continuation*, as it may be used to extend an analytic function to a larger interval.

Exercises

1. Find the interval of convergence of $\sum_{n=1}^{\infty} f_n(x)$, where $f_n(x) =$

   (a) $\dfrac{(-1)^n n}{2^n}(x-1)^n$ (b) $\dfrac{2^{3n} n^3 x^n}{\sqrt{n!}}$ (c) $\dfrac{n^2 n!}{(2n)!}(x-2)^n$

   (d) $\dfrac{(-1)^n n x^n}{(n+1)\ln(n+2)}$ (e) $\dfrac{n! x^n}{n^{n+2-1/n}}$ (f) $\dfrac{(1+2/n)^n}{3^n}(x+1)^n$

   (g) $[3+(-1)^n]^n \sin\left[\dfrac{1}{n}\right] x^n$ (h) $\dfrac{2^n+5^n}{3^n+4^n}x^{2n}$ (i) $\dfrac{(1.5)(2.5)\cdots(n+.5)}{n!}x^n$.

2. Use (7.5) to represent the following functions as power series about the given point a. In each case, find the representation interval.

   (a) $\dfrac{x}{(x+1)^2}$, $a=0$ (b) $\dfrac{x^3}{2-3x}$, $a=0$ (c) $\dfrac{x}{3+2x}$, $a=1$.

3. Use (7.12) to find power series representations for (a) $x \ln x$, (b) $x^2 \ln x$ about the point $a = 1$.

4. Without using 7.4.16, find the Maclaurin series and representation interval for the following functions:

(a) $\ln\left(\dfrac{1+2x}{1-3x}\right)$ (b) $(1+x^2)\arctan x$ (c) $x^3 e^{-3x^2}$

(d) $\dfrac{e^x-1}{x}$ (e) $\dfrac{\sin\sqrt{x}}{\sqrt{x}}$ (f) $\dfrac{\cos x-1}{x^2}$

(g) $\sin x\cos x$ (h) $\dfrac{1}{\sqrt{9-x^2}}$ (i) $\sin(x+\pi/3)$.

5. Use an identity and 7.4.9 to find the Maclaurin series for $\arccos x$.

6. Verify the identity (7.17).

7. Without using 7.4.16, show that

(a) $\sin x = \sum_{n=0}^{\infty} a_n(x-a)^n,\quad a_{2n} = \dfrac{(-1)^n \sin a}{(2n)!},\quad a_{2n+1} = \dfrac{(-1)^n \cos a}{(2n+1)!}.$

(b) $\cos x = \sum_{n=0}^{\infty} b_n(x-a)^n,\quad b_{2n} = \dfrac{(-1)^n \cos a}{(2n)!},\quad b_{2n+1} = \dfrac{(-1)^{n+1} \sin a}{(2n+1)!}.$

8. Prove that $\displaystyle\sum_{n=0}^{\infty}\frac{4(-1)^n}{2n+1} = \sum_{n=0}^{\infty}\frac{2(2n)!}{(n!)^2(2n+1)4^n} = \pi.$

9. Find a power series representation for $\int_0^x f(t)\,dt$ if $f(t) =$

(a) $\dfrac{\sin t - t}{t^3}$ (b) $\cos t\sqrt{t}$ (c) $\dfrac{\cos t - 1}{t}$ (d) $\dfrac{e^{t^2}-1}{t^2}$.

10. Find a closed form for the series $\sum_{n=0}^{\infty} f_n(x)$, $|x|<1$, where $f_n(x) =$

(a) n^2x^n (b) $(-1)^n(2n+1)x^{2n+1}$ (c) $\dfrac{x^{n+1}}{(n+1)(n+2)}$ (d) $\dfrac{n^2x^n}{n+1}$.

11. Sum the series $\sum_{n=0}^{\infty}(n^2+n+1)3^{-n}$.

12. Use 7.4.13 to find a series representation and representation interval for

(a) $\dfrac{\ln(1-x)}{1+x}$ (b) $\dfrac{e^{-x}}{\sqrt{1-x}}$ (c) $\dfrac{\sin x}{1-x^2}$ (d) $e^{x^2}\sin x$ (e) $\dfrac{\arctan x}{x(1+x^2)}$.

13. By calculating the Maclaurin series of the function $\sin^2 x$ in two ways, establish the identity

$$\frac{2^{2n+1}}{(2n+2)!} = \sum_{k=0}^{n}\frac{1}{(2k+1)!(2n-2k+1)!}.$$

14. By calculating the Maclaurin series of the function $\cos^2 x$ in two ways, establish the identity

$$\frac{2^{2n-1}}{(2n)!} = \sum_{k=0}^{n} \frac{1}{(2k)!(2n-2k)!}.$$

15. By calculating the Maclaurin series of the function $(1-x)^{-3/2}$ in two ways, establish the identity

$$\frac{(2n+1)!}{(n!)^2 4^n} = \sum_{k=0}^{n} \frac{(2k)!}{(k!)^2 4^k}.$$

16. Show that the Fibonacci power series $s(x)$ (7.4.3) has the closed form

$$(1-x-x^2)^{-1}, \quad |x| < (\sqrt{5}-1)/2.$$

Conclude from Abel's continuity theorem (7.4.4) that $s(x)$ cannot converge at the endpoint $(\sqrt{5}-1)/2$.

17. Let $a_n \to L \in \mathbb{R}$ and set $s(x) := \sum_{n=0}^{\infty} a_n x^n$, $|x| < 1$. For $m \in \mathbb{N}$, define

$$\varphi_m(x) := \sum_{k=0}^{2m-1} (-1)^k x^k.$$

Prove that $\lim_{x \to 1^-} \varphi_m(x)s(x) = mL$. *Hint.* Use Abel's continuity theorem.

18. Use the method of 7.4.9 to establish the representation

$$\ln\left(\sqrt{1+x^2}+x\right) = \sum_{n=0}^{\infty} \frac{(-1)^n (2n)!}{(2n+1)(n!)^2 4^n} x^{2n+1}, \quad |x| < 1.$$

19. Let R be the radius of convergence of $\sum_n c_n(x-a)^n$ and let $p \geq 0$. Prove:

(a) If $\liminf_n |c_n| n^p > 0$, then $R \leq 1$.

(b) If $\limsup_n |c_n|/n^p < +\infty$, then $R \geq 1$.

20. Let R_a and R_b denote the radii of convergence of $A(x) := \sum_n a_n x^n$ and $B(x) := \sum_n b_n x_n$, respectively. Suppose that

$$\limsup(|a_n|/|b_n|) < +\infty.$$

Prove that $R_a \geq R_b$.

21. Let R_s and R_t denote the radii of convergence of

$$s(x) := \sum_n c_n (x-a)^n \quad \text{and} \quad t(x) := \sum_n c_{n^2} (x-a)^n,$$

respectively. Prove:

(a) If $R_s > 1$, then $R_t = +\infty$.

(b) If $R_s \le 1$, then no conclusion is possible.

22. Let R_s and R_t denote the radii of convergence of

$$s(x) := \sum_n c_n (x-a)^n \quad \text{and} \quad t(x) := \sum_n c_n (x-a)^{n^2},$$

respectively. Prove:

(a) If $0 < R_s < +\infty$, then $R_t = 1$.

(b) If $R_s = 0$, then $R_t \le 1$, and any value of $R_t \le 1$ is possible.

(c) If $R_s = +\infty$, then $R_t \ge 1$, and value of $R_t \ge 1$ is possible.

23. Suppose that the series

$$A := \sum_{n=0}^{\infty} a_n, \quad B := \sum_{n=0}^{\infty} b_n, \quad \text{and} \quad C := \sum_{n=0}^{\infty} c_n$$

converge, where $c_n = \sum_{k=0}^{n} a_n b_{n-k}$. Use 7.4.12 and 7.4.4 to prove that $AB = C$.

24. Prove that for any $a, b \in \mathbb{R}$ and $n \in \mathbb{N}$,

$$\binom{a}{0}\binom{b}{n} + \binom{a}{1}\binom{b}{n-1} + \cdots + \binom{a}{n}\binom{b}{0} = \binom{a+b}{n}.$$

25. Let $n \in \mathbb{Z}^+$. The *Bessel function of order n* may be defined as the power series

$$J_n(x) = \sum_{k=0}^{\infty} \frac{(-1)^k}{(n+k)!k!} \left(\frac{x}{2}\right)^{n+2k}.$$

Prove:

(a) The radius of convergence of $J_n(x)$ is $+\infty$.

(b) J_n satisfies *Bessel's differential equation* $x^2 y'' + xy' + (x^2 - n^2)y = 0$.

(c) $\dfrac{d}{dx} x^n J_n(x) = x^n J_{n-1}(x), \ n \ge 1$.

26. Prove that $\displaystyle\lim_{x \to 1^-} \sum_{n=1}^{\infty} \frac{x^n}{n^p} \begin{cases} = +\infty & \text{if } p \le 1, \\ < +\infty & \text{if } p > 1. \end{cases}$

27. Let $\{c_n\}$ tend monotonically to 0. Prove that $\sum_n c_n x^n$ is continuous on $[-1, 1)$.

28. Let $f(x)$ be bounded on $[0, 1]$.

    (a) Prove that $t(x) := \sum_n nx^n f(x)$ converges pointwise on $[0, 1)$ and uniformly on $[0, r]$ for $0 < r < 1$.

    (b) Suppose that $L := \lim_{x \to 1^-}(1 - x)^{-2} f(x)$ exists. Prove that the convergence of $t(x)$ in (a) is uniform on $[0, 1)$ iff $L = 0$. (Compare with Exercise 7.3.7.)

29. Show that the function

$$f(x) = \begin{cases} e^{-1/x^2} & \text{if } x \neq 0, \\ 0 & \text{otherwise} \end{cases}$$

is not analytic at 0. (See Exercise 4.6.1.)

30. Prove 7.4.20.

31. Prove: If $f(x)$ is analytic at a, then $f'(x)$ and $g(x) := \int_a^x f(t)\,dt$ are analytic at a.

32. Let f be analytic at a and let $\{a_n\}$ be a sequence of distinct real numbers such that $a_n \to a$ and $f(a_n) = 0$ for all n. Prove that f is identically zero in a neighborhood of a. *Hint.* Assume that $a_n \uparrow a$ (how?). Construct, by induction, sequences $\{a_n^{(k)}\}_n$ such that $\lim_n a_n^{(k)} = a$ and $f^{(k)}(a_n^{(k)}) = 0$ for all n and k.

33. Let f be analytic at a and $f(a) \neq 0$. Carry out the following steps to show that $1/f$ is analytic at a.

    (a) Assume that $f(a) = 1$ and that

$$f(x) = \sum_{n=0}^{\infty} a_n (x - a)^n \neq 0, \ \ |x - a| < r$$

for some r. Define a series g formally by

$$g(x) = \sum_{n=0}^{\infty} b_n (x - a)^n,$$

where the sequence $\{b_n\}$ is given recursively by

$$b_0 = 1 \ \text{ and } \ b_n = -\sum_{k=1}^{n} a_k b_{n-k}, \quad n \geq 1.$$

Show that if $g(x)$ converges for $|x - a| < r_1$ for some $0 < r_1 < r$, then $f(x)g(x) = 1$ for $|x - a| < r_1$.

    (b) Show that if $|a_n| \leq M^n$ for all n, then $|b_n| \leq (2M)^n$ for all n.

    (c) Conclude that g is analytic at a and that $g = 1/f$.

Part II

Functions of Several Variables

Chapter 8

Metric Spaces

The essential feature in the notion of limit of a function is the idea of nearness. This is made precise by a distance function, which, in the real number system, is derived from the absolute value function. It turns out that there are many other important mathematical structures equipped with a distance function and therefore admitting a definition of limit. In this chapter, we examine the general properties of these structures.

8.1 Definitions and Examples

8.1.1 Definition. A *metric on a nonempty set X* is a function $d : X \times X \to \mathbb{R}$ such that, for all x, y, $z \in X$,

(a) $d(x, y) \geq 0$ (nonnegativity),

(b) $d(x, y) = 0$ iff $x = y$ (coincidence),

(c) $d(x, y) = d(y, x)$ (symmetry),

(d) $d(x, y) \leq d(x, z) + d(y, z)$ (triangle inequality).

The ordered pair (X, d) is called a *metric space.* A nonempty subset E of X with the metric $d\big|_{E\times E}$ is called a *subspace* of X and is denoted by (E, d). ◊

The real number system is a metric space under the *usual metric* $d(x, y) = |x - y|$. The following example shows that any nonempty set may be given a metric.

8.1.2 Example. (Discrete metric space). On a nonempty set X define $d(x, x) = 0$ for all $x \in X$, and $d(x, y) = 1$ if $x \neq y$. Then d is easily seen to be a metric, called the *discrete metric on X*. For example, the triangle inequality $d(x, y) \leq d(x, z) + d(y, z)$ holds because the left side of the inequality is at most 1, in which case either $x \neq z$ or $y \neq z$, implying that the right side must be at least 1. ◊

8.1.3 Definition. A subset E of a metric space X is said to be *bounded* if for some $x_0 \in X$ and $M > 0$, $d(x, x_0) \leq M$ for all $x \in E$. ◊

DOI: 10.1201/9781003629382-8

The point x_0 in the definition may be replaced by any other point $y_0 \in X$ since for $x \in E$,

$$d(x, y_0) \le d(x, x_0) + d(x_0, y_0) \le M + d(x_0, y_0).$$

The notions of sequential convergence and completeness in $\mathbb{R}$ readily carry over to general metric spaces:

8.1.4 Definition. A sequence $\{x_n\}$ in a metric space (X, d) is said to *converge to a member x of X* if $\lim_n d(x_n, x) = 0$. In this case we write $x_n \to x$ or $\lim_n x_n = x$. A *cluster point* of a sequence in X is the limit of a convergent subsequence. ◊

The limit of a sequence $\{x_n\}$ in X, if it exists, must be unique. Indeed, if $x_n \to x$ and $x_n \to y$, then, by the triangle inequality,

$$0 \le d(x, y) \le d(x, x_n) + d(y, x_n) \to 0,$$

hence $d(x, y) = 0$ and so $x = y$.

8.1.5 Definition. A sequence $\{x_n\}$ in a metric space (X, d) is said to be *Cauchy* if $\lim_{m,n} d(x_m, x_n) = 0$. A metric space (X, d) is said to be *complete* if every Cauchy sequence in X converges to a member of X. A subset E of X is *complete* if it is complete as a subspace of X, that is, every Cauchy sequence in E converges to a member of E. ◊

The real number system is complete under the usual metric. The subspace $\mathbb{Q}$ of $\mathbb{R}$ is not complete: a sequence of rational numbers converging to an irrational number is Cauchy but does not converge in $\mathbb{Q}$. A discrete metric space is complete, since every Cauchy sequence is eventually constant and therefore trivially converges.

8.1.6 Proposition.

(a) *Every Cauchy sequence is bounded.*

(b) *Every convergent sequence is Cauchy, hence bounded.*

Proof. (a) If $\{x_n\}$ is Cauchy, we may choose an index N such that $d(x_m, x_n) < 1$ for all m, $n \ge N$. Then, for all $n \in \mathbb{N}$,

$$d(x_N, x_n) < 1 + \max\{d(x_N, x_1), d(x_N, x_2), \ldots, d(x_N, x_{N-1})\}.$$

(b) If $x_n \to x$, then the inequality $d(x_m, x_n) < d(x_m, x) + d(x_n, x)$ implies that $\{x_n\}$ is Cauchy. □

The notions of pointwise convergence and uniform convergence of a sequence of real-valued functions easily extend to general metric spaces:

8.1.7 Definition. Let S be a nonempty set and let (X, d) be a metric space. A sequence of functions $f_n : S \to X$ is said to *converge pointwise to a function* $f : S \to X$ if $f_n(s) \to f(s)$ for each $s \in S$. In this case we write $f = \lim_n f$ or $f_n \to f$ (on S). The sequence *converges uniformly to* f *on* S if for each $\varepsilon > 0$ there exists $N \in \mathbb{N}$ such that $d\big(f_n(s), f(s)\big) < \varepsilon$ for all $n \geq N$ and $s \in S$. ◊

The following definition describes an important class of metric spaces.

8.1.8 Definition. Let $\mathcal{X}$ be a vector space. A *norm* on $\mathcal{X}$ is a function $\|\cdot\|$ from $\mathcal{X}$ to $\mathbb{R}$ such that for all $\boldsymbol{x}$, $\boldsymbol{y} \in \mathcal{X}$ and $t \in \mathbb{R}$

(a) $\|\boldsymbol{x}\| \geq 0$ (nonnegativity),

(b) $\|\boldsymbol{x}\| = 0$ iff $\boldsymbol{x} = \boldsymbol{0}$ (coincidence),

(c) $\|t\boldsymbol{x}\| = |t|\,\|\boldsymbol{x}\|$ (absolute homogeneity),

(d) $\|\boldsymbol{x} + \boldsymbol{y}\| \leq \|\boldsymbol{x}\| + \|\boldsymbol{y}\|$ (triangle inequality).

The pair $(\mathcal{X}, \|\cdot\|)$ is then called a *normed vector space.* ◊

The proof of the following proposition is left to the reader.

8.1.9 Proposition. *If* $(\mathcal{X}, \|\cdot\|)$ *is a normed vector space, then the function* $d(\boldsymbol{x}, \boldsymbol{y}) := \|\boldsymbol{x} - \boldsymbol{y}\|$ *is a metric on* $\mathcal{X}$.

From 1.5.4 and Exercise 1.5.4, we see that $\|\cdot\|_2$, $\|\cdot\|_1$, and $\|\cdot\|_\infty$ are norms on $\mathbb{R}^n$, hence, according to 8.1.9, give rise to metrics. We denote these, respectively, by d_2, d_1, and d_∞. In Exercise 17 the reader is asked to show that $\mathbb{R}^n$ is complete in each of these metrics. The metric d_2 is called the *Euclidean* metric on $\mathbb{R}^n$. The metric d_1 is the *ℓ^1 metric on $\mathbb{R}^n$* and d_∞ the *max metric on $\mathbb{R}^n$*. For $n = 1$, all three metrics reduce to absolute value on $\mathbb{R}$.

8.1.10 Example. Let S be a nonempty set and let $\mathcal{B}(S)$ denote the set of all bounded real-valued functions on S. Then $\mathcal{B}(S)$ is a vector space under the operations of addition $f + g$ and scalar multiplication cf defined by

$$(f + g)(s) = f(s) + g(s) \quad \text{and} \quad (cf)(s) = cf(s), \quad s \in S.$$

The *supremum norm of* $f \in \mathcal{B}(S)$ is defined by

$$\|f\|_\infty = \sup\{|f(s)| \,:\, s \in S\}.$$

It is easy to check that $\|\cdot\|_\infty$ is indeed a norm. For example, the triangle inequality follows by taking the supremum over $s \in S$ in the inequality

$$|f(s) + g(s)| \leq |f(s)| + |g(s)| \leq \|f\|_\infty + \|g\|_\infty.$$

Note that convergence of a sequence of functions in $\mathcal{B}(S)$ is simply uniform convergence on S. For this reason, $\|\cdot\|_\infty$ is also called the *uniform norm.*

The space $\mathcal{B}(S)$ is complete in the metric $d_\infty(f,g) := \|f-g\|_\infty$ induced by the norm. To see this, let $\{f_n\}$ be a Cauchy sequence in $\mathcal{B}(S)$ and let $\varepsilon > 0$. Choose N such that $d_\infty(f_n, f_m) < \varepsilon$ for all m, $n \ge N$. For such m, n

$$|f_n(s) - f_m(s)| < \varepsilon \ \text{ for all } s \in S. \tag{8.1}$$

Thus $\{f_n(s)\}$ is a Cauchy sequence in $\mathbb{R}$ for every $s \in S$. Since $\mathbb{R}$ is complete, $f_n(s) \to f(s)$ for some $f(s) \in \mathbb{R}$. Fixing n in (8.1) and letting $m \to +\infty$ yields

$$|f_n(s) - f(s)| \le \varepsilon \ \text{ for all } s \in S \text{ and all } n \ge N.$$

In particular, $f(s) = f(s) - f_n(s) + f_N(s)$, the difference of two bounded functions, hence f is bounded, Therefore, $f_n \to f$ in $\mathcal{B}(S)$. ◇

In the case $S = \mathbb{N}$, $\mathcal{B}(S)$ may be identified with the set of all bounded sequences and as such is denoted by ℓ^∞.

8.1.11 Example. Let ℓ^1 denote the set of all sequences $\boldsymbol{a} = \{a_n\}$ in $\mathbb{R}$ such that $\sum_n |a_n| < +\infty$. Clearly, ℓ^1 is a vector subspace of ℓ^∞. It is easy to check that $\|\boldsymbol{a}\|_1 := \sum_n |a_n|$ defines a norm on ℓ^1. We show that $\big(\ell^1, \|\cdot\|_1\big)$ is complete in this norm.

Let $\{\boldsymbol{a}_n := (a_{1,n}, a_{2,n}, \ldots)\}_{n=1}^\infty$ be a Cauchy sequence in ℓ^1, and let $\varepsilon > 0$. Choose N so that

$$\|\boldsymbol{a}_n - \boldsymbol{a}_m\|_1 = \sum_{k=1}^\infty |a_{k,n} - a_{k,m}| < \varepsilon \ \text{ for all } \ n,\, m \ge N. \tag{8.2}$$

Since $|a_{k,n} - a_{k,m}| \le \|\boldsymbol{a}_n - \boldsymbol{a}_m\|_1$, the sequence $\{a_{k,n}\}_n$ is Cauchy for each k, hence converges. Let $a_k = \lim_n a_{k,n}$. Fix $K \in \mathbb{N}$ and $n \ge N$. From (8.2),

$$\sum_{k=1}^K |a_{k,n} - a_{k,m}| < \varepsilon \ \text{ for all } m \ge N.$$

Letting $m \to +\infty$, we obtain $\sum_{k=1}^K |a_{k,n} - a_k| \le \varepsilon$. Since K was arbitrary,

$$\|\boldsymbol{a}_n - \boldsymbol{a}\|_1 = \sum_{k=1}^\infty |a_{k,n} - a_k| \le \varepsilon \ \text{ for all } \ n \ge N.$$

It follows that $\boldsymbol{a} \in \ell^1$ and $\boldsymbol{a}_n \to \boldsymbol{a}$. ◇

8.1.12 Definition. Let (X,d) and (Y,ρ) be metric spaces. The *product metric* $d \times \rho$ on $X \times Y$ is defined by

$$(d\times\rho)\big((x,y),(a,b)\big) := d(x,a) + \rho(y,b), \quad x,\, a \in X, y,\, b \in Y.$$

The pair $(X \times Y, d \times \rho)$ is called the *product of the metric spaces X and Y*. ◇

In Exercise 13 the reader is asked to prove, among other things, that $d \times \rho$ is indeed a metric and that a sequence $\{(x_n, y_n)\}$ converges to (a,b) in $X \times Y$ in this metric iff $x_n \to a$ in X and $y_n \to b$ in Y.

Exercises

1. Determine whether d is a metric on $\mathbb{R}^2$, where $d((x_1, x_2), (y_1, y_2)) =$

   (a) $2|x_1 - y_1| + 3|x_2 - y_2|$ (b) $|x_1^2 - y_1^2| + |x_2^2 - y_2^2|$

   (c) $|x_1^3 - y_1^3| + |x_2^3 - y_2^3|$ (d) $|x_1 - x_2| + |y_1 - y_2|$

   (e) $\dfrac{|x_1 - y_1| + |x_2 - y_2|}{2 + |x_1 - y_1| + |x_2 - y_2|}$ (f) $|e^{x_1} - e^{y_1}| + |e^{x_2} - e^{y_2}|$.

2. (*p-adic metric*). Let p be a fixed prime number. Define $\rho_p(n, n) = 0$, and for $m \neq n \in \mathbb{Z}$ define $\rho_p(m, n) = 1/p^\alpha$, where α is the power of p in the unique prime factorization of $|m - n|$. (For example, $\rho_2(42, 2) = 1/8$, $\rho_5(42, 2) = 1/5$, and $\rho_3(42, 2) = 1$.) Show that ρ_p is a metric on $\mathbb{Z}$.

3. (*Hamming distance*). Let A be a nonempty set and $X := A^n$. For $x = (x_1, \ldots, x_n)$ and $y = (y_1, \ldots, y_n)$ in X define $d(x, y)$ to be the number of indices j for which $x_j \neq y_j$. Show that d is a metric on X. (The metric is named after Richard Hamming, who pioneered the field of error correcting codes.)

4. Let X be as in Exercise 3. Define $\rho(x, x) = 0$, and for distinct points $x = (x_1, \ldots, x_n)$ and $y = (y_1, \ldots, y_n)$ in X define $\rho(x, y) = 2^{-j}$, where j is the smallest index for which $x_j \neq y_j$. Show that ρ is a metric on X.

5. Prove that a metric d satisfies

   $$|d(x, y) - d(a, b)| \leq d(x, a) + d(y, b).$$

   Conclude that if $x_n \to a$ and $y_n \to b$, then $d(x_n, y_n) \to d(a, b)$.

6. Let X and Y be nonempty sets and let $f : X \to Y$ be one-to-one. Show that if ρ is a metric on Y, then $d(x, y) := \rho(f(x), f(y))$ defines a metric on X.

7. Prove that a finite union of bounded sets in a metric space is bounded.

8. Prove 8.1.9.

9. ($\Downarrow$ 8.5.8.) Prove that a Cauchy sequence with a cluster point converges.

10. Let $E_1, \ldots, E_m$ be complete subspaces of (X, d). Prove that the finite union $E_1 \cup \cdots \cup E_m$ is complete. Does the analogous assertion hold for a countable union of complete subspaces?

11. Let $X := [1, +\infty)$ have the metric

    $$d(x, y) = |x^{-1} - y^{-1}|$$

    (see Exercise 6). Show that $x_n \to x$ with respect to the usual metric on X iff $x_n \to x$ with respect to d. Is (X, d) complete?

12. Same as Exercise 11 but with the metric

$$\rho(x, y) = \left|x(1 + x^2)^{-1} - y(1 + y^2)^{-1}\right|.$$

13. Let (X, d) and (Y, ρ) be metric spaces and let $(Z, \eta) = (X \times Y, d \times \rho)$ be the product space. Prove:

    (a) η is a metric on Z.
    (b) A sequence $\{(x_n, y_n)\}$ is Cauchy in Z iff $\{x_n\}$ is Cauchy in X and $\{y_n\}$ is Cauchy in Y.
    (c) A sequence $\{(x_n, y_n)\}$ converges to (x, y) in Z iff $x_n \to x$ in X and $y_n \to y$ in Y.
    (d) Z is complete iff X and Y are complete.

14. Metrics d, ρ on a set X are said to be *metrically equivalent* if there exist positive constants a and b such that

$$d(x, y) \le a\,\rho(x, y) \quad \text{and} \quad \rho(x, y) \le b\,d(x, y) \quad \text{for all} \quad x,\ y \in X.$$

For example, by Exercise 1.5.6, the metrics d_1, d_2, and d_∞ are metrically equivalent. Suppose that d and ρ are metrically equivalent. Let $\{x_n\}$ be a sequence in X and let $x \in X$. Prove the following:

    (a) $x_n \to x$ in (X, d) iff $x_n \to x$ in (X, ρ).
    (b) $\{x_n\}$ is Cauchy in (X, d) iff $\{x_n\}$ is Cauchy in (X, ρ).
    (c) (X, d) is complete iff (X, ρ) is complete.

15. Let d be a metric on a set X and $a > 0$. Define

$$\rho(x, y) := \min\{d(x, y), a\}.$$

Prove:

    (a) ρ is a metric on X.
    (b) A sequence is Cauchy in (X, ρ) iff it is Cauchy in (X, d).
    (c) A sequence converges in (X, ρ) iff it converges in (X, d).
    (d) (X, ρ) is complete iff (X, d) is complete.

    Are d and ρ metrically equivalent? Does $\sigma(x, y) := \max\{d(x, y), a\}$ define a metric on X?

16. Let ρ_1 and ρ_2 be metrics on X. Prove that $\max\{\rho_1, \rho_2\}$ is a metric. Is $\min\{\rho_1, \rho_2\}$ a metric?

17. Let $\boldsymbol{x} := (x_1, \ldots, x_n)$, $\boldsymbol{x}_k := (x_{1,k}, \ldots, x_{n,k}) \in \mathbb{R}^n$, $k = 1, 2, \ldots$. Prove:

    (a) $\boldsymbol{x}_k \to \boldsymbol{x}$ in $(\mathbb{R}^n, d_2)$ iff $x_{j,k} \to x_j$ for $j = 1, \ldots, n$.
    (b) $\{\boldsymbol{x}_k\}$ is Cauchy in $(\mathbb{R}^n, d_2)$ iff $\{x_{j,k}\}_{k=1}^\infty$ is Cauchy in $\mathbb{R}$ for each $j = 1, \ldots, n$.
    (c) $\mathbb{R}^n$ is complete in each of the metrics d_1, d_2, d_∞. (Use Exercise 14.)

18. Let d be a metric on a set X and define

$$\rho(x, y) := \frac{d(x, y)}{1 + d(x, y)}.$$

Verify that (a)–(d) of Exercise 15 hold. Are d and ρ metrically equivalent?

19. Let ρ_1 and ρ_2 be metrics on a set X and let α, $\beta > 0$. Define

$$\rho(x, y) := \alpha\rho_1(x, y) + \beta\rho_2(x, y).$$

Prove:

(a) ρ is a metric on X.

(b) A sequence $\{x_n\}$ converges to x in (X, ρ) iff it converges to x in both (X, ρ_1) and (X, ρ_2).

20. Let $\{d_k\}_{k=1}^{\infty}$ be a sequence of metrics on a set X. For x, $y \in X$ define

$$\rho_k(x, y) = \frac{d_k(x, y)}{1 + d_k(x, y)} \quad \text{and} \quad \rho(x, y) = \sum_{k=1}^{\infty} 2^{-k} \rho_k(x, y).$$

Prove:

(a) ρ is a metric on X (See Exercise 18)

(b) $\rho(x_n, x) \to 0$ iff $d_k(x_n, x) \to 0$ for every k.

21. Let $\mathcal{C}(\mathbb{R})$ denote the set of continuous, real-valued functions on $\mathbb{R}$. For f, $g \in \mathcal{C}(\mathbb{R})$ define

$$\rho(f, g) := \sum_{k=1}^{\infty} 2^{-k} \rho_k(f, g),$$

where

$$\rho_k(f, g) = \frac{d_k(f, g)}{1 + d_k(f, g)}, \quad d_k(f, g) = \sup_{-k \le x \le k} |f(x) - g(x)|.$$

Prove:

(a) ρ is a metric on $\mathcal{C}(\mathbb{R})$.

(b) $f_n \to f$ in this metric iff $f_n \to f$ uniformly on each bounded subset of $\mathbb{R}$.

(c) $\mathcal{C}(\mathbb{R})$ is complete in this metric.

22. For $f \in \mathcal{C}([a, b])$ define $\|f\|_1 = \int_a^b |f|$. Show that $\|\cdot\|_1$ is a norm on $\mathcal{C}([a, b])$ and that $\mathcal{C}([a, b])$ is not complete in the metric induced by this norm.

23. Show that the sequence of functions

$$f_n(x, y) = (1 + x^n)^{1/n}(1 + y^n)^{-1/n}$$

converges uniformly to $f(x, y) = x/y$ on $[1, b] \times [1, b]$ for any $b > 1$.

8.2 Open and Closed Sets

Throughout this section, (X, d) denotes an arbitrary metric space.

It is frequently useful to formulate assertions regarding a metric space X in terms of subsets of X rather than in terms of the metric. The subsets of greatest interest in this regard are described in the next two definitions.

8.2.1 Definition. Let $x \in X$ and $r > 0$. The sets

$$B_r(x) := \{y \in X : d(x,y) < r\} \quad \text{and} \quad C_r(x) := \{y \in X : d(x,y) \le r\}$$

are called, respectively, the *open and closed balls with center x and radius r*. The set

$$S_r(x) := C_r(x) \setminus B_r(x) = \{y \in X : d(x,y) = r\}$$

is called the *sphere with center x and radius r*. The ball $B_r(x)$ is also called a *neighborhood of x*. ♢

The open (closed) balls in $\mathbb{R}$ with the usual metric are simply the bounded open (closed) intervals. The spheres are the endpoints of these intervals. The open (closed) balls in Euclidean space $\mathbb{R}^2$ are open (closed) disks and the spheres are circles. The open and closed balls in a discrete metric space X are the sets X and $\{x\}$; the spheres are $X \setminus \{x\}$ and the empty set.

8.2.2 Definition. A subset U of X is said to be *open* if either $U = \emptyset$ or else U has the following property:

$$\text{For each } x \in U \text{ there exists } \varepsilon > 0 \text{ such that } B_\varepsilon(x) \subseteq U.$$

A subset of X is *closed* if its complement is open. The collection of all open sets is called the *(metric) topology* of (X, d). ♢

In any metric space, X and $\emptyset$ are both open and closed. There are many metric spaces for which these are the only subsets that are both open and closed; Euclidean space $\mathbb{R}^n$ is an important example (see Section 8.7). The sets $\mathbb{Q}$ and $\mathbb{I}$ are neither open nor closed in $\mathbb{R}$ since every open ball (that is, open interval) contains members of both sets. A finite set F is always closed. Indeed, if $x \in F^c$, then $B_r(x) \subseteq F^c$, where $r = \min\{d(x,y) : y \in F\}$, hence F^c is open.

8.2.3 Proposition. *An open ball is open, a closed ball is closed, and a sphere is closed.*

Proof. Let $x \in B_r(x_0)$. We claim that $B_\varepsilon(x) \subseteq B_r(x_0)$, where $\varepsilon := r - d(x, x_0)$. Indeed, if $y \in B_\varepsilon(x)$ then

$$d(y, x_0) \le d(y, x) + d(x, x_0) < \varepsilon + d(x, x_0) = r,$$

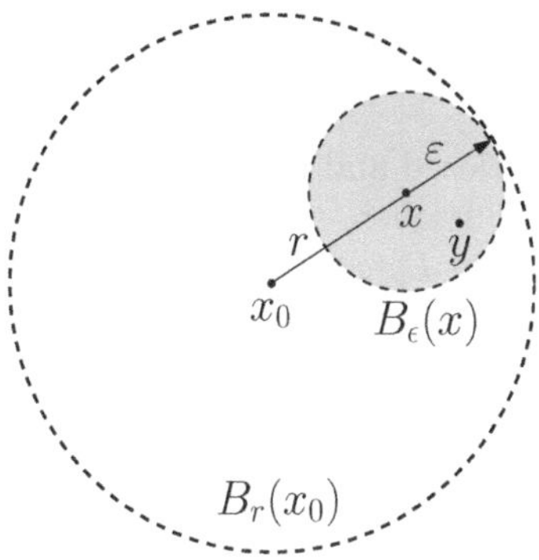

FIGURE 8.1: An open ball is open.

hence $y \in B_r(x_0)$ (Figure 8.1). Since x was arbitrary, $B_r(x_0)$ is open. A similar argument shows that $C_r(x_0)^c$ and $S_r(x_0)^c$ are open, hence $C_r(x_0)$ and $S_r(x_0)$ are closed (See Exercise 2) That $S_r(x_0)$ is closed also follows from 8.2.6 below. □

8.2.4 Theorem. *Open sets in (X, d) have the following properties:*

(a) *If $U_\mathfrak{i}$ is open for each $\mathfrak{i}$ in an index set $\mathfrak{I}$, then $\bigcup_{\mathfrak{i}\in\mathfrak{I}} U_\mathfrak{i}$ is open.*

(b) *If $V_1, \ldots, V_n$ are open, then $V_1 \cap \cdots \cap V_n$ is open.*

Proof. (a) Let U denote the union. If $x \in U$, then $x \in U_\mathfrak{i}$ for some $\mathfrak{i}$, hence there exists $r > 0$ such that $B_r(x) \subseteq U_\mathfrak{i} \subseteq U$. Therefore, U is open.

(b) Let V denote the intersection and let $x \in V$. For each $j = 1, \ldots, n$, there exists $r_j > 0$ such that $B_{r_j}(x) \subseteq V_j$. Then $B_r(x) \subseteq V$, where $r = \min\{r_1, \ldots, r_n\}$. Therefore, V is open. □

8.2.5 Corollary. *A nonempty subset U is open iff it is the union of open balls.*

For example, in a discrete metric space every subset is a union of open balls $\{x\} = B_1(x)$ and hence is open. It follows that every subset is also closed.

8.2.6 Corollary. *Closed sets in (X, d) have the following properties:*

(a) *If $C_\mathfrak{i}$ is closed for each $\mathfrak{i}$ in an index set $\mathfrak{I}$, then $\bigcap_{\mathfrak{i}\in\mathfrak{I}} C_\mathfrak{i}$ is closed.*

(b) *If $C_1, \ldots, C_n$ are closed, then $C_1 \cup \cdots \cup C_n$ is closed.*

Proof. In (a), each set $C_\mathfrak{i}^c$ is open, hence, using DeMorgan's law and 8.2.4,

$$\left(\bigcap_{\mathfrak{i}\in\mathfrak{I}} C_\mathfrak{i}\right)^c = \bigcup_{\mathfrak{i}\in\mathfrak{I}} C_\mathfrak{i}^c$$

is open, that is, $\bigcap_{\mathfrak{i}\in\mathfrak{I}} C_\mathfrak{i}$ is closed. Part (b) is proved in a similar manner, using DeMorgan's law for complements of finite unions. □

8.2.7 Theorem. *A subset C of X is closed iff C contains the limit of each convergent sequence in C.*

Proof. Assume that C is closed and let $\{x_n\}$ be a sequence in C with $x_n \to x$. If $x \notin C$, then, because C^c is open, there exists $\varepsilon > 0$ such that $B_\varepsilon(x) \cap C = \emptyset$. But then x_n is eventually in $B_\varepsilon(x) \subseteq C^c$, impossible. Therefore, $x \in C$.

Now suppose C is not closed. Then C^c is not open, hence there exists $x \in C^c$ such that $B_{1/n}(x) \not\subseteq C^c$, that is, $B_{1/n}(x) \cap C \neq \emptyset$, for every $n \in \mathbb{N}$. Choosing a point x_n in this intersection, we obtain a sequence $\{x_n\}$ in C that converges to a point not in C. Thus the sequential criterion fails. □

8.2.8 Corollary. *Let (X, d) be a metric space and let Y be a subspace of X.*

(a) *If X is complete and Y is closed, then Y is complete.*

(b) *If Y is complete, then Y is closed.*

Proof. (a) Let $\{y_n\}$ be a Cauchy sequence in Y. Since X is complete, there exists $x \in X$ such that $y_n \to x$. Since Y is closed, $x \in Y$. Therefore, Y is complete.

(b) Let $\{y_n\}$ be a sequence in Y such that $y_n \to x \in X$. Then $\{y_n\}$ is Cauchy and hence converges to some $y \in Y$. Since limits are unique, $x = y$. Therefore, $x \in Y$, hence Y is closed. □

8.2.9 Example. Let $\mathcal{C}([a, b])$ denote the set of all continuous real-valued functions on the interval $[a, b]$. Each such function is bounded, hence $\mathcal{C}([a, b])$ is a vector subspace of $\mathcal{B}([a, b])$ (8.1.10). Since the uniform limit of continuous functions is continuous (7.2.2), $\mathcal{C}([a, b])$ is closed in the uniform metric. Since $\mathcal{B}([a, b])$ is complete, 8.2.8(a) shows that $\mathcal{C}([a, b])$ is complete. ◊

8.2.10 Example. The subspace $\mathcal{D}([a, b])$ of $\mathcal{C}([a, b])$ consisting of all differentiable functions is not complete in the uniform metric. To see this take $[a, b] = [0, 1]$ and define a sequence of continuous functions $g_n(x)$, $n \geq 2$, on $[0, 1]$ such that $g_n = 1$ on $[0, 1/2]$, $g_n = 0$ on $[1/2 + 1/n, 1]$, and g_n is linear on $[1/2, 1/2 + 1/n]$. Also, define $g(t)$ on $[0, 1]$ by $g = 1$ on $[0, 1/2]$ and $g = 0$ on $(1/2, 1]$ (See Figure 8.3)

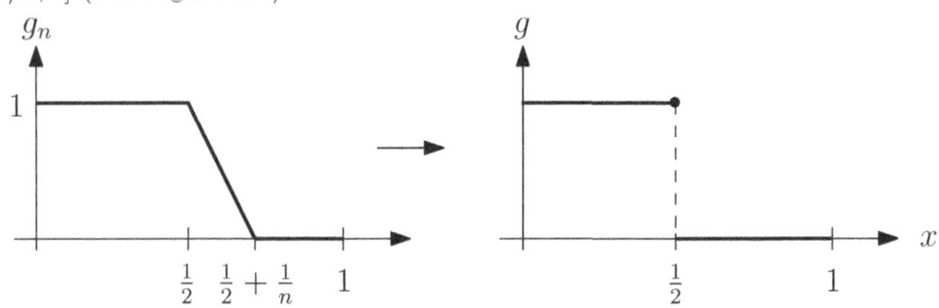

FIGURE 8.2: The functions g_n and g.

Now set

$$f_n(x) = \int_0^x g_n(t)\,dt \text{ and } f(x) = \int_0^x g(t)\,dt, \qquad x \in [0,1].$$

Then $f_n \in \mathcal{D}([0,1])$, $f \in \mathcal{C}([0,1])$, and

$$|f_n(x) - f(x)| \le \int_0^1 |g_n - g| = \int_{1/2}^{1/2+1/n} g_n = \frac{1}{2n}.$$

Therefore, $f_n \to f$ uniformly on $[0,1]$. Since f is not differentiable at $1/2$, $\mathcal{D}([0,1])$ is not closed. ◊

8.2.11 Definition. Let Y be a subset of X. A subset $A \subseteq Y$ is said to be *relatively open (relatively closed) in* Y if A is open (closed) in the subspace (Y,d) of (X,d). ◊

8.2.12 Theorem. *Let $A \subseteq Y \subseteq X$. Then A is relatively open (relatively closed) in Y iff $A = Y \cap B$ for some open (closed) subset B of X.*

Proof. By definition, a nonempty open set A in the subspace Y is a union of open balls in Y. The latter are of the form $Y \cap B_r(y)$, where $y \in Y$ and $B_r(y)$ is an open ball of X. Therefore, $A = Y \cap B$, where B is the corresponding union of the open balls $B_r(y)$.

From the first paragraph, the closed sets of Y are of the form $Y \setminus A = Y \cap B^c$, where B is open in X. Since B^c is closed in X, the assertion regarding closed sets follows. □

8.2.13 Definition. Let $\mathcal{X}$ be a vector space and let $\boldsymbol{a}, \boldsymbol{b} \in \mathcal{X}$. The *line segment from* $\boldsymbol{a}$ *to* $\boldsymbol{b}$ is defined by

$$[\boldsymbol{a} : \boldsymbol{b}] = \{(1-t)\boldsymbol{a} + t\boldsymbol{b} \,:\, 0 \le t \le 1\}.$$

A subset E of $\mathcal{X}$ is said to be *convex* if $\boldsymbol{a}, \boldsymbol{b} \in E$ implies $[\boldsymbol{a} : \boldsymbol{b}] \subseteq E$. ◊

FIGURE 8.3: Convex and non-convex sets.

Recall that, by definition, the convex subsets of $\mathbb{R}$ are the intervals. The reader may easily check that if $D \subseteq \mathbb{R}^p$ and $E \subseteq \mathbb{R}^q$ are convex, then $D \times E$, as a subset of $\mathbb{R}^{p+q}$, is convex. In particular, Cartesian products $I_1 \times \cdots \times I_n$ of intervals I_j are convex in $\mathbb{R}^n$. Other examples are given in Exercise 5.

Exercises

1. Sketch $B_1(0,0) \subseteq \mathbb{R}^2$ for the metrics d_1 and d_∞ derived from the norms $\|\cdot\|_1$ and $\|\cdot\|_\infty$.

2. Prove that a closed ball is closed.

3. Let x, y be distinct points in a metric space (X, d). Find the largest number r such that $B_r(x) \cap B_r(y) = \emptyset$.

4. Show that every open subset U of $\mathbb{R}^n$ is a countable union of open balls as well as a countable union of bounded open n-dimensional intervals $(a_1, b_1) \times \cdots \times (a_n, b_n)$.

5. Prove that open and closed balls in a normed vector space are convex. Are spheres convex?

6. Show by example that arbitrary intersections of open sets may not be open and that arbitrary unions of closed sets may not be closed.

7. Metrics d and ρ on a set X are said to be *topologically equivalent* if they have the property that a sequence $\{x_n\}$ converges to x in (X, d) iff it converges to x in (X, ρ).

   (a) Prove that metrically equivalent metrics are topologically equivalent (See Exercise 8.1.14)

   (b) Prove that d and ρ are topologically equivalent iff (X, d) and (X, ρ) have the same topologies, that is, the metrics produce the same open sets.

   (c) Are topologically equivalent metrics necessarily metrically equivalent?

8. Prove that the metric $\rho(x, y) = |e^x - e^y|$ on $\mathbb{R}$ is topologically equivalent to the usual metric. Is $\mathbb{R}$ complete in this metric? Is ρ metrically equivalent to the usual metric on $\mathbb{R}$?

9. Let Y be a subspace of (X, d) with the property that for some $r > 0$, $d(x, y) \geq r$ for all x, $y \in Y$ with $x \neq y$. Prove that Y is complete, hence closed. Conclude that finite metric spaces, discrete metric spaces, and the subspaces $\mathbb{N}$ and $\mathbb{Z}$ of $\mathbb{R}$ are complete.

10. Let $x_n \to x_0$ in (X, d). Prove that the set $C := \{x_0, x_1, x_2, \ldots\}$ is closed in X.

11. Let Y be open (closed) in (X, d). Prove that a subset U of Y is relatively open (relatively closed) in Y iff it is open (closed) in X.

12. Prove that the set

$$C := \{f \in \mathcal{C}([0,1]) \ : \ f(x) = f(1-x) \text{ for all } x \in [0,1]\}$$

is closed in the supremum metric (8.1.10) but not in the metric of Exercise 8.1.22.

13. Prove that the subspaces

$$\mathcal{V} := \left\{ f \in \mathcal{B}([0,+\infty)) \;:\; \lim_{x\to+\infty} f(x) \text{ exists in } \mathbb{R} \right\} \quad \textit{and}$$

$$\mathcal{W} := \left\{ f \in \mathcal{V} : \lim_{x\to+\infty} f(x) = 0 \right\}$$

are closed in the supremum metric.

8.3 Closure, Interior, and Boundary

Throughout this section, (X, d) denotes an arbitrary metric space.

8.3.1 Definition. Let $E \subseteq X$.

- The *closure* $\mathrm{cl}(E) = \mathrm{cl}_X(E)$ of E in X is the intersection of all closed subsets of X containing E.
- The *interior* $\mathrm{int}(E) = \mathrm{int}_X(E)$ of E is the union of all open subsets of X contained in E.
- The *boundary* $\mathrm{bd}(E) = \mathrm{bd}_X(E)$ of E is the set $\mathrm{cl}(E) \setminus \mathrm{int}(E)$. ♢

8.3.2 Examples. (a) Since every nonempty open set of $\mathbb{R}$ (with the usual metric) contains rational and irrational points,

$$\mathrm{int}(\mathbb{Q}) = \mathrm{int}(\mathbb{I}) = \emptyset \text{ and } \mathrm{cl}(\mathbb{Q}) = \mathrm{cl}(\mathbb{I}) = \mathbb{R}, \text{ hence } \mathrm{bd}(\mathbb{Q}) = \mathrm{bd}(\mathbb{I}) = \mathbb{R}.$$

For bounded intervals we have

$$\mathrm{cl}((a,b)) = [a,b], \quad \mathrm{int}([a,b]) = (a,b), \quad \text{and} \quad \mathrm{bd}((a,b)) = \mathrm{bd}([a,b]) = \{a,b\}.$$

(b) In a discrete metric space, a subset E is both open and closed, hence $\mathrm{cl}(E) = \mathrm{int}(E) = E$ and $\mathrm{bd}(E) = \emptyset$. ♢

By 8.2.4 and 8.2.6, $\mathrm{int}(E)$ is open and $\mathrm{cl}(E)$ is closed, hence $\mathrm{bd}(E)$ is closed. The following proposition asserts that $\mathrm{int}(E)$ is the largest open set contained in E and $\mathrm{cl}(E)$ is the smallest closed set containing E.

8.3.3 Proposition. *If U is open, C is closed, and $U \subseteq E \subseteq C$, then*

$$U \subseteq \mathrm{int}(E) \subseteq E \subseteq \mathrm{cl}(E) \subseteq C.$$

Proof. Simply note that U is one of the open sets in the definition of $\mathrm{int}(E)$ and that C is one of the closed sets in the definition of $\mathrm{cl}(E)$. □

8.3.4 Corollary. *Let $E \subseteq X$. Then*

(a) *E is open in X iff* $\text{int}(E) = E$ (b) $\text{int}\big(\text{int}(E)\big) = \text{int}(E)$

(c) *E is closed in X iff* $\text{cl}(E) = E$ (d) $\text{cl}\big(\text{cl}(E)\big) = \text{cl}(E)$

Proof. If E is open, take $U = E$ in the proposition. If E is closed, take $C = E$. This proves (a) and (c). Parts (b) and (d) follow from these. □

8.3.5 Proposition. *For any subset E of X,*

(a) $\text{cl}(E^c) = \big(\text{int}(E)\big)^c$ (b) $\text{int}(E^c) = \big(\text{cl}(E)\big)^c$

(c) $\text{bd}(E) = \text{cl}(E) \cap \text{cl}(E^c) = \text{bd}(E^c)$

Proof. For (a) we have

$$\big(\text{int}(E)\big)^c = \Bigg[\bigcup_{\substack{U \subseteq E \\ U \text{ open}}} U\Bigg]^c = \bigcap_{\substack{C \supseteq E^c \\ C \text{ closed}}} C = \text{cl}(E^c).$$

Parts (b) and (c) follow from (a). □

8.3.6 Proposition. *Let $E \subseteq X$. Then $x \in \text{cl}(E)$ iff there exists a sequence $\{a_n\}$ in E such that $a_n \to x$.*

Proof. Let C be the set of all limits of convergent sequences in E, including constant sequences, so $E \subseteq C$. We show that $C = \text{cl}(E)$, which will establish the proposition.

First, C is closed. If not, then C^c is not open, hence there exists $y \in C^c$ and for each n a point $y_n \in B_{1/n}(y) \cap C$. By definition of C, each y_n is the limit of a sequence in E, hence there exists $a_n \in E$ such that $d(y_n, a_n) < 1/n$. By the triangle inequality, $d(a_n, y) < 2/n$, hence $a_n \to y$. But then $y \in C$, a contradiction. Therefore, C must be closed.

It follows that $\text{cl}(E) \subseteq C$. Since $\text{cl}(E)$ contains the limit of all convergent sequences in E (8.2.7), $C \subseteq \text{cl}(E)$. Therefore, $C = \text{cl}(E)$. □

8.3.7 Example. (Topologist's sine curve). Let

$$A = \{(x, \sin(1/x)) \,:\, 0 < x \le 2/\pi\} \quad \text{and} \quad B = \{0\} \times [-1, 1].$$

We show that $\text{cl}(A) = A \cup B$.

For the inclusion $A \cup B \subseteq \text{cl}(A)$, note first that

$$\left\{\sin\frac{1}{x} \,:\, \frac{2}{(4n+3)\pi} \le x \le \frac{2}{(4n+1)\pi}\right\} = [-1, 1], \quad n \in \mathbb{Z}^+.$$

It follows from the intermediate value theorem that for each $y \in [-1, 1]$ and $n \in \mathbb{N}$ there exists $x_n \in \mathbb{R}$ such that

$$0 < x_n \le \frac{2}{(4n+1)\pi} \quad \text{and} \quad \sin(1/x_n) = y.$$

Since $(x_n, y) \in A$ and $(x_n, y) \to (0, y)$, we see that $(0, y) \in \mathrm{cl}(A)$. Therefore, $B \subseteq \mathrm{cl}(A)$, hence $A \cup B \subseteq \mathrm{cl}(A)$.

The reverse inclusion will follow if we show that $A \cup B$ is closed. For this we use 8.2.7. Let $\{(x_n, y_n)\}$ be a sequence in $A \cup B$ with $(x_n, y_n) \to (x, y)$.

Case 1. There exists a subsequence $\{(x_{n_k}, y_{n_k})\}$ that lies in B. Then, since B is closed, $(x, y) \in B$.

Case 2. $\{(x_n, y_n)\}$ eventually lies in A, so $y_n = \sin(1/x_n)$ for all sufficiently large n. Since $\lim_{t\to 0} \sin(1/t)$ does not exist, x cannot be zero, hence $y = \sin(1/x)$, that is, $(x, y) \in A$.

In each case $(x, y) \in A \cup B$, hence $A \cup B$ is closed. ◊

8.3.8 Definition. A subset E of X is said to be *dense in* X if $\mathrm{cl}(E) = X$; equivalently, every $x \in X$ is the limit of a sequence in E. ◊

By 8.3.2, $\mathbb{Q}$ and $\mathbb{I}$ are dense in $\mathbb{R}$. The set of all points in $\mathbb{R}^2$ with rational coordinates is dense in $\mathbb{R}^2$. A discrete space has no proper dense subsets. In Section 8.8 we show that the set of polynomials on $[a, b]$ is dense in $\mathcal{C}([a, b])$ in the uniform norm.

8.3.9 Example. (Dirichlet). If ξ is irrational, then the set

$$E := \{n\xi + m : m \in \mathbb{Z},\ n \in \mathbb{N}\}$$

is dense in $\mathbb{R}$. To verify this we show that for any $x \in \mathbb{R}$ and $k \in \mathbb{N}$ there exists $z \in E$ such that $|z - x| < 1/k$.

To this end, let

$$y_j = j\xi - \lfloor j\xi \rfloor, \quad j = 1, \ldots, k+1.$$

Because ξ is irrational, $0 < y_j < 1$, hence y_j must be in one of the intervals $(0, 1/k), (1/k, 2/k), \ldots, ((k-1)/k, 1)$. Since there are only k intervals, one of these must contain y_i and y_j for some $i \neq j$.[1] By the irrationality of ξ, $y_j \neq y_i$. Hence one of the quantities $\pm(y_j - y_i)$, call it y, is in E and $|y| < 1/k$.

We now consider two cases. If $y > 0$, choose $m \in \mathbb{Z}$ such that $x + m > 0$ and let n be the smallest integer such that $ny > x + m$. Then $n \in \mathbb{N}$ and $(n-1)y \le x + m$, hence $z := ny - m \in E$ and

$$0 < z - x = ny - m - x \le y < 1/k.$$

On the other hand, if $y < 0$, choose $m \in \mathbb{Z}$ such that $x + m < 0$ and let n be the smallest integer such that $n(-y) > -(x + m)$, that is, $ny < x + m$. Again, $z := ny - m \in E$, and in this case, since $(n-1)y \ge x + m$,

$$-1/k < y \le ny - m - x = z - x < 0.$$

In either case, $|z - x| < 1/k$, as required. ◊

[1]This is an instance of the so-called *pigeon hole principle*.

8.3.10 Example. We show that the set $S := \{\sin n : n \in \mathbb{N}\}$ is dense in the interval $[-1, 1]$. Let $x \in \mathbb{R}$ and take $\xi = 1/2\pi$ in the preceding example. Then $n_k/2\pi + m_k \to x$ for some integer sequences $\{n_k\}$ and $\{m_k\}$ with $n_k > 0$, hence

$$\sin n_k = \sin\left[2\pi(n_k/2\pi + m_k)\right] \to \sin(2\pi x).$$

Since x was arbitrary, every member of $[-1, 1]$ is the limit of a sequence in S. A similar argument shows that $\{\cos n : n \in \mathbb{N}\}$ is dense in $[-1, 1]$. ◊

8.3.11 Definition. A metric space is said to be *separable* if it has a countable dense subset. ◊

For example, $\mathbb{R}^n$ is separable (consider all points with rational coordinates). An uncountable discrete space is not separable. The space $\mathcal{C}([a, b])$ is separable in the supremum norm (Exercise 19).

Exercises

1. Let (X, d) be a metric space and $A, B \subseteq X$. Prove the following:

   (a) $\text{cl}(A \cup B) = \text{cl}(A) \cup \text{cl}(B)$ (b) $\text{cl}(A \cap B) \subseteq \text{cl}(A) \cap \text{cl}(B)$
   (c) $\text{int}(A \cap B) = \text{int}(A) \cap \text{int}(B)$ (d) $\text{int}(A \cup B) \supseteq \text{int}(A) \cup \text{int}(B)$
   (e) $\text{bd}(A \cup B) \subseteq \text{bd}(A) \cup \text{bd}(B)$ (f) $\text{bd}(\text{cl}(A)) \subseteq \text{bd}(A)$
   (g) $\text{bd}(\text{int}(A)) \subseteq \text{bd}(A)$ (h) $\text{cl}(A) = A \cup \text{bd}(A)$.

   Show by examples that the inclusions may be strict.

2. Prove: $\text{bd}(A \cap B) \subseteq \big(A \cap \text{bd}(B)\big) \cup \big(B \cap \text{bd}(A)\big) \cup \big(\text{bd}(A) \cap \text{bd}(B)\big)$. Show that the inclusion may be strict.

3. Find $\text{cl}(A) \setminus A$ for $A =$

   (a) $\{(1/n, 1/m) : m, n \in \mathbb{N}\}$ (b) $\{(\cos t, \sin t, e^{-t}) : t > 0\}$
   (c) $\left\{\left(\cos t, \sin t, \dfrac{t}{1+|t|}\right) : t \in \mathbb{R}\right\}$ (d) $\{(t\cos t, t\sin t, t) : t > 0\}$
   (e) $\left\{\left(\dfrac{\cos t}{1+t}, \dfrac{\sin t}{1+t}\right) : t > 0\right\}$ (f) $\left\{\left(\dfrac{t\cos t}{1+t}, \dfrac{t\sin t}{1+t}\right) : t > 0\right\}$.

4. An induction argument shows that parts (a) and (c) of Exercise 1 hold for any finite number of sets. Show, by example, that the analogous statements for infinitely many sets are false.

5. Prove that if $\text{cl}(A) \cap \text{cl}(B) = \emptyset$, then $\text{int}(A \cup B) = \text{int}(A) \cup \text{int}(B)$.

6. Let Y be a subspace of (X, d) and $A \subseteq Y$. Prove:
   (a) $\text{cl}_Y(A) = \text{cl}_X(A) \cap Y$ (b) $\text{int}_X(A) \cap Y \subseteq \text{int}_Y(A)$
   (c) $\text{bd}_Y(A) \subseteq \text{bd}_X(A)$.
   Show by examples that the inclusions in (b) and (c) may be strict.

7. Let $x_n \to x_0$ in X. Show that $\text{cl}(\{x_1, x_2, \ldots\}) = \{x_0, x_1, x_2, \ldots\}$.

8. Let $f_n(x) = x^n$, $0 \le x \le 1$. Show that the set $\{f_1, f_2, \ldots\}$ is closed in $(\mathcal{C}([0,1]), \|\cdot\|_\infty)$. Is it closed in $(\mathcal{C}([0,1]), \|\cdot\|_1)$?

9. Let $B = B_r(x_0)$ and $C = C_r(x_0)$. Prove that

   (a) $B \subseteq \text{int}(C)$ (b) $\text{cl}(B) \subseteq C$ (c) $\text{bd}(B) \subseteq C \setminus B$.

   Show, by example, that the inclusions may be strict.

10. Prove that in a normed vector space the inclusions in Exercise 9 are equalities.

11. Prove that the set $E = \{(x, y) : x, y \in \mathbb{Q} \text{ and } x \ne y\}$ is neither open nor closed and is dense in Euclidean space $\mathbb{R}^2$.

12. Let $x \in \mathbb{R}$, $r \in \mathbb{Q}$, $r \ne 0$. In each case, find the largest interval in which the given set is dense.

    (a) $\{\sin(rn) : n \in \mathbb{N}\}$ (b) $\{\sin(x + n) : n \in \mathbb{N}\}$

    (c) $\{\sin n \cos n : n \in \mathbb{N}\}$ (d) $\{\tan^2 n : n \in \mathbb{N}\}$.

13. Show that $\lim_n \sin(\pi n x)$ does not exist for any irrational number x. Conclude that $\lim_n \sin(nr)$ does not exist for any nonzero rational number r.

14. (a) Let E be dense in X and let F be a proper finite subset of E. Show that $E \setminus F$ is dense in $X \setminus F$. Is $E \setminus F$ is necessarily dense in X?

    (b) Let $\mathcal{X}$ be a normed vector space with $\{\boldsymbol{a}_1, \boldsymbol{a}_2, \ldots\}$ dense in $\mathcal{X}$. Show that $\{\boldsymbol{a}_n : n \ge N\}$ is dense in $\mathcal{X}$ for every $N \in \mathbb{N}$. Is the corresponding result true for general metric spaces?

    Conclude, for example, that $\{\sin n : n \ge N\}$ is dense in $[-1, 1]$.

15. Show that $\liminf_n \sin n = -1$ and $\limsup_n \sin n = 1$.

16. Let Y be dense in X and $U \subseteq X$ open. Show that $U \cap Y$ is dense in U. What if U is not open?

17. Let $X = \mathbb{R}^n$ with the Euclidean metric and let $Y \subseteq X$ have the property of Exercise 8.2.9. Prove that Y^c is open and dense in X. Conclude that $\mathbb{N}^c$ and $\mathbb{Z}^c$ are open and dense in $\mathbb{R}$.

18. Show that in a separable space, every nonempty open set U is a countable union of open balls.

19. Use the Weierstrass approximation theorem (8.8.5, below) to show that $(\mathcal{C}([a,b]), \|\cdot\|_\infty)$ is separable

20. (a) Let $\{I_\mathfrak{i} : \mathfrak{i} \in \mathfrak{I}\}$ be a family of open intervals in $\mathbb{R}$ with the property that each pair has a nonempty intersection. Show that $\bigcup_{\mathfrak{i} \in \mathfrak{I}} I_\mathfrak{i}$ is an open interval.

    (b) Prove that every nonempty open set in $\mathbb{R}$ is a countable union of disjoint open intervals.

8.4 Limits and Continuity

> *In this section, (X, d), (Y, ρ), and (Z, μ) denote arbitrary metric spaces.*

8.4.1 Definition. Let $E \subseteq X$. A member a of X is said to be an *accumulation point* of E if $E \cap \big(B_r(a) \setminus \{a\}\big) \neq \emptyset$ for each $r > 0$. A member of E that is not an accumulation point is called an *isolated point* of E. ◊

It follows from the definition that a is an accumulation point of E iff there exists a sequence of *distinct* points of E converging to a. No subset of a discrete metric space has an accumulation point. The set of functions $x \mapsto x^n$ in $\mathcal{C}([0,1])$, $n \in \mathbb{N}$, has no accumulation points in the uniform norm, but the zero function is an accumulation point of this sequence in the norm $\|f\|_1 = \int_0^1 f(x)\,dx$.

8.4.2 Definition. Let $E \subseteq X$, $f : E \to Y$, and let $a \in X$ be either a member of E or an accumulation point of E. If $b \in Y$, we write

$$b = \lim_{\{x \to a,\, x \in E\}} f(x)$$

if for each $\varepsilon > 0$ there exists $\delta > 0$ such that

$$x \in E \quad \text{and} \quad d(x, a) < \delta \text{ implies } \rho(f(x), b) < \varepsilon. \tag{8.3}$$

In the special case $E = X \setminus \{a\}$, we write simply $b = \lim_{x \to a} f(x)$. ◊

Note that condition (8.3) may be written

$$f\big(E \cap B_\delta(a)\big) \subseteq B_\varepsilon(b).$$

This observation will be useful later in proving a global characterization of continuity.

Many of the results in Chapter 3 on limits of functions on subsets of $\mathbb{R}$ hold for real-valued functions defined on a metric space. These include the theorems on limits of sums, products, and quotients of functions, the comparison theorem, the squeeze principle, and the sequential characterization of limit. The statements and proofs are essentially the same: simply replace $|x - y|$ by the metric $d(x, y)$. For future reference, we explicitly state:

8.4.3 Sequential Characterization of Limit. *Let a be an accumulation point of $E \subseteq X$ and let $f : E \to Y$. Then $\lim_{\{x \to a,\, x \in E\}} f(x)$ exists and equals $b \in Y$ iff $f(a_n) \to b$ for all sequences $\{a_n\}$ in E with $a_n \to a$.*

The following theorem gives sufficient conditions for a double limit to equal an iterated limit.

8.4.4 Iterated Limit Theorem. *Let $X \times Y$ have the product metric $\eta := d \times \rho$, and let a and b be accumulation points of $X \setminus \{a\}$ and $Y \setminus \{b\}$, respectively. Suppose that $f : X \times Y \setminus \{(a,b)\} \to Z$ has the following properties:*

(a) $\lim_{y\to b} f(x,y)$ *exists in Z for each $x \in X$, and*

(b) $\lim_{(x,y)\to(a,b)} f(x,y)$ *exists in Z.*

Then $\lim_{x\to a}\lim_{y\to b} f(x,y)$ *exists and*

$$\lim_{x\to a}\lim_{y\to b} f(x,y) = \lim_{(x,y)\to(a,b)} f(x,y),$$

Proof. Let $g(x)$ denote the limit in (a) and z the limit in (b). Given $\varepsilon > 0$ choose $\delta > 0$ such that

$$\mu\big(f(x,y), z\big) < \varepsilon \quad \text{for all } (x,y) \in X \times Y \text{ with } 0 < \eta\big((x,y),(a,b)\big) < \delta.$$

Let $0 < d(x,a) < \delta$. Then, for all y sufficiently near b, $\eta\big((x,y),(a,b)\big) < \delta$, hence

$$\mu\big(g(x), z\big) \le \mu\big(g(x), f(x,y)\big) + \mu\big(f(x,y), z\big) < \mu\big(g(x), f(x,y)\big) + \varepsilon.$$

Letting $y \to b$ in this inequality, noting that $f(x,y) \to g(x)$, we obtain $\mu\big(g(x), z\big) \le \varepsilon$. This shows that $\lim_{x\to a} g(x) = z$. □

We can state a slightly expanded version of the theorem as

$$\lim_{(x,y)\to(a,b)} f(x,y) = \lim_{x\to a}\lim_{y\to b} f(x,y) = \lim_{y\to b}\lim_{x\to a} f(x,y)$$

whenever the limit on the left exists and inner limits on the right exist for each x and y, respectively. The limits on the right are called *iterated limits* and the limit on the left is sometimes called a *double limit*. In particular, if the iterated limits exist and are unequal, then the double limit cannot exist.

In many cases, the iterated limit theorem (suitably modified) still holds if f is defined on subsets E of $X \times Y$ more general than $X \times Y \setminus \{(a,b)\}$. This is the case in Examples **(c)** and **(d)** that follow.

8.4.5 Examples. In (a)–(e), $X = Y = Z = \mathbb{R}$. Note that in this case the product metric η is equivalent to the Euclidean metric on $\mathbb{R}^2$.

(a) Let $E = (0, +\infty) \times (0, +\infty)$. To calculate the limit

$$\lim_{\substack{(x,y)\to(0,0)\\(x,y)\in E}} \frac{\sin(x+2y)}{2x+y}$$

we write the function as

$$\frac{\sin(x+2y)}{x+2y}\,\frac{x+2y}{2x+y}.$$

As $(x, y) \to (0, 0)$ along E, the first factor tends to 1 but the second factor has no limit. Indeed, along a path $y = mx$, $m > 0$, $x > 0$,

$$\frac{x + 2y}{2x + y} = \frac{x + 2mx}{2x + mx} = \frac{1 + 2m}{2 + m}.$$

Therefore, the double limit cannot exist. Note that the iterated limits exist and are unequal:

$$\lim_{y\to 0+} \lim_{x\to 0+} f(x, y) = \lim_{y\to 0+} \frac{\sin(2y)}{y} = 2,$$
$$\lim_{x\to 0+} \lim_{y\to 0+} f(x, y) = \lim_{x\to 0+} \frac{\sin x}{2x} = \frac{1}{2}.$$

(b) Let E be as in **(a)** and let p, $q > 0$. The limit

$$L := \lim_{\substack{(x,y)\to(0,0)\\(x,y)\in E}} \frac{x^p + y^q}{x^2 + y^2}$$

exists iff $p, q > 2$ or $p = q = 2$. In the former case, $L = 0$ and in the latter, $L = 1$. This is best seen by converting to polar coordinates $x = r\cos\theta$, $y = r\sin\theta$, $0 < \theta < \pi/2$, so that the above limit may be written

$$L = \lim_{r\to 0+} \left(r^{p-2}\cos^p\theta + r^{q-2}\sin^q\theta\right).$$

Note that both iterated limits exist iff $p, q \geq 2$.

(c) Let $E = \{(x, y) : x > 0, y > 0, x \neq y\}$. Then the limit

$$\lim_{\substack{(x,y)\to(0,0)\\(x,y)\in E}} \frac{x^p - y^p}{x - y}$$

exists iff $p \geq 1$ and has zero limit if $p > 1$. Indeed, if $0 < x < y$, then, by the mean value theorem, there exists $t \in (x, y)$ such that $x^p - y^p = pt^{p-1}(x - y)$, hence

$$px^{p-1} < \frac{x^p - y^p}{x - y} < py^{p-1},$$

and the assertion follows from the squeeze principle. Clearly, the iterated limits exist (hence equal the double limit) iff $p \geq 1$.

(d) Let E be as in **(c)**. The limit

$$\lim_{\substack{(x,y)\to(0,0)\\(x,y)\in E}} \frac{x^p + y^p}{y - x}$$

does not exist for any value of p. Indeed, along the path $y = mx$, $m, x > 0$, $m \neq 1$, the function has values

$$\frac{x^p + (mx)^p}{mx - x} = \frac{x^{p-1}(1 + m^p)}{1 - m}$$

so the limit cannot exist if $p \leq 1$. Now assume $p > 1$ and set $\theta_r = mr^{p-1} + \pi/4$. Along the path given by

$$\begin{aligned} x = r\cos\theta_r &= \frac{r}{\sqrt{2}}\left[\cos\left(mr^{p-1}\right) - \sin\left(mr^{p-1}\right)\right] \\ y = r\sin\theta_r &= \frac{r}{\sqrt{2}}\left[\cos\left(mr^{p-1}\right) + \sin\left(mr^{p-1}\right)\right], \end{aligned}$$

where $r \downarrow 0$, the function has values

$$\frac{x^p + y^p}{y - x} = \frac{1}{\sqrt{2}m} \frac{mr^{p-1}}{\sin(mr^{p-1})}\left(\cos^p\theta_r + \sin^p\theta_r\right),$$

which tends to $2^{(1-p)/2}/m$ as $r \to 0$.

Note that neither of the iterated limits exists if $p < 1$. On the other hand, if $p > 1$ then clearly

$$\lim_{x\to 0}\lim_{y\to 0}\frac{x^p + y^p}{y - x} = \lim_{y\to 0}\lim_{x\to 0}\frac{x^p + y^p}{y - x} = 0,$$

and if $p = 1$, then

$$\lim_{x\to 0}\lim_{y\to 0}\frac{x^p + y^p}{y - x} = -1, \quad \text{while} \quad \lim_{y\to 0}\lim_{x\to 0}\frac{x^p + y^p}{y - x} = 1.$$

(e) Let $E = \{(x, y) : x > 0, y > 0\}$. Then the limit

$$\lim_{\substack{(x,y)\to(0,0)\\(x,y)\in E}} \frac{x^p y}{x + y}$$

exists iff $p > 0$, in which case the limit is zero. Indeed, along the path $y = mx$ the function has values $mx^p/(1 + m)$, so the limit cannot exist if $p \leq 0$. If $p > 0$, one can introduce polar coordinates as in (b). Both iterated limits exist iff $p \geq 0$, but are unequal if $p = 0$. ◊

8.4.6 Definition. A function $f : X \to Y$ is said to be *continuous at a point* $a \in X$ if $\lim_{x\to a} f(x) = f(a)$. Also, f is said to be *continuous on a set* $E \subseteq X$ if f is continuous at each point of E. If $E = X$, then f is simply said to be *continuous.* If f is one-to-one and onto Y and if $f^{-1} : Y \to X$ is continuous, then f is called a *homeomorphism.* ◊

From the sequential characterization of limit we have

8.4.7 Sequential Characterization of Continuity. *Let $f : X \to Y$ and $a \in X$. Then f is continuous at a iff $f(a_n) \to f(a)$ for all sequences $\{a_n\}$ in X with $a_n \to a$.*

The next theorem gives an important global characterization of continuity.

8.4.8 Theorem. *Let $f : X \to Y$. The following statements are equivalent:*

(a) *f is continuous.*

(b) *$f^{-1}(V)$ is open in X for each open subset V of Y.*

(c) *$f^{-1}(C)$ is closed in X for each closed subset C of Y.*

Proof. That (b) and (c) are equivalent follows from the general set-theoretic identity $f^{-1}(B^c) = \left(f^{-1}(B)\right)^c$.

(a) $\Rightarrow$ (b): Let $V \subseteq Y$ be open. If $x \in f^{-1}(V)$, then $f(x) \in V$ so there exists $\varepsilon > 0$ such that $B_\varepsilon\big(f(x)\big) \subseteq V$. By continuity there exists $\delta > 0$ such that $f\big(B_\delta(x)\big) \subseteq B_\varepsilon\big(f(x)\big)$. Therefore, $f\big(B_\delta(x)\big) \subseteq V$, hence $B_\delta(x) \subseteq f^{-1}(V)$.

(b) $\Rightarrow$ (a): Let $x \in X$ and $\varepsilon > 0$. Since $U := f^{-1}\left(B_\varepsilon\big(f(x)\big)\right)$ is open in X and contains x, we may choose $\delta > 0$ such that $B_\delta(x) \subseteq U$. Then

$$f\big(B_\delta(x)\big) \subseteq f(U) \subseteq B_\varepsilon\big(f(x)\big),$$

which shows that f is continuous at x. □

8.4.9 Definition. A function $f : X \to Y$ is said to be *uniformly continuous on a set* $E \subseteq X$ if for each $\varepsilon > 0$ there exists a $\delta > 0$ such that

$$\rho(f(u), f(v)) < \varepsilon \quad \text{for all } u, v \in E \text{ with } d(u,v) < \delta. \qquad \Diamond$$

8.4.10 Example. The function

$$f(x,y) = \frac{1}{2.1 + \sin x + \sin y}$$

is uniformly continuous on $\mathbb{R}^2$. Indeed, for all (x,y), $(a,b) \in \mathbb{R}^2$,

$$\begin{aligned} |f(x,y) - f(a,b)| &= \frac{|\sin x + \sin y) - (\sin a + \sin b)|}{(2.1 + \sin x + \sin y)(2.1 + \sin a + \sin b)} \\ &\le \frac{|\sin x - \sin a| + |\sin y - \sin b|}{(2.1 + \sin x + \sin y)(2.1 + \sin a + \sin b)} \\ &\le 100|\sin x - \sin a| + 100|\sin y - \sin b)| \\ &\le 100(|x-a| + |y-b|) \\ &\le 200\sqrt{(x-a)^2 + (y-b)^2}. \qquad \Diamond \end{aligned}$$

The proof of the following theorem is entirely analogous to that of 3.5.2. The details are left to the reader.

8.4.11 Sequential Characterization of Uniform Continuity. *A function $f : X \to Y$ is uniformly continuous on $E \subseteq X$ iff $\rho\big(f(u_n), f(v_n)\big) \to 0$ for all sequences $\{u_n\}$ and $\{v_n\}$ in E with $d(u_n, v_n) \to 0$.*

For example, every function on a discrete metric space is uniformly continuous, since eventually $u_n = v_n$. The indefinite integral function $F(f)(x) = \int_a^x f(t)\,dt$ on the space $\mathcal{C}([a,b])$ is uniformly continuous with respect to the uniform norm, since $\|f_n - g_n\|_\infty \to 0 \Rightarrow \|F(f_n) - F(g_n)\|_\infty \to 0$. The addition function $(x,y) \mapsto x+y$ is uniformly continuous on $\mathbb{R}^2$ since $(x_n, y_n) - (a_n, b_n) \to (0,0)$ obviously implies that $(x_n + y_n) - (a_n + b_n) \to 0$. On the other hand, the multiplication function $(x,y) \mapsto xy$ is not uniformly continuous on $\mathbb{R}^2$, since $(n+1/n, n+1/n) - (n,n) \to 0$ but $(n+1/n)^2 - n^2 \to 2$.

Exercises

1. For each of the functions $f(\boldsymbol{x})$ below, find $\lim_{\{\boldsymbol{x}\to\boldsymbol{0},\, \boldsymbol{x}\in E\}} f(\boldsymbol{x})$ and the corresponding iterated limits or show that the limits fail to exist. In each case take E to be the natural domain of the function.

   (a) $\dfrac{y^2 + \sin^2 x}{3x^2 + 2y^2}$ (b) $\dfrac{x^2 y}{5x^2 + 2y^2}$ (c) $\dfrac{x^2 y^2}{x^4 + 7y^4}$.

   (d) $\dfrac{\sin x \sin y}{\sqrt{x^2 + y^2}}$ (e) $\dfrac{x^4}{x^4 - xy^2 + y^4}$ (f) $(x+y)\sin\dfrac{1}{x^2 + y^2}$

   (g) $\dfrac{\sin(3xy^2 + 2xy^3)}{xy^2}$ (h) $\dfrac{xy^2\cos(xy)}{x^2 + y^2}$ (i) $\dfrac{\sqrt{(1+x^2)(1+y^2)} - 1}{x^2 + y^2}$

   (j) $\dfrac{3x + 2y}{(x^2 + y^2)^{1/3}}$ (k) $\dfrac{1 - \cos(xy)}{\sin x \sin y}$ (l) $\dfrac{x^2 + |y|^{2.1}}{x^2 + y^2}$

   (m) $\dfrac{1 - \cos\sqrt{|xy|}}{|x|^p}$ (n) $\dfrac{\sin x \pm \sin y}{x - y}$ (o) $\dfrac{x - y}{\ln x - \ln y}$

   (p) $\dfrac{x|y|^{1.1}}{\sin^2\sqrt{x^2 + y^2}}$ (q) $\dfrac{3x^2 + 2y^2 + z^2}{x^2 + y^2 + z^2}$ (r) $\dfrac{xy + yz + xz}{\sqrt{x^2 + y^2 + z^2}}$.

2. Let $a > 0$, $p > 1$. Evaluate the limit
$$\lim_{\substack{(x,y)\to(0,0)\\(x,y)\in E}} \frac{x^2 - 5y^2}{x^2 + 3y^2}$$
for the sets

   (a) $E = \{(x,y) : |y| \le a|x|^p|\}$ (b) $E = \{(x,y) : |y| < |x|\}$.

3. Let f be continuously differentiable on $(-\pi/2, \pi/2)$. Define g on the set
$$E := \{(x,y) \in (-\pi/2, \pi/2)^2 : x \ne y\}$$
by
$$g(x,y) = \frac{f(x) - f(y)}{\sin x - \sin y}.$$
Show that g has a continuous extension to $(-\pi/2, \pi/2) \times (-\pi/2, \pi/2)$.

4. Let f and g be continuously differentiable on some open interval (a, b) and suppose that g' is never 0. Define h on the set
$$E := \{(x, y) \in (a, b)^2 \ : \ x \neq y\}$$
by
$$h(x, y) = \frac{f^2(x) - f^2(y)}{g(x) - g(y)}.$$
Prove that h has a continuous extension to $(a, b) \times (a, b)$.

5. Let $f : X \to Y$. Prove that the following statements are equivalent:
   (a) f is continuous.
   (b) $f\big(\mathrm{cl}(A)\big) \subseteq \mathrm{cl}\big(f(A)\big)$ for each subset A of X.
   (c) $\mathrm{cl}\big(f^{-1}(B)\big) \subseteq f^{-1}(\mathrm{cl}(B))$ for each subset B of Y.
   (d) $f^{-1}\big(\mathrm{int}(B)\big) \subseteq \mathrm{int}\big(f^{-1}(B)\big)$ for each subset B of Y.

6. Show that $d : X \times X \to \mathbb{R}$ is uniformly continuous with respect to the product metric $\eta := d \times d$ on $X \times X$.

7. Let $f : [0, a) \to \mathbb{R}$ and
$$g(x, y) := f\Big(\sqrt{x^2 + y^2}\Big), \quad \sqrt{x^2 + y^2} < a.$$
   (a) Prove that g is uniformly continuous iff f is uniformly continuous.
   (b) Use (a) to show that the functions
$$\sqrt{x^2 + y^2}, \quad \frac{1}{\sqrt{x^2 + y^2 + 1}}, \quad \text{and} \quad \sin\sqrt{x^2 + y^2}$$
   are uniformly continuous on $\mathbb{R}^2$ but $\sin(x^2 + y^2)$ is not.

8. Let $f(x)$ be uniformly continuous on $\mathbb{R}$. Prove that the function $g(x, y) := f(\alpha x + \beta y)$ is uniformly continuous on $\mathbb{R}^2$. Give an example of a bounded uniformly continuous function f on $\mathbb{R}$ such that the function $h(x, y) := f(xy)$ is not uniformly continuous on $\mathbb{R}^2$.

9. Show that the function $f(x, y) = (1 - \sin x \sin y)^{-1}$ is uniformly continuous on the set
$$E_r := [-\pi/2 + r, \pi/2 - r] \times [-\pi/2 + r, \pi/2 - r]$$
for any $0 < r < \pi/2$, but is not uniformly continuous on
$$E := (-\pi/2, \pi/2) \times (-\pi/2, \pi/2).$$

10. Let $f : (X, d) \to (Y, \rho)$ and $g : (Y, \rho) \to (Z, \mu)$ be (uniformly) continuous. Prove that $g \circ f : (X, d) \to (Z, \mu)$ is (uniformly) continuous.

11. Let $f : X \to \mathbb{R}^k$, say $f(x) = \big(f_1(x), \dots, f_k(x)\big)$. Prove that f is (uniformly) continuous iff each f_j is (uniformly) continuous.

12. Let $f_n : (X, d) \to (Y, \rho)$ converge uniformly to f on X. Prove that if each f_n is (uniformly) continuous, then f is (uniformly) continuous.

8.5 Compact Sets

Throughout this section, (X, d) and (Y, ρ) denote arbitrary metric spaces.

Compactness is one of the most important concepts in analysis. For example, it allows the formulation of results such as the extreme value theorem and the uniform continuity theorem in the context of general metric spaces. It is also the key feature that distinguishes the finite dimensional space $\mathbb{R}^n$ from its infinite dimensional counterparts ℓ^∞ and ℓ^1.

8.5.1 Definition. Let $E \subseteq X$. A collection $\mathcal{U} = \{U_\mathfrak{i} : \mathfrak{i} \in \mathfrak{I}\}$ of subsets of X is called a *cover* of E if E is contained in the union of the sets $U_\mathfrak{i}$. If each $U_\mathfrak{i}$ is open, then $\mathcal{U}$ is called an *open cover of* E. A cover $\mathcal{U}$ of E is said to have a *finite subcover* if there exists a finite subset $\mathfrak{I}_0$ of $\mathfrak{I}$ such that $\{U_\mathfrak{i} : \mathfrak{i} \in \mathfrak{I}_0\}$ is a cover of E. If every open cover of E has a finite subcover, then E is said to be *compact*. ◊

Finite subsets of a metric space are compact. In a discrete metric space, these are the only compact sets. Indeed, if E is an infinite subset of a discrete space, then $\big\{\{x\} : x \in E\big\}$ is an open cover of E with no finite subcover.

8.5.2 Proposition. *A compact subset of a metric space is closed and bounded.*

Proof. Let E be compact and let $a \in E^c$. For each $x \in E$, let U_x and V_x denote disjoint open balls with centers x and a, respectively (see Figure 8.4). Then $\{U_x : x \in E\}$ is an open cover of E, hence there exists a finite subset E_0 of E such that $\{U_x : x \in E_0\}$ covers E. Set $V = \bigcap_{x \in E_0} V_x$. Then V is an open ball with center a, and since $V \cap U_x = \emptyset$ for each $x \in E_0$, $V \subseteq E^c$. Therefore, E^c is open.

To show that E is bounded, choose any $x \in X$ and consider the open cover $\{B_n(x) : n \in \mathbb{N}\}$ of E. Let F be a finite subset of $\mathbb{N}$ such that $\{B_n(x) : n \in F\}$ covers E. Then $E \subseteq B_m(x)$, where m is the largest member of F. □

The converse of 8.5.2 is false. For example, the set $\mathbb{Q} \cap [-\sqrt{2}, \sqrt{2}]$ is closed and bounded in $\mathbb{Q}$ but not compact. Indeed, if $\{r_n\}$ is a sequence in $\mathbb{Q}$ with $r_n \uparrow \sqrt{2}$, then $\{(-r_n, r_n) : n \in \mathbb{N}\}$ is an open cover of $\mathbb{Q} \cap [-\sqrt{2}, \sqrt{2}]$ with no finite subcover. For another example, consider a discrete metric space. Here, the entire metric space is closed and bounded but only finite sets are compact.

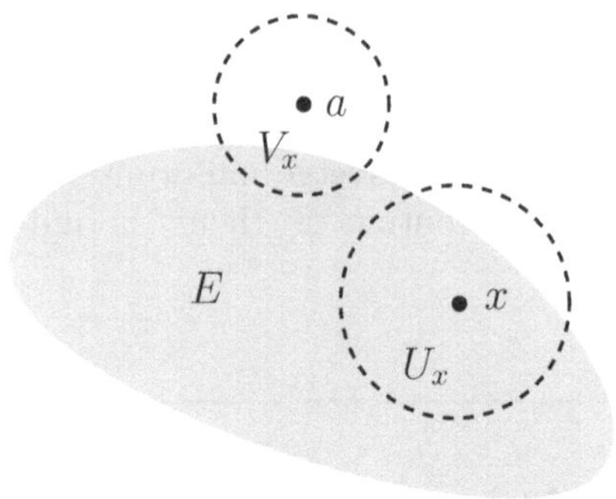

FIGURE 8.4: The neighborhoods U_x and V_x.

8.5.3 Proposition. *A closed subset of a compact metric space is compact.*

Proof. Let X be compact, $E \subseteq X$ closed, and let $\mathcal{U} = \{U_\mathfrak{i} : \mathfrak{i} \in \mathfrak{I}\}$ be an open cover of E. Then $\mathcal{U} \cup \{E^c\}$ is an open cover of X, hence there exists a finite subset $\mathfrak{I}_0$ of $\mathfrak{I}$ such that $X = E^c \cup \bigcup_{\mathfrak{i}\in\mathfrak{I}_0} U_\mathfrak{i}$. It follows that $E \subseteq \bigcup_{\mathfrak{i}\in\mathfrak{I}_0} U_\mathfrak{i}$. □

Closely related to compactness is the notion of *total boundedness.*

8.5.4 Definition. Let $E \subseteq X$ and $\varepsilon > 0$. An *ε-net for E* is a set $F \subseteq X$ such that $\{B_\varepsilon(x) : x \in F\}$ covers E. E is said to be *totally bounded* if for each $\varepsilon > 0$ there exists a finite ε-net for E. ◊

An ε-net F for E has the property that every member of E is within ε of a member of F. For example, $\mathbb{Q}$ is an ε-net for $\mathbb{R}$ for any ε, and $\mathbb{Z}$ is a .51-net for $\mathbb{R}$.

The following proposition shows that the set F in the definition of total boundedness may be taken to be a subset of E.

8.5.5 Proposition. *If E has a finite ε-net F, then E has a finite 2ε-net contained in E.*

Proof. For each $x \in F$, apply the following procedure: If $E \cap B_\varepsilon(x) = \emptyset$, remove x from F. Otherwise, choose any $a \in E \cap B_\varepsilon(x)$ and replace $B_\varepsilon(x)$ by $B_{2\varepsilon}(a)$ and x in F by a. Since $B_\varepsilon(x) \subseteq B_{2\varepsilon}(a)$, the revised set is a finite 2ε-net for E contained in E. □

Since a finite union of open balls is bounded (Exercise 8.1.3), every totally bounded set is bounded. The converse is false. For example, in a discrete space all sets are bounded but no infinite set can be totally bounded.

Open and closed balls in $\mathcal{C}([0,1])$ with the supremum norm are bounded but not totally bounded (Exercise 8). Contrast this with the following example:

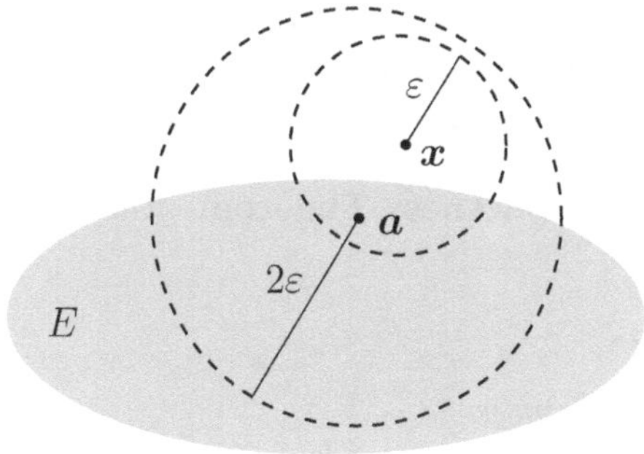

FIGURE 8.5: A 2ε-net.

8.5.6 Example. Every bounded subset E of $\mathbb{R}^n$ is totally bounded. To see this, let $\varepsilon > 0$ and $0 < \delta < 2\varepsilon/\sqrt{n}$, and choose $k \in \mathbb{N}$ so large that $E \subseteq [-k\delta, k\delta]^n$. Subdividing, we see that I is a finite union of sets of the form

$$J := [a_1, b_1] \times [a_2, b_2] \times \cdots \times [a_n, b_n], \quad \text{where} \quad b_j - a_j = \delta.$$

(See Figure 8.6.) The largest diagonal in J has length $\sqrt{n}\,\delta < 2\varepsilon$, hence J may

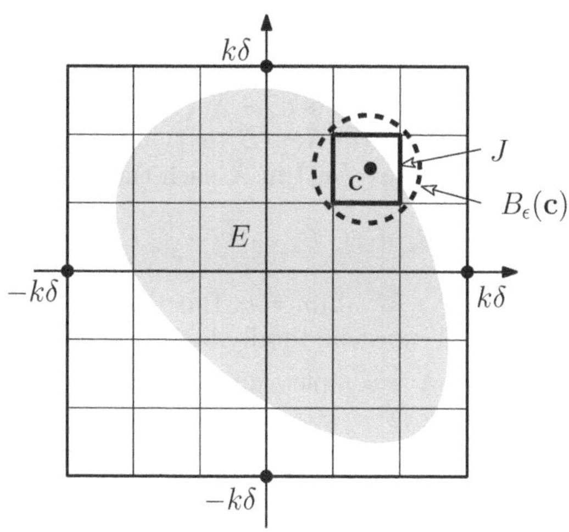

FIGURE 8.6: A bounded set in $\mathbb{R}^n$ is totally bounded.

be enclosed in an open ball with radius ε and center $\mathbf{c} = (c_1, \ldots c_n)$, where $c_j = (a_j + b_j)/2$. The resulting collection of balls is a finite ε-cover of E. ◊

8.5.7 Definition. A subset E of X is said to be *sequentially compact* if every sequence in E has a cluster point in E. ◊

By the Bolzano–Weierstrass theorem, closed and bounded intervals in $\mathbb{R}$ are sequentially compact. The same is not true in $\mathbb{Q}$; for example, the set

$\mathbb{Q} \cap [0, \sqrt{2}]$ is not sequentially compact. In a discrete space, no infinite set can be sequentially compact since sequences with distinct terms cannot converge.

The following theorem is one of the fundamental results of real analysis.

8.5.8 Sequential Compactness Theorem. *The following statements are equivalent:*

(a) *X is compact.*

(b) *X is sequentially compact.*

(c) *X is complete and totally bounded.*

Proof. (a) $\Rightarrow$ (b): We prove the contrapositive $\sim$(b) $\Rightarrow$ $\sim$(a). Let $\{a_n\}$ be a sequence in X with no cluster point. Then for each $x \in X$ there must exist an open ball $B(x)$ with center x that contains only finitely many terms of the sequence. This implies that every finite subcollection of the open cover $\{B(x) : x \in X\}$ of X contains only finitely many terms of the sequence and hence cannot cover X. Therefore, X is not compact.

(b) $\Rightarrow$ (c): Let X be sequentially compact and let $\{a_n\}$ be a Cauchy sequence in X. By hypothesis, $\{a_n\}$ has a convergent subsequence, say $a_{n_k} \to a \in X$. By Exercise 8.1.9, $a_n \to a$. Therefore, X is complete.

Suppose that X is not totally bounded. Then there exists $\varepsilon > 0$ such that no finite collection of open balls of radius ε covers X. Choose any $a_1 \in X$. Since $B_\varepsilon(a_1)$ does not cover X, there exists $a_2 \in X \setminus B_\varepsilon(a_1)$. Since $B_\varepsilon(a_1) \cup B_\varepsilon(a_2)$ does not cover X, there exists $a_3 \in X \setminus \big(B_\varepsilon(a_1) \cup B_\varepsilon(a_2)\big)$. Continuing in this fashion, we construct a sequence $\{a_n\}$ in X such that

$$a_n \in X \setminus \big[B_\varepsilon(a_1) \cup B_\varepsilon(a_2) \cup \cdots \cup B_\varepsilon(a_{n-1})\big].$$

It follows that $d(a_n, a_m) \geq \varepsilon$ for all $m \neq n$. But then no subsequence of $\{a_n\}$ can converge. Therefore, X must be totally bounded.

(c) $\Rightarrow$ (a): Assume that X is complete and totally bounded but not compact. Then X has an open cover $\mathcal{U} = \{U_\mathfrak{i} : \mathfrak{i} \in \mathfrak{I}\}$ with no finite subcover. For each k let F_k be a finite set of points in X such that $\{B_{1/k}(x) : x \in F_k\}$ is a cover of X. Consider the case $k = 1$. If for each $x \in F_1$ the ball $B_1(x)$ could be covered by finitely many members of $\mathcal{U}$, then X itself would have such a cover, contradicting our assumption. Thus there exists $x_1 \in F_1$ such that $E_1 := B_1(x_1)$ cannot be covered by finitely many members of $\mathcal{U}$. Since $\{B_{1/2}(x) : x \in F_2\}$ covers X, $\{E_1 \cap B_{1/2}(x) : x \in F_2\}$ covers E_1, so by similar reasoning there exists $x_2 \in F_2$ such that $E_2 := E_1 \cap B_{1/2}(x_2)$ cannot be covered by finitely many members of $\mathcal{U}$. In this manner we construct a sequence of points x_n in X and decreasing sets

$$E_n = B_1(x_1) \cap B_{1/2}(x_2) \cap \cdots \cap B_{1/n}(x_n) = E_{n-1} \cap B_{1/n}(x_n) \tag{8.4}$$

that cannot be covered by finitely many members of $\mathcal{U}$. In particular, $E_n \neq \emptyset$. Choose a point $y_n \in E_n$. If $n > m$, then $y_n \in E_m$, hence from (8.4)

$$d(x_m, x_n) \leq d(x_m, y_n) + d(y_n, x_n) < 1/m + 1/n.$$

It follows that $\{x_n\}$ is a Cauchy sequence. Since X is complete, $x_n \to x$ for some $x \in X$. Choose $\mathfrak{i} \in \mathfrak{I}$ such that $x \in U_\mathfrak{i}$. Since $U_\mathfrak{i}$ is open, there exists $r > 0$ such that $B_r(x) \subseteq U_\mathfrak{i}$. Next, choose $n > 2/r$ such that $d(x_n, x) < r/2$. By the triangle inequality, $B_{1/n}(x_n) \subseteq B_r(x)$. But then $E_n \subseteq U_\mathfrak{i}$, contradicting the noncovering property of E_n. Therefore, X must be compact, completing the proof. □

8.5.9 Corollary. (Heine–Borel Theorem) *A subset of $\mathbb{R}^n$ is compact iff it is closed and bounded.*

Proof. We have already seen that a compact set in a metric space is closed and bounded. Conversely, let $C \subseteq \mathbb{R}^n$ be closed and bounded. Since $\mathbb{R}^n$ is complete (Exercise 8.1.17), C is complete (8.2.8). Since C is bounded, it is totally bounded (8.5.6). By the theorem, C is compact. □

The validity of the preceding corollary ultimately rests on the finite dimensionality of $\mathbb{R}^n$. For infinite dimensional normed spaces such as $\mathcal{C}([0,1])$, a closed and bounded set need not be compact (Exercise 8). In the next section, we characterize the compact subsets of spaces like $\mathcal{C}([0,1])$.

8.5.10 Theorem. *If $f : X \to Y$ is continuous and X is compact, then $f(X)$ is compact.*

Proof. Let $\{V_\mathfrak{i} : \mathfrak{i} \in \mathfrak{I}\}$ be an open cover of $f(X)$ in Y. For each $\mathfrak{i} \in \mathfrak{I}$, set $U_\mathfrak{i} = f^{-1}(V_\mathfrak{i})$. Then $\{U_\mathfrak{i} : \mathfrak{i} \in \mathfrak{I}\}$ is an open cover of X, hence there exists a finite subset $\mathfrak{I}_0$ of $\mathfrak{I}$ such that $\{U_\mathfrak{i} : \mathfrak{i} \in \mathfrak{I}_0\}$ is a cover of X. It follows that $\{V_\mathfrak{i} : \mathfrak{i} \in \mathfrak{I}_0\}$ is a finite cover of $f(X)$. □

8.5.11 Corollary. *Let $f : X \to Y$ be continuous, one-to-one, and onto Y. If X is compact then $f^{-1} : Y \to X$ is continuous, hence f is a homeomorphism.*

Proof. Let $g = f^{-1}$ and let C be a closed subset of X. Then C is compact (8.5.3), hence, by the theorem, $g^{-1}(C) = f(C)$ is compact and therefore closed in Y (8.5.2). By 8.4.8, g is continuous. □

Corollary 8.5.11 is false for noncompact X (Exercise 19).

8.5.12 Extreme Value Theorem. *If $f : X \to \mathbb{R}$ is continuous and X is compact, then there exist points x_m and x_M in X such that*

$$f(x_m) \leq f(x) \leq f(x_M) \quad \textit{for all} \quad x \in X.$$

Proof. By 8.5.10 and 8.5.2, $f(X)$ is closed and bounded in $\mathbb{R}$ and therefore contains its supremum and infimum. □

8.5.13 Theorem. *If $f : X \to Y$ is continuous and X is compact, then f is uniformly continuous.*

Proof. Let $\varepsilon > 0$. By continuity, for each $x \in X$ there exists $\gamma_x > 0$ such that

$$f\big(B_{\gamma_x}(x)\big) \subseteq B_{\varepsilon/2}\big(f(x)\big). \tag{8.5}$$

Set $\delta_x = \gamma_x/2$. The collection $\{B_{\delta_x}(x) : x \in X\}$ is an open cover of X, hence there exists a finite set $F \subseteq X$ such that the collection $\{B_{\delta_x}(x) : x \in F\}$ covers X. Let $\delta := \min_{x\in F}\delta_x$ and let a, $b \in X$ with $d(a,b) < \delta$. Choose $x \in F$ such that $a \in B_{\delta_x}(x)$. Then

$$d(x,a) < \delta_x < \gamma_x \quad \text{and} \quad d(x,b) \le d(a,b) + d(x,a) < \delta_x + \delta_x = \gamma_x,$$

so a, $b \in B_{\gamma_x}(x)$. By (8.5),

$$\rho\big(f(a), f(b)\big) \le \rho\big(f(a), f(x)\big) + \rho\big(f(x), f(b)\big) < \varepsilon/2 + \varepsilon/2 = \varepsilon.$$

Therefore, f is uniformly continuous. □

The following is a generalization of 3.5.9.

8.5.14 Corollary. *Let X be compact, Y complete, E a dense subset of X, and $f : E \to Y$ continuous. The following statements are equivalent:*

(a) $\lim_{\{x\to a,\, x\in E\}} f(x)$ *exists for each $a \in X$.*

(b) *f has a continuous extension to X; that is, there exists a continuous function $g : X \to Y$ such that $g|_E = f$.*

(c) *f is uniformly continuous on E.*

Proof. (a) $\Rightarrow$ (b): For each $a \in X$ define $g(a) = \lim_{\{x\to a,\, x\in E\}} f(x)$. Since f is continuous, $g|_E = f$. If g is not continuous at $a \in X$, then there exist $\varepsilon > 0$ and a sequence $\{x_n\}$ in X such that $x_n \to a$ and $\rho\big(g(x_n), g(a)\big) \ge \varepsilon$ for all n. By definition of $g(x_n)$, for each n we may choose $a_n \in E$ such that $d(x_n, a_n) < 1/n$ and $\rho\big(g(x_n), f(a_n)\big) < \varepsilon/2$. Then $a_n \to a$ but

$$\rho\big(f(a_n), g(a)\big) \ge \rho\big(g(x_n), g(a)\big) - \rho\big(g(x_n), f(a_n)\big) > \varepsilon/2,$$

contradicting the definition of $g(a)$. Therefore, g is continuous.

(b) $\Rightarrow$ (c): By 8.5.13, g is uniformly continuous on X, hence f is uniformly continuous on E.

(c) $\Rightarrow$ (a): Let $a \in X$ and let $\{x_n\}$ be a sequence in E such that $x_n \to a$. Since f is uniformly continuous, $\{f(x_n)\}$ is Cauchy and therefore converges to some $b \in Y$. If $\{y_n\}$ is another sequence in E such that $y_n \to a$, then $d(y_n, x_n) \to 0$ so, by uniform continuity again, $\rho\big(f(y_n), f(x_n)\big) \to 0$, hence $f(y_n) \to b$. By the sequential criterion for limits, $\lim_{\{x\to a,\, x\in E\}} f(x)$ exists and equals b. □

Exercises

1. Determine which of the following subsets of $\mathbb{R}^2$ are closed, bounded, or compact.

   (a) $\{(x, y) : 2x^2 + y^2 + 6y \leq 8x\}$ (b) $\{(x, y) : 3x^2 + 2y \leq 6x\}$

   (c) $\{(x, y) : xy = 1\}$ (d) $\{(x, y) : x^{1/3} + y^{1/3} = 1\}$

   (e) $\{(x, y) : x^{2/3} + y^{2/3} = 1\}$ (f) $\left\{\left[\dfrac{x\cos x}{1+x}, \dfrac{x\sin x}{1+x}\right] : x \geq 0\right\}$

   (g) $\{(e^{-x}\cos x, e^{-x}\sin x) : x \geq 0\}$ (h) $\{(x, y) : x^3/y + y^3/x > 0\}$.

2. Let $\{x_n\}$ be a convergent sequence in X with $x_n \to x_0$. Prove that the set $\{x_0, x_1, x_2, \ldots\}$ is compact.

3. Prove that a finite union of totally bounded (compact) sets is totally bounded (compact).

4. Prove that the intersection of an arbitrary family of compact subsets of a metric space X is compact.

5. Prove that $X \times Y$ is compact in the product metric $\eta := d \times \rho$ iff X and Y are compact.

6. Prove that the closure of a totally bounded subset of a metric space is totally bounded.

7. Prove that a subset E of a complete metric space X is totally bounded iff every sequence in E has a cluster point in X.

8. Prove that in $\big(\mathcal{C}([0, 1]), \|\cdot\|_\infty\big)$ the closed ball with radius 1 and center the zero function is not compact.

9. Let $\mathcal{C}_0([0, +\infty))$ be the vector subspace of $\mathcal{B}([0, +\infty))$ consisting of all real-valued continuous functions f on $[0, +\infty)$ such that $\lim_{x\to+\infty} f(x) = 0$. Prove that $\mathcal{C}_0([0, +\infty))$ is closed in the uniform norm and that the closed ball $C_1(0)$ in $\mathcal{C}_0([0, +\infty))$ with radius 1 and center the zero function is not compact and therefore is not totally bounded.

10. For $n \in \mathbb{N}$, define $f_n \in \mathcal{B}([0, +\infty))$ by $f_n = 1$ on $[n, n + 1]$ and zero elsewhere. Prove that the set $E := \{f_1, f_2, \ldots\}$ is bounded but not totally bounded in the sup metric.

11. (Cantor's intersection theorem). Let $C_1, C_2, \ldots$ be a sequence of nonempty compact subsets of a metric space X such that $C_{n+1} \subseteq C_n$ for all n. Prove that $\bigcap_{n=1}^{\infty} C_n \neq \emptyset$.

12. A collection of subsets of a metric space X is said to have the *finite intersection property* if every finite subcollection has a nonempty intersection. Prove that X is compact iff every collection of closed subsets of X with the finite intersection property has a nonempty intersection.

13. The *diameter* of a nonempty subset A of (X, d) is defined by

$$d(A) := \sup\{d(a, b) \,:\, a,\, b \in A\}.$$

(a) Prove that if A is compact, then there exist points a, $b \in A$ such that $d(A) = d(a, b)$.

(b) Give an example of a closed and bounded set A in a metric space such that $d(A) > d(a, b)$ for all a, $b \in A$.

14. ($\Downarrow$ 8.7.2.) The *distance between nonempty subsets A and B of* (X, d) is defined as

$$d(A, B) := \inf\{d(a, b) \,:\, a \in A,\, b \in B\}.$$

(a) Prove that if A and B are disjoint with A closed and B compact, then $d(A, B) > 0$.

(b) Show by example that the conclusion in (a) is false if B is merely closed.

(c) Show that if both sets are compact, then there exist $a \in A$ and $b \in B$ such that $d(A, B) = d(a, b)$.

15. ($\Downarrow$ 11.2.17.) Let A be a nonempty subset of X and define $d(A, \cdot) : X \to \mathbb{R}$ by $d(A, x) = d(A, \{x\})$ (see Exercise 14). Prove the following:

(a) $|d(A, x) - d(A, y)| \le d(x, y)$, hence $d(A, \cdot)$ is uniformly continuous.

(b) $d(A, x) = 0$ iff $x \in \mathrm{cl}(A)$.

(c) If A and B are disjoint closed sets, then the function

$$F_{AB}(x) = \frac{d(x, A)}{d(x, A) + d(x, B)}, \quad x \in X,$$

is well-defined and continuous, $0 \le F_{AB} \le 1$ on X, and

$$A = \{x \,:\, F_{AB}(x) = 0\}, \quad B = \{x \,:\, F_{AB}(x) = 1\}.$$

(d) If A and B are disjoint closed sets of X, then there exist disjoint open sets U and V such that $A \subseteq U$ and $B \subseteq V$. (U and V are then said to *separate A and B*.)

16. Referring to 8.1.10, show that the set $\{f \in \ell^\infty \,:\, |f(n)| \le e^{-n}\}$ is compact. Is $\{f \in \mathcal{B}([1, +\infty)) \,:\, |f(x)| \le e^{-x}\}$ compact?

17. (Lebesgue's number). Let X be compact and let $\mathcal{U} = \{U_\mathfrak{i} \,:\, \mathfrak{i} \in \mathfrak{I}\}$ be an open cover of X. Prove that there exists a number $r > 0$ such that every set with diameter $< r$ (Exercise 13) is contained in some $U_\mathfrak{i}$.

18. (Dini's Theorem). Let X be compact and let f_n, $g : X \to \mathbb{R}$ be continuous such that either $f_n \downarrow g$ or $f_n \uparrow g$ on X. Prove that the convergence is uniform. (See 7.1.12.)

19. Let $f : [0, 2\pi) \to \mathbb{R}^2$ be defined by $f(t) = (\cos t, \sin t)$. Show that f is continuous, one-to-one, and maps $[0, 2\pi)$ onto the circle $x^2 + y^2 = 1$ but has a discontinuous inverse.

20. Let $f : \mathbb{R}^2 \to \mathbb{R}$ be defined by $f(x, y) = x$. Prove or disprove:

    (a) If $E \subseteq \mathbb{R}^2$ is closed, then $f(E)$ is closed.

    (b) If $E \subseteq \mathbb{R}^2$ is open, then $f(E)$ is open.

21. Let A and B be compact subsets of $\mathbb{R}$. Prove that the sets

$$AB := \{ab : a \in A, b \in B\} \quad \text{and} \quad A + B := \{a + b : a \in A, b \in B\}$$

    are compact.

22. (⇓ 13.6.5.) Let a sequence of continuous functions $f_n : (X, d) \to (Y, \rho)$ converge uniformly to f on X, let $C \subseteq X$ be compact and $U \subseteq Y$ open. Prove that if $f(C) \subseteq U$, then $f_n(C) \subseteq U$ for all sufficiently large n.

---

*8.6 The Arzelà–Ascoli Theorem

Throughout this section, (X, d) and (Y, ρ) denote arbitrary metric spaces and $\mathcal{C}(X, Y)$ denotes the set of all continuous functions from X to Y.

As noted in the preceding section, closed and bounded subsets in infinite-dimensional spaces such as $\mathcal{C}([0, 1])$ need not be compact. The additional property of *equicontinuity* is needed to give a complete characterization of compact subsets of such spaces.

8.6.1 Definition. A family $\mathcal{F}$ of functions in $\mathcal{C}(X, Y)$ is said to be

- *equicontinuous at* $a \in X$ if for each $\varepsilon > 0$ there exists a $\delta > 0$ such that $\rho(f(x), f(a)) < \varepsilon$ for all $x \in X$ with $d(x, a) < \delta$ and all $f \in \mathcal{F}$.
- *equicontinuous on* $E \subseteq X$ if $\mathcal{F}$ is equicontinuous at each point of E.
- *uniformly equicontinuous on* E if for each $\varepsilon > 0$ there exists a $\delta > 0$ such that $\rho(f(x), f(y)) < \varepsilon$ for all $f \in \mathcal{F}$ and $x, y \in E$ with $d(x, y) < \delta$. ♢

The distinguishing feature of equicontinuity is that, while δ may vary with the point a, it is independent of the functions f. With uniform equicontinuity, δ is independent of both f and a.

8.6.2 Example. For each $x, t \in \mathbb{R}$, define $f_t(x) = tx$. Let $I = (c, d)$ be a bounded interval and set $M = \max\{|c|, |d|\}$. The inequality

$$|f_t(x) - f_t(y)| = |t|\,|x - y| \le M|x - y|, \quad t \in I,$$

shows that the collection of functions $\{f_t : t \in I\}$ is uniformly equicontinuous on $\mathbb{R}$. On the other hand, the larger collection $\{f_t : t \in \mathbb{R}\}$ is not equicontinuous at any $a \in \mathbb{R}$. Indeed, no δ can be chosen so that $|tx - ta| < 1$ for all $t \in \mathbb{R}$ and all $x \in \mathbb{R}$ with $|x - a| < \delta$. ◊

A straightforward modification of the proof of 8.5.13 yields

8.6.3 Theorem. *If X is compact and $\mathcal{F}$ is equicontinuous on X, then $\mathcal{F}$ is uniformly equicontinuous.*

For the desired characterization of equicontinuity we shall need another definition.

8.6.4 Definition. A metric space is said to have the *Bolzano–Weierstrass property* if every bounded sequence has a cluster point. ◊

Compact metric spaces and the space $\mathbb{R}^n$ have the Bolzano–Weierstrass property. Infinite discrete metric spaces, the space $\mathbb{Q}$, and the infinite dimensional space $\big(C([0,1]), \|\cdot\|_\infty\big)$ do not.

8.6.5 Proposition. (a) *A metric space with the Bolzano–Weierstrass property is complete.*

(b) *A metric space has the Bolzano–Weierstrass property iff every closed and bounded set is compact.*

Proof. For (a), use the fact that a Cauchy sequence is bounded and apply Exercise 8.1.9. Part (b) follows from 8.5.8. □

The following lemma may be proved using familiar ideas such as those found in 8.1.10. The details are left to the reader.

8.6.6 Lemma. *Let (X, d) be compact and (Y, ρ) complete. For $f, g \in \mathcal{C}(X, Y)$ define*

$$\sigma(f, g) = \sup_{x \in X} \rho(f(x), g(x)).$$

Then σ is a metric on $\mathcal{C}(X, Y)$ and $\mathcal{C}(X, Y)$ is complete in this metric.

8.6.7 Lemma. *A compact metric space X has a countable dense subset of the form $D = \bigcup_{k=1}^{\infty} F_k$, where F_k is a finite $(1/k)$-net for X.*

Proof. For each $k \in \mathbb{N}$, the collection $\{B_{1/k}(x) : x \in X\}$ is an open cover of X, hence has a finite subcover $\{B_{1/k}(x) : x \in F_k\}$. By definition of ε-net, $\bigcup_{k=1}^{\infty} F_k$ is dense in X. □

8.6.8 Arzelà–Ascoli Theorem. *Let X be compact and let Y have the Bolzano–Weierstrass property. Then a set $\mathcal{F}$ is compact in $\big(\mathcal{C}(X,Y),\sigma\big)$ iff it is closed, bounded, and equicontinuous.*

Proof. Suppose $\mathcal{F}$ is compact in $\mathcal{C}(X,Y)$, hence closed and bounded. If $\mathcal{F}$ is not equicontinuous at some point $a \in X$, then there exists an $\varepsilon > 0$ and for every n members x_n of X and f_n of $\mathcal{F}$ such that $d(x_n, a) < 1/n$ and

$$\rho(f_n(x_n), f_n(a)) \geq \varepsilon. \tag{8.6}$$

By compactness of $\mathcal{F}$, we may assume that $\{f_n\}$ converges uniformly to some $f \in \mathcal{F}$ (otherwise, take a subsequence). Since $x_n \to a$, the uniform convergence of $\{f_n\}$ implies that $f_n(x_n) \to f(a)$. But this contradicts (8.6). Therefore, $\mathcal{F}$ is equicontinuous.

Conversely, assume that $\mathcal{F}$ is closed, bounded, and equicontinuous and let $\{f_n\}$ be any sequence in $\mathcal{F}$. We show that $\{f_n\}$ has a convergent subsequence. The compactness of $\mathcal{F}$ will then follow from 8.5.8. Let F_k and $D = \{x_1, x_2, \ldots\}$ be as in 8.6.7. We show first that $\{f_n\}$ has a subsequence that converges pointwise on D. For this we use the Bolzano–Weierstrass property of Y and the following diagonalization argument: Because $\{f_n\}$ is bounded, we may choose a subsequence $\{f_n^{(1)}\}$ of $\{f_n^{(0)} := f_n\}$ such that the sequence $\{f_n^{(1)}(x_1)\}$ converges to some $y_1 \in Y$. We may then choose a subsequence $\{f_n^{(2)}\}$ of $\{f_n^{(1)}\}$ such that $\{f_n^{(2)}(x_2)\}$ converges to some $y_2 \in Y$. Continuing in this way, we obtain for each k a sequence $\{f_n^{(k)}\}$ such that $\{f_n^{(k+1)}\}$ is a subsequence of $\{f_n^{(k)}\}$ and $\lim_n f_n^{(k)}(x_k) = y_k$. Now take the *diagonal sequence* $\{g_n := f_n^{(n)}\}$, which is a subsequence of $\{f_n\}$ and for each k, except for the first $k-1$ terms, is a subsequence of $\{f_n^{(k)}\}$. It follows that $\lim_n g_n(x_k) = y_k$ for each k. The scheme may be depicted as follows:

$$\begin{array}{lll}
\underline{f_1^{(1)}}, f_2^{(1)}, \ldots, f_n^{(1)}, \ldots & \to y_1 & \text{at } x_1 \\
f_1^{(2)}, \underline{f_2^{(2)}}, \ldots, f_n^{(2)}, \ldots & \to y_2 & \text{at } x_2 \\
\qquad\quad \vdots & & \\
f_1^{(n)}, f_2^{(n)}, \ldots, \underline{f_n^{(n)}}, \ldots & \to y_n & \text{at } x_n \\
\qquad\quad \vdots \quad \searrow & & \\
\qquad\qquad\qquad y_k & & \text{at each } x_k
\end{array}$$

Having obtained a subsequence $\{g_n\}$ of $\{f_n\}$ that converges pointwise on the dense set D, we now show that $\{g_n\}$ converges uniformly on X, which will complete the proof. By the uniform equicontinuity of $\{g_n\}$, given $\varepsilon > 0$ we may choose a $\delta > 0$ such that

$$\rho\big(g_n(x), g_n(y)\big) < \varepsilon/3, \text{ for all } n \in \mathbb{N} \text{ and } x,\, y \in X \text{ with } d(x,y) < \delta. \tag{8.7}$$

Let $k > 1/\delta$. Since $\{g_n\}$ converges pointwise on F_k and F_k is finite, we may choose N_k so that

$$\rho\big(g_n(y), g_m(y)\big) < \varepsilon/3, \quad \text{for all } n,\, m \geq N_k \text{ and all } y \in F_k. \tag{8.8}$$

Since F_k is a δ-net, given $x \in X$ there exists $y \in F_k$ such that $d(x, y) < \delta$. It follows from (8.7) and (8.8) that for $m,\, n \geq N_k$,

$$\begin{aligned}\rho\big(g_n(x), g_m(x)\big) &\leq \rho\big(g_n(x), g_n(y)\big) + \rho\big(g_n(y), g_m(y)\big) + \rho\big(g_m(y), g_m(x)\big)\\ &< \varepsilon/3 + \varepsilon/3 + \varepsilon/3 = \varepsilon.\end{aligned}$$

Since x was arbitrary, $\{g_n\}$ is a Cauchy sequence in $\mathcal{C}(X, Y)$. Since $\mathcal{C}(X, Y)$ is complete, $\{g_n\}$ converges in $\mathcal{C}(X, Y)$. □

Remark. The proof of the sufficiency of the theorem did not require that $\mathcal{F}$ be uniformly bounded. All that was used was the property of *pointwise boundedness*, that is, $\{f(x) : f \in \mathcal{F}\}$ bounded in Y for each $x \in X$. Uniform boundedness is then a consequence of equicontinuity. ◊

8.6.9 Example. Let X be compact. Then any convergent sequence of functions f_n in $\mathcal{C}(X, \mathbb{R})$, say $f_n \to f$, is equicontinuous. This may be verified directly, but a quick proof uses 8.6.8 applied the set $\{f, f_1, f_2, \ldots\}$, whose compactness is readily established. ◊

Exercises

1. Let $X \times Y$ have the product metric $\eta := d \times \rho$ and let $f : X \to Y$. The *graph* of f is the set

   $$G(f) = \{(x, y) : x \in X \text{ and } y = f(x)\}.$$

   Prove that if f is continuous, then $G(f)$ is closed in $X \times Y$. Conversely, prove that if $G(f)$ is closed, $f(X)$ is bounded, and Y has the Bolzano–Weierstrass property, then f is continuous. Give an example of a real-valued discontinuous function on $[0, 1]$ with a closed graph.

2. Let X have the Bolzano–Weierstrass property and let $\{x_n\}$ be a bounded sequence in X with only finitely many cluster points $y_1, \ldots, y_k$. Prove that the set $C := \{y_1, \ldots, y_k, x_1, x_2, \ldots\}$ is compact.

3. Prove that a subset $\mathcal{F}$ of $\mathcal{C}(X, Y)$ is equicontinuous at $a \in X$ iff for any sequences $\{f_n\}$ in $\mathcal{F}$ and $\{x_n\}$ in X with $x_n \to a$, $\rho\big(f_n(x_n), f_n(a)\big) \to 0$.

4. Prove that a subset $\mathcal{F}$ of $\mathcal{C}(X, Y)$ is uniformly equicontinuous on $E \subseteq X$ iff for any sequences $\{f_n\}$ in $\mathcal{F}$ and $\{x_n\}$, $\{a_n\}$ in E with $d(x_n, a_n) \to 0$, $\rho\big(f_n(x_n), f_n(a_n)\big) \to 0$.

5. Prove that a finite set of uniformly continuous functions $f : X \to Y$ is uniformly equicontinuous.

6. Prove that the uniform closure of a set $\mathcal{F} \subseteq \mathcal{C}(X,Y)$ of uniformly equicontinuous functions is uniformly equicontinuous.

7. Let $c, p > 0$ and define $f_n(x) = (nx)^{-p}$, $x \geq c$. Show that the sequence $\{f_n\}$ is uniformly equicontinuous.

8. Define $f_n(x) = \ln(n + x)$. Show that the sequence $\{f_n\}$ is uniformly equicontinuous on $(0, +\infty)$.

9. Define $f_n(x) = \sin(nx)$. Use Exercise 3 and Exercise 8.3.13 to show that the sequence $\{f_n\}$ is not equicontinuous at any nonzero rational number r.

10. Let $M > 0$ and define
$$\mathcal{R}_M := \{f \,:\, f \text{ is locally integrable on } [0, +\infty) \text{ and } \|f\|_\infty \leq M\}.$$
For $f \in \mathcal{R}_M$ define
$$F_f(x) = \int_0^x f, \quad x \geq 0.$$
Prove that the set $\mathcal{F} := \{F_f \,:\, f \in \mathcal{R}_M\}$ is uniformly equicontinuous on $[0, +\infty)$.

11. Let $M > 0$ and define
$$\mathcal{D}_M := \{f : (a,b) \to \mathbb{R} \,:\, |f'(x)| \leq M \text{ for all } a < x < b\}.$$
Show that $\mathcal{D}_M$ is uniformly equicontinuous. Conclude that if g has a bounded derivative on $\mathbb{R}$, then the set of functions $\{g_t \,:\, t \in \mathbb{R}\}$ is uniformly equicontinuous on I, where $g_t(x) = g(t + x)$.

12. Let $f : X \times Y \to \mathbb{R}$ have the property that $f(x, y)$ is continuous in y for each fixed x and continuous in x for each fixed y. Define
$$\mathcal{F} := \{f(\,\cdot\,, y) \,:\, y \in Y\}.$$
Prove:

    (a) If $\mathcal{F}$ is equicontinuous, then f is continuous.

    (b) If f is continuous and Y is compact, then $\mathcal{F}$ is equicontinuous.

13. Let X be compact. Show that a totally bounded subset of $\mathcal{C}(X,Y)$ is uniformly equicontinuous.

14. Let $\{f_\mathfrak{i} \,:\, \mathfrak{i} \in \mathfrak{I}\}$ be a uniformly bounded subset of $\mathcal{R}_a^b$. Define
$$F_\mathfrak{i}(x) := \int_a^x f_\mathfrak{i}(t)\, dt, \qquad a \leq x \leq b.$$
Show that $\{F_\mathfrak{i} \,:\, \mathfrak{i} \in \mathfrak{I}\}$ is a totally bounded subset of $\mathcal{C}([a,b])$.

15. Let $f(t,x,y)$ be continuous on $[a,b]\times[a,b]\times[a,b]$ and set $f_t(x,y) = f(t,x,y)$. Prove that the family $\{f_t : t\in[a,b]\}$ is uniformly equicontinuous on $[a,b]\times[a,b]$. Apply this to the function

$$f(t,x,y) = \frac{1+t\sin x}{2+t\sin y} \quad \text{on} \quad [0,1]\times[0,1]\times[0,1].$$

8.7 Connected Sets

Throughout this section, (X,d) and (Y,ρ) denote arbitrary metric spaces.

8.7.1 Definition. A pair (U,V) of open sets in X is said to *separate* X if

$$X = U\cup V, \quad U\neq\emptyset, \quad V\neq\emptyset, \quad \text{and } U\cap V=\emptyset.$$

The pair (U,V) is then called a *separation of* X. The space X is said to be *connected* if it has no separation, and *disconnected* otherwise. A subset E of X is *connected* if it is connected as a subspace of X. ◊

It follows from the definition that if E is disconnected, then there exist sets U, V open in X such that $(E\cap U, E\cap V)$ is a separation of E. The sets U and V need not be disjoint here, but the next theorem shows that this useful state of affairs may always be achieved. In this case we shall call (U,V) a *separation of* E.

8.7.2 Theorem. *A subset E of X is disconnected iff there exists a separation $(E\cap U, E\cap V)$ of E such that $U\cap V=\emptyset$.*

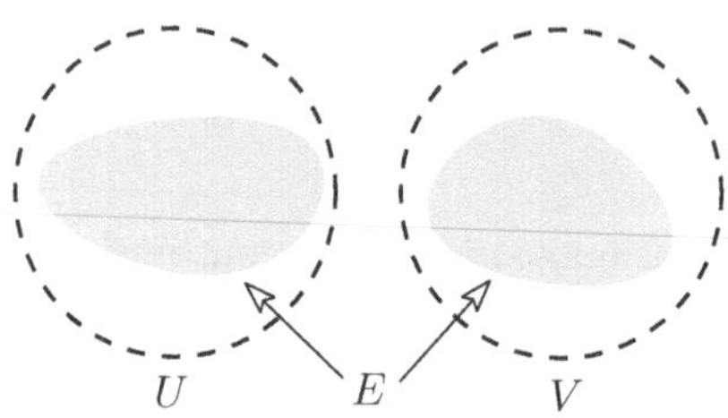

FIGURE 8.7: A separation (U,V) of E.

Proof. The sufficiency is clear. For the necessity, assume that E is disconnected and that $(E\cap U_1, E\cap V_1)$ is a separation of E. Here, U_1 and V_1 are open in X but may not be disjoint. However, since $E\cap U_1$ and $E\cap V_1$ are disjoint, $\mathrm{cl}_E(E\cap U_1)\cap V_1=\emptyset$. Indeed, if, to the contrary, $x\in \mathrm{cl}_E(E\cap U_1)\cap V_1$ for some x, then there would be a sequence $\{x_n\}$ in $E\cap U_1$ converging to x, which would imply that eventually $x_n \in E \cap V_1$, impossible. Recalling that $\mathrm{cl}_E(E \cap U_1) = E \cap \mathrm{cl}_X(U_1)$ we now see that

$$v \notin \mathrm{cl}_X(U_1) \quad \text{for each } v \in E \cap V_1.$$

Similarly,

$$u \notin \mathrm{cl}_X(V_1) \quad \text{for each } u \in E \cap U_1.$$

By Exercise 8.5.14, it follows that for $u \in E \cap U_1$ and $v \in E \cap V_1$ the distances

$$r(u) := \inf\{d(u,x) : x \in \mathrm{cl}_X(V_1)\} \quad \text{and} \quad s(v) := \inf\{d(v,x) : x \in \mathrm{cl}_X(U_1)\}$$

are positive. Now define

$$U = \bigcup_{u \in E \cap U_1} B_{r(u)/2}(u), \quad \text{and} \quad V = \bigcup_{v \in E \cap V_1} B_{s(v)/2}(v).$$

Clearly, U and V are open in X and contain $E \cap U_1$ and $E \cap V_1$, respectively. To prove that (U, V) is a separation of E, it remains to show that $U \cap V = \emptyset$.

Suppose to the contrary that there exists a point $x \in U \cap V$. Then, by the above, $d(x,u) < r(u)/2$ for some $u \in U_1$ and $d(x,v) < s(v)/2$ for some $v \in V_1$. Adding and using the triangle inequality, we have

$$d(u,v) < r(u)/2 + s(v)/2.$$

But by definition of $r(u)$ and $s(v)$, $d(u,v) \geq r(u)$ and $d(u,v) \geq s(v)$ and so

$$d(u,v) \geq \big[r(u) + s(v)\big]/2.$$

This contradiction shows that $U \cap V = \emptyset$ and completes the proof of the theorem. □

In any metric space, the empty set and the singletons $\{x\}$ are trivially connected, but no other finite subsets are connected. In a discrete space the only connected sets are the empty set and the singletons. The set $\mathbb{Q}$ is not connected in $\mathbb{R}$, since the open sets $(-\infty, \sqrt{2})$ and $(\sqrt{2}, +\infty)$ separate $\mathbb{Q}$. Here is a characterization of connectedness that is frequently useful.

8.7.3 Theorem. *X is not connected iff there exists a continuous function from X onto $\{0,1\}$. Equivalently, X is connected iff every continuous function from X into $\{0,1\}$ is constant.*

Proof. Assume that X is not connected and let (U, V) separate X. Define

$$g(x) = \begin{cases} 0 & \text{if } x \in U, \\ 1 & \text{if } x \in V. \end{cases}$$

Then g maps X onto $\{0,1\}$. Let W be any open set in $\mathbb{R}$. Then $g^{-1}(W)$ is one of the sets $\emptyset$, U, V, or X, each of which is open in X. Therefore, g is continuous. Conversely, if a continuous function g from X onto $\{0,1\}$ exists, then the open sets $g^{-1}((-1,1/2))$ and $g^{-1}((1/2,2))$ separate X. □

8.7.4 Corollary. *The nonempty connected subsets of* $\mathbb{R}$ *are the intervals.*

Proof. By the intermediate value theorem, there can be no continuous function from an interval onto $\{0,1\}$. Hence intervals must be connected.

Now let E be a nonempty subset of $\mathbb{R}$ that is not an interval. Choose real numbers $a < c < b$ with a, $b \in E$ but $c \notin E$. Then $(-\infty, c)$ and $(c, +\infty)$ separate E, hence E is not connected. □

The following is a generalization of the intermediate value theorem.

8.7.5 Corollary. *If* $f : X \to Y$ *is continuous and* X *is connected, then* $f(X)$ *is connected.*

Proof. Let $g : f(X) \to \{0,1\}$ be continuous. Then $g \circ f : X \to \{0,1\}$ is continuous and hence must be constant. It follows that g itself must be constant. □

8.7.6 Corollary. *If* $A \subseteq X$ *is connected and* $A \subseteq B \subseteq \mathrm{cl}(A)$, *then* B *is connected. In particular, the closure of a connected set is connected.*

Proof. Let $g : B \to \{0,1\}$ be continuous. Then $g|_A$ is continuous, hence must be constant. Since $B \subseteq \mathrm{cl}(A)$, g itself must be constant. Therefore, B is connected. □

The converse of 8.7.6 is false. For example, $\mathrm{cl}(\mathbb{Q}) = \mathbb{R}$ is connected but $\mathbb{Q}$ is not.

8.7.7 Definition. A *path in* X *from* x *to* y is a continuous function φ from an interval $[a,b]$ to X such that $\varphi(a) = x$, the *initial point* of the path, and $\varphi(b) = y$, the *terminal point* of the path.. ◊

Note that if $\varphi : [a,b] \to X$ is a path from x to y, then

$$-\varphi(t) := \varphi(-t), \quad -b \le t \le -a,$$

defines a path from y to x. Moreover, if $\vartheta : [c,d] \to X$ is a path from y to z, then the *sum* or *concatenation* $\varphi + \vartheta : [0,2] \to X$ of the paths φ and ϑ is a path from x to z, where

$$(\varphi + \vartheta)(t) = \begin{cases} \varphi\big(a + (b-a)t\big) & \text{if } 0 \le t \le 1, \\ \vartheta\big(c + (d-c)(t-1)\big) & \text{if } 1 \le t \le 2. \end{cases}$$

8.7.8 Definition. X is said to be *path connected* if for each pair of points x, $y \in X$ there exists a path in X from x to y. A subset E of X is *path connected* if it is path connected as a subspace of X. ◊

For example, a convex subset C of a normed vector $\mathcal{X}$ is path connected. Indeed, if $\boldsymbol{x}$, $\boldsymbol{y} \in C$, then the line segment $\varphi(t) := (1-t)\boldsymbol{x} + t\boldsymbol{y}$, $\quad 0 \le t \le 1$, joins $\boldsymbol{x}$ to $\boldsymbol{y}$ and lies in C. In particular, open and closed balls in $\mathcal{X}$ are path connected.

8.7.9 Theorem. *If X is path connected, then it is connected.*

Proof. Let $g : X \to \{0,1\}$ be a continuous function, let $x, y \in X$, and let $\varphi[a,b] \to X$ be a path from x to y. Then $g \circ \varphi : [a,b] \to \{0,1\}$ is continuous and, because $[a,b]$ is connected, must be constant. In particular,

$$g(x) = (g \circ \varphi)(a) = (g \circ \varphi)(b) = g(y).$$

Since x and y were arbitrary, g is constant. □

8.7.10 Example. A sphere in $\mathbb{R}^n$, $n > 1$, is path connected, hence connected. For example, consider the sphere

$$S = \{\boldsymbol{x} \in \mathbb{R}^n : \|\boldsymbol{x}\|_2 = 1\}.$$

We show that there is a path from the point $\boldsymbol{a} = (1, 0, \ldots, 0)$ to any point $\boldsymbol{b} = (b_1, b_2, \ldots, b_n)$. It will then follow that any pair of points in S may be joined by a path in S through $\boldsymbol{a}$.

If $\boldsymbol{b} = (-1, 0, \ldots, 0)$, then $(\cos t, \sin t, 0, \ldots, 0)$, $0 \le t \le \pi$, is such a path. Suppose $\boldsymbol{b} \ne (-1, 0, \ldots, 0)$. Then the line segment

$$\varphi(t) = (1-t)\boldsymbol{a} + t\boldsymbol{b} = (1 - t + tb_1, tb_2, \ldots, tb_n), \quad 0 \le t \le 1,$$

is never zero, hence $\|\varphi(t)\|_2^{-1}\varphi(t)$ is a path from $\boldsymbol{a}$ to $\boldsymbol{b}$ in S. ◊

8.7.11 Example. The subset $B_1(-1,0) \cup B_1(1,0)$ of $\mathbb{R}^2$ is not connected, hence not path connected. However, its closure $C_1(-1,0) \cup C_1(1,0)$ is path connected, as can be seen from the figure, hence is connected. ◊

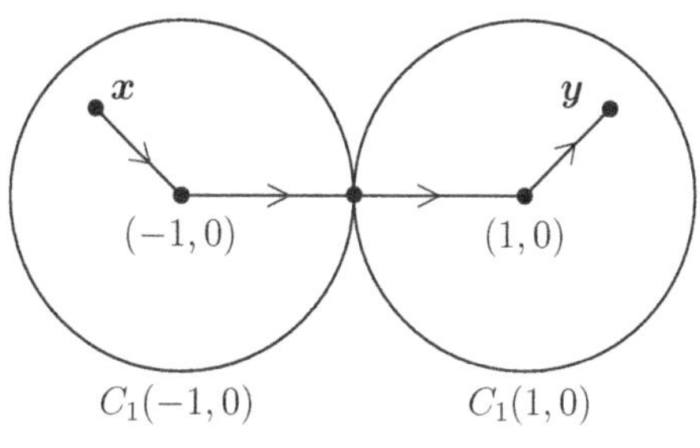

FIGURE 8.8: $C_1(-1,0) \cup C_1(1,0)$ is path connected.

The converse of 8.7.9 is false, as the following example—the topologist's sine curve (8.3.7)—demonstrates.

8.7.12 Example. Let

$$A = \{(x, \sin(1/x)) : 0 < x < 2/\pi\}, \quad B = \{0\} \times [-1,1], \quad \text{and} \quad E = A \cup B.$$

Since A is connected and $E = \text{cl}(A)$, 8.7.6 shows that E is connected. However, no point in A can be joined to a point in B by a path in E. Indeed, suppose such a path existed, say $\varphi : [a, b] \to E$, where

$$\varphi(t) = \big(x(t), y(t)\big), \quad \varphi(a) \in A, \quad \text{and} \quad \varphi(b) \in B.$$

Let

$$S := \big\{t \in [a, b] \,:\, \varphi([a, t]) \subseteq A\big\}.$$

Since S is nonempty and bounded, $c := \sup S$ exists and $c \in [a, b]$. Note that $x(t) > 0$ on S. If $x(c) > 0$, then $c < b$, hence, by continuity, $x(s)$ is positive on $[a, c + \delta]$ for some $\delta > 0$, contradicting the definition of c. Therefore, $x(c) = 0$ and $x(t) > 0$ on $[a, c)$. This implies that $\varphi(t) = \big(x(t), \sin(1/x(t))\big)$ on $[a, c)$ and $\lim_{t\to c^-} x(t) = 0$. By continuity, for each $\delta > 0$ the set $x([c-\delta, c])$ is an interval of the form $[0, d]$, $d > 0$. Therefore, $y(t) = \sin(1/x(t))$ takes on all values in $[-1, 1]$ on each interval $[c - \delta, c)$, which implies that $\lim_{t\to c^-} y(t)$ cannot exist. But this contradicts the continuity of φ at c. ♢

While there is no strict converse to 8.7.9, the next theorem provides a partial converse.

8.7.13 Theorem. *An open connected subset E of a normed vector space $\mathcal{X}$ is path connected.*

Proof. Fix a point $\boldsymbol{x} \in E$ and let U denote the set of all points $\boldsymbol{u} \in E$ for which there exists a path in E from $\boldsymbol{x}$ to $\boldsymbol{u}$. We claim that U is open. Let

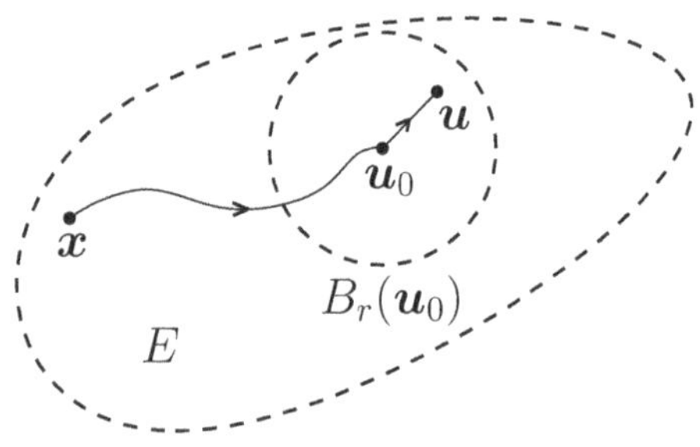

FIGURE 8.9: E is path connected.

$\boldsymbol{u}_0 \in U$ and choose $r > 0$ such that $B_r(\boldsymbol{u}_0) \subseteq E$. By definition of U, there exists a path in E from $\boldsymbol{x}$ to $\boldsymbol{u}_0$. Since $B_r(\boldsymbol{u}_0)$ is convex, there exists a line segment in $B_r(\boldsymbol{u}_0)$ from $\boldsymbol{u}_0$ to any point $\boldsymbol{u} \in B_r(\boldsymbol{u}_0)$. The sum of these paths is then a path in E from $\boldsymbol{x}$ to $\boldsymbol{u}$. Therefore, $B_r(\boldsymbol{u}_0) \subseteq U$, which shows that U is open. A similar argument shows that $V := E \setminus U$ is open. Since E is connected, and $\boldsymbol{x} \in U$, $V = \emptyset$, and so $E = U$. □

Exercises

1. Determine which sets are connected in $\mathbb{R}^2$:

(a) $B_1(-1,0) \cup \{(0,0)\} \cup B_1(1,0)$
(b) $\mathbb{R}^2 \setminus \{(1/m, 1/n) : m, n \in \mathbb{N}\}$
(c) $\mathbb{Q}^2$
(d) $\mathbb{R}^2 \setminus \mathbb{Q}^2$
(e) $\{(x, \sin(1/x)) : x \neq 0\} \cup \{(0, a)\}$
(f) $\mathbb{R}^2 \setminus G$, where G is the graph of a bounded function $f : [a, b] \to \mathbb{R}$
(g) $\mathbb{R}^2 \setminus G$, where G is the graph of an equation $F(x, y) = 0$
(h) $\{(x, y, z) : x^2 + y^2 - z^2 = 1\}$
(i) $\{(x, y, z) : x^2 + y^2 - z^2 = -1\}$
(j) $\{(x, y, z) : x^2 + y^2 - z^2 = 0,\ 0 < x^2 + y^2 \leq 1\}$.

2. Prove that a metric space X is connected iff it has no proper nonempty subset that is both open and closed.

3. Prove that X is connected iff it cannot be expressed as the union of nonempty sets A and B such that $A \cap \mathrm{cl}_X(B) = \mathrm{cl}_X(A) \cap B = \emptyset$. *Hint.* Use 8.7.3 and the sequential characterization of continuity.

4. Prove that $X \times Y$ is connected in the product metric $d \times \rho$ iff X and Y are connected.

5. Let X be connected and $f : X \to \mathbb{R}$ continuous. Suppose there exist $u, v \in X$ such that $f(u)f(v) < 0$. Show that the equation $f(x) = 0$ has a solution.

6. Let X be connected and $f : X \to Y$ continuous. Suppose f has the property that for each $x \in X$ there exists $\varepsilon > 0$, possibly depending on x, such that f is constant on $B_\varepsilon(x)$. Prove that f is constant on X.

7. Let X be connected and let $g, h : X \to \mathbb{R}$ be continuous such that $g(x) \neq h(x)$ for all $x \in X$. Prove that $g > h$ or $h > g$ on X.

8. ($\Downarrow$ 12.2.10.) Let $\mathcal{X}$ be a normed vector space and $\boldsymbol{u}, \boldsymbol{v} \in \mathcal{X}$. A *polygonal path P from $\boldsymbol{u}$ to $\boldsymbol{v}$* is a finite sequence of line segments $L_k = [\boldsymbol{x}_k : \boldsymbol{x}_{k+1}]$, $k = 1, \ldots, n-1$, where $\boldsymbol{x}_1 = \boldsymbol{u}$ and $\boldsymbol{x}_n = \boldsymbol{v}$. The path P is *non-overlapping* if $L_j \cap L_k = \emptyset$ unless $j = k - 1$, in which case $L_j \cap L_k = \boldsymbol{x}_k$. A subset E of a normed vector space $\mathcal{X}$ is *polygonally connected* if for each pair of points $\boldsymbol{u}$ and $\boldsymbol{v}$ in E there exists a polygonal path from $\boldsymbol{u}$ to $\boldsymbol{v}$ contained in E. For example, a convex set is polygonally connected. Prove that every open connected subset E of $\mathcal{X}$ is polygonally connected. Show also that it is always possible to choose P to be non-overlapping

9. Show that for $n > 1$ the complement of an open ball or a closed ball in $\mathbb{R}^n$ is path connected, hence connected.

10. Suppose $A \subseteq X$ is connected. By 8.7.6, $\text{cl}(A)$ is connected. Prove or disprove: (a) $\text{int}(A)$ is connected. (b) $\text{bd}(A)$ is connected.

11. The *exterior* $\text{ext}(E)$ of a subset E of a metric space X is defined as the interior of E^c. Show that $X = \text{int}(E) \cup \text{bd}(E) \cup \text{ext}(E)$. Conclude that X is connected iff every subset of X with nonempty interior and nonempty exterior also has a nonempty boundary.

12. Let $\{A_n\}$ be a finite or infinite sequence of connected subsets of X such that $A_n \cap A_{n+1} \neq \emptyset$ for each n. Prove that $\bigcup_n A_n$ is connected.

13. Let $\{A_\mathfrak{i} : \mathfrak{i} \in \mathfrak{I}\}$ be a collection of nonempty connected sets and $\mathfrak{i}_0 \in \mathfrak{I}$ such that $A_\mathfrak{i} \cap A_{\mathfrak{i}_0} \neq \emptyset$ for all $\mathfrak{i}$. Prove that $\bigcup_\mathfrak{i} A_\mathfrak{i}$ is connected.

14. Let $\{A_n\}$ be an infinite sequence of compact connected subsets of X such that $A_{n+1} \subseteq A_n$. Prove that $\bigcap_n A_n$ is connected.

15. Let $X = A_1 \cup \cdots \cup A_p$ and $Y = B_1 \cup \cdots \cup B_q$, $p < q$, where A_j and B_j are connected, the A_j's are pairwise disjoint, and the B_j's are pairwise disjoint and closed. Show that no continuous function $f : X \to Y$ can map X onto Y.

16. Prove that no one-to-one continuous function can map a closed line segment L onto a circle C. Show, however, that there are continuous functions that can do this.

17. Suppose closed line segments L_1, L_2, L_3 in the plane meet at a single endpoint P. Show that no one-to-one continuous function can map a closed line segment L onto $L_1 \cup L_2 \cup L_3$. Show, however, that there are continuous functions that can do this.

18. Let C_1 and C_2 be tangent circles in the plane. Show that no one-to-one continuous function can map $C_1 \cup C_2$ onto a circle C. Show, however, that there are continuous functions that can do this.

19. Show that no one-to-one continuous function can map the set

$$E := \{(x, y, z) : x^2 + y^2 = z^2,\ x^2 + y^2 \leq 1\}$$

onto a closed disk D. Show, however, that there are continuous functions that can do this.

20. Let $\mathcal{X}$ be a normed vector space and $f : \mathcal{X} \to \mathbb{R}$ continuous. Let

$$A := \{\boldsymbol{x} \in \mathcal{X} : f(\boldsymbol{x}) \geq c\} \quad \text{and} \quad B := \{\boldsymbol{x} \in \mathcal{X} : f(\boldsymbol{x}) = c\}.$$

Prove that $\text{bd}(A) \subseteq B$ and that the inclusion may be strict.

21. Let X be connected and have at least two points. Show that X is uncountable. *Hint.* For all sufficiently small $r > 0$, $X \neq B_r(x) \cup C_r^c(x)$.

22. Let U be an open subset of a normed vector space $\mathcal{X}$ and let $\boldsymbol{x} \in U$. The *component of* U *containing* $\boldsymbol{x}$ is the union $C_{\boldsymbol{x}}$ of all connected subsets of U containing $\boldsymbol{x}$.

    (a) Prove that $C_{\boldsymbol{x}}$ is open and connected and that U is a union of pairwise disjoint components.

    (b) Show that the number of components is countable if $\mathcal{X}$ is a Euclidean space $\mathbb{R}^n$.

23. Let (X, d) be complete, (Y, ρ) connected, $c > 0$, and $f : X \to Y$ a continuous mapping such that $f(X)$ is open and $\rho\big(f(u), f(v)\big) \geq c\, d(u, v)$ for all u, $v \in X$. Prove that Y is complete.

8.8 The Stone–Weierstrass Theorem

Let (X, d) be a compact metric space and let $\mathcal{C}(X)$ denote the space of all continuous real-valued functions of X with the supremum norm $\|f\|_\infty = \sup_{x \in X} |f(x)|$. A member f of $\mathcal{C}(X)$ is said to be *uniformly approximated by members of a subset* $\mathcal{S}$ *of* $\mathcal{C}(X)$ if $f \in \operatorname{cl}(\mathcal{S})$. This is equivalent to the existence of a sequence $\{f_n\}$ in $\mathcal{S}$ converging uniformly to f on X.

Weierstrass's approximation theorem asserts that any function in $\mathcal{C}\big([a, b]\big)$ may be uniformly approximated by polynomials. Stone's generalization of Weierstrass's theorem replaces $[a, b]$ by a compact metric space[2] and the set of polynomials by a more general class of functions. We first prove Weierstrass's theorem and then use it to prove Stone's theorem.

The proof of Weierstrass's theorem given below is due to Lebesgue. The idea is to show that every continuous function may be uniformly approximated by piecewise linear functions, which in turn may be uniformly approximated by polynomials.

8.8.1 Definition. Let $a = x_0 < x_1 < \cdots < x_k = b$. A function g on $[a, b]$ is said to be *piecewise linear with vertices* (x_j, y_j) if, for $j = 0, 1, \ldots, k-1$,

$$g(x) = y_j + m_j(x - x_j), \quad m_j = \frac{y_{j+1} - y_j}{x_{j+1} - x_j}, \quad x_j \leq x \leq x_{j+1}. \qquad \diamond$$

Note that a piecewise linear function is necessarily continuous and that its graph consists of a sequence of line segments joined at the vertices. (See Figure 8.10.)

8.8.2 Lemma. *Every continuous function* f *on* $[a, b]$ *may be uniformly approximated by a piecewise linear function.*

---

[2]more generally, by a compact, Hausdorff topological space.

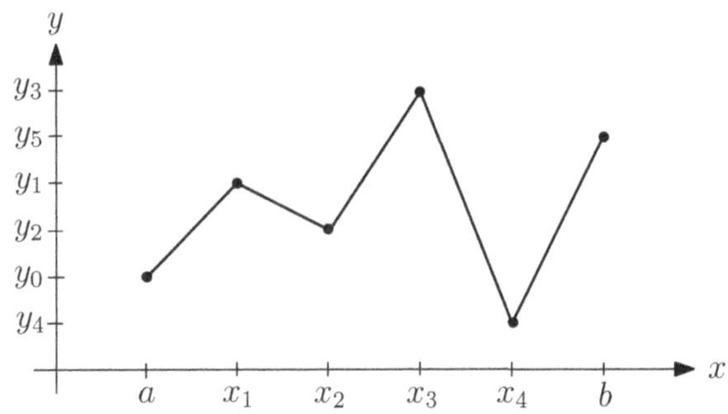

FIGURE 8.10: A piecewise linear function.

Proof. By uniform continuity, given $\varepsilon > 0$ we may choose $\delta > 0$ such that $|f(x) - f(y)| < \varepsilon/2$ whenever $|x - y| \le \delta$. Let $\{x_0 = a < x_1 < \cdots < x_k = b\}$ have mesh $< \delta$ and let g be as in 8.8.1 with $y_j = f(x_j)$. If $x_j \le x \le x_{j+1}$, then

$$|m_j|(x - x_j) = |f(x_{j+1}) - f(x_j)| \frac{x - x_j}{x_{j+1} - x_j} \le |f(x_{j+1}) - f(x_j)| < \varepsilon/2,$$

hence

$$|f(x) - g(x)| \le |f(x) - f(x_j)| + |m_j|(x - x_j) < \varepsilon. \qquad \square$$

8.8.3 Lemma. *The function g in 8.8.1 may be written*

$$g(x) = y_0 + \sum_{j=0}^{k-1} c_j (x - x_j)^+, \quad a \le x \le b,$$

for suitably chosen constants c_j.

Proof. For $0 \le j \le k-1$ and $x_j \le x \le x_{j+1}$, the desired equation reduces to

$$y_j + m_j(x - x_j) = y_0 + \sum_{i=0}^{j} c_i(x - x_i) = y_0 - \sum_{i=0}^{j} c_i x_i + x \sum_{i=0}^{j} c_i.$$

This holds iff

$$m_j = \sum_{i=0}^{j} c_i, \quad \text{and} \quad y_j - m_j x_j = y_0 - \sum_{i=0}^{j} c_i x_i. \tag{8.9}$$

The first equation in (8.9) is satisfied by taking $c_0 = m_0$ and $c_j = m_j - m_{j-1}$,

$j \geq 1$. For this choice,

$$\begin{aligned} y_0 - \sum_{i=0}^{j} c_i x_i &= y_0 + \sum_{i=1}^{j} m_{i-1} x_i - \sum_{i=0}^{j} m_i x_i \\ &= y_0 - m_j x_j + \sum_{i=0}^{j-1} m_i (x_{i+1} - x_i) \\ &= y_0 - m_j x_j + \sum_{i=0}^{j-1} (y_{i+1} - y_i) \\ &= y_j - m_j x_j, \end{aligned}$$

which shows that the second equation in (8.9) is also satisfied. □

8.8.4 Lemma. *The functions $|x|$ and x^+ may be uniformly approximated by polynomials on any bounded interval I.*

Proof. By 7.4.10, the binomial series

$$\sum_{n=0}^{\infty} \binom{1/2}{n} (-t)^n$$

converges uniformly to $\sqrt{1-t}$ on $[-1, 1]$. Setting $t = 1 - x^2$ we see that

$$\sum_{n=0}^{\infty} \binom{1/2}{n} (x^2 - 1)^n$$

converges uniformly to $\sqrt{x^2} = |x|$ on $[-1, 1]$. Thus if $s_n(x)$ denotes the nth partial sum of the last series and m is chosen so that $I \subseteq [-m, m]$, then $Q_n(x) := m s_n(x/m)$ defines a sequence of polynomials converging uniformly to $|x|$ on I. Since $x^+ = \frac{1}{2}(x + |x|)$, the polynomials $P_n(x) := \frac{1}{2}(x + Q_n(x))$ converge uniformly to x^+ on I. □

8.8.5 Weierstrass Approximation Theorem. *The set of all polynomials on $[a, b]$ is dense in $\mathcal{C}([a, b])$. That is, every member of $\mathcal{C}([a, b])$ may be uniformly approximated by polynomials.*

Proof. Let $f \in \mathcal{C}([a, b])$ and $\varepsilon > 0$. By 8.8.2, there exists a piecewise linear function g on $[a, b]$ such that $\|f - g\|_\infty < \varepsilon/2$. By 8.8.3 and 8.8.4, there exists a polynomial P such that $\|P - g\|_\infty < \varepsilon/2$. Then, by the triangle inequality, $\|f - P\|_\infty < \varepsilon$. □

For the statement of the Stone–Weierstrass theorem, we need the following definitions.

8.8.6 Definition. A collection $\mathcal{A}$ of real-valued functions on a set S is said to be an *algebra* if $\mathcal{A}$ is closed under addition, multiplication, and scalar multiplication; that is,

$$f, g \in \mathcal{A} \text{ and } \alpha \in \mathbb{R} \Rightarrow f+g,\ fg,\ \alpha f \in \mathcal{A}.$$

$\mathcal{A}$ is said to *separate points of* S if for each pair of distinct points s and t in S there exists $f \in \mathcal{A}$ such that $f(s) \neq f(t)$. ◊

For example, the collection of all polynomials on $[a, b]$ is an algebra that separates points of $[a, b]$.

8.8.7 Stone–Weierstrass Theorem. *Let X be a compact metric space and let $\mathcal{A}$ be an algebra in $\mathcal{C}(X)$ that contains the constant functions and separates points of X. Then $\mathcal{A}$ is dense in $\mathcal{C}(X)$.*

Proof. Set $\mathcal{B} := \text{cl}(\mathcal{A})$. The proof that $\mathcal{B} = \mathcal{C}(X)$ consists of the following sequence of steps.

I. $\mathcal{B}$ is an algebra in $\mathcal{C}(X)$.

⟦ If $f_n, g_n \in \mathcal{A}$, $f_n \to f$, $g_n \to g$, and $\alpha \in \mathbb{R}$, then

(a) $\|\alpha f_n - \alpha f\|_\infty = |\alpha|\,|f_n - f\|_\infty \to 0,$

(b) $\|(f_n + g_n) - (f+g)\|_\infty \leq \|f_n - f\|_\infty + \|g_n - g\|_\infty \to 0,$ and

(c) $$\begin{aligned}\|f_n g_n - fg\|_\infty &\leq \|f_n g_n - f g_n\|_\infty + \|f g_n - fg\|_\infty \\ &\leq \|g_n\|_\infty \|f_n - f\|_\infty + \|f\|_\infty \|g_n - g\|_\infty \\ &\to 0,\end{aligned}$$

the convergence in (c) holding because $\{g_n\}$ is uniformly bounded. (Each g_n is bounded and g_n converges uniformly to a bounded function.) Thus $\mathcal{B}$ is closed under addition, multiplication, and scalar multiplication. ⟧

II. $f \in \mathcal{B} \Rightarrow |f| \in \mathcal{B}$.

⟦ Let $M = \|f\|_\infty$. By 8.8.4 there exists a sequence of polynomials $P_n(x)$ converging uniformly to $|x|$ on $[-M, M]$. It follows that $P_n \circ f$ converges uniformly to $|f|$ on X. Because $\mathcal{B}$ is an algebra containing the constants, $P_n \circ f \in \mathcal{B}$. Indeed, if $P_n(x) = \sum_{j=0}^k a_j x^j$, then $P_n \circ f = \sum_{j=0}^k a_j f^j$. Since $\mathcal{B}$ is closed, $|f| \in \mathcal{B}$. ⟧

III. $f_1, \ldots, f_k \in \mathcal{B} \Rightarrow \max\{f_1, \ldots, f_k\}, \min\{f_1, \ldots, f_k\} \in \mathcal{B}$.

⟦ By induction, it suffices to consider the case $k = 2$. This follows from step II and the identities

$$\max\{f_1, f_2\} = \tfrac{1}{2}\big(f_1 + f_2 + |f_1 - f_2|\big),$$
$$\min\{f_1, f_2\} = \tfrac{1}{2}\big(f_1 + f_2 - |f_1 - f_2|\big). \quad ⟧$$

IV. Let $f \in \mathcal{C}(X)$. Then for each pair of distinct points x, y in X there exists a function $g_{xy} \in \mathcal{A}$ such that $g_{xy}(x) = f(x)$ and $g_{xy}(y) = f(y)$.

⟦ Choose a function $h \in \mathcal{A}$ such that $h(x) \neq h(y)$ ($\mathcal{A}$ separates points). Define

$$g_{xy}(z) = f(x) + \frac{f(x) - f(y)}{h(x) - h(y)} \big(h(z) - h(x)\big), \quad z \in X.$$

Because $\mathcal{A}$ contains the constant functions, $g_{xy} \in \mathcal{A}$. Clearly, $g_{xy}(x) = f(x)$ and $g_{xy}(y) = f(y)$. ⟧

V. If $f \in \mathcal{C}(X)$, $x \in X$, and $\varepsilon > 0$, then there exists a function $g_x \in \mathcal{B}$ such that

$$g_x(x) = f(x) \quad \text{and} \quad g_x(z) < f(z) + \varepsilon \quad \text{for all} \quad z \in X.$$

⟦ By continuity, for each $y \in X$ the set

$$U_y := \{z \in X \,:\, g_{xy}(z) < f(z) + \varepsilon\}$$

is open in X, where g_{xy} is the function in step IV. Moreover, U_y contains both x and y. Since X is compact, there exist $y_1, \dots y_k \in X$ such that

$$X = U_{y_1} \cup \cdots \cup U_{y_k}.$$

Set $g_x := \min\{g_{xy_1}, \dots, g_{xy_k}\}$. Then g_x clearly has the required properties and, by step III, $g_x \in \mathcal{B}$. ⟧

VI. If $f \in \mathcal{C}(X)$ and $\varepsilon > 0$, then there exists a function $g \in \mathcal{B}$ such that

$$f(z) - \varepsilon < g(z) < f(z) + \varepsilon, \quad \text{for all} \quad z \in X.$$

⟦ By continuity, for each $x \in X$ the set

$$V_x := \{z \in X \,:\, g_x(z) > f(z) - \varepsilon\}$$

is open in X, where g_x is the function in step V. Moreover, V_x clearly contains x and

$$f(z) - \varepsilon < g_x(z) < f(z) + \varepsilon, \quad \text{for all} \quad z \in V_x.$$

Since X is compact, there exist $x_1, \dots, x_m \in X$ such that

$$X = V_{x_1} \cup \cdots \cup V_{x_k}.$$

Set $g := \max\{g_{x_1}, \dots, g_{x_m}\}$. By step III, $g \in \mathcal{B}$, and g clearly satisfies the desired inequality. ⟧

To complete the proof of the theorem, observe that step VI asserts that $\mathcal{C}(X) = \mathrm{cl}(\mathcal{B})$. Since $\mathcal{B}$ is closed, $\mathcal{C}(X) = \mathcal{B}$. □

8.8.8 Example. A *trigonometric polynomial* is a function on $\mathbb{R}$ of the form

$$T(x) = a_0 + \sum_{j=1}^{m} a_j \cos(jx) + b_j \sin(jx), \quad a_j,\, b_j \in \mathbb{R}.$$

The collection $\mathcal{T}([a,b])$ of all trigonometric polynomials on the interval $[a,b]$ clearly contains the constant functions and is closed under addition and scalar multiplication. Since

$$\sin jx \,\sin kx = \tfrac{1}{2}\big[\sin(j-k)x + \sin(j+k)x\big],$$

with similar identities holding for $\sin jx \,\cos kx$ and $\cos jx\, \cos kx$, $\mathcal{T}([a,b])$ is an algebra.

If $0 < b - a < 2\pi$, then $\{\cos x, \sin x\}$, and hence $\mathcal{T}([a,b])$, separate points of $[a,b]$. By the Stone–Weierstrass theorem, every member of $\mathcal{C}([a,b])$ may be uniformly approximated by trigonometric polynomials on $[a,b]$.

If $b - a = 2\pi$, then $\mathcal{T}([a,b])$ no longer separates points of $[a,b]$. However, in this case every member f of $\mathcal{C}([a,b])$ with $f(a) = f(b)$ may be uniformly approximated by a trigonometric polynomial. We verify this for the interval $[0,2\pi]$. Let $\mathcal{E}$ denote the algebra of continuous functions $f : [0,2\pi] \to \mathbb{R}$ with $f(0) = f(2\pi)$, and let X denote the circle $x^2 + y^2 = 1$ with the Euclidean $\mathbb{R}^2$ metric. For each $f \in \mathcal{E}$, define $F_f : X \to \mathbb{R}$ by

$$F_f(\cos t, \sin t) = f(t), \quad 0 \le t \le 2\pi.$$

It is straightforward to verify that F_f is continuous. For example, if $(\cos t_n, \sin t_n) \to (1,0)$, then every convergent subsequence $\{t_{n_k}\}$ converges either to 0 or to 2π, hence

$$F_f(\cos t_{n_k}, \sin t_{n_k}) = f(t_{n_k}) \to f(0) = f(1) = F_f(1,0).$$

The set $\mathcal{A} := \{F_T \,:\, T \in \mathcal{T}([0,2\pi])\}$ is easily seen to be an algebra that contains the constant functions. Moreover, $\mathcal{A}$ separates points of X. Indeed, if $\boldsymbol{x} := (\cos s, \sin s)$ and $\boldsymbol{y} := (\cos t, \sin t)$ with $\boldsymbol{x} \ne \boldsymbol{y}$, then, say, $\cos s \ne \cos t$ hence $F_T(\boldsymbol{x}) \ne F_T(\boldsymbol{y})$, where $T(x) = \cos x$. Therefore, each F_f may be uniformly approximated on X by members of $\mathcal{A}$. It follows that each member of $\mathcal{E}$ may be uniformly approximated on $[0,2\pi]$ by trigonometric polynomials. ◊

Exercises

1. Give an example of a bounded continuous function that cannot be approximated uniformly by polynomials on $(0,1)$.

2. Let f be continuous on $[a,+\infty)$ such that $\lim_{x\to+\infty} f^{(m)}(x) \ne 0$ for all sufficiently large $m \in \mathbb{N}$. Prove that f cannot be uniformly approximated by polynomials on $[a,+\infty)$. Give an example of such a function.

3. Let $f \in \mathcal{C}([a,b])$ have the property that $\int_a^b x^n f(x)\,dx = 0$ for all $n \in \mathbb{Z}^+$. Prove that $f = 0$ on $[a,b]$. Show that if $a \geq 0$, then it is enough that the given property holds for even integers n in $\mathbb{Z}^+$.

4. Let $f : [a,b] \to \mathbb{R}$ have continuous derivatives up to order k such that

$$\int_a^b x^n f^{(k)}(x)\,dx = 0 \quad \text{for all } n \in \mathbb{Z}^+.$$

Prove that f is a polynomial.

5. Let $f : [a,b] \to \mathbb{R}$ have continuous derivatives up to order k. Prove that there exists a sequence of polynomials P_n such that $\lim_n P_n^{(j)} = f^{(j)}$ uniformly on $[a,b]$ for $j = 0, 1, \ldots, k$.

6. Let X be compact and let $\mathcal{A}$ be an algebra in $\mathcal{C}(X)$ that contains the constant functions and separates the points of X. Let $x_0 \in X$ and let $f \in \mathcal{C}(X)$ satisfy $f(x_0) = 0$. Prove that there exists a sequence $f_n \in \mathcal{A}$ converging uniformly to f such that $f_n(x_0) = 0$ for all n.

7. Show that there exists a sequence of polynomials P_n converging uniformly to $\sin x$ on $[0,\pi]$ such that $P_n(0) = P_n(\pi) = 0$ for all n.

8. Let f be an odd (even) continuous function on $[-a,a]$, $a > 0$. Prove that there is a sequence of odd (even) polynomials that converges uniformly to f on $[-a,a]$.

9. Let $f \in \mathcal{C}([0,2\pi])$ have the properties $f(0) = f(2\pi)$ and

$$\int_0^{2\pi} f(x)(\sin^m x)(\cos^n x)\,dx = 0 \quad \text{for all } m,\, n \in \mathbb{Z}^+.$$

Prove that f is identically zero on $[0,2\pi]$.

10. Let $f : \mathbb{R} \to \mathbb{R}$ be continuous and periodic with period 2π. Prove that there exists a sequence of trigonometric polynomials that converges uniformly to f on $\mathbb{R}$.

11. Let $f \in \mathcal{C}([-\pi/2,\pi/2])$ with $f(0) = 0$. Prove that f can be uniformly approximated on $[-\pi/2,\pi/2]$ by functions of the form $\sum_{j=1}^m b_j \sin(jx)$.

12. Let g be continuous and one-to-one on $[a,b]$. Prove that any function in $\mathcal{C}([a,b])$ may be uniformly approximated by functions of the form $\sum_{j=0}^m a_j g^j$.

13. Prove the following version of the Stone–Weierstrass theorem: If $\mathcal{V}$ is a linear subspace of $\mathcal{C}(X)$ that contains the constant functions, separates points of X, and contains $|f|$ for all $f \in \mathcal{V}$, then $\mathcal{V}$ is dense in $\mathcal{C}(X)$.

14. Show that for any $f \in C([0,2\pi])$ there exists a sequence of trigonometric polynomials T_n such that $\int_0^{2\pi} |f - T_n| \to 0$.

15. Let X and Y be compact metric spaces and let $f(x,y) \in \mathcal{C}(X \times Y)$ be a continuous real-valued function on $X \times Y$. Show that for every $\varepsilon > 0$ there exist $g_1, \ldots, g_n \in \mathcal{C}(X)$ and $h_1, \ldots, h_n \in \mathcal{C}(Y)$ such that

$$\left| f(x,y) - \sum_{i=1}^{n} g_i(x) h_i(y) \right| < \varepsilon \text{ for all } (x,y) \in X \times Y.$$

16. Let $\mathcal{E}_0$ denote the algebra of all continuous functions $f[a,b] :\to \mathbb{R}$ such that $f(a) = f(b) = 0$. If $\mathcal{A}_0$ is an algebra in $\mathcal{E}_0$ that separates points of (a,b), show that $\mathcal{A}_0$ is dense in $\mathcal{E}_0$ in the uniform norm. *Hint.* Use ideas of 8.8.8 by considering the algebra generated by $\mathcal{A}_0$ and the constant functions.

17. Let $C_0(\mathbb{R})$ denote the algebra of all continuous functions f on $\mathbb{R}$ such that $\lim_{t \to \pm\infty} f(t) = 0$. Let $\mathcal{B}_0$ be an algebra in $C_0(\mathbb{R})$ that separates points of $\mathbb{R}$. Show that $\mathcal{B}_0$ is dense in $C_0(\mathbb{R})$ in the uniform norm. *Hint.* Consider $\theta(t) = \tan^{-1}[(t-\pi)/2]$, $0 < t < 2\pi$ and use Exercise 16.

*8.9 Baire's Theorem

Let (X,d) be a metric space. The *diameter* $d(E)$ of a nonempty subset E of X is defined by

$$d(E) = \sup_{x,y \in E} d(x,y).$$

8.9.1 Lemma. *If X is complete, then the intersection C of any decreasing sequence of nonempty closed sets C_n in X with $d(C_n) \to 0$ contains a single point.*

Proof. For each n choose a point $x_n \in C_n$. If $m > n$, then $x_m \in C_n$, hence $d(x_m, x_n) \le d(C_n)$. Since $d(C_n) \to 0$, $\{x_n\}$ is Cauchy. Let $x_n \to x$. Since $x_n, x_{n+1}, \ldots \in C_n$ and C_n is closed, $x \in C_n$ for all n, that is, $x \in C$. Since $d(C) \le d(C_n) \to 0$, $C = \{x\}$. □

8.9.2 Baire Category Theorem. *Let X be a complete metric space. Then the following statements hold:*

(a) *If $U_n \subseteq X$ is open and dense in X for all n, then $G := \bigcap_n U_n$ is dense in X.*

(b) *If $C_n \subseteq X$ is closed and has empty interior for all n, then $F := \bigcup_n C_n$ has empty interior.*

Proof. To prove (a), we show that $B \cap G \neq \emptyset$ for any open ball B. Since $B \cap U_1$ is open and nonempty, $C_1 := C_{r_1}(x_1) \subseteq B \cap U_1$ for some $x_1 \in X$ and $0 < r_1 \leq 1$. Since $B_{r_1}(x_1) \cap U_2$ is open and nonempty, $C_2 := C_{r_2}(x_2) \subseteq B_{r_1}(x_1) \cap U_2$ for some $x_2 \in X$ and $0 < r_2 < 1/2$. Continuing in this manner, we obtain a decreasing sequence of closed balls $C_n \subseteq B \cap U_n$ with diameters tending to zero. By 8.9.1, $\bigcap_n C_n$ contains a point x. Then $x \in B \cap U_n$ for all n, hence $x \in B \cap G$.

Part (b) follows from (a). Indeed, suppose $\operatorname{int}(C_n) = \emptyset$ for all n. Then $U_n := C_n^c$ is dense in X, hence $\bigcap_n U_n$ is dense in X. It follows that the interior of $(\bigcap_n U_n)^c = F$ is empty. □

We give three applications of Baire's theorem. The first is known as the *principle of uniform boundedness.*

8.9.3 Theorem. *Let $\mathcal{X}$ and $\mathcal{Y}$ be complete normed vector spaces and let $\mathcal{L}$ be a family of continuous linear transformations from $\mathcal{X}$ to $\mathcal{Y}$ such that*

$$\sup_{T \in \mathcal{L}} \|T\boldsymbol{x}\| < \infty \quad \textit{for each } \boldsymbol{x} \in \mathcal{X}.$$

Then there exists $M > 0$ such that $\|T\boldsymbol{x}\| \leq M\|\boldsymbol{x}\|$ for all $\boldsymbol{x} \in \mathcal{X}$ and $T \in \mathcal{L}$.

Proof. For each n, set

$$C_n = \{\boldsymbol{x} \in \mathcal{X} : \|T\boldsymbol{x}\| \leq n \text{ for all } T \in \mathcal{L}\}.$$

By hypothesis, $\mathcal{X} = \bigcup_n C_n$. By continuity of the transformations T, each C_n is closed. Therefore, Baire's theorem shows that $\operatorname{int}(C_n) \neq \emptyset$ for some n. Thus there exists $\boldsymbol{x}_0$ and $r > 0$ such that $\|T\boldsymbol{y}\| \leq n$ for all $T \in \mathcal{L}$ and $\boldsymbol{y} \in \mathcal{X}$ with $\|\boldsymbol{y} - \boldsymbol{x}_0\| \leq r$. If $\|\boldsymbol{x}\| \leq r$, then, taking $\boldsymbol{y} = \boldsymbol{x} + \boldsymbol{x}_0$, we have

$$\|T\boldsymbol{x}\| \leq \|T\boldsymbol{x} + T\boldsymbol{x}_0\| + \|T\boldsymbol{x}_0\| = \|T\boldsymbol{y}\| + \|T\boldsymbol{x}_0\| \leq n + \|T\boldsymbol{x}_0\|.$$

It follows that for all $\boldsymbol{x} \neq \mathbf{0}$ and $T \in \mathcal{L}$

$$\left\| T\left(\frac{r\boldsymbol{x}}{\|\boldsymbol{x}\|}\right) \right\| \leq n + \|T\boldsymbol{x}_0\|$$

hence

$$\|T\boldsymbol{x}\| \leq r^{-1}\big(n + \|T\boldsymbol{x}_0\|\big)\|\boldsymbol{x}\|.$$ □

The following corollary is one of the few instances in analysis (Dini's theorem being another) when pointwise convergence of a sequence of continuous functions is sufficient to convey the continuity property to the limit function.

8.9.4 Corollary. *Let $\mathcal{X}$ and $\mathcal{Y}$ be complete normed vector spaces and let $\{T_n\}$ be a sequence of continuous linear transformations from $\mathcal{X}$ to $\mathcal{Y}$ converging pointwise on $\mathcal{X}$ to a function T. Then T is linear and continuous.*

Proof. Linearity of T is clear. For continuity, note that $\sup_n \|T_n\boldsymbol{x}\| < +\infty$ for each $\boldsymbol{x} \in \mathcal{X}$, hence, by the theorem, there exists $M > 0$ such that $\|T_n\boldsymbol{x}\| \le M\|\boldsymbol{x}\|$ for all n and $\boldsymbol{x}$. Letting $n \to +\infty$ yields $\|T\boldsymbol{x}\| \le M\|\boldsymbol{x}\|$, hence T is continuous. □

For the second application of Baire's theorem, recall that there exist functions $f : \mathbb{R} \to \mathbb{R}$ whose set of discontinuity points is precisely $\mathbb{Q}$ (3.3.3). The obvious question raised by this fact is answered in the following theorem.

8.9.5 Theorem. *There is no function $f : \mathbb{R} \to \mathbb{R}$ whose set of continuity points is precisely $\mathbb{Q}$.*

Proof. For each n, let U_n denote the union of all intervals (a, b) such that $|f(x) - f(y)| < 1/n$ for all x, $y \in (a, b)$. Then U_n is open and the set of continuity points of f is precisely $C := \bigcap_{n=1}^{\infty} U_n$. Suppose that $C = \mathbb{Q}$. Then each U_n contains $\mathbb{Q}$ and hence is dense in $\mathbb{R}$. Let $\{r_1, r_2, \ldots\}$ be an enumeration of $\mathbb{Q}$. Then the open sets $V_m := \mathbb{R} \setminus \{r_m\}$ are also dense in $\mathbb{R}$ and have intersection $\mathbb{I}$. By Baire's theorem, the collection of sets $\{U_n, V_m : m, n \in \mathbb{N}\}$ has a nonempty intersection. But this intersection is $\mathbb{Q} \cap \mathbb{I} = \emptyset$. Therefore, C cannot equal $\mathbb{Q}$. □

The last application of Baire's theorem shows that there is a rich supply of continuous, nowhere differentiable functions. For the proof we need the following lemma.

8.9.6 Lemma. *If g is piecewise linear on $[a, b]$, then there exists $M > 0$ such that*

$$|g(x) - g(y)| \le M|x - y| \quad \textit{for all } x,\ y \in [a, b].$$

Proof. Let g be as in 8.8.1 and set $M = \max_j\{|m_j|\}$. If

$$x_i \le x \le x_{i+1} \le x_j \le y \le x_{j+1}$$

then

$$\begin{aligned}|g(x) - g(y)| &\le |g(x) - g(x_{i+1})| + |g(x_{i+1}) - g(x_{i+2})| + \cdots + |g(x_j) - g(y)| \\ &\le |m_i|(x_{i+1} - x) + |m_{i+1}|(x_{i+2} - x_{i+1}) + \cdots + |m_j|(y - x_j) \\ &\le M(y - x).\end{aligned}$$

□

8.9.7 Theorem. *The set of all continuous, nowhere differentiable functions on an interval $[a, b]$ is dense in $\mathcal{C}([a, b])$ in the uniform norm.*

Proof. For each $n \in \mathbb{N}$ and $f \in \mathcal{C}([a, b])$ define

$$E_n(f) = \{x \in [a, b] \ : \ |f(y) - f(x)| \le n|x - y| \ \text{ for all } \ y \in [a, b]\}.$$

we break the proof into several steps:

I. $\bigcup_{n=1}^{\infty} E_n(f)$ contains all points at which f is differentiable.

⟦ Let x be such a point and choose $\delta > 0$ such that

$$\left|\frac{f(y)-f(x)}{y-x} - f'(x)\right| < 1 \text{ for all } y \in [a,b] \text{ with } 0 < |x-y| < \delta.$$

Then

$$|f(y)-f(x)| \le \begin{cases} (1+|f'(x)|)\,|y-x| & \text{if } |x-y| < \delta, \\ 2\|f\|_\infty \le 2\delta^{-1}\|f\|_\infty|y-x| & \text{if } |x-y| \ge \delta, \end{cases}$$

which shows that $x \in E_n(f)$ for all $n > 1 + |f'(x)| + 2\delta^{-1}\|f\|_\infty$. ⟧

II. $\mathcal{E}_n := \{f \in \mathcal{C}([a,b]) : E_n(f) \ne \emptyset\}$ is closed in $\mathcal{C}([a,b])$.

⟦ Let $\{f_k\}$ be a sequence in $\mathcal{E}_n$ converging uniformly to $f \in \mathcal{C}([a,b])$. For each k, choose a point $x_k \in E_n(f_k)$. We may assume that $x_k \to x$ for some $x \in [a,b]$ (otherwise, take a subsequence). Then for all $y \in [a,b]$,

$$\begin{aligned} |f(y)-f(x)| &\le |f(y)-f_k(y)| + |f_k(y)-f_k(x_k)| \\ &\qquad + |f_k(x_k)-f_k(x)| + |f_k(x)-f(x)| \\ &\le 2\|f-f_k\|_\infty + n|y-x_k| + n|x_k-x|. \end{aligned}$$

Letting $k \to \infty$ shows that $|f(y)-f(x)| \le n|y-x|$, that is, $x \in E_n(f)$. Therefore, $f \in \mathcal{E}_n$. ⟧

III. $\mathcal{E}_n^c$ is dense in $\mathcal{C}([a,b])$.

⟦ Let $f \in \mathcal{C}([a,b])$ and $\varepsilon > 0$. We construct a function $h \in B_\varepsilon(f) \cap \mathcal{E}_n^c$. By 8.8.2, there exists a piecewise linear function g such that $\|f-g\|_\infty < \varepsilon/2$. By 8.9.6, there exists $M > 0$ such that

$$|g(x)-g(y)| \le M|x-y| \quad \text{for all } x,\ y \in [a,b].$$

Let $r > 0$ and let $x_0 = a < x_1 < \cdots < x_{2p} = b$ be a partition of $[a,b]$ with mesh $< r$. Construct a "sawtooth" piecewise linear function h_r with vertices

$$(x_0,1),(x_2,1),\ldots,(x_{2p},1) \quad \text{and} \quad (x_1,-1),(x_3,-1),\ldots,(x_{2p-1},-1),$$

and set $h := g + \varepsilon h_r/2$. Then

$$\|h-f\|_\infty \le \|h-g\|_\infty + \|g-f\|_\infty = \frac{\varepsilon}{2}\|h_r\|_\infty + \|g-f\|_\infty < \frac{\varepsilon}{2} + \frac{\varepsilon}{2} = \varepsilon,$$

so $h \in B_\varepsilon(f)$. To show that $h \in \mathcal{E}_n^c$, let x be an arbitrary member of $[a,b]$. If $h_r(x) \le 0$ (≥ 0) choose x_j such that $|x-x_j| < r$ and $h_r(x_j) = 1$

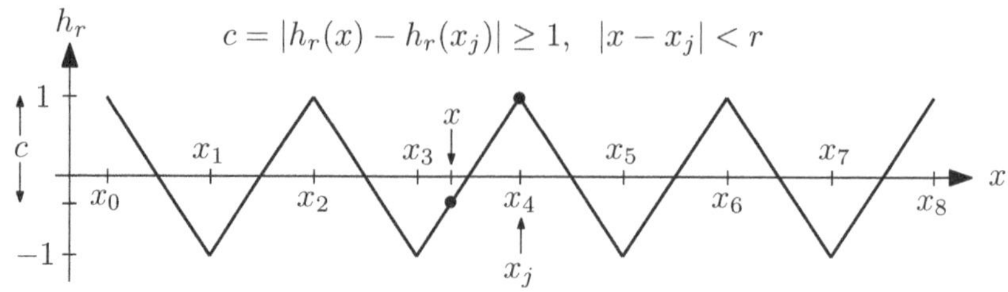

FIGURE 8.11: The sawtooth function h_r.

$(= -1)$ (see Figure 8.11). Then $|h_r(x) - h_r(x_j)| \ge 1$, hence

$$\begin{aligned} |h(x) - h(x_j)| &\ge \frac{\varepsilon}{2}|h_r(x) - h_r(x_j)| - |g(x) - g(x_j)| \\ &\ge \frac{\varepsilon}{2} - M|x - x_j| \\ &\ge \left(\frac{\varepsilon}{2r} - M\right)|x - x_j|. \end{aligned}$$

If r is chosen so that $\dfrac{\varepsilon}{2r} - M > n$, then $x \notin E_n(h)$, hence $h \notin \mathcal{E}_n$.]]

To complete the proof note that by step III and Baire's theorem, $\mathcal{F} := \bigcap_{n=1}^{\infty} \mathcal{E}_n^c$ is dense in $\mathcal{C}([a,b])$. Since $f \in \mathcal{F}$ implies that $E_n(f) = \emptyset$ for every n, and since a point at which f is differentiable must lie in some $E_n(f)$, no member of $\mathcal{F}$ can be differentiable at any point of $[a,b]$. □

Exercises

1. Prove the converse of 8.9.1: If the intersection of any decreasing sequence of nonempty closed sets C_n in X with $d(C_n) \to 0$ contains a single point, then X is complete.

2. Let $\mathbb{Q}$ have the usual metric. Find a decreasing sequence of closed sets C_n in $\mathbb{Q}$ with $d(C_n) \to 0$ and $\bigcap_n C_n = \emptyset$.

3. Show that 8.9.2 does not hold in $\mathbb{Q}$ with the usual metric.

4. Let $D = \{x_1, x_2, \ldots\}$ be a proper subset of a complete metric space X. Show that (a) and (b) of 8.9.2 hold for $Y := X \setminus D$. Conclude that the set of irrationals $\mathbb{I}$ with the usual metric satisfies (a) and (b) of the theorem.

Chapter 9

Differentiation on $\mathbb{R}^n$

For the remainder of the book, the Euclidean norm $\|\cdot\|_2$ on the spaces $\mathbb{R}^n$ will be denoted simply by $\|\cdot\|$.

In this chapter we extend the ideas of Chapter 4 to vector-valued functions of several variables. This will require some notions from linear algebra, a brief review of which may be found in Appendix B.

9.1 Definition of the Derivative

To motivate the general definition of the derivative of a function on $\mathbb{R}^n$, we begin with two important special cases.

Derivative of a Vector-Valued Function of a Real Variable

The definition of derivative in this case is a natural extension of the definition of the derivative of a scalar-valued function:

9.1.1 Definition. Let $I \subseteq \mathbb{R}$ be an interval and $a \in I$. A function $f : I \to \mathbb{R}^m$ is said to be *differentiable at* a if the (vector) limit

$$f'(a) := \lim_{h\to 0} \frac{f(a+h) - f(a)}{h} = \lim_{t\to a} \frac{f(t) - f(a)}{t - a}$$

exists in $\mathbb{R}^m$. (The limit is one-sided if a is an endpoint of I.) The vector $f'(a)$ is called the *derivative of* f *at* a. If f is differentiable at each point in I, then f is said to be *differentiable on* I and the resulting function $f' : I \to \mathbb{R}^m$ is called the *derivative of* f *on* I. ◊

The function f may be viewed as a parametrization of a curve C in $\mathbb{R}^m$. The vector $f'(a)$ is then called the *tangent vector* to C at the point $f(a)$. If the variable t is interpreted as time, then C may be viewed as the path of a particle in $\mathbb{R}^m$. In this context, $f'(a)$ is called the *velocity* of the particle at time a and $\|f'(a)\|$ the *speed*. The curve is said to be *smooth* if f' is continuous and nonzero on I. Parameterized curves will be examined in detail in Chapter 12.

DOI: 10.1201/9781003629382-9

Note that the function $f : I \to \mathbb{R}^m$ may be written $f = (f_1, \ldots, f_m)$, where $f_j : I \to \mathbb{R}$ is the *jth component function* of f.

9.1.2 Proposition. *Let I be an interval and $f = (f_1, \ldots, f_m) : I \to \mathbb{R}^m$. Then f is differentiable at $a \in I$ iff each f_j is differentiable at a, in which case $f'(a) = (f_1'(a), \ldots, f_m'(a))$. In particular, if f is differentiable at a, then f is continuous at a.*

Proof. The assertions follow directly from the inequalities

$$\left|\frac{f_j(a+h) - f_j(a)}{h} - x_j\right|^2 \le \left\|\frac{f(a+h) - f(a)}{h} - (x_1, \ldots, x_m)\right\|^2$$
$$\le \sum_{i=1}^m \left|\frac{f_i(a+h) - f_i(a)}{h} - x_i\right|^2. \qquad \square$$

The *differential of f at a* is the linear transformation $df_a : \mathbb{R} \to \mathbb{R}^m$ that takes a real number h to the vector $hf'(a)$:

$$df_a(h) = hf'(a), \quad h \in \mathbb{R}.$$

Definition 9.1.1 may then be rephrased as follows: f is differentiable at a iff there exists a linear transformation $T : \mathbb{R} \to \mathbb{R}^m$ such that

$$\lim_{h \to 0} \frac{f(a+h) - f(a) - Th}{|h|} = 0,$$

in which case $T = df_a$

Derivative of a Real-Valued Function of Several Variables

The derivative of a scalar-valued function of n variables is defined as follows:

9.1.3 Definition. Let $U \subseteq \mathbb{R}^n$ be open and $\boldsymbol{a} \in U$. Then $f : U \to \mathbb{R}$ is said to be *differentiable at* $\boldsymbol{a}$ if there exists a vector $f'(\boldsymbol{a})$ in $\mathbb{R}^n$ such that

$$\lim_{\boldsymbol{h} \to 0} \frac{f(\boldsymbol{a}+\boldsymbol{h}) - f(\boldsymbol{a}) - f'(\boldsymbol{a}) \cdot \boldsymbol{h}}{\|\boldsymbol{h}\|} = 0. \tag{9.1}$$

The vector $f'(\boldsymbol{a})$ is called the derivative of f at $\boldsymbol{a}$. The *differential of f at $\boldsymbol{a}$* is the linear transformation $df_{\boldsymbol{a}} \in \mathcal{L}(\mathbb{R}^n, \mathbb{R})$ defined by

$$df_{\boldsymbol{a}}(\boldsymbol{h}) = f'(\boldsymbol{a}) \cdot \boldsymbol{h}, \quad \boldsymbol{h} \in \mathbb{R}^n. \qquad \diamond$$

Now let

$$\boldsymbol{e}^j = (0, \ldots, 0, \overset{j}{1}, 0, \ldots, 0), \quad j = 1, \ldots, n,$$

denote the standard basis vectors in $\mathbb{R}^n$. If $f'(\boldsymbol{a})$ exists, then, taking $\boldsymbol{h} = t\boldsymbol{e}^j$ in (9.1), we have

$$\lim_{t \to 0} \frac{f(\boldsymbol{a} + t\boldsymbol{e}^j) - f(\boldsymbol{a}) - tf'(\boldsymbol{a}) \cdot \boldsymbol{e}^j}{t} = 0,$$

or, equivalently,

$$\lim_{t\to 0}\frac{f(\boldsymbol{a}+t\boldsymbol{e}^j)-f(\boldsymbol{a})}{t}=f'(\boldsymbol{a})\cdot\boldsymbol{e}^j. \tag{9.2}$$

The expression on the right is just the jth component of $f'(\boldsymbol{a})$. The limit on the left is called the jth *partial derivative of* f *at* $\boldsymbol{a}$ and is denoted variously by

$$\partial_j f = f_{x_j} = \frac{\partial f}{\partial x_j}.$$

We have proved the following result.

9.1.4 Proposition. *If* f *is differentiable at* $\boldsymbol{a}$*, then the partial derivatives* $\partial_j f(\boldsymbol{a})$ *of* f *exist at* $\boldsymbol{a}$ *and*

$$f'(\boldsymbol{a}) = \big(\partial_1 f(\boldsymbol{a}), \partial_2 f(\boldsymbol{a}), \ldots, \partial_n f(\boldsymbol{a})\big). \tag{9.3}$$

In particular, the derivative is unique.

The vector on the right in (9.3) is called the *gradient of* f *at* $\boldsymbol{a}$ and is denoted by ∇f or $\operatorname{grad} f$. The linear transformation $df_{\boldsymbol{a}} \in \mathcal{L}(\mathbb{R}^n, \mathbb{R})$ may now be written

$$df_{\boldsymbol{a}}(\boldsymbol{h}) = \nabla f(\boldsymbol{a})\cdot\boldsymbol{h}, \quad \boldsymbol{h}\in\mathbb{R}^n. \tag{9.4}$$

For an alternate notation, let $dx_j : \mathbb{R}^n \to \mathbb{R}$ be the linear function defined by

$$dx_j(\boldsymbol{h}) = h_j, \quad \boldsymbol{h} = (h_1, \ldots, h_n).$$

Then $df_{\boldsymbol{a}}$ may be expressed as

$$df_{\boldsymbol{a}}(\boldsymbol{h}) = \sum_{j=1}^{n}\frac{\partial f(\boldsymbol{a})}{\partial x_j}dx_j(\boldsymbol{h}).$$

If the partial derivatives of f exist at each point of U, we write simply

$$df = \sum_{j=1}^{n}\frac{\partial f}{\partial x_j}dx_j.$$

For example,

$$d\sin(x^2y) = 2xy\cos(x^2y)\,dx + x^2\cos(x^2y)\,dy.$$

We show below that if f has continuous partial derivatives on U, then f is differentiable on U. The continuity hypothesis cannot be removed: There are functions f that are not differentiable on U but whose partial derivatives exist throughout U. This is the case for the function in the following example.

9.1.5 Example. Let $m \in \mathbb{N}$. The function

$$f(x,y) = \begin{cases} \dfrac{x^m y}{x^2+y^2} & \text{if } (x,y) \neq (0,0), \\ 0 & \text{otherwise} \end{cases}$$

exhibits a variety of behavior depending on the values of m. The partial derivatives of f are

$$f_x(x,y) = \begin{cases} \dfrac{mx^{m+1}y + mx^{m-1}y^3 - 2x^{m+1}y}{(x^2+y^2)^2}, & \text{if } \boldsymbol{x} \neq (0,0), \\ 0 & \text{otherwise,} \end{cases}$$

$$f_y(x,y) = \begin{cases} \dfrac{x^m(x^2-y^2)}{(x^2+y^2)^2}, & \text{if } \boldsymbol{x} \neq (0,0), \\ 0 & \text{otherwise.} \end{cases}$$

If $m = 1$, f is not continuous at $(0,0)$, hence is not differentiable there (see 9.1.11, below). If $m = 1$ or 2, the partial derivatives exist at $(0,0)$ but are not continuous there. If $m = 2$, the function is continuous at $(0,0)$, with zero partial derivatives at $(0,0)$, but is not differentiable there since in this case the limit

$$\lim_{\boldsymbol{x}\to\boldsymbol{0}} \frac{f(\boldsymbol{x}) - f(\boldsymbol{0}) - \boldsymbol{0}\cdot\boldsymbol{x}}{\|\boldsymbol{x}\|}, = \lim_{(x,y)\to(0,0)} \frac{x^2 y}{(x^2+y^2)^{3/2}}$$

fails to exist. If $m \geq 3$, f has continuous partial derivatives and is differentiable on $\mathbb{R}^2$. ◊

The definition of the jth partial derivative of f at $\boldsymbol{a}$ may be written explicitly as

$$\partial_j f(\boldsymbol{a}) = \lim_{h\to 0} \frac{f(a_1,\ldots,a_j+h,\ldots,a_n) - f(a_1,\ldots,a_j,\ldots,a_n)}{h}.$$

This is simply the derivative at a_j of the one-variable function

$$t \mapsto f(a_1,\ldots,a_{j-1},t,a_{j+1},\ldots,a_n).$$

Thus, to find the jth partial derivative of $f(x_1,\ldots,x_j,\ldots,x_n)$, one simply differentiates f with respect to x_j while holding the other variables fixed. It follows that the standard formulas for derivatives of functions of one variable hold for partial derivatives of functions of several variables. For example, the product rule takes the form

$$\partial_j(fg)(\boldsymbol{a}) = f(\boldsymbol{a})\partial_j g(\boldsymbol{a}) + g(\boldsymbol{a})\partial_j f(\boldsymbol{a}),$$

and the quotient rule becomes

$$\partial_j\left(\frac{f}{g}\right)(\boldsymbol{a}) = \frac{g(\boldsymbol{a})\partial_j f(\boldsymbol{a}) - f(\boldsymbol{a})\partial_j g(\boldsymbol{a})}{g^2(\boldsymbol{a})}, \quad g(\boldsymbol{a}) \neq 0.$$

Derivative of a Vector-Valued Function of Several Variables

We now consider the general case. The following definition includes the two special cases discussed before.

9.1.6 Definition. Let $U \subseteq \mathbb{R}^n$ be open. A function $f : U \to \mathbb{R}^m$ is said to be *differentiable at* $\boldsymbol{a} \in U$ if there exists a linear transformation $df_{\boldsymbol{a}} : \mathbb{R}^n \to \mathbb{R}^m$, called the *differential of* f *at* $\boldsymbol{a}$, such that

$$\lim_{\boldsymbol{h}\to 0} \frac{f(\boldsymbol{a}+\boldsymbol{h}) - f(\boldsymbol{a}) - df_{\boldsymbol{a}}(\boldsymbol{h})}{\|\boldsymbol{h}\|} = 0.$$

The $m \times n$ matrix $[df_{\boldsymbol{a}}]$ is called the *derivative of* f *at* $\boldsymbol{a}$, or the *Jacobian matrix of* f *at* $\boldsymbol{a}$, and is denoted by $f'(\boldsymbol{a})$. ◊

9.1.7 Example. If $T \in \mathcal{L}(\mathbb{R}^n, \mathbb{R}^n)$, then, by the linearity of T,

$$T(\boldsymbol{x}+\boldsymbol{h}) - T(\boldsymbol{x}) - T\boldsymbol{h} = 0 \quad \text{for all } \boldsymbol{h}.$$

It follows that $dT_{\boldsymbol{x}} = T$ for all $\boldsymbol{x}$. This is the n-dimensional version of the familiar result that the derivative of the function $x \to tx$ is the constant t. ◊

9.1.8 Theorem. *Let* $U \subseteq \mathbb{R}^n$ *be open,* $f = (f_1, \ldots, f_m) : U \to \mathbb{R}^m$*, and let* $\boldsymbol{a} \in U$*. Then* f *is differentiable at* $\boldsymbol{a}$ *iff each function* $f_i : U \to \mathbb{R}$ *is differentiable at* $\boldsymbol{a}$*. In this case,* $\partial_j f_i(\boldsymbol{a})$ *exists and equals* $df_{\boldsymbol{a}}(\boldsymbol{e}^j) \cdot \boldsymbol{e}^i$*, and*

$$df_{\boldsymbol{a}}(\boldsymbol{h}) = \big(\nabla f_1(\boldsymbol{a}) \cdot \boldsymbol{h}, \ldots, \nabla f_m(\boldsymbol{a}) \cdot \boldsymbol{h}\big), \quad \boldsymbol{h} \in \mathbb{R}^n. \tag{9.5}$$

In particular, if the differential exists, it is unique.

Proof. Let f be differentiable at $\boldsymbol{a}$. For $i = i, \ldots, m$ and $j = 1, \ldots, n$, let $b_{ij} = df_{\boldsymbol{a}}(\boldsymbol{e}^j) \cdot \boldsymbol{e}^i$, the ith component of $df_{\boldsymbol{a}}(\boldsymbol{e}^j)$ and the (i,j)th entry of the matrix $[df_{\boldsymbol{a}}]$. Then

$$df_{\boldsymbol{a}}(\boldsymbol{h}) = \big(\boldsymbol{b}_1 \cdot \boldsymbol{h}, \ldots, \boldsymbol{b}_m \cdot \boldsymbol{h}\big), \quad \text{where } \boldsymbol{b}_i := (b_{i1}, \ldots, b_{in}).$$

Thus for each i,

$$|f_i(\boldsymbol{a}+\boldsymbol{h}) - f_i(\boldsymbol{a}) - \boldsymbol{b}_i \cdot \boldsymbol{h}| \le \|f(\boldsymbol{a}+\boldsymbol{h}) - f(\boldsymbol{a}) - df_{\boldsymbol{a}}(\boldsymbol{h})\|,$$

from which it follows that

$$\lim_{\boldsymbol{h}\to 0} \frac{f_i(\boldsymbol{a}+\boldsymbol{h}) - f_i(\boldsymbol{a}) - \boldsymbol{b}_i \cdot \boldsymbol{h}}{\|\boldsymbol{h}\|} = 0.$$

Therefore, the derivative of f_i at $\boldsymbol{a}$ exists and equals $\boldsymbol{b}_i$. By 9.1.4, $\boldsymbol{b}_i = \nabla f_i(\boldsymbol{a})$, that is, $b_{ij} = \partial_j f_i(\boldsymbol{a})$.

Conversely, suppose each f_j is differentiable at $\boldsymbol{a}$. Then $\nabla f_j(\boldsymbol{a})$ exists and by (9.4),

$$\lim_{\boldsymbol{h}\to 0} \frac{|f_i(\boldsymbol{a}+\boldsymbol{h}) - f_i(\boldsymbol{a}) - \nabla f_i(\boldsymbol{a}) \cdot \boldsymbol{h}|}{\|\boldsymbol{h}\|} = 0, \quad i = 1, \ldots, m.$$

Let $T(\boldsymbol{h})$ denote the right side of (9.5). Then T is linear and

$$\frac{\|f(\boldsymbol{a}+\boldsymbol{h}) - f(\boldsymbol{a}) - T(\boldsymbol{h})\|^2}{\|\boldsymbol{h}\|^2} = \sum_{i=1}^{m} \frac{|f_i(\boldsymbol{a}+\boldsymbol{h}) - f_i(\boldsymbol{a}) - \nabla f_i(\boldsymbol{a}) \cdot \boldsymbol{h}|^2}{\|\boldsymbol{h}\|^2} \to 0$$

as $\boldsymbol{h} \to 0$. Therefore, $df_{\boldsymbol{a}}$ exists and equals T. □

By the theorem, the (i,j) entry of $f'(\boldsymbol{a})$ is $\partial_j f_i(\boldsymbol{a})$. The effect of $df_{\boldsymbol{a}}$ on a vector $\boldsymbol{h} \in \mathbb{R}^n$ may therefore be expressed in matrix form as

$$f'(\boldsymbol{a})\boldsymbol{h}^t = \begin{bmatrix} \partial_1 f_1(\boldsymbol{a}) & \cdots & \partial_n f_1(\boldsymbol{a}) \\ \vdots & & \vdots \\ \partial_1 f_m(\boldsymbol{a}) & \cdots & \partial_n f_m(\boldsymbol{a}) \end{bmatrix} \begin{bmatrix} h_1 \\ \vdots \\ h_n \end{bmatrix} = \begin{bmatrix} \nabla f_1(\boldsymbol{a}) \cdot \boldsymbol{h} \\ \vdots \\ \nabla f_m(\boldsymbol{a}) \cdot \boldsymbol{h} \end{bmatrix},$$

where $\boldsymbol{h}^t$ denotes the transpose of the vector $\boldsymbol{h}$. In the special case $m = n$, the determinant of $f'(\boldsymbol{a})$ is called the *Jacobian of f at* $\boldsymbol{a}$ and is denoted variously by

$$\det f'(\boldsymbol{a}) = J_f(\boldsymbol{a}) = \frac{\partial(f_1, \ldots, f_n)}{\partial(x_1, \ldots, x_n)}(\boldsymbol{a}).$$

9.1.9 Example. The transformation $(x, y, z) = (r\cos\theta, r\sin\theta, z)$ from cylindrical coordinates to rectangular coordinates in $\mathbb{R}^3$ has Jacobian

$$\frac{\partial(x, y, z)}{\partial(r, \theta, z)} = \begin{vmatrix} \cos\theta & \sin\theta & 0 \\ -r\sin\theta & r\cos\theta & 0 \\ 0 & 0 & 1 \end{vmatrix} = r. \qquad \Diamond$$

The following characterization of differentiability will be useful.

9.1.10 Theorem. *Let $f : U \to \mathbb{R}^m$, where $U \subseteq \mathbb{R}^n$ is open. Then f is differentiable at $\boldsymbol{a} \in U$ iff there exists $T \in \mathcal{L}(\mathbb{R}^n, \mathbb{R}^m)$ and, for sufficiently small r, a function $\eta : B_r(\mathbf{0}) \to \mathbb{R}^m$ such that*

$$f(\boldsymbol{a}+\boldsymbol{h}) = f(\boldsymbol{a}) + T\boldsymbol{h} + \|\boldsymbol{h}\|\,\eta(\boldsymbol{h}), \quad \textit{and} \quad \lim_{\boldsymbol{h}\to\mathbf{0}} \eta(\boldsymbol{h}) = \mathbf{0}. \tag{9.6}$$

In this case, $T = df_{\boldsymbol{a}}$.

Proof. Assume that f is differentiable at $\boldsymbol{a}$. Choose $r > 0$ such that $B_r(\boldsymbol{a}) \subseteq U$ and define $\eta : B_r(\mathbf{0}) \to \mathbb{R}^m$ by $\eta(\mathbf{0}) = \mathbf{0}$ and

$$\eta(\boldsymbol{h}) = \frac{f(\boldsymbol{a}+\boldsymbol{h}) - f(\boldsymbol{a}) - df_{\boldsymbol{a}}(\boldsymbol{h})}{\|\boldsymbol{h}\|} \quad \text{if } \boldsymbol{h} \neq \mathbf{0}.$$

Then (9.6) holds with $T = df_{\boldsymbol{a}}$.

Conversely, if (9.6) holds for some η and T, then

$$\lim_{\boldsymbol{h}\to\mathbf{0}} \frac{\|f(\boldsymbol{a}+\boldsymbol{h}) - f(\boldsymbol{a}) - T\boldsymbol{h}\|}{\|\boldsymbol{h}\|} = \lim_{\boldsymbol{h}\to\mathbf{0}} \|\eta(\boldsymbol{h})\| = 0,$$

hence f is differentiable at $\boldsymbol{a}$ with $df_{\boldsymbol{a}} = T$. □

9.1.11 Corollary. *If f is differentiable at $\boldsymbol{a}$, then f is continuous at $\boldsymbol{a}$.*

Proof. By (9.6) and the continuity of linear transformations,

$$\lim_{\boldsymbol{h}\to\boldsymbol{0}} \left[f(\boldsymbol{a}+\boldsymbol{h}) - f(\boldsymbol{a})\right] = \lim_{\boldsymbol{h}\to\boldsymbol{0}} \|\boldsymbol{h}\|\eta(\boldsymbol{h}) + \lim_{\boldsymbol{h}\to\boldsymbol{0}} df_{\boldsymbol{a}}(\boldsymbol{h}) = \boldsymbol{0}. \qquad \square$$

Exercises

1. Find the differential df for each of the functions $f(x, y)$:

   (a) $\dfrac{x-y}{x+y}$ (b) $\ln(x^2+y^3)$ (c) $\arctan(xy^2)$

   (d) $\cos\dfrac{x}{y}$ (e) $\sin(x^2y)$ (f) $\arcsin\dfrac{y}{x}$, $0 < y < x$

   (g) $\sec(ye^x)$ (h) e^{xy^2} (i) $\tan\dfrac{3x+2y}{2x+3y}$.

2. Find $f'(\boldsymbol{x})$ where $f(\boldsymbol{x}) =$

   (a) (x^3-y^3, x^2y^2) (b) $(e^x\sin y, e^y \sin x)$ (c) $\left(\dfrac{xy}{x^2+y^2}, \dfrac{x^2-y^2}{x^2+y^2}\right)$

   (d) $\left(\ln(x^2+y^2+z^2+1), xyz\right)$ (e) $\left(\arctan(x-y), e^{xy}, x/y\right)$.

3. For each of the functions $f(x, y)$ below, find all values of p, $q \in \mathbb{N}$ for which on $\mathbb{R}^2$

   (i) f_x, f_y exist, (ii) f_x, f_y are continuous, (iii) f' exists.

   (a) $\begin{cases} \dfrac{x^p+y^q}{x^2+y^2} & \text{if } (x,y)\neq \boldsymbol{0},\\ 0 & \text{otherwise}\end{cases}$ (b) $\begin{cases} \dfrac{x^py^q}{x^2+y^2} & \text{if } (x,y)\neq \boldsymbol{0},\\ 0 & \text{otherwise}\end{cases}$

   (c) $\begin{cases} (x-y)^p\sin(x-y)^{-1} & \text{if } x\neq y,\\ 0 & \text{otherwise}\end{cases}$ (d) $\begin{cases} x^p\sin\dfrac{1}{x}+y^q & \text{if } x\neq 0,\\ y^q & \text{otherwise}\end{cases}$

   (e) $\begin{cases} \dfrac{x^p+y^q}{x-y} & \text{if } x\neq y,\\ 0 & \text{otherwise}\end{cases}$ (f) $\begin{cases} \dfrac{x^py^q}{x-y} & \text{if } x\neq y,\\ 0 & \text{otherwise.}\end{cases}$

4. Find all values of p, q, $s \in (0, +\infty)$ for which on $\mathbb{R}^2$

   (i) f_x, f_y exist, (ii) f_x, f_y are continuous, (iii) f' exists,

   where $f(0,0) = 0$ and, for $(x, y) \neq (0,0)$, $f(x,y) =$

   (a) $|x|^p|y|^q\ln(x^2+y^2)$ (b) $\dfrac{\sin(x^2+y^2)^p}{(x^2+y^2)^q}$ (c) $\dfrac{\tan(x^2+y^2)^p}{(x^2+y^2)^q}$

   (d) $\dfrac{\sin|x|^p|y|^q}{(x^2+y^2)^s}$ (e) $\dfrac{\sin^{-1}|x|^p|y|^q}{(x^2+y^2)^s}$.

5. *Spherical coordinates* (ρ, ϕ, θ) *in* $\mathbb{R}^3$ are defined by

$$x = \rho \sin\phi \cos\theta, \quad y = \rho \sin\phi \sin\theta, \quad z = \rho \cos\phi,$$

where $\rho \geq 0$, $0 \leq \phi \leq \pi$, and $0 \leq \theta < 2\pi$. Show that

$$\frac{\partial(x, y, z)}{\partial(\rho, \phi, \theta)} = \rho^2 \sin\phi.$$

6. Let $(u, v) = \big(\sin f(x, y), \cos f(x, y)\big)$. Find $\dfrac{\partial(u, v)}{\partial(x, y)}$.

7. Let $(u, v, w) = (y/z, z/x, x/y)$, where $xyz \neq 0$. Find $\dfrac{\partial(u, v, w)}{\partial(x, y, z)}$.

8. Let

$$f(\boldsymbol{x}) = \sum_{i=1}^{n} x_i^{a_i} \quad \text{and} \quad g(\boldsymbol{x}) = \prod_{i=1}^{n} x_i^{a_i},$$

where x_i, $a_i > 0$ and $\sum_i a_i = 1$. Find

(a) $\boldsymbol{x} \cdot \nabla f(\boldsymbol{x})$. (b) $\boldsymbol{x} \cdot \nabla g(\boldsymbol{x})$.

9. Let $f(\boldsymbol{x})$ be defined implicitly by the equation

$$\frac{1}{f(\boldsymbol{x})} = \sum_{i=1}^{n} \frac{1}{x_i}.$$

Express $\nabla f(\boldsymbol{x})$ in terms of f.

10. Let $f(\boldsymbol{x}) = \ln\big(\sum_{i=1}^{n} e^{x_i}\big)$. Express $\nabla f(\boldsymbol{x})$ in terms of f.

11. Let the equation $\alpha x_n - x_1 x_2 \cdots x_{n-1} = 0$, $\alpha \neq 0$, define each of the variables $x_1, \ldots, x_{n-1}$ as a differentiable function of x_n. Show that

$$x_n^{n-2} \frac{\partial x_1}{\partial x_n} \frac{\partial x_2}{\partial x_n} \cdots \frac{\partial x_{n-1}}{\partial x_n} = \alpha.$$

12. Let $\boldsymbol{x} = (x_1, \ldots, x_n)$. Find ∂_i of

(a) $\|x\|$ (b) $\dfrac{1}{\|x\|}$ (c) $\dfrac{x_i}{\|x\|}$ (d) $\dfrac{x_i}{\|x\|^2}$.

13. Let $f : \mathbb{R} \to \mathbb{R}$ be differentiable and $p > 0$. Show that for $\boldsymbol{x} \neq \boldsymbol{0}$,

$$\boldsymbol{x} \cdot \nabla \|\boldsymbol{x}\|^p = p\|\boldsymbol{x}\|^p \quad \text{and} \quad \boldsymbol{x} \cdot \nabla f\big(\|\boldsymbol{x}\|^p\big) = pf'\big(\|\boldsymbol{x}\|^p\big)\|\boldsymbol{x}\|^p.$$

9.2 Properties of the Differential

In this section we consider analogs of differentiation rules for single-variable functions. Deeper properties of the differential are taken up in later sections.

Linearity of the Differential

9.2.1 Theorem. *Let* $U \subseteq \mathbb{R}^n$ *be open, let* f, $g : U \to \mathbb{R}^m$ *be differentiable at* $\boldsymbol{a} \in U$, *and let* α, $\beta \in \mathbb{R}$. *Then* $\alpha f + \beta g$ *is differentiable at* $\boldsymbol{a}$ *and*

$$d(\alpha f + \beta g)_{\boldsymbol{a}} = \alpha df_{\boldsymbol{a}} + \beta dg_{\boldsymbol{a}}.$$

Proof. By 9.1.10, there exist functions $\eta(\boldsymbol{h})$, $\mu(\boldsymbol{h})$, defined for $\boldsymbol{h} \in \mathbb{R}^n$ with sufficiently small norm, such that

$$\begin{aligned} f(\boldsymbol{a}+\boldsymbol{h}) &= f(\boldsymbol{a}) + df_{\boldsymbol{a}}(\boldsymbol{h}) + \|\boldsymbol{h}\|\eta(\boldsymbol{h}), \quad \lim_{\boldsymbol{h}\to\boldsymbol{0}} \eta(\boldsymbol{h}) = \boldsymbol{0}, \text{ and} \\ g(\boldsymbol{a}+\boldsymbol{h}) &= g(\boldsymbol{a}) + dg_{\boldsymbol{a}}(\boldsymbol{h}) + \|\boldsymbol{h}\|\mu(\boldsymbol{h}), \quad \lim_{\boldsymbol{h}\to\boldsymbol{0}} \mu(\boldsymbol{h}) = \boldsymbol{0}. \end{aligned}$$

Then

$$(\alpha f + \beta g)(\boldsymbol{a}+\boldsymbol{h}) = (\alpha f + \beta g)(\boldsymbol{a}) + (\alpha df_{\boldsymbol{a}} + \beta dg_{\boldsymbol{a}})(\boldsymbol{h}) + \|\boldsymbol{h}\|\big(\alpha\eta + \beta\mu\big)(\boldsymbol{h})$$

and

$$\lim_{\boldsymbol{h}\to\boldsymbol{0}} \big(\alpha\eta + \beta\mu\big)(\boldsymbol{h}) = \boldsymbol{0}.$$

Another application of 9.1.10 completes the proof. □

The Norm of a Linear Transformation

For additional properties of the differential, including product rules, we need the notion of *operator norm* on the space $\mathcal{L}(\mathbb{R}^n, \mathbb{R}^m)$ of linear transformations from $\mathbb{R}^n$ to $\mathbb{R}^m$.

9.2.2 Definition. Let $T \in \mathcal{L}(\mathbb{R}^n, \mathbb{R}^m)$. The *operator norm* of T is defined as

$$\|T\| = \sup\big\{\|T\boldsymbol{x}\| \,:\, \boldsymbol{x} \in \mathbb{R}^n,\ \|\boldsymbol{x}\| = 1\big\}. \qquad \Diamond$$

The following proposition justifies the use of the term "norm."

9.2.3 Proposition. $\|T\|$ *defines a norm on* $\mathcal{L}(\mathbb{R}^n, \mathbb{R}^m)$ *such that*

$$\|T\boldsymbol{x}\| \le \|T\|\,\|\boldsymbol{x}\| \quad \text{for all} \quad \boldsymbol{x} \in \mathbb{R}^n. \tag{9.7}$$

Moreover, if $[a_{ij}]_{m\times n}$ *is the matrix of* T, *then for all* k, ℓ

$$|a_{k\ell}| \le \|T\| \le \left(\sum_{i=1}^{m}\sum_{j=1}^{n} a_{ij}^2\right)^{1/2}. \tag{9.8}$$

Proof. Inequality (9.7) is clear if $\boldsymbol{x} = \mathbf{0}$. If $\boldsymbol{x} \neq \mathbf{0}$, then $\|\boldsymbol{x}\|^{-1}\boldsymbol{x}$ has norm 1 hence

$$\|\boldsymbol{x}\|^{-1}\|T\boldsymbol{x}\| = \big\|T(\|\boldsymbol{x}\|^{-1}\boldsymbol{x})\big\| \leq \|T\|.$$

To verify (9.8), let $\boldsymbol{a}_i = (a_{i1}, \ldots, a_{in})$. Since $T\boldsymbol{x} = \big(\boldsymbol{a}_1 \cdot \boldsymbol{x}, \ldots, \boldsymbol{a}_n \cdot \boldsymbol{x}\big)$, by the Cauchy–Schwarz inequality,

$$\|T\boldsymbol{x}\|^2 = \sum_{i=1}^{m}(\boldsymbol{a}_i \cdot \boldsymbol{x})^2 \leq \sum_{i=1}^{m}\|\boldsymbol{a}_i\|^2\|\boldsymbol{x}\|^2 = \sum_{i,j} a_{ij}^2, \quad \|\boldsymbol{x}\| = 1,$$

which verifies the second inequality in (9.8). The first inequality follows from

$$|a_{k\ell}|^2 \leq \sum_{i=1}^{m}|a_{i\ell}|^2 = \|T\boldsymbol{e}^\ell\|^2 \leq \|T\|^2.$$

To see that $\|T\|$ defines a norm, note that homogeneity follows directly from the definition, and the triangle inequality $\|T_1 + T_2\| \leq \|T_1\| + \|T_2\|$ is a consequence of

$$\|(T_1 + T_2)\boldsymbol{x}\| \leq \|T_1\boldsymbol{x}\| + \|T_2\boldsymbol{x}\| \leq \|T_1\| + \|T_2\|, \quad \|\boldsymbol{x}\| = 1.$$

The property of coincidence follows directly from (9.7). □

9.2.4 Corollary. *A linear transformation $T : \mathbb{R}^n \to \mathbb{R}^m$ is uniformly continuous.*

Proof. This follows from

$$\|T\boldsymbol{x} - T\boldsymbol{y}\| = \|T(\boldsymbol{x} - \boldsymbol{y})\| \leq \|T\|\,\|\boldsymbol{x} - \boldsymbol{y}\|,$$

using the linearity of T. □

Since $\mathcal{L}(\mathbb{R}^n, \mathbb{R}^m)$ is a normed vector space, it is a metric space under the distance function $\rho(T_1, T_2) := \|T_1 - T_2\|$. Thus the methods of Chapter 8 apply. In particular, we have the following consequence of 9.2.3.

9.2.5 Corollary. *Let (X, d) be a metric space and let F be a function from X to $\mathcal{L}(\mathbb{R}^n, \mathbb{R}^m)$. For each $x \in X$, let $[a_{ij}(x)]_{m\times n}$ denote the matrix of $F(x)$. Then F is (uniformly) continuous with respect to the metric ρ iff each function $a_{ij}(x)$ is (uniformly) continuous on X.*

Proof. The matrix of $F(x) - F(y)$ is $[a_{ij}(x) - a_{ij}(y)]_{m\times n}$, hence, by (9.8),

$$|a_{k\ell}(x) - a_{k\ell}(y)|^2 \leq \|F(x) - F(y)\|^2 \leq \sum_{i,j}[a_{ij}(x) - a_{ij}(y)]^2.$$

The assertion follows. □

Product Rules

We consider two product rules; additional product rules, as well as a quotient rule, are given in the exercises.

9.2.6 Theorem (Scalar Product Rule). *Let U be open in $\mathbb{R}^n$ and $f : U \to \mathbb{R}^m$ and $\psi : U \to \mathbb{R}$ differentiable at $\boldsymbol{a} \in U$. Then*

$$d(\psi f)_{\boldsymbol{a}}(\boldsymbol{h}) = \psi(\boldsymbol{a})df_{\boldsymbol{a}}(\boldsymbol{h}) + \big(\nabla\psi(\boldsymbol{a}) \cdot \boldsymbol{h}\big)f(\boldsymbol{a}), \quad \boldsymbol{h} \in \mathbb{R}^n. \tag{9.9}$$

Proof. By 9.1.10, there exist functions $\eta(\boldsymbol{h})$ and $\mu(\boldsymbol{h})$, defined for $\boldsymbol{h} \in \mathbb{R}^n$ with sufficiently small norm, such that

$$\begin{aligned} f(\boldsymbol{a}+\boldsymbol{h}) - f(\boldsymbol{a}) - df_{\boldsymbol{a}}(\boldsymbol{h}) &= \|\boldsymbol{h}\|\eta(\boldsymbol{h}), \quad \lim_{\boldsymbol{h}\to\boldsymbol{0}} \eta(\boldsymbol{h}) = \boldsymbol{0}, \\ \psi(\boldsymbol{a}+\boldsymbol{h}) - \psi(\boldsymbol{a}) - \nabla\psi(\boldsymbol{a}) \cdot \boldsymbol{h} &= \|\boldsymbol{h}\|\mu(\boldsymbol{h}), \quad \lim_{\boldsymbol{h}\to\boldsymbol{0}} \mu(\boldsymbol{h}) = 0. \end{aligned}$$

Let $T\boldsymbol{h}$ denote the right side of (9.9) and set

$$\begin{aligned} \nu(\boldsymbol{h}) &:= (\psi f)(\boldsymbol{a}+\boldsymbol{h}) - (\psi f)(\boldsymbol{a}) - T\boldsymbol{h} \\ &= \psi(\boldsymbol{a}+\boldsymbol{h})f(\boldsymbol{a}+\boldsymbol{h}) - \psi(\boldsymbol{a})f(\boldsymbol{a}) - \psi(\boldsymbol{a})df_{\boldsymbol{a}}(\boldsymbol{h}) - \big(\nabla\psi(\boldsymbol{a}) \cdot \boldsymbol{h}\big)f(\boldsymbol{a}). \end{aligned}$$

Then T is linear and

$$\begin{aligned} \nu(\boldsymbol{h}) = \psi(\boldsymbol{a}+\boldsymbol{h})&\big[f(\boldsymbol{a}+\boldsymbol{h}) - f(\boldsymbol{a}) - df_{\boldsymbol{a}}(\boldsymbol{h})\big] \\ &+ \big[\psi(\boldsymbol{a}+\boldsymbol{h}) - \psi(\boldsymbol{a}) - \nabla\psi(\boldsymbol{a}) \cdot \boldsymbol{h}\big]f(\boldsymbol{a}) + \big[\psi(\boldsymbol{a}+\boldsymbol{h}) - \psi(\boldsymbol{a})\big]df_{\boldsymbol{a}}(\boldsymbol{h}) \\ &= \psi(\boldsymbol{a}+\boldsymbol{h})\|\boldsymbol{h}\|\eta(\boldsymbol{h}) + \|\boldsymbol{h}\|\mu(\boldsymbol{h})f(\boldsymbol{a}) + \big[\psi(\boldsymbol{a}+\boldsymbol{h}) - \psi(\boldsymbol{a})\big]df_{\boldsymbol{a}}(\boldsymbol{h}). \end{aligned}$$

Since $\|df_{\boldsymbol{a}}(\boldsymbol{h})\| \le \|df_{\boldsymbol{a}}\|\,\|\boldsymbol{h}\|$,

$$\frac{\|\nu(\boldsymbol{h})\|}{\|\boldsymbol{h}\|} \le |\psi(\boldsymbol{a}+\boldsymbol{h})|\,\|\eta(\boldsymbol{h})\| + |\mu(\boldsymbol{h})|\,\|f(\boldsymbol{a})\| + |\psi(\boldsymbol{a}+\boldsymbol{h}) - \psi(\boldsymbol{a})\|, \|df_{\boldsymbol{a}}\|.$$

By continuity of ψ at $\boldsymbol{a}$, the right side of the last inequality tends to zero as $\boldsymbol{h} \to \boldsymbol{0}$, proving the theorem. □

9.2.7 Theorem (Dot Product Rule). *Let U be open in $\mathbb{R}^n$ and $f,\ g : U \to \mathbb{R}^m$ differentiable at $\boldsymbol{a} \in U$. Then*

$$d(f \cdot g)_{\boldsymbol{a}}(\boldsymbol{h}) = f(\boldsymbol{a}) \cdot dg_{\boldsymbol{a}}(\boldsymbol{h}) + g(\boldsymbol{a}) \cdot df_{\boldsymbol{a}}(\boldsymbol{h}), \quad \boldsymbol{h} \in \mathbb{R}^n. \tag{9.10}$$

Proof. Let $\eta(\boldsymbol{h})$ and $\mu(\boldsymbol{h})$ be functions defined for sufficiently small $\|\boldsymbol{h}\|$ such that

$$\begin{aligned} f(\boldsymbol{a}+\boldsymbol{h}) - f(\boldsymbol{a}) - df_{\boldsymbol{a}}(\boldsymbol{h}) &= \|\boldsymbol{h}\|\eta(\boldsymbol{h}), \quad \lim_{\boldsymbol{h}\to\boldsymbol{0}} \eta(\boldsymbol{h}) = \boldsymbol{0}, \\ g(\boldsymbol{a}+\boldsymbol{h}) - g(\boldsymbol{a}) - dg_{\boldsymbol{a}}(\boldsymbol{h}) &= \|\boldsymbol{h}\|\mu(\boldsymbol{h}), \quad \lim_{\boldsymbol{h}\to\boldsymbol{0}} \mu(\boldsymbol{h}) = \boldsymbol{0}. \end{aligned}$$

Let $T\boldsymbol{h}$ denote the right side of (9.10) and define

$$\begin{aligned}\nu(\boldsymbol{h}) &:= (f \cdot g)(\boldsymbol{a}+\boldsymbol{h}) - (f \cdot g)(\boldsymbol{a}) - T\boldsymbol{h}, \quad \boldsymbol{h} \in \mathbb{R}^n \\ &= f(\boldsymbol{a}+\boldsymbol{h}) \cdot g(\boldsymbol{a}+\boldsymbol{h}) - f(\boldsymbol{a}) \cdot g(\boldsymbol{a}) - f(\boldsymbol{a}) \cdot dg_{\boldsymbol{a}}(\boldsymbol{h}) - g(\boldsymbol{a}) \cdot df_{\boldsymbol{a}}(\boldsymbol{h}).\end{aligned}$$

Then T is linear and

$$\begin{aligned}\nu(\boldsymbol{h}) &= f(\boldsymbol{a}+\boldsymbol{h}) \cdot \big[g(\boldsymbol{a}+\boldsymbol{h}) - g(\boldsymbol{a}) - dg_{\boldsymbol{a}}(\boldsymbol{h})\big] \\ &\qquad + g(\boldsymbol{a}) \cdot \big[f(\boldsymbol{a}+\boldsymbol{h}) - f(\boldsymbol{a}) - df_{\boldsymbol{a}}(\boldsymbol{h})\big] + \big[f(\boldsymbol{a}+\boldsymbol{h}) - f(\boldsymbol{a})\big] \cdot dg_{\boldsymbol{a}}(\boldsymbol{h}) \\ &= \|\boldsymbol{h}\| f(\boldsymbol{a}+\boldsymbol{h}) \cdot \mu(\boldsymbol{h}) + \|\boldsymbol{h}\| g(\boldsymbol{a}) \cdot \eta(\boldsymbol{h}) + \big[df_{\boldsymbol{a}}(\boldsymbol{h}) + \|\boldsymbol{h}\|\eta(\boldsymbol{h})\big] \cdot dg_{\boldsymbol{a}}(\boldsymbol{h}).\end{aligned}$$

By the Cauchy–Schwarz and operator norm inequalities,

$$\begin{aligned}\frac{|\nu(\boldsymbol{h})|}{\|\boldsymbol{h}\|} &\le \|f(\boldsymbol{a}+\boldsymbol{h})\|\,\|\mu(\boldsymbol{h})\| + \|g(\boldsymbol{a})\|\,\|\eta(\boldsymbol{h})\| \\ &\qquad + \|df_{\boldsymbol{a}}\|\,\|dg_{\boldsymbol{a}}\|\,\|\boldsymbol{h}\| + \|\eta(\boldsymbol{h})\|\,\|dg_{\boldsymbol{a}}\|\,\|\boldsymbol{h}\|.\end{aligned}$$

Since the right side of this inequality tends to zero as $\boldsymbol{h} \to \boldsymbol{0}$ so does the left, completing the proof. □

Continuity of the Differential

If U is an open subset of $\mathbb{R}^n$ and $f : U \to \mathbb{R}^m$ is differentiable, then the mapping $\boldsymbol{x} \mapsto df_{\boldsymbol{x}}$ is a function from U to $\mathcal{L}(\mathbb{R}^n, \mathbb{R}^m)$. Since $\mathcal{L}(\mathbb{R}^n, \mathbb{R}^m)$ is a metric space in the operator norm, the notion of continuity of this mapping is meaningful.

9.2.8 Definition. Let $U \subseteq \mathbb{R}^n$ be open. A function $f : U \to \mathbb{R}^m$ is said to be *continuously differentiable on* U if $df_{\boldsymbol{x}}$ exists and is continuous as a function of $\boldsymbol{x}$ on U. In this case, f is also said to be of *class* C^1 *on* U. A function g is *continuously differentiable on a subset* E *of* $\mathbb{R}^n$ if g is the restriction to E of a continuously differentiable function f on an open set $U \supseteq E$. ◊

9.2.9 Theorem. *Let* $f = (f_1, \ldots, f_m) : U \to \mathbb{R}^m$, *where* $U \subseteq \mathbb{R}^n$ *is open. Then* f *is continuously differentiable on* U *iff the partial derivatives* $\partial_j f_i$, $1 \le i \le m$, $1 \le j \le n$, *exist and are continuous on* U.

Proof. If f is continuously differentiable on U then, by 9.2.5, the matrix $f'(\boldsymbol{x})$ has continuous entries. By 9.1.8, these entries are the partial derivatives of the components of f.

For the sufficiency, by 9.1.8 we may assume that $m = 1$, that is, f is real-valued. Suppose then that the partial derivatives $\partial_j f$ exist and are continuous on U. Let $\boldsymbol{a} \in U$ and $\varepsilon > 0$. Choose $r > 0$ such that $B_r(\boldsymbol{a}) \subseteq U$ and fix $\boldsymbol{h} = (h_1, \ldots, h_n)$ such that $\|\boldsymbol{h}\| < r$. For $1 \le j \le n$ set

$$g_j(t) := f\big(\boldsymbol{a} + \boldsymbol{h}_j(t)\big), \quad \boldsymbol{h}_j(t) := (h_1, \ldots, h_{j-1}, th_j, 0, \ldots, 0), \quad 0 \le t \le 1.$$

Then

$$g_j(1) - g_j(0) = f\big(\boldsymbol{a} + \boldsymbol{h}_j(1)\big) - f\big(\boldsymbol{a} + \boldsymbol{h}_j(0)\big).$$

Also, by the mean value theorem and the chain rule, there exists $t_j \in (0,1)$ such that

$$g_j(1) - g_j(0) = g_j'(t_j) = h_j \partial_j f\big(\boldsymbol{a} + \boldsymbol{h}_j(t_j)\big).$$

Therefore,

$$f(\boldsymbol{a}+\boldsymbol{h}) - f(\boldsymbol{a}) = \sum_{j=1}^{n} \big[g_j(1) - g_j(0)\big] = \sum_{j=1}^{n} h_j \partial_j f\big(\boldsymbol{a} + \boldsymbol{h}_j(t_j)\big),$$

hence

$$f(\boldsymbol{a}+\boldsymbol{h}) - f(\boldsymbol{a}) - \nabla f(\boldsymbol{a}) \cdot \boldsymbol{h} = \sum_{j=1}^{n} \big[\partial_j f\big(\boldsymbol{a} + \boldsymbol{h}_j(t_j)\big) - \partial_j f(\boldsymbol{a})\big] h_j = \nu(\boldsymbol{h}) \cdot \boldsymbol{h},$$

where

$$\nu(\boldsymbol{h}) := \sum_{j=1}^{n} \big[\partial_j f\big(\boldsymbol{a} + \boldsymbol{h}_j(t_j)\big) - \partial_j f(\boldsymbol{a})\big] \boldsymbol{e}^i.$$

Since $\lim_{\boldsymbol{h}\to\boldsymbol{0}} \boldsymbol{h}_j(t_j) = \boldsymbol{0}$, the continuity of $\partial_j f$ at $\boldsymbol{a}$ implies that $\lim_{\boldsymbol{h}\to\boldsymbol{0}} \nu(\boldsymbol{h}) = \boldsymbol{0}$. Since $|\nu(\boldsymbol{h}) \cdot \boldsymbol{h}| \le \|\nu(\boldsymbol{h})\| \, \|\boldsymbol{h}\|$,

$$\frac{|f(\boldsymbol{a}+\boldsymbol{h}) - f(\boldsymbol{a}) - \nabla f(\boldsymbol{a}) \cdot \boldsymbol{h}|}{\|\boldsymbol{h}\|} \le \|\nu(\boldsymbol{h})\| \to 0,$$

completing the proof. □

Exercises

1. Prove that for $T \in \mathcal{L}(\mathbb{R}^n, \mathbb{R}^m)$,

$$\|T\| = \sup\big\{\|T\boldsymbol{x}\| \,:\, \boldsymbol{x} \in \mathbb{R}^n,\ \|\boldsymbol{x}\| \le 1\big\}.$$

2. Let $T_1 \in \mathcal{L}(\mathbb{R}^m, \mathbb{R}^k)$ and $T_2 \in \mathcal{L}(\mathbb{R}^n, \mathbb{R}^m)$. Prove that

$$\|T_1 T_2\| \le \|T_1\| \, \|T_2\|.$$

   (We use the standard notation $T_1 T_2$ for composition of linear operators.)

3. (Quotient rule) Let U, f, and ψ be as in 9.2.6. If $\psi(\boldsymbol{a}) \ne 0$, prove that

$$d\left[\frac{f}{\psi}\right]_{\boldsymbol{a}}(\boldsymbol{h}) = \frac{\psi(\boldsymbol{a}) df_{\boldsymbol{a}}(\boldsymbol{h}) - \big(\nabla \psi(\boldsymbol{a}) \cdot \boldsymbol{h}\big) f(\boldsymbol{a})}{\psi^2(\boldsymbol{a})}.$$

4. Find $dg_{\boldsymbol{x}}(\boldsymbol{x})$, $\|\boldsymbol{x}\| \ne \boldsymbol{0}$, if $g(\boldsymbol{x}) =$

   (a) $\|\boldsymbol{x}\|\boldsymbol{x}$ (b) $\|\boldsymbol{x}\|^{-2}\boldsymbol{x}$ (c) $\|\boldsymbol{x}\|^{-1}\boldsymbol{x}$.

5. The *cross product* of vectors $\boldsymbol{a} = (a_1, a_2, a_3)$ and $\boldsymbol{b} = (b_1, b_2, b_3)$ is defined by

$$\boldsymbol{a} \times \boldsymbol{b} = \begin{vmatrix} a_2 & a_3 \\ b_2 & b_3 \end{vmatrix} \boldsymbol{e}^1 - \begin{vmatrix} a_1 & a_3 \\ b_1 & b_3 \end{vmatrix} \boldsymbol{e}^2 + \begin{vmatrix} a_1 & a_2 \\ b_1 & b_2 \end{vmatrix} \boldsymbol{e}^3.$$

(See Exercise 1.5.9.) Let $f : U \to \mathbb{R}^3$ and $g : U \to \mathbb{R}^3$, where $U \subseteq \mathbb{R}^n$ is open. Define $f \times g$ on U by

$$(f \times g)(\boldsymbol{x}) = f(\boldsymbol{x}) \times g(\boldsymbol{x}).$$

Prove that

$$d(f \times g)_{\boldsymbol{a}}(\boldsymbol{h}) = f(\boldsymbol{a}) \times dg_{\boldsymbol{a}}(\boldsymbol{h}) + df_{\boldsymbol{a}}(\boldsymbol{h}) \times g(\boldsymbol{a}).$$

6. Let $V \subseteq \mathbb{R}^p$ and $W \subseteq \mathbb{R}^q$ be open, $f : V \to \mathbb{R}^k$, $g : W \to \mathbb{R}^k$, and $\alpha, \beta \in \mathbb{R}$. Define F on $V \times W \subseteq \mathbb{R}^{p+q}$ by

$$F(\boldsymbol{x}, \boldsymbol{y}) = \alpha f(\boldsymbol{x}) + \beta g(\boldsymbol{y}), \quad \boldsymbol{x} \in V, \quad \boldsymbol{y} \in W.$$

If f is differentiable at $\boldsymbol{a} \in V$ and g is differentiable at $\boldsymbol{b} \in W$, prove that F is differentiable at $\boldsymbol{c} := (\boldsymbol{a}, \boldsymbol{b})$ and

$$dF_{\boldsymbol{c}}(\boldsymbol{h}, \boldsymbol{k}) = \alpha df_{\boldsymbol{a}}(\boldsymbol{h}) + \beta dg_{\boldsymbol{b}}(\boldsymbol{k}), \quad \boldsymbol{h} \in \mathbb{R}^p, \ \boldsymbol{k} \in \mathbb{R}^q.$$

7. Let $V \subseteq \mathbb{R}^p$ and $W \subseteq \mathbb{R}^q$ be open and $f : V \to \mathbb{R}^k$, $g : W \to \mathbb{R}^k$. Define F on $V \times W \subseteq \mathbb{R}^{p+q}$ by

$$F(\boldsymbol{x}, \boldsymbol{y}) = f(\boldsymbol{x}) \cdot g(\boldsymbol{y}), \quad \boldsymbol{x} \in V, \quad \boldsymbol{y} \in W.$$

If f is differentiable at $\boldsymbol{a} \in V$ and g is differentiable at $\boldsymbol{b} \in W$, prove that F is differentiable at $\boldsymbol{c} := (\boldsymbol{a}, \boldsymbol{b})$ and

$$dF_{\boldsymbol{c}}(\boldsymbol{h}, \boldsymbol{k}) = g(\boldsymbol{b}) \cdot df_{\boldsymbol{a}}(\boldsymbol{h}) + f(\boldsymbol{a}) \cdot dg_{\boldsymbol{b}}(\boldsymbol{k}), \quad \boldsymbol{h} \in \mathbb{R}^p, \ \boldsymbol{k} \in \mathbb{R}^q.$$

8. Formulate and prove the analog of Exercise 7 for cross products.

9. Let $f : I \to \mathbb{R}^m$ be differentiable and $\|f\| = 1$ on an open interval I. Prove that $f(t)$ and $f'(t)$ are perpendicular for all t, that is, $f \cdot f' = 0$ on I.

10. Let $f : [a, b] \to \mathbb{R}^m$ be differentiable and $\boldsymbol{v} \notin A := f[a, b]$. Referring to Exercise 8.5.15 with $d(\boldsymbol{x}, \boldsymbol{y}) = \|\boldsymbol{x} - \boldsymbol{y}\|$, show that

    (a) $d(A, \boldsymbol{v}) = \|f(t_0) - \boldsymbol{v}\|$ for some $t_0 \in [a, b]$.

    (b) $\big(f(t_0) - \boldsymbol{v}\big) \cdot f'(t_0) = 0$ if $t_0 \in (a, b)$.

11. A path $\varphi : [a, b] \to \mathbb{R}^n$ is *piecewise smooth* if there exists a partition $a_0 = a < a_1 < \cdots < a_n = b$ of $[a, b]$ such that φ' exists and is continuous on each subinterval $[a_{j-1}, a_j]$. Let $U \subseteq \mathbb{R}^n$ be nonempty open and connected. Show that if $\varepsilon > 0$, then any pair of points can be joined by a piecewise smooth path φ in U such that $\sup_{a_{j-1} \le t \le a_j} \|\varphi'(t)\| < \varepsilon$ for each j.

9.3 Further Properties of the Differential

In this section we prove two important theorems, the first of which is an n-dimensional version of the chain rule.

9.3.1 Chain Rule. *Let* $U \subseteq \mathbb{R}^n$ *and* $V \subseteq \mathbb{R}^m$ *be open and* $f : U \to \mathbb{R}^m$, $g : V \to \mathbb{R}^k$ *with* $f(U) \subseteq V$. *If* f *is differentiable at* $\boldsymbol{a} \in U$ *and* g *is differentiable at* $\boldsymbol{b} := f(\boldsymbol{a})$, *then* $g \circ f : U \to \mathbb{R}^k$ *is differentiable at* $\boldsymbol{a}$ *and the linear transformation* $d(g \circ f)_{\boldsymbol{a}} : \mathbb{R}^n \to \mathbb{R}^k$ *is the composition of the linear transformations* $dg_{\boldsymbol{b}} : \mathbb{R}^m \to \mathbb{R}^k$ *and* $df_{\boldsymbol{a}} : \mathbb{R}^n \to \mathbb{R}^m$:

$$d(g \circ f)_{\boldsymbol{a}} = dg_{\boldsymbol{b}} \circ df_{\boldsymbol{a}}.$$

Proof. Choose r, $s > 0$ such that $B_r(\boldsymbol{a}) \subseteq U$ and $B_s(\boldsymbol{b}) \subseteq V$. By 9.1.10 there exist functions $\eta : B_r(\boldsymbol{0}) \to \mathbb{R}^m$ and $\nu : B_s(\boldsymbol{0}) \to \mathbb{R}^k$ such that

$$f(\boldsymbol{a}+\boldsymbol{h}) = f(\boldsymbol{a}) + df_{\boldsymbol{a}}(\boldsymbol{h}) + \|\boldsymbol{h}\|\eta(\boldsymbol{h}), \qquad \lim_{\boldsymbol{h}\to\boldsymbol{0}} \eta(\boldsymbol{h}) = \boldsymbol{0}, \quad \text{and} \tag{9.11}$$

$$g(\boldsymbol{b}+\boldsymbol{k}) = g(\boldsymbol{b}) + dg_{\boldsymbol{b}}(\boldsymbol{k}) + \|\boldsymbol{k}\|\nu(\boldsymbol{k}), \qquad \lim_{\boldsymbol{k}\to\boldsymbol{0}} \nu(\boldsymbol{k}) = \boldsymbol{0}. \tag{9.12}$$

Set

$$\boldsymbol{k} = f(\boldsymbol{a}+\boldsymbol{h}) - f(\boldsymbol{a}) = df_{\boldsymbol{a}}(\boldsymbol{h}) + \|\boldsymbol{h}\|\eta(\boldsymbol{h}). \tag{9.13}$$

By the continuity of f at $\boldsymbol{a}$, $\boldsymbol{k} \in B_s(\boldsymbol{0})$ for all sufficiently small $\|\boldsymbol{h}\|$. For such $\boldsymbol{h}$ set

$$\mu(\boldsymbol{h}) = (g \circ f)(\boldsymbol{a}+\boldsymbol{h}) - (g \circ f)(\boldsymbol{a}) - (dg_{\boldsymbol{b}} \circ df_{\boldsymbol{a}})(\boldsymbol{h}).$$

To complete the proof we show that

$$\lim_{\boldsymbol{h}\to 0} \frac{\mu(\boldsymbol{h})}{\|\boldsymbol{h}\|} = 0. \tag{9.14}$$

From (9.11), (9.12), and (9.13),

$$\begin{aligned}\mu(\boldsymbol{h}) &= g(\boldsymbol{b}+\boldsymbol{k}) - g(\boldsymbol{b}) - dg_{\boldsymbol{b}}(\boldsymbol{k}) + dg_{\boldsymbol{b}}[\boldsymbol{k} - df_{\boldsymbol{a}}(\boldsymbol{h})] \\ &= \|\boldsymbol{k}\|\nu(\boldsymbol{k}) + \|\boldsymbol{h}\|dg_{\boldsymbol{b}}(\eta(\boldsymbol{h})),\end{aligned}$$

hence

$$\frac{\|\mu(\boldsymbol{h})\|}{\|\boldsymbol{h}\|} \le \frac{\|\boldsymbol{k}\|}{\|\boldsymbol{h}\|}\|\nu(\boldsymbol{k})\| + \|dg_{\boldsymbol{b}}(\eta(\boldsymbol{h})\|$$

Since

$$\|\boldsymbol{k}\| = \big\|df_{\boldsymbol{a}}(\boldsymbol{h}) + \|\boldsymbol{h}\|\eta(\boldsymbol{h})\big\| \le \big(\|df_{\boldsymbol{a}}\| + \|\eta(\boldsymbol{h})\|\big)\|\boldsymbol{h}\|,$$

we have

$$\frac{\|\mu(\boldsymbol{h})\|}{\|\boldsymbol{h}\|} \le \big(\|df_{\boldsymbol{a}}\| + \|\eta(\boldsymbol{h})\|\big)\|\nu(\boldsymbol{k})\| + \|dg_{\boldsymbol{b}}(\eta(\boldsymbol{h}))\|.$$

Since $\boldsymbol{h} \to \boldsymbol{0}$ implies $\boldsymbol{k} \to \boldsymbol{0}$, (9.14) follows. □

9.3.2 Remark. Let f be differentiable on U and g differentiable on V. Set $\boldsymbol{y} = f(\boldsymbol{x})$ and $\boldsymbol{z} = (g \circ f)(\boldsymbol{x}) = g(\boldsymbol{y})$. Then the chain rule may be written in matrix form as $(g \circ f)'(\boldsymbol{x}) = g'(\boldsymbol{y})f'(\boldsymbol{x})$ or

$$\begin{bmatrix} \dfrac{\partial z_1}{\partial x_1} & \cdots & \dfrac{\partial z_1}{\partial x_n} \\ \vdots & & \vdots \\ \dfrac{\partial z_k}{\partial x_1} & \cdots & \dfrac{\partial z_k}{\partial x_n} \end{bmatrix} = \begin{bmatrix} \dfrac{\partial z_1}{\partial y_1} & \cdots & \dfrac{\partial z_1}{\partial y_m} \\ \vdots & & \vdots \\ \dfrac{\partial z_k}{\partial y_1} & \cdots & \dfrac{\partial z_k}{\partial y_m} \end{bmatrix} \begin{bmatrix} \dfrac{\partial y_1}{\partial x_1} & \cdots & \dfrac{\partial y_1}{\partial x_n} \\ \vdots & & \vdots \\ \dfrac{\partial y_m}{\partial x_1} & \cdots & \dfrac{\partial y_m}{\partial x_n} \end{bmatrix}.$$

From this we obtain the familiar formulas

$$\frac{\partial z_\ell}{\partial x_j} = \sum_{i=1}^{m} \frac{\partial z_\ell}{\partial y_i}\frac{\partial y_i}{\partial x_j} \qquad j = 1, \ldots, n, \quad \ell = 1, \ldots, k. \qquad \diamond$$

9.3.3 Example. Let the partial derivatives of $u = f(x, y)$ and $v = g(x, y)$ exist on $\mathbb{R}$. If $x = r\cos\theta$ and $y = r\sin\theta$, we may use the chain rule to find f_x, f_y, g_x, and g_y in terms of u_r, v_r, u_θ, and v_θ. Indeed, from 9.3.2,

$$\begin{bmatrix} u_r & u_\theta \\ v_r & v_\theta \end{bmatrix} = \begin{bmatrix} f_x & f_y \\ g_x & g_y \end{bmatrix} \begin{bmatrix} \cos\theta & -r\sin\theta \\ \sin\theta & r\cos\theta \end{bmatrix},$$

hence

$$\begin{bmatrix} f_x & f_y \\ g_x & g_y \end{bmatrix} = \begin{bmatrix} u_r & u_\theta \\ v_r & v_\theta \end{bmatrix} \begin{bmatrix} \cos\theta & -r\sin\theta \\ \sin\theta & r\cos\theta \end{bmatrix}^{-1} = \frac{1}{r}\begin{bmatrix} u_r & u_\theta \\ v_r & v_\theta \end{bmatrix} \begin{bmatrix} r\cos\theta & r\sin\theta \\ -\sin\theta & \cos\theta \end{bmatrix}.$$

Thus, for example, $f_x = (\cos\theta)u_r - r^{-1}(\sin\theta)u_\theta$. $\diamond$

9.3.4 Remark. The chain rule may be used to suggest a definition of the tangent plane to a smooth surface. Let $f : U \to \mathbb{R}$ be differentiable on the open subset U of $\mathbb{R}^n$ and let $c \in \mathbb{R}$. The set

$$S = \{\boldsymbol{x} \in U : f(\boldsymbol{x}) = c \quad \text{and} \quad \nabla f(\boldsymbol{x}) \neq \boldsymbol{0}\}$$

is called a *level surface of f in $\mathbb{R}^n$*. Let $\boldsymbol{a} \in S$ and let $\varphi : (-r, r) \to \mathbb{R}^n$ be a smooth path in S such that $\varphi(0) = \boldsymbol{a}$. The existence of such paths may be justified by the implicit function theorem, proved in the next section. Applying the chain rule to the identity $f\big(\varphi(t)\big) = c$, we see that

$$0 = (f \circ \varphi)'(0) = \nabla f(\boldsymbol{a}) \cdot \varphi'(0).$$

Since $\varphi'(0)$ is tangent to the curve at $\boldsymbol{a}$, $\nabla f(\boldsymbol{a})$ is perpendicular to S at $\boldsymbol{a}$. The *tangent hyperplane to S at $\boldsymbol{a}$* is then defined as the set of all points $\boldsymbol{x} \in \mathbb{R}^n$ such that $\boldsymbol{x} - \boldsymbol{a}$ is perpendicular to $\nabla f(\boldsymbol{a})$, that is,

$$(\boldsymbol{x} - \boldsymbol{a}) \cdot \nabla f(\boldsymbol{a}) = 0.$$

For example, the hyperplane tangent at $\boldsymbol{a}$ to the $(n-1)$-dimensional sphere $\{\boldsymbol{x} \in \mathbb{R}^n : \|\boldsymbol{x}\|^2 = 1\}$ is the set of all $\boldsymbol{x}$ such that

$$\sum_{i=1}^{n} 2a_i(x_i - a_i) = 0 \quad \text{or} \quad \boldsymbol{a} \cdot \boldsymbol{x} = 1.$$

The tangent hyperplane at $\boldsymbol{a}$ to a surface S may be seen as the best linear approximation to S near $\boldsymbol{a}$. ◊

The second main result of this section is an n-dimensional version of the mean value theorem of Chapter 4. While such a theorem is not generally available for vector-valued functions (Exercise 14), there is a version for scalar-valued functions. For its statement, we recall that the line segment in $\mathbb{R}^n$ from $\boldsymbol{a}$ to $\boldsymbol{b}$ is defined by

$$[\boldsymbol{a} : \boldsymbol{b}] = \{(1-t)\boldsymbol{a} + t\boldsymbol{b} : 0 \le t \le 1\}.$$

9.3.5 Mean Value Theorem. *Let $U \subseteq \mathbb{R}^n$ be open and let $f : U \to \mathbb{R}$ be differentiable on U. For each pair of points $\boldsymbol{a}, \boldsymbol{b} \in U$ with $[\boldsymbol{a} : \boldsymbol{b}] \subseteq U$, there exists $\boldsymbol{c} \in [\boldsymbol{a} : \boldsymbol{b}]$ such that*

$$f(\boldsymbol{b}) - f(\boldsymbol{a}) = df_{\boldsymbol{c}}(\boldsymbol{b} - \boldsymbol{a}) = \nabla f(\boldsymbol{c}) \cdot (\boldsymbol{b} - \boldsymbol{a}).$$

Proof. Set $\varphi(t) = (1-t)\boldsymbol{a} + t\boldsymbol{b}$, $0 \le t \le 1$, and $g = f \circ \varphi$. Since $\varphi'(t) = \boldsymbol{b} - \boldsymbol{a}$, the chain rule and one-variable mean value theorem imply that

$$f(\boldsymbol{b}) - f(\boldsymbol{a}) = g(1) - g(0) = g'(c) = df_{\varphi(c)}(\boldsymbol{b} - \boldsymbol{a})$$

for some $c \in (0,1)$. Setting $\boldsymbol{c} = \varphi(c)$ completes the proof. □

We conclude this section with two applications of the mean value theorem.

9.3.6 Theorem. *Let $U \subseteq \mathbb{R}^n$ be open and let $f : U \to \mathbb{R}^m$ be continuously differentiable on U. Let $C \subseteq U$ be compact and convex and define $c := \sup_{\boldsymbol{z} \in C} \|df_{\boldsymbol{z}}\|$. Then $c < +\infty$ and $\|f(\boldsymbol{x}) - f(\boldsymbol{y})\| \le c\|\boldsymbol{x} - \boldsymbol{y}\|$, $\boldsymbol{x}, \boldsymbol{y} \in C$.*

Proof. Since $\boldsymbol{z} \mapsto df_{\boldsymbol{z}}$ is continuous and C is compact, $c < +\infty$. Let $\boldsymbol{x}, \boldsymbol{y} \in C$ and $\boldsymbol{u} \in \mathbb{R}^m$. By 9.3.5 applied to the scalar function $g := \boldsymbol{u} \cdot f$, there exists a point $\boldsymbol{c} \in [\boldsymbol{x} : \boldsymbol{y}] \subseteq C$ such that

$$\boldsymbol{u} \cdot \left[f(\boldsymbol{x}) - f(\boldsymbol{y})\right] = g(\boldsymbol{x}) - g(\boldsymbol{y}) = dg_{\boldsymbol{c}}(\boldsymbol{x} - \boldsymbol{y}) = \boldsymbol{u} \cdot df_{\boldsymbol{c}}(\boldsymbol{x} - \boldsymbol{y}).$$

Taking $\boldsymbol{u} = f(\boldsymbol{x}) - f(\boldsymbol{y})$ and using the Cauchy–Schwarz and the operator norm inequalities, we have

$$\|f(\boldsymbol{x}) - f(\boldsymbol{y})\|^2 = \left[f(\boldsymbol{x}) - f(\boldsymbol{y})\right] \cdot \left[df_{\boldsymbol{c}}(\boldsymbol{x} - \boldsymbol{y})\right] \le c\|f(\boldsymbol{x}) - f(\boldsymbol{y})\|\,\|\boldsymbol{x} - \boldsymbol{y}\|.$$

Dividing by $\|f(\boldsymbol{x}) - f(\boldsymbol{y})\|$ completes the proof. □

9.3.7 Corollary. *Let $U \subseteq \mathbb{R}^n$ be open and connected, and let $f : U \to \mathbb{R}^m$ be differentiable on U. If $df_{\boldsymbol{x}} = \mathbf{0}$ for all $\boldsymbol{x} \in U$, then f is constant.*

Proof. Let $\boldsymbol{x} \in U$ and choose $r > 0$ such that $C_r(\boldsymbol{x}) \subseteq U$. Since $C_r(\boldsymbol{x})$ is compact and convex, 9.3.6 implies that

$$\|f(\boldsymbol{x}) - f(\boldsymbol{y})\| \le c\|\boldsymbol{x} - \boldsymbol{y}\|, \quad \boldsymbol{y} \in C_r(\boldsymbol{x}), \quad c := \sup_{\boldsymbol{z} \in C_r(\boldsymbol{x})} \|df_{\boldsymbol{z}}\|.$$

By hypothesis, $c = 0$, hence $f(\boldsymbol{y}) = f(\boldsymbol{x})$ for all $\boldsymbol{y} \in C_r(\boldsymbol{x})$. Thus f is constant on any ball contained in U.

Now let $\boldsymbol{a} \in U$ and define

$$U_{\boldsymbol{a}} = \{\boldsymbol{x} \in U : f(\boldsymbol{x}) = f(\boldsymbol{a})\} \quad \text{and} \quad V_{\boldsymbol{a}} = \{\boldsymbol{x} \in U : f(\boldsymbol{x}) \ne f(\boldsymbol{a})\}.$$

By the first paragraph, if $\boldsymbol{x} \in U_{\boldsymbol{a}}$, then a ball with center $\boldsymbol{x}$ is contained in $U_{\boldsymbol{a}}$. Therefore, $U_{\boldsymbol{a}}$ is open. A similar argument shows that $V_{\boldsymbol{a}}$ is open. Since U is connected and $U_{\boldsymbol{a}} \ne \emptyset$, $U_{\boldsymbol{a}} = U$, that is, $f(\boldsymbol{x}) = f(\boldsymbol{a})$ for all $\boldsymbol{x} \in U$. □

Exercises

1. Let g, φ, $\psi : \mathbb{R} \to \mathbb{R}$ be differentiable and let $f(x, y) = g\big(\varphi(x)\psi(y)\big)$. Find $\nabla f(x, y)$ in terms of g, φ, and ψ.

2. Let $\varphi : \mathbb{R} \to \mathbb{R}$ and $g : \mathbb{R}^3 \to \mathbb{R}$ be differentiable and set $f(x, y) := g\big(x, \varphi(x + 2y), \varphi(x - 3y)\big)$. Find f_y in terms of g and φ.

3. Let $g : \mathbb{R}^2 \to \mathbb{R}$ be differentiable, $\boldsymbol{a}$, $\boldsymbol{b} \in \mathbb{R}^n$, and set $f(\boldsymbol{x}) = g\big(\boldsymbol{a} \cdot \boldsymbol{x}, \boldsymbol{b} \cdot \boldsymbol{x}\big)$. Find ∇f.

4. Let the partial derivatives of $f : \mathbb{R}^2 \to \mathbb{R}$ of f exist and let

$$z = f(x, y) = f(r\cos\theta, r\sin\theta).$$

Prove that
$$\left[r\frac{\partial z}{\partial r}\right]^2 + \left[\frac{\partial z}{\partial \theta}\right]^2 = \left[r\frac{\partial z}{\partial x}\right]^2 + \left[r\frac{\partial z}{\partial y}\right]^2.$$

5. Let $F : \mathbb{R}^n \to \mathbb{R}$ be differentiable and set $f(x) = F(x, \ldots, x)$. Prove that $f'(x) = (1, \ldots, 1) \cdot \nabla F(x, \ldots, x)$.

6. Let $f(x, y)$ be continuously differentiable. Prove that

$$f(x, y) = \int_0^1 (x, y) \cdot \nabla f(tx, ty)t\,dt + \int_0^1 f(tx, ty)\,dt.$$

7. Let $f : U \to \mathbb{R}^m$ be differentiable on an open set $U \subseteq \mathbb{R}^n$. Find $(T \circ f)'(\boldsymbol{x})$ for $T \in \mathcal{L}(\mathbb{R}^m, \mathbb{R}^k)$.

8. Let $f : \mathbb{R}^n \to \mathbb{R}$ be differentiable and $\boldsymbol{a} = (a_1, \ldots, a_n) \in \mathbb{R}^n$ with $a_n \neq 0$. Prove that $\boldsymbol{a} \cdot \nabla f(\boldsymbol{x}) = 0$ for all $\boldsymbol{x} \in \mathbb{R}^n$ iff there exists a differentiable function $g : \mathbb{R}^{n-1} \to \mathbb{R}$ such that

$$f(x_1, x_2, \ldots, x_n) = g\big(x_1 - b_1 x_n, x_2 - b_2 x_n, \ldots, x_{n-1} - b_{n-1} x_n\big),$$

where $b_j = a_j / a_n$, $1 \leq j \leq n-1$.

9. Let $U \subseteq \mathbb{R}^n$ be open and $f : U \to \mathbb{R}$ smooth. Let $\alpha, \beta : I \to \mathbb{R}^n$ be smooth paths in U such that $(\nabla f) \circ \alpha = \alpha'$, $\alpha(t_1) = \beta(t_1)$, and $\|\alpha'(t_1)\| = \|\beta'(t_1)\| = 1$ for some $t_1 \in I$ (that is, α and β both have unit speed at the intersection). Show that $(f \circ \alpha)'(t_1) \geq (f \circ \beta)'(t_1)$.

10. Let $U \subseteq \mathbb{R}^n$ be open and $f : U \to \mathbb{R}$ differentiable at $\boldsymbol{a} \in U$. If $\boldsymbol{u} \in \mathbb{R}^n$ with $\|\boldsymbol{u}\| = 1$, define the *directional derivative of f in the direction of $\boldsymbol{u}$* by

$$D_{\boldsymbol{u}} f(\boldsymbol{a}) = \lim_{t \to 0} \frac{f(\boldsymbol{a} + t\boldsymbol{u}) - f(\boldsymbol{a})}{t}.$$

   (a) Show that if f is differentiable at $\boldsymbol{a}$, then $D_{\boldsymbol{u}} f(\boldsymbol{a})$ exists and equals $\boldsymbol{u} \cdot \nabla f(\boldsymbol{a})$.

   (b) Show that if $D_{\boldsymbol{u}} f$ exists, then $D_{-\boldsymbol{u}} f$ exists and $D_{-\boldsymbol{u}} f = -D_{\boldsymbol{u}} f$.

   (c) Define

$$f(x, y) = \begin{cases} \dfrac{xy^2}{x^2 + y^4} & \text{if } (x, y) \neq (0, 0), \\ 0 & \text{otherwise.} \end{cases}$$

   Show that $D_{\boldsymbol{u}} f(0, 0)$ exists for each $\boldsymbol{u}$ but f is not even continuous at $(0, 0)$.

   (d) Find all unit vectors $\boldsymbol{u}$ such that $D_{\boldsymbol{u}}(xy)^{1/3}$ exists at $(0, 0)$.

   (e) Find all unit vectors $\boldsymbol{u}$ such that $D_{\boldsymbol{u}}|x + y|$ exists at $(x_0, -x_0)$.

   (f) Find all unit vectors $\boldsymbol{u}$ such that $D_{\boldsymbol{u}}(x + y)^{1/3}$ exists at $(0, 0)$.

11. Let $z = F(x, y)$, where $x = x(u, v)$, $y = y(u, v)$, $z = z(u, v)$, and the partial derivatives of these functions exist on $\mathbb{R}^2$. Suppose that $x_u y_v - y_u x_v \neq 0$. Find z_x and z_y in terms of z_u, z_v, x_u, x_v, y_u, and y_v.

12. Let f and f_x be continuous on $[a, b] \times [c, d]$. Use the mean value theorem to prove that

$$\frac{d}{dx} \int_a^b f(t, x)\, dt = \int_a^b f_x(t, x)\, dt, \quad c \leq x \leq d.$$

13. Let f and f_x be continuous on $\mathbb{R}^2$ and $u(x)$, $v(x)$ differentiable on $\mathbb{R}$. Use Exercises 5 and 12 to prove that

$$\frac{d}{dx} \int_{u(x)}^{v(x)} f(t, x)\, dt = \int_{u(x)}^{v(x)} f_x(t, x)\, dt + f\big(v(x), x\big) v'(x) - f\big(u(x), x\big) u'(x).$$

14. Show that the mean value theorem does not generally hold for vector-valued functions.

15. A function $f : \mathbb{R}^n \setminus \{\mathbf{0}\} \to \mathbb{R}$ is *homogeneous of degree* $p > 0$ if $f(t\boldsymbol{x}) = t^p f(\boldsymbol{x})$ for all $t > 0$ and all $\boldsymbol{x} \neq \mathbf{0}$. Prove that a differentiable function f is homogeneous of degree p iff $\boldsymbol{x} \cdot \nabla f(\boldsymbol{x}) = pf(\boldsymbol{x})$ for every $\boldsymbol{x} \neq \mathbf{0}$.

16. Prove the following generalization of the Cauchy mean value theorem: Let $U \subseteq \mathbb{R}^n$ be open and convex and let $f,\, g : U \to \mathbb{R}$ be differentiable on U. Then, for each pair of points $\boldsymbol{a},\, \boldsymbol{b} \in U$, there exists $\boldsymbol{c} \in [\boldsymbol{a} : \boldsymbol{b}]$ such that
$$\big(f(\boldsymbol{b}) - f(\boldsymbol{a})\big)\nabla g(\boldsymbol{c}) \cdot (\boldsymbol{b} - \boldsymbol{a}) = \big(g(\boldsymbol{b}) - g(\boldsymbol{a})\big)\nabla f(\boldsymbol{c}) \cdot (\boldsymbol{b} - \boldsymbol{a}).$$

17. Let $f : U \to \mathbb{R}^m$ be continuously differentiable on the open set $U \subseteq \mathbb{R}^n$ and let C be a compact convex subset of U. Prove that
$$\|f(\boldsymbol{x}) - f(\boldsymbol{y}) - df_{\boldsymbol{y}}(\boldsymbol{x} - \boldsymbol{y})\| \le \sup_{\boldsymbol{z} \in C} \|df_{\boldsymbol{z}} - df_{\boldsymbol{y}}\|\, \|\boldsymbol{x} - \boldsymbol{y}\|, \quad \boldsymbol{x},\, \boldsymbol{y} \in C,$$
and that the supremum is finite.

18. Let $f(x, y) = \big(x^2 - y^2, 2xy\big)$ and $(a, b) \neq (0, 0)$. Show that if the functions $\varphi,\, \psi : (-1, 1) \to \mathbb{R}^2$ are differentiable and $\varphi(0) = \psi(0) = (a, b)$, then
$$\frac{(f \circ \varphi)'(0) \cdot (f \circ \psi)'(0)}{\|(f \circ \varphi)'(0)\|\, \|(f \circ \psi)'(0)\|} = \frac{\varphi'(0) \cdot \psi'(0)}{\|\varphi'(0)\|\, \|\psi'(0)\|}.$$
Thus the angle between the curves φ and ψ at their intersection is preserved under the transformation f.

9.4 Inverse Function Theorem

The one-dimensional inverse function theorem of Section 4.4 has the following n-dimensional extension.

9.4.1 Inverse Function Theorem. *Let $U \subseteq \mathbb{R}^n$ be open and let $f : U \to \mathbb{R}^n$ be continuously differentiable on U. If $J_f(\boldsymbol{a}) \neq 0$ for some $\boldsymbol{a} \in U$, then there exist open sets $U_{\boldsymbol{a}} \subseteq U$ and $V_{\boldsymbol{a}} = f(U_{\boldsymbol{a}})$ with $\boldsymbol{a} \in U_{\boldsymbol{a}}$ such that f is one-to-one on $U_{\boldsymbol{a}}$ and $f^{-1} : V_{\boldsymbol{a}} \to U_{\boldsymbol{a}}$ is continuously differentiable. Moreover,*
$$\big(df_{\boldsymbol{x}}\big)^{-1} = d(f^{-1})_{\boldsymbol{y}}, \qquad \boldsymbol{x} \in U_{\boldsymbol{a}},\ \boldsymbol{y} := f(\boldsymbol{x}). \tag{9.15}$$

The conclusion of the theorem may be summarized by saying that f has a continuously differentiable *local inverse* at $\boldsymbol{a}$. Of course, since f need not be one-to-one on U, f may not have a "global" inverse.

The proof of the theorem requires two lemmas. The first is of some independent interest.

9.4.2 Lemma (Contraction Mapping Principle). *Let* (X, d) *be a complete metric space and let* $\varphi : X \to X$ *be a continuous function such that, for some* $0 \le c < 1$,

$$d\big(\varphi(x), \varphi(y)\big) \le c\, d(x, y) \quad \textit{for all } x,\, y \in X.$$

Then there exists a unique point $x \in X$ *such that* $\varphi(x) = x$.

Proof. Choose any point x_0 in X and define a sequence $\{x_n\}$ recursively by $x_n = \varphi(x_{n-1})$, $n \ge 1$. By hypothesis,

$$d(x_{k+1}, x_k) \le c\, d(x_k, x_{k-1}) \le c^2 d(x_{k-1}, x_{k-2}) \le \cdots \le c^k d(x_1, x_0).$$

Thus, by the triangle inequality, for $m > n$

$$d(x_n, x_m) \le \sum_{k=n}^{m-1} d(x_k, x_{k+1}) \le d(x_1, x_0) \sum_{k=n}^{\infty} c^k.$$

Since $c < 1$, the series $\sum_{k=1}^{\infty} c^k$ converges, hence the sum on the right tends to zero as $n \to \infty$. It follows that $\{x_n\}$ is a Cauchy sequence and therefore converges to some $x \in X$. Letting $n \to +\infty$ in the equation $x_n = \varphi(x_{n-1})$ yields $\varphi(x) = x$. If also $\varphi(y) = y$, then $d(x, y) = d\big(\varphi(x), \varphi(y)\big) \le c\, d(x, y)$, which is possible only if $x = y$. □

9.4.3 Lemma. *Let* $U \subseteq \mathbb{R}^n$ *be open and* $f : U \to \mathbb{R}^n$ *continuously differentiable. If* $\boldsymbol{a} \in U$ *with* $J_f(\boldsymbol{a}) \ne 0$, *then there exists* $r > 0$ *such that the linear transformation* $df_{\boldsymbol{x}}$ *is invertible for each* $\boldsymbol{x} \in B_r(\boldsymbol{a})$.

Proof. Since f' is continuous, its entries are continuous, hence $J_f(\boldsymbol{x})$ is a continuous function of $\boldsymbol{x}$. Since $J_f(\boldsymbol{a}) \ne 0$, there exists $r > 0$ such that $J_f(\boldsymbol{x}) \ne 0$ on $B_r(\boldsymbol{a}) \subseteq U$. Since a linear transformation on $\mathbb{R}^n$ is invertible iff the determinant of its matrix is not zero, $df_{\boldsymbol{x}}$ is invertible for $\boldsymbol{x} \in B_r(\boldsymbol{a})$. □

Proof of the inverse function theorem.

By 9.4.3, there exists an $r > 0$ such that $C_r(\boldsymbol{a}) \subseteq U$ and $df_{\boldsymbol{x}}$ is invertible for each $\boldsymbol{x}$ in an open set W_r containing $C_r(\boldsymbol{a})$. Let $T = df_{\boldsymbol{a}}$ and define $g = T^{-1} \circ f$ on W_r. Then $dg_{\boldsymbol{a}} = T^{-1} \circ df_{\boldsymbol{a}} = I_n$, the identity transformation on $\mathbb{R}^n$. Now apply 9.3.6 to the function $g(\boldsymbol{x}) - \boldsymbol{x}$ on $C_r(\boldsymbol{a})$. The constant c in that theorem is

$$\sup\{\|dg_z - dg_{\boldsymbol{a}}\| \,:\, z \in C_r(\boldsymbol{a})\},$$

which we can make less than $1/2$ by taking r sufficiently small, using the continuity of the function $\boldsymbol{z} \mapsto dg_{\boldsymbol{z}}$ at $\boldsymbol{a}$. Thus

$$\|g(\boldsymbol{x}) - g(\boldsymbol{y}) - (\boldsymbol{x} - \boldsymbol{y})\| \le \tfrac{1}{2}\|\boldsymbol{x} - \boldsymbol{y}\|, \quad \boldsymbol{x},\, \boldsymbol{y} \in C_r(\boldsymbol{a}). \tag{9.16}$$

Since

$$\|\boldsymbol{x} - \boldsymbol{y}\| - \|g(\boldsymbol{x}) - g(\boldsymbol{y})\| \le \|g(\boldsymbol{x}) - g(\boldsymbol{y}) - (\boldsymbol{x} - \boldsymbol{y})\|,$$

we see from (9.16) that

$$\tfrac{1}{2}\|\boldsymbol{x}-\boldsymbol{y}\| \le \|g(\boldsymbol{x})-g(\boldsymbol{y})\|, \quad \boldsymbol{x},\, \boldsymbol{y} \in B_r(\boldsymbol{a}).$$

In particular, g is one-to-one on $B_r(\boldsymbol{a})$.

Next, we use 9.4.2 to show that $g\big(B_r(\boldsymbol{a})\big)$ is open. Let $\boldsymbol{c} \in B_r(\boldsymbol{a})$, $\boldsymbol{d} = g(\boldsymbol{c})$ and choose $s > 0$ so that $C_s(\boldsymbol{c}) \subseteq B_r(\boldsymbol{a})$. We claim that

$$B_{s/2}(\boldsymbol{d}) \subseteq g\big(C_s(\boldsymbol{c})\big) \subseteq g\big(B_r(\boldsymbol{a})\big). \tag{9.17}$$

The second inclusion is clear. For the first, let $\boldsymbol{u} \in B_{s/2}(\boldsymbol{d})$. To show that $\boldsymbol{u} \in g\big(B_r(\boldsymbol{a})\big)$ define

$$\varphi(\boldsymbol{x}) = \boldsymbol{x} - g(\boldsymbol{x}) + \boldsymbol{u}, \quad \boldsymbol{x} \in C_s(\boldsymbol{c}).$$

Then

$$\begin{aligned}
\|\boldsymbol{c}-\varphi(\boldsymbol{x})\| &= \|g(\boldsymbol{x})-g(\boldsymbol{c})-(\boldsymbol{x}-\boldsymbol{c})+\boldsymbol{d}-\boldsymbol{u}\| \\
&\le \|g(\boldsymbol{x})-g(\boldsymbol{c})-(\boldsymbol{x}-\boldsymbol{c})\| + \|\boldsymbol{d}-\boldsymbol{u}\| \\
&\le \tfrac{1}{2}\|\boldsymbol{x}-\boldsymbol{c}\| + \|\boldsymbol{d}-\boldsymbol{u}\| && \text{by (9.16)} \\
&< s/2 + s/2 = s,
\end{aligned}$$

so $\varphi\big(C_s(\boldsymbol{c})\big) \subseteq B_s(\boldsymbol{c})$. Moreover, using (9.16) again we have

$$\|\varphi(\boldsymbol{x})-\varphi(\boldsymbol{y})\| \le \tfrac{1}{2}\|\boldsymbol{x}-\boldsymbol{y}\|, \quad \boldsymbol{x},\, \boldsymbol{y} \in C_s(\boldsymbol{c}).$$

By Lemma 9.4.2, $\varphi(\boldsymbol{x}) = \boldsymbol{x}$ for some $\boldsymbol{x} \in B_s(\boldsymbol{c})$, hence $\boldsymbol{u} = g(\boldsymbol{x}) \in g\big(B_s(\boldsymbol{c})\big)$. Since $\boldsymbol{u}$ was arbitrary, (9.17) holds. Since $\boldsymbol{d} \in g\big(B_r(\boldsymbol{a})\big)$ was arbitrary, $g\big(B_r(\boldsymbol{a})\big)$ is open.

Next, we show that $g^{-1} : g\big(B_r(\boldsymbol{a})\big) \to B_r(\boldsymbol{a})$ is differentiable at $\boldsymbol{b} := g(\boldsymbol{a})$. Since $\boldsymbol{b} \in g\big(B_r(\boldsymbol{a})\big)$ and $g\big(B_r(\boldsymbol{a})\big)$ is open, $\boldsymbol{b}+\boldsymbol{k} \in g\big(B_r(\boldsymbol{a})\big)$ for sufficiently small $\|\boldsymbol{k}\|$, that is, for each such $\boldsymbol{k}$,

$$\boldsymbol{b}+\boldsymbol{k} = g(\boldsymbol{a}+\boldsymbol{h}) \text{ for some } \|\boldsymbol{h}\| < r.$$

By (9.16),

$$\|\boldsymbol{h}\| - \|\boldsymbol{k}\| \le \|\boldsymbol{h}-\boldsymbol{k}\| = \|g(\boldsymbol{a}+\boldsymbol{h})-g(\boldsymbol{a})-\boldsymbol{h}\| \le \tfrac{1}{2}\|\boldsymbol{h}\|,$$

hence $\|\boldsymbol{k}\| \ge \frac{1}{2}\|\boldsymbol{h}\|$. Since $g^{-1}(\boldsymbol{b}+\boldsymbol{k}) = \boldsymbol{a}+\boldsymbol{h}$ and $g^{-1}(\boldsymbol{b}) = \boldsymbol{a}$, recalling that $dg_{\boldsymbol{a}} = I_n$ we have

$$\frac{\|g^{-1}(\boldsymbol{b}+\boldsymbol{k})-g^{-1}(\boldsymbol{b})-I_n\boldsymbol{k}\|}{\|\boldsymbol{k}\|} = \frac{\|\boldsymbol{h}-\boldsymbol{k}\|}{\|\boldsymbol{k}\|} \le 2\,\frac{\|g(\boldsymbol{a}+\boldsymbol{h})-g(\boldsymbol{a})-dg_{\boldsymbol{a}}(\boldsymbol{h})\|}{\|\boldsymbol{h}\|}.$$

Since $\boldsymbol{k} \to \boldsymbol{0}$ implies that $\boldsymbol{h} \to \boldsymbol{0}$, which in turn implies that the right side of the above inequality tends to zero, we see that g^{-1} is differentiable at $\boldsymbol{b}$ with derivative I_n.

Now set $U_{\boldsymbol{a}} = B_r(\boldsymbol{a})$ and $V_{\boldsymbol{a}} = (T \circ g)(U_{\boldsymbol{a}})$. Since T is invertible, it is a homeomorphism, hence $V_{\boldsymbol{a}}$ is open. Moreover, since g is one-to-one on $U_{\boldsymbol{a}}$ and maps $U_{\boldsymbol{a}}$ onto $g(U_{\boldsymbol{a}})$, $f = T \circ g$ is one-to-one on $U_{\boldsymbol{a}}$ and maps $U_{\boldsymbol{a}}$ onto $V_{\boldsymbol{a}}$. Since $f^{-1} = g^{-1} \circ T^{-1}$, the chain rule implies that f^{-1} is differentiable at $f(\boldsymbol{a}) = T\boldsymbol{b}$.

Now observe that the entire above argument may be used at any point $\boldsymbol{x}$ of $U_{\boldsymbol{a}}$, since all that is needed is the invertibility of $df_{\boldsymbol{x}}$. Therefore, f^{-1} is differentiable on $V_{\boldsymbol{a}}$.

To verify (9.15) apply the chain rule to $f^{-1} \circ f = I_n$:

$$d(f^{-1})_{\boldsymbol{y}} \circ df_{\boldsymbol{x}} = d(f^{-1} \circ f)_{\boldsymbol{x}} = d(I_n)_{\boldsymbol{x}} = I_n, \quad \boldsymbol{y} = f(\boldsymbol{x}) \in V_{\boldsymbol{a}}. \qquad \square$$

9.4.4 Corollary. *Let $U \subseteq \mathbb{R}^n$ be open and $f : U \to \mathbb{R}^n$ continuously differentiable with $J_f(\boldsymbol{x}) \neq 0$ for each $\boldsymbol{x} \in U$. Then f is an open map, that is, if $E \subseteq U$ is open, then $f(E)$ is open. If particular, $f(U)$ is open.*

Proof. In the notation of the theorem, $f(E)$ is the union of the open sets $f(U_{\boldsymbol{a}} \cap E)$, $\boldsymbol{a} \in E$. $\square$

Since continuous differentiability is a local property, we have

9.4.5 Global Inverse Function Theorem. *Under the conditions of the preceding corollary, if f is also one-to-one on U, then $f^{-1} : f(U) \to U$ is continuously differentiable.*

9.4.6 Example. The function $(x, y) = f(r, \theta) = (r\cos\theta, r\sin\theta)$, $r > 0$, $\theta \in \mathbb{R}$, has Jacobian r, hence is locally invertible at each point of its domain. Since the function is not one-to-one, it has no global inverse. However, if the domain of f is suitably restricted, say by requiring $\theta_0 < \theta < \theta_0 + 2\pi$, then f is one-to-one on the resulting open set $U_{\theta_0} := (0, +\infty) \times (\theta_0, \theta_0 + 2\pi)$.

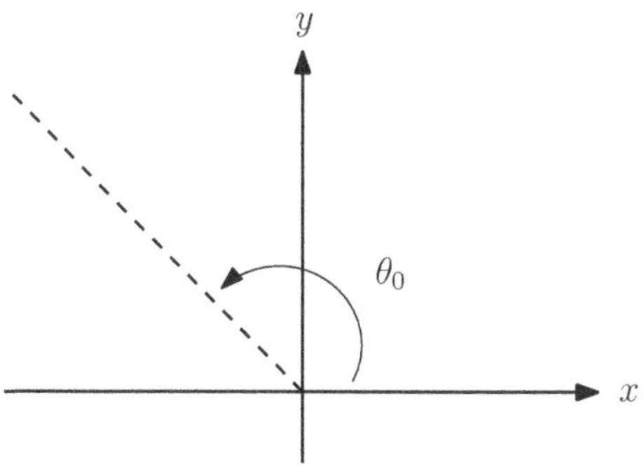

FIGURE 9.1: The domain of $\arg_{\theta_0}$.

By 9.4.5, the restriction g of f to U_{θ_0} has a continuously differentiable inverse

$$(r(x,y), \theta(x,y)) = g^{-1}(x,y)$$

on the open set $V_{\theta_0} = f(U_{\theta_0})$, obtained by removing the ray (r, θ_0), $r \geq 0$, from $\mathbb{R}^2$. Clearly, $r(x,y) = \sqrt{x^2+y^2}$. The function $\theta(x,y)$ is called the *argument of* (x,y) (*determined by* θ_0) and is denoted by $\arg_{\theta_0}(x,y)$. Thus

$$g^{-1}(x,y) = \left(\sqrt{x^2+y^2}, \arg_{\theta_0}(x,y)\right) \text{ on } V_{\theta_0}.$$

For example, if $\theta_0 = -\pi$, then $\arg_{\theta_0}(x,y) = \arctan(y/x)$ for $x > 0$. $\Diamond$

If a function f has a nonzero Jacobian on an open set U and if f is one-to-one on an open subset U_0 of U, then the inverse of the restriction of f to U_0 is called a *branch of* f^{-1} (even though a global f^{-1} may not exist). In the preceding example, g^{-1} is one of infinitely many branches of f^{-1}.

9.4.7 Example. The function $(x,y) = f(u,\theta) = (e^u \cos\theta, e^u \sin\theta)$, where $(u,\theta) \in \mathbb{R}^2$, has Jacobian e^u, hence is locally invertible at each point of $\mathbb{R}^2$. The set $U_{\theta_0} = \mathbb{R} \times (\theta_0, \theta_0 + 2\pi)$ is open, and f restricted to U_{θ_0} is one-to-one. Therefore, the corresponding branch of f^{-1} is continuously differentiable on $f(U_{\theta_0})$, which is the set V_{θ_0} of 9.4.6. The inverse may be given explicitly by

$$u = \ln\sqrt{x^2+y^2}, \quad \theta = \arg_{\theta_0}(x,y). \qquad \Diamond$$

9.4.8 Example. Let $(u,v) = f(x,y) = \left(2x^2 - 3y^2, 3x^2 - 2y^2\right)$. The Jacobian is nonzero on the open set $U = \{(x,y) : xy \neq 0\}$. Solving the equations for x^2 and y^2 yields

$$x^2 = \frac{3v - 2u}{5} \text{ and } y^2 = \frac{2v - 3u}{5}.$$

Restricting f to each of the open quadrants of $\mathbb{R}^2$, we obtain four natural branches of f^{-1}, each defined on the open set

$$V := \{(u,v) : 3v > 2u \text{ and } 2v > 3u\} = \{(u,v) : v > \max\{2u/3, 3u/2\}\},$$

and each of the form

$$f^{-1}(u,v) = \left(\pm\sqrt{\frac{3v-2u}{5}}, \pm\sqrt{\frac{2v-3u}{5}}\right), \quad (u,v) \in V,$$

For example, in the open second quadrant of the x, y plane, one chooses the minus sign in the first coordinate and the plus sign in the second. $\Diamond$

Exercises

1. Find the largest set at each point of which the inverse function theorem guarantees a local C^1 inverse of f, where $f(\boldsymbol{x}) =$

(a) $(x+y, xy)$

(b) $(\sin x + \cos y, \cos x + \sin y)$

(c) $\left(ye^{-x^2}, xe^{-y^2}\right)$

(d) $(\sin x + \sin y, \cos x - \cos y)$

(e) $\left(ye^{-2x}, ye^{3x}\right)$

(f) $\left(\ln\sqrt{xy}, \dfrac{1}{x^2+y^2}\right)$, $x,\ y > 0$

(g) $(xy, x^2 - y^2)$

(h) $\left(\dfrac{x}{1+x^2+y^2}, \dfrac{y}{1+x^2+y^2}\right)$

(i) (x^2+y^2, xy)

(j) (xy^2, x^2z, yz^2)

(k) $\left(ye^{-x^2}, ye^{x^2}\right)$

(l) $(x/y, y/z, z/x)$, $xyz \neq 0$.

2. Find a local inverse of the function in the specified part below of Exercise 1 about the point (a, b) and find $df^{-1}_{(u,v)}$.

(i) (a), $a > b > 0$ (ii) (e), $ab \neq 0$ (iii) (f), $a > b > 0$
(iv) (g), $a > b > 0$ (v) (i), $a,\ b > 0$ (vi) (k), $a,\ b > 0$.

Show that for part (a) in Exercise 1, no inverse is possible on $(0, +\infty)^2$.

3. Let $f(\rho,\phi,\theta) = \big(x(\rho,\phi,\theta), y(\rho,\phi,\theta), z(\rho,\phi,\theta)\big)$ be the spherical coordinate transformation of Exercise 9.1.5. Find an explicit formula for the branch of f^{-1} on the set $\{(\rho,\phi,\theta) : \rho > 0,\ 0 < \phi < \pi,\ 0 < \theta < \pi\}$.

4. Let
$$f(x,y) := \left(\frac{x}{x^2+y^2}, \frac{y}{x^2+y^2}\right), \quad (x,y) \neq (0,0).$$
Show that $f = f^{-1}$ and find J_f.

5. By considering the function
$$f(x) = \begin{cases} x + x^2\sin(1/x) & \text{if } x \neq 0, \\ 0 & \text{otherwise,} \end{cases}$$
show that the hypothesis in the statement of the inverse function theorem that df be continuous on U cannot be removed.

6. Let $U \subseteq \mathbb{R}^n$ be open and $f : U \to \mathbb{R}^n$ of class C^1 such that for some $c > 0$, $\|f(\boldsymbol{x}) - f(\boldsymbol{y})\| \geq c\|\boldsymbol{x} - \boldsymbol{y}\|$ for all $\boldsymbol{x},\ \boldsymbol{y} \in U$, where $c > 0$. Prove that $df_{\boldsymbol{x}}$ is invertible for each $\boldsymbol{x} \in U$. Conclude that $f : U \to f(U)$ is a homeomorphism.

9.5 Implicit Function Theorem

The implicit function theorem is one of the most important applications of the inverse function theorem. The theorem gives conditions under which an equation of the form $F(\boldsymbol{x}, \boldsymbol{y}) = \mathbf{0}$ may be solved *locally* for $\boldsymbol{y}$ in terms of $\boldsymbol{x}$. The resulting function is then said to be *implicitly defined by the equation* $F(\boldsymbol{x}, \boldsymbol{y}) = 0$. The following simple example illustrates the basic idea.

9.5.1 Example. Let $F(x, y, z) = x^2 + y^2 + z^2 - 1$. Consider the problem of finding all points (a, b, c) with $F(a, b, c) = 0$ such that the equation $F(x, y, z) = 0$ has a continuously differentiable solution $z = z(x, y)$ satisfying $z(a, b) = c$. The key fact here is that such a solution is possible if $F_z(a, b, c)(= 2c) \neq 0$. Indeed, in this case $a^2 + b^2 = 1 - c^2 < 1$, hence $x^2 + y^2 < 1$ for all (x, y, z) sufficiently near (a, b, c) that satisfy $F(x, y, z) = 0$. For such points the solution

$$z(x, y) = \pm\sqrt{1 - x^2 - y^2}$$

is continuously differentiable, and if the sign chosen is that of c, then $z(x, y)$ is the unique solution satisfying $z(a, b) = c$. $\lozenge$

Notation. For the statement and proof of the implicit function theorem, we use the following conventions: For points $\boldsymbol{z} \in \mathbb{R}^{n+m}$ we write

$$\boldsymbol{z} = (\boldsymbol{x}, \boldsymbol{y}) = (x_1, \ldots x_n, y_1, \ldots y_m), \quad \boldsymbol{x} \in \mathbb{R}^n, \ \boldsymbol{y} \in \mathbb{R}^m.$$

For a differentiable function

$$F(\boldsymbol{z}) = F(\boldsymbol{x}, \boldsymbol{y}) = (F_1(\boldsymbol{x}, \boldsymbol{y}), \ldots, F_m(\boldsymbol{x}, \boldsymbol{y})),$$

we denote by $F_{\boldsymbol{y}}(\boldsymbol{x}, \boldsymbol{y})$ the $m \times m$ matrix with (i, j)th entry $\dfrac{\partial F_i}{\partial y_j}(\boldsymbol{x}, \boldsymbol{y})$. $\lozenge$

9.5.2 Implicit Function Theorem. *Let U be an open subset of $\mathbb{R}^{n+m}$, let $F = (F_1, \ldots, F_m) : U \to \mathbb{R}^m$ be continuously differentiable, and let $F(\boldsymbol{a}, \boldsymbol{b}) = \mathbf{0}$ for some $(\boldsymbol{a}, \boldsymbol{b}) \in U$. If*

$$\frac{\partial(F_1, \ldots, F_m)}{\partial(y_1, \ldots, y_m)} = \det F_{\boldsymbol{y}}(\boldsymbol{a}, \boldsymbol{b}) \neq 0,$$

then there is an open set $V_{\boldsymbol{a}} \subseteq \mathbb{R}^n$ containing $\boldsymbol{a}$ and a unique continuously differentiable mapping $f : V_{\boldsymbol{a}} \to \mathbb{R}^m$ such that

$$f(\boldsymbol{a}) = \boldsymbol{b} \ \text{ and } \ F\big(\boldsymbol{x}, f(\boldsymbol{x})\big) = \mathbf{0} \ \text{ for every } \ \boldsymbol{x} \in V_{\boldsymbol{a}}.$$

Proof. Define $G : U \to \mathbb{R}^{n+m}$ by

$$G(\boldsymbol{x}, \boldsymbol{y}) = \big(\boldsymbol{x}, F(\boldsymbol{x}, \boldsymbol{y})\big) = \big(\boldsymbol{x}, F_1(\boldsymbol{x}, \boldsymbol{y}), \ldots, F_m(\boldsymbol{x}, \boldsymbol{y})\big).$$

Then G is continuously differentiable, and

$$G'(\boldsymbol{x},\boldsymbol{y}) = \begin{bmatrix} I_{n\times n} & O_{n\times m} \\ A & F_{\boldsymbol{y}} \end{bmatrix},$$

where $I_{n\times n}$ is the $n \times n$ identity matrix, $O_{m\times n}$ is the $m \times n$ zero matrix, and A is an $m \times n$ matrix of partial derivatives of the components of F with respect to $\boldsymbol{x}$. Therefore, $J_G = \det F_{\boldsymbol{y}}$. Since $\det F_{\boldsymbol{y}}(\boldsymbol{a},\boldsymbol{b}) \neq 0$, by the inverse function theorem there exists an open set $W \subseteq U$ containing $(\boldsymbol{a},\boldsymbol{b})$ and an open set $V \subseteq \mathbb{R}^{n+m}$ containing $G(\boldsymbol{a},\boldsymbol{b}) = (\boldsymbol{a},\boldsymbol{0})$ such that $G(W) = V$ and

$$H = (H_1,\ldots,H_n,H_{n+1},\ldots,H_{n+m}) := G^{-1} : V \to W$$

is continuously differentiable. Note that the identities $H\big(G(\boldsymbol{x},\boldsymbol{y})\big) = (\boldsymbol{x},\boldsymbol{y})$ and $G\big(H(\boldsymbol{x},\boldsymbol{y})\big) = (\boldsymbol{x},\boldsymbol{y})$ imply, respectively, that

$$\big(H_{n+1}(G(\boldsymbol{x},\boldsymbol{y})),\ldots,H_{n+m}(G(\boldsymbol{x},\boldsymbol{y}))\big) = \boldsymbol{y}, \quad (\boldsymbol{x},\boldsymbol{y}) \in W \tag{9.18}$$

and

$$F\big(\boldsymbol{x},H_{n+1}(\boldsymbol{x},\boldsymbol{y}),\ldots,H_{n+m}(\boldsymbol{x},\boldsymbol{y})\big) = \boldsymbol{y}, \quad (\boldsymbol{x},\boldsymbol{y}) \in V. \tag{9.19}$$

Now let $V_{\boldsymbol{a}} = \{\boldsymbol{x} \in \mathbb{R}^n : (\boldsymbol{x},\boldsymbol{0}) \in V\}$. Then $V_{\boldsymbol{a}}$ is open and contains $\boldsymbol{a}$. Define f on $V_{\boldsymbol{a}}$ by

$$f(\boldsymbol{x}) = \big(H_{n+1}(\boldsymbol{x},\boldsymbol{0}),\ldots,H_{n+m}(\boldsymbol{x},\boldsymbol{0})\big).$$

Then f is continuously differentiable, and since $(\boldsymbol{a},\boldsymbol{0}) = G(\boldsymbol{a},\boldsymbol{b})$, (9.18) implies that

$$f(\boldsymbol{a}) = \big(H_{n+1}(\boldsymbol{a},\boldsymbol{0}),\ldots,H_{n+m}(\boldsymbol{a},\boldsymbol{0})\big) = \boldsymbol{b}.$$

Furthermore, (9.19) implies that $F(\boldsymbol{x}, f(\boldsymbol{x})) = \boldsymbol{0}$ on $V_{\boldsymbol{a}}$. This establishes the existence of f.

To show uniqueness, assume that $F(\boldsymbol{x}, g(\boldsymbol{x})) = \boldsymbol{0}$ for some function $g : V_{\boldsymbol{a}} \to \mathbb{R}^m$. Then

$$G(\boldsymbol{x}, f(\boldsymbol{x})) = \big(\boldsymbol{x}, F(\boldsymbol{x}, f(\boldsymbol{x}))\big) = \big(\boldsymbol{x}, F(\boldsymbol{x}, g(\boldsymbol{x}))\big) = G(\boldsymbol{x}, g(\boldsymbol{x})).$$

Since G is one-to-one, $f(\boldsymbol{x}) = g(\boldsymbol{x})$. □

9.5.3 Example. The point $(x,y,u,v) = (-1,1,1,1)$ is a solution of the system

$$\begin{aligned} F(x,y,u,v) &:= xu^2 + y = 0 \\ G(x,y,u,v) &:= xy^2 + u^2v^2 = 0, \end{aligned}$$

and at that point

$$\frac{\partial(F,G)}{\partial(u,v)} = 4xu^3v \neq 0.$$

By the implicit function theorem, there are C^1 functions $u(x,y)$ and $v(x,y)$ defined on a ball $B_r(-1,1)$ that satisfy the above system with $u(-1,1) = v(-1,1) = 1$. If $r < 1$, then $(x,y) \in B_r(-1,1)$ implies that $x < 0 < y$ and we have the explicit solution

$$u = \sqrt{-y/x}, \quad v = -x\sqrt{y}.$$ ◊

9.5.4 Remark. Let $f = (f_1, \ldots, f_m)$ be the function in the statement of the implicit function theorem. Set $\boldsymbol{y} = f(\boldsymbol{x})$ and $\boldsymbol{w} = F(\boldsymbol{x}, \boldsymbol{y}) = F(\boldsymbol{z})$. Applying the chain rule to the identity $F\big(\boldsymbol{x}, f(\boldsymbol{x})\big) = \mathbf{0}$ yields

$$\frac{\partial w_i}{\partial x_j} + \frac{\partial w_i}{\partial y_1}\frac{\partial y_1}{\partial x_j} + \cdots + \frac{\partial w_i}{\partial y_m}\frac{\partial y_m}{\partial x_j} = 0, \quad i = 1, \ldots, m, \ j = 1, \ldots, n.$$

This may be written in matrix form as

$$\begin{bmatrix} \dfrac{\partial w_1}{\partial y_1} & \cdots & \dfrac{\partial w_1}{\partial y_m} \\ \vdots & & \vdots \\ \dfrac{\partial w_m}{\partial y_1} & \cdots & \dfrac{\partial w_m}{\partial y_m} \end{bmatrix} \begin{bmatrix} \dfrac{\partial y_1}{\partial x_1} & \cdots & \dfrac{\partial y_1}{\partial x_n} \\ \vdots & & \\ \dfrac{\partial y_m}{\partial x_1} & \cdots & \dfrac{\partial y_m}{\partial x_n} \end{bmatrix} = - \begin{bmatrix} \dfrac{\partial w_1}{\partial x_1} & \cdots & \dfrac{\partial w_1}{\partial x_n} \\ \vdots & & \\ \dfrac{\partial w_m}{\partial x_1} & \cdots & \dfrac{\partial w_m}{\partial x_n} \end{bmatrix},$$

or, in the above notation, as $F_{\boldsymbol{y}}(\boldsymbol{z}) f'(\boldsymbol{x}) = -F_{\boldsymbol{x}}(\boldsymbol{z})$. Therefore,

$$f'(\boldsymbol{x}) = -F_{\boldsymbol{y}}(\boldsymbol{z})^{-1} F_{\boldsymbol{x}}(\boldsymbol{z}),$$

which shows that the partial derivatives of the solution f in the implicit function theorem may be calculated by carrying out a matrix inversion. However, this is practical only for small dimensions, and even in this case it is often easier to apply the chain rule directly and then use Cramer's rule. The next example illustrates the latter approach. ◊

9.5.5 Example. Suppose (x_0, y_0, u_0, v_0) satisfies the system

$$F(x, y, u, v) = G(x, y, u, v) = 0, \tag{9.20}$$

where F and G are C^1 in a neighborhood of (x_0, y_0, u_0, v_0) and

$$\frac{\partial(F, G)}{\partial(u, v)}(x_0, y_0, u_0, v_0) \neq 0.$$

Then (9.20) has a C^1 solution $u = u(x, y)$, $v = v(x, y)$ near (x_0, y_0) such that $u_0 = u(x_0, y_0)$, $v_0 = v(x_0, y_0)$. Differentiating each equation in (9.20) with respect to x and y, we obtain the two systems

$$\begin{aligned} F_u u_x + F_v v_x &= -F_x \\ G_u u_x + G_v v_x &= -G_x \end{aligned} \qquad \begin{aligned} F_u u_y + F_v v_y &= -F_y \\ G_u u_y + G_v v_y &= -G_y \end{aligned}$$

Cramer's rule gives the following solutions near (x_0, y_0, u_0, v_0):

$$u_x = -\frac{\dfrac{\partial(F, G)}{\partial(x, v)}}{\dfrac{\partial(F, G)}{\partial(u, v)}}, \quad v_x = -\frac{\dfrac{\partial(F, G)}{\partial(u, x)}}{\dfrac{\partial(F, G)}{\partial(u, v)}}, \quad u_y = -\frac{\dfrac{\partial(F, G)}{\partial(y, v)}}{\dfrac{\partial(F, G)}{\partial(u, v)}}, \quad v_y = -\frac{\dfrac{\partial(F, G)}{\partial(u, y)}}{\dfrac{\partial(F, G)}{\partial(u, v)}}. \ \lozenge$$

Exercises

1. What does the implicit function theorem tell us about solving the equation $x + y^2 + e^{xy} = 1$ near $(0, 0)$ for one of the variables in terms of the other?

2. Suppose (x_0, y_0, z_0) satisfies the equation $F(x, y, z) = 0$, where F is C^1 in a neighborhood of (x_0, y_0, z_0) and $F_z(x_0, y_0, z_0) \neq 0$. By the implicit function theorem, $F(x, y, z) = 0$ has a C^1 solution $z = z(x, y)$ near (x_0, y_0) with $z_0 = z(x_0, y_0)$. Show that near (x_0, y_0, z_0),
$$z_x = -\frac{F_x}{F_z} \quad \text{and} \quad z_y = -\frac{F_y}{F_z}.$$

3. Show that for each of the functions F below, the equation $F(x, y, z) = 0$ has a local C^1 solution $z = z(x, y)$ on some ball $B_r(a, b)$ such that $z(a, b) = c$. Calculate z_x in a neighborhood of (a, b, c).

   (a) $\sin(xyz) + \cos(xyz) - 1$, $\quad (a, b, c) = (1, \pi, 0)$
   (b) $e^{xyz} + x + y + z - 1$, $\quad (a, b, c) = (0, 0, 0)$
   (c) $z \sin(x + y + z) - \pi\sqrt{3}/6$, $\quad (a, b, c) = (\pi/6, \pi/6, \pi/3)$
   (d) $xyz + \ln(x + y + z) - 1 - \ln 3$, $\quad (a, b, c) = (1, 1, 1)$
   (e) $x \ln z + y \ln x + z \ln y$, $\quad (a, b, c) = (1, 1, 1)$
   (f) $x \sin z + y \sin x + z \sin y - 3\pi/2$, $\quad (a, b, c) = (\pi/2, \pi/2, \pi/2)$
   (g) $z^{2n} + xz^{2n-1} + xy - 1,\ n \in \mathbb{N}$, $\quad (a, b, c) = (1, 1, -1)$
   (h) $\cos(xyz) + \cos(xz) + \cos(yz)$, $\quad (a, b, c) = (0, 1, \pi/2)$.

4. Suppose (x_0, y_0, z_0) satisfies the system $F(x, y, z) = G(x, y, z) = 0$, where F and G are C^1 in a neighborhood of (x_0, y_0, z_0) and
$$\frac{\partial(F, G)}{\partial(x, y)}(x_0, y_0, z_0) \neq 0.$$
By the implicit function theorem, the system has a C^1 solution $(x, y) = (x(z), y(z))$ near (x_0, y_0) with $(x_0, y_0) = (x(z_0), y(z_0))$. Show that near (x_0, y_0, z_0),
$$x'(z) = -\frac{\dfrac{\partial(F, G)}{\partial(z, y)}}{\dfrac{\partial(F, G)}{\partial(x, y)}}, \quad \text{and} \quad y'(z) = -\frac{\dfrac{\partial(F, G)}{\partial(x, z)}}{\dfrac{\partial(F, G)}{\partial(x, y)}}.$$

5. Show that each pair of variables in the system
$$\begin{aligned} \sin(x + z) + \ln(y + z) &= \sqrt{2}/2 \\ e^{xz} + \sin(\pi y + z) &= 1 \end{aligned}$$

are C^1 functions of the other variable near $(x, y, z) = (\pi/4, 1, 0)$. In the case $(x, y) = (x(z), y(z))$, calculate $x'(z)$ and $y'(z)$ in a neighborhood of $(\pi/4, 1, 0)$.

6. Show that each pair of variables in the system

$$\begin{aligned} xy + yz + xz &= 11 \\ xyz + x + y &= 9 \end{aligned}$$

are C^1 functions of the other variable near $(x, y, z) = (1, 2, 3)$. In the case $(x, y) = (x(z), y(z))$, calculate $x'(z)$ and $y'(z)$ in a neighborhood of $(1, 2, 3)$.

7. Show that each pair of the variables (u, v), (x, y), and (x, v) in the system

$$\begin{aligned} x^2 - y^2 + uv - v^2 &= 0 \\ x^2 + y^2 + uv + u^2 &= 4 \end{aligned}$$

are C^1 functions of the remaining variables near $(x, y, u, v) = (1, 1, 1, 1)$. In the case $u(x, y)$, $v(x, y)$, calculate u_x in a neighborhood of $(1, 1)$.

8. Show that the system

$$\begin{aligned} x - y + z + u^2 &= 2 \\ -x + 2z + u^3 &= 2 \\ -y + 3z + u^4 &= 3 \end{aligned}$$

cannot be solved for x, y, and z in terms of u near the point $(x, y, z, u) = (1, 1, 1, 1)$, but for any other group of three variables, a local C^1 solution in terms of the fourth variable is possible.

9. Let $f(x, y)$ be continuously differentiable with $f(0, 0) = 0$. Give conditions on f_x and f_y such that each of the equations below has a C^1 solution $y = y(x)$ on some interval $(-r, r)$ with $y(0) = 0$. Calculate $y'(x)$ in each case.

   (a) $f(2y, 2x - 3y) = 0$. (b) $f(f(x, y), y) = 0$. (c) $f(f(x, y), f(x, y)) = 0$.

10. Let $f(x, y)$ be continuously differentiable with $f(0, 0) = 0$. Give conditions on f_x and f_y under which each of the equations below has a C^1 solution $z = z(x, y)$ on some open ball $B_r(0, 0)$ with $z(0, 0) = 0$.

    (a) $f(2y + 3z, 3x - 2z) = 0$

    (b) $f(f(x, -z), z\ln(e^2 + x + y)) = 0$

    (c) $f(e^{2z} f(x, 2z), f(y, \sin 3z)) = 0$

    (d) $f(f(z, x), f(y, z)) = 0$.

11. Let $f(x, y)$ be continuously differentiable with $f(0, 0) = 0$. For each system below, give conditions on f_x and f_y under which the system has a C^1 solution $x = g(z)$, $y = h(z)$ on some interval $(-r, r)$ with $g(0) = h(0) = 0$.

(a) $f\big(f(x,y), f(z,y)\big) = 0$
$f\big(f(y,z), f(x,z)\big) = 0$

(b) $f\big(f(z,z), f(x,y)\big) = 0$
$f\big(f(x,y), f(y,z)\big) = 0.$

For each system, calculate $g'(z)$.

12. Let $f(x, y)$ be continuously differentiable with $f(0, 0) = 0$. What does the implicit function theorem tell us about the possibility of solving the system

$$f\big(f(u,x), f(v,y)\big) = f\big(f(y,u), f(x,v)\big) = 0$$

(a) for (x, y) in terms of (u, v) such that $x(0, 0) = y(0, 0) = 0$?

(b) for (u, v) in terms of (x, y) such that $u(0, 0) = v(0, 0) = 0$?

13. Let f, g, and h be continuously differentiable and $f(1) = g(1) = h(1) = 0$. Give conditions on f', g', and h' so that the system

$$f(xu) + g(yu) + h(zu) = 0$$
$$f(xv) + g(yv) + h(zv) = 0$$

has a C^1 solution $u = u(x, y, z)$, $v = v(x, y, z)$ on some ball $B_r(1, 1, 1)$ such that $u(1, 1, 1) = v(1, 1, 1) = 1$. Calculate u_x.

14. Let $D \subseteq \mathbb{R}^2$ be compact and let $F(x, y, z)$ be continuous on the set $E := D \times [a, b]$ such that for each $(x, y) \in D$ there exists a unique $z = z(x, y) \in [a, b]$ for which $F\big(x, y, z(x, y)\big) = 0$. Prove that $z(x, y)$ is continuous on D.

15. Suppose the equation $F(x_1, \dots, x_n) = 0$ may be solved for each variable x_j in terms of the others. Show that under suitable conditions

$$\frac{\partial x_2}{\partial x_1}\frac{\partial x_3}{\partial x_2}\cdots\frac{\partial x_n}{\partial x_{n-1}}\frac{\partial x_1}{\partial x_n} = (-1)^n.$$

Verify this for each of the functions

(a) $F(x_1, x_2, x_3) = x_1x_2x_3 - 1$,

(b) $F(x_1, x_2, x_3, x_4) = x_1x_2x_3x_4 - 1$.

16. Let $p(x, y)$ and $q(x, y)$ be C_1 on an open set U containing $(0, 0)$ such that $p(0, 0) = q(0, 0) = 0$ and for $(x, y) \in U \setminus \{(0, 0)\}$

$$p(x, y) > 0, \quad \text{and} \quad -1 \le q(x, y) \le 1.$$

Let

$$f(x, y, z) = z^3 + p(x, y)z + q(x, y), \quad (x, y) \in U, \quad z \in \mathbb{R}.$$

Prove that there is a unique solution $z = z(x, y)$ to $f(x, y, z) = 0$ on all of U which is C^1 on $U \setminus \{(0, 0)\}$ and satisfies $z(0, 0) = 0$.

9.6 Higher-Order Partial Derivatives

Let f be a real-valued function defined on an open subset of $\mathbb{R}^2$ with first partial derivatives f_x and f_y. The *higher-order partial derivatives* are defined inductively by

$$\begin{aligned}
f_{xx} &= \frac{\partial^2 f}{\partial x^2} := \frac{\partial}{\partial x}\frac{\partial f}{\partial x}, & f_{yy} &= \frac{\partial^2 f}{\partial y^2} := \frac{\partial}{\partial y}\frac{\partial f}{\partial y}, \\
f_{xy} &= \frac{\partial^2 f}{\partial y \partial x} := \frac{\partial}{\partial y}\frac{\partial f}{\partial x}, & f_{yx} &= \frac{\partial^2 f}{\partial x \partial y} := \frac{\partial}{\partial x}\frac{\partial f}{\partial y}, \\
f_{xxx} &= \frac{\partial^3 f}{\partial x^3} := \frac{\partial}{\partial x}\frac{\partial^2 f}{\partial x^2}, & f_{xxy} &= \frac{\partial^3 f}{\partial y \partial x^2} := \frac{\partial}{\partial y}\frac{\partial^2 f}{\partial x^2} \\
&\vdots & &\vdots
\end{aligned}$$

Analogous definitions are given for functions of n variables. For such a function f, integers $m_i \in \mathbb{Z}^+$ and a permutation $(i_1, \ldots, i_n)$ of $(1, \ldots, n)$,

$$\frac{\partial^m f}{\partial x_{i_1}^{m_1} \cdots \partial x_{i_n}^{m_n}}, \qquad m := m_1 + \cdots + m_n,$$

is called a *partial derivative of order* m.

The following result will allow some simplifications in calculating higher-order partial derivatives.

9.6.1 Theorem. *Let $U \subseteq \mathbb{R}^2$ be open and let $f : U \to \mathbb{R}$ have continuous first partial derivatives f_x and f_y on U. If f_{xy} exists on U and is continuous at $(a, b) \in U$, then $f_{yx}(a, b)$ exists and equals $f_{xy}(a, b)$.*

Proof. Choose $r > 0$ such that $(a-r, a+r) \times (b-r, b+r) \subseteq U$. For $|h|, |k| < r$, define

$$\begin{aligned}
\varphi_k(x) &= f(x, b+k) - f(x, b), \quad x \in (a-r, a+r), \\
\psi_h(y) &= f(a+h, y) - f(a, y), \quad y \in (b-r, b+r), \\
\Delta(h, k) &= \varphi_k(a+h) - \varphi_k(a) \\
&= \psi_h(b+k) - \psi_h(b) \\
&= f(a+h, b+k) - f(a, b+k) + f(a, b) - f(a+h, b).
\end{aligned}$$

By the mean value theorem applied twice, there exist s, $t \in (0, 1)$ such that

$$\begin{aligned}
\Delta(h, k) &= \varphi_k'(a+sh)h \\
&= \big[f_x(a+sh, b+k) - f_x(a+sh, b)\big]h \\
&= f_{xy}(a+sh, b+tk)hk.
\end{aligned}$$

By continuity of f_{xy} at (a, b),

$$\lim_{(h,k)\to(0,0)} \frac{\Delta(h,k)}{hk} = \lim_{(h,k)\to(0,0)} f_{xy}(a+sh, b+tk) = f_{xy}(a,b).$$

On the other hand, for each h,

$$\lim_{k\to 0} \frac{\Delta(h,k)}{k} = \lim_{k\to 0} \frac{\psi_h(b+k) - \psi_h(b)}{k} = \psi_h'(b) = f_y(a+h,b) - f_y(a,b),$$

so by the iterated limit theorem (8.4.4),

$$\lim_{h\to 0} \frac{f_y(a+h,b) - f_y(a,b)}{h} = \lim_{h\to 0}\lim_{k\to 0} \frac{\Delta(h,k)}{hk} = \lim_{(h,k)\to(0,0)} \frac{\Delta(h,k)}{hk}.$$

Therefore, $f_{yx}(a,b) = f_{xy}(a,b)$. □

The following example shows that continuity of at least one of the second partial derivatives in the theorem is essential.

9.6.2 Example. Let $f(0,0) = 0$ and define

$$f(x,y) = \frac{x^3y - y^3x}{x^2+y^2} \quad \text{if} \quad (x,y) \neq (0,0).$$

Then the first partial derivatives exist and are continuous on $\mathbb{R}^2$, the second partial derivatives exist on $\mathbb{R}^2$, but $f_{xy}(0,0) \neq f_{yx}(0,0)$. Indeed, since $f_x(0,0) = 0$,

$$f_x(0,y) = \lim_{h\to 0} \frac{f(h,y) - f(0,y)}{h} = \lim_{h\to 0} \frac{h^2y - y^3}{h^2+y^2} = -y,$$

and similarly $f_y(x,0) = x$. Therefore, $f_{xy}(0,0) = -1$ and $f_{yx}(0,0) = 1$. ◊

Theorem 9.6.1 may be extended to functions f of n variables. Indeed, if $1 \le i < j \le n$, then under suitable continuity conditions one has

$$\frac{\partial^2 f}{\partial x_i \partial x_j} = \frac{\partial^2 f}{\partial x_j \partial x_i},$$

since the only "active" variables in this identity are x_i and x_j. Combining this observation with an induction argument leads to the following result.

9.6.3 Corollary. *Let f be a real-valued function defined on an open subset of $\mathbb{R}^n$ and let $m = m_1 + m_2 + \cdots + m_n$, $m_i \in \mathbb{Z}^+$. Then, for any permutation $(i_1, \ldots, i_n)$ of $(1, \ldots, n)$,*

$$\frac{\partial^m f}{\partial x_{i_1}^{m_{i_1}} \cdots \partial x_{i_n}^{m_{i_n}}} = \frac{\partial^m f}{\partial x_1^{m_1} \cdots \partial x_n^{m_n}},$$

provided that all partial derivatives of f up to order m are continuous on U.

9.6.4 Definition. Let $r \in \mathbb{N}$. A real-valued function f on an open set $U \subseteq \mathbb{R}^n$ is said to be of *class* C^r *on* U (or simply C^r *on* U) if all partial derivatives up to order r exist and are continuous on U. Also, f is of *class* C^∞ *on* U if it is of class C^r on U for every $r \in \mathbb{N}$. A vector-valued function is C^r if each component function is C^r. Continuous functions are said to be of *class* C^0. A function is of *class* C^r *on a set* if it is the restriction of a C^r function on a larger open set. ◊

9.6.5 Remarks. (a) A function of class $r+1$ is of class r. The function

$$f(x_1, \ldots, x_n) = \begin{cases} x_1^{r+1} & \text{if } x_1 \geq 0, \\ 0 & \text{otherwise} \end{cases}$$

is C^r on $\mathbb{R}^n$ but not C^{r+1}.

(b) The standard rules of differentiation show that if f and g are real-valued functions of class C^r, then so are αf, $f+g$, fg, and f/g. For example, if $f(x,y)$ and $g(x,y)$ are of class C^2, then

$$(fg)_{xx} = f_{xx}g + fg_{xx} + 2f_x g_x,$$

with similar formulas holding for $(fg)_{xy}$ and $(fg)_{yy}$. Since the terms on the right are continuous, fg is C^2. In particular, polynomials and rational functions of several variables are of class C^∞.

(c) The composite $f = g \circ h$ of real-valued C^r functions is again C^r. This follows from the chain rule: The matrix equation

$$f'(\boldsymbol{x}) = g'\big(h(\boldsymbol{x})\big)h'(\boldsymbol{x})$$

shows that the entries of $f'(\boldsymbol{x})$ are sums of products of C^{r-1} functions, hence the entries of $f(\boldsymbol{x})$ are C^r.

(d) If the function f in the statement of the inverse function theorem is C^r on U, then the local inverse of f is also C^r. This is proved by induction on r as follows. Assume that the assertion holds for $r-1$, and let f be C^r on U. Then the entries of the matrix $f'(\boldsymbol{x})$ are C^{r-1}, hence, near $\boldsymbol{a}$, the entries of

$$(f^{-1})'(\boldsymbol{y}) = \left[f'(f^{-1}(\boldsymbol{y}))\right]^{-1}$$

are C^{r-1}, as these are rational functions of the entries of f'. Therefore, the entries of f^{-1} are C^r.

(e) If the function F in the statement of the implicit function theorem is C^r, then the solution $\boldsymbol{y} = f(\boldsymbol{x})$ to the equation $F(\boldsymbol{x}, \boldsymbol{y}) = \boldsymbol{0}$ is C^r. This follows from **(d)**, since f is constructed using the inverse function theorem. ◊

The following example illustrates how the chain rule may be used to calculate higher-order partial derivatives of composite functions.

9.6.6 Example. Let $u = f(x, y)$ be C^2 on $\mathbb{R}^2$ and let $x = r\cos\theta$, $y = r\sin\theta$. Then

$$\begin{aligned}
u_r &= (\cos\theta)u_x + (\sin\theta)u_y, \quad u_\theta = -(r\sin\theta)u_x + (r\cos\theta)u_y, \\
u_{rr} &= (\cos\theta)u_{xr} + (\sin\theta)u_{yr} = (\cos\theta)^2 u_{xx} + (2\sin\theta\cos\theta)u_{xy} + (\sin\theta)^2 u_{yy}, \\
u_{\theta\theta} &= -(r\cos\theta)u_x - (r\sin\theta)u_{x\theta} - (r\sin\theta)u_y + (r\cos\theta)u_{y\theta}, \\
&= (r\sin\theta)^2 u_{xx} - (2r^2\sin\theta\cos\theta)u_{xy} + (r\sin\theta)^2 u_{yy} - ru_r.
\end{aligned}$$

Calculations like these are useful for changing coordinates in differential operators. For example, the above equations imply that

$$\frac{\partial^2}{\partial x^2} + \frac{\partial^2}{\partial y^2} = \frac{\partial^2}{\partial r^2} + \frac{1}{r}\frac{\partial}{\partial r} + \frac{1}{r^2}\frac{\partial^2}{\partial\theta^2}. \tag{9.21}$$

The operator on the left is called the *Laplacian*. The equation expresses the Laplacian in polar coordinates. ◊

Exercises

1. Let $z = f(x, y)$ be C^2 on $\mathbb{R}^2$. Show that the following equations hold for the given functions $x = x(r, t)$ and $y = y(r, t)$.

   (a) $z_{xx} + z_{yy} = \dfrac{z_{rr} + z_{tt}}{a^2 + b^2}, \quad x = ar + bt, \ y = at - br$

   (b) $xz_{xx} - z_{yy} = \dfrac{rz_{rr} - tz_{tt}}{t - r}, \quad x = rt, \ y = r + t$

   (c) $z_{xx} + 4z_{yy} = \dfrac{z_{rr} + z_{tt}}{r^2 + t^2}, \quad x = rt, \ y = r^2 - t^2$

   (d) $x^2 z_{xx} + y^2 z_{yy} = \frac{1}{2}[r^2 z_{rr} + z_{tt} - rz_r], \quad x = re^t, \ y = re^{-t}$

   (e) $z_{xx} + z_{yy} = e^{-2r}[z_{rr} + z_{tt}], \quad x = e^r\sin t, \ y = e^r\cos t$

   (f) $a^2x^2 z_{xx} + b^2y^2 z_{yy} = z_{rr} + z_{tt} - az_r - bz_t, \quad x = e^{ar}, \ y = e^{bt}$.

2. Let $z = f(x, y)$ be C^2 on $\mathbb{R}^2$, $x = ar + bs$, and $y = cr + ds$. Show that

$$\begin{bmatrix} z_{rr} \\ z_{ss} \\ z_{rs} \end{bmatrix} = \begin{bmatrix} a^2 & c^2 & 2ac \\ b^2 & d^2 & 2bd \\ ab & cd & ad + bc \end{bmatrix} \begin{bmatrix} z_{xx} \\ z_{yy} \\ z_{xy} \end{bmatrix}.$$

   In particular, if $x = r - s$ and $y = r + s$, show that

$$\begin{bmatrix} z_{xx} \\ z_{yy} \\ z_{xy} \end{bmatrix} = \frac{1}{4}\begin{bmatrix} 1 & 1 & -2 \\ 1 & 1 & 2 \\ 1 & -1 & 0 \end{bmatrix} \begin{bmatrix} z_{rr} \\ z_{ss} \\ z_{rs} \end{bmatrix}.$$

3. Let $z = f(x, y)$, $x = g(r, s)$, $y = h(r, s)$ be C^2 on $\mathbb{R}^2$. Show that

$$\frac{\partial^2 z}{\partial r^2} = \frac{\partial z}{\partial x}\frac{\partial^2 x}{\partial r^2} + \frac{\partial z}{\partial y}\frac{\partial^2 y}{\partial r^2} + \frac{\partial^2 z}{\partial x^2}\left(\frac{\partial x}{\partial r}\right)^2 + \frac{\partial^2 z}{\partial y^2}\left(\frac{\partial r}{\partial r}\right)^2 + 2\frac{\partial^2 z}{\partial x\partial y}\frac{\partial x}{\partial r}\frac{\partial y}{\partial r}.$$

4. Let $F(x, y, z)$ be C^2 on an open set U and assume that the equation $F(x, y, z) = 0$ defines z implicitly as a function of x and y. Express z_{xx} in terms of partial derivatives of F.

5. Show that each of the following functions $u = u(t, x)$ satisfies the *one-dimensional heat equation* $u_t = k^2 u_{xx}$.

   (a) $u = (a \sin x + b \cos x) \exp(-k^2 t)$. (b) $u = t^{-1/2} \exp\left(-x^2/4k^2 t\right)$.

6. Let $f(x)$ and $g(x)$ be twice differentiable.

   (a) Show that the function $u(t, x) = \big[f(x - ct) + g(x + ct)\big]$ satisfies the *one-dimensional wave equation* $u_{tt} = c^2 u_{xx}$.

   (b) Show that the function $v(t, x) = \dfrac{1}{x}\,[f(x - ct) + g(x + ct)]$, $x > 0$, satisfies the equation $v_{tt} = c^2\left(1 + \dfrac{1}{x}\right) v_{xx} + \dfrac{c^2}{x} v_x$.

7. (Spherical coordinate analog of (9.21)). Let $w = f(x, y, z)$ be of class C^2 on $\mathbb{R}^3$, where

$$x = \rho \sin\phi \cos\theta, \quad y = \rho \sin\phi \sin\theta, \quad \text{and} \quad z = \rho \cos\phi.$$

   Show that

$$\frac{\partial^2 w}{\partial x^2} + \frac{\partial^2 w}{\partial y^2} + \frac{\partial^2 w}{\partial z^2} = \frac{\partial^2 w}{\partial \rho^2} + \frac{2}{\rho}\frac{\partial w}{\partial \rho} + \frac{1}{\rho^2}\frac{\partial^2 w}{\partial \phi^2} + \frac{\cos\phi}{\rho^2 \sin\phi}\frac{\partial w}{\partial \phi} + \frac{1}{\rho^2 \sin^2\phi}\frac{\partial^2 w}{\partial \theta^2}.$$

8. Show that if $f(x, y)$ is C^2 and homogeneous of degree $n \geq 2$ (Exercise 9.3.15), then

$$x^2 f_{xx} + 2xy f_{xy} + y^2 f_{yy} = n(n - 1) f(x, y).$$

9. Let g be C^2 on $(0, +\infty)$, $p \neq 0$, and $f(\boldsymbol{x}) = g\left(\|\boldsymbol{x}\|^p\right)$, $\boldsymbol{x} \in \mathbb{R}^n \setminus \{\boldsymbol{0}\}$. Show that

$$\frac{1}{p}\sum_{i=1}^{n} f_{x_i x_i} = (n + p - 2)\|\boldsymbol{x}\|^{p-2} g'\left(\|\boldsymbol{x}\|^p\right) + p\|\boldsymbol{x}\|^{2(p-1)} g''\left(\|\boldsymbol{x}\|^p\right) \quad \text{and}$$

$$\frac{1}{p}\sum_{i<j} f_{x_i x_j} = \Big[(p - 2)\|\boldsymbol{x}\|^{p-4} g'\left(\|\boldsymbol{x}\|^p\right) + \|\boldsymbol{x}\|^{2(p-2)} g''\left(\|\boldsymbol{x}\|^p\right)\Big] \sum_{i<j} x_i x_j.$$

10. Let $r > 0$. Show that the substitutions

$$s = e^x, \quad t = T - \frac{2\tau}{\sigma^2}, \quad \text{and} \quad v(t, s) = u(\tau, x)$$

   transform the partial differential equation

$$v_t(t, s) + rsv_s(t, s) + \tfrac{1}{2}\sigma^2 s^2 v_{ss}(t, s) - rv(t, s) = 0, \quad s > 0,\ 0 \leq t \leq T$$

into

$$u_\tau(\tau, x) = (k-1)u_x(\tau, x) + u_{xx}(x, \tau) - ku(x, \tau), \quad k := 2r/\sigma^2.$$

The first equation arises in the Black–Scholes theory of option pricing. The second is an example of a *diffusion equation.*

11. Show that the substitutions

$$u(\tau, x) = e^{ax+b\tau}w(\tau, x),\ a := \tfrac{1}{2}(1-k),\ b := a(k-1)+a^2-k = -\tfrac{1}{4}(k+1)^2,$$

reduce the diffusion equation in Exercise 10 to the *heat equation* $w_\tau(\tau, x) = w_{xx}(\tau, x)$.

9.7 Higher-Order Differentials and Taylor's Theorem

Higher-order differentials of a function f of several variables are analogs of higher-order derivatives of functions of a single variable. These may be conveniently expressed in terms of higher-order partial derivatives of f. An important consequence is Taylor's theorem in n-dimensions, which is used to establish convergence of power series in several variables.

We begin by giving an alternate description of the space $\mathcal{L}(\mathbb{R}^n, \mathcal{L}(\mathbb{R}^n, \mathbb{R}))$. For a member B of this space and each $\boldsymbol{h} \in \mathbb{R}^n$, $B\boldsymbol{h} \in \mathcal{L}(\mathbb{R}^n, \mathbb{R})$ has matrix

$$\left[(B\boldsymbol{h})\boldsymbol{e}^1 \quad \cdots \quad (B\boldsymbol{h})\boldsymbol{e}^n\right],$$

which we identify with the *vector* $\left((B\boldsymbol{h})\boldsymbol{e}^1, \ldots, (B\boldsymbol{h})\boldsymbol{e}^n\right)$, so that $(B\boldsymbol{h})\boldsymbol{k}$ may be written $(B\boldsymbol{h}) \cdot \boldsymbol{k}$. Now define

$$\tilde{B}(\boldsymbol{h}, \boldsymbol{k}) := (B\boldsymbol{h}) \cdot \boldsymbol{k}, \quad \boldsymbol{h},\ \boldsymbol{k} \in \mathbb{R}^n.$$

Clearly, $\tilde{B}$ is linear in $\boldsymbol{h}$ for each fixed $\boldsymbol{k}$ and linear in $\boldsymbol{k}$ for each fixed $\boldsymbol{h}$. Such a function is called a *bilinear functional on* $\mathbb{R}^n$. Using the bilinearity, we have

$$\tilde{B}(\boldsymbol{h}, \boldsymbol{k}) = \tilde{B}\left(\sum_{i=1}^n h_i\boldsymbol{e}^i, \sum_{j=1}^n k_j\boldsymbol{e}^i\right) = \sum_{i=1}^n\sum_{j=1}^n B_{ij}h_ik_j, \tag{9.22}$$

where $B_{ij} := \tilde{B}(\boldsymbol{e}^i, \boldsymbol{e}^j) = (B\boldsymbol{e}^i) \cdot \boldsymbol{e}^j$. In matrix notation,

$$\tilde{B}(\boldsymbol{h}, \boldsymbol{k}) = [h_1 \ \cdots \ h_n]\left[B_{ij}\right]\begin{bmatrix} k_1 \\ \vdots \\ k_n \end{bmatrix}.$$

Conversely, given any bilinear functional $\tilde{B}$ on $\mathbb{R}^n$, the equation $(B\boldsymbol{h})\boldsymbol{k} := \tilde{B}(\boldsymbol{h}, \boldsymbol{k})$ defines a member B of $\mathcal{L}(\mathbb{R}^n, \mathcal{L}(\mathbb{R}^n, \mathbb{R}))$. Thus, identifying B with $\tilde{B}$, we see that $\mathcal{L}(\mathbb{R}^n, \mathcal{L}(\mathbb{R}^n, \mathbb{R}))$ may be viewed as the vector space of all bilinear functionals on $\mathbb{R}^n$.

Now let $U \subseteq \mathbb{R}^n$ be open and let $f : U \to \mathbb{R}$ be C^2 on U. Then df is a function on U taking values in $\mathcal{L}(\mathbb{R}^n, \mathbb{R})$. Identifying df with the vector function $\nabla f = (\partial_1 f, \ldots, \partial_n f)$, we define $d^2 f_{\boldsymbol{x}} \in \mathcal{L}(\mathbb{R}^n, \mathcal{L}(\mathbb{R}^n, \mathbb{R}))$ by

$$d^2 f_{\boldsymbol{x}} = d(df)_{\boldsymbol{x}} = d(\nabla f)_{\boldsymbol{x}}$$

that is, by the above identification,

$$d^2 f_{\boldsymbol{x}}(\boldsymbol{h}, \boldsymbol{k}) = d(\nabla f)_{\boldsymbol{x}}(\boldsymbol{h}) \cdot \boldsymbol{k}, \quad \boldsymbol{x} \in U, \quad \boldsymbol{h}, \boldsymbol{k} \in \mathbb{R}^n.$$

The matrix of $d(\nabla f)_{\boldsymbol{x}}$ has (i, j) entry $\partial_j \partial_i f(\boldsymbol{x}) = \partial_i \partial_j f(\boldsymbol{x})$, since f is C^2. Thus

$$d^2 f_{\boldsymbol{x}}(\boldsymbol{h}, \boldsymbol{k}) = \sum_{i,j=1}^{n} \frac{\partial^2 f(\boldsymbol{x})}{\partial x_i \partial x_j} h_i k_j, \quad \boldsymbol{h}, \boldsymbol{k} \in \mathbb{R}^n.$$

The bilinear function $d^2 f_{\boldsymbol{x}}$ is called the *second-order differential of f at $\boldsymbol{x}$*.

For higher-order differentials, we need the following generalization of a bilinear functional:

9.7.1 Definition. An *m-multilinear functional* on $\mathbb{R}^n$ is a real-valued function $M(\boldsymbol{h}^1, \ldots, \boldsymbol{h}^m)$ of vectors $\boldsymbol{h}^j = (h_1^j, \ldots, h_n^j) \in \mathbb{R}^n$ that is linear in each variable $\boldsymbol{h}_j$ when the other variables are held fixed. ◊

Analogous to (9.22) we have

$$M(\boldsymbol{h}^1, \ldots, \boldsymbol{h}^m) = \sum_{j_1=1}^{n} \cdots \sum_{j_m=1}^{n} M_{j_1,\ldots,j_m} h_{j_1}^1 \cdots h_{j_m}^m \tag{9.23}$$

where $M_{j_1,\ldots,j_m} := M(\boldsymbol{e}^{j_1}, \ldots, \boldsymbol{e}^{j_m})$.

Now let f be C^m on U, $m \geq 2$. The *mth-order differential of f at $\boldsymbol{x}$* is defined inductively by

$$d^m f_{\boldsymbol{x}} = d(d^{m-1} f)_{\boldsymbol{x}}.$$

As in the case $m = 2$, we may interpret $d^m f_{\boldsymbol{x}}$ as the m-multilinear functional

$$d^m f_{\boldsymbol{x}}(\boldsymbol{h}^1, \ldots, \boldsymbol{h}^m) = \sum_{j_1=1}^{n} \cdots \sum_{j_m=1}^{n} \frac{\partial^m f(\boldsymbol{x})}{\partial x_{j_1} \cdots \partial x_{j_m}} h_{j_1}^1 \cdots h_{j_m}^m.$$

The mth *total differential $D^m f_{\boldsymbol{x}}$ of f at $\boldsymbol{x}$* is then defined by

$$\begin{aligned} D^m f_{\boldsymbol{x}}(\boldsymbol{h}) &:= d^m f_{\boldsymbol{x}}(\boldsymbol{h}, \ldots, \boldsymbol{h}), \quad \boldsymbol{h} := (h_1, \ldots, h_n) \\ &= \sum_{j_1=1}^{n} \cdots \sum_{j_m=1}^{n} \frac{\partial^m f(\boldsymbol{x})}{\partial x_{j_1} \cdots \partial x_{j_m}} h_{j_1} \cdots h_{j_m}, \end{aligned} \tag{9.24}$$

which is frequently written

$$D^m f_{\boldsymbol{x}} = \sum_{j_1, j_2, \ldots, j_m = 1}^{n} \frac{\partial^m f(\boldsymbol{x})}{\partial x_{j_1} \partial x_{j_2} \ldots \partial x_{j_m}} dx_{j_1} dx_{j_2} \cdots dx_{j_m}, \quad dx_j(\boldsymbol{h}) := h_j.$$

By 9.6.3, each partial derivative in (9.24) may be expressed as

$$\frac{\partial^m f}{\partial x_1^{m_1} \ldots \partial x_n^{m_n}}, \quad m := m_1 + \cdots + m_n, \quad m_j \in \mathbb{Z}^+.$$

Similarly, the corresponding product of h's in (9.24) may be written in the form $h_1^{m_1} \ldots h_n^{m_n}$. For a fixed *multi-index* $(m_1, \ldots, m_n) \in \mathbb{Z}^+ \times \cdots \times \mathbb{Z}^+$, the number of terms in (9.24) of the form

$$\frac{\partial^m f}{\partial x_1^{m_1} \ldots \partial x_n^{m_n}} h_1^{m_1} \cdots h_n^{m_n}$$

is given by the *multinomial coefficient*

$$\binom{m}{m_1, m_2, \ldots, m_n} = \frac{m}{m_1! m_2! \cdots m_n!}, \tag{9.25}$$

which is the number of distinct ways of arranging m objects, where m_1 are alike, m_2 are alike, etc. With this notation, (9.24) may be written

$$D^m f_{\boldsymbol{x}}(\boldsymbol{h}) = \sum \binom{m}{m_1, \ldots, m_n} \frac{\partial^m f(\boldsymbol{x})}{\partial x_1^{m_1} \ldots \partial x_n^{m_n}} h_1^{m_1} \cdots h_n^{m_n}, \tag{9.26}$$

or, in differential notation,

$$D^m f_{\boldsymbol{x}} = \sum \binom{m}{m_1, \ldots, m_n} \frac{\partial^m f(\boldsymbol{x})}{\partial x_1^{m_1} \ldots \partial x_n^{m_n}} (dx_1)^{m_1} \cdots (dx_n)^{m_n},$$

where the sums are taken over all multi-indices $(m_1, \ldots, m_n) \in \mathbb{Z}^+ \times \cdots \times \mathbb{Z}^+$ for which $m_1 + \cdots + m_n = m$.

We may go a step further by appealing to the following generalization of the binomial theorem.

9.7.2 Multinomial Theorem. *Let* $h_1, \ldots, h_n \in \mathbb{R}$ *and* $m \in \mathbb{N}$. *Then*

$$(h_1 + \cdots + h_n)^m = \sum \binom{m}{m_1, \ldots, m_n} h_1^{m_1} \cdots h_n^{m_n}, \tag{9.27}$$

where the summation is taken over all multi-indices $(m_1, \ldots, m_n)$ *for which* $m_1 + \cdots + m_n = m$.

Proof. The theorem may be proved by induction, but we give a combinatorial argument instead. The left side of (9.27) expands into a sum of products of the form $x_1 \cdots x_m$, where each x_i is one of the terms in the sum $h_1 + \cdots + h_n$.

Each such product may be written uniquely as $h_1^{m_1} \cdots h_n^{m_n}$, where $m_j \geq 0$ and $m_1 + \cdots + m_n = m$. For each fixed $(m_1, \ldots, m_n)$, the number of products of this form is the number of ways m_1 factors in the product $x_1 \cdots x_n$ may be chosen to be h_1, m_2 factors may be chosen to be h_2, etc. This number is precisely the multinomial coefficient (9.25). $\square$

Now consider the operator $h_i \frac{\partial}{\partial x_i}$, which takes a C^1 function f to the function $h_i \frac{\partial f}{\partial x_i}$. If multiplication of such operators is defined as operator composition, then the usual laws of algebra hold. For example, the operator

$$\left(h_1 \frac{\partial}{\partial x_1} + h_2 \frac{\partial}{\partial x_2}\right) h_2 \frac{\partial}{\partial x_2}$$

applied to a C^2 function f yields

$$h_1 h_2 \frac{\partial^2 f}{\partial x_1 \partial x_2} + h_2^2 \frac{\partial^2 f}{\partial x_2^2} = \left(h_1 h_2 \frac{\partial^2}{\partial x_1 \partial x_2} + h_2^2 \frac{\partial^2}{\partial x_2^2}\right) f,$$

hence we may write

$$\left(h_1 \frac{\partial}{\partial x_1} + h_2 \frac{\partial}{\partial x_2}\right) h_2 \frac{\partial}{\partial x_2} = \left(h_1 h_2 \frac{\partial^2}{\partial x_1 \partial x_2} + h_2^2 \frac{\partial^2}{\partial x_2^2}\right).$$

Similarly,

$$\left(h_1 \frac{\partial}{\partial x_1} + h_2 \frac{\partial}{\partial x_2}\right)^2 = h_1^2 \frac{\partial^2}{\partial x_1^2} + 2h_1 h_2 \frac{\partial^2}{\partial x_1 \partial x_2} + h_2^2 \frac{\partial^2}{\partial x_2^2}.$$

The last example suggests that the multinomial theorem is valid in this setting. This is indeed the case (a similar proof works). It follows from (9.26) that the mth total differential may be written in operator form as

$$D^m f_{\boldsymbol{x}}(\boldsymbol{h}) = \left(h_1 \frac{\partial}{\partial x_1} + h_2 \frac{\partial}{\partial x_2} + \cdots + h_n \frac{\partial}{\partial x_n}\right)^m = (\boldsymbol{h} \cdot \nabla)^m.$$

We may now state the n-dimensional version of Taylor's theorem.

9.7.3 Taylor's Theorem. *Let $U \subseteq \mathbb{R}^n$ be open, $m \in \mathbb{N}$, and let $f : U \to \mathbb{R}$ be C^{m+1} on U. Then for each pair of distinct points $\boldsymbol{a}$, $\boldsymbol{x} \in U$ for which $[\boldsymbol{a} : \boldsymbol{x}] \subseteq U$, there exists a point $\mathbf{c} \in [\boldsymbol{a} : \boldsymbol{x}]$ depending on $\boldsymbol{x}$ and $\boldsymbol{a}$ such that*

$$f(\boldsymbol{x}) = \sum_{p=0}^{m} \frac{1}{p!} (\boldsymbol{h} \cdot \nabla)^p f(\boldsymbol{a}) + \frac{1}{(m+1)!} (\boldsymbol{h} \cdot \nabla)^{m+1} f(\boldsymbol{c}), \quad \boldsymbol{h} := \boldsymbol{x} - \boldsymbol{a}. \tag{9.28}$$

Proof. Let the line segment $[\boldsymbol{a} : \boldsymbol{x}]$ be described by

$$\varphi(t) := (1 - t)\boldsymbol{a} + t\boldsymbol{x} = \boldsymbol{a} + t\boldsymbol{h}, \quad 0 \leq t \leq 1.$$

Since U is open, there exists an $r > 0$ such that $\varphi((-r, 1+r)) \subseteq U$. Let $F = f \circ \varphi$. By the chain rule,

$$F'(t) = \sum_{j=1}^{n} \frac{\partial f(\varphi(t))}{\partial x_j} h_j \quad \text{and} \quad \frac{d}{dt} \frac{\partial f(\varphi(t))}{\partial x_j} = \sum_{i=1}^{n} \frac{\partial^2 f(\varphi(t))}{\partial x_i \partial x_j} h_i,$$

hence

$$F''(t) = \sum_{i,j=1}^{n} \frac{\partial^2 f(\varphi(t))}{\partial x_i \partial x_j} h_i h_j.$$

An induction argument shows that

$$F^{(p)}(t) = \sum_{j_1,\ldots,j_p=1}^{n} \frac{\partial^p f(\varphi(t))}{\partial x_{j_1} \cdots \partial x_{j_p}} h_{j_1} \ldots h_{j_p} = (\boldsymbol{h} \cdot \nabla)^p f(\varphi(t)).$$

By Taylor's theorem in one variable, there exists $c \in (0, 1)$ such that

$$f(\boldsymbol{x}) = F(1) = \sum_{p=0}^{m} \frac{F^{(p)}(0)}{p!} + \frac{F^{(m+1)}(c)}{(m+1)!}.$$

Setting $\boldsymbol{c} = \varphi(c)$ completes the proof. □

The summation in (9.28) is called an *mth-order Taylor polynomial about* $\boldsymbol{a}$ and is denoted by $T_m(\boldsymbol{x}, \boldsymbol{a})$. For example, the second-order Taylor polynomial of a C^2 function $f(x_1, x_2)$ is

$$f + h_1 \frac{\partial f}{\partial x_1} + h_2 \frac{\partial f}{\partial x_2} + \frac{1}{2} h_1^2 \frac{\partial^2 f}{\partial x_1^2} + h_1 h_2 \frac{\partial^2 f}{\partial x_1 \partial x_2} + \frac{1}{2} h_2^2 \frac{\partial^2 f}{\partial x_2^2},$$

where $h_j = x_j - a_j$ and the terms are evaluated at (a_1, a_2). The last term in (9.28) is called the *remainder* term and is denoted by $R_m(\boldsymbol{x}, \boldsymbol{a})$.

The following theorem gives a sufficient condition for a C^∞ function to be expressed as a multi-variable Taylor's series.

9.7.4 Taylor Series Representation. *Let* $U \subseteq \mathbb{R}^n$ *be open and convex and let* $f : U \to \mathbb{R}$ *be* C^∞ *on* U. *Suppose that for some* $M < +\infty$

$$\left| \frac{\partial^p f(\boldsymbol{x})}{\partial x_1^{p_1} \partial x_2^{p_2} \ldots \partial x_n^{p_n}} \right| \leq M$$

for all $\boldsymbol{x} \in U$, $p \in \mathbb{N}$, *and all* $p_j \in \mathbb{Z}^+$, *where* $p = p_1 + \cdots + p_n$. *Then*

$$f(\boldsymbol{x}) = \sum_{p=0}^{\infty} \frac{1}{p!} (\boldsymbol{h} \cdot \nabla)^p f(\boldsymbol{a}), \quad \boldsymbol{a}, \boldsymbol{x} \in U, \quad \boldsymbol{h} := \boldsymbol{x} - \boldsymbol{a}. \tag{9.29}$$

Proof. By 9.7.3, the theorem will follow if we show that the remainder term

$$R_m(\boldsymbol{x}, \boldsymbol{a}) = \frac{1}{(m+1)!}(\boldsymbol{h} \cdot \nabla)^{m+1} f(\boldsymbol{c})$$

tends to zero as $m \to \infty$. By (9.26),

$$|R_m(\boldsymbol{x}, \boldsymbol{a})| \leq \frac{M}{(m+1)!} \sum \binom{m+1}{m_1, \ldots, m_n} |h_1|^{m_1} |h_2|^{m_2} \ldots |h_n|^{m_n},$$

where the summation is taken over all multi-indices $(m_1, \ldots, m_n)$ for which $m_1 + \cdots + m_n = m + 1$. By the multinomial theorem, this sum is h^{m+1}, where $h = |h_1| + |h_2| + \cdots + |h_n|$. Therefore,

$$|R_m(\boldsymbol{x}, \boldsymbol{a})| \leq \frac{Mh^{m+1}}{(m+1)!},$$

which implies $\lim_m R_m(\boldsymbol{x}, \boldsymbol{a}) = 0$. □

The series on the right in (9.29) is called the *Taylor series for f about* $\boldsymbol{a}$. While the theorem may be applied directly, in many cases it is easier to make use of single-variable series. For example, from the series expansion for e^x we have

$$e^{xy} = \sum_{n=0}^{\infty} \frac{x^n y^n}{n!} \quad \text{and} \quad e^{x+y} = e^x e^y = \sum_{n=0}^{\infty} \Big(\sum_{j=0}^{n} \frac{x^j}{j!} \frac{y^{n-j}}{(n-j)!} \Big).$$

Exercises

1. Let f be of class C^3. Write out explicitly

   (a) $D^2 f(x, y)$ (b) $D^3 f(x, y)$ (c) $D^2 f(x, y, z)$.

2. Calculate $D^2 f$ for the functions $f(x, y) =$

   (a) $x^3 y^2 + x^2 y^3$ (b) $\dfrac{1}{x^2 y}$ (c) $\sin(xy)$ (d) $e^{x^2+y^2}$ (e) $\ln(x^2 + y)$.

3. Find $D^{m+n+1}(x^m y^n)$.

4. Let $f(x, y)$ be C^n. Show that for $1 \leq k \leq n$,

   $$\frac{\partial^k}{\partial t^k} f(tx, ty)\Big|_{t=1} = \big((x, y) \cdot \nabla\big)^k f(x, y).$$

   Conclude that if f is homogeneous of degree n (Exercise 9.3.15), then

   $$\big((x, y) \cdot \nabla\big)^k f(x, y) = n(n-1) \cdots (n-k+1) f(x, y).$$

5. Write out explicitly

   (a) the first-order Taylor polynomial for a C^1 function $f(x_1, x_2, x_3)$.

   (b) the third-order Taylor polynomial for a C^3 function $f(x_1, x_2)$,

6. A *polynomial of degree* $m+n$ *in two variables* x *and* y is a function of the form

$$\sum_{i=0}^{m}\sum_{j=0}^{n} a_{ij}x^i y^j, \quad \text{where } a_{ij} \in \mathbb{R} \text{ and } a_{mn} \neq 0.$$

   Prove that $f(x, y)$ is a polynomial of degree $\leq p$ on $B_r(a, b)$ iff $D^{p+1}f(x, y) = 0$ for all $(x, y) \in B_r(a, b)$.

7. Let $P(x, y)$ be a polynomial in x, y. Prove that the polynomials $P(x \pm 1, y \pm 1)$ may be written as linear combinations of derivatives

$$\frac{\partial^k P(x, y)}{\partial x^i \partial y^j}, \quad k = i + j.$$

8. Let $\varphi(t)$ be of class C^m on an interval $(-r, r)$ and let $f(\boldsymbol{x}) = \varphi(\boldsymbol{b} \cdot \boldsymbol{x})$ where $\boldsymbol{b}$, $\boldsymbol{x} \in \mathbb{R}^n$. Show that the Taylor polynomial for f of order m about $\mathbf{0}$ is

$$\sum_{p=0}^{m} \frac{\varphi^{(p)}(0)}{p!}(\boldsymbol{b} \cdot \boldsymbol{x})^p.$$

9. Let $U \subseteq \mathbb{R}^2$ be open and connected, and let f be C^∞ on U such that for each $(x, y) \in U$ there exists $r > 0$ and $p \in \mathbb{N}$ depending on (x, y) such that $D^p f = 0$ on $B_r(x, y)$. Prove that there exists a single $p \in \mathbb{N}$ such that $D^p f = 0$ on U. *Hint.* Use Exercise 6.

10. Let $U \subseteq \mathbb{R}^n$ be open and let f be C^p on U such that all partial derivatives of f of order $r < p$ vanish throughout U. Let C be a compact convex subset of U. Prove that there exists $c < +\infty$ such that

$$\|f(\boldsymbol{x}) - f(\boldsymbol{y})\| \leq c\|\boldsymbol{x} - \boldsymbol{y}\|^p, \quad \boldsymbol{x}, \boldsymbol{y} \in C.$$

11. Use the one-variable Taylor series to find third-order Taylor polynomials with $\boldsymbol{a} = (0, 0)$ for the functions

   (a) $\sin(x + y)$ (b) $\cos\sqrt{xy}$ (c) $\ln(1 - x - y)^{-1}$

   (d) $\arctan(x + y)$ (e) e^{2x+3y} (f) $\dfrac{y}{1 + xy}$.

*9.8 Optimization

Throughout the section, $f : U \to \mathbb{R}$ denotes a C^1 function on an open subset U of $\mathbb{R}^n$.

In this section we use differential theory to find the maximum and minimum values of f on subsets E of U. The first step is to find all local extrema.

Local Extrema and Critical Points

9.8.1 Definition. Let $\boldsymbol{a} \in U$. If $f(\boldsymbol{a})$ is the maximum (minimum) value of f on some ball in U with center $\boldsymbol{a}$ then f is said to have a *local maximum* (*local minimum*) *at* $\boldsymbol{a} \in U$ In either case, f is said to have a *local extremum at* $\boldsymbol{a}$. ♢

The following theorem gives a necessary condition for the existence of a local extremum.

9.8.2 Local Extremum Theorem. *If f has a local extremum at $\boldsymbol{a}$, then* $df_{\boldsymbol{a}} = 0$.

Proof. The function $g(t) := f(a_1, \ldots a_{j-1}, t, a_{j+1}, \ldots, a_n)$ has a local extremum at $t = a_j$, hence, by the single-variable local extremum theorem (4.2.2), $\partial_j f(a_1, \ldots, a_n) = g'(a_j) = 0$. □

9.8.3 Definition. A point $\boldsymbol{a} \in U$ is called a *critical point of f* if $df_{\boldsymbol{a}} = 0$. A critical point $\boldsymbol{a}$ is a *local maximum* (*local minimum*) *point* if f has a local maximum (local minimum) at $\boldsymbol{a}$. If $\boldsymbol{a}$ is neither a local maximum nor a local minimum point, then $\boldsymbol{a}$ is called a *saddle point*. ♢

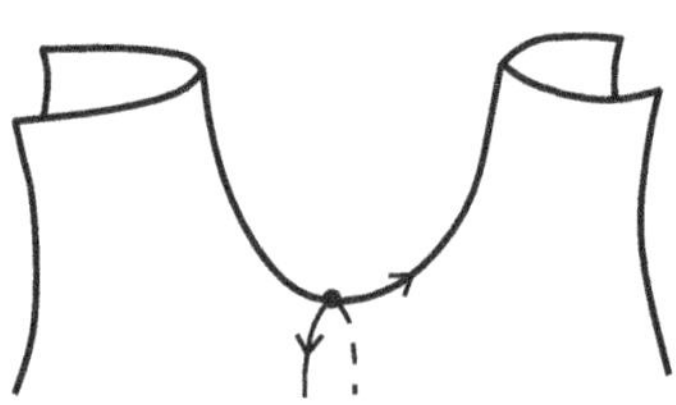

FIGURE 9.2: Saddle point.

By definition, a critical point $\boldsymbol{a}$ is a saddle point iff in each ball $B_r(\boldsymbol{a})$ there exist points $\boldsymbol{x}$ and $\boldsymbol{y}$ such that $f(\boldsymbol{x}) < f(\boldsymbol{a}) < f(\boldsymbol{y})$. This means that the graph of f rises in some directions from $\boldsymbol{a}$ and falls in others. A familiar example is $f(x, y) = y^2 - x^2$ at $(0, 0)$ (Figure 9.2).

Second Derivative Test

The following theorem gives sufficient conditions for a critical point of a function f to be a local maximum point, a local minimum point, or a saddle point. It may be seen as an extension of the second derivative test for functions of one variable.

9.8.4 Second Derivative Test. *Let f be C^2 on U and let $\boldsymbol{a} \in U$ be a critical point of f.*

(a) *If $D^2 f_{\boldsymbol{a}}(\boldsymbol{h}) > 0$ for all $\boldsymbol{h} \neq \boldsymbol{0}$, then $\boldsymbol{a}$ is a local minimum point.*

(b) *If $D^2 f_{\boldsymbol{a}}(\boldsymbol{h}) < 0$ for all $\boldsymbol{h} \neq \boldsymbol{0}$, then $\boldsymbol{a}$ is a local maximum point.*

(c) *If $D^2 f_{\boldsymbol{a}}(\boldsymbol{h}) > 0$ for some $\boldsymbol{h}$ and $D^2 f_{\boldsymbol{a}}(\boldsymbol{k}) < 0$ for some $\boldsymbol{k}$, then $\boldsymbol{a}$ is a saddle point of f.*

Proof. Choose $r > 0$ such that $B_r(\boldsymbol{a}) \subseteq U$. By 9.28 with $m = 1$, for each $\boldsymbol{h}$ with $\|\boldsymbol{h}\| < r$ there exists $\boldsymbol{c} \in [\boldsymbol{a} : \boldsymbol{a} + \boldsymbol{h}]$ such that

$$f(\boldsymbol{a}+\boldsymbol{h}) - f(\boldsymbol{a}) = \tfrac{1}{2} D^2 f_{\boldsymbol{c}}(\boldsymbol{h}) = \tfrac{1}{2}\big[D^2 f_{\boldsymbol{a}}(\boldsymbol{h}) + \eta(\boldsymbol{h})\big], \tag{9.30}$$

where

$$\eta(\boldsymbol{h}) = D^2 f_{\boldsymbol{c}}(\boldsymbol{h}) - D^2 f_{\boldsymbol{a}}(\boldsymbol{h}) = \sum_{i,j=1}^{n} h_i h_j \left[\frac{\partial^2 f(\boldsymbol{c})}{\partial x_i \partial x_j} - \frac{\partial^2 f(\boldsymbol{a})}{\partial x_i \partial x_j}\right].$$

Set

$$\varepsilon(\boldsymbol{h}) = \begin{cases} \|\boldsymbol{h}\|^{-2}\eta(\boldsymbol{h}) & \text{if } \|\boldsymbol{h}\| \neq 0, \\ 0 & \text{if } \|\boldsymbol{h}\| = 0. \end{cases}$$

Since $|h_i h_j| \leq \|\boldsymbol{h}\|^2$,

$$|\varepsilon(\boldsymbol{h})| \leq \sum_{i,j=1}^{n} \left|\frac{\partial^2 f(\boldsymbol{c})}{\partial x_i \partial x_j} - \frac{\partial^2 f(\boldsymbol{a})}{\partial x_i \partial x_j}\right|.$$

Since f is C^2, $\lim_{\boldsymbol{h}\to\boldsymbol{0}} \varepsilon(\boldsymbol{h}) = 0$.

With these preliminaries out of the way, assume that the hypothesis in (a) holds. Since the function $D^2 f_{\boldsymbol{a}}(\boldsymbol{h})$ is continuous in $\boldsymbol{h}$, it has a positive minimum m on the sphere $S_1(\boldsymbol{0})$ in $\mathbb{R}^n$. Thus

$$D^2 f_{\boldsymbol{a}}(\boldsymbol{h}) = \|\boldsymbol{h}\|^2 D^2 f_{\boldsymbol{a}}\left(\frac{\boldsymbol{h}}{\|\boldsymbol{h}\|}\right) \geq m\|\boldsymbol{h}\|^2, \ \boldsymbol{h} \neq \boldsymbol{0},$$

so from (9.30)

$$f(\boldsymbol{a}+\boldsymbol{h}) - f(\boldsymbol{a}) \geq \tfrac{1}{2}\big(m\|\boldsymbol{h}\|^2 + \eta(\boldsymbol{h})\big) = \tfrac{1}{2}\big(m + \varepsilon(\boldsymbol{h})\big)\|\boldsymbol{h}\|^2.$$

Since $m > 0$ and $\varepsilon(\boldsymbol{h}) \to 0$, $f(\boldsymbol{a}+\boldsymbol{h}) - f(\boldsymbol{a}) > 0$ for all $\boldsymbol{h} \neq \boldsymbol{0}$ with sufficiently small norm. This proves (a). Part (b) follows from (a) by considering $-f$.

To prove (c), suppose for some $\boldsymbol{h}$, $\boldsymbol{k}$ that $D^2 f_{\boldsymbol{a}}(\boldsymbol{h}) > 0$ and $D^2 f_{\boldsymbol{a}}(\boldsymbol{k}) < 0$. By (9.30),

$$f(\boldsymbol{a}+t\boldsymbol{h}) - f(\boldsymbol{a}) = \frac{t^2}{2}\left[D^2 f_{\boldsymbol{a}}(\boldsymbol{h}) + \|\boldsymbol{h}\|^2 \varepsilon(t\boldsymbol{h})\right],$$

for all $t > 0$. Therefore, $f(\boldsymbol{a}+t\boldsymbol{h}) - f(\boldsymbol{a}) > 0$ for all sufficiently small $t > 0$. Similarly, $f(\boldsymbol{a}+t\boldsymbol{k}) - f(\boldsymbol{a}) < 0$ for all sufficiently small $t > 0$. □

9.8.5 Example. Let $f(x,y,z) = x^2 + y^2 + xy + 3x + \sin^2 z$. The system

$$f_x = 2x + y + 3 = 0, \quad f_y = x + 2y = 0, \quad f_z = \sin(2z) = 0$$

has solutions $\boldsymbol{a}_n = (-2, 1, n\pi/2)$, $n \in \mathbb{Z}$. From

$$f_{xx} = f_{yy} = 2, \quad f_{zz} = 2\cos(2z), \quad f_{xy} = 1, \quad \text{and} \quad f_{xz} = f_{yz} = 0,$$

we have

$$\begin{aligned} D^2 f(h,k,\ell) &= \left(h\frac{\partial}{\partial x} + k\frac{\partial}{\partial y} + \ell\frac{\partial}{\partial z}\right)^2 f \\ &= h^2 f_{xx} + k^2 f_{yy} + \ell^2 f_{zz} + 2(hk f_{xy} + h\ell f_{xz} + k\ell f_{yz}) \\ &= 2\big(h^2 + k^2 + hk + \ell^2 \cos(2z)\big). \end{aligned}$$

Therefore,

$$D^2 f_{\boldsymbol{a}_n}(h,k,\ell) = \begin{cases} 2(h^2 + k^2 + hk + \ell^2) & \text{if } n = 2k, \\ 2(h^2 + k^2 + hk - \ell^2) & \text{if } n = 2k+1. \end{cases}$$

Since $h^2 + k^2 + hk \geq 0$ for all h, k, $\boldsymbol{a}_{2k}$ is a local minimum point and $\boldsymbol{a}_{2k+1}$ a saddle point. ◊

The second derivative test gives no information if $D^2 f_{\boldsymbol{a}} = 0$. For example, the critical point $(0,0)$ of the function $f(x,y) = x^n + y^2$, $n \geq 3$, is a saddle point if n is odd and a local minimum point if n is even.

For $n = 2$, there is a simpler version of the second derivative test:

9.8.6 Corollary. *Let $U \subseteq \mathbb{R}^2$ be open and let $f : U \to \mathbb{R}$ be C^2 on U. For a critical point (a,b) of f, set*

$$\Delta = \Delta(a,b) = \begin{vmatrix} f_{xx}(a,b) & f_{xy}(a,b) \\ f_{yx}(a,b) & f_{yy}(a,b) \end{vmatrix} = f_{xx}(a,b) f_{yy}(a,b) - f_{xy}^2(a,b).$$

(a) *If $\Delta > 0$ and $f_{xx}(a,b) > 0$, then (a,b) is a local minimum point.*

(b) *If $\Delta > 0$ and $f_{xx}(a,b) < 0$, then (a,b) is a local maximum point.*

(c) *If $\Delta < 0$, then (a,b) is a saddle point of f.*

Proof. Let

$$\alpha = f_{xx}(a,b), \quad \beta = f_{xy}(a,b), \quad \text{and} \quad \gamma = f_{yy}(a,b).$$

Then $\Delta = \alpha\gamma - \beta^2$ and

$$D^2 f_{(a,b)}(h,k) = \alpha h^2 + 2\beta hk + \gamma k^2, \quad h,\, k \in \mathbb{R}. \tag{9.31}$$

If $\alpha \neq 0$, completing the square yields

$$D^2 f_{(a,b)}(h,k) = \alpha\left[h + \frac{k\beta}{\alpha}\right]^2 + \frac{k^2(\alpha\gamma - \beta^2)}{\alpha} = \alpha\left[h + \frac{k\beta}{\alpha}\right]^2 + \frac{k^2\Delta}{\alpha}.$$

Thus if $\Delta > 0$, $\alpha > 0$, and $(h,k) \neq (0,0)$, then $D^2 f_{\boldsymbol{a}}(h,k) > 0$, hence, by the theorem, (a) holds. A similar argument proves (b).

Now suppose $\Delta < 0$. If $\alpha \neq 0$, then from (9.31)

$$D^2 f_{(a,b)}(1,0) = \alpha \quad \text{and} \quad D^2 f(a,b)(-\beta\alpha^{-1}, 1) = \frac{\Delta}{\alpha},$$

which have opposite signs. If $\gamma \neq 0$, then completing the square yields

$$D^2 f_{(a,b)}(h,k) = \gamma\left[k + \frac{h\beta}{\gamma}\right]^2 + \frac{h^2\Delta}{\gamma},$$

and one may argue similarly. (This also shows that (a) and (b) hold with f_{xx} in the statement replaced by f_{yy}.) Finally, if $\alpha = \gamma = 0$, then $\beta \neq 0$, and (9.31) shows that, again, $D^2 f_{\boldsymbol{a}}(h,k)$ has positive and negative values. This proves (c). □

9.8.7 Example. Let $f(x,y) = 3x^2y + 2xy^2 - 6xy$. Since

$$f_x(x,y) = 2y(3x + y - 3) \quad \text{and} \quad f_y(x,y) = x(3x + 4y - 6),$$

the critical points are $(0,0)$, $(2,0)$, $(0,3)$, and $(2/3,1)$.

TABLE 9.1: Values of Δ.

(a,b)	$(0,0)$	$(2,0)$	$(0,3)$	$(2/3,1)$
$f_{xx}(a,b)$	0	0	18	6
$f_{yy}(a,b)$	0	8	0	$8/3$
$f_{xy}(a,b)$	-6	6	6	2
$\Delta(a,b)$	-36	-36	-36	12

Table 9.1 shows that f has three saddle points and one local minimum point. ♢

9.8.8 Example. Let

$$f(x,y) = (cx^2 + y^2)e^{-x^2-y^2}, \quad c \neq 0,\ 1.$$

The system

$$f_x = 2xe^{-x^2-y^2}(c - cx^2 - y^2) = 0, \quad f_y = 2ye^{-x^2-y^2}(1 - cx^2 - y^2) = 0$$

has solutions $(0,0)$, $(0,\pm 1)$, and $(\pm 1, 0)$. The second partial derivatives are

$$\begin{aligned} f_{xx} &= 2e^{-x^2-y^2}\big[c - 3cx^2 - y^2 + 2x^2(cx^2 + y^2 - c)\big], \\ f_{yy} &= 2e^{-x^2-y^2}\big[1 - cx^2 - 3y^2 + 2y^2(cx^2 + y^2 - 1)\big], \quad \text{and} \\ f_{xy} &= 4xye^{-x^2-y^2}\big[cx^2 + y^2 + -c - 1\big]. \end{aligned}$$

TABLE 9.2: Values of Δ.

(a,b)	$(0,0)$	$(0,1))$	$(0,-1)$	$(1,0)$	$(-1,0)$
$f_{xx}(a,b)$	$2c$	$2(c-1)/e$	$2(c-1)/e$	$-4c/e$	$-4c/e$
$f_{yy}(a,b)$	2	$-4/e$	$-4/e$	$2(1-c)/e$	$2(1-c)/e$
$f_{xy}(a,b)$	0	0	0	0	0
$\Delta(a,b)$	$4c$	$8(1-c)/e$	$8(1-c)/e^2$	$8(c-1)/e$	$8(c-1)/e^2$

The values of Δ at the critical points (a,b) are given in Table 9.2. Assigning values to c produces a variety of local extreme points. For example, if $c > 1$, then $(0,\pm 1)$ are saddle points and the remaining critical points are local minimum points of f. ◊

Global Extrema

We now turn to the problem described at the beginning of the section, namely, to find the points in a subset E of U at which f has a maximum or a minimum. Such points, called *global extrema*, will always exist if E is closed and bounded. The following examples illustrate a common technique for finding them.

9.8.9 Example. Let

$$f(x,y) = 2x^3 - x^2 + 3y^2, \quad E := \big\{(x,y) \,:\, x^2 + y^2 \leq 1\big\}.$$

By 9.8.2, the extreme values of f occur at points on $\text{bd}(E)$ or at critical points of f in $\text{int}(E)$. Solving the system

$$f_x = 6x^2 - 2x = 0, \quad f_y = 6y = 0$$

yields the critical points $(0,0)$ and $(1/3, 0)$, which are candidates for extrema in $\text{int}(E)$. To find possible extrema on $\text{bd}(E)$ we substitute $1 - x^2$ for y^2 in the expression for f to obtain the function

$$F(x) = 2x^3 - 4x^2 + 3, \quad -1 \le x \le 1.$$

Since the only zero of $F'(x)$ in $[-1, 1]$ is $x = 0$, single-variable optimization theory gives us the additional extrema candidates $(0, \pm 1)$ and $(\pm 1, 0)$. Calculating the values of f at these six points shows that $f(0, \pm 1) = 3$ is the maximum value of f on E and $f(-1, 0) = -3$ is the minimum. ◊

9.8.10 Example. Let

$$f(x, y, z) = (x-1)^2 + (y-2)^2 + z^2, \quad E := \{(x, y, z) : x^2 + y^2 + z^2 \le 6\}.$$

The solution of the system $f_x = f_y = f_z = 0$ is $(1, 2, 0)$, at which f has minimum value zero. The maximum of f must then occur on $\text{bd}(E)$. Substituting the expression $6 - x^2 - y^2$ for z^2 in the definition of f, we obtain the function

$$F(x, y) = (x-1)^2 + (y-2)^2 + 6 - x^2 - y^2 = 11 - 2x - 4y, \quad x^2 + y^2 \le 6.$$

The system $F_x = F_y = 0$ has no solution, hence the extreme values of F must lie on the boundary $x^2 + y^2 = 6$. To find these values, let $x = \sqrt{6}\cos\theta$ and $y = \sqrt{6}\sin\theta$, so

$$F(x, y) = G(\theta) := 11 - 2\sqrt{6}\cos\theta - 4\sqrt{6}\sin\theta, \quad 0 \le \theta \le 2\pi.$$

Applying single-variable optimization techniques to G, we see that possible extreme values of F on $x^2 + y^2 = 6$ occur at points (x, y) for which $\theta = 0$ and $\theta = \arctan 2$, that is, $(x, y) = (\sqrt{6}, 0)$ and $\pm\left(\sqrt{\frac{6}{5}}, 2\sqrt{\frac{6}{5}}\right)$. Calculating the values of F at these points shows that the maximum value of f on E is

$$f\left(-\sqrt{\frac{6}{5}}, -2\sqrt{\frac{6}{5}}, 0\right) \approx 22. \qquad \lozenge$$

In the above examples, E was the closure of an open set whose boundary is a smooth surface. In many important cases, however, E itself is a surface. The surfaces we shall consider are of the form

$$E = \{\boldsymbol{x} \in U : g_1(\boldsymbol{x}) = \cdots = g_m(\boldsymbol{x}) = 0\},$$

where $U \subseteq \mathbb{R}^n$ is open, $m < n$, and the functions g_j are C^1 on U. The equations $g_j(\boldsymbol{x}) = 0$ are then called *constraints*, and E is the *constraint set.* If $f(\boldsymbol{a})$ is the maximum or minimum value of f on E, then f is said to have an *extremum at* $\boldsymbol{a}$ *subject to the constraints* $g_j = 0$.

9.8.11 Example. We find the points on the surface $z^2 - x^2y = 1$ closest to the origin. This is equivalent to minimizing $f(x, y, z) = x^2 + y^2 + z^2$ subject to the constraint $z^2 - x^2y - 1 = 0$. Since the surface is unbounded, it suffices to consider that part of the surface inside a ball with center $\mathbf{0}$. To find the minimum, we substitute $z^2 = x^2y + 1$ into f to obtain a function $F(x, y) = x^2(1+y) + y^2 + 1$ defined on an open disk containing a point at which f is minimum. The critical points of F, solutions of the system

$$F_x = 2x(1 + y) = 0, \quad F_y = x^2 + 2y = 0,$$

are $(0, 0)$, and $(\pm\sqrt{2}, -1)$. The last two are easily seen to be saddle points, while $(0, 0)$ is a local minimum point. Therefore, the minimum of f occurs at $(0, 0, \pm 1)$, hence the distance from the surface to the origin is 1. ◊

Lagrange Multipliers

In 9.8.11, it was possible to solve the constraint equation for one of the variables in terms of the others, reducing the dimension by one, thereby simplifying the problem. This is not always possible, but the implicit function theorem may be used to solve the constraint equation locally. This is the method used in the proof of the next theorem. For its statement, we use the following notational conventions, similar to those used in the proof of the implicit function theorem.

Notation. Let $m < n$ and $p := n - m$. For points $\boldsymbol{z} \in \mathbb{R}^n = \mathbb{R}^{m+p}$ we write

$$\boldsymbol{z} = (\boldsymbol{x}, \boldsymbol{y}) = (x_1, \ldots x_m, y_1, \ldots y_p), \quad \boldsymbol{x} \in \mathbb{R}^m, \ \boldsymbol{y} \in \mathbb{R}^p.$$

If $G := (g_1, \ldots, g_m) : U \to \mathbb{R}^m$, then $G(\boldsymbol{z})$ may be written as $G(\boldsymbol{x}, \boldsymbol{y})$. If G is differentiable, we define

$$G_{\boldsymbol{x}} = \begin{bmatrix} \dfrac{\partial g_1}{\partial x_1} & \cdots & \dfrac{\partial g_1}{\partial x_m} \\ \vdots & \cdots & \vdots \\ \dfrac{\partial g_m}{\partial x_1} & \cdots & \dfrac{\partial g_m}{\partial x_m} \end{bmatrix} \text{ and } G_{\boldsymbol{y}} = \begin{bmatrix} \dfrac{\partial g_1}{\partial y_1} & \cdots & \dfrac{\partial g_1}{\partial y_p} \\ \vdots & \cdots & \vdots \\ \dfrac{\partial g_m}{\partial y_1} & \cdots & \dfrac{\partial g_m}{\partial y_p} \end{bmatrix}. \quad ◊$$

9.8.12 Lagrange Multipliers. *Let $U \subseteq \mathbb{R}^n$ be open and let f, $g_j : U \to \mathbb{R}$, $j = 1, \ldots, m < n$ be C^1 functions. Set $G := (g_1, \ldots, g_m)$. Suppose that f has a global extremum at $\mathbf{c} = (\boldsymbol{a}, \mathbf{b}) \in U$ subject to the constraint $G = \mathbf{0}$. If* $\det G_{\boldsymbol{x}}(\mathbf{c}) \neq 0$, *then there exist constants $\lambda_1, \ldots, \lambda_m$ such that*

$$\nabla f(\mathbf{c}) = \sum_{i=1}^{m} \lambda_i \nabla g_i(\mathbf{c}). \tag{9.32}$$

Proof. Equation (9.32) is the system

$$\partial_j f(\mathbf{c}) = \lambda_1 \frac{\partial g_1}{\partial x_j} + \cdots + \lambda_m \frac{\partial g_m}{\partial x_j}, \quad j = 1, \ldots, m,$$

$$\partial_{j+m} f(\mathbf{c}) = \lambda_1 \frac{\partial g_1}{\partial y_j} + \cdots + \lambda_m \frac{\partial g_m}{\partial y_j}, \quad j = 1, \ldots, p,$$

which may be written in matrix form as

$$\begin{bmatrix}\lambda_1 & \cdots & \lambda_m\end{bmatrix} G_{\boldsymbol{x}}(\boldsymbol{c}) = \begin{bmatrix}\partial_1 f(\boldsymbol{c}) & \cdots & \partial_m f(\boldsymbol{c})\end{bmatrix} \tag{9.33}$$

$$\begin{bmatrix}\lambda_1 & \cdots & \lambda_m\end{bmatrix} G_{\boldsymbol{y}}(\boldsymbol{c}) = \begin{bmatrix}\partial_{m+1} f(\boldsymbol{c}) & \cdots & \partial_n f(\boldsymbol{c})\end{bmatrix}. \tag{9.34}$$

Equation (9.33) is satisfied by defining

$$\begin{bmatrix}\lambda_1 & \cdots & \lambda_m\end{bmatrix} := \begin{bmatrix}\partial_1 f(\boldsymbol{c}) & \cdots & \partial_m f(\boldsymbol{c})\end{bmatrix} G_{\boldsymbol{x}}^{-1}(\boldsymbol{c}). \tag{9.35}$$

It remains to show that (9.34) is satisfied for this choice of $\begin{bmatrix}\lambda_1 & \cdots & \lambda_m\end{bmatrix}$.

By the implicit function theorem applied to G, there is an open set $V_{\boldsymbol{b}} \subseteq \mathbb{R}^p$ containing $\boldsymbol{b}$ and a continuously differentiable mapping

$$h = (h_1, \ldots, h_m) : V_{\boldsymbol{b}} \to \mathbb{R}^m$$

such that

$$h(\boldsymbol{b}) = \boldsymbol{a} \text{ and } G\big(h(\boldsymbol{y}), \boldsymbol{y}\big) = \boldsymbol{0} \text{ for every } \boldsymbol{y} \in V_{\boldsymbol{b}}.$$

Applying the chain rule to each component equation $g_i\big(h(\boldsymbol{y}), \boldsymbol{y}\big) = 0$ yields

$$\frac{\partial g_i}{\partial x_1}\frac{\partial h_1}{\partial y_j} + \cdots + \frac{\partial g_i}{\partial x_m}\frac{\partial h_m}{\partial y_j} + \frac{\partial g_i}{\partial y_j} = 0, \quad i = 1, \ldots, m, \ j = 1, \ldots, p,$$

which may be written in matrix form as

$$\begin{bmatrix} \dfrac{\partial g_1}{\partial x_1} & \cdots & \dfrac{\partial g_1}{\partial x_m} \\ \vdots & & \vdots \\ \dfrac{\partial g_m}{\partial x_1} & \cdots & \dfrac{\partial g_m}{\partial x_m} \end{bmatrix} \begin{bmatrix} \dfrac{\partial h_1}{\partial y_1} & \cdots & \dfrac{\partial h_1}{\partial y_p} \\ \vdots & & \vdots \\ \dfrac{\partial h_m}{\partial y_1} & \cdots & \dfrac{\partial h_m}{\partial y_p} \end{bmatrix} = - \begin{bmatrix} \dfrac{\partial g_1}{\partial y_1} & \cdots & \dfrac{\partial g_1}{\partial y_p} \\ \vdots & & \vdots \\ \dfrac{\partial g_m}{\partial y_1} & \cdots & \dfrac{\partial g_1}{\partial y_p} \end{bmatrix}$$

or in the above notation as

$$G_{\boldsymbol{x}}(\boldsymbol{c}) h'(\boldsymbol{b}) = -G_{\boldsymbol{y}}(\boldsymbol{c}).$$

Multiplying the last equation on the left by $\begin{bmatrix}\partial_1 f(\boldsymbol{c}) & \cdots & \partial_m f(\boldsymbol{c})\end{bmatrix} G_{\boldsymbol{x}}^{-1}(\boldsymbol{c})$ and using (9.35), we obtain

$$\begin{bmatrix}\partial_1 f(\boldsymbol{c}) & \cdots & \partial_m f(\boldsymbol{c})\end{bmatrix} h'(\boldsymbol{b}) = - \begin{bmatrix}\lambda_1 & \cdots & \lambda_m\end{bmatrix} G_{\boldsymbol{y}}(\boldsymbol{c}). \tag{9.36}$$

Since $f\big(h(\boldsymbol{y}), \boldsymbol{y}\big)$ has a local extremum at $\boldsymbol{b}$, its partial derivatives must vanish there:

$$\frac{\partial f(\boldsymbol{c})}{\partial x_1}\frac{\partial h_1(\boldsymbol{b})}{\partial y_j} + \cdots + \frac{\partial f(\boldsymbol{c})}{\partial x_m}\frac{\partial h_m(\boldsymbol{b})}{\partial y_j} + \frac{\partial f(\boldsymbol{c})}{\partial y_j} = 0, \ j = 1, 2, \ldots, p.$$

In matrix form,

$$\begin{bmatrix}\partial_1 f(\boldsymbol{c}) & \cdots & \partial_m f(\boldsymbol{c})\end{bmatrix} h'(\boldsymbol{b}) = - \begin{bmatrix}\partial_{m+1} f(\boldsymbol{c}) & \cdots & \partial_n f(\boldsymbol{c})\end{bmatrix}. \tag{9.37}$$

Equation (9.34) now follows from (9.36) and (9.37). □

9.8.13 Example. Let $\mathbf{c}$, $\boldsymbol{x} \in \mathbb{R}^n$, $\mathbf{c} \neq \mathbf{0}$. We find the extreme values of $f(\boldsymbol{x}) := \mathbf{c} \cdot \boldsymbol{x}$ on the sphere $\|\boldsymbol{x}\| = 1$, that is, with constraint $g(\boldsymbol{x}) := \|\boldsymbol{x}\|^2 - 1 = 0$. By Lagrange multipliers, the extreme values occur at points $\boldsymbol{x}$ for which $\nabla f(\boldsymbol{x}) = \lambda \nabla g(\boldsymbol{x})$ for some $\lambda \in \mathbb{R}$. This yields the system $c_i = 2\lambda x_i$, $1 \leq i \leq n$. Squaring and adding, $\|\mathbf{c}\|^2 = 4\lambda^2 \|\boldsymbol{x}\|^2 = 4\lambda^2$, hence $2\lambda = \pm\|\mathbf{c}\|$ and $\boldsymbol{x} = \mathbf{c}/2\lambda = \pm\mathbf{c}/\|\mathbf{c}\|$. Therefore, the extreme values of f are $f\big(\pm\mathbf{c}/\|\mathbf{c}\|\big) = \pm\|\mathbf{c}\|$. ◊

The following example has an important application to directional derivatives: Let h be differentiable on $B_r(\boldsymbol{a})$. From Exercise 9.3.10, the directional derivative $D_{\boldsymbol{x}} h(\boldsymbol{a})$ of h at $\boldsymbol{a}$ in the direction of a unit vector $\boldsymbol{x}$ is $\mathbf{c} \cdot \boldsymbol{x}$, where $\mathbf{c} = \nabla h(\boldsymbol{a})$. Thus, by the example, $D_{\boldsymbol{x}} h(\boldsymbol{a})$ is maximum when $\boldsymbol{x} = \mathbf{c}/\|\mathbf{c}\|$, that is, when $\boldsymbol{x}$ is in the direction of the gradient of h.

9.8.14 Example. Let $\boldsymbol{x} = (x_1, \ldots, x_n)$, $\boldsymbol{a} = (a_1, \ldots, a_n)$, and $\mathbf{c} = (c_1, \ldots, c_n)$, where $x_j \geq 0$, $a_j > 0$, and $c_j > 0$. We find the maximum value of $f(\boldsymbol{x}) = x_1^{a_1} x_2^{a_2} \cdots x_n^{a_n}$ subject to the constraint $\mathbf{c} \cdot \boldsymbol{x} = 1$. Note that the conditions $x_j \geq 0$ and $c_j > 0$ imply that the constraint set is closed and bounded. Set $g(\boldsymbol{x}) = \mathbf{c} \cdot \boldsymbol{x} - 1$. The maximum of f occurs at points $\boldsymbol{x}$ for which $\nabla f(\boldsymbol{x}) = \lambda \nabla g(\boldsymbol{x})$ for some $\lambda \in \mathbb{R}$. This leads to the equations

$$a_j f(\boldsymbol{x}) = \lambda c_j x_j, \quad j = 1, \ldots, n. \tag{9.38}$$

Adding and using the constraint yields $a f(\boldsymbol{x}) = \lambda$, or $f(\boldsymbol{x}) = \lambda/a$, where $a = \sum_{j=1}^n a_j$. From (9.38), $a_j = a c_j x_j$ so the maximum occurs at the point

$$\left(\frac{a_1}{ac_1}, \frac{a_2}{ac_2}, \ldots, \frac{a_n}{ac_n}\right).$$

In particular, if $a_1 = \cdots = a_n = 1$ and $c_1 = \cdots = c_n = 1/c$, $c > 0$, then $f(x_1, x_2, \ldots, x_n) = x_1 x_2 \cdots x_n$ has maximum $f(c/n, \ldots, c/n) = (c/n)^n$. Thus $x_1 x_2 \cdots x_n \leq (c/n)^n$, or equivalently $(x_1 x_2 \cdots x_n)^{1/n} \leq c/n$ for all $x_j > 0$ satisfying $x_1 + \cdots + x_n = c$. Since c is arbitrary, we obtain the classic result

$$(x_1 x_2 \ldots x_n)^{1/n} \leq \frac{x_1 + x_2 + \cdots + x_n}{n}, \quad x_j \geq 0,$$

which asserts that the geometric mean of nonnegative data does not exceed the arithmetic mean. ◊

Exercises

1. In each case classify the critical point $\boldsymbol{a} := (\pi/2, \pi/2, \pi/2)$ of the function.
   (a) $(\sin x)(\sin y)(\sin z)$ (b) $(\sin x)(\cos y)(\cos z)$.
2. Show that the function $x^2 + 2y^2 + 3z^2 - xy - yz - xz$ on $\mathbb{R}^3$ has minimum value zero.

3. Find and classify the critical points of the following functions.

(a) $x^3 + 2xy + 3x^2 + y^2$ (b) $x^3 + 3x^2y^2 - 6x^2 - 12y^2$
(c) $x^2y^2 + 2/x + 2/y$ (d) $x^4 + 2y^2 - 4xy$
(e) $x^{-1} + y^{-1} + \ln(x^2 + y^2)$ (f) $x^{-1} + y^{-1} + \arctan(y/x)$
(g) $x^3 - xy^2 + x^2 - y^2$ (h) $x^4 - 2x^2 + 4y^3 - 12y$
(i) $xy - x^2y - xy^2$ (j) $x^4 - 4x^3 + 4x^2 + y^2$.

4. Find the maximum and minimum values of each of the following functions f on $\mathbb{R}^2 \setminus \{(0,0)\}$.

(a) $\dfrac{x+y}{\sqrt{x^2+y^2}}$ (b) $\dfrac{x+\sqrt{3}y}{\sqrt{x^2+y^2}}$ (c) $\dfrac{x+2y}{\sqrt{x^2+y^2}}$ (d) $\dfrac{x^2+xy}{x^2+y^2}$.

5. Show that the point (x, x^2) on the curve $y = x^2$ nearest the point $(1, 2)$ satisfies the equation $2x^3 - 3x - 1 = 0$.

In Exercises 6–9, use the method of 9.8.9 and 9.8.10.

6. Find the extreme values of the following functions on the disk $D := \{(x,y) : x^2 + y^2 \le 1\}$:
(a) $3x^2 + 2y^2 - x$ (b) $x^2 + xy - x + y^2$ (c) $\cos(xy)$ (d) $\sin(xy)$.

7. Prove that the maximum of $f(x,y) = x^2 + ay^2 + (a-1)y$ on the disk $D := \{(x,y) : x^2 + y^2 \le 1\}$ occurs on $\mathrm{bd}(D)$.

8. Let $f(x,y) = x^2 + y^2 + axy$ on the disk $D := \{(x,y) : x^2 + y^2 \le 1\}$. Prove that a maximum of f occurs on $\mathrm{bd}(D)$, and that a minimum of f occurs on $\mathrm{bd}(D)$ iff $|a| \ge 2$.

9. Show that the maximum (minimum) value of $f_n(x_1, \ldots, x_n) = \sum_{i=1}^n x_i$ on $C_1(\mathbf{0})$ is $\sqrt{n}$ $(-\sqrt{n})$.

10. Let $f(x,y) = ax^{-1} + by^{-1} + xy$, $a, b > 0$. Prove that f has a minimum on $(0, +\infty) \times (0, +\infty)$ and that the minimum value is $3(ab)^{1/3}$.

11. Consider the data points (x_i, y_i), $1 \le i \le n$, where $x_i \ne x_j$ for at least one pair of points. The *linear least squares fit* is the line $y = mx + b$ with the property that the sum of squares of the vertical distances from the data points to the line, namely, $\sum_{i=1}^n (y_i - mx_i - b)^2$, is minimum. Show that

$$m = \frac{\boldsymbol{x} \cdot \boldsymbol{y} - n\overline{x}\,\overline{y}}{\|\boldsymbol{x}\|^2 - n\overline{x}^2}, \quad \text{and} \quad b = \overline{y} - m\overline{x}, \quad \text{where}$$

$$\boldsymbol{x} = (x_1, \ldots, x_n), \quad \boldsymbol{y} = (y_1, \ldots, y_n), \quad \overline{x} := \frac{1}{n}\sum_{i=1}^n x_i, \quad \text{and} \quad \overline{y} := \frac{1}{n}\sum_{i=1}^n y_i.$$

12. Let $\boldsymbol{x}$, $\boldsymbol{a}$, $\boldsymbol{b}$, $\boldsymbol{c} \in \mathbb{R}^n$. Prove that $f(\boldsymbol{x}) := \|\boldsymbol{x}-\boldsymbol{a}\|^2 + \|\boldsymbol{x}-\boldsymbol{b}\|^2 + \|\boldsymbol{x}-\mathbf{c}\|^2$ has a minimum value and find the point at which it occurs.

In Exercises 13–28, use Lagrange multipliers.

13. Show that the maximum value of $f(\boldsymbol{x}) = x_1 x_2 \cdots x_n$ on the set $E := \{\boldsymbol{x} : x_i \geq 0 \text{ and } \sum_{i=1}^n x_i \leq 1\}$ is n^{-n}.

14. Find the maximum and minimum of $2x - 3y$ subject to the constraint $(x+1)^2 + (y-1)^2 = 1$.

15. Find the maximum and minimum of $ax^2 + 2bxy + y^2$ subject to the constraint $x^2 + y^2 = c^2$, where $abc \neq 0$ and $(a+1)^2 + 4(a - b^2) \geq 0$.

16. Show that the point (x, y, z) on the surface $x^2 + y^2 + z^2 = 1$ nearest the point $(1, 2, 3)$ is $(1/\sqrt{14}, 2/\sqrt{14}, 3/\sqrt{14})$.

17. Show that the point (x, y, z) on the surface $z = x^2 + y^2$ nearest the point $(1, 2, 3)$ satisfies the equations
$$10x^3 - 5x - 1 = 0, \quad y = 2x, \quad \text{and} \quad z = x^2 + y^2 = 5x^2.$$

18. Show that the point on the surface $x^2 + y^2 - z^2 = 1$ nearest to the point $(1, 2, 3)$ is
$$(x, 2x, 3x/(2x-1)), \quad \text{where} \quad 20x^4 - 20x^3 - 8x^2 + 4x - 1 = 0.$$

19. Show that the point on the surface $z^2 - x^2 - y^2 = 1$ nearest $(1, 2, 3)$ is
$$(x, 2x, 3x/(2x-1)), \quad \text{where} \quad 20x^4 - 20x^3 - 4x + 1 = 0.$$

20. The intersection of the surfaces $z = x^2 + y^2$ and $x + y + z = 1$ is an ellipse lying above the xy plane. Find the highest and lowest points of the ellipse.

21. Let $a, b, c > 0$. Show that the maximum and minimum values of the function $f(x, y, z) = ax + by + cz$ subject to the constraints $x^2 + z^2 = 1$, $y^2 + z^2 = 1$, $x, y, z \geq 0$, are, respectively, the maximum and minimum of the quantities c, $a + b$, and $(a + b + cd)/\sqrt{1 + d^2}$, where $d := c/(a + b)$.

22. Find the maximum and minimum values of $x + 2y + 3z$ subject to the constraints $x + y + z = 1$ and $x^2 + y^2 + z^2 = 1$.

23. Let $a > 1/3$. Show that the maximum value of xyz subject to the constraints $x + y + z = 1$ and $x^2 + y^2 + z^2 = a$ is
$$xyz = \frac{1}{27}(1 - 3t^2 + 2t^3), \quad \text{where} \quad t = \sqrt{\frac{3a-1}{2}}.$$

24. Let $\boldsymbol{x} = (x_1, \cdots, x_n)$, $\boldsymbol{a} = (a_1, \cdots, a_n) \neq \boldsymbol{0}$, and $\boldsymbol{b} = (b_1, \cdots, b_n)$. Show that the shortest distance from $\boldsymbol{b}$ to the hyperplane $\boldsymbol{a} \cdot \boldsymbol{x} = c$ is $\sqrt{2}\,|c - \boldsymbol{a} \cdot \boldsymbol{b}|\,\|\boldsymbol{a}\|^{-1}$.

25. Let $p \geq 2$. Show that the largest distance from the surface $\sum_{i=1}^{n} |x_i|^p = 1$ to the origin is $n^{(p-2)/2p}$ and the smallest distance is 1.

26. Find the distance from the point $\boldsymbol{a} = (a_1, \ldots, a_n)$ to the $(n-1)$-dimensional sphere $\|\boldsymbol{x}\| = 1$ in $\mathbb{R}^n$, where $a_j > 0$, and $\|\boldsymbol{a}\| \neq 1$.

27. Let $\boldsymbol{a} = (a_1, \ldots, a_n)$ and $\boldsymbol{b} = (b_1, \ldots, b_n)$, where $a_i,\ b_i > 0$.

    (a) Show that the minimum value of the function $\boldsymbol{a} \cdot \boldsymbol{x}$ subject to the constraint $\displaystyle\sum_{i=1}^{n} b_i/x_i = 1$, where $x_i > 0$, is $\displaystyle\left(\sum_{i=1}^{n} \sqrt{a_i b_i}\right)^2$.

    (b) Show that the minimum value of $\sum_{i=1}^{n} a_i x_i^2$ subject to the constraint in (a) is $\displaystyle\left(\sum_{i=1}^{n} a_i^{1/3} b_i^{2/3}\right)^3$.

28. Let $\boldsymbol{a} = (a_1, \ldots, a_n)$ and $\boldsymbol{b} = (b_1, \ldots, b_n)$, where $a_i,\ b_i > 0$ and $\sum_{i=1}^{n} b_i = 1$. Find the minimum value of $\boldsymbol{a} \cdot \boldsymbol{x}$ subject to the constraint $\prod_{i=1}^{n} x_j^{b_j} = 1$, where $x_i > 0$.

29. Let $U \subseteq \mathbb{R}^n$ be open and let $f : U \to \mathbb{R}$ be C^2 on U. Show that if f has a local maximum (minimum) at $\boldsymbol{a} \in U$, then $D^2 f_{\boldsymbol{a}}(\boldsymbol{h}) \leq 0\ (\geq 0)$ for all $\boldsymbol{h} \in \mathbb{R}^n$.

30. Prove the following generalization of Rolle's theorem: Let $U \subseteq \mathbb{R}^n$ be bounded and open and let $f : U \to \mathbb{R}$ be differentiable on U, continuous on $\mathrm{cl}(U)$, and constant on $\mathrm{bd}(U)$. Then $f'(\boldsymbol{u}) = \boldsymbol{0}$ for some $\boldsymbol{u} \in U$.

31. Let $U \subseteq \mathbb{R}^n$ be open and $f : U \to \mathbb{R}^n$ C^1 on U such that $J_f \neq 0$ on U. Let $\boldsymbol{a} \in U$ and let $C := C_r(\boldsymbol{a}) \subseteq U$, $r > 0$. Prove that if $\sup_C \|f(\boldsymbol{x}) - \boldsymbol{x}\| < r/2$, then the equation $f(\boldsymbol{x}) = \boldsymbol{a}$ has a solution in C.

Chapter 10

Lebesgue Measure on $\mathbb{R}^n$

The methods of Chapter 5 may be modified in a natural way to construct the Riemann integral of a function of several variables. In Section 11.1, we briefly describe how this is done. However, the main goal of the present chapter and the next is to construct the more general Lebesgue integral. The choice to develop the n-dimensional Lebesgue integral rather than the n-dimensional Riemann integral is motivated by the fact that, as an analytical tool, the former has several distinct advantages over the latter. For example, the Lebesgue theory allows the interchange of limit and integral in more general settings. Furthermore, the collection of Lebesgue integrable functions, which includes unbounded functions on unbounded domains, is significantly larger than the set of Riemann integrable functions. These advantages make the Lebesgue theory better suited for many applications, both pure and applied.

The key idea in Riemann integration on $\mathbb{R}^n$ is the partitioning of the domain of the integrand f into n-dimensional subintervals. The Riemann integral is then obtained as a limit of Riemann sums, that is, sums of function values times the volumes of the subintervals. In Lebesgue integration, it is the *range* of f rather than the domain that is partitioned into subintervals (see Figure 10.7). This still produces a partition of the domain of f; however, the sets in this partition are generally more complicated than subintervals. The Lebesgue integral is constructed by multiplying the *measure* of these sets by function values, adding the results, and then taking limits. In this chapter we construct the measure, and in the next chapter we construct the integral. The precise connection between the Riemann and Lebesgue integrals is made in Section 11.4.

10.1 General Measure Theory

In this section we give a brief description of those aspects of measure theory that will be needed to construct Lebesgue measure on $\mathbb{R}^n$. For a comprehensive treatment, see, for example, [4].

DOI: 10.1201/9781003629382-10

Sigma Fields

10.1.1 Definition. A *σ-field on a nonempty set* S is a collection $\mathcal{F}$ of subsets of S such that

(a) $S, \emptyset \in \mathcal{F}$;

(b) $A \in \mathcal{F}$ implies $A^c \in \mathcal{F}$;

(c) $A_k \in \mathcal{F}$, $k \in \mathbb{N}$, implies $\bigcup_k A_k \in \mathcal{F}$. ◊

Part (c) of the definition says that $\mathcal{F}$ is closed under countable unions. By DeMorgan's law,

$$\bigcap_k A_k = \left(\bigcup_k A_k^c\right)^c,$$

hence part (b) implies that $\mathcal{F}$ is also closed under countable intersections.

The collection of *all* subsets of S and the collection $\{\emptyset, S\}$ are simple examples of σ-fields. The following examples are somewhat more interesting.

10.1.2 Example. If $\mathcal{A}$ is an arbitrary collection of subsets of S, then the *σ-field generated by* $\mathcal{A}$ is the intersection $\sigma(\mathcal{A})$ of all σ-fields containing $\mathcal{A}$. It is the smallest σ-field containing $\mathcal{A}$ in the sense that if $\mathcal{F}$ is a σ-field containing $\mathcal{A}$ then $\mathcal{F}$ contains $\sigma(\mathcal{A})$. In the special case where $\mathcal{A} = \{A_1, A_2, \ldots\}$ is a countable partition of S, $\sigma(\mathcal{A})$ is simply the collection $\mathcal{F}$ of all unions of members of $\mathcal{A}$. Indeed, $\mathcal{F}$ is clearly closed under countable unions, and the calculation

$$\left(\bigcup_{k\in F} A_k\right)^c = \bigcup_{k\in F^c} A_k, \quad F \subseteq \mathbb{N}$$

shows that $\mathcal{F}$ is closed under complements. Thus, by minimality, $\sigma(\mathcal{A}) = \mathcal{F}$. ◊

10.1.3 Example. If $\mathcal{F}$ is a σ-field on S and $E \subseteq S$, then the collection

$$\mathcal{F}_E := \{A \cap E : A \in \mathcal{F}\}$$

is a σ-field of subsets of E. Moreover, $\mathcal{F}_E \subseteq \mathcal{F}$ iff $E \in \mathcal{F}$. (See Exercise 2.) ◊

Measure on a Sigma Field

10.1.4 Definition. A *measure* on a σ-field $\mathcal{F}$ of subsets of S is a function $\mu : \mathcal{F} \to [0, +\infty]$ such that $\mu(\emptyset) = 0$ and μ has the *additivity property*

$$\mu\left(\bigcup_k A_k\right) = \sum_k \mu(A_k)$$

for any finite or infinite sequence of pairwise disjoint sets $A_k \in \mathcal{F}$. The extended real number $\mu(A)$ is called the measure of A. ◊

10.1.5 Example. Let $\{p_k\}$ be a sequence of nonnegative real numbers. Define

$$\mu(E) = \sum_{k \in E} p_k, \quad E \subseteq \mathbb{N},$$

where the sum may be infinite. (By convention, the sum over the empty set is zero.) It is not difficult to show that μ is a measure on the σ-field of all subsets of $\mathbb{N}$.

In the special case $p_k = 1$ for all k, $\mu(E)$ counts the number of elements in E if E is a finite set, and $\mu(E) = +\infty$ otherwise. In this case, μ is called a *counting measure*. ◊

10.1.6 Proposition. *Let μ be a measure on a σ-field $\mathcal{F}$ and $A_1, A_2, \cdots \in \mathcal{F}$.*

(a) *If $A_1 \subseteq A_2$, then $\mu(A_1) \le \mu(A_2)$* (monotonicity).

(b) $\mu(\bigcup_k A_k) \le \sum_k \mu(A_k)$ (subadditivity).

(c) $\mu(A_1) + \mu(A_2) = \mu(A_1 \cup A_2) + \mu(A_1 \cap A_2)$ (inclusion-exclusion).

(d) *If $A_k \uparrow A$, then $\mu(A_k) \uparrow \mu(A)$* (continuity from below).

(e) *If $A_k \downarrow A$ and $\mu(A_1) < +\infty$, then $\mu(A_k) \downarrow \mu(A)$* (continuity from above).

Proof. (a) By additivity, $\mu(A_2) = \mu(A_2 \setminus A_1) + \mu(A_1) \ge \mu(A_1)$.

(b) Write

$$\bigcup_k A_k = A_1 \cup (A_2 \cap A_1^c) \cup \cdots \cup (A_m \cap A_1^c \cap \cdots \cap A_{m-1}^c) \cup \cdots .$$

Since the sets in the union on the right are pairwise disjoint, by countable additivity and monotonicity

$$\mu\bigg(\bigcup_k A_k\bigg) = \mu(A_1) + \sum_{m \ge 2} \mu(A_1^c \cap \cdots \cap A_{m-1}^c \cap A_m) \le \sum_{m \ge 1} \mu(A_m).$$

(c) Since $A_1 \cup A_2$ is the union of the pairwise disjoint sets $A_1 \cap A_2^c$, $A_1 \cap A_2$, and $A_2 \cap A_1^c$, additivity implies that

$$\mu(A_1 \cup A_2) = \mu(A_1 \cap A_2^c) + \mu(A_1 \cap A_2) + \mu(A_2 \cap A_1^c).$$

Similarly,

$$\mu(A_1) + \mu(A_2) = \mu(A_1 \cap A_2^c) + 2\mu(A_2 \cap A_1) + \mu(A_2 \cap A_1^c).$$

It follows that $\mu(A_1 \cup A_2) = +\infty$ iff $\mu(A_1) + \mu(A_2) = +\infty$, which proves (c) in the infinite case. In the finite case, simply subtract the above equations to get (c).

(d) This is clear if some A_k has infinite measure, so assume $\mu(A_k) < +\infty$ for all k. Set $A_0 = \emptyset$ and $E_k = A_k \setminus A_{k-1}$. The sets E_k are pairwise disjoint, $A = \bigcup_{k=1}^{\infty} E_k$, and $\mu(E_k) = \mu(A_k) - \mu(A_{k-1})$, hence by additivity

$$\mu(A) = \sum_{k=1}^{\infty} \mu(E_k) = \lim_n \sum_{k=1}^{n} \big[\mu(A_k) - \mu(A_{k-1})\big] = \lim_n \mu(A_n).$$

(e) Note that $A_1 \setminus A_k \uparrow A_1 \setminus A$, hence, by (d),

$$\mu(A_1) - \mu(A) = \mu(A_1 \setminus A) = \lim_k \mu(A_1 \setminus A_k) = \mu(A_1) - \lim_k \mu(A_k). \qquad \square$$

Exercises

For the following exercises, $\mathcal{F}$ is a σ-field of subsets of a set S and μ is a measure on $\mathcal{F}$.

1. Find an example which shows that the hypothesis $\mu(A_1) < +\infty$ in 10.1.6(e) cannot be removed.

2. Verify that the collection $\mathcal{F}_E$ in 10.1.3 is a σ-field.

3. Let $A, B \in \mathcal{F}$ with $\mu(B) = 0$. Show that $\mu(A \cup B) = \mu(A \setminus B) = \mu(A)$.

4. Let $A_k, B_k \in \mathcal{F}$ and let s denote the sum $\sum_k \mu(A_k \setminus B_k)$. Prove that

   (a) $\mu\Big(\bigcup_k A_k \setminus \bigcup_k B_k\Big) \le s$ (b) $\mu\Big(\bigcap_k A_k \setminus \bigcap_k B_k\Big) \le s$.

5. (General inclusion-exclusion principle). Let $\mu(A_1 \cup \cdots \cup A_n) < +\infty$. Prove that for $n \ge 2$

$$\mu(A_1 \cup \cdots \cup A_n) = \sum_{i=1}^{n} \mu(A_i) - \sum_{1 \le i < j \le n}^{n} \mu(A_i \cap A_j)$$
$$+ \sum_{1 \le i < j < k \le n}^{n} \mu(A_i \cap A_j \cap A_k) - \cdots + (-1)^{n-1}\mu(A_1 \cap \cdots \cap A_n).$$

   Hint. Use induction on n. The case $n = 2$ is 10.1.6(c).

6. For a sequence of sets $A_k \in \mathcal{F}$, define

$$\liminf_k A_k = \bigcup_{k=1}^{\infty} \bigcap_{j=k}^{\infty} A_j \quad \text{and} \quad \limsup_k A_k = \bigcap_{k=1}^{\infty} \bigcup_{j=k}^{\infty} A_j.$$

   Prove the following:

   (a) $\liminf_k A_k \subseteq \limsup_k A_k$

(b) $\mu\big(\liminf_k A_k\big) \le \liminf_k \mu(A_k)$

(c) $\mu\big(\limsup_k A_k\big) \ge \limsup_k \mu(A_k)$ if $\mu\left(\bigcup_k A_k\right) < +\infty$

(d) $\mu\big(\limsup_k A_k\big) = 0$ if $\sum_k \mu(A_k) < +\infty$.

7. Let $\{E_k\}$ be a sequence in $\mathcal{F}$, $m \in \mathbb{N}$, and let A denote the set of all $x \in S$ such that $x \in E_k$ for finitely many and at least m values of k. Prove that $A \in \mathcal{F}$ and

$$\mu(A) \le \frac{1}{m} \sum_{k=1}^{\infty} \mu(A \cap E_k).$$

8. Let $\{E_k\}$ be a sequence in $\mathcal{F}$, $m \in \mathbb{N}$, and let B denote the set of all $x \in S$ such that $x \in E_k$ for at most m values of k. Prove that $B \in \mathcal{F}$ and

$$\mu(B) \ge \frac{1}{m} \sum_{k=1}^{\infty} \mu(B \cap E_k).$$

9. Let $\mathcal{E}$ be a collection of pairwise disjoint members of $\mathcal{F}$, and let $A \in \mathcal{F}$. Show that $\mu(A \cap E) > 0$ for at most countably many members of $\mathcal{E}$. *Hint.* Consider

$$\mathcal{E}_m := \{E \in \mathcal{E} : \mu(A \cap E) \ge 1/m\}, \quad m \in \mathbb{N}.$$

10.2 Lebesgue Outer Measure

10.2.1 Definition. An *n-dimensional interval* in $\mathbb{R}^n$ is a Cartesian product

$$I = A_1 \times A_2 \times \cdots \times A_n,$$

where each A_j is an interval in $\mathbb{R}$. If I is bounded, then the n-dimensional volume $|I|$ of I is defined by

$$|I| := \prod_{j=1}^{n} |A_j| = \prod_{j=1}^{n} (b_j - a_j),$$

where $a_j \le b_j$ are the endpoints of A_j. If $A_j = [a_j, b_j)$ for each j, then I is said to be *half-open*. We denote by

- $\mathcal{I}$ the collection of all bounded intervals in $\mathbb{R}^n$,
- $\mathcal{H}$ the collection of all bounded half-open intervals in $\mathbb{R}^n$,
- $\mathcal{O}$ the collection of all bounded open intervals in $\mathbb{R}^n$,

- $\mathcal{C}$ the collection of all bounded closed intervals in $\mathbb{R}^n$. ◊

Note that each of the above collections is closed under the formation of nonempty finite intersections.

10.2.2 Lemma. *Let $I, I_1, \ldots, I_m \in \mathcal{H}$.*

(a) *If $I_1, \ldots, I_m$ are pairwise disjoint and $I = \bigcup_{j=1}^m I_j$, then $|I| = \sum_{j=1}^m |I_j|$.*

(b) *If $I \subseteq \bigcup_{j=1}^m I_j$, then $|I| \le \sum_{j=1}^m |I_j|$.*

(c) *If $I_1, \ldots, I_m$ are pairwise disjoint and $I \supseteq \bigcup_{j=1}^m I_j$, then $|I| \ge \sum_{j=1}^m |I_j|$.*

Proof. For ease of notation, we prove the lemma for the case $n = 2$, in which case the intervals are half-open rectangles. Let $I = [a, b) \times [c, d)$. We may

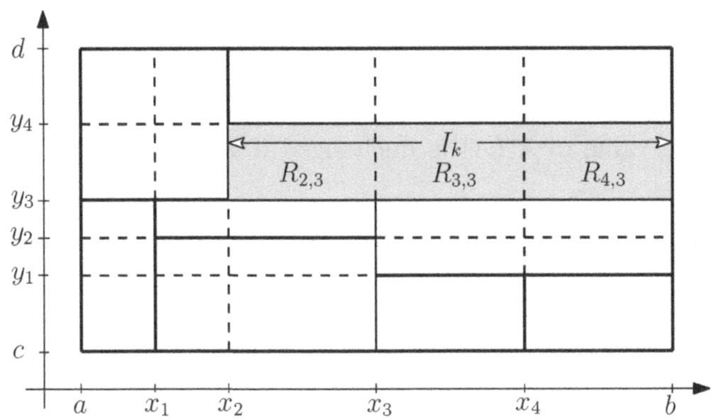

FIGURE 10.1: Pairwise disjoint interval grid.

assume in (b) that $I = \bigcup_{j=1}^m I_j$, otherwise we could replace I_j by $I_j \cap I$. Thus in each case the rectangles I_j are contained in I, hence the coordinates of their vertices form partitions

$$\{x_0 := a \le x_1 \le \ldots \le x_p := b\} \text{ and } \{y_0 := c \le y_1 \le \ldots \le y_q := d\}$$

of $[a, b]$ and $[c, d]$, respectively. These partitions generate a grid of subrectangles $R_{i,j} = [x_i, x_{i+1}) \times [y_j, y_{j+1})$ with union I such that each rectangle I_k is a union of subrectangles $R_{i,j}$. Case (a) is depicted in Figure 10.1; the rectangles I_k in the figure are shown with solid boundaries, and the dashed lines are the extensions of these boundaries. Since

$$b - a = \sum_{i=0}^{p-1} (x_{i+1} - x_i) \text{ and } d - c = \sum_{j=1}^{q-1} (y_{j+1} - y_j),$$

we have

$$|I| = \left[\sum_{i=0}^{p-1} (x_{i+1} - x_i)\right]\left[\sum_{j=0}^{q-1} (y_{j+1} - y_j)\right] = \sum_{i=0}^{p-1} \sum_{j=0}^{q-1} |R_{ij}|. \tag{10.1}$$

Similarly, $|I_k| = \sum_{\{(i,j):R_{i,j}\subseteq I_k\}} |R_{ij}|$, hence

$$\sum_k |I_k| = \sum_k \sum_{\{(i,j):R_{i,j}\subseteq I_k\}} |R_{ij}|. \tag{10.2}$$

We now compare (10.1) and (10.2). In part (a), every $R_{i,j}$ is contained in exactly one I_k, hence $|I| = \sum_{k=1}^m |I_k|$. In (b), a rectangle $R_{i,j}$ could be contained in more than one I_k, so $|I| \le \sum_{k=1}^m |I_k|$. Finally, in (c) not every $R_{i,j}$ is necessarily contained in an I_k, hence $|I| \ge \sum_{k=1}^m |I_k|$. □

10.2.3 Definition. The *Lebesgue outer measure* of a subset A of $\mathbb{R}^n$ is defined by

$$\lambda^*(A) = \lambda_n^*(A) := \inf\left\{\sum\nolimits_j |I_j| \,:\, I_j \in \mathcal{I} \text{ and } \bigcup\nolimits_j I_j \supseteq A\right\}. \qquad \Diamond$$

10.2.4 Remark. The number of intervals I_j covering A in the definition of $\lambda^*(A)$ may be finite or infinite. Of course, every bounded subset of $\mathbb{R}^n$ has a covering by a finite many I_j's. By slightly adjusting endpoints, one may show that the value of $\lambda^*(A)$ is unchanged if $\mathcal{I}$ is replaced by $\mathcal{H}$, $\mathcal{O}$, or $\mathcal{C}$. (Exercise 1.) ◊

10.2.5 Proposition. *Lebesgue outer measure on $\mathbb{R}^n$ has the following properties:*

(a) $0 \le \lambda^*(A) \le +\infty$.

(b) $\lambda^*(\emptyset) = 0$.

(c) $\lambda^*(I) = |I|$ *for each* $I \in \mathcal{I}$.

(d) *If* $A \subseteq B$, *then* $\lambda^*(A) \le \lambda^*(B)$ (monotonicity).

(e) $\lambda^*\left(\bigcup_k A_k\right) \le \sum_k \lambda^*(A_k)$ (subadditivity).

(f) *If* $I, J \in \mathcal{H}$ *and* $I \cap J = \emptyset$, *then* $\lambda^*(I \cup J) = |I| + |J|$.

Proof. Parts (a) and (d) follow directly from the definition, and (b) follows from the observation that $\emptyset$ may be covered by a single interval of arbitrarily small volume.

To prove (c), note first that, because $\{I\}$ is a covering, $\lambda^*(I) \le |I|$. For the reverse inequality, let $\varepsilon > 0$ and choose a closed, bounded interval $J \subseteq I$ such that $|J| > |I| - \varepsilon$. Let $\{I_k\}$ be any sequence of intervals covering I. By 10.2.4, we may take $I_k \in \mathcal{O}$. Let $\{J_k\}$ be a sequence in $\mathcal{H}$ such that $I_k \subseteq J_k$ and $|J_k| < |I_k| + \varepsilon/2^k$ (Figure 10.2). Since J is compact, there exists an m such that

$$J \subseteq I_1 \cup \cdots \cup I_m \subseteq J_1 \cup \cdots \cup J_m.$$

Therefore,

$$|I| - \varepsilon < |J| \le |J_1| + \cdots + |J_m| \le \varepsilon + \sum_{k=1}^{\infty} |I_k|,$$

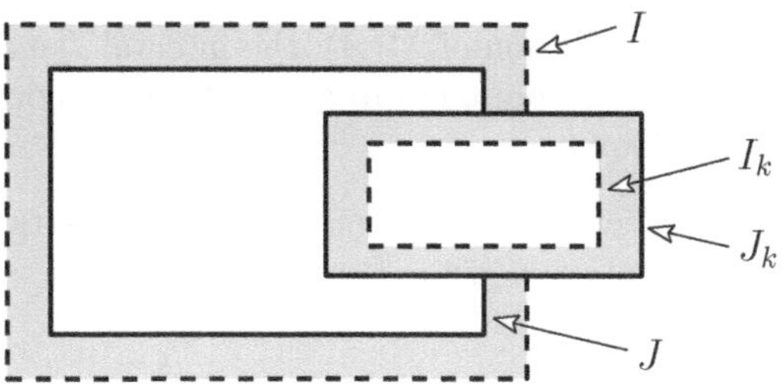

FIGURE 10.2: The coverings $\{I_k\}$ and $\{J_k\}$.

the second inequality by 10.2.2(b). Letting $\varepsilon \to 0$, we have $|I| \le \sum_{k=1}^{\infty} |I_k|$. Therefore, $|I| \le \lambda^*(I)$.

For (e), we may assume that $\lambda^*(A_k) < +\infty$ for all k. Let $\varepsilon > 0$ and for each k choose a sequence $\{I_{k,j}\}_{j=1}^{\infty}$ in $\mathcal{I}$ such that

$$A_k \subseteq \bigcup_{j=1}^{\infty} I_{k,j} \quad \text{and} \quad \sum_{j=1}^{\infty} \lambda^*(I_{k,j}) \le \lambda^*(A_k) + \frac{\varepsilon}{2^k}.$$

Since the countable collection $\{I_{k,j} : k, j = 1, 2, \ldots\}$ covers $\bigcup_{k=1}^{\infty} A_k$,

$$\lambda^*\left(\bigcup_{k=1}^{\infty} A_k\right) \le \sum_{k=1}^{\infty}\sum_{j=1}^{\infty} \lambda^*(I_{k,j}) \le \sum_{k=1}^{\infty} \lambda^*(A_k) + \varepsilon.$$

Since ε was arbitrary, (e) follows.

For (f), let $I \cup J \subseteq \bigcup_{k=1}^{m} I_k$, where $I_k \in \mathcal{H}$. Since $I_k \supseteq (I_k \cap I) \cup (I_k \cap J)$, 10.2.2(c) shows that $|I_k| \ge |I_k \cap I| + |I_k \cap J|$. Therefore, by (c),

$$\sum_{k=1}^{m} |I_k| \ge \sum_{k=1}^{m} |I_k \cap I| + \sum_{k=1}^{m} |I_k \cap J| \ge \lambda^*(I) + \lambda^*(J) = |I| + |J|,$$

Taking the infimum we have $\lambda^*(I \cup J) \ge |I| + |J|$. The reverse inequality follows from (e). □

Exercises

1. Prove the assertions in 10.2.4. More generally prove the following:

   Let $\mathcal{J}$ be a collection of bounded intervals with the property that for each bounded interval I and each $\varepsilon > 0$, there exists $J \in \mathcal{J}$ containing I such that $|J| < |I| + \varepsilon$. For $A \subseteq \mathbb{R}^n$, define

   $$\alpha(A) := \inf\left\{\sum_k |J_k| \ : \ J_k \in \mathcal{J} \text{ and } \bigcup_k J_k \supseteq A\right\}.$$

   Then $\lambda^*(A) = \alpha(A)$.

2. Prove that in the definition of $\lambda^*(A)$, the interval $\mathcal{I}$ may be replaced by the collection $\mathcal{I}_r$ of all bounded intervals I whose coordinate intervals have rational endpoints.

3. Let $\mathcal{U}$ be the collection of all bounded open subsets of $\mathbb{R}$ and $\mathcal{K}$ the collection of all compact sets. Prove that $\lambda^*(A)$ equals each of the following:

   (a) $\inf\left\{\sum_j \lambda^*(U_j) : U_j \in \mathcal{U} \text{ and } \bigcup_j U_j \supseteq A\right\}$

   (b) $\inf\left\{\sum_j \lambda^*(K_j) : K_j \in \mathcal{K} \text{ and } \bigcup_j K_j \supseteq A\right\}$.

4. Show that Lebesgue outer measure is *translation invariant*, that is,

$$\lambda^*(A + \boldsymbol{x}) = \lambda^*(A) \text{ for every } A \subseteq \mathbb{R}^n \text{ and } \boldsymbol{x} \in \mathbb{R}^n,$$

   where $A + \boldsymbol{x} := \{\boldsymbol{a} + \boldsymbol{x} : \boldsymbol{a} \in A\}$.

5. Show that Lebesgue outer measure has the *reflection property*

$$\lambda^*(-A) = \lambda^*(A) \text{ for every } A \subseteq \mathbb{R}^n,$$

   where $-A := \{\boldsymbol{x} : -\boldsymbol{x} \in A\}$.

6. Show that Lebesgue outer measure has the *dilation property*

$$\lambda^*(rA) = |r|^n \lambda^*(A) \text{ for every } A \subseteq \mathbb{R}^n \text{ and } r \in \mathbb{R},$$

   where $rA := \{r\boldsymbol{x} : \boldsymbol{x} \in A\}$.

10.3 Lebesgue Measure

By subadditivity of outer measure,

$$\lambda^*(C) \leq \lambda^*(C \cap E) + \lambda^*(C \cap E^c)$$

for all subsets E and C of $\mathbb{R}^n$. The following definition singles out those sets E that also satisfy the reverse inequality for all sets C.

10.3.1 Definition. A subset E of $\mathbb{R}^n$ is said to be *Lebesgue measurable* if

$$\lambda^*(C) \geq \lambda^*(C \cap E) + \lambda^*(C \cap E^c) \tag{10.3}$$

for all subsets C of $\mathbb{R}^n$. The collection of all Lebesgue measurable subsets of $\mathbb{R}^n$ is denoted by $\mathcal{M} = \mathcal{M}(\mathbb{R}^n)$. The restriction of λ^* to $\mathcal{M}$ is called *Lebesgue measure on* $\mathbb{R}^n$ and is denoted by $\lambda = \lambda_n$. Any particular set C satisfying (10.3) is called a *test set* for E. ◊

Thus if C is a test set for E, then $\lambda^*(C) = \lambda^*(C \cap E) + \lambda^*(C \cap E^c)$): the set E "splits" the outer measure of C.

10.3.2 Theorem. *$\mathcal{M}$ is a sigma field containing all sets of outer measure zero and λ is a measure on $\mathcal{M}$.*

Proof. Clearly, $\emptyset$, $\mathbb{R}^n \in \mathcal{M}$, and since E and E^c appear symmetrically in (10.3), $E^c \in \mathcal{M}$ iff $E \in \mathcal{M}$. If $\lambda^*(E) = 0$, then, by monotonicity,

$$\lambda^*(C \cap E) + \lambda^*(C \cap E^c) \le \lambda^*(E) + \lambda^*(C \cap E^c) = \lambda^*(C \cap E^c) \le \lambda^*(C),$$

hence $E \in \mathcal{M}$. Therefore, $\mathcal{M}$ contains all sets of Lebesgue outer measure 0.

It remains to show that, for a sequence $\{E_k\}$ in $\mathcal{M}$, $\bigcup_{k=1}^\infty E_k \in \mathcal{M}$ and furthermore that $\lambda^*\left(\bigcup_{k=1}^\infty E_k\right) = \sum_{k=1}^\infty \lambda^*(E_k)$ if the sets E_k are pairwise disjoint. This is accomplished in the following four steps:

I. *If E, $F \in \mathcal{M}$, then $E \cup F$, $E \cap F \in \mathcal{M}$.*

⟦ To show that $E \cup F \in \mathcal{M}$, take any set C as a test set for E and take $C \cap E^c$ as a test set for F to obtain

$$\lambda^*(C) = \lambda^*(C \cap E) + \lambda^*(C \cap E^c) \text{ and}$$
$$\lambda^*(C \cap E^c) = \lambda^*(C \cap E^c \cap F) + \lambda^*(C \cap E^c \cap F^c).$$

Combining these and using subadditivity,

$$\begin{aligned}\lambda^*(C) &= \lambda^*(C \cap E) + \lambda^*(C \cap E^c \cap F) + \lambda^*(C \cap E^c \cap F^c) \\ &\ge \lambda^*\big[(C \cap E) \cup (C \cap E^c \cap F)\big] + \lambda^*(C \cap E^c \cap F^c).\end{aligned} \qquad (10.4)$$

Since $\big(C \cap E\big) \cup \big(C \cap E^c \cap F\big) = C \cap (E \cup F)$, by (10.4)

$$\begin{aligned}\lambda^*(C) &\ge \lambda^*\big[C \cap (E \cup F)\big] + \lambda^*\big[C \cap E^c \cap F^c\big] \\ &= \lambda^*\big[C \cap (E \cup F)\big] + \lambda^*\big[C \cap (E \cup F)^c\big].\end{aligned}$$

This shows that $E \cup F \in \mathcal{M}$. That $E \cap F \in \mathcal{M}$ follows from De Morgan's law $E \cap F = (E^c \cup F^c)^c$. ⟧

II. *If $C \subseteq \mathbb{R}^n$ and E, $F \in \mathcal{M}$ with $E \cap F = \emptyset$, then*

$$\lambda^*\big(C \cap (E \cup F)\big) = \lambda^*(C \cap E) + \lambda^*(C \cap F).$$

⟦ Use $C \cap (E \cup F)$ as a test set for E to obtain

$$\begin{aligned}\lambda^*\big[C \cap (E \cup F)\big] &= \lambda^*\big[C \cap (E \cup F) \cap E\big] + \lambda^*\big[C \cap (E \cup F) \cap E^c\big] \\ &= \lambda^*(C \cap E) + \lambda^*(C \cap F).\end{aligned}$$ ⟧

III. *If the sets E_k are pairwise disjoint and $F := \bigcup_k E_k$, then $F \in \mathcal{M}$ and $\lambda(F) = \sum_k \lambda(E_k)$.*

⟦ Set $F_k = \bigcup_{j=1}^k E_j$ and let $C \subseteq \mathbb{R}^n$. By steps I and II and induction, $F_k \in \mathcal{M}$ and

$$\lambda^*(C \cap F_k) = \sum_{j=1}^{k} \lambda^*(C \cap E_j).$$

Thus, by monotonicity,

$$\lambda^*(C) = \lambda^*(C \cap F_k) + \lambda^*(C \cap F_k^c) \geq \sum_{j=1}^{k} \lambda^*(C \cap E_j) + \lambda^*(C \cap F^c).$$

Since k was arbitrary, by subadditivity

$$\lambda^*(C) \geq \sum_{j=1}^{\infty} \lambda^*(C \cap E_j) + \lambda^*(C \cap F^c) \geq \lambda^*(C \cap F) + \lambda^*(C \cap F^c) \geq \lambda^*(C).$$

The inequalities are therefore equalities, which shows that $F \in \mathcal{M}$. Taking $C = F$ verifies the second assertion of III. ⟧

IV. $\displaystyle\bigcup_{k=1}^{\infty} E_k \in \mathcal{M}$.

⟦ Use I, III and $\bigcup_{k=1}^{\infty} E_k = E_1 \cup (E_2 \cap E_1^c) \cup (E_3 \cap E_1^c \cap E_2^c) \cup \ldots$. ⟧ □

10.3.3 Definition. A set E is said to have *(Lebesgue) measure zero* if $\lambda(E) = 0$. A property $P(\boldsymbol{x})$ depending on points $\boldsymbol{x} \in \mathbb{R}^n$ is said to hold *almost everywhere (a.e.)* or *for almost all* $\boldsymbol{x}$ if the set of all $\boldsymbol{x}$ for which $P(\boldsymbol{x})$ is false has measure zero. ◊

For example, the Dirichlet function is zero a.e. More generally, if $E \in \mathcal{M}$ then $\mathbf{1}_E = 0$ a.e. iff $\lambda(E) = 0$.

By subadditivity, a countable union of sets of measure zero has measure zero. Since a point has measure zero, it follows that every countable set has measure zero. In particular, $\mathbb{Q}^n$ has measure zero. The following is an example of an uncountable set with measure zero.

10.3.4 Example. (Cantor ternary set). Remove from $I_{0,1} := [0, 1]$ the "middle third" open interval $(1/3, 2/3)$, leaving closed intervals $I_{1,1}$ and $I_{1,2}$ with union E_1 and total length $2/3$. Next, remove from each of $I_{1,1}$ and $I_{1,2}$ the middle third open interval, leaving closed intervals $I_{2,1}$, $I_{2,2}$, $I_{2,3}$, and $I_{2,4}$ with union E_2 and total length $4/9 = (2/3)^2$. By induction, one obtains a decreasing sequence of closed sets $E_k = \bigcup_{j=1}^{2^k} I_{k,j}$ such that, by subadditivity, $\lambda^*(E_k) \leq (2/3)^k$. If E denotes the intersection of these sets, then E is closed and, by monotonicity, $\lambda^*(E) \leq (2/3)^k$ for all k. Therefore, $\lambda^*(E) = 0$.

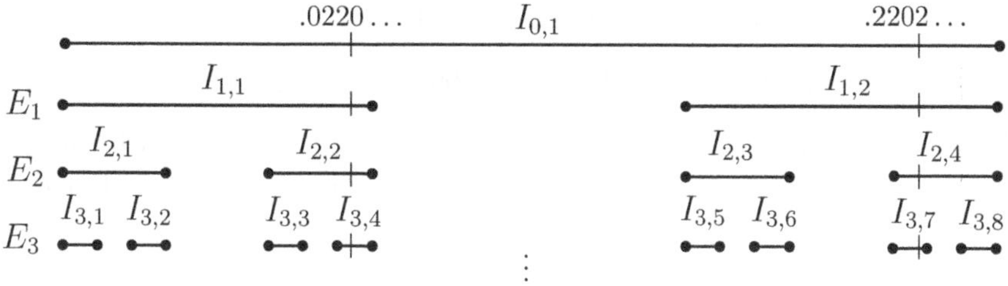

FIGURE 10.3: Middle thirds construction.

To show that E is uncountable, we use the fact that every real number $x \in [0,1]$ has both *ternary* and *binary* representations

$$x = .d_1d_2\ldots(\text{ternary}) = \sum_{k=1}^{\infty} d_k 3^{-k}, \quad \text{where} \quad d_k \in \{0,1,2\},$$

$$x = .e_1e_2\ldots(\text{binary}) \ = \sum_{k=1}^{\infty} e_k 2^{-k}, \quad \text{where} \quad e_k \in \{0,1\}.$$

These are obvious analogs of the decimal representation of a real number (see Exercise 6.1.14). As with decimal representations, there is some ambiguity; for example, $1/3 = .1000\ldots = .0222\ldots$ (ternary). Now observe that if $d_k = 0$ or

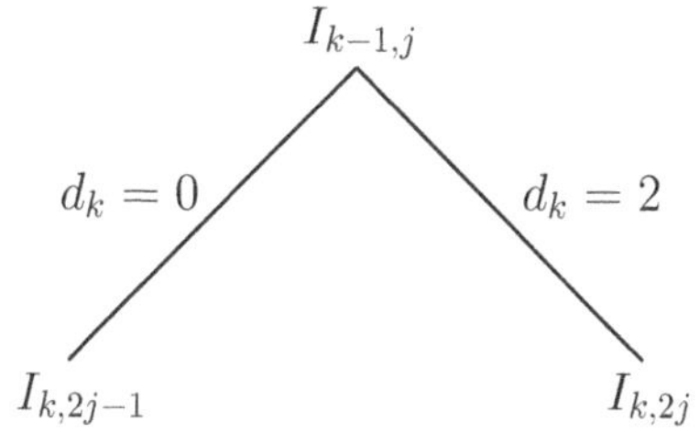

FIGURE 10.4: $\boldsymbol{x \in I_{k-1,j} \Rightarrow x \in I_{k,2j-1+d_k/2}}$.

2 for all k in the above ternary representation, then $x \in E$. For example,

$$.0220\ldots \in I_{1,1} \cap I_{2,2} \cap I_{3,4} \cap I_{4,7} \cap \cdots$$

and

$$.2202\ldots \in I_{1,2} \cap I_{2,4} \cap I_{3,7} \cap I_{4,14} \cap \cdots$$

(see Figure 10.3). In general, if $x \in I_{k-1,j}$, then $x \in I_{k,2j-1+d_k/2}$. Conversely, let $x \in E$. Since $x \in E_1$, we may choose $d_1 = 0$ or 2. Similarly, since $x \in E_2$, we may choose $d_2 = 0$ or 2, etc. Continuing in this manner, we see that every member of E has a (unique) ternary representation with digits 0 or 2.

Now define $\varphi : E \to [0,1]$ by

$$\varphi\big(.d_1d_2\ldots(\text{ternary})\big) = .e_1e_2\ldots(\text{binary}), \quad \text{where } d_k \in \{0,2\} \text{ and } e_k = d_k/2.$$

The function φ is not one-to-one; for example,

$$\varphi(.0222\ldots) = .0111\ldots = .1000\ldots = \varphi(.2000\ldots).$$

However, by removing from E the countable set of all numbers with ternary representations having a tail end of zeros, these being necessarily rational, we obtain a set F on which φ is one-to-one. Since $\varphi(F) = (0, 1)$, it follows that E is uncountable. ◊

We show in the next section that intervals, open sets, and closed sets are Lebesgue measurable. It follows that countable unions and intersections of these sets are also Lebesgue measurable. The reader may well ask if there are any subsets of $\mathbb{R}^n$ that are *not* Lebesgue measurable. The answer is that there are many, but their construction is surprisingly intricate. The following is an example for the case $n = 1$.

10.3.5 Example. (A non-measurable set). Consider sets of the form $x + \mathbb{Q}$, $x \in \mathbb{R}$. We claim that if $(x + \mathbb{Q}) \cap (y + \mathbb{Q}) \neq \emptyset$, then $x + \mathbb{Q} = y + \mathbb{Q}$. To see this, choose $z \in (x + \mathbb{Q}) \cap (y + \mathbb{Q})$, say $z = x + r_1 = y + r_2$, $r_1, r_2 \in \mathbb{Q}$. Then, for any $r \in \mathbb{Q}$,

$$x + r = y + r_2 - r_1 + r \in y + \mathbb{Q} \quad \text{and} \quad y + r = x + r_1 - r_2 + r \in x + \mathbb{Q},$$

hence $x + \mathbb{Q} = y + \mathbb{Q}$. It follows that every real number is in exactly one of the sets $x + \mathbb{Q}$. Now form a set E by choosing exactly one number in each of the distinct sets $x + \mathbb{Q}$.[1] For each $x \in \mathbb{R}$, the set $E \cap (x + \mathbb{Q})$ has a single member, hence $x = y + r$ for unique $y \in E$ and $r \in \mathbb{Q}$. Thus $\mathbb{R}$ may be expressed as a disjoint union

$$\mathbb{R} = \bigcup_{k=1}^{\infty} (r_k + E), \tag{10.5}$$

where $\{r_1, r_2, \ldots\}$ is an enumeration of $\mathbb{Q}$.

Suppose, for a contradiction, that E is Lebesgue measurable. Then $\lambda(E) > 0$; otherwise, by (10.5), translation invariance (Exercise 1), and countable additivity, $\mathbb{R}$ would have measure zero. On the other hand, let I be an arbitrary bounded interval and set $J = \mathbb{Q} \cap (0, 1)$. Since I is measurable (Section 10.4, below), the set

$$F := \bigcup_{r \in J} (r + E \cap I)$$

is measurable. Also, since I and J are bounded, so is F. Thus, by countable additivity and translation invariance,

$$+\infty > \lambda(F) = \sum_{r \in J} \lambda(r + E \cap I) = \sum_{r \in J} \lambda(E \cap I).$$

[1] The existence of E requires the axiom of choice, one of the axioms of Zermelo–Fraenkel set theory.

Since J is an infinite set, $\lambda(E \cap I) = 0$. But then

$$\lambda(E) = \sum_{k=0}^{\infty} \lambda(E \cap [k, k+1)) + \sum_{k=0}^{\infty} \lambda(E \cap [-k-1, -k)) = 0.$$

This contradiction shows that E cannot be Lebesgue measurable. ◊

Exercises

1. ($\Downarrow$ 11.2.18.) Show that $E \in \mathcal{M}$ and $\boldsymbol{x} \in \mathbb{R}^n$ imply that $\boldsymbol{x} + E \in \mathcal{M}$. Conclude from Exercise 10.2.4 that $\lambda(\boldsymbol{x} + E) = \lambda(E)$.

2. Show that $E \in \mathcal{M}$ implies that $-E \in \mathcal{M}$. Conclude from Exercise 10.2.5 that $\lambda(-E) = \lambda(E)$.

3. Show that $E \in \mathcal{M}$ and $r \neq 0$ imply that $rE \in \mathcal{M}$. Conclude from Exercise 10.2.6 that $\lambda(rE) = |r|^n \lambda(E)$.

4. Show that for any $\varepsilon > 0$ there exists an open set D dense in $\mathbb{R}^n$ such that $\lambda(D) < \varepsilon$.

5. Prove that if f and g are continuous real-valued functions on $\mathbb{R}^n$ which are equal a.e., then $f = g$. Does the same result hold if only one of the functions is continuous?

6. Let A be the subset of $[0, 1]$ whose members are missing the digit three in their decimal expansions. Prove that A is uncountable and $\lambda(A) = 0$.

10.4 Borel Sets

Recall that the σ-field generated by a collection $\mathcal{A}$ of sets is the intersection of all σ-fields containing $\mathcal{A}$ (10.1.2). The following special case is of particular importance.

10.4.1 Definition. The *Borel σ-field* $\mathcal{B} = \mathcal{B}(\mathbb{R}^n)$ is the σ-field generated by the open sets of $\mathbb{R}^n$. A member of $\mathcal{B}$ is called a *Borel set.* ◊

10.4.2 Remark. Since open sets and closed sets are complements of one another, $\mathcal{B}$ is also generated by the closed sets. Furthermore, since an open set is a countable union of n-dimensional open intervals (Exercise 8.2.4), $\mathcal{B}$ is also generated by $\mathcal{O}$. Since every open interval is a countable union of closed and bounded intervals and every closed interval is a countable intersection of open intervals, $\mathcal{B}$ is also generated by $\mathcal{C}$. Similar considerations show that $\mathcal{B}$ is generated by $\mathcal{H}$ as well. ◊

10.4.3 Theorem. $\mathcal{B}(\mathbb{R}^n) \subseteq \mathcal{M}(\mathbb{R}^n)$.

Proof. By 10.4.2, it suffices to show that $\mathcal{H} \subseteq \mathcal{M}$. Note first that if $I,\ J \in \mathcal{H}$ then, using partitions as in the proof of 10.2.2, $I \setminus J$ may be expressed (usually in several ways) as a disjoint union of members of $\mathcal{H}$. (See Figure 10.5.)

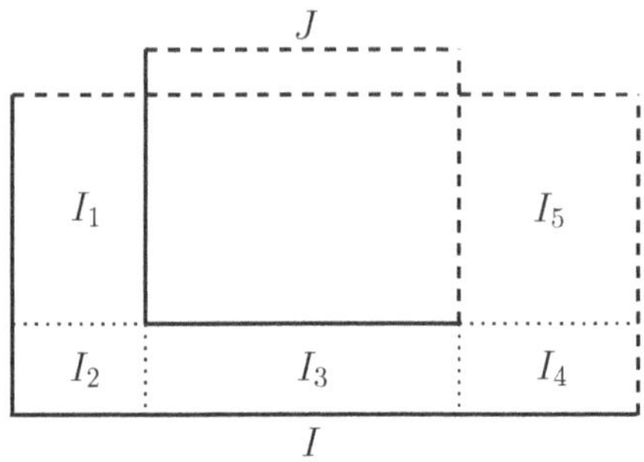

FIGURE 10.5: $I \setminus J = I_1 \cup I_2 \cup I_3 \cup I_4 \cup I_5$.

Now let $I \in \mathcal{H}$, $C \subseteq \mathbb{R}^n$, and let $\{I_k\}$ be any sequence in $\mathcal{H}$ that covers C. We show that

$$\lambda^*(C \cap I) + \lambda^*(C \cap I^c) \le \sum_k \lambda^*(I_k). \tag{10.6}$$

Taking the infimum over all such sequences $\{I_k\}$ produces the inequality $\lambda^*(C \cap I) + \lambda^*(C \cap I^c) \le \lambda^*(C)$, proving that $I \in \mathcal{M}$.

To verify (10.6), we may assume that $\sum_{k=1}^{\infty} \lambda^*(I_k) < +\infty$. For each k there exist, according to the observation at the beginning of the proof, intervals $J_{j,k} \in \mathcal{H}$ such that $I_k \setminus I = \bigcup_{j=1}^{m_k} J_{j,k}$ (disjoint union). Then

$$I_k = (I_k \cap I) \cup (I_k \setminus I) = (I_k \cap I) \cup \bigcup_{j=1}^{m_k} J_{j,k} \quad \text{(disjoint union)},$$

hence, by 10.2.5(f) and induction,

$$\lambda^*(I_k) = \lambda^*(I_k \cap I) + \sum_{j=1}^{m_k} \lambda^*(J_{j,k}).$$

Since $\{I_k \cap I\}_k$ covers $C \cap I$ and $\{J_{j,k}\}_{j,k}$ covers $C \cap I^c$,

$$\sum_k \lambda^*(I_k) = \sum_k \lambda^*(I_k \cap I) + \sum_k \sum_{j=1}^{m_k} \lambda^*(J_{j,k}) \ge \lambda^*(C \cap I) + \lambda^*(C \cap I^c). \quad \square$$

It may be shown that the inclusion $\mathcal{B} \subseteq \mathcal{M}$ is proper.[2] The importance of

[2]See, for example, [4].

Borel sets is that they are closely linked to the topology of $\mathbb{R}^n$ and hence are better suited for contexts involving continuous functions.

The remainder of the section demonstrates the precise connection between $\mathcal{B}$ and $\mathcal{M}$.

10.4.4 Lemma. *For any bounded $E \in \mathcal{M}$, there exists a decreasing sequence of bounded open sets $U_k \supseteq E$ such that*

$$\lim_k \lambda(U_k) = \lim_k \lambda\big(\mathrm{cl}(U_k)\big) = \lambda(E).$$

Proof. By definition of $\lambda(E)$, for each k we may choose a sequence of open intervals $I_{j,k}$ with union V_k containing E such that

$$\lambda(E) \le \lambda(V_k) \le \lambda\big(\mathrm{cl}(V_k)\big) \le \sum_j |\mathrm{cl}(I_{j,k})| < \lambda(E) + 1/k.$$

The sequence of open sets $U_k := V_1 \cap \cdots \cap V_k$ is decreasing, contains E, and satisfies

$$\lambda(E) \le \lambda(U_k) \le \lambda\big(\mathrm{cl}(U_k)\big) \le \lambda\big(\mathrm{cl}(V_k)\big) \le \lambda(E) + 1/k.$$

Letting $k \to +\infty$ proves the assertion. □

10.4.5 Lemma. *For any $E \in \mathcal{M}$, there exists an increasing sequence of compact sets $C_k \subseteq E$ such that* $\lim_k \lambda(C_k) = \lambda(E)$.

Proof. Suppose first that E is bounded. Let I be a bounded open interval containing $\mathrm{cl}(E)$ and let $\varepsilon > 0$. Choose a sequence of open intervals I_k with

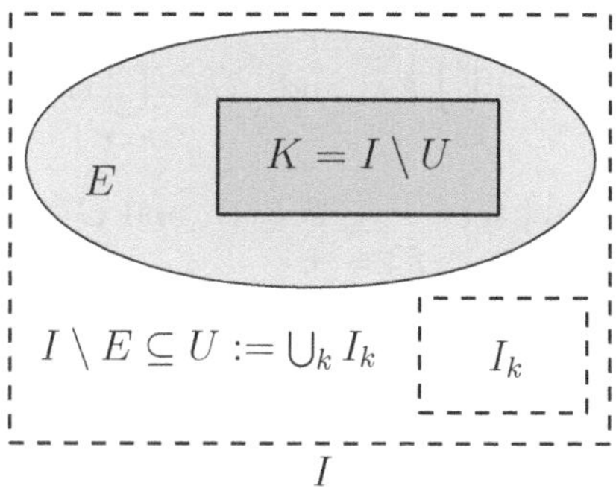

FIGURE 10.6: $K = \mathrm{cl}(E) \setminus U$.

union $U \supseteq I \setminus E$ such that $\sum_{k=1}^{\infty} |I_k| < \lambda(I \setminus E) + \varepsilon$. Since I is open, we may assume that $I_k \subseteq I$ (otherwise, replace I_k by $I_k \cap I$). Then $I \setminus E \subseteq U \subseteq I$ and

$$\lambda(U) \le \lambda(I \setminus E) + \varepsilon = \lambda(I) - \lambda(E) + \varepsilon.$$

Set $K = I \setminus U$. Then $K \subseteq E \subseteq \text{cl}(E) \subseteq I$, hence $K = \text{cl}(E) \setminus U$. Therefore, K is compact and

$$\lambda(K) = \lambda(I) - \lambda(U) \geq \lambda(I) - \big(\lambda(I) - \lambda(E) + \varepsilon\big) = \lambda(E) - \varepsilon.$$

Now let $E \in \mathcal{M}$ be arbitrary and let $\{E_k\}$ be a sequence of bounded measurable sets such that $E_k \uparrow E$. By the first paragraph, for each k we may choose a compact set $K_k \subseteq E_k$ such that $\lambda(K_k) > \lambda(E_k) - 1/k$. These conditions still hold if K_k is replaced by the compact set $C_k = K_1 \cup \cdots \cup K_k$. The sequence $\{C_k\}$ is increasing, contained in E, and $\lambda(C_k) \to \lambda(E)$. □

10.4.6 Lemma. *If $E \in \mathcal{M}$ is bounded, then there exists an increasing sequence of compact sets C_k and a decreasing sequence of bounded open sets U_k such that*

$$C_k \subseteq E \subseteq U_k \quad \textit{and} \quad \lim_k \lambda(U_k \setminus C_k) = 0.$$

Proof. If C_k and U_k are as in 10.4.4 and 10.4.5 with U_k bounded, then

$$\lambda(U_k \setminus C_k) = \lambda(U_k \setminus E) + \lambda(E \setminus C_k) \to 0. \qquad \square$$

10.4.7 Theorem. *If $E \in \mathcal{M}$, then there exist Borel sets F and G such that $F \subseteq E \subseteq G$ and $\lambda(G \setminus F) = 0$.*

Proof. Suppose first that E is bounded. Set $F = \bigcup_{k=1}^{\infty} C_k$ and $G = \bigcap_{k=1}^{\infty} U_k$, where C_k and U_k are the sets in 10.4.6. Then $F \subseteq E \subseteq G$ and $G \setminus F \subseteq U_k \setminus C_k$ for all k, hence $\lambda(G \setminus F) \leq \lambda(U_k \setminus C_k) \to 0$.

In the general case, there exists a sequence of bounded Borel sets $E_k \uparrow E$. By the first paragraph, there exist Borel sets F_k and G_k such that $F_k \subseteq E_k \subseteq G_k$ and $\lambda(G_k \setminus F_k) = 0$. Let

$$F = \bigcup_{k=1}^{\infty} F_k \quad \text{and} \quad G = \bigcup_{k=1}^{\infty} G_k.$$

Then F and G are Borel sets, $F \subseteq E \subseteq G$, and $G \setminus F \subseteq \bigcup_{k=1}^{\infty} G_k \setminus F_k$. By countable subadditivity, $\lambda(G \setminus F) = 0$. □

10.4.8 Corollary. *Every $E \in \mathcal{M}$ is the disjoint union of a Borel set and a set of Lebesgue measure zero.*

Proof. By the theorem, $E = F \cup (E \setminus F)$, where $F \in \mathcal{B}$ and $\lambda(E \setminus F) = 0$. □

Exercises

1. Let $\varepsilon > 0$. Construct an explicit compact subset $C \subseteq E := [0,1] \cap \mathbb{I}$ such that $\lambda(E \setminus C) < \varepsilon$.

2. Show that the graph $G := \{(x, y) : y = f(x)\}$ of a continuous function $f : \mathbb{R} \to \mathbb{R}$ is a Borel set with two-dimensional Lebesgue measure zero.

3. Let E denote the Cantor set (10.3.4). Show that $E + \mathbb{Q}$ and $E + E$ are Borel sets and find their measures.

4. Let $B \in \mathcal{B}(\mathbb{R}^n)$, $\boldsymbol{y} \in \mathbb{R}^n$, and $r \in \mathbb{R}$. Prove that $B + \boldsymbol{y} := \{\boldsymbol{x} + \boldsymbol{y} : \boldsymbol{x} \in B\}$, $rB := \{r\boldsymbol{x} : \boldsymbol{x} \in B\}$ and $-B := \{\boldsymbol{x} : -\boldsymbol{x} \in B\}$ are Borel sets.

10.5 Measurable Functions

In this section, $\mathcal{F}$ denotes a σ-field of subsets of a set S.

Definition and Basic Properties

10.5.1 Lemma. *Let $f : S \to \overline{\mathbb{R}}$. The following statements are equivalent:*

(a) $f^{-1}(\{+\infty\}), f^{-1}(\{-\infty\}) \in \mathcal{F}$, *and* $f^{-1}(U) \in \mathcal{F}$ *for all open sets* $U \subseteq \mathbb{R}$.

(b) $f^{-1}(\{+\infty\}), f^{-1}(\{-\infty\}) \in \mathcal{F}$, *and* $f^{-1}(F) \in \mathcal{F}$ *for all closed sets* $F \subseteq \mathbb{R}$.

(c) $f^{-1}(\{+\infty\}), f^{-1}(\{-\infty\}) \in \mathcal{F}$, *and* $f^{-1}(B) \in \mathcal{F}$ *for all Borel sets* $B \subseteq \mathbb{R}$.

(d) $\{x : f(x) \le t\} \in \mathcal{F}$ *for all* $t \in \mathbb{R}$.

(e) $\{x : f(x) < t\} \in \mathcal{F}$ *for all* $t \in \mathbb{R}$.

(f) $\{x : f(x) \ge t\} \in \mathcal{F}$ *for all* $t \in \mathbb{R}$.

(g) $\{x : f(x) > t\} \in \mathcal{F}$ *for all* $t \in \mathbb{R}$.

Proof. The equivalence of (a) and (b) follows from the general set theoretic relation $f^{-1}(A^c) = \left[f^{-1}(A)\right]^c$. Clearly, (c) implies (b). For the converse, denote by $\mathcal{G}$ the collection of all Borel subsets B of $\mathbb{R}$ such that $f^{-1}(B) \in \mathcal{F}$. Then $\mathcal{G}$ is a σ-field. If (b) holds, then $\mathcal{G}$ contains the closed sets, hence, by minimality, $\mathcal{G} = \mathcal{B}$. This proves (c) and hence shows that (a)–(c) are equivalent.

The implications (c) $\Rightarrow$ (d) $\Rightarrow$ (e) $\Rightarrow$ (f) $\Rightarrow$ (g) $\Rightarrow$ (d) are proved using the following set relations:

$$\begin{aligned}
&(c) \Rightarrow (d): \quad \{x : f(x) \le t\} = f^{-1}(\{-\infty\}) \cup f^{-1}((-\infty, t]).\\
&(d) \Rightarrow (e): \quad \{x : f(x) < t\} = \bigcup_{n=1}^{\infty} \{x : f(x) \le t - 1/n\}.\\
&(e) \Rightarrow (f): \quad \{x : f(x) \ge t\} = \{x : f(x) < t\}^c.\\
&(f) \Rightarrow (g): \quad \{x : f(x) > t\} = \bigcup_{n=1}^{\infty} \{x : f(x) \ge t + 1/n\}.\\
&(g) \Rightarrow (d): \quad \{x : f(x) \le t\} = \{x : f(x) > t\}^c.
\end{aligned}$$

Thus (d)–(g) are equivalent and are implied by (a)–(c).

Now assume that (d)–(g) hold. Then the sets

$$f^{-1}(+\infty) = \bigcap_{k=1}^{\infty} \{x : f(x) > k\}, \quad f^{-1}(-\infty) = \bigcap_{k=1}^{\infty} \{x : f(x) < -k\}$$

are members of $\mathcal{F}$, and for $-\infty < a < b < +\infty$,

$$f^{-1}((a, b)) = \{x : f(x) > a\} \cap \{x : f(x) < b\} \in \mathcal{F}.$$

Since every open subset of $\mathbb{R}$ is a countable union of open intervals, (a) holds, completing the proof. □

10.5.2 Definition. A function $f : S \to \overline{\mathbb{R}}$ is said to be *measurable with respect to* $\mathcal{F}$, or simply $\mathcal{F}$*-measurable*, if any (hence all) of the conditions in Lemma 10.5.1 hold. ◊

The following theorem shows that the collection of all measurable functions is closed under the standard ways of combining functions. The functions f^+, f^-, $\sup_n f_n$, $\inf_n f_n$, $\limsup_n f_n$, and $\liminf_n f_n$ in the statement of the theorem are defined by

$$\begin{aligned}
f^+(x) &:= \max\{f(x), 0\}, & f^-(x) &:= \max\{-f(x), 0\},\\
(\sup_k f_k)(x) &:= \sup_k f_k(x), & (\inf_k f_k)(x) &:= \inf_k f_k(x),\\
(\limsup_k f_k)(x) &:= \limsup_k f_k(x), & (\liminf_k f_k)(x) &:= \liminf_k f_k(x).
\end{aligned}$$

10.5.3 Theorem. *Let f, g, f_k be measurable with respect to a σ-field $\mathcal{F}$ on S. If $\alpha \in \mathbb{R}$ and $p > 0$, then $f + g$, αf, f^2, fg, $|f|^p$, f^+, f^-, $\sup_k f_k$, $\inf_k f_k$, $\limsup_k f_k$, and $\liminf_k f_k$ are measurable.*

Proof. The proof is based on the following equalities. The details are left to the reader.

- $\{x : (f + g)(x) < t\} = \bigcup_{r \in \mathbb{Q}} \{x : f(x) < r\} \cap \{x : g(x) < t - r\}$.

- $\{x : \alpha f(x) < t\} = \{x : f(x) < t/\alpha\}$ for $\alpha > 0$.
- $\{x : f^2(x) < t\} = \{x : -\sqrt{t} < f(x) < \sqrt{t}\}$ for $t > 0$.
- $fg = \frac{1}{2}[(f+g)^2 - f^2 - g^2]$.
- $\{x : |f|^p(x) < t\} = \{x : -t^{1/p} < f(x) < t^{1/p}\}$ for $t > 0$.
- $f^+ = \frac{1}{2}(|f| + f), \quad f^- = \frac{1}{2}(|f| - f)$.
- $\left\{x : \sup_k f_k(x) \le t\right\} = \bigcap_k \{x : f_k(x) \le t\}$.
- $\inf_k f_k = -\sup_k(-f_k)$.
- $\liminf_k f_k = \sup_k \inf_{j\ge k} f_j; \quad \limsup_k f_k = -\liminf_k(-f_k)$. □

10.5.4 Corollary. *If $f_k : S \to \overline{\mathbb{R}}$ is $\mathcal{F}$-measurable for every k and if $f_k \to f$ on S, then f is $\mathcal{F}$-measurable.*

Simple Functions

10.5.5 Definition. The *indicator function* of a set $A \subseteq S$ is the function $\mathbf{1}_A$ on S defined by

$$\mathbf{1}_A(x) = \begin{cases} 1 & \text{if } x \in A, \text{ and} \\ 0 & \text{if } x \notin A. \end{cases} \qquad \Diamond$$

For example, the Dirichlet function may be expressed as $\mathbf{1}_{\mathbb{Q}}$.

10.5.6 Definition. A function $f : S \to \mathbb{R}$ with finite range is called a *simple function*. The collection of all nonnegative $\mathcal{F}$-measurable simple functions is denoted by $\mathcal{S}_+(\mathcal{F})$. ◊

10.5.7 Remarks. (a) A linear combination of indicator functions is a simple function. Conversely, a simple function f may be expressed in many ways as a linear combination of indicator functions. The most important of these is the *standard form*

$$f = \sum_{j=1}^{m} a_j \mathbf{1}_{A_j}, \quad A_j := \{x \in S : f(x) = a_j\}, \tag{10.7}$$

where $a_1, \ldots, a_m \in \mathbb{R}$ are the distinct values of f. Note that the sets A_j form a partition of $\mathbb{R}^n$. By 10.5.3 and Exercise 8, f is $\mathcal{F}$-measurable iff $A_j \in \mathcal{F}$ for each j.

(b) If $f_1, f_2 \in \mathcal{S}_+(\mathcal{F})$, $\alpha \ge 0$, and $p > 0$, then the functions

$$\alpha f_1, \quad f_1 + f_2, \quad f_1 f_2, \quad f_1^p, \quad \max\{f_1, f_2\}, \quad \min\{f_1, f_2\}$$

are nonnegative, measurable, and have finite ranges, hence are in $\mathcal{S}_+(\mathcal{F})$. ◊

The following theorem shows that the collection $\mathcal{S}_+(\mathcal{F})$ generates all measurable functions. It is a key ingredient in the development of the Lebesgue theory.

10.5.8 Theorem. *For each nonnegative $\mathcal{F}$-measurable function f on S, there exists a sequence $\{f_k\}$ in $\mathcal{S}_+(\mathcal{F})$ such that $f_k \uparrow f$ on S.*

Proof. Let $f_0 = 0$, and for each $k \in \mathbb{N}$ define

$$f_k = \sum_{j=1}^{k2^k} \frac{j-1}{2^k} \mathbf{1}_{A_{k,j}} + k\mathbf{1}_{A_k}, \quad \text{where } A_k = \{x : f(x) \geq k\} \text{ and}$$
$$A_{k,j} = \left\{x : (j-1)2^{-k} \leq f(x) < j2^{-k}\right\}, \quad j = 1, 2, \ldots, k2^k.$$

(See Figure 10.7.) We show that $f_k(x) \uparrow f(x)$ for each $x \in S$. This is

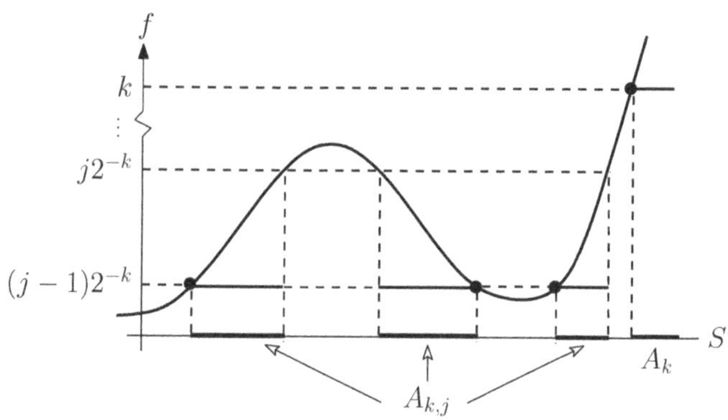

FIGURE 10.7: The components of f_k.

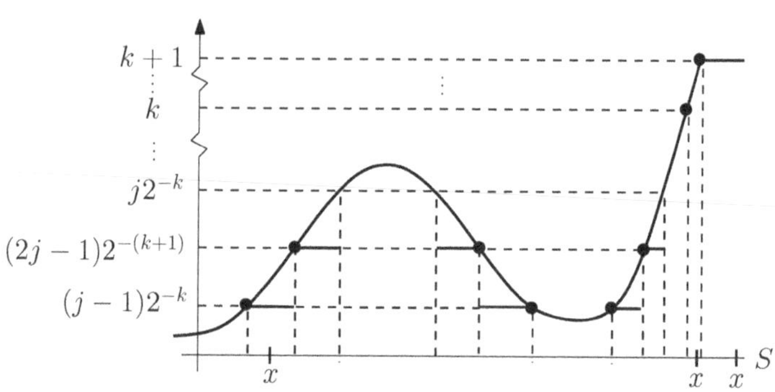

FIGURE 10.8: The components of f_{k+1}.

clear if $f(x) = +\infty$, since then $f_k(x) = k$ for all k. Suppose $f(x) \in \mathbb{R}$ and let $k \in \mathbb{N}$. If $f(x) \geq k+1$, then $f_{k+1}(x) = k+1 > k = f_k(x)$. If $k \leq f(x) < k+1$, then $f_{k+1}(x) \geq k = f_k(x)$. Finally, suppose that $f(x) < k$.

Then $(j-1)2^{-k} \le f(x) < j2^{-k}$ for some $1 \le j \le k2^k$, hence

$$\frac{2j-2}{2^{k+1}} \le f(x) < \frac{2j-1}{2^{k+1}} \quad \text{or} \quad \frac{2j-1}{2^{k+1}} \le f(x) < \frac{2j}{2^{k+1}}.$$

(See Figure 10.8.) In either case,

$$f_{k+1}(x) \ge \frac{2j-2}{2^{k+1}} = \frac{j-1}{2^k} = f_k(x).$$

Thus $f_k \uparrow$ on S. Since $0 \le f(x) - f_k(x) < 2^{-k}$ for all sufficiently large k, $f_k(x) \to f(x)$. □

Lebesgue and Borel Measurable Functions

10.5.9 Definition. A function $f : \mathbb{R}^n \to \overline{\mathbb{R}}$ is said to be *Borel (Lebesgue) measurable* if f is measurable with respect to the σ-field $\mathcal{B}(\mathbb{R}^n)$ $(\mathcal{M}(\mathbb{R}^n))$. ◊

10.5.10 Proposition. *If f is Lebesgue measurable and $f = g$ a.e., then g is Lebesgue measurable.*

Proof. Let $A = \{\boldsymbol{x} : f(\boldsymbol{x}) \ne g(\boldsymbol{x})\}$. By hypothesis, A has Lebesgue measure zero, hence A^c and $\{\boldsymbol{x} : g(\boldsymbol{x}) < t\} \cap A \in \mathcal{M}$. Therefore,

$$\{\boldsymbol{x} : g(\boldsymbol{x}) < t\} = \big(\{\boldsymbol{x} : f(\boldsymbol{x}) < t\} \cap A^c\big) \cup \big(\{\boldsymbol{x} : g(\boldsymbol{x}) < t\} \cap A\big) \in \mathcal{M}. \quad \square$$

If f is Borel measurable and $f = g$ a.e., then g need not be Borel measurable. Indeed, there exist sets $E \in \mathcal{M} \setminus \mathcal{B}$ with measure zero, hence $\mathbf{1}_E = 0$ a.e. but $\mathbf{1}_E$ is not Borel measurable.[3]

Clearly, a Borel measurable function is Lebesgue measurable. The preceding paragraph shows that the converse is false. However, we have

10.5.11 Proposition. *If $f : \mathbb{R}^n \to \overline{\mathbb{R}}$ is Lebesgue measurable, then there exists a Borel measurable function $g : \mathbb{R}^n \to \overline{\mathbb{R}}$ such that $g = f$ a.e.*

Proof. Consider first the case $f = \mathbf{1}_E$, $E \in \mathcal{M}$. By 10.4.8, E is the disjoint union of a Borel set F and a set A of Lebesgue measure zero. Thus $g := \mathbf{1}_F$ is Borel measurable and $f = g + \mathbf{1}_A = g$ a.e. The assertion therefore holds for indicator functions.

If f is a simple function, then each term in the standard form of f is a.e. equal to a Borel function. Therefore, the assertion holds for simple functions.

If $f \ge 0$, then, by 10.5.8, there exists a sequence of nonnegative Lebesgue measurable simple functions f_k such that $f_k \to f$ on $\mathbb{R}^n$. By the previous paragraph, for each k there exists a Borel measurable function g_k such that $f_k = g_k$ a.e. Let

$$A_k := \{\boldsymbol{x} : f_k(\boldsymbol{x}) \ne g_k(\boldsymbol{x})\} \quad \text{and} \quad A := \bigcup_{k=1}^{\infty} A_k.$$

[3] See, for example, [4].

Then $A \in \mathcal{M}$, $\lambda(A) = 0$, and $f_k(\boldsymbol{x}) = g_k(\boldsymbol{x})$ for all $\boldsymbol{x} \in A^c$ and all k. Let B denote the set of all $\boldsymbol{x}$ such that the sequence $\{g_k(\boldsymbol{x})\}$ does not converge. Then $B \subseteq A$ and, by 10.5.3, $B \in \mathcal{B}$. Let $g = \lim_k g_k \mathbf{1}_{B^c}$. Then g is Borel measurable and $\{\boldsymbol{x} : g(\boldsymbol{x}) \neq f(\boldsymbol{x})\} \subseteq A$ so $g = f$ a.e. Therefore, the assertion holds for nonnegative f. The general case follows from the identity $f = f^+ - f^-$. □

Part (a) of 10.5.1 implies that a continuous function $f : \mathbb{R}^n \to \mathbb{R}$ is Borel measurable. In a similar vein,

10.5.12 Proposition. *If $f : \mathbb{R}^n \to \mathbb{R}$ is continuous except on a set E of Lebesgue measure zero, then f is Lebesgue measurable.*

Proof. Let $U \subseteq \mathbb{R}$ be open. Then

$$f^{-1}(U) = A \cup B, \quad \text{where} \quad A := f^{-1}(U) \cap E \quad \text{and} \quad B := f^{-1}(U) \cap E^c.$$

Since $A \subseteq E$ and $\lambda(E) = 0$, $A \in \mathcal{M}$. Since f is continuous on E^c, B is open in E^c, hence $B = V \cap E^c$ for some open subset V of $\mathbb{R}^n$. Therefore, $B \in \mathcal{M}$, so $f^{-1}(U) \in \mathcal{M}$. By 10.5.1, f is Lebesgue measurable. □

Proposition 10.5.12 implies that a function with at most countably many discontinuities is Lebesgue measurable. An examination of the proof shows that such a function is in fact Borel measurable. In particular, monotone functions on $\mathbb{R}$, hence also functions of bounded variation, are Borel measurable (see 3.3.6 and 5.9.7).

Note that a function that is continuous except on a set of measure zero is not necessarily equal a.e. to a continuous function (Exercise 12). Conversely, a function equal a.e. to a continuous function need not be continuous anywhere; the Dirichlet function is an obvious example.

Exercises

In Exercises 1–8, $\mathcal{F}$ denotes a σ-field of subsets of a set S.

1. Let $f : S \to \overline{\mathbb{R}}$ have the property that $\mathbf{1}_{A_k} f$ is $\mathcal{F}$-measurable for every k, where $A_k \in \mathcal{F}$ and $\bigcup_k A_k = S$. Prove that f is $\mathcal{F}$-measurable.

2. Prove that if $f : S \to \mathbb{R}$ is $\mathcal{F}$-measurable and never zero, then $1/f$ is $\mathcal{F}$-measurable.

3. Let $f : S \to \overline{\mathbb{R}}$ have the property that $\{x : f(x) < r\} \in \mathcal{F}$ for all $r \in \mathbb{Q}$. Prove that f is $\mathcal{F}$-measurable.

4. Let $f : S \to \mathbb{R}$ be $\mathcal{F}$-measurable and let $g : \mathbb{R} \to \mathbb{R}$ be continuous. Show that $g \circ f$ is $\mathcal{F}$-measurable.

5. Let $g,\ h : S \to \mathbb{R}$ be $\mathcal{F}$-measurable functions. Prove that the following sets are $\mathcal{F}$-measurable:

   (a) $\{x \in S : g(x) > h(x)\}$ (b) $\{x \in S : g(x) \geq h(x)\}$

   (c) $\{x \in S : g(x) = h(x)\}$ (d) $\{x \in S : g(x)h(x) = 1\}$.

6. Let $\{f_k : S \to \mathbb{R}\}$ be a sequence of $\mathcal{F}$-measurable functions. Prove that the set $\{x \in S : \lim_k f_k(x) \text{ exists in } \overline{\mathbb{R}}\}$ is $\mathcal{F}$-measurable.

7. Let $f : S \to \overline{\mathbb{R}}$ have range consisting of the distinct values $a_k,\ k \in \mathbb{N}$. Show that f is $\mathcal{F}$-measurable iff $\{x \in S : f(x) = a_k\} \in \mathcal{F}$ for every k.

8. Let $E \subseteq S$. Prove that $\mathbf{1}_E$ is $\mathcal{F}$-measurable iff $E \in \mathcal{F}$.

9. Let A, B, and C be subsets of S. Prove:

   (a) $\mathbf{1}_{AB} = \mathbf{1}_A \mathbf{1}_B$ (b) $\mathbf{1}_{A\cup B} = \mathbf{1}_A + \mathbf{1}_B - \mathbf{1}_A\mathbf{1}_B$

   (c) $\mathbf{1}_{A^c} = 1 - \mathbf{1}_A$ (d) $\mathbf{1}_A \leq \mathbf{1}_B$ iff $A \subseteq B$.

10. Define the *symmetric difference* $A\Delta B$ of sets A and B by

$$A \,\Delta\, B = (A \setminus B) \cup (B \setminus A) = (A \cup B) \setminus (A \cap B).$$

    Prove that $\mathbf{1}_{A\Delta B} = |\mathbf{1}_A - \mathbf{1}_B|$.

11. Let $A_k \subseteq S$ and set $B = \liminf_k A_k$ and $C = \limsup_k A_k$. (see Exercise 10.1.6). Prove that

    (a) $\mathbf{1}_B = \liminf_k \mathbf{1}_{A_k}$ (b) $\mathbf{1}_C = \limsup_k \mathbf{1}_{A_k}$.

12. Prove that $\mathbf{1}_{[0,1]}$ is not equal a.e. to a continuous function on $\mathbb{R}$.

13. Let $f : \mathbb{R} \to \mathbb{R}$. Prove that if f' exists on $\mathbb{R}$, then f' is Borel measurable.

14. Let $f(x) = \lfloor x^{-1} \rfloor^{-1}$, $0 < x \leq 1$. Show that f is Borel measurable on $(0, 1]$.

15. Let $f(x) = 1 + r(\lfloor x^{-1} \rfloor)$, $0 < x \leq 1$, where $r(k)$ denotes the remainder on division of an integer k by 3. Show that f is Borel measurable.

16. Define $f : [0, 1] \to \mathbf{R}$ by $f(x) = 0$ if x is rational and $f(x) = 1/\sqrt{d}$ if x is irrational, where d is the first nonzero digit in the decimal expansion of x. Prove that f is Borel measurable.

17. Prove that if the function f in 10.5.8 is bounded, then the convergence of the sequence is uniform.

18. Let $f : \mathbb{R}^2 \to \mathbb{R}$ have the property that $f(x, y)$ is continuous in x for each y and Borel measurable in y for each x. Let $g : \mathbb{R} \to \mathbb{R}$ be Borel measurable. Prove that the function $h(y) := f(g(y), y)$ is Borel measurable. *Hint.* Start with indicator functions g.

19. ($\Downarrow$ 11.5.4.) Let $f = (f_1, \ldots, f_m) : \mathbb{R}^n \to \mathbb{R}^m$, where each $f_j : \mathbb{R}^n \to \mathbb{R}$ is Borel measurable. Prove:

    (a) $\mathcal{F} := \{B \in \mathcal{B}(\mathbb{R}^m) : f^{-1}(B) \in \mathcal{B}(\mathbb{R}^n)\}$ is a σ-field.

    (b) $\mathcal{F} = \mathcal{B}(\mathbb{R}^m)$, that is, $f^{-1}(B) \in \mathcal{B}(\mathbb{R}^n)$ for every $B \in \mathcal{B}(\mathbb{R}^m)$.

    (c) If $F : \mathbb{R}^m \to \mathbb{R}$ is Borel measurable, then the function $g := F \circ f$ is Borel measurable.

20. (a) Show that $B \times \mathbb{R} \in \mathcal{B}(\mathbb{R}^2)$ for all $B \in \mathcal{B}(\mathbb{R})$.

    (b) Let $f : \mathbb{R} \to \overline{\mathbb{R}}$ be Borel measurable and define $g : \mathbb{R}^2 \to \overline{\mathbb{R}}$ by $g(x, y) = f(x)$. Show that g is Borel measurable.

21. Let $\mathbf{0} \in A \subseteq \mathbb{R}^n$. Define the "radius function" $f_A : \mathbb{R}^n \to \overline{\mathbb{R}}$ by

$$f_A(\boldsymbol{x}) := \sup\{t \geq 0 : t\boldsymbol{x} \in A\}, \quad \boldsymbol{x} \in \mathbb{R}^n.$$

    (a) Let $\mathbf{0} \in A_k$ for all k and $A_k \uparrow A$. Show that $f_{A_k} \uparrow f_A$.

    (b) Show that if A is open, then f_A is positive and Borel measurable.

    (c) Use (b) to show that if A is compact, then f_A is Borel measurable.

    (d) Conclude from 10.4.5 that f_A is Borel measurable for any Borel set A containing $\mathbf{0}$.

Chapter 11

Lebesgue Integration on $\mathbb{R}^n$

In this chapter we use the measure theory developed in Chapter 10 to construct the Lebesgue integral of a measurable function of several variables. For comparison purposes, we begin with a brief description of the Riemann integral on compact subintervals of $\mathbb{R}^n$.

11.1 Riemann Integration on $\mathbb{R}^n$

The n-dimensional Riemann integral is constructed in essentially the same way as the one-dimensional integral: Let f be a bounded real-valued function on an n-dimensional interval

$$[\mathbf{a}, \mathbf{b}] := [a_1, b_1] \times [a_2, b_2] \times \cdots \times [a_n, b_n],$$

where $\boldsymbol{a} := (a_1, \ldots, a_n)$ and $\boldsymbol{b} := (b_1, \ldots, b_n)$.

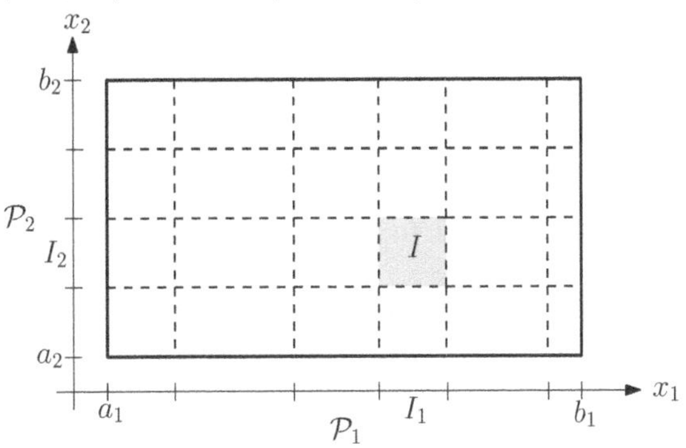

FIGURE 11.1: Partition of $[a, b] \times [c, d]$.

For each j, let $\mathcal{P}_j$ be a partition of the coordinate interval $[a_j, b_j]$. The collection of all Cartesian products of the resulting coordinate subintervals produces a partition $\mathcal{P}$ of $[\mathbf{a}, \mathbf{b}]$ consisting of n-dimensional subintervals $I = I_1 \times I_2 \times \cdots \times I_n$ with volume $\Delta V_I := |I_1|\,|I_2| \cdots |I_n|$ (see Figure 11.1). The *lower and upper*

DOI: 10.1201/9781003629382-11

sums of f over $\mathcal{P}$ are defined by

$$\underline{S}(f,\mathcal{P}) = \sum_{I\in\mathcal{P}} m_I \Delta V_I, \qquad m_I := \inf_{\boldsymbol{x}\in I} f(\boldsymbol{x}),$$
$$\overline{S}(f,\mathcal{P}) = \sum_{I\in\mathcal{P}} M_I \Delta V_I, \qquad M_I := \sup_{\boldsymbol{x}\in I} f(\boldsymbol{x}).$$

The *lower and upper integrals* on $[\mathbf{a},\mathbf{b}]$ are defined by

$$\underline{\int_{\boldsymbol{a}}^{\boldsymbol{b}}} f := \sup_{\mathcal{P}} \underline{S}(f,\mathcal{P}) \quad \text{and} \quad \overline{\int_{\boldsymbol{a}}^{\boldsymbol{b}}} f := \inf_{\mathcal{P}} \overline{S}(f,\mathcal{P}),$$

where the supremum and infimum are taken over all partitions $\mathcal{P}$ of $[\mathbf{a},\mathbf{b}]$. If the two integrals are equal, then f is said to be *Riemann–Darboux integrable on* $[\mathbf{a},\mathbf{b}]$. The common value of these integrals is then denoted by $\int_{\boldsymbol{a}}^{\boldsymbol{b}} f$. As in the one-variable case,

$$\int_{\boldsymbol{a}}^{\boldsymbol{b}} f = \lim_{\|\mathcal{P}\|\to 0} S(f,\mathcal{P},\{\boldsymbol{\xi}_I\}_I),$$

where $\|\mathcal{P}\| = \max_j \|\mathcal{P}_j\|$ and $S(f,\mathcal{P},\{\boldsymbol{\xi}_I\}_I)$ is the *Riemann sum*

$$S(f,\mathcal{P},\{\boldsymbol{\xi}_I\}_I) := \sum_I f(\boldsymbol{\xi}_I)\Delta V_I, \quad \boldsymbol{\xi}_I \in I,\ I \in \mathcal{P}.$$

The n-dimensional Riemann integral has properties analogous to those of the one-dimensional integral. Moreover, as is shown in Section 11.5, if f is continuous, then $\int_{\boldsymbol{a}}^{\boldsymbol{b}} f$ may be expressed as an *iterated integral*

$$\int_{a_1}^{b_1} \cdots \int_{a_n}^{b_n} f(x_1,\ldots,x_n)\,dx_n \cdots dx_1,$$

effectively reducing the theory to the one-dimensional case. Integrals over regions bounded by "nice" surfaces may be similarly evaluated.

11.2 The Lebesgue Integral

The Lebesgue integral on $\mathbb{R}^n$ is defined first for nonnegative Lebesgue measurable simple functions and is then extended to a larger class of functions, including all nonnegative Lebesgue measurable functions. The identity $f = f^+ - f^-$ is then used to define the integral for general measurable functions.

The Integral of a Simple Function

11.2.1 Definition. Let $f \in \mathcal{S}_+(\mathcal{M})$ have standard form

$$f = \sum_{j=1}^{m} a_j \mathbf{1}_{A_j}, \quad A_j := \{\boldsymbol{x} : f(\boldsymbol{x}) = a_j\},$$

where $\{A_1, \ldots, A_m\}$ is a (measurable) partition of $\mathbb{R}^n$. The *Lebesgue integral of f on $\mathbb{R}^n$* is defined by

$$\int f\, d\lambda := \sum_{j=1}^{m} a_j \lambda(A_j). \qquad \Diamond$$

Note that the above sum may contain a term of the form $0 \cdot (+\infty)$. While this expression was heretofore undefined, it is now necessary to make the definition

$$0 \cdot (+\infty) := 0.$$

In particular, the integral of the identically zero function is $0 \cdot \lambda(\mathbb{R}^n) = 0$.

11.2.2 Lemma. *If $f, g \in \mathcal{S}_+(\mathcal{M})$ and $\alpha \geq 0$, then*

(a) $\displaystyle\int \alpha f\, d\lambda = \alpha \int f\, d\lambda$; (b) $\displaystyle\int (f+g)\, d\lambda = \int f\, d\lambda + \int g\, d\lambda$;

(c) $\displaystyle\int f\, d\lambda \leq \int g\, d\lambda$ *if* $f \leq g$ *a.e.* (d) $\displaystyle\int f\, d\lambda = \int g\, d\lambda$ *if* $f = g$ *a.e.*

Proof. Part (a) is immediate from the definition, and (d) follows from (c). To prove (b) and (c), let f and g have standard representations

$$f = \sum_{i=1}^{m} a_i \mathbf{1}_{A_i} \quad \text{and} \quad g = \sum_{j=1}^{k} b_j \mathbf{1}_{B_j},$$

so

$$\int f = \sum_{i=1}^{m} a_i \lambda(A_i) \quad \text{and} \quad \int g = \sum_{j=1}^{k} b_j \lambda(B_j).$$

Since $\mathbb{R}^n = \bigcup_{i=1}^m A_i = \bigcup_{j=1}^k B_j$ and the unions are disjoint,

$$\lambda(A_i) = \sum_{j=1}^{k} \lambda(A_i \cap B_j) \quad \text{and} \quad \lambda(B_j) = \sum_{i=1}^{m} \lambda(A_i \cap B_j),$$

hence

$$\int f\, d\lambda = \sum_{i=1}^{m}\sum_{j=1}^{k} a_i \lambda(A_i \cap B_j) \quad \text{and} \quad \int g\, d\lambda = \sum_{j=1}^{k}\sum_{i=1}^{m} b_j \lambda(A_i \cap B_j). \quad (11.1)$$

Now let $c_1, \ldots, c_p$ be the distinct values of $f+g$, and set

$$C_\ell = \{\boldsymbol{x} : (f+g)(\boldsymbol{x}) = c_\ell\}, \quad \ell = 1, \ldots, p.$$

Then

$$f+g = \sum_{\ell=1}^{p} c_\ell \mathbf{1}_{C_\ell} \text{ and } C_\ell = \bigcup_{\{(i,j):a_i+b_j=c_\ell\}} A_i \cap B_j \text{ (disjoint)},$$

so

$$\begin{aligned}\int (f+g)\,d\lambda &= \sum_{\ell=1}^{p} c_\ell \lambda(C_l) = \sum_{\ell=1}^{p} c_\ell \sum_{\{(i,j):a_i+b_j=c_\ell\}} \lambda(A_i \cap B_j) \\ &= \sum_{i=1}^{m}\sum_{j=1}^{k} (a_i+b_j)\lambda(A_i \cap B_j).\end{aligned}$$

By (11.1), the last sum is $\int f\,d\lambda + \int g\,d\lambda$, proving (b).

For (c), suppose $f \le g$ a.e. and let $E = \{\boldsymbol{x} : f(\boldsymbol{x}) \le g(\boldsymbol{x})\}$. Then $\lambda(E^c) = 0$ and $a_i \le b_j$ for all i, j for which $A_i \cap B_j \cap E \ne \emptyset$. From

$$\lambda(A_i \cap B_j) = \lambda(A_i \cap B_j \cap E) + \lambda(A_i \cap B_j \cap E^c) = \lambda(A_i \cap B_j \cap E)$$

and (11.1), we have

$$\int f\,d\lambda = \sum_{i=1}^{m}\sum_{j=1}^{k} a_i\lambda(A_i \cap B_j \cap E) \le \sum_{j=1}^{k}\sum_{i=1}^{m} b_j\lambda(A_i \cap B_j \cap E) = \int g\,d\lambda. \qquad \square$$

The Integral of a Measurable Function

11.2.3 Definition. Let $f : \mathbb{R}^n \to \mathbb{R}$ be Lebesgue measurable. If $f \ge 0$, define

$$\int f\,d\lambda = \int f(\boldsymbol{x})\,d\lambda(\boldsymbol{x}) := \sup\left\{\int f_s\,d\lambda : f_s \le f,\ f_s \in \mathcal{S}_+(\mathcal{M})\right\}. \tag{11.2}$$

In general, define *the Lebesgue integral on* $\mathbb{R}^n$ by

$$\int f\,d\lambda := \int f^+\,d\lambda - \int f^-\,d\lambda,$$

provided at least one of the terms on the right is finite. For $E \in \mathcal{M}$ define the *Lebesgue integral on* E by

$$\int_E f\,d\lambda := \int f \cdot \mathbf{1}_E\,d\lambda$$

whenever the right side is defined. If both $\int_E f^+\,d\lambda$ and $\int_E f^-\,d\lambda$ are finite, then f is said to be *(Lebesgue) integrable on* E. The collection of all integrable functions on E is denoted by $\mathcal{L}^1(E)$. Finally, f is said to be *integrable* if it is integrable on $\mathbb{R}^n$. $\lozenge$

Note that from the definition, $f \geq 0 \Rightarrow \int f \geq 0$. More generally,

11.2.4 Proposition. *If* $f, g : \mathbb{R}^n \to \overline{\mathbb{R}}$ *are Lebesgue measurable,* $f \leq g$ *a.e., and* $\int f\,d\lambda$, $\int g\,d\lambda$ *are defined, then* $\int f\,d\lambda \leq \int g\,d\lambda$. *In particular, if* $f \geq 0$ *and* g *is integrable, then* f *is integrable*

Proof. Assume first that $f, g \geq 0$. Let $f_s \in \mathcal{S}_+(\mathcal{M})$ with $f_s \leq f$ and set $g_s := \mathbf{1}_E f_s$, where $E := \{\boldsymbol{x} : f(\boldsymbol{x}) \leq g(\boldsymbol{x})\}$. Then $g_s \in \mathcal{S}_+(\mathcal{M})$, $f_s = g_s$ a.e., and $g_s \leq \mathbf{1}_E f \leq \mathbf{1}_E g \leq g$. By 11.2.2, $\int f_s\,d\lambda = \int g_s\,d\lambda \leq \int g\,d\lambda$. Since f_s was arbitrary, $\int f\,d\lambda \leq \int g\,d\lambda$.

In the general case, $f^+ \leq g^+$ and $f^- \geq g^-$ a.e., hence, by the first part of the proof,

$$\int f\,d\lambda = \int f^+\,d\lambda - \int f^-\,d\lambda \leq \int g^+\,d\lambda - \int g^-\,d\lambda = \int g\,d\lambda. \qquad \square$$

11.2.5 Corollary. *If* $f : \mathbb{R}^n \to \overline{\mathbb{R}}$ *is integrable and* $f = g$ *a.e., then* g *is integrable and* $\int f\,d\lambda = \int g\,d\lambda$.

Proof. By 10.5.10, g is Lebesgue measurable. Moreover, $f^+ = g^+$ a.e., so by 11.2.4, $\int f^+\,d\lambda = \int g^+\,d\lambda$. Similarly, $\int f^-\,d\lambda = \int g^-\,d\lambda$. $\square$

11.2.6 Proposition. *If* $f : \mathbb{R}^n \to \overline{\mathbb{R}}$ *is integrable, then* f *is finite a.e.*

Proof. Suppose first that $f \geq 0$. Let

$$A = \{\boldsymbol{x} : f(\boldsymbol{x}) = +\infty\} \quad \text{and} \quad A_k = \{\boldsymbol{x} : f(\boldsymbol{x}) \geq k\}.$$

Since $f \geq f\mathbf{1}_{A_k} \geq k\mathbf{1}_{A_k} \geq k\mathbf{1}_A$,

$$0 \leq \lambda(A) \leq \frac{1}{k}\int f\,d\lambda < +\infty.$$

Letting $k \to +\infty$ shows that $\lambda(A) = 0$.

In the general case, apply the result of the first paragraph to f^+ and f^- to obtain

$$\lambda\big(\{\boldsymbol{x} : f^+(\boldsymbol{x}) = +\infty\}\big) = \lambda\big(\{\boldsymbol{x} : f^-(\boldsymbol{x}) = +\infty\}\big) = 0,$$

hence $\lambda\big(\{\boldsymbol{x} : |f(\boldsymbol{x})| = +\infty\}\big) = 0$. $\square$

11.2.7 Proposition. *Let* $f : \mathbb{R}^n \to [0, +\infty]$ *be Lebesgue measurable. Then* $\int f\,d\lambda = 0$ *iff* $f = 0$ *a.e.*

Proof. The sufficiency follows from 11.2.5. For the necessity, suppose $\int f\,d\lambda = 0$ and let

$$B = \{\boldsymbol{x} : f(\boldsymbol{x}) > 0\} \quad \text{and} \quad B_k = \{\boldsymbol{x} : f(\boldsymbol{x}) > 1/k\}.$$

Then $B = \bigcup_{k=1}^\infty B_k$ and $f \geq f\mathbf{1}_{B_k} \geq k^{-1}\mathbf{1}_{B_k}$ so

$$0 \leq \lambda(B_k) \leq k\int f\,d\lambda = 0.$$

Therefore, $\lambda(B_k) = 0$. By countable subadditivity, $\lambda(B) = 0$. $\square$

By 11.2.4, $f \geq 0$ implies that $\int_A f\,d\lambda \geq 0$ for all $A \in \mathcal{M}$. The following is a converse:

11.2.8 Proposition. *Let $f : \mathbb{R}^n \to \overline{\mathbb{R}}$ be Lebesgue measurable and suppose that $\int_A f\,d\lambda$ is defined for all $A \in \mathcal{M}$.*

(a) *If $\int_A f\,d\lambda \geq 0$ for all $A \in \mathcal{M}$, then $f \geq 0$ a.e.*

(b) *If $\int_A f\,d\lambda = 0$ for all $A \in \mathcal{M}$, then $f = 0$ a.e.*

Proof. Part (b) follows from part (a). To prove (a), let

$$A_k = \{\boldsymbol{x} : f(\boldsymbol{x}) \leq -k^{-1}\} \quad \text{and} \quad A = \{\boldsymbol{x} : f(\boldsymbol{x}) < 0\}.$$

Then $f\mathbf{1}_{A_k} \leq -k^{-1}\mathbf{1}_{A_k}$ or $\mathbf{1}_{A_k} \leq -kf\mathbf{1}_{A_k}$, hence, since $\int_{A_k} f \geq 0$,

$$0 \leq \lambda(A_k) \leq -k \int_{A_k} f\,d\lambda \leq 0.$$

Therefore, $\lambda(A_k) = 0$. Since $A = \bigcup_{k=1}^{\infty} A_k$, $\lambda(A) = 0$. □

11.2.9 Remark. The above properties of integrals on $\mathbb{R}^n$ also hold for integrals on $E \in \mathcal{M}$. For example, if f is integrable on E, then f is finite a.e. on E: simply replace f in 11.2.6 by $f \cdot \mathbf{1}_E$. This observation applies to most of the results that follow. We shall usually refrain from making this explicit, but the reader is invited to formulate and verify such generalizations. ◊

Linearity of the Integral

The following lemma is a special case of the monotone convergence theorem proved in the next section.

11.2.10 Lemma (Beppo–Levi). *If $\{f_k\}$ is a sequence of nonnegative Lebesgue measurable functions such that $f_k \uparrow f$ on $\mathbb{R}^n$, then*

$$\int f\,d\lambda = \lim_k \int f_k\,d\lambda.$$

Proof. By 10.5.3, f is Lebesgue measurable, hence $\int f\,d\lambda$ is defined. It follows from $0 \leq f_k \leq f_{k+1} \leq f$ and 11.2.4 that

$$\int f_k\,d\lambda \leq \int f_{k+1}\,d\lambda \leq \int f\,d\lambda.$$

Therefore, $L := \lim \int f_k\,d\lambda$ exists in $\overline{\mathbb{R}}$ and $L \leq \int f\,d\lambda$. For the reverse inequality, it suffices to show that $\int g\,d\lambda \leq L$ for any $g \in \mathcal{S}_+(\mathcal{M})$ with $g \leq f$. Let $0 < r < 1$ and set $E_k = \{\boldsymbol{x} : f_k(\boldsymbol{x}) \geq rg(\boldsymbol{x})\}$. Since the sequence $\{f_k\}$ is increasing, $E_k \subseteq E_{k+1}$. Since $f_k(\boldsymbol{x}) \geq rg(\boldsymbol{x})$ for all large k, $E_k \uparrow \mathbb{R}^n$. If g has standard form $\sum_{j=1}^m a_j \mathbf{1}_{A_j}$, then

$$f_k \geq f_k \mathbf{1}_{E_k} \geq r \sum_{j=1}^m a_j \mathbf{1}_{E_k \cap A_j},$$

hence

$$\int f_k \, d\lambda \geq r \sum_{j=1}^m a_j \lambda(E_k \cap A_j).$$

Letting $k \to +\infty$, noting that $E_k \cap A_j \uparrow_k A_j$, we then obtain

$$L \geq r \sum_{j=1}^m a_j \lambda(A_j) = r \int g \, d\lambda.$$

Letting $r \uparrow 1$ yields $L \geq \int g \, d\lambda$, as required. □

11.2.11 Theorem. *If f, $g : \mathbb{R}^n \to [0, +\infty]$ are Lebesgue measurable, then*

$$\int (\alpha f + \beta g) \, d\lambda = \alpha \int f \, d\lambda + \beta \int g \, d\lambda \quad \alpha, \beta \in \mathbb{R}^+.$$

In particular, if f and g are integrable then so is $\alpha f + \beta g$.

Proof. By 10.5.8, there exist sequences $\{f_k\}$ and $\{g_k\}$ in $\mathcal{S}_+(\mathcal{M})$ such that $f_k \uparrow f$ and $g_k \uparrow g$. Then $\alpha f_k + \beta g_k \uparrow f + g$ and, by 11.2.10 and 11.2.2,

$$\begin{aligned}\int (\alpha f + \beta g) \, d\lambda &= \lim_k \int (\alpha f_k + \beta g_k) \, d\lambda \\ &= \alpha \lim_k \int f_k \, d\lambda + \beta \lim_k \int g_k \, d\lambda \\ &= \alpha \int f \, d\lambda + \beta \int g \, d\lambda.\end{aligned}$$

□

11.2.12 Corollary. *Let f, $g : \mathbb{R}^n \to \overline{\mathbb{R}}$ be Lebesgue measurable.*

(a) *f is integrable iff $|f|$ is integrable.*

(b) *If f is integrable and $|g| \leq |f|$, then g is integrable.*

(c) *If f and g are integrable, then $f + g$ is integrable.*

(d) *If f is integrable and $E \in \mathcal{M}$, then f is integrable on E.*

Proof. (a) If f is integrable then, by definition, both f^+ and f^- are integrable, hence, by the theorem, $|f| = f^+ + f^-$ is integrable. Conversely, if $|f|$ is integrable, then the inequalities $0 \leq \int f^\pm \, d\lambda \leq \int |f| \, d\lambda$ show that both f^+ and f^- are integrable, hence f is integrable.

(b) By (a), $|f|$ is integrable. The inequality $|g| \le |f|$ then implies that $|g|$ is integrable. By (a) again, g is integrable.

(c) If f and g are integrable, then so are $|f|$ and $|g|$. The inequality $|f+g| \le |f| + |g|$ then shows that $|f+g|$ is integrable. By (a), $f+g$ is integrable.

(d) This follows from (b) since $|f\mathbf{1}_E| \le |f|$. □

The following theorem complements 11.2.11.

11.2.13 Theorem. *Let $f, g : \mathbb{R}^n \to \overline{\mathbb{R}}$ be Lebesgue measurable with g integrable, and let $c \in \mathbb{R}$. Then the following hold:*

(a) *cg is integrable and* $\displaystyle\int cg\,d\lambda = c\int g\,d\lambda.$

(b) *If f is integrable, then $f+g$ is integrable and*

$$\int (f+g)\,d\lambda = \int f\,d\lambda + \int g\,d\lambda. \tag{11.3}$$

(c) *If* $\displaystyle\int f\,d\lambda$ *is defined, then* $\displaystyle\int (f+g)\,d\lambda$ *is defined and* (11.3) *holds.*[1]

Proof. (a) If $c \ge 0$, then $(cg)^+ = cg^+$ and $(cg)^- = cg^-$, hence, by 11.2.11, the functions $(cg)^\pm$ are integrable and

$$\int cg\,d\lambda = \int (cg)^+\,d\lambda - \int (cg)^-\,d\lambda = c\int g^+\,d\lambda - c\int g^-\,d\lambda = c\int g\,d\lambda.$$

Next, observe that $(-g)^+ = g^-$ and $(-g)^- = g^+$ so

$$\int (-g)\,d\lambda = \int (-g)^+\,d\lambda - \int (-g)^-\,d\lambda = \int g^-\,d\lambda - \int g^+\,d\lambda = -\int g\,d\lambda.$$

Therefore, if $c < 0$,

$$\int cg\,d\lambda = \int (-c)(-g)\,d\lambda = -c\int (-g)\,d\lambda = c\int g\,d\lambda.$$

(b) By 11.2.12, $f+g$ is integrable. By 11.2.6, there exists a set A of Lebesgue measure zero such that $f(\boldsymbol{x}), g(\boldsymbol{x}) \in \mathbb{R}$ for $\boldsymbol{x} \in A^c$. Then on the set A^c,

$$(f+g)^+ - (f+g)^- = f+g = f^+ - f^- + g^+ - g^-, \quad \text{hence}$$
$$(f+g)^+ + f^- + g^- = (f+g)^- + f^+ + g^+.$$

---

[1]To avoid undefined expressions such as $\infty - \infty$ in the integrand $f+g$ in (b) and (c), it must be assumed that g is finite-valued. This is no real loss of generality since g is integrable, hence finite-valued a.e. (11.2.6).

By 11.2.11 and 11.2.5,

$$\int (f+g)^+ \, d\lambda + \int f^- \, d\lambda + \int g^- \, d\lambda = \int (f+g)^- \, d\lambda + \int f^+ \, d\lambda + \int g^+ \, d\lambda.$$

Since the integrals in this equation are finite, rearranging yields

$$\begin{aligned} \int (f+g)\, d\lambda &= \int (f+g)^+ \, d\lambda - \int (f+g)^- \, d\lambda \\ &= \int f^+ \, d\lambda - \int f^- \, d\lambda + \int g^+ \, d\lambda - \int g^- \, d\lambda \\ &= \int f \, d\lambda + \int g \, d\lambda. \end{aligned}$$

(c) The cases to be considered are

(i) $\int f^- \, d\lambda < +\infty$ and $\int f^+ \, d\lambda = +\infty$;

(ii) $\int f^+ \, d\lambda < +\infty$ and $\int f^- \, d\lambda = +\infty$.

Suppose that (i) holds. We may assume that both f^- and g are finite-valued. Since

$$\int (f+g)^- \, d\lambda \le \int (f^- + g^-) \, d\lambda < +\infty,$$

$\int (f+g)\, d\lambda$ is defined. If $\int (f+g)^+ \, d\lambda < +\infty$, then $(f+g)$ would be integrable, hence, by part (b,) so would $f^+ = (f+g) + f^- - g$, contrary to our assumption. Therefore,

$$\int (f+g)\, d\lambda = +\infty = \int f \, d\lambda + \int g \, d\lambda.$$

Case (ii) is similar (or apply Case (i) to $-f$). □

11.2.14 Corollary. *If f is integrable, then* $\left| \int f \, d\lambda \right| \le \int |f| \, d\lambda.$

Proof. Since $\pm f \le |f|$, $\pm \int f \, d\lambda = \int \pm f \, d\lambda \le \int |f| \, d\lambda.$ □

Approximation of Integrable Functions

11.2.15 Definition. For $E \in \mathcal{M}$ and $f \in \mathcal{L}^1(E)$ define the $\mathcal{L}^1$ *seminorm of* f by

$$\|f\|_1 := \int_E |f| \, d\lambda.$$

11.2.16 Theorem. *$\mathcal{L}^1(E)$ is a linear space and $\|\cdot\|_1$ has all the properties of a norm except the coincidence property.*

Proof. That $\mathcal{L}^1(E)$ is a linear space follows from 11.2.13. Coincidence may fail since $\|f\|_1 = 0$ only implies that $f = 0$ a.e. (Consider the Dirichlet function.) The other properties of a norm are easily established. □

11.2.17 Theorem. *Let $f \in \mathcal{L}^1(\mathbb{R}^n)$ and $\varepsilon > 0$. Then there exists a simple function g and a continuous function h, each vanishing outside a bounded interval, such that $\|f - g\|_1 < \varepsilon$ and $\|f - h\|_1 < \varepsilon$.*

Proof. By considering positive and negative parts, we may assume that $f \geq 0$. By definition of $\int f\,d\lambda$, there exists $f_s \in \mathcal{S}_+(\mathcal{M})$ with $f_s \leq f$ such that

$$\|f - f_s\|_1 = \int f\,d\lambda - \int f_s\,d\lambda < \varepsilon/4.$$

Let $f_s = \sum_{i=1}^m a_i \mathbf{1}_{A_i}$, where $a_i > 0$. Since

$$\sum_{i=1}^m a_i \lambda(A_i) = \int f_s\,d\lambda \leq \int f\,d\lambda < +\infty,$$

$\lambda(A_i) < +\infty$ for each i. Let $M = \max_i a_i$. By 10.1.6(d), there exists a bounded interval I such that

$$\lambda(A_i) - \lambda(I \cap A_i) < \varepsilon/(4Mm), \quad i = 1, \ldots, m.$$

Set $B_i := A_i \cap I$ and $g := \sum_{i=1}^m a_i \mathbf{1}_{B_i}$. Then

$$\|g - f_s\|_1 = \sum_{i=1}^m a_i \big[\lambda(A_i) - \lambda(B_i)\big] < \varepsilon/4,$$

hence

$$\|f - g\|_1 \leq \|f - f_s\|_1 + \|f_s - g\|_1 < \varepsilon/2.$$

To obtain h, for each i choose a compact set C_i and a bounded open set U_i such that $C_i \subseteq B_i \subseteq U_i$ and $\lambda(U_i \backslash C_i) < \varepsilon/(4mM)$ (10.4.6). By Exercise 8.5.15, there exists a continuous function $h_i : \mathbb{R}^n \to [0,1]$ such that $h_i = 1$ on C_i and $h_i = 0$ on U_i^c. Since $h_i - \mathbf{1}_{B_i} = 0$ on $C_i \cup U_i^c = (U_i \setminus C_i)^c$,

$$\|\mathbf{1}_{B_i} - h_i\|_1 = \int_{U_i \setminus C_i} |\mathbf{1}_{B_i} - h_i|\,d\lambda \leq 2\lambda(U_i \setminus C_i) < \varepsilon/2mM.$$

The function $h := \sum_{i=1}^m a_i h_i$ is continuous and by the triangle inequality $\|g - h\|_1 < \varepsilon/2$. Therefore, $\|f - h\|_1 < \varepsilon$, completing the proof. □

Translation Invariance of the Integral

11.2.18 Theorem. *If* $f : \mathbb{R}^n \to \overline{\mathbb{R}}$ *is Lebesgue measurable and* $\boldsymbol{y} \in \mathbb{R}^n$, *then*

$$\int f(\boldsymbol{x} + \boldsymbol{y})\, d\lambda(\boldsymbol{x}) = \int f(\boldsymbol{x})\, d\lambda(\boldsymbol{x}) \tag{11.4}$$

in the sense that if one side is defined, then so is the other and the integrals are then equal.

Proof. If $E \in \mathcal{M}$, then $E - \boldsymbol{y} \in \mathcal{M}$ and $\lambda(E - \boldsymbol{y}) = \lambda(E)$ (Exercise 10.3.1), hence

$$\int \mathbf{1}_E(\boldsymbol{y} + \boldsymbol{x})\, d\lambda(\boldsymbol{x}) = \int \mathbf{1}_{E-\boldsymbol{y}}\, d\lambda = \lambda(E - \boldsymbol{y}) = \int \mathbf{1}_E\, d\lambda.$$

Therefore, (11.4) holds for indicator functions.

For a function h, define $h_{\boldsymbol{y}}(\boldsymbol{x}) := h(\boldsymbol{y} + \boldsymbol{x})$. Let $f \geq 0$ and let $g \in \mathcal{S}_+(\mathcal{M})$ with $g \leq f$. Then $g_{\boldsymbol{y}} \leq f_{\boldsymbol{y}}$ and, by the first paragraph and linearity, $\int g = \int g_{\boldsymbol{y}}$, hence $\int g \leq \int f_{\boldsymbol{y}}$. Taking the supremum over g yields $\int f \leq \int f_{\boldsymbol{y}}$. Replacing $\boldsymbol{y}$ by $-\boldsymbol{y}$ and f by $f_{\boldsymbol{y}}$ in this inequality produces the reverse inequality. Therefore, (11.4) holds for $f \geq 0$. The general case follows from this and the identities $(f^{\pm})_{\boldsymbol{y}} = (f_{\boldsymbol{y}})^{\pm}$. □

Exercises

1. Let f and g be integrable. Prove:
   (a) If $\int_E f\, d\lambda \leq \int_E g\, d\lambda$ for all $E \in \mathcal{M}$, then $f \leq g$ a.e.
   (b) If $\int_E f\, d\lambda = \int_E g\, d\lambda$ for all $E \in \mathcal{M}$, then $f = g$ a.e.

2. Let $f(x) = 1 + r(\lfloor x^{-1} \rfloor)$ for $0 < x \leq 1$, where $r(k)$ is the remainder on division of the positive integer k by 3. (Cf. Exercise 10.5.15.) Show that

$$\int_{(0,1]} f\, d\lambda = \frac{3}{2} + \sum_{k=1}^{\infty} \left[\frac{1}{3k(3k+1)} + \frac{2}{(3k+1)(3k+2)} + \frac{3}{(3k+2)(3k+3)}\right].$$

3. Define $f : [0, 1] \to \mathbf{R}$ by $f(x) = 0$ if x is rational, and $f(x) = d^2$ if x is irrational, where d is the first nonzero digit in the decimal expansion of x. (See Exercise 10.5.16.) Show that $\int_{[0,1]} f\, d\lambda = 95/3$.

4. (a) Prove the following mean value theorem for integrals: Let f be continuous on a compact connected set $K \subseteq \mathbb{R}^n$. Then there exists $\boldsymbol{x}_K \in K$ such that

$$\int_K f\, d\lambda = f(\boldsymbol{x}_K)\lambda(K).$$

   (b) Let f be continuous on $C_1(\boldsymbol{x}_0)$. Prove that

$$\lim_{r \to 0} \frac{1}{\lambda(C_r(\boldsymbol{x}_0))} \int_{C_r(\boldsymbol{x}_0)} f\, d\lambda = f(\boldsymbol{x}_0).$$

5. Let f be Lebesgue measurable on $\mathbb{R}$ and let $m \le f \le M$ on $E \in \mathcal{M}(\mathbb{R})$.

   (a) Prove that if g is integrable on E, then there exists $a \in [m, M]$ such that
$$\int_E f|g|\,d\lambda = a\int_E |g|\,d\lambda$$

   (b) Show that part (a) may be false if $|g|$ is replaced by g.

   (c) Use (a) to show that at each point x where f is continuous,
$$\lim_{y\to x}\frac{1}{y-x}\left[\int_{[a,y]} f\,d\lambda - \int_{[a,x]} f\,d\lambda\right] = f(x).$$

6. (Cauchy–Schwarz inequality) Let f and g be Lebesgue measurable on $\mathbb{R}^n$. Prove that
$$\left(\int |fg|\,d\lambda\right)^2 \le \int f^2\,d\lambda \cdot \int g^2\,d\lambda.$$
   (See 5.7.19.)

7. Prove that if f is integrable on $[0,1]$ and $\varepsilon > 0$, then there exists a polynomial P on $[0,1]$ such that $\int_0^1 |f - P|\,d\lambda < \varepsilon$.

8. (Absolute continuity of the integral). Let $f \ge 0$ be integrable on $\mathbb{R}^n$. Prove that for each $\varepsilon > 0$ there exists a $\delta > 0$ such that
$$\int_E f\,d\lambda < \varepsilon \quad \text{for all } E \in \mathcal{M}(\mathbb{R}^n) \text{ with } \lambda(E) < \delta.$$
   Conclude that if $\{E_k\}$ is a sequence in $\mathcal{M}(\mathbb{R}^n)$ with $\lambda(E_k) \to 0$, then $\int_{E_k} f\,d\lambda \to 0$. *Hint.* Begin with simple functions.

9. Let f be integrable. Prove that $\lim_k \int_{[k,k+1]} f d\lambda = 0$. (A quick proof uses the dominated convergence theorem. For now, give a proof starting with simple functions.)

10. Suppose $f : I = [0,1] \to [-1,1]$ is integrable. Prove that
$$\int_I f^2\,d\lambda \le \varepsilon^2 + \lambda\big(\{x : |f(x)| > \varepsilon\}\big) \quad \text{for every } \varepsilon > 0.$$

11. Let $f : I = [0,1] \to \mathbb{R}$ be integrable. Prove that
$$\int_I f^2\,d\lambda \ge \varepsilon^2\lambda\big(\{x \in I : |f(x)| > \varepsilon\}\big) \quad \text{for every } \varepsilon > 0.$$

12. Prove: If f is integrable and $f < 1$ a.e. on $I = [0,1]$, then $\int_I f < 1$.

13. Let f be Lebesgue integrable on $E \in \mathcal{M}$ with $0 < \lambda(E) < +\infty$ and $\int_E f\,d\lambda \geq \lambda(E)$. Prove that $\lambda\{\boldsymbol{x} \in E : f(\boldsymbol{x}) \geq 1\} > 0$.

14. Let f be integrable on $\mathbb{R}^n$. Show that for each $r > 0$ the function $f_r(\boldsymbol{x}) := f(r\boldsymbol{x})$ is integrable and $\int f_r\,d\lambda = r^{-n}\int f\,d\lambda$.

15. Let $f : [0,1] \to \mathbf{R}$ be a bounded Lebesgue measurable function such that $\int_{[0,1]} x^{2k} f(x)\,d\lambda(x) = 0$ for all $k \in \mathbb{Z}^+$. Prove that $f = 0$ a.e.

16. Let f be Lebesgue integrable and g, g' bounded and continuous on $\mathbb{R}$. Carry out the following steps to show that

$$\lim_k \int f(x)g'(kx)\,d\lambda(x) = 0. \tag{11.5}$$

    (a) Prove (11.5) for $f = \mathbf{1}_{[a,b]}$. (Use the fact that the Riemann and Lebesgue integrals of a continuous function on a closed bounded interval are equal (Section 11.4)

    (b) Use (a) to show that (11.5) holds for $f = \mathbf{1}_U$, where U is bounded and open.

    (c) Use (b) and 10.4.7 to show that (11.5) holds for $f = \mathbf{1}_E$, where $E \in \mathcal{M}$ is bounded.

    (d) Use (c) and 11.2.17 to complete the proof.

    If $g'(x) = \sin x$ or $\cos x$, then (11.5) is known as the *Riemann–Lebesgue lemma.*

11.3 Convergence Theorems

In this section we state and prove three pointwise convergence theorems for the Lebesgue integral. The first of these is a generalization of 11.2.10.

11.3.1 Monotone Convergence Theorem. *Let* $f_k : \mathbb{R}^n \to \overline{\mathbb{R}}$ *be Lebesgue measurable with* $f_k \uparrow f$ *on* $\mathbb{R}^n$ *and let* $\int f_1^-\,d\lambda < +\infty$. *Then*

$$\int f\,d\lambda = \lim_k \int f_k\,d\lambda. \tag{11.6}$$

Proof. From $0 \leq f^- \leq f_k^- \leq f_1^-$ we have $\int f_k^-\,d\lambda < +\infty$ and $\int f^-\,d\lambda < +\infty$, hence the integrals in the assertion of the theorem are defined. If $\int f_1^+\,d\lambda = +\infty$, then from $f_1^+ \leq f_k^+ \leq f^+$ we see that each side of (11.6) is $+\infty$. If $\int f_1^+\,d\lambda < +\infty$, then f_1 is integrable and we may apply 11.2.10 to $f_k - f_1$ (≥ 0) to obtain

$$\int f_k\,d\lambda = \int (f_k - f_1)\,d\lambda + \int f_1\,d\lambda \to \int (f - f_1)\,d\lambda + \int f_1\,d\lambda = \int f\,d\lambda. \ \square$$

11.3.2 Remark. Equation (11.6) is still true if the inequalities $f_k \le f_{k+1} \le f$ and the convergence $f_k \uparrow f$ hold only almost everywhere. To see this, let A denote the set on which $f_k \le f_{k+1}$ for all k and $f_k \uparrow f$. Set $\tilde{f}_k = f_k \mathbf{1}_A$ and $\tilde{f} = f\mathbf{1}_A$. Then (11.6) holds for the new functions. Since $\lambda(A^c) = 0$, 11.2.5 shows that the equation holds for the original functions. Analogous remarks apply to the other convergence theorems in this section. ◊

11.3.3 Corollary. *If g_k is Lebesgue measurable and nonnegative for every k, then*

$$\int \Big(\sum_{k=1}^{\infty} g_k\Big)\, d\lambda = \sum_{k=1}^{\infty} \int g_k \, d\lambda.$$

Proof. Let $f_k = \sum_{j=1}^{k} g_j$ and $f = \sum_{j=1}^{\infty} g_j$. Then $0 \le f_k \uparrow f$ on $\mathbb{R}^n$, hence, by the theorem and linearity,

$$\int f \, d\lambda = \lim_k \int f_k \, d\lambda = \lim_k \sum_{j=1}^{k} \int g_j \, d\lambda. \qquad \square$$

11.3.4 Corollary. *Let $f \ge 0$ be Lebesgue measurable. Define a function μ on $\mathcal{M}(\mathbb{R}^n)$ by*

$$\mu(E) := \int_E f \, d\lambda, \quad E \in \mathcal{M}(\mathbb{R}^n).$$

Then μ is a measure on $\mathcal{M}(\mathbb{R}^n)$.

Proof. For countable additivity, apply 11.3.3 to $g_k = \mathbf{1}_{E_k}$. □

11.3.5 Fatou's Lemma. *If f_k is nonnegative and Lebesgue measurable for every k, then*

$$\int \liminf_k f_k \, d\lambda \le \liminf_k \int f_k \, d\lambda. \tag{11.7}$$

Proof. Let $g_k = \inf_{j \ge k} f_j$ and $g = \liminf_k f_k$. Then $g_k \le f_k$, $g_k \uparrow g$, and g_k and g are Lebesgue measurable (10.5.3). By the monotone convergence theorem,

$$\int \liminf_k f_k \, d\lambda = \int g \, d\lambda = \lim_k \int g_k \, d\lambda = \liminf_k \int g_k \, d\lambda \le \liminf_k \int f_k \, d\lambda. \quad \square$$

The inequality in (11.7) may be strict. For example, if $f_k = k\mathbf{1}_{[0,1/k]}$, then the left side is zero while the right side is one.

11.3.6 Dominated Convergence Theorem. *Let $g : \mathbb{R}^n \to [0, +\infty]$ be integrable and let $\{f_k : \mathbb{R}^n \to \overline{\mathbb{R}}\}$ be a sequence of Lebesgue measurable functions such that $|f_k| \le g$ for all k. If $f_k \to f$ on $\mathbb{R}^n$, then f is integrable and $\int f_k \, d\lambda \to \int f \, d\lambda$.*

Proof. Since $|f| \le g$, f_k and f are integrable (11.2.12). Fatou's lemma applied to $g \pm f_k$ (≥ 0) shows that

$$\int g\,d\lambda + \int f\,d\lambda \le \liminf_k \int (g + f_k)\,d\lambda = \int g\,d\lambda + \liminf_k \int f_k\,d\lambda$$

and

$$\int g\,d\lambda - \int f\,d\lambda \le \liminf_k \int (g - f_k)\,d\lambda = \int g\,d\lambda - \limsup_k \int f_k\,d\lambda.$$

Subtracting $\int g\,d\lambda$ in each inequality yields

$$\int f\,d\lambda \le \liminf_k \int f_k\,d\lambda \le \limsup_k \int f_k\,d\lambda \le \int f\,d\lambda. \qquad \square$$

The following example illustrates that care must be taken when applying the dominated convergence theorem.

11.3.7 Example. Let $p > 0$ and define

$$f_k(x) := \frac{k}{1 + k^2 x^{2p}}, \quad 0 < x \le 1, \quad \text{and} \quad I_k := \int_{(0,1]} f_k\,d\lambda, \quad k \in \mathbb{N}.$$

Clearly, $f_k \to 0$ for all $p > 0$. We show that $\lim_k I_k = 0$ iff $0 < p < 1$. By Section 11.4, below, the integrals are Riemann, hence, making the substitution $t = kx^p$ and setting $q = p^{-1} - 1$, we obtain

$$I_k = \int_0^1 \frac{k}{1 + k^2 x^{2p}}\,dx = \frac{1}{pk^q} \int_0^k \frac{t^q}{1 + t^2}\,dt = \int g_k\,d\lambda,$$

where

$$g_k(t) = \frac{1}{pk^q} \frac{t^q}{1 + t^2} \mathbf{1}_{[0,k]}.$$

If $p = 1$, then $q = 0$ and $I_k = \arctan k \to \pi/2$. If $0 < p < 1$, then $g_k \to 0$ and $g_k(t) \le p^{-1}(1 + t^2)^{-1}$ for all $t \ge 0$ and all k, so $I_k \to 0$ by the dominated convergence theorem. Finally, if $p > 1$, then $-1 < q < 0$ and

$$I_k \ge \frac{1}{pk^q} \int_0^1 \frac{1}{1 + t^2}\,dt \to +\infty. \qquad \lozenge$$

The following theorem gives general conditions under which one may "differentiate under the integral sign."

11.3.8 Theorem. *Let $f(x, y)$ be Lebesgue measurable on $I := (a, b) \times (c, d)$ such that for each y in (c, d) the function $f(\cdot, y)$ is Lebesgue integrable on (a, b) and the derivative f_y exists on I. If there exists an integrable function g on (a, b) such that $|f_y(x, y)| \le g(x)$ for all $(x, y) \in I$, then*

$$\frac{d}{dy} \int_{(a,b)} f(x, y)\,d\lambda(x) = \int_{(a,b)} \frac{\partial f}{\partial y}(x, y)\,d\lambda(x).$$

Proof. We prove the right-hand derivative version. Let $y \in (c, d)$ and $y_k \downarrow y$. Set

$$G(y) = \int_{(a,b)} f(x, y)\, d\lambda(x) \quad \text{and} \quad g_k(x) = \frac{f(x, y_k) - f(x, y)}{y_k - y}.$$

By the mean value theorem, $g_k(x) = f_y(x, t_k)$ for some $t_k \in (y, y_k)$, hence $|g_k| \le g$. Since $g_k(x) \to f_y(x, y)$, the dominated convergence theorem implies that

$$\frac{G(y_k) - G(y)}{y_k - y} = \int_{(a,b)} g_k(x)\, d\lambda(x) \to \int_{(a,b)} f_y(x, y)\, d\lambda(x).$$

Since $\{y_k\}$ was arbitrary, $G_r'(y)$ exists and equals $\int_{(a,b)} f_y(x, y)\, d\lambda(x)$. □

Exercises

1. Prove the following:

   (a) $\lim_k \int_{[0,\pi]} (\sin^k x)\,(1 - \sin x)^k \, d\lambda(x) = 0$

   (b) $\lim_k \int_{[0,+\infty)} \dfrac{k \sin\left(x^{3/2}\right)}{1 + k^2x^2}\, d\lambda(x) = 0.$

2. Let $f : \mathbb{R}^n \to (0, +\infty)$ be integrable. Prove that

   (a) $\int k \ln(1 + k^{-1}f)\, d\lambda \to \int f\, d\lambda$ (b) $\int k \ln(1 + k^{-2}f)\, d\lambda \to 0$

   (c) $\int k \sin\left(k^{-1}f\right) d\lambda \to \int f\, d\lambda$ (d) $\int_E f^{1/k}\, d\lambda \to \lambda(E),\ E \in \mathcal{M}.$

3. Let $f,\ g : \mathbb{R}^n \to (0, +\infty)$ be Lebesgue measurable with g integrable. Prove:

   $$\int g(1 + k^{-1}f)^k \exp(-f)\, d\lambda \to \int g\, d\lambda.$$

4. Let $f : \mathbb{R}^n \to [1, +\infty)$ be Lebesgue measurable and $g : \mathbb{R}^n \to [0, +\infty)$ integrable. Prove that

   $$\int k^2 g \exp(-kf)\, d\lambda \to 0.$$

5. Let f be integrable on $(0, \infty)$. Show that for each $t \in \mathbb{R}$ the function $f(x)\sin(tx)/x$ is integrable on $(0, \infty)$ and prove that the integral $\int_{(0,\infty)} f(x)x^{-1}\sin(tx)\, d\lambda(x)$ is continuous in t.

6. Prove that the derivative of the gamma function (5.7.8) is

   $$\Gamma'(x) = \int_0^\infty t^{x-1}e^{-t}\ln t\, dt, \quad x > 0.$$

(Use the fact, proved in the next section, that the improper Riemann integral and the Lebesgue integral of a nonnegative continuous function are equal.)

7. Let $f : [0, +\infty) \to \mathbb{R}$ be bounded and Lebesgue measurable and suppose that $\lim_{x \to +\infty} f(x) = r$. Show that

$$\lim_k \int_{[0,a]} f(kx)\, d\lambda(x) = ar \quad \text{for every } a > 0.$$

8. Let $f : \mathbb{R}^n \to \mathbb{R}$ be Lebesgue measurable and have countable range $\{a_1, a_2, \ldots\}$. Set $A_k = \{\boldsymbol{x} \in \mathbb{R}^n : f(\boldsymbol{x}) = a_k\}$. Prove that f is integrable iff the series $\sum_{k=1}^{\infty} a_k \lambda(A_k)$ converges absolutely, in which case the value of the series is $\int f\, d\lambda$.

9. Let $p > 1$ and $f(x) := \lfloor x^{-1} \rfloor^{-p}$, $0 < x < 1$. Find $\int_{(0,1)} f\, d\lambda$.

10. Let f_k, f be integrable and E_k, $E \in \mathcal{M}(\mathbb{R}^n)$ such that

$$\lim_k \|f_k - f\|_1 = 0 \quad \text{and} \quad \lim_k \lambda(E_k \Delta E) = 0$$

(see Exercise 10.5.10). Prove that

$$\lim_k \int_{E_k} f_k\, d\lambda = \int_E f\, d\lambda.$$

11. Let $f : \mathbb{R}^n \to \overline{\mathbb{R}}$ be integrable and $\varepsilon > 0$.

(a) Prove that the set $A = \{\boldsymbol{x} : |f(\boldsymbol{x})| \geq \varepsilon\}$ has finite measure.

(b) Show that there exists $B \in \mathcal{M}$ with $\lambda(B) < +\infty$ such that

$$\left| \int f\, d\lambda - \int_B f\, d\lambda \right| < \varepsilon.$$

12. Let $\{f_k\}$ be a sequence of integrable functions on $\mathbb{R}^n$ such that $\displaystyle\sum_{k=1}^{\infty} \|f_k\|_1 < +\infty$. Prove that $\lim_k f_k(\boldsymbol{x}) = 0$ a.e.

13. Let f be Lebesgue integrable on $\mathbb{R}$ and $p > 0$. Prove that the series $\sum_{k=1}^{\infty} k^{-p} f(kx)$ converges absolutely a.e. on $\mathbb{R}$.

14. Let $T : \mathcal{C}([a,b]) \to \mathcal{C}([a,b])$ be linear and continuous in the $\mathcal{L}^1$ norm. If $f : [a,b] \times [c,d] \to \mathbb{R}$ is continuous, prove that

$$T \int_c^d f(\cdot, x)\, dx = \int_c^d T f(\cdot, x)\, dx$$

where the integrals may be taken to be Riemann.

15. Prove the following extension of Fatou's lemma: If f_k, g are Lebesgue integrable on $\mathbb{R}^n$ and $f_k \geq g$ for all k, then

$$\int \liminf_k f_k \, d\lambda \leq \liminf_k \int f_k \, d\lambda.$$

16. Let g be integrable on $\mathbb{R}^n$ and let $\{f_k\}$ be a sequence of Lebesgue measurable functions on $\mathbb{R}^n$ such that $|f_k| \leq g$. Show that

$$\int \liminf_k f_k \leq \liminf_k \int f_k \leq \limsup_k \int f_k \leq \int \limsup_k f_k.$$

17. Let f, f_k be nonnegative Lebesgue integrable functions on $\mathbb{R}^n$ such that $f_k \to f$. Prove that

$$\int f_k \, d\lambda \to \int f \, d\lambda \text{ iff } \|f_k - f\|_1 \to 0.$$

Hint. For the necessity, note that $(f_k - f)^- \leq f$.

18. Let f, f_k be integrable on $\mathbb{R}^n$ with $f_k \to f$. Prove that

$$\|f_k - f\|_1 \to 0 \text{ iff } \int |f_k| \, d\lambda \to \int |f| \, d\lambda.$$

Hint. For the sufficiency use Fatou's lemma.

19. Let f be Lebesgue integrable on $\mathbb{R}$ such that $\int_{[a,b]} f \, d\lambda = 0$ for all intervals $[a, b]$. Prove that $f = 0$ a.e. *Hint.* Use 11.3.4, 10.4.4, and Exercise 11.2.8.

20. Let $f : \mathbb{R}^2 \to \mathbb{R}$ have the property that $f(x, y)$ is Lebesgue measurable in y for each x and continuous in x for each y. Suppose there exists an integrable function $g : \mathbb{R} \to \mathbb{R}$ such that $|f(x, y)| \leq g(y)$ for all x and y. Prove that the function

$$F(x) := \int f(x, y) \, d\lambda(y)$$

is continuous.

21. Let $f \geq 0$ be integrable on $[1, +\infty)$. Prove that $\sum_{k=1}^{\infty} f(x+k)$ is integrable on $[0, 1]$. Conclude that the series converges a.e. on $[1, +\infty)$.

22. Let f be Lebesgue measurable on $I = [0, 1]$ and set

$$A_k = \{x \in I : |f(x)| \geq k\}.$$

Prove:

(a) f is integrable on I iff $\sum_{k=0}^{\infty} \lambda(A_k)$ converges.

(b) If f is integrable on $I = [0, 1]$, then $\lim_k k\lambda(A_k) = 0$.

11.4 Connections with Riemann Integration

Throughout the section, f denotes an arbitrary bounded real-valued function on a closed and bounded interval $[a, b]$.

In this section we show that f is Riemann integrable if and only if its set of discontinuities has Lebesgue measure zero. The first step is to show that the upper and lower integrals of f may be expressed as integrals of Borel measurable functions.

By 5.2.1 there exists a sequence of partitions $\{\mathcal{P}_k\}$ of $[a, b]$ such that $\mathcal{P}_{k+1}$ is a refinement of $\mathcal{P}_k$, $\|\mathcal{P}_k\| \to 0$, and

$$\underline{\int_a^b} f = \lim_k \underline{S}(f, \mathcal{P}_k), \quad \overline{\int_a^b} f = \lim_k \overline{S}(f, \mathcal{P}_k).$$

Define

$$h_k = \sum_j m_j \mathbf{1}_{[x_{j-1}, x_j]} \quad \text{and} \quad g_k = \sum_j M_j \mathbf{1}_{[x_{j-1}, x_j]},$$

where

$$m_j := \inf_{x_{j-1} \le x \le x_j} f(x), \quad M_j := \sup_{x_{j-1} \le x \le x_j} f(x),$$

and the intervals $[x_{j-1}, x_j]$ are those generated by the partition $\mathcal{P}_k$. Then g_k and h_k are Borel measurable simple functions and

$$\underline{S}(f, \mathcal{P}_k) = \int_{[a,b]} h_k \, d\lambda, \quad \overline{S}(f, \mathcal{P}_k) = \int_{[a,b]} g_k \, d\lambda.$$

Moreover,

$$h_1 \le h_2 \le \cdots \le f \le \ldots \le g_2 \le g_1,$$

hence $h(x) := \lim_k h_k(x)$ and $g(x) := \lim_k g_k(x)$ exist in $\mathbb{R}$ for each $x \in [a, b]$, $h \le f \le g$, and h and g are Borel measurable. If M is a bound for $|f|$, then $|h_k|, |g_k| \le M$ a.e., hence by the dominated convergence theorem

$$\underline{\int_a^b} f = \lim_k \underline{S}(f, \mathcal{P}_k) = \lim_k \int_{[a,b]} h_k \, d\lambda = \int_{[a,b]} h \, d\lambda \tag{11.8}$$

and

$$\overline{\int_a^b} f = \lim_k \overline{S}(f, \mathcal{P}_k) = \lim_k \int_{[a,b]} g_k \, d\lambda = \int_{[a,b]} g \, d\lambda. \tag{11.9}$$

11.4.1 Lemma. *$f \in \mathcal{R}_a^b$ iff $g = h$ a.e. In this case, f is Lebesgue measurable and $\int_a^b f = \int_{[a,b]} f \, d\lambda$.*

Proof. From (11.8) and (11.9), $f \in \mathcal{R}_a^b$ iff $\int_{[a,b]}(g-h)\,d\lambda = 0$, which, by 11.2.7, is equivalent to $g = h$ a.e. If this holds, then $h = f = g$ a.e. so f is Lebesgue measurable and $\int_a^b f = \int_{[a,b]} f\,d\lambda$ by (11.8) and (11.9). □

11.4.2 Lemma. *Suppose that $x \in [a,b]$ is not a member of any of the partitions $\mathcal{P}_k$. Then f is continuous at x iff $h(x) = g(x)$.*

Proof. Suppose f is continuous at x. Given $\varepsilon > 0$, choose $\delta > 0$ such that $y \in [a,b]$ and $|x-y| < \delta$ implies $|f(x)-f(y)| < \varepsilon$. Choose N so that $\|\mathcal{P}_k\| < \delta$ for all $k \ge N$ and fix $k \ge N$. Since x is in some subinterval (x_{j-1}, x_j) of $\mathcal{P}_k$,

$$f(x) - \varepsilon < f(y) < f(x) + \varepsilon \ \text{ for all } y \in [x_{j-1}, x_j],$$

hence

$$f(x) - \varepsilon \le h_k(x) = m_j \le M_j = g_k(x) \le f(x) + \varepsilon.$$

Letting $k \to +\infty$ yields

$$f(x) - \varepsilon \le h(x) \le g(x) \le f(x) + \varepsilon,$$

and since ε was arbitrary, $g(x) = h(x)$.

Conversely, let $g(x) = h(x)$. Given $\varepsilon > 0$, choose k such that

$$|g_k(x) - g(x)| < \varepsilon \ \text{ and } \ |h_k(x) - h(x)| < \varepsilon.$$

Suppose that x is in the open subinterval (x_{i-1}, x_i) of $\mathcal{P}_k$. Choose $\delta > 0$ so that $(x-\delta, x+\delta) \subseteq (x_{i-1}, x_i)$. Then for all $y \in (x-\delta, x+\delta)$,

$$h(x) - \varepsilon \le h_k(x) \le f(y) \le g_k(x) \le g(x) + \varepsilon = h(x) + \varepsilon,$$

which implies that $|f(x) - f(y)| < 2\varepsilon$. Therefore, f is continuous at x. □

Here is the main result of the section.

11.4.3 Theorem. *Let $f : [a,b] \to \mathbb{R}$ be bounded. Then $f \in \mathcal{R}_a^b$ iff the set D of discontinuities of f has Lebesgue measure zero. In this case, f is Lebesgue measurable and*

$$\int_a^b f(x)\,dx = \int_{[a,b]} f\,d\lambda.$$

Proof. Let A denote the union of the partitions $\mathcal{P}_k$ and set $B = \{x : g(x) \ne h(x)\}$. By 11.4.2,

$$B \cap A^c \subseteq D \subseteq A \cup B$$

Since A is countable, $\lambda(A) = 0$, hence $\lambda(B \cap A^c) = \lambda(A \cup B) = \lambda(B)$. It follows that $\lambda(B) = \lambda(D)$. Thus, by 11.4.1, $f \in \mathcal{R}_a^b$ iff $\lambda(D) = 0$. □

11.4.4 Example. Let $A := (0,1) \setminus E$, where E is the Cantor ternary set (10.3.4). Since A is open, the function $f(x) = \mathbf{1}_A(x)\sin(\pi x)$ is continuous on A. Since $\lambda(E) = 0$, f is both Riemann and Lebesgue integrable on $[0,1]$ and

$$\int_0^1 f(x)\,dx = \int_0^1 \sin(\pi x)\,dx = \frac{2}{\pi}. \qquad \Diamond$$

11.4.5 Remark. Theorem 11.4.3 readily extends to n-dimensional Riemann integrals; the statement and proof are essentially the same. Note that in this case, a Riemann integrable function f may be discontinuous on m-dimensional hyperplanes, $m < n$, as these have Lebesgue measure zero (see 11.6.8). $\Diamond$

Here is the connection between improper integrals and Lebesgue integrals.

11.4.6 Corollary. *Let g be locally Riemann integrable on $[a,b)$ (where b could be infinite). Then g is Lebesgue measurable on $[a,b)$. Moreover:*

(a) *If $g \geq 0$, then g is improperly integrable on $[a,b)$ iff g is Lebesgue integrable on $[a,b)$, in which case*

$$\int_a^b g = \int_{[a,b)} g\,d\lambda. \tag{11.10}$$

(b) *If g is Lebesgue integrable on $[a,b)$, then g is improperly integrable on $[a,b)$ and (11.10) holds.*

(c) *If g is improperly integrable on $[a,b)$, then g need not be Lebesgue integrable on $[a,b)$.*

Proof. (a) Let $b_k \uparrow b$ and let D denote the set of discontinuities of g on $[a,b)$. Since g is Riemann integrable on $[a,b_k]$, $\lambda([a,b_k] \cap D) = 0$. By the theorem, $\mathbf{1}_{[a,b_k]}g$ is Lebesgue measurable for every k and

$$\int_a^{b_k} g = \int_{[a,b_k]} g\,d\lambda = \int \mathbf{1}_{[a,b_k]} g\,d\lambda.$$

Taking limits we see that g is Lebesgue measurable and, by the monotone convergence theorem, 11.10 holds.

(b) If g is Lebesgue integrable on $[a,b)$, then, by (a), g^+ and g^- are improperly integrable on $[a,b)$ hence (b) holds.

(c) The function

$$g(x) = x^{-1}\sin x$$

is improperly integrable but not absolutely improperly integrable on $[1,+\infty)$ (5.7.18). Since a Lebesgue integrable function is absolutely integrable, g cannot be Lebesgue integrable on $[1,+\infty)$. $\square$

11.5 Iterated Integrals

For the remainder of the text, we also use the notation $d\boldsymbol{x}$ for $d\lambda(\boldsymbol{x})$ and $\int_a^b f(x)\,dx$ for $\int_{[a,b]} f(x)\,d\lambda(x)$, etc.

In this section we state and prove a result that gives general conditions under which the Lebesgue integral of a function on $\mathbb{R}^n$ may be expressed as an iterated integral, a useful tool for evaluating integrals.

11.5.1 Fubini–Tonelli Theorem. *Let f be Borel measurable on $\mathbb{R}^n$ and let $p, q \in \mathbb{N}$ with $p + q = n$.*

(a) *If $f \geq 0$, then the functions*

$$\int_{\mathbb{R}^q} f(\boldsymbol{x}, \mathbf{z})\,d\boldsymbol{z} \quad \text{and} \quad \int_{\mathbb{R}^p} f(\mathbf{z}, \boldsymbol{y})\,d\boldsymbol{z}$$

are Borel measurable in $\boldsymbol{x} \in \mathbb{R}^p$ and $\boldsymbol{y} \in \mathbb{R}^q$, respectively, and

$$\int_{\mathbb{R}^n} f(\mathbf{z}, \boldsymbol{y})\lambda(\,d\boldsymbol{z},\, d\boldsymbol{y}) = \int_{\mathbb{R}^p}\int_{\mathbb{R}^q} f(\boldsymbol{x}, \mathbf{z})\,d\boldsymbol{z}\,d\boldsymbol{x} = \int_{\mathbb{R}^q}\int_{\mathbb{R}^p} f(\mathbf{z}, \boldsymbol{y})\,d\boldsymbol{z}\,d\boldsymbol{y}. \tag{11.11}$$

(b) *If either of the iterated integrals*

$$\int_{\mathbb{R}^p}\int_{\mathbb{R}^q} |f(\boldsymbol{x}, \mathbf{z})|\,d\boldsymbol{z}\,d\boldsymbol{x} \quad \text{or} \quad \int_{\mathbb{R}^q}\int_{\mathbb{R}^p} |f(\mathbf{z}, \boldsymbol{y})|\,d\boldsymbol{z}\,d\boldsymbol{y}$$

is finite, then both are finite, f is integrable, and (11.11) holds.

By induction we have

11.5.2 Corollary. *Let f be Borel measurable on $\mathbb{R}^n$ such that*

$$\int_{-\infty}^{\infty}\cdots\int_{-\infty}^{\infty} |f(x_1, \ldots, x_n)|\,dx_{i_1}\cdots dx_{i_n} < +\infty$$

for some *permutation $(i_1, \ldots, i_n)$ of $(1, \ldots, n)$. Then f is integrable and*

$$\int_{\mathbb{R}^n} f\,d\lambda = \int_{-\infty}^{\infty}\cdots\int_{-\infty}^{\infty} f(x_1, \ldots, x_n)\,dx_{j_1}\cdots dx_{j_n}.$$

for every permutation $(j_1, \ldots, j_n)$ of $(1, \ldots, n)$.

11.5.3 Example. We prove the *Gaussian density formula*

$$\int_{-\infty}^{\infty} \varphi(t)\,dt = 1, \quad \text{where} \quad \varphi(t) := \frac{e^{-t^2/2}}{\sqrt{2\pi}}. \tag{11.12}$$

By 11.4.6, the integral may be interpreted either as a Lebesgue integral or as an improper Riemann integral. The function φ is called the *standard normal* (or *Gaussian*) *density.* It plays an important role in probability and statistics. For example, $\sigma^{-1}\int_a^b \varphi\big[(x-\mu)/\sigma\big]\,dx$ is the probability that randomly chosen data from a normally distributed population with mean μ and standard deviation σ lies between a and b, and $\sigma^{-1}\int_{-\infty}^{\infty}\varphi\big[(x-\mu)/\sigma\big]x\,dx$ is the data average.

To verify (11.12) note that because the integrand is an even function, the left side is $2\int_0^\infty \varphi(t)\,dt$. By a change of variable,

$$2\int_0^\infty \varphi(t)\,dt = \frac{2}{\sqrt{2\pi}}\int_0^\infty e^{-t^2/2}\,dt = \frac{2}{\sqrt{\pi}}\int_0^\infty e^{-t^2}\,dt.$$

Thus it suffices to show that

$$\frac{2}{\sqrt{\pi}}\int_0^\infty e^{-t^2}\,dt = 1.$$

Let I denote the integral on the left. Then

$$\begin{aligned}
I^2 &= \int_0^\infty e^{-y^2}\int_0^\infty e^{-t^2}\,dt\,dy \\
&= \int_0^\infty e^{-y^2}\int_0^\infty ye^{-x^2y^2}\,dx\,dy, && \text{by the substitution } t = xy \\
&= \int_0^\infty\int_0^\infty ye^{-y^2(1+x^2)}\,dy\,dx, && \text{by 11.5.1} \\
&= \frac{1}{2}\int_0^\infty (1+x^2)^{-1}\int_0^\infty e^{-u}\,du\,dx, && \text{by the substitution } u = y^2(1+x^2) \\
&= \frac{1}{2}\arctan x\Big|_0^\infty && \text{because } \textstyle\int_0^\infty e^{-u}\,du = 1 \\
&= \frac{\pi}{4}
\end{aligned}$$

◇

11.5.4 Example. Let $f,\, g : \mathbb{R}^n \to \overline{\mathbb{R}}$ be Borel measurable and integrable. By Exercise 10.5.19, the function $F(\boldsymbol{x},\boldsymbol{y}) := f(\boldsymbol{x}-\boldsymbol{y})g(\boldsymbol{y})$ is Borel measurable in $(\boldsymbol{x},\boldsymbol{y})$. By the Fubini–Tonelli theorem and translation invariance of the integral,

$$\int_{\mathbb{R}^n\times\mathbb{R}^n} |F(\boldsymbol{x},\boldsymbol{y})|\,d\lambda(\boldsymbol{x},\boldsymbol{y}) = \int_{\mathbb{R}^n} |g(\boldsymbol{y})|\int_{\mathbb{R}^n} |f(\boldsymbol{x}-\boldsymbol{y})|d\boldsymbol{x}\,d\boldsymbol{y} = \|g\|_1\|f\|_1 < +\infty.$$

Therefore, F is integrable, hence the function

$$(f * g)(\boldsymbol{x}) := \int_{\mathbb{R}^n} f(\boldsymbol{x}-\boldsymbol{y})g(\boldsymbol{y})d\boldsymbol{y},$$

called the *convolution of f and g*, is finite a.e. and integrable on $\mathbb{R}^n$. Convolutions are useful in calculating the probability distribution of a sum of independent random variables. $\Diamond$

11.5.5 Example. (Volume of a simplex). Let $a > 0$ and let $\boldsymbol{e}^j$, $1 \le j \le n$, be the standard basis in $\mathbb{R}^n$. Define the *n-dimensional simplex in $\mathbb{R}^n$* by

$$S(a, n) = \Big\{\boldsymbol{x} : \sum_{j=1}^{n} x_j \le a \text{ and } x_j \ge 0\Big\}.$$

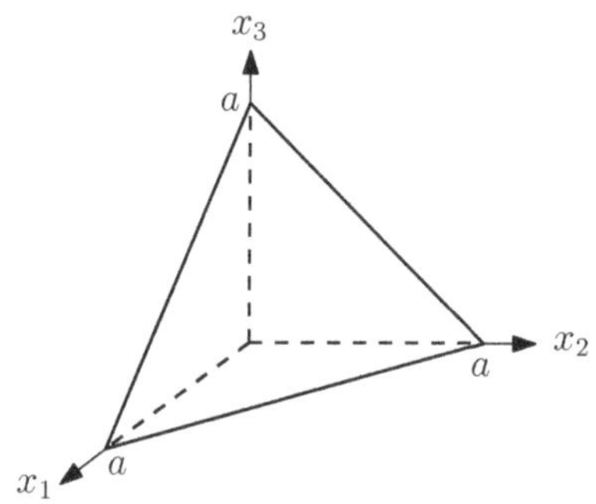

FIGURE 11.2: Three-dimensional simplex.

We use the Fubini–Tonelli theorem and induction to show that

$$\lambda_n\big(S(a, n)\big) = \frac{a^n}{n!}.$$

The formula holds for $n = 1$ since $S(a, 1) = [0, a]$. Assume the formula holds for $n - 1$ and all $a > 0$. Then

$$\begin{aligned}
\lambda_n\big(S(a, n)\big) &= \int \mathbf{1}_{S(a,n)}(x_1, \ldots, x_n)\, d(x_1, \ldots, x_n) \\
&= \int_{[0,a]} \mathbf{1}_{S(a-x_n, n-1)}(x_1, \ldots, x_{n-1})\, d(x_1, \ldots, x_{n-1})\, dx_n \\
&= \frac{1}{(n-1)!} \int_{[0,a]} (a - x_n)^{n-1}\, dx_n.
\end{aligned}$$

The last integral evaluates to a^n/n, completing the proof. $\Diamond$

11.5.6 Example. Let $C_r^n(\boldsymbol{x})$ denote the closed ball in $\mathbb{R}^n$ with center $\boldsymbol{x}$ and radius r. We show that $\lambda\big(C_r^n(\boldsymbol{x})\big) = r^n \alpha_n$, where

$$\alpha_n = \begin{cases} \dfrac{(2\pi)^{n/2}}{n(n-2)\cdots 4\cdot 2} & \text{if } n \text{ is even,} \\[2ex] \dfrac{2(2\pi)^{(n-1)/2}}{n(n-2)\cdots 3\cdot 1} & \text{if } n \text{ is odd.} \end{cases}$$

For ease of notation we write C_r^n for $C_r^n(\mathbf{0})$ and denote by $\mathbf{1}_r$ the indicator function of C_r^n. By the translation and dilation properties of Lebesgue measure, $\lambda(C_r^n(\boldsymbol{x})) = r^n\lambda(C_1^n)$, hence, it suffices to establish the formula for $r = 1$ and $\boldsymbol{x} = \mathbf{0}$.

If $n = 1$, then $C_1^n = [-1, 1]$ and $\alpha_n = 2$, so the formula holds in this case. By a simple integration, $\lambda(C_1^2) = \pi$, hence the formula holds for $n = 2$ as well. Now assume that $n > 2$. From

$$\begin{aligned} C_1^n &= \{(x_1, \ldots, x_n) : x_1^2 + \cdots + x_n^2 \le 1\} \\ &= \{(x_1, \ldots, x_n) : x_3^2 + \cdots + x_n^2 \le 1 - x_1^2 - x_2^2, \ \ (x_1, x_2) \in C_1^2\} \end{aligned}$$

we have

$$\mathbf{1}_1(x_1, \ldots, x_n) = \mathbf{1}_{\sqrt{1-x_1^2-x_2^2}}(x_3, \ldots, x_n)\mathbf{1}_1(x_1, x_2),$$

hence, by the Fubini–Tonelli theorem,

$$\lambda(C_1^n) = \int_{\mathbb{R}^2} \mathbf{1}_1(x_1, x_2) \int_{\mathbb{R}^{n-2}} \mathbf{1}_{\sqrt{1-x_1^2-x_2^2}}(x_3, \ldots, x_n)\, d\lambda(x_3, \ldots, x_n)\, dx_1\, dx_2.$$

The inner integral is

$$\lambda\left(C^{n-2}_{\sqrt{1-x_1^2-x_2^2}}\right) = (1 - x_1^2 - x_2^2)^{(n-2)/2}\lambda(C_1^{n-2}),$$

hence, changing to polar coordinates,[2]

$$\begin{aligned} \lambda(C_1^n) &= \lambda(C_1^{n-2}) \int_{x_1^2+x_2^2 \le 1} (1 - x_1^2 - x_2^2)^{(n-2)/2}\, dx_1\, dx_2 \\ &= \lambda(C_1^{n-2}) \int_0^{2\pi} \int_0^1 (1 - r^2)^{(n-2)/2} r\, dr\, d\theta \\ &= \frac{2\pi}{n}\lambda(C_1^{n-2}). \end{aligned}$$

Iterating, we obtain

$$\begin{aligned} \lambda(C_1^n) &= \frac{2\pi}{n}\lambda(C_1^{n-2}) = \frac{(2\pi)^2}{n(n-2)}\lambda(C_1^{n-4}) = \cdots \\ &= \frac{(2\pi)^{m-1}}{n(n-2)\cdots(n-2(m-2))}\lambda(C_1^{n-2(m-1)}). \end{aligned}$$

Thus

$$\lambda(C_1^{2m}) = \frac{(2\pi)^{m-1}}{2m(2m-2)\cdots 4}\lambda(C_1^2) = \frac{(2\pi)^m}{2m(2m-2)\cdots 2}$$

and

$$\lambda(C_1^{2m-1}) = \frac{(2\pi)^{m-1}}{(2m-1)(2m-3)\cdots 3}\lambda(C_1^1) = \frac{2(2\pi)^{m-1}}{(2m-1)(2m-3)\cdots 3}. \quad \Diamond$$

[2]The general change of variables theorem for Lebesgue integrals is proved in the next section.

Proof of the Fubini–Tonelli theorem.

We show first that part (b) of the theorem is a consequence of part (a). Indeed, if one of the iterated integrals in (b) is finite, then, by part (a) applied to $|f|$, so is the other and f is integrable. Applying part (a) to $f^{\pm}$, we see that (11.11) holds.

Next, observe that if part (a) of the theorem holds for indicator functions then, by linearity of the integrals, it holds for nonnegative simple functions. By 10.5.8 and the monotone convergence theorem, (a) holds for all nonnegative Borel measurable functions.

It remains then to prove (a) for indicator functions. The proof consists of several lemmas, the first of which is a special case of a theorem due to E.B. Dynkin.

11.5.7 Lemma. *Let $\mathcal{F}$ denote the intersection of all collections $\mathcal{G}$ of subsets of $\mathbb{R}^n$ with the following properties:*

(a) *If $A, B \in \mathcal{G}$ and $A \subseteq B$, then $B \setminus A \in \mathcal{G}$.*

(b) *If $A_k \in \mathcal{G}$ and $A_k \uparrow A$, then $A \in \mathcal{G}$.*

(c) *$\mathcal{G}$ contains every bounded interval.*

Then $\mathcal{F}$ is a σ-field containing $\mathcal{B}(\mathbb{R}^n)$.

Proof. It is easy to see that $\mathcal{F}$ itself has properties (a)–(c). Moreover, from (b) and (c), $\mathcal{F}$ contains every interval. In particular, $\mathbb{R}^n \in \mathcal{F}$.

We show first that $\mathcal{F}$ is closed under finite intersections. To see this, fix $A \in \mathcal{F}$ and define

$$\mathcal{F}_A := \{B \in \mathcal{F} : A \cap B \in \mathcal{F}\}.$$

One easily checks that $\mathcal{F}_A$ has properties (a) and (b). Furthermore, if A is an interval, then $\mathcal{F}_A$ has property (c) so by minimality $\mathcal{F} \subseteq \mathcal{F}_A$. This shows that if $B \in \mathcal{F}$, then $A \cap B \in \mathcal{F}$ for all intervals A; in other words, $\mathcal{F}_B$ contains all intervals. Thus $\mathcal{F}_B$ has properties (a)–(c). By minimality, $\mathcal{F} \subseteq \mathcal{F}_B$, that is, $A, B \in \mathcal{F} \Rightarrow A \cap B \in \mathcal{F}$. By induction, $\mathcal{F}$ is closed under finite intersections.

Now observe that property (a), together with the fact that $\mathbb{R}^n \in \mathcal{F}$, implies that $\mathcal{F}$ is closed under complements. Thus if $\{E_k\}$ is a sequence in $\mathcal{F}$; then, by the result of the preceding paragraph,

$$A_k := \bigcup_{j=1}^{k} E_k = \left(\bigcap_{j=1}^{k} E_k^c\right)^c \in \mathcal{F}.$$

By (b), $\bigcup_{k=1}^{\infty} E_k = \bigcup_{k=1}^{\infty} A_k \in \mathcal{F}$. This shows that $\mathcal{F}$ is a σ-field. Since $\mathcal{F}$ contains all intervals, it must contain $\mathcal{B}(\mathbb{R}^n)$. □

11.5.8 Lemma. *Let $p, q \in \mathbb{N}$ with $p + q = n$. If $A \in \mathcal{B}(\mathbb{R}^p)$ and $B \in \mathcal{B}(\mathbb{R}^q)$, then*

$$A \times B \in \mathcal{B}(\mathbb{R}^n) \ \text{ and } \ \lambda(A \times B) = \lambda(A)\lambda(B). \tag{11.13}$$

Proof. For fixed bounded intervals $I \subseteq \mathbb{R}^p$ and $J \subseteq \mathbb{R}^q$, define

$$\mathcal{G}_{I,J} = \{B \in \mathcal{B}(\mathbb{R}^q) : I \times (B \cap J) \in \mathcal{B}(\mathbb{R}^n) \ \& \ \lambda(I \times (B \cap J)) = \lambda(I)\lambda(B \cap J)\}.$$

We show that $\mathcal{G}_{I,J}$ has properties (a)–(c) of 11.5.7. Clearly, (c) holds. If $B \in \mathcal{G}_{I,J}$, then

$$\begin{aligned} I \times (B^c \cap J) &= (I \times J) \setminus (I \times (B \cap J)) \in \mathcal{B}(\mathbb{R}^n) \ \text{ and} \\ \lambda(I \times (B^c \cap J)) &= \lambda(I \times J) - \lambda(I \times (B \cap J)) \\ &= \lambda(I)\left[\lambda(J) - \lambda(B \cap J)\right] \\ &= \lambda(I)\lambda(J \cap B^c), \end{aligned}$$

hence $B^c \in \mathcal{G}$. Therefore, $\mathcal{G}_{I,J}$ is closed under complements. Now let $B_k \in \mathcal{G}_{I,J}$ and $B_k \uparrow B$. Then

$$I \times (J \cap B) = \bigcup_{k=1}^{\infty} I \times (J \cap B_k) \in \mathcal{B}(\mathbb{R}^n)$$

and, by 10.1.6,

$$\lambda(I \times J \cap B)) = \lim_k \lambda(I \times (J \cap B_k)) = \lambda(I) \lim_k \lambda(J \cap B_k) = \lambda(I)\lambda(J \cap B),$$

which shows that $B \in \mathcal{G}_{I,J}$. Therefore, $\mathcal{G}_{I,J}$ has properties (a)–(c) of 11.5.7, so $\mathcal{B}(\mathbb{R}^q) = \mathcal{G}_{I,J}$. We have shown that for all bounded intervals $I \subseteq \mathbb{R}^p$, $J \subseteq \mathbb{R}^q$ and all $B \in \mathcal{B}(\mathbb{R}^q)$,

$$I \times (B \cap J) \in \mathcal{B}(\mathbb{R}^n) \ \text{ and } \ \lambda(I \times (B \cap J)) = \lambda(I)\lambda(B \cap J).$$

Taking a sequence of bounded intervals $J_k \uparrow \mathbb{R}^n$, we see that

$$I \times B \in \mathcal{B}(\mathbb{R}^n) \ \text{ and } \ \lambda(I \times B)) = \lambda(I)\lambda(B). \tag{11.14}$$

Now fix $B \in \mathcal{B}(\mathbb{R}^q)$ and let $I \subseteq \mathbb{R}^p$ be a bounded interval. Define

$$\mathcal{H}_{B,I} = \{A \in \mathcal{B}(\mathbb{R}^p) : (A \cap I) \times B \in \mathcal{B}(\mathbb{R}^n) \ \& \ \lambda((A \cap I) \times B) = \lambda(A \cap I)\lambda(B)\}.$$

By (11.14), $\mathcal{H}_{B,I}$ contains all intervals. Arguing as above, we see that $\mathcal{H}_{B,I} = \mathcal{B}(\mathbb{R}^p)$. Thus for all $A \in \mathcal{B}(\mathbb{R}^p)$, $B \in \mathcal{B}(\mathbb{R}^q)$, and all bounded intervals $I \subseteq \mathbb{R}^p$,

$$(A \cap I) \times B \in \mathcal{B}(\mathbb{R}^n) \ \text{ and } \ \lambda((A \cap I) \times B) = \lambda(A \cap I)\lambda(B).$$

Taking a sequence of bounded intervals $I_k \uparrow \mathbb{R}^n$ in the last equation yields (11.13). $\square$

The following lemma asserts that part (a) of the Fubini–Tonelli theorem holds for indicator functions of Borel sets and hence completes the proof of the theorem.

11.5.9 Lemma. *Let $p,\ q \in \mathbb{N}$ with $p + q = n$ and let $C \in \mathcal{B}(\mathbb{R}^n)$. Then*

$$\int_{\mathbb{R}^q} \mathbf{1}_C(\boldsymbol{x}, \mathbf{z})\, d\boldsymbol{z} \quad \text{and} \quad \int_{\mathbb{R}^p} \mathbf{1}_C(\mathbf{z}, \boldsymbol{y})\, d\boldsymbol{z}$$

are Borel measurable functions of $\boldsymbol{x} \in \mathbb{R}^p$ and $\boldsymbol{y} \in \mathbb{R}^q$, respectively, and

$$\lambda(C) = \int_{\mathbb{R}^p} \int_{\mathbb{R}^q} \mathbf{1}_C(\boldsymbol{x}, \mathbf{z})\, d\boldsymbol{z}\, d\boldsymbol{x} = \int_{\mathbb{R}^q} \int_{\mathbb{R}^p} \mathbf{1}_C(\mathbf{z}, \boldsymbol{y})\, d\boldsymbol{z}\, d\boldsymbol{y}.$$

Proof. Let $\mathcal{G}$ denote the collection of all $C \in \mathcal{B}(\mathbb{R}^n)$ for which the assertions of the lemma hold. We show that $\mathcal{G} = \mathcal{B}(\mathbb{R}^n)$. The first step is to show that $\mathcal{G}$ has properties (b) and (c) of 11.5.7.

For property (b), let $C_k \in \mathcal{G}$ and $C_k \uparrow C$. Then $\mathbf{1}_{C_k}(\boldsymbol{x}, \mathbf{z}) \uparrow \mathbf{1}_C(\boldsymbol{x}, \mathbf{z})$, hence, by the monotone convergence theorem,

$$\int_{\mathbb{R}^q} \mathbf{1}_{C_k}(\boldsymbol{x}, \mathbf{z})\, d\boldsymbol{z} \uparrow \int_{\mathbb{R}^q} \mathbf{1}_C(\boldsymbol{x}, \mathbf{z})\, d\boldsymbol{z}, \quad \boldsymbol{x} \in \mathbb{R}^p.$$

Thus $\int_{\mathbb{R}^q} \mathbf{1}_C(\boldsymbol{x}, \mathbf{z})\, d\boldsymbol{z}$ is Borel measurable in $\boldsymbol{x}$. Applying the monotone convergence theorem again, we see that

$$\lambda(C) = \lim_k \lambda(C_k) = \lim_k \int_{\mathbb{R}^p} \int_{\mathbb{R}^q} \mathbf{1}_{C_k}(\boldsymbol{x}, \mathbf{z})\, d\boldsymbol{z}\, d\boldsymbol{x} = \int_{\mathbb{R}^p} \int_{\mathbb{R}^q} \mathbf{1}_C(\boldsymbol{x}, \mathbf{z})\, d\boldsymbol{z}\, d\boldsymbol{x},$$

and similarly for the other iterated integral. Therefore, $\mathcal{G}$ has property (b).

For property (c), let $A \in \mathcal{B}(\mathbb{R}^p)$, $B \in \mathcal{B}(\mathbb{R}^q)$, and $C = A \times B$. Then

$$\int_{\mathbb{R}^q} \mathbf{1}_C(\boldsymbol{x}, \mathbf{z})\, d\boldsymbol{z} = \int_{\mathbb{R}^q} \mathbf{1}_A(\boldsymbol{x}) \mathbf{1}_B(\mathbf{z})\, d\boldsymbol{z} = \mathbf{1}_A(\boldsymbol{x}) \lambda(B),$$

which is Borel measurable in $\boldsymbol{x}$ and, together with 11.5.8, implies that

$$\int_{\mathbb{R}^p} \int_{\mathbb{R}^q} \mathbf{1}_C(\boldsymbol{x}, \mathbf{z})\, d\boldsymbol{z}\, d\boldsymbol{x} = \lambda(A)\lambda(B) = \lambda(C).$$

Similar assertions hold for the other iterated integral. Thus, $\mathcal{G}$ contains Cartesian products of Borel sets and, in particular, all intervals.

Now let I be a bounded interval in $\mathbb{R}^n$ and let $\mathcal{G}_I = \{B \in \mathcal{B} : B \cap I \in \mathcal{G}\}$. Since $\mathcal{G}$ has properties (b) and (c) of 11.5.7, so does $\mathcal{G}_I$. We claim that $\mathcal{G}_I$ also has property (a). To see this, let $C,\ D \in \mathcal{G}_I$ with $C \subseteq D$ and let $E = D \setminus C$. Since $\mathbf{1}_{E \cap I} = \mathbf{1}_{D \cap I} - \mathbf{1}_{C \cap I}$,

$$\int_{\mathbb{R}^q} \mathbf{1}_{E \cap I}(\boldsymbol{x}, \mathbf{z})\, d\boldsymbol{z} = \int_{\mathbb{R}^q} \mathbf{1}_{D \cap I}(\boldsymbol{x}, \mathbf{z})\, d\boldsymbol{z} - \int_{\mathbb{R}^q} \mathbf{1}_{C \cap I}(\boldsymbol{x}, \mathbf{z})\, d\boldsymbol{z},$$

which, because $C \cap I$ and $D \cap I \in \mathcal{G}$, is Borel measurable in $\boldsymbol{x}$ and implies that

$$\begin{aligned}\int_{\mathbb{R}^p}\int_{\mathbb{R}^q} \mathbf{1}_{E\cap I}(\boldsymbol{x},\mathbf{z})\,d\boldsymbol{z}\,d\boldsymbol{x} &= \int_{\mathbb{R}^p}\int_{\mathbb{R}^q} \mathbf{1}_{D\cap I}(\boldsymbol{x},\mathbf{z})\,d\boldsymbol{z}\,d\boldsymbol{x} - \int_{\mathbb{R}^p}\int_{\mathbb{R}^q} \mathbf{1}_{C\cap I}(\boldsymbol{x},\mathbf{z})\,d\boldsymbol{z}\,d\boldsymbol{x}\\ &= \lambda(D\cap I) - \lambda(C\cap I)\\ &= \lambda(E\cap I).\end{aligned}$$

Here we have used the fact that, because I is bounded, the calculations take place in $\mathbb{R}$, hence subtraction is legitimate. The other iterated integral is treated similarly. Therefore $E \in \mathcal{G}_I$, as required.

Since $\mathcal{G}_I$ has properties (a)–(c) of 11.5.7, $\mathcal{G}_I$ contains all Borel sets. This means that for any $C \in \mathcal{B}(\mathbb{R}^n)$ and bounded interval $I \subseteq \mathbb{R}^n$, the functions

$$\int_{\mathbb{R}^q} \mathbf{1}_{C\cap I}(\boldsymbol{x},\mathbf{z})\,d\boldsymbol{z} \quad\text{and}\quad \int_{\mathbb{R}^p} \mathbf{1}_{C\cap I}(\mathbf{z},\boldsymbol{y})\,d\boldsymbol{z}$$

are Borel measurable in $\boldsymbol{x}$ and $\boldsymbol{y}$, respectively, and

$$\int_{\mathbb{R}^p}\int_{\mathbb{R}^q} \mathbf{1}_{C\cap I}(\boldsymbol{x},\mathbf{z})\,d\boldsymbol{z}\,d\boldsymbol{x} = \lambda(C\cap I) = \int_{\mathbb{R}^q}\int_{\mathbb{R}^p} \mathbf{1}_{C\cap I}(\mathbf{z},\boldsymbol{y})\,d\boldsymbol{z}\,d\boldsymbol{y}.$$

Taking an increasing sequence of bounded intervals I tending to $\mathbb{R}^n$ and using the monotone convergence theorem shows that $C \in \mathcal{G}$. Therefore, $\mathcal{G} = \mathcal{B}(\mathbb{R}^n)$, as required. □

Exercises

1. Prove that $\lambda_n\big(\{(x_1,\ldots,x_n) : x_j \in \mathbb{Q} \text{ for some } j\}\big) = 0$.

2. Evaluate $\int_{[0,+\infty)^n} f$, where

   (a) $f(\boldsymbol{x}) = x_1\cdots x_n e^{-\|\boldsymbol{x}\|^2}$ (b) $f(\boldsymbol{x}) = x_1\cdots x_n(1+\|\boldsymbol{x}\|^2)^{-n-1}$.

3. (Cavalieri's principle). For $E \in \mathcal{M}(\mathbb{R}^n)$ and $t \in \mathbb{R}$, define

$$E_t := \big\{\boldsymbol{x} = (x_1,\ldots,x_{n-1}) \in \mathbb{R}^{n-1} : (\boldsymbol{x},t) \in E\big\}.$$

   Suppose that $E_t \in \mathcal{M}(\mathbb{R}^n)$ for all $t \in [a,b]$. Prove that

$$\lambda_n\Big[E \cap \big(\mathbb{R}^{n-1} \times [a,b]\big)\Big] = \int_a^b \lambda_{n-1}(E_t)\,dt.$$

   Thus the "volume" of the portion of E between the hyperplanes $x_n = a$ and $x_n = b$ is the integral from a to b of the "cross-sectional areas" $\lambda_{n-1}(E_t)$.

4. Let f and g be Riemann integrable on $[0,1]$. Prove that

$$\int_0^1 \int_0^x g(x-y)f(y)\,dy\,dx = \int_0^1 \int_0^{1-y} g(x)f(y)\,dx\,dy$$
$$= \int_0^1 \int_0^{1-x} g(x)f(y)\,dy\,dx.$$

5. Evaluate $\displaystyle\int_{0\le x\le x_1\le\cdots\le x_m\le 1} x\,d\lambda(x,x_1,\ldots,x_m)$.

6. Show that

$$\int_0^1 \int_0^1 \frac{x^2-y^2}{(x^2+y^2)^2}\,dy\,dx = -\int_0^1 \int_0^1 \frac{x^2-y^2}{(x^2+y^2)^2}\,dx\,dy = \frac{\pi}{4}.$$

Why does this not contradict the Fubini–Tonelli theorem?

7. Let f be integrable on $(0,1)$, $p>0$, and define

$$g(x) = \int_{[x^{1/p},1)} t^{-p} f(t)\,dt, \quad 0<x<1.$$

Prove that g is integrable on $(0,1)$ and that

$$\int_{(0,1)} g\,d\lambda = \int_{(0,1)} f\,d\lambda.$$

8. Let f' be continuous on $[-1,1]$. Show that

(a) $$\int_0^{2\pi} \int_0^1 f'(r\cos\theta) r\cos^2\theta\,dr\,d\theta$$
$$= \int_0^{2\pi} f(\cos\theta)\cos\theta\,d\theta - \int_0^{2\pi}\int_0^{\cos\theta} f(x)\,dx\,d\theta$$

(b) $$\int_0^{2\pi} \int_0^1 f'(r\cos\theta) r\sin^2\theta\,dr\,d\theta = \int_0^{2\pi}\int_0^{\cos\theta} f(x)\,dx\,d\theta$$

(c) $$\int_0^{2\pi} \int_0^1 f'(r\cos\theta) r\,dr\,d\theta = \int_0^{2\pi} f(\cos\theta)\cos\theta\,d\theta.$$

9. Let $a, b>0$. Use the Fubini–Tonelli theorem, the dominated convergence theorem, and the identity $1/x = \int_0^\infty e^{-xt}\,dt$, $x>0$, to prove that

(a) $$\int_0^\infty \frac{\sin x}{x}\,dx = \frac{\pi}{2}$$ (b) $$\int_0^\infty \frac{e^{-ax}-e^{-bx}}{x}\,dx = \ln b - \ln a.$$

10. Show that $\varphi * \varphi(x) = \dfrac{1}{\sqrt{2}}\varphi\left(\dfrac{x}{\sqrt{2}}\right)$.

11. Let $f,\, g : \mathbb{R}^n \to \overline{\mathbb{R}}$ be Borel measurable and integrable. Prove:

    (a) $f * g = g * f$.

    (b) If f and g are continuous, then

$$\frac{d}{dz}\int_{A_z} f(x)g(y)\,dx\,dy = f * g(z), \quad \text{where} \quad A_z = \{(x,y) \,:\, x+y \le z\}.$$

12. Let $f : [0,1] \to (0,+\infty]$ be Lebesgue measurable. Use the Fubini–Tonelli theorem to prove that

$$\int_{[0,1]} f\,d\lambda \int_{[0,1]} 1/f\,d\lambda \ge 1.$$

    (A simpler but less interesting proof uses the Cauchy–Schwarz inequality.)

13. Let f and g be positive Lebesgue measurable functions on $[0,1]$ such that $fg \ge 1$. Use the preceding exercise to prove that

$$\int_{[0,1]} f\,d\lambda \int_{[0,1]} g\,d\lambda \ge 1.$$

    (The Cauchy–Schwarz inequality may be used here as well.)

14. Let f and g be Lebesgue integrable on $[a,b]$ and for $x \in [a,b]$ let

$$F(x) = F(a) + \int_{[a,x]} f(t)\,d\lambda(t) \;\text{ and }\; G(x) = G(a) + \int_{[a,x]} g(t)\,d\lambda(t),$$

    where $F(a)$ and $G(a)$ are arbitrary. Prove that

$$\int_{[a,b]} F(x)g(x)\,d\lambda(x) + \int_{[a,b]} G(x)f(x)\,d\lambda(x) = F(b)G(b) - F(a)G(a).$$

15. (a) Verify that the function

$$\kappa(t,x) = \frac{1}{2\sqrt{\pi t}}e^{-x^2/4t} = \frac{1}{\sqrt{2t}}\varphi\left(\frac{x}{\sqrt{2t}}\right)$$

    is a solution of the *heat equation*

$$w_t(t,x) = w_{xx}(t,x), \quad x \in \mathbb{R},\; t > 0.$$

    (b) Let $w_0(x)$ be integrable on $\mathbb{R}$ and define

$$w(t,x) = \int_{-\infty}^{\infty} w_0(y)\kappa(t,x-y)\,dy,$$

    the convolution of κ with w_0. Show that $w(t,x)$ satisfies the heat equation.

(c) Verify that

$$w(t,x) = \int_{-\infty}^{\infty} w_0\big(x + z\sqrt{2t}\big)\varphi(z)\,dz.$$

(d) Use (c) and the dominated convergence theorem to show that if w_0 is continuous and satisfies $|w_0(x)| \le ae^{b|x|}$ for some positive constants a, b and for all x, then $\lim_{t\to 0+} w(t,x) = w_0(x)$. Conclude that the solution $w(t,x)$ may be continuously extended to $[0,+\infty) \times \mathbb{R}$ and consequently satisfies the *boundary condition* $w(0,x) = w_0(x)$.

16. For a Borel measurable function $f : \mathbb{R} \to [0,+\infty)$, define

$$A := \{(x,y) \,:\, 0 \le y \le f(x)\} \quad \text{and} \quad A_y := \{x \,:\, f(x) > y\}\,, \quad y \in \mathbb{R}.$$

Prove:

(a) $A \in \mathcal{B}(\mathbb{R}^2)$.

(b) The function $y \mapsto \lambda(A_y)$ is Borel measurable and

$$\int f(x)\,d\lambda(x) = \int_{(0,+\infty)} \lambda(A_y)\,d\lambda(y) = \lambda(A).$$

(c) Part (b) holds if A and A_y are replaced, respectively, by

$$B = \{(x,y) \,:\, 0 \le y < f(x)\} \quad \text{and} \quad B_y = \{x \,:\, f(x) \ge y\}\,.$$

(d) $\lambda\big(\{(x,y) \,:\, f(x) = y\}\big) = 0$. (The graph of a Borel function has measure zero.)

11.6 Change of Variables

In Chapter 5 we proved that if $\varphi : [a,b] \to \mathbb{R}$ is continuously differentiable with everywhere nonzero derivative and if f is Riemann integrable on $[c,d] := \varphi([a,b])$, then

$$\int_c^d f(y)\,dy = \int_a^b f(\varphi(x))|\varphi'(x)|\,dx.$$

In this section we prove the following n-dimensional version of this result.

11.6.1 Change of Variables Theorem. *Let U and V be open subsets of $\mathbb{R}^n$ and let $\varphi : U \to V$ be C^1 on U with C^1 inverse $\varphi^{-1} : V \to U$. If f is Lebesgue measurable on V and either $f \ge 0$ or f is integrable, then*

$$\int_V f(\boldsymbol{y})\,d\boldsymbol{y} = \int_U (f \circ \varphi)(\boldsymbol{x})|J_\varphi(\boldsymbol{x})|\,d\boldsymbol{x}, \tag{11.15}$$

where J_φ is the Jacobian of φ on U.

11.6.2 Example. *Spherical coordinates* $(r, \theta_1, \theta_2, \ldots, \theta_{n-1})$ in $\mathbb{R}^n$ are defined by the transformation formulas

$$\begin{aligned} x_1 &= r\cos\theta_1 \\ x_2 &= r\sin\theta_1\cos\theta_2 \\ x_3 &= r\sin\theta_1\sin\theta_2\cos\theta_3 \\ &\vdots \\ x_{n-1} &= r\sin\theta_1\sin\theta_2\cdots\sin\theta_{n-2}\cos\theta_{n-1} \\ x_n &= r\sin\theta_1\sin\theta_2\cdots\sin\theta_{n-2}\sin\theta_{n-1}, \end{aligned}$$

where

$$r > 0, \quad 0 < \theta_j < \pi, \ j = 1, \ldots, n-2, \quad \text{and} \quad 0 < \theta_{n-1} < 2\pi.$$

Note that $\sin\theta_j > 0$ for $j \le n-2$ and $\sum_{j=1}^n x_j^2 = r^2$. Let

$$U := (0, +\infty) \times (0, \pi)^{n-2} \times (0, 2\pi) \ \text{ and } \ V := \mathbb{R}^n \setminus \big(\mathbb{R}^{n-2} \times [0, +\infty) \times \{0\}\big)$$

and define φ on U by

$$\varphi(r, \theta_1, , \ldots, \theta_{n-1}) = (x_1, \ldots, x_n),$$

where the x_j are as above. Clearly U and V are open and φ is C^∞ on U. We claim that φ maps U onto V and has a C^∞ inverse on U.

The inclusion $\varphi(U) \subseteq V$ is established as follows: If $(r, \theta_1, \ldots, \theta_{n-1}) \in U$ and $(x_1, \ldots, x_n) = \varphi(r, \theta_1, \ldots, \theta_{n-1}) \notin V$, then $x_{n-1} \ge 0$ and $x_n = 0$. But the latter implies that $\theta_{n-1} = \pi$, which gives the contradiction $x_{n-1} < 0$.

For the reverse inclusion, we show that for each $(x_1, \ldots, x_n) \in V$ there exists a unique solution $(r, \theta_1, \theta_2, \ldots, \theta_{n-1})$ to the above system. Clearly, r and θ_1 have the unique solutions

$$r = \left(\sum_{j=1}^n x_j^2\right)^{1/2} \quad \text{and} \quad \theta_1 = \arccos(x_1/r).$$

In particular, the system has a unique solution if $n = 2$. Now set

$$y_j = x_j/(r\sin\theta_1), \quad 2 \le j \le n.$$

By induction, we may assume that the reduced system

$$\begin{aligned} y_2 &= \cos\theta_2 \\ y_3 &= \sin\theta_2\cos\theta_3 \\ &\vdots \\ y_{n-1} &= \sin\theta_2\cdots\sin\theta_{n-2}\cos\theta_{n-1} \\ y_n &= \sin\theta_2\cdots\sin\theta_{n-2}\sin\theta_{n-1} \end{aligned}$$

has a unique solution $(\theta_2, \dots, \theta_{n-1})$. Then the original system has the unique solution $(r, \theta_1, \dots, \theta_{n-1})$. Therefore, φ is one-to-one and $\varphi(U) = V$.

By standard properties of determinants and a reduction argument,

$$J_\varphi(r, \theta_1, \theta_2, \dots, \theta_{n-1}) = r^{n-1} \sin^{n-2}\theta_1 \sin^{n-3}\theta_2 \cdots \sin^2\theta_{n-3} \sin\theta_{n-2}.$$

Since $J_\varphi > 0$ on U, the inverse function theorem implies that φ has a global C^∞ inverse on U. Hence, by the change of variables theorem, if f is Lebesgue measurable on $\mathbb{R}^n$ and either $f \geq 0$ or f is integrable, then

$$\int_V f \, d\lambda = \int_U (f \circ \varphi) J_\varphi \, d\lambda.$$

Since V differs from $\mathbb{R}^n$ by a set of measure zero, we may write the last equation as

$$\int_{-\infty}^{\infty} \cdots \int_{-\infty}^{\infty} f(x_1, \dots, x_n) \, dx_1 \cdots dx_n \tag{11.16}$$

$$= \int_0^\infty \int_0^\pi \cdots \int_0^\pi \int_0^{2\pi} f\big(r\cos\theta_1, r\sin\theta_1\cos\theta_2, \dots, r\sin\theta_1 \cdots \sin\theta_{n-1}\big)$$
$$\big(r^{n-1}\sin^{n-2}\theta_1 \sin^{n-3}\theta_2 \cdots \sin^2\theta_{n-3}\sin\theta_{n-2}\big)\, d\theta_{n-1}\, d\theta_{n-2} \cdots d\theta_1\, dr.$$

In particular, taking f to be the indicator function of $C_1^n(0)$ and using 11.5.6, we see that the left side of (11.16) is $\lambda\big(C_1^n(0)\big) = \alpha_n$ and the right side is

$$\int_0^1 \int_0^\pi \cdots \int_0^\pi \int_0^{2\pi} \big(r^{n-1}\sin^{n-2}\theta_1 \cdots \sin^2\theta_{n-3}\sin\theta_{n-2}\big)\, d\theta_{n-1}\, d\theta_{n-2} \cdots d\theta_1\, dr$$
$$= \frac{2\pi}{n} \int_0^\pi \cdots \int_0^\pi \big(\sin^{n-2}\theta_1 \cdots \sin^2\theta_{n-3}\sin\theta_{n-2}\big)\, d\theta_{n-2} \cdots d\theta_1.$$

In particular,

$$\int_0^\pi \cdots \int_0^\pi \big(\sin^{n-2}\theta_1 \sin^{n-3}\theta_2 \cdots \sin^2\theta_{n-3}\sin\theta_{n-2}\big)\, d\theta_{n-2} \cdots d\theta_1 = \frac{n\alpha_n}{2\pi}. \quad \Diamond$$

Proof of the change of variables theorem.

Before we begin the proof proper, we make some reductions. First, by considering f^+ and f^-, we need only prove the case $f \geq 0$. Second, since a Lebesgue measurable function is equal a.e. to a Borel measurable function, we may assume that f is Borel measurable. Note that in this case $f \circ \varphi$ is also Borel measurable. To prove (11.15) it then suffices to verify that

$$\int_V f \, d\lambda \leq \int_U (f \circ \varphi)|J_\varphi| \, d\lambda \tag{11.17}$$

for all Borel measurable functions $f : V :\to [0, +\infty]$. Indeed, if (11.17) holds for all f and φ, then, switching the roles of U and V it must also be the case that

$$\int_U g \, d\lambda \leq \int_V (g \circ \varphi^{-1})|J_{\varphi^{-1}}| \, d\lambda$$

for all Borel measurable $g : U :\to [0, +\infty]$. Taking $g = (f \circ \varphi)|J_\varphi|$ and recalling that $J_\varphi J_{\varphi^{-1}} = 1$, we obtain the reverse of inequality (11.17). Finally, by considering simple functions and using linearity, 10.5.8, and the monotone convergence theorem, it suffices to prove (11.17) for indicator functions $f = \mathbf{1}_B$, where $B \in \mathcal{B}(\mathbb{R}^n)$ and $B \subseteq V$. Equation (11.17) then reduces to

$$\lambda(B) \le \int_{\varphi^{-1}(B)} |J_\varphi|\, d\lambda, \quad B \subseteq V,\ B \in \mathcal{B}(\mathbb{R}^n),$$

or, equivalently, (taking $B = \varphi(E)$),

$$\lambda(\varphi(E)) \le \int_E |J_\varphi|\, d\lambda, \quad E \subseteq U,\ E \in \mathcal{B}(\mathbb{R}^n). \tag{11.18}$$

The proof of (11.18) is a sequence of lemmas, the first of which treats the case of a linear change of variable.

11.6.3 Lemma. *If $T \in \mathcal{L}(\mathbb{R}^n, \mathbb{R}^n)$ is nonsingular, then*

$$\lambda(T(E)) = |\det T|\lambda(E), \quad E \in \mathcal{B}(\mathbb{R}^n). \tag{11.19}$$

Proof. Since T is nonsingular, $T(E) \in \mathcal{B}(\mathbb{R}^n)$ so the left side of (11.19) is defined. Furthermore, if (11.19) holds for T_1 and T_2, then it holds for T_1T_2:

$$\lambda(T_1T_2(E)) = |\det T_1|\lambda(T_2(E)) = |\det T_1|\,|\det T_2|\lambda(E) = |\det(T_1T_2)|\lambda(E).$$

Now observe that a nonsingular linear transformation T may be expressed as a product of elementary linear transformations, that is, linear transformations whose matrices are obtained from the identity matrix by one of the following operations:

(a) Interchange of two rows.

(b) Multiplication of a row by a nonzero constant.

(c) Addition of one row to another.

This is simply the assertion that a matrix may be put into reduced row echelon form by a sequence of elementary row operations. (See Appendix B.) We claim that (11.19) holds for elementary linear transformations T and bounded intervals $E = I_1 \times \cdots \times I_n$.

In case (a), $\det T = -1$ and $T(E)$ is the interval obtained from E by interchanging a pair of intervals I_i and I_j, hence (11.19) holds in this case.

In (b), $T(E)$ is the interval obtained from E by multiplying one of the coordinate intervals by a nonzero constant a, hence $\lambda(T(E)) = |a|\lambda(E)$. Since $|\det T| = |a|$, (11.19) holds in this case as well.

For case (c), assume for definiteness that the matrix of T is the result of adding row two of the identity matrix to row one, so

$$T(x_1, x_2, x_3, \ldots, x_n) = (x_1 + x_2, x_2, x_3, \ldots, x_n).$$

Then $\det T = 1$ and

$$\lambda(T(E)) = \int \mathbf{1}_{T(E)}(\boldsymbol{x})\,d\boldsymbol{x} = \int \mathbf{1}_E(x_1 - x_2, x_2, \ldots, x_n)\,d\boldsymbol{x}.$$

By the Fubini–Tonelli theorem and translation invariance, the last integral evaluates to

$$\begin{aligned}\iint \cdots \int \mathbf{1}_{I_1}(x_1 - x_2)\mathbf{1}_{I_2}(x_2)\cdots\mathbf{1}_{I_n}(x_n)\,dx_n \cdots dx_2\,dx_1 \\ &= |I_n|\cdots|I_3| \int \mathbf{1}_{I_2}(x_2) \int \mathbf{1}_{I_1}(x_1 - x_2)\,dx_1\,dx_2 \\ &= |I_n|\cdots|I_3|\,|I_2|\,|I_1| \\ &= \lambda(E).\end{aligned}$$

Therefore, (c) holds.

It now follows that (11.19) holds for all nonsingular T and all intervals E. To verify (11.19) for all Borel sets E, we use 11.5.7. For a fixed bounded interval I, let $\mathcal{G}_I$ denote the collection of all $E \in \mathcal{B}(\mathbb{R}^n)$ for which

$$\lambda(T(E \cap I)) = |\det T|\lambda(E \cap I). \tag{11.20}$$

By the first part of the proof, $\mathcal{G}_I$ contains all intervals. Let $A, B \in \mathcal{G}_I$ with $A \subseteq B$, and set $C = A \cap I$ and $D = B \cap I$. Then $(B \setminus A) \cap I = D \setminus C$ and

$$\lambda(T(D\backslash C)) = \lambda(T(D)) - \lambda(T(C)) = |\det T|(\lambda(D) - \lambda(C)) = |\det T|\lambda(D\backslash C),$$

hence $B \setminus A \in \mathcal{G}_I$. (The operation of substraction is legitimate because C and D are bounded.) Now let $A_k \in \mathcal{G}_I$, $A_k \uparrow A$. Letting $k \to +\infty$ in

$$\lambda(T(A_k \cap I)) = |\det T|\lambda(A_k \cap I)$$

shows that $A \in \mathcal{G}_I$. Therefore, $\mathcal{G}_I$ satisfies (a)–(c) of 11.5.7, hence (11.20) holds for every $E \in \mathcal{B}(\mathbb{R}^n)$. Taking a sequence of bounded intervals $I_k \uparrow \mathbb{R}^n$ in (11.20) yields (11.19). □

For the remaining lemmas, the following terminology and notation will be useful. The *cube with center* $\boldsymbol{y} \in \mathbb{R}^n$ *and edge* $r > 0$ is the semi-closed interval

$$Q = Q_r(\boldsymbol{y}) := \{\boldsymbol{x} \in \mathbb{R}^n : y_j - r/2 \le x_j < y_j + r/2,\ j = 1, \ldots, n\}.$$

Note that $|Q| = r^n$ and the diameter of Q is $r\sqrt{n}$. Thus

$$B_{r/2}(\boldsymbol{y}) \subseteq Q_r(\boldsymbol{y}) \subseteq B_{r\sqrt{n}/2}(\boldsymbol{y}).$$

A *paving* of a subset A of $\mathbb{R}^n$ is a finite collection $\mathcal{Q}_r$ of pairwise disjoint cubes with edge r that covers A. Two pavings $\mathcal{Q}_r = \{Q_r(\boldsymbol{x}_j) : 1 \le j \le m\}$ and $\mathcal{Q}_s = \{Q_s(\boldsymbol{x}_j) : 1 \le j \le m\}$ with the same centers are said to be *concentric*. Any bounded set A has a paving $\mathcal{Q}_r$ with arbitrarily small r. Indeed, if $A \subseteq [a, b)^n$, one need only subdivide $[a, b)$ into subintervals of size $(b - a)/k$ for sufficiently large k and form Cartesian products of these subintervals.

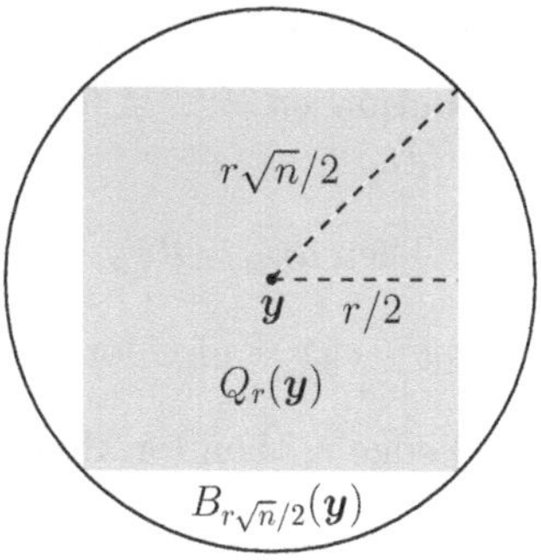

FIGURE 11.3: Concentric cube and ball.

11.6.4 Lemma. *Let* $K \subseteq U$ *be compact.*

(a) *For each sufficiently small* $\delta > 0$, *there exists a compact set* K_δ *with* $K \subseteq K_\delta \subseteq U$.

(b) *For each* $r < \delta$, *there exists a paving* $\mathcal{Q}_r$ *of* K *contained in* K_δ.

Proof. For subsets $A, B \subseteq \mathbb{R}^n$, denote by $d(A, B)$ the distance between A and B:

$$d(A, B) = \inf \{\|\boldsymbol{a} - \boldsymbol{b}\| : \boldsymbol{a} \in A,\ \boldsymbol{b} \in B\}.$$

Since K is compact and U^c is closed, $\delta_0 := d(U^c, K) > 0$. For $0 < \delta < \delta_0/\sqrt{n}$, let

$$K_\delta = \left\{\boldsymbol{x} : d(\boldsymbol{x}, K) \le \delta\sqrt{n}\right\}.$$

Then K_δ is compact and $K \subseteq K_\delta \subseteq U$. Let Q be a cube with edge r. If

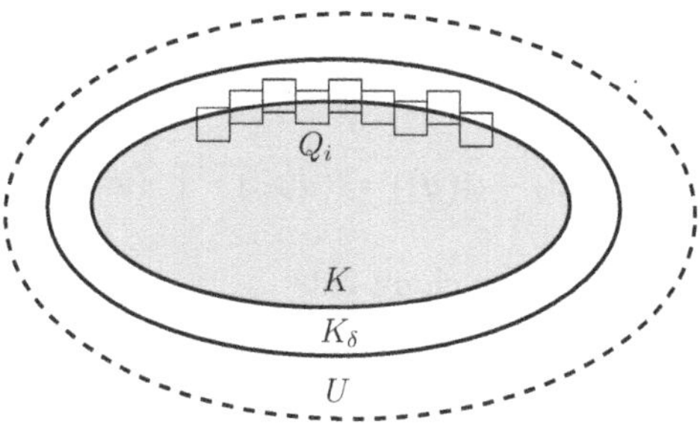

FIGURE 11.4: The paving $\mathcal{Q}_r$.

$\boldsymbol{x} \in Q \cap K$ and $\boldsymbol{y} \in Q \cap K_\delta^c$, then

$$\delta\sqrt{n} < d(\boldsymbol{y}, K) \le \|\boldsymbol{x} - \boldsymbol{y}\| \le r\sqrt{n}.$$

Therefore, if $r < \delta$ and $Q \cap K \ne \emptyset$, then $Q \cap K_\delta^c = \emptyset$, that is, $Q \subseteq K_\delta$. Since K is bounded, there exists a paving $\mathcal{Q}_r$ of K. Removing those members of $\mathcal{Q}_r$ that do not meet K produces a paving of K contained in K_δ. □

11.6.5 Lemma. *Let ψ be C^1 on U, Q a cube contained in U, and let I_n denote the identity transformation on $\mathbb{R}^n$. If $\|d\psi'_{\boldsymbol{x}} - I_n\| \le c$ for all $\boldsymbol{x} \in Q$, then $\lambda(\psi(Q)) \le [(1+c)n]^n \lambda(Q)$.*

Proof. Let $\tilde{\psi}(\boldsymbol{x}) = \psi(\boldsymbol{x}) - \boldsymbol{x}$. Then $d\tilde{\psi}_{\boldsymbol{x}} = d\psi_{\boldsymbol{x}} - I_n$. By 9.3.6,

$$\|\tilde{\psi}(\boldsymbol{x}) - \tilde{\psi}(\boldsymbol{y})\| \le c\|\boldsymbol{x} - \boldsymbol{y}\|, \quad \text{for all } \boldsymbol{x},\ \boldsymbol{y} \in Q.$$

Thus, if Q has center $\boldsymbol{x}_0$ and edge r, then for all $\boldsymbol{x} \in Q$

$$\|\psi(\boldsymbol{x}) - \psi(\boldsymbol{x}_0)\| \le \|\tilde{\psi}(\boldsymbol{x}) - \tilde{\psi}(\boldsymbol{x}_0)\| + \|\boldsymbol{x} - \boldsymbol{x}_0\| \le (c+1)\|\boldsymbol{x} - \boldsymbol{x}_0\| \le (c+1)\sqrt{n}r/2,$$

that is, $\psi(Q)$ is contained in the closed ball C with center $\psi(\boldsymbol{x}_0)$ and radius $\sqrt{n}(c+1)r/2$. Since C is contained in the cube with center $\psi(\boldsymbol{x}_0)$ and edge $(c+1)nr$,

$$\lambda(\psi(Q)) \le [(c+1)nr]^n = [(c+1)n]^n \lambda(Q). \qquad \square$$

11.6.6 Lemma. *Let $\psi : U \to \mathbb{R}^n$ be C^1 on U and let $K \subseteq U$ be compact. Then, for each $\varepsilon > 0$, there exists $\delta > 0$, a compact set K_δ with $K \subseteq K_\delta \subseteq U$, and concentric pavings $\mathcal{Q}_r$, $\mathcal{Q}_{nr}$ of K contained in K_δ with arbitrarily small r such that for any $Q_r(\boldsymbol{y}) \in \mathcal{Q}_r$,*

$$\lambda\Big(\varphi(Q_r(\boldsymbol{y}))\Big) \le (1+\varepsilon)^n |J_\varphi(\boldsymbol{y})| \lambda(Q_{nr}(\boldsymbol{y})) \tag{11.21}$$

Moreover, δ may be chosen so that

$$\int_{K_\delta} |J_\varphi(\boldsymbol{x})|\, d\boldsymbol{x} < \int_K |J_\varphi(\boldsymbol{x})|\, d\boldsymbol{x} + \varepsilon. \tag{11.22}$$

Proof. Let $M = \sup\big\{\|(d\varphi_{\boldsymbol{y}})^{-1}\| : \boldsymbol{y} \in K_\delta\big\}$, where K_δ is chosen as in 11.6.4. For $\boldsymbol{x},\ \boldsymbol{y} \in U$ define

$$\psi^{\boldsymbol{y}}(\boldsymbol{x}) = \big(d\varphi_{\boldsymbol{y}}\big)^{-1}\big(\varphi(\boldsymbol{x}) - \varphi(\boldsymbol{y})\big) = \big(d\varphi_{\boldsymbol{y}}\big)^{-1}\big(\varphi(\boldsymbol{x})\big) - \big(d\varphi_{\boldsymbol{y}}\big)^{-1}\big(\varphi(\boldsymbol{y})\big).$$

Since $\big(d\varphi_{\boldsymbol{y}}\big)^{-1}$ is linear, by the chain rule

$$d(\psi^{\boldsymbol{y}})_{\boldsymbol{x}} = \big(d\varphi_{\boldsymbol{y}}\big)^{-1} \circ d\varphi_{\boldsymbol{x}}.$$

Thus for all $\boldsymbol{x} \in U$, $\boldsymbol{y} \in K_\delta$, and $\mathbf{z} \in \mathbb{R}^n$,

$$\|d(\psi^{\boldsymbol{y}})_{\boldsymbol{x}}(\mathbf{z}) - \mathbf{z}\| = \Big\|\big(d\varphi_{\boldsymbol{y}}\big)^{-1}\big(d\varphi_{\boldsymbol{x}}(\mathbf{z}) - d\varphi_{\boldsymbol{y}}(\mathbf{z})\big)\Big\| \le M\|d\varphi_{\boldsymbol{x}} - d\varphi_{\boldsymbol{y}}\|\,\|\mathbf{z}\|.$$

Therefore, by definition of the operator norm,

$$\|d(\psi^{\boldsymbol{y}})_{\boldsymbol{x}} - I_n\| \le M\|d\varphi_{\boldsymbol{x}} - d\varphi_{\boldsymbol{y}}\|. \tag{11.23}$$

Now, by the uniform continuity of $d\varphi$ on K_δ, there exists $0 < \delta_1 < \delta$ such that $\|d\varphi_{\boldsymbol{x}} - d\varphi_{\boldsymbol{y}}\| \le \varepsilon/M$ for all $\boldsymbol{x}, \boldsymbol{y} \in K_\delta$ with $\|\boldsymbol{x} - \boldsymbol{y}\| < \delta_1\sqrt{n}$. Let

$r < \delta_1/n$ and let $\mathcal{Q}_r$ $\mathcal{Q}_{nr}$ be concentric pavings of K contained in K_δ. If $\boldsymbol{x} \in Q := Q_r(\boldsymbol{y}) \in \mathcal{Q}_r$, then $\|\boldsymbol{x} - \boldsymbol{y}\| < r\sqrt{n} < \delta_1\sqrt{n}$, hence, from (11.23), $\|d(\psi^{\boldsymbol{y}})_{\boldsymbol{x}} - I_n\| < \varepsilon$. By 11.6.5,

$$\lambda(\psi^{\boldsymbol{y}}(Q)) \le [(1+\varepsilon)n]^n \lambda(Q) = (1+\varepsilon)^n \lambda(Q_{nr}(\boldsymbol{y})). \tag{11.24}$$

On the other hand, since $\psi^{\boldsymbol{y}}(Q) = (d\varphi_{\boldsymbol{y}}))^{-1}(\varphi(Q)) - (d\varphi_{\boldsymbol{y}})^{-1}(\varphi(\boldsymbol{y}))$, by translation invariance and 11.6.3,

$$\lambda(\psi^{\boldsymbol{y}}(Q)) = \lambda\big[(d\varphi_{\boldsymbol{y}})^{-1}(\varphi(Q))\big] = |J_\varphi(\boldsymbol{y})|^{-1}\lambda(\varphi(Q)). \tag{11.25}$$

Inequality (11.21) now follows from (11.24) and (11.25).

For (11.22), note that since $K_{1/k} \downarrow K$ and $\mu(A) := \int_A |J_\varphi|\, d\lambda$ is a measure on the Borel sets (11.3.4), $\mu(K_{1/k}) \downarrow \mu(K)$. Thus there exists k such that $\mu(K_{1/k}) < \mu(K) + \varepsilon$. Taking $\delta < 1/k$ completes the proof. □

11.6.7 Lemma. *If $K \subseteq U$ is compact, then*

$$\lambda(\varphi(K)) \le \int_K |J_\varphi(\boldsymbol{y})|\, d\boldsymbol{y}.$$

Proof. Let $\varepsilon > 0$ and choose $\delta > 0$ as in 11.6.6. By uniform continuity of $J_\varphi(\boldsymbol{x})$ on K_δ, there exists $\delta_1 < \delta$ such that

$$|J_\varphi(\boldsymbol{x}) - J_\varphi(\boldsymbol{y})| < \varepsilon \quad \text{for all } \boldsymbol{x}, \boldsymbol{y} \in K_\delta \text{ with } \|\boldsymbol{x} - \boldsymbol{y}\| < \delta_1.$$

Choose pavings $\mathcal{Q}_r = \{Q_r(\boldsymbol{y})\}_{\boldsymbol{y}}$ and $\mathcal{Q}_{nr} = \{Q_{nr}(\boldsymbol{y})\}_{\boldsymbol{y}}$ as in 11.6.6. Then for $\boldsymbol{x} \in Q_{nr}(\boldsymbol{y})$

$$|J_\varphi(\boldsymbol{y})| \le |J_\varphi(\boldsymbol{x}) - J_\varphi(\boldsymbol{y})| + |J_\varphi(\boldsymbol{x})| < \varepsilon + |J_\varphi(\boldsymbol{x})|,$$

hence, by (11.21),

$$(1+\varepsilon)^{-n}\lambda(\varphi(Q_r(\boldsymbol{y}))) \le |J_\varphi(\boldsymbol{y})|\lambda(Q_{nr}(\boldsymbol{y})) \le \int_{Q_{nr}(\boldsymbol{y})} \big(|J_\varphi(\boldsymbol{x})| + \varepsilon\big)\, d\boldsymbol{x},$$

so

$$\begin{aligned}(1+\varepsilon)^{-n}\lambda(\varphi(K)) &\le \sum_{\boldsymbol{y}} (1+\varepsilon)^{-n}\lambda(\varphi(Q_r(\boldsymbol{y}))) \\ &\le \int_{K_\delta} \big(|J_\varphi(\boldsymbol{x})| + \varepsilon\big)\, d\boldsymbol{x} \\ &\le \int_K |J_\varphi(\boldsymbol{x})|\, d\boldsymbol{x} + \varepsilon\big(1 + \lambda(K_\delta)\big). \qquad \text{by (11.22)}\end{aligned}$$

Letting $\varepsilon \to 0$ verifies the lemma. □

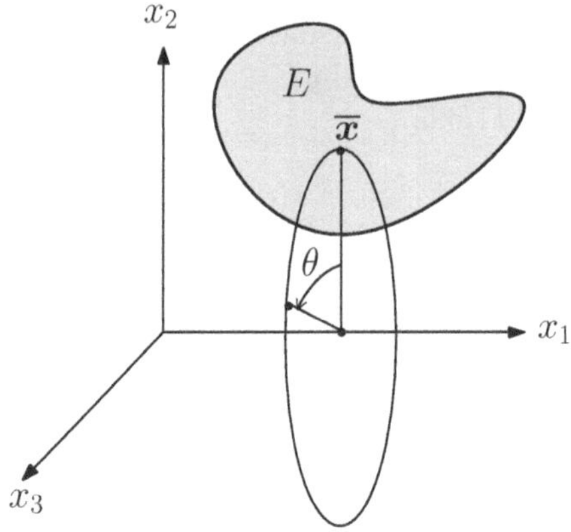

FIGURE 11.5: Theorem of Pappus.

Now use 10.4.5 to obtain an increasing sequence of compact sets $K_k \subseteq E$ such that $\lambda(K_k) \uparrow \lambda(E)$. Then $\lambda\big(\varphi(K_k)\big) \uparrow \lambda\big(\varphi(E)\big)$ and, by 11.6.7,

$$\lambda(\varphi(K_k)) \le \int_{K_k} |J_\varphi(\boldsymbol{y})|\, d\boldsymbol{y} \le \int_E |J_\varphi(\boldsymbol{y})|\, d\boldsymbol{y}.$$

Letting $k \to +\infty$ yields (11.18), completing the proof of the change of variables theorem.

11.6.8 Remark. If $\mathcal{V}$ is a linear subspace of $\mathbb{R}^n$ of dimension $m < n$, then $\lambda_n(\mathcal{V}) = 0$. To see this, let $\mathbf{v}_1, \ldots, \mathbf{v}_m, \ldots, \mathbf{v}_n$, be an orthonormal basis for $\mathbb{R}^n$, where the first m vectors form a basis for $\mathcal{V}$.[3] Define $T_{\mathcal{V}} \in \mathcal{L}(\mathbb{R}^n, \mathbb{R}^n)$ such that $T_{\mathcal{V}}(\mathbf{v}_j) = \mathbf{e}_j$, $1 \le j \le n$. Then $T_{\mathcal{V}}$ is an orthogonal transformation and $T_{\mathcal{V}}(\mathcal{V}) = \mathbb{R}^m \times \{\mathbf{0}\}$. By 11.6.3

$$\lambda_n(\mathcal{V}) = |\det(T_{\mathcal{V}})|\lambda_n(\mathbb{R}^m \times \{\mathbf{0}\}) = 0,$$

as claimed. This also shows that (11.19) holds for singular transformations T as well, since then both sides of that equation are zero.

While the n-dimensional volume of a subset E of $\mathcal{V}$ is zero, E may still have positive m-dimensional measure. This is defined as

$$\lambda_{\mathcal{V}}(E) := \lambda_m\big(T_{\mathcal{V}}(E)\big) \quad \text{for } E \in T_{\mathcal{V}}^{-1}\big(\mathcal{B}(\mathbb{R}^m)\big).$$

From a geometric point of view, this is a reasonable definition, since an orthogonal transformation is either a rotation or a rotation combined with a reflection and therefore does not change volumes or areas. To see that the definition does not depend on the particular choice of the orthonormal basis, let $\mathbf{w}_1, \ldots, \mathbf{w}_n$ be another orthonormal basis for $\mathbb{R}^n$ whose first m members

[3]This is always possible by the Gram–Schmidt process.

form a basis for $\mathcal{V}$ and let $\tilde{T}_\mathcal{V} \in \mathcal{L}(\mathbb{R}^n, \mathbb{R}^n)$ satisfy $\tilde{T}_\mathcal{V}(\mathbf{w}_j) = \mathbf{e}_j$, $1 \le j \le n$. Set $T = \tilde{T}_\mathcal{V} T_\mathcal{V}^{-1}$. Then, by (11.19),

$$\lambda_m\big(\tilde{T}_\mathcal{V}(E)\big) = \lambda_m\big(TT_\mathcal{V}(E)\big) = |\det T|\lambda_m\big(T_\mathcal{V}(E)\big) = \lambda_m\big(T_\mathcal{V}(E)\big),$$

the last equality because T is orthogonal and hence has determinant ± 1. $\Diamond$

Exercises

1. Define the *n-dimensional ellipsoid*

$$E = \left\{(x_1, \ldots, x_n) : \left(\frac{x_1}{a_1}\right)^2 + \cdots + \left(\frac{x_n}{a_n}\right)^2 \le 1\right\},$$

where $a_j > 0$. Prove that $\lambda_n(E) = a_1 \cdots a_n\, \lambda_n\big(C_1(\mathbf{0})\big)$.

2. Show that the volume of the solid with surface $\sqrt{|x|} + \sqrt{|y|} + \sqrt{|z|} = 1$ is given by

$$64 \int_0^1 \int_0^{1-u} \int_0^{1-u-v} uvw\, dw\, dv\, du.$$

3. ($\Downarrow$ 13.2.5 and 13.4.2.) Let h be Lebesgue integrable on $[0, +\infty)$. Use 11.6.2 to prove that

$$\int_{\mathbb{R}^n} h(\|\boldsymbol{x}\|)\, d\boldsymbol{x} = n\alpha_n \int_0^\infty h(r) r^{n-1}\, dr.$$

4. Use Exercise 3 to show that for $n \ge 2$

(a) $\displaystyle\int_{\mathbb{R}^n} \exp(-\|\boldsymbol{x}\|)\, d\boldsymbol{x} = n!\, \alpha_n$ (b) $\displaystyle\int_{\mathbb{R}^n} \exp(-\|\boldsymbol{x}\|^2)\, d\boldsymbol{x} = \pi^{n/2}$.

5. A hole of radius $R \in (0,1)$ is drilled in the $(n+1)$-dimensional ball $C_1^{n+1}(\mathbf{0})$ from the north pole $(0, 0, \ldots, 1)$ to the south pole $(0, 0, \ldots, -1)$. Use Exercise 3 to show that the amount removed from the ball is

$$n\alpha_n\big[R\sqrt{1-R^2} - \arcsin\sqrt{1-R^2} + \pi/2\big].$$

6. (Theorem of Pappus) Let $E \in \mathcal{M}(\mathbb{R}^n)$ be bounded with positive n-dimensional Lebesgue measure such that $x_n > 0$ for all $\boldsymbol{x} = (x_1, \ldots, x_n) \in E$. Define

$$E_r = \{(x_1, \ldots, x_{n-1}, x_n \cos\theta, x_n \sin\theta) : \boldsymbol{x} \in E,\ 0 < \theta < 2\pi\}.$$

Prove that

$$\lambda_{n+1}(E_r) = 2\pi \overline{x}_n \lambda_n(E),$$

where

$$\overline{x}_n := \frac{1}{\lambda_n(E)} \int_E x_n\, d\lambda_n(x_1, \ldots, x_n),$$

the nth coordinate of the *centroid* $\overline{\boldsymbol{x}}$ of E. Thus if $n = 2$, then E_r is the rotation of E about the x_1-axis, and the theorem of Pappus asserts that the volume of E_r is equal to the area of E times the distance the centroid of E travels around the x_1-axis.

Chapter 12

Curves and Surfaces in $\mathbb{R}^n$

12.1 Parameterized Curves

A *parameterized curve* $\mathbb{R}^n$ is a continuous function $\varphi : I \to \mathbb{R}^n$, where I is an interval in $\mathbb{R}$. We shall usually refer to φ as simply a *curve*. The range $\varphi(I)$ of φ is called the *trace* of φ and is denoted by $\operatorname{trace}(\varphi)$. The curve is said to *lie in a set* $E \subseteq \mathbb{R}^n$ if $\operatorname{trace}(\varphi) \subseteq E$. The curve is called *simple* if φ is one-to-one. If $I = [a, b]$, the point $\varphi(a)$ is the *initial point* of the curve and $\varphi(b)$ the *terminal point.* The curve φ is then said to be *closed* if $\varphi(a) = \varphi(b)$, and *simple closed* if it is closed and φ is one-to-one on (a, b), that is, the curve intersects itself only at the initial and terminal points. For example, the curve $(\cos(2k\pi t), \sin(2k\pi t))$, $t \in [0, 1]$, $k \in \mathbb{N}$, is a simple closed curve iff $k = 1$; its trace is the circle $x^2 + y^2 = 1$.

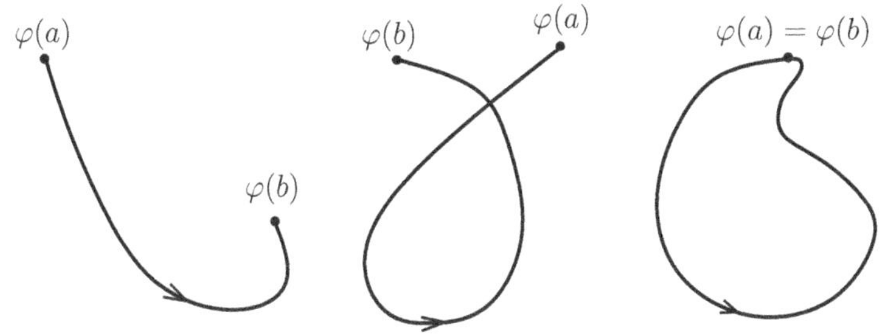

Simple curve Non-simple curve Simple closed curve

FIGURE 12.1: Curves in $\mathbb{R}^2$.

A curve $\varphi : I \to \mathbb{R}^n$ is said to be of *class* C^r if φ is C^r on an open interval containing I. A C^1 curve φ is *smooth* if $\varphi'(t) \neq \mathbf{0}$ for all $t \in I$. For example, on $[-1, 1]$ the curve $\varphi(t) = (t, t^2)$ is smooth but the curve $\psi(t) = (t^3, t^6)$, which has the same trace as φ, is not.

A curve $\varphi : [a, b] \to \mathbb{R}^n$ is said to be *piecewise smooth* if, for some partition $a = a_0 < a_1 < \cdots < a_m = b$, φ is smooth on each interval $[a_{j-1}, a_j]$. This implies that φ' is uniformly continuous on each *interval of smoothness* (a_{j-1}, a_j) and has right-hand and left-hand limits at the left and right endpoints, respec-

DOI: 10.1201/9781003629382-12

tively. Thus a piecewise smooth curve may be viewed as a concatenation (*sum*) of smooth curves, as shown in Figure 12.2. Note that at junctions that are *corners* there are two tangent vectors, and at junctions which are *cusps* there is one. A point on a smooth portion of the curve will be called a *smooth point.* A piecewise smooth curve therefore consists of smooth points and finitely many corner or cusp points.

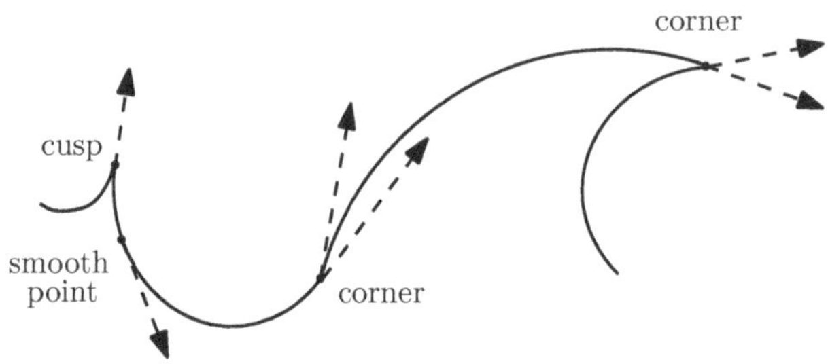

FIGURE 12.2: A piecewise smooth curve with tangent vectors.

A *reparametrization of a curve* $\varphi : I \to \mathbb{R}^n$ is a curve $\psi = \varphi \circ \alpha : J \to \mathbb{R}^n$, where $\alpha : J \to I$ is continuous, strictly increasing, and $\alpha(J) = I$ (hence $\operatorname{trace}(\psi) = \operatorname{trace}(\varphi)$). If φ is smooth, then α is required to be smooth with positive Jacobian. If ψ is a reparametrization of φ, then φ and ψ are said to be *equivalent.* For example, the smooth curve (t, t^2, t^3) $(t > 0)$ is equivalent to the curve (e^t, e^{2t}, e^{3t}) $(t \in \mathbb{R})$.

A curve $\varphi : I \to \mathbb{R}^n$ has a *positive direction*, namely, the direction that $\varphi(t)$ moves as t increases. An equivalent curve $\psi = \varphi \circ \alpha$ has the same direction since α is strictly increasing. The curve $-\varphi$, defined by

$$(-\varphi)(t) := \varphi(-t), \quad -t \in I,$$

has the opposite (negative) direction. If φ is piecewise smooth, then the positive direction is given by the tangent vectors $\varphi'(t)$, defined at smooth points. At corners and cusps the tangent vectors are right- and left-hand limits. The set of tangent vectors to a curve is called the *tangent vector field* (defined more precisely later).

12.1.1 Proposition. . *Let* $\varphi_j : [a_j, b_j] \to \mathbb{R}^n$, $j = 1, \ldots, k$, *be piecewise* C^1 *curves such that* $\varphi_j(b_j) = \varphi_{j+1}(a_{j+1})$, $j = 1, \ldots, k-1$. *Then there exists a piecewise* C^1 *curve* $\varphi : [0, 1] \to \mathbb{R}^n$, *denoted by*

$$\varphi = \varphi_1 + \varphi_2 + \cdots + \varphi_k$$

and called the sum of the curves φ_j, *such that* $\varphi\big|_{[(j-1)/k, j/k]}$ *is equivalent to* φ_j.

Proof. Define $\alpha_j : [(j-1)/k, j/k] \to [a_j, b_j]$ by

$$\alpha_j(t) = b_j + (b_j - a_j)(kt - j), \quad (j-1)/k \le t \le j/k,$$

and $\varphi : [0, 1] \to \mathbb{R}^n$ by $\varphi = \varphi_j \circ \alpha_j$ on $[(j-1)/k, j/k]$. □

Exercises

1. Prove that the notion of equivalent smooth curves is an equivalence relation.

2. Show that if φ is smooth and $\psi = \varphi \circ \alpha$ is an equivalent curve, then

$$\frac{\varphi'(t)}{\|\varphi'(t)\|} = \frac{\psi'(t)}{\|\psi'(t)\|}.$$

   Thus the unit tangent vector field is invariant under a reparametrization.

3. Sketch the trace of the curve $\varphi(t) = (t^2, t^3 - t)$ on the interval $[-2, 2]$. Find all points on the trace where there are two tangent vectors and express these vectors in terms of the standard basis.

4. Find the tangent vector field of the given curve φ on the interval $[0, 2\pi]$. Sketch the trace and find all points on the trace at which there are two tangent vectors. Express these vectors in terms of the standard basis.

   (a) $\varphi(t) = \big(\sin t, \cos(2t)\big)$ (b) $\varphi(t) = \big(\cos t, \sin(2t)\big)$

   (c) $\varphi(t) = \big(\cos t, \cos(2t)\big)$ (d) $\varphi(t) = \big(\sin t, \sin(2t)\big)$.

5. In (a)–(d) below, find a smooth simple curve or a smooth simple closed curve $\varphi : I \to C$ with trace C.

   (a) C is the intersection of the elliptic cylinder $\dfrac{x^2}{a^2} + \dfrac{y^2}{b^2} = 1$ and the plane $x + y + z = 1$.

   (b) C is the intersection of the elliptic cylinder $\dfrac{x^2}{a^2} + \dfrac{y^2}{b^2} = 1$ and the surface $z = 2xy$.

   (c) C is the intersection in the first octant of the paraboloid $z = x^2 + y^2$ and the plane $x + y + z = 1$.

   (d) C is the intersection in the first octant of the cone $z = x^2 + y^2$ and the plane $x + y + z = 1$.

6. Let $\varphi : [a, b] \to \mathbb{R}^n$ be a C^1 curve with the property that for some $\boldsymbol{x} \in \mathbb{R}^n$, $\varphi(t) = \boldsymbol{x}$ for infinitely many $t \in [a, b]$. Prove that φ is not smooth.

7. Let f be C^1 on an open set U and let φ be a C^1 curve in U. Suppose that $\varphi'(t) = \nabla f\big(\varphi(t)\big)$ for all $t > a$ and that the limit $\boldsymbol{x} := \lim_{t\to+\infty} \varphi(t)$ exists in U. Prove that $\nabla f(\boldsymbol{x}) = \mathbf{0}$.

   Hint. Assume $\nabla f(\boldsymbol{x}) \neq \mathbf{0}$. Let $g = f \circ \varphi$ and show that $g'(t) > \|\nabla f(\boldsymbol{x})\|^2/2$ for all sufficiently large t.

12.2 Integration on Curves

Rectifiable Curves

Let $\varphi : I \to \mathbb{R}^n$ be a parameterized curve. Assume first that $I = [a, b]$. For a partition $\mathcal{P} = \{t_0 = a < t_1 < \cdots < t_{k-1} < t_k = b\}$ of $[a, b]$, define

$$L_{\mathcal{P}}(\varphi) = \sum_{j=1}^{k} \|\varphi(t_j) - \varphi(t_{j-1})\|,$$

which is the length of the inscribed polygonal line with segments joining the points $\varphi(t_{j-1})$ and $\varphi(t_j)$.

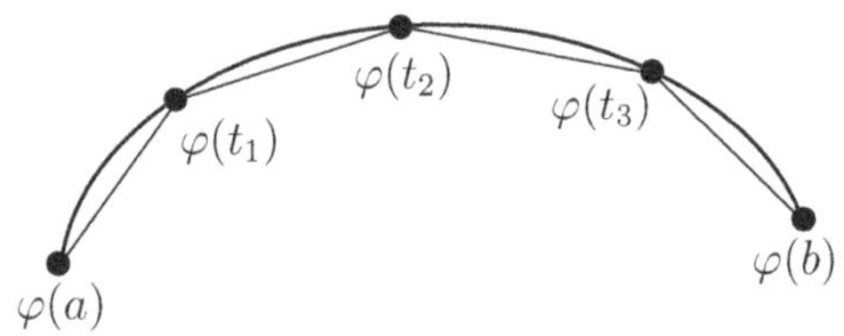

FIGURE 12.3: Inscribed polygonal line.

The *(arc) length* of φ is defined as

$$\text{length}(\varphi) := \sup_{\mathcal{P}} L_{\mathcal{P}}(\varphi),$$

where the supremum is taken over all partitions $\mathcal{P}$ of $[a, b]$. If $\text{length}(\varphi) < +\infty$, then φ is said to be *rectifiable.* Note that if $\psi = \varphi \circ \alpha$ is equivalent to φ, then $\text{length}(\psi) = \text{length}(\varphi)$, since $\alpha : [c, d] \to [a, b]$ induces a one-to-one correspondence between partitions of $[c, d]$ and $[a, b]$.

If $I = [a, b)$ (where b could be infinite), define

$$\text{length}(\varphi) := \sup_{a<t<b} \text{length}\left(\varphi\big|_{[a,t]}\right).$$

A similar definition is given for intervals $I = (a, b]$. For an open interval $I = (a, b)$, define

$$\text{length}(\varphi) := \text{length}\left(\varphi\big|_{(a,c]}\right) + \text{length}\left(\varphi\big|_{[c,b)}\right),$$

where $a < c < b$. By 12.2.3 below, the expression on the right does not depend on the intermediate point c.

12.2.1 Example. Let $\alpha > 0$. The curve

$$\varphi(t) = \big(x(t), y(t)\big) = \big(t, t^\alpha \sin(1/t)\big), \quad 0 < t \le b, \quad \varphi(0) = 0,$$

is rectifiable iff $\alpha > 1$. This follows from the inequalities

$$\sum_{j=1}^{k} |y(t_j) - y(t_{j-1})| \leq \sum_{j=1}^{k} \|\varphi(t_j) - \varphi(t_{j-1})\| \leq 2(b-a) + 2\sum_{j=1}^{k} |y(t_j) - y(t_{j-1})|$$

and 5.9.3. ◊

We prove in 12.2.4 below that piecewise C^1 curves on $[a, b]$ are rectifiable. For this, we require two lemmas. The proof of the first is similar to that of the corresponding result for lower Darboux sums and is left as an exercise.

12.2.2 Lemma. *Let* $\varphi : [a, b] \to \mathbb{R}^n$ *be a curve and let* $\mathcal{P}$ *and* $\mathcal{Q}$ *be partitions of* $[a, b]$*. If* $\mathcal{P}$ *is a refinement of* $\mathcal{Q}$*, then* $L_{\mathcal{Q}}(\varphi) \leq L_{\mathcal{P}}(\varphi)$*.*

12.2.3 Lemma. *Let* $\varphi : [a, b] \to \mathbb{R}^n$ *be a curve and* $c \in (a, b)$*. Then*

$$\text{length}(\varphi) = \text{length}\big(\varphi|_{[a,c]}\big) + \text{length}\big(\varphi|_{[c,b]}\big).$$

In particular, φ *is rectifiable iff* $\varphi|_{[a,c]}$ *and* $\varphi|_{[c,b]}$ *are rectifiable.*

Proof. Let $\mathcal{P}'$ and $\mathcal{P}''$ be partitions of $[a, c]$ and $[c, b]$, respectively, and set $\mathcal{P} = \mathcal{P}' \cup \mathcal{P}''$. Then $\mathcal{P}$ is a partition of $[a, b]$ and

$$\text{length}(\varphi) \geq L_{\mathcal{P}}(\varphi) = L_{\mathcal{P}'}\big(\varphi|_{[a,c]}\big) + L_{\mathcal{P}''}\big(\varphi|_{[c,b]}\big).$$

Taking suprema over $\mathcal{P}'$ and then $\mathcal{P}''$ yields

$$\text{length}(\varphi) \geq \text{length}\big(\varphi|_{[a,c]}\big) + \text{length}\big(\varphi|_{[c,b]}\big).$$

For the reverse inequality, let $\mathcal{P} = \{t_0 = a < t_1 < \cdots < t_k = b\}$ be a partition of $[a, b]$ and suppose $c \in (t_{i-1}, t_i]$. If $\mathcal{P}' = \{t_0 = a < t_1 < \cdots < t_{i-1} < c\}$ and $\mathcal{P}'' = \{c \leq t_i < \cdots < t_k = b\}$, then an application of the triangle inequality shows that

$$L_{\mathcal{P}}(\varphi) \leq L_{\mathcal{P}'}\big(\varphi|_{[a,c]}\big) + L_{\mathcal{P}''}\big(\varphi|_{[c,b]}\big) \leq \text{length}\big(\varphi|_{[a,c]}\big) + \text{length}\big(\varphi|_{[c,b]}\big).$$

Since $\mathcal{P}$ was arbitrary,

$$\text{length}(\varphi) \leq \text{length}\big(\varphi|_{[a,c]}\big) + \text{length}\big(\varphi|_{[c,b]}\big). \qquad \square$$

12.2.4 Theorem. *Let* $\varphi : [a, b] \to \mathbb{R}^n$ *be piecewise* C^1*. Then* φ *is rectifiable and*

$$\text{length}(\varphi) = \sum_{j=1}^{m} \int_{a_{j-1}}^{a_j} \|\varphi'(t)\|\, dt,$$

where φ *is smooth on the intervals* $[a_{j-1}, a_j]$*,* $a = a_0 < a_1 < \cdots < a_m = b$*.*

Proof. By 12.2.3 we may assume that $\varphi = (\varphi_1, \ldots, \varphi_n)$ is C^1 on $[a, b]$. Given $\varepsilon > 0$, choose $\delta > 0$ so that

$$\left| \int_a^b \|\varphi'(t)\|\, dt - \sum_{k=1}^m \|\varphi'(t_k)\|\Delta t_k \right| < \varepsilon, \quad \Delta t_k := t_k - t_{k-1} \tag{12.1}$$

for all partitions $\mathcal{P} = \{t_0 = a < t_1 < \cdots < t_{m-1} < t_m = b\}$ with $\|\mathcal{P}\| < \delta$. For such a partition $\mathcal{P}$, choose $s_{j,k} \in (t_{k-1}, t_k)$ such that

$$\varphi_j(t_k) - \varphi_j(t_{k-1}) = \varphi_j'(s_{j,k})\Delta t_k, \quad k = 1, \ldots, m, \ j = 1, \ldots, n.$$

Then

$$L_{\mathcal{P}}(\varphi) = \sum_{k=1}^m \|\varphi(t_k) - \varphi(t_{k-1})\| = \sum_{k=1}^m \left(\sum_{j=1}^n |\varphi_j'(s_{j,k})|^2 \right)^{1/2} \Delta t_k,$$

hence

$$\begin{aligned} &\left| L_{\mathcal{P}}(\varphi) - \sum_{k=1}^m \|\varphi'(t_k)\|\Delta t_k \right| \\ &\qquad = \left| \sum_{k=1}^m \left\{ \left(\sum_{j=1}^n |\varphi_j'(s_{j,k})|^2 \right)^{1/2} - \left(\sum_{j=1}^n |\varphi_j'(t_k)|^2 \right)^{1/2} \right\} \Delta t_k \right|. \end{aligned}$$

Taking a smaller δ if necessary, we may assume that the absolute value of the term in braces is less than $\varepsilon/(b-a)$. This is possible by the uniform continuity of φ'. It follows that

$$\left| L_{\mathcal{P}}(\varphi) - \sum_{k=1}^m \|\varphi'(t_k)\|\Delta t_k \right| < \varepsilon. \tag{12.2}$$

From (12.1) and (12.2) we now have

$$\int_a^b \|\varphi'(t)\|\, dt - 2\varepsilon < L_{\mathcal{P}}(\varphi) < \int_a^b \|\varphi'(t)\|\, dt + 2\varepsilon \tag{12.3}$$

for all $\mathcal{P}$ with $\|\mathcal{P}\| < \delta$. Since $L_{\mathcal{P}}(\varphi) \le \text{length}(\varphi)$ and ε was arbitrary, the first inequality in (12.3) implies that

$$\int_a^b \|\varphi'(t)\|\, dt \le \text{length}(\varphi).$$

For the reverse inequality, let $\mathcal{Q}$ be any partition of $[a, b]$. Refine $\mathcal{Q}$ to obtain a partition $\mathcal{P}$ with $\|\mathcal{P}\| < \delta$. Then, from 12.2.2 and the second inequality in (12.3),

$$L(\varphi, \mathcal{Q}) < \int_a^b \|\varphi'(t)\|\, dt + 2\varepsilon.$$

Since $\mathcal{Q}$ and ε are arbitrary, $\text{length}(\varphi) \le \int_a^b \|\varphi'(t)\|\, dt$. □

The proof of the following corollary is left to the reader.

12.2.5 Corollary. *If* $\varphi : [a, b) \to \mathbb{R}^n$ *is* C^1, *then* $\text{length}(\varphi)$ *is the improper integral* $\int_a^b \|\varphi'(t)\|\, dt$.

12.2.6 Example. Let $\varphi(t) = \big(e^{-t}\cos t, e^{-t}\sin t\big)$, where $0 \le t < +\infty$. Then $\|\varphi'(t)\| = e^{-t}$, hence $\text{length}(\varphi) = \int_0^\infty e^{-t} = 1$. ◊

Line Integrals

Let $\varphi : [a, b] \to \mathbb{R}^n$ be a C^1 curve with trace C and let $f : C \to \mathbb{R}$ be continuous. The *line integral of f over φ* is defined by

$$\int_\varphi f\, ds = \int_C f\, ds = \int_a^b f\big(\varphi(t)\big)\|\varphi'(t)\|\, dt.$$

Note that if $\psi = \varphi \circ \alpha$ is an equivalent parametrization, where $\alpha : [c, d] \to [a, b]$ is C^1, then, by the chain rule and the change of variables theorem,

$$\begin{aligned}\int_c^d f\big(\psi(t)\big)\|\psi'(t)\|\, dt &= \int_c^d f\big(\varphi(\alpha(t))\big)\|\varphi'\big(\alpha(t)\big)\|\alpha'(t)\, dt\\ &= \int_a^b f\big(\varphi(u)\big)\|\varphi'(u)\|\, du.\end{aligned}$$

The value of a line integral is therefore independent of the choice of parametrization.

If $\varphi : [a, b] \to \mathbb{R}^n$ is piecewise C^1, then the line integral is defined as

$$\int_\varphi f\, ds = \sum_j \int_{\varphi_j} f\, ds,$$

where φ_j is the restriction of φ to $[a_j, a_{j+1}]$ and φ is C^1 on $[a_j, a_{j+1}]$. If $\varphi : I \to \mathbb{R}^n$ is C^1, where I is an arbitrary interval, then the line integral is defined as an improper integral, as in the case of arc length.

12.2.7 Remark. Theorem 12.2.4 shows that arc length is the line integral of the constant function 1. Using techniques similar to those found in the proof of that theorem, one may show that if φ is C^1, then $\int_\varphi f$ is the limit of sums of the form

$$\sum_{j=1}^k (f \circ \varphi)(t_j^*)\|\varphi(t_j) - \varphi(t_{j-1})\|, \quad t_j^* \in (t_{j-1}, t_j),$$

as $\max_j \|\varphi(t_j) - \varphi(t_{j-1})\| \to 0$. This interpretation is useful in applications. For example, if $f(\boldsymbol{x})$ is the mass per unit length at the point $\boldsymbol{x}$ of a wire C, then $(f \circ \varphi)(t_j^*)\|\varphi(t_j) - \varphi(t_{j-1})\|$ is approximately the mass of a small piece of the wire. Summing and taking the limit gives the mass of the wire as the line integral $\int_C f\, ds$. ◊

Vector Fields

12.2.8 Definition. A *vector field on a set* $E \subseteq \mathbb{R}^n$ is a function

$$\vec{F} = (f_1, \dots, f_n) : E \to \mathbb{R}^n.$$

The vector field is said to be of *class* C^r if each f_j is C^r. ◊

Geometrically, a vector field assigns to each point of E a unique vector in $\mathbb{R}^n$, as illustrated in Figure 12.4.

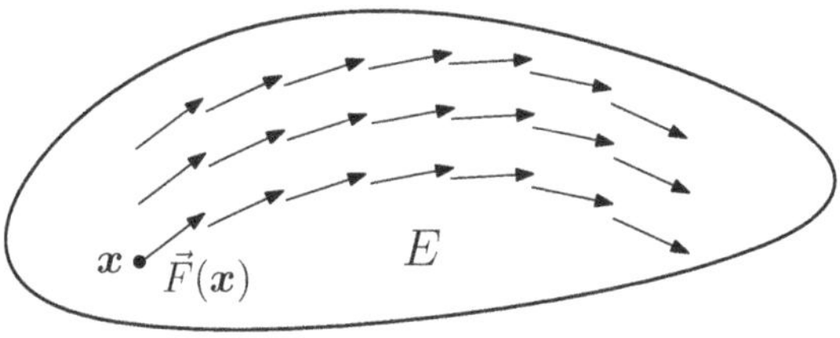

FIGURE 12.4: Vector field on E.

If φ is a simple smooth curve and $\boldsymbol{x} = \varphi(t)$, then

$$\vec{v}_\varphi(\boldsymbol{x}) := \varphi'(t) \quad \text{and} \quad \vec{T}_\varphi(\boldsymbol{x}) := \frac{\varphi'(t)}{\|\varphi'(t)\|}$$

denote, respectively, the *tangent vector field* and *unit tangent vector field* along φ. If φ denotes the position of a particle at time t, then the tangent vector field is called the *velocity vector field* of the particle.

Vector fields that describe forces, such as gravitation or electromagnetism, are called *force fields*. Line integrals may then be used to calculate the work done by the force in moving a particle along a curve. Specifically, suppose the particle moves along a simple smooth curve $\varphi : [a, b] \to \mathbb{R}^3$ under the action of a continuous force field $\vec{F} = (f_1, f_2, f_3)$ on $C := \text{trace}(\varphi)$. The work $\Delta_j W$ done by the force in moving the particle from a point $\boldsymbol{x}_j = \varphi(t_j)$ on C to a nearby point $\boldsymbol{x}_{j+1} = \varphi(t_{j+1})$ is approximately the component of the force in the direction of the tangent to the curve at $\boldsymbol{x}_j$ multiplied by the distance the particle travels:

$$\Delta_j W \approx \left[\vec{F}(\boldsymbol{x}_j) \cdot \vec{T}_\varphi(\boldsymbol{x}_j)\right] \|\boldsymbol{x}_j - \boldsymbol{x}_{j+1}\|.$$

The total work W done by the force is then approximately $\sum_j \Delta_j W$. Since F is continuous, the approximation gets better by taking smaller intervals. It is therefore reasonable to *define* the total work done by the force in moving the particle along the curve as the limit of $\sum_j \Delta_j W$ as $\max_j \|\boldsymbol{x}_j - \boldsymbol{x}_{j+1}\| \to 0$. By 12.2.7, we are therefore led to the definition

$$W := \int_\varphi \vec{F} \cdot \vec{T}\, ds = \int_a^b \vec{F}(\varphi(t)) \cdot \vec{T}_\varphi(\varphi(t)) \|\varphi'(t)\|\, dt.$$

Since $\vec{T}_\varphi(\varphi(t)) = \|\varphi'(t)\|^{-1}\varphi'(t)$, we see that

$$W = \int_a^b \vec{F}(\varphi(t)) \cdot \varphi'(t)\, dt = \int_a^b \left[f_1(\boldsymbol{x})\frac{dx_1}{dt} + f_2(\boldsymbol{x})\frac{dx_2}{dt} + f_3(\boldsymbol{x})\frac{dx_3}{dt} \right] dt,$$

where $\boldsymbol{x} = \varphi(t)$. The last integral is frequently written

$$\int_\varphi (f_1\, dx_1 + f_2\, dx_2 + f_3\, dx_3).$$

The integrand is called a (*differential*) 1-*form on* C in $\mathbb{R}^3$.

Differential 1-Forms in $\mathbb{R}^n$

Let f_j be defined on a set $S \subseteq \mathbb{R}^n$. The symbol

$$\omega := f_1\, dx_1 + \cdots + f_n\, dx_n$$

is called a (*differential*) 1-*form on* S. The form is said to be C^r on S if each f_j is C^r on S, where $r \in \mathbb{N} \cup \{+\infty\}$. Given another 1-form

$$\eta = g_1\, dx_1 + \cdots + g_n\, dx_n$$

on S and a, $b \in \mathbb{R}$, the 1-form $a\omega + b\eta$ on S is defined by

$$a\omega + b\eta := (af_1 + bg_1)\, dx_1 + \cdots + (af_n + bg_n)\, dx_n.$$

If $\vec{H} = (h_1, \dots, h_n)$ is a vector field on S, we define the *inner product* $\omega \cdot \vec{H}$ *of* ω *and* $\vec{H}$ *on* S by

$$\omega \cdot \vec{H}(\boldsymbol{x}) := \sum_{j=1}^n f_j(\boldsymbol{x})h_j(\boldsymbol{x}), \quad \boldsymbol{x} \in S.$$

The *integral* of a continuous (that is, C^0) 1-form ω over a C^1 curve $\varphi : [a, b] \to S$ is defined as

$$\int_\varphi \omega = \int_a^b [f_1(\varphi(t))\varphi_1'(t) + \cdots + f_n(\varphi(t))\varphi_n'(t)]\, dt = \int_a^b [\vec{F}(\varphi(t))] \cdot \varphi'(t)\, dt,$$

where $\vec{F} := (f_1, \dots, f_n)$. If φ is only piecewise C^1, then $\int_\varphi \omega$ is defined to be the sum of the integrals over the intervals on which φ is C^1.

The following properties of the integral are easily established:

- $\displaystyle\int_\varphi (a\omega + b\eta) = a\int_\varphi \omega + b\int_\varphi \eta,$
- $\displaystyle\int_{-\varphi} \omega = -\int_\varphi \omega,$ and
- $\displaystyle\int_{\varphi_1+\cdots+\varphi_k} \omega = \sum_{j=1}^k \int_{\varphi_j} \omega.$

A continuous 1-form $\omega = f_1\,dx_1 + \cdots + f_n\,dx_n$ on an open set $U \subseteq \mathbb{R}^n$ is said to be *exact* if there exists a C^1 function f on U such that $f_j = \partial_j f$ on U for each j. We then write

$$\omega = df = \sum_{i=1}^{n} \frac{\partial f}{\partial x_j}\,dx_j.$$

The following proposition shows that the integral of an exact form over a curve depends only on f and the endpoints of the curve.

12.2.9 Proposition. *If $\varphi : [a, b] \to U$ is piecewise C^1, then*

$$\int_\varphi df = f(\varphi(b)) - f(\varphi(a)).$$

Proof. If φ is C^1, then, by the chain rule and the fundamental theorem of calculus,

$$\int_\varphi df = \int_a^b \sum_{i=1}^{n} (\partial_j f)(\varphi(t))\varphi_i'(t)\,dt = \int_a^b (f \circ \varphi)'(t)\,dt = f(\varphi(b)) - f(\varphi(a)).$$

If φ is only piecewise C^1, subdivide the interval $[a, b]$ into intervals on which φ is smooth, apply the above result to each subinterval, and sum the results. □

12.2.10 Theorem. *Let $U \subseteq \mathbb{R}^n$ be open and connected and let ω be a continuous 1-form on U. The following statements are equivalent:*

(a) *ω is exact.*

(b) *$\int_\varphi \omega = 0$ for every closed piecewise C^1 curve φ in U.*

(c) *$\int_\phi \omega = \int_\psi \omega$ for every pair of piecewise C^1 curves ϕ, $\psi : [a, b] \to \mathbb{R}^n$ in U with $\phi(a) = \psi(a)$ and $\phi(b) = \psi(b)$.*

Proof. That (a) implies (b) follows from 12.2.9.

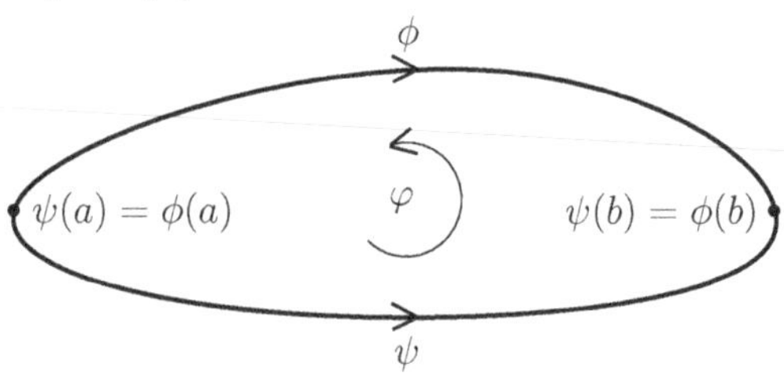

FIGURE 12.5: $\varphi = \psi - \phi$.

For (b) implies (c), define a closed, piecewise smooth curve $\varphi : [a, b+1] \to \mathbb{R}^n$ by

$$\varphi(t) = \psi(t), \quad a \le t \le b, \quad \varphi(t) = \phi\left(b + (b - t)(b - a)\right), \quad b \le t \le b + 1$$

(See Figure 12.5) Then $\varphi|_{[b,b+1]}$ is equivalent to $-\phi$, hence if (b) holds,

$$0 = \int_{\varphi} \omega = \int_{\psi} \omega - \int_{\phi} \omega,$$

proving (c).

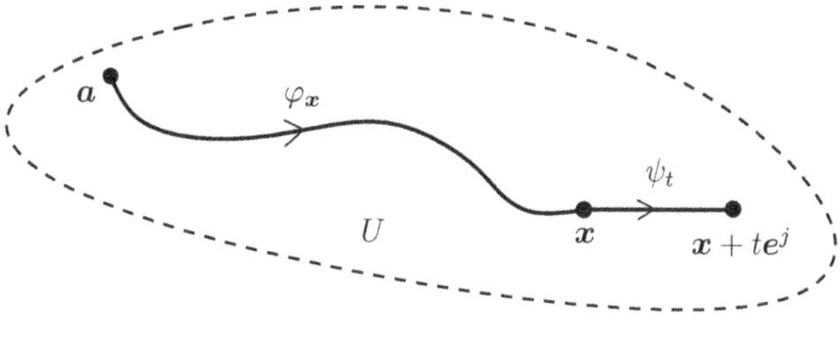

FIGURE 12.6: $\varphi_{\boldsymbol{x}+t\boldsymbol{e}^j} = \varphi_{\boldsymbol{x}} + \psi_t$.

Now assume that (c) holds and let $\omega = \sum_{j=1}^n f_j\, dx_j$. To establish (a), we construct a function f on U such that $\partial_j f = f_j$. Choose any point $\boldsymbol{a} \in U$. By Exercise 8.7.8, for each $\boldsymbol{x} \in U$ there exists a piecewise C^1 curve $\varphi_{\boldsymbol{x}}$ in U with initial point $\boldsymbol{a}$ and terminal point $\boldsymbol{x}$. Define $f(\boldsymbol{x}) = \int_{\varphi_{\boldsymbol{x}}} \omega$. By (c), $f(\boldsymbol{x})$ is independent of the path and hence is well-defined. Fix j, let $t > 0$, and denote by ψ_t the line segment $\boldsymbol{x} + u\boldsymbol{e}^j$, $0 \le u \le t$. Then ψ_t lies in U for sufficiently small $t > 0$, and by continuity of f_j,

$$\frac{1}{t}\left[f(\boldsymbol{x} + t\boldsymbol{e}^j) - f(\boldsymbol{x})\right] = \frac{1}{t}\int_{\psi_t} \omega = \frac{1}{t}\int_0^t f_j(\boldsymbol{x} + u\boldsymbol{e}^j)\, du \to f_j(\boldsymbol{x})$$

as $t \to 0^+$. A similar argument works for the case $t \to 0^-$. Therefore, $\partial_j f(\boldsymbol{x}) = f_j(\boldsymbol{x})$, as required. □

Exercises

1. Determine which of the following curves are rectifiable.
   (a) $\varphi(t) = (t, t^{-p})$, $0 < t \le 1$, where $p > 0$
   (b) $\varphi(t) = (e^{-t}, e^{-t^2}, e^{-t^3})$, $0 \le t < +\infty$
   (c) $\varphi(t) = \left(t^{-1}, e^{-t}\right)$, $t \ge 1$
   (d) $\varphi(t) = \left(t^{-1}, e^{-t}\right)$, $0 < t \le 1$.
2. Evaluate $\int_{\varphi} f$ for
   (a) $\varphi(t) = (t^3/3, t^4/4)$, $1 \le t \le 2$, $f(x, y) = x/y$
   (b) $\varphi(t) = \big(t, \sin(2t), \cos(2t)\big)$, $0 \le t \le \pi/4$, $f(x, y, z) = xz$
   (c) $\varphi(t) = \big(t, t^2/2, t^3/3\big)$, $0 \le t \le 1$, $f(x, y, z) = x + 6z$
   (d) $\varphi(t) = \big(\sin t, \sqrt{2}\cos t, \sin t\big)$, $0 \le t \le \pi/2$, $f(x, y, z) = xyz$.

3. Set up, but do not evaluate, the integral that gives the circumference of the ellipse $\frac{x^2}{a^2} + \frac{y^2}{b^2} = 1$. (Your answer should involve $\sin^2 t$.)

4. In each case below, find a smooth simple curve or a smooth simple closed curve with trace C. Use the parametrization to find an integral that gives the length of the curve. (Do not evaluate the integral.)

   (a) $C = \{(x, y) : x^3 - 7y^2 = 1,\ 1 < x < 2, y > 0\}$

   (b) $C = \{(x, y) : 9(x-1)^2 + 4(y-2)^2 = 36\}$

   (c) $C = \{(x, y) : x^2 - y^2 = 4,\ x > 2,\ 0 < x + y < a\}$.

5. Let $\varphi(x) = (x, g(x))$, $a \le x \le b$, where g is continuously differentiable, and let $f(x, y)$ be continuous on the graph of g. Show that

$$\int_\varphi f = \int_a^b f(x, g(x)) \sqrt{1 + [g'(x)]^2}\, dx.$$

   Use this to find

   (a) $\int_\varphi f$ if $g(x) = (2/5)x^{5/2}$ and $f(x, y) = x^2$, $0 \le x \le 1$.

   (b) the length of the graph of the equation $x^{2/3} + y^{2/3} = 1$.

   (c) the length of the graph of the function $g(x) = \frac{x^p}{2p} + \frac{x^{2-p}}{2(p-2)}$, where $0 < a \le x \le b$ and $p > 2$.

6. Prove 12.2.5.

7. Let a smooth curve $\varphi : [a, b] \to \mathbb{R}^2$ be described in polar coordinates by

$$\varphi(t) = \big(r(t)\cos\theta(t), r(t)\sin\theta(t)\big), \quad r(t) \ge 0.$$

   Show that

$$\text{length}(\varphi) = \int_a^b \sqrt{r(t)\big[\theta'(t)\big]^2 + \big[r'(t)\big]^2}\, dt.$$

8. Let $\vec{F} = (F_1, F_2, F_3)$ be a force field in $\mathbb{R}^3$ that moves a particle of mass m along a smooth curve $\varphi : [a, b] \to \mathbb{R}^3$. The *kinetic energy* of the particle at time t is defined as $\frac{1}{2}m\|\varphi'(t)\|^2$. Use Newton's second law $\vec{F} = m\varphi''$ to show that the work done by the force in moving the particle from $\varphi(a)$ to $\varphi(b)$ is the change in kinetic energy

$$\tfrac{1}{2}m\|\varphi'(b)\|^2 - \tfrac{1}{2}m\|\varphi'(a)\|^2.$$

9. A force field $\vec{F}$ in $\mathbb{R}^3$ is said to be *conservative* if there exists a function $P(x,y,z)$ such that $\vec{F} = -\nabla P$. $P(x,y,z)$ is called the *potential energy* of an object at the point (x,y,z).

(a) Show that the work done by a conservative force in moving the object along a curve φ from $\varphi(a)$ to $\varphi(b)$ is

$$P(\varphi(a)) - P(\varphi(b)).$$

(b) Deduce the *Law of Conservation of Energy*

$$P(\varphi(b)) + \tfrac{1}{2}m\|\varphi'(b)\|^2 = P(\varphi(a)) + \tfrac{1}{2}m\|\varphi'(a)\|^2,$$

that is, the sum of the potential and kinetic energies is constant.

(c) Find a potential function for the gravitational force field

$$F(\boldsymbol{x}) = -mMG\|\boldsymbol{x}\|^{-3}\boldsymbol{x},$$

where M is the mass of the earth (concentrated at the origin, the center of the earth), m is the mass of the particle at point $\boldsymbol{x}$, and G is the gravitation constant.

10. For a smooth curve $\varphi : [a,b] :\to \mathbb{R}^n$, define the *arc length* function $s = s(t)$ by

$$s(t) = \int_a^t \|\varphi'(\tau)\|\, d\tau, \quad a \le t \le b.$$

Show that s has a smooth inverse $t = t(s)$, $0 \le s \le \ell := \text{length}(\varphi)$. The curve $\psi(s) = \varphi(t(s))$ is called a *reparametrization of* φ *by arc length.* Show that, for a continuous vector field $\vec{F}$ on $\text{trace}(\varphi) = \text{trace}(\psi)$,

$$\int_\varphi \vec{F} \cdot \vec{T}_\varphi = \int_0^\ell \vec{F}(\psi(s)) \cdot \psi'(s)\, ds.$$

11. Let $\mathcal{P} = \{a = t_0 < t_1 < \cdots < t_k = b\}$ be a partition of $[a,b]$. For $f : [a,b] \to \mathbb{R}$, define

$$V_{\mathcal{P}}(f) = \sum_{j=1}^{k} |f(t_j) - f(t_{j-1})|.$$

Then f is said to have *bounded variation on the interval* $[a,b]$ if $\sup_{\mathcal{P}} V_{\mathcal{P}}(f) < +\infty$. (Section 5.9.) Show that a curve $\varphi = (\varphi_1, \ldots, \varphi_n) : [a,b] \to \mathbb{R}^n$ is rectifiable iff each component function φ_i has bounded variation on $[a,b]$.

12.3 Parameterized Surfaces

12.3.1 Definition. Let $1 \le m \le n$. A *smooth parameterized m-surface in* $\mathbb{R}^n$ is a C^1 function $\varphi = (\varphi_1, \ldots, \varphi_n) : U \to \mathbb{R}^n$, where $U \subseteq \mathbb{R}^m$ is open and the derivative $\varphi'(\boldsymbol{u})$ has rank m at each point $\boldsymbol{u} \in U$. A *reparametrization of* φ is a smooth parameterized m-surface $\psi = \varphi \circ \alpha : V \to \mathbb{R}^n$, where $V \subseteq \mathbb{R}^m$ is open and $\alpha : V \to U$ is C^1 with C^1 inverse $\alpha^{-1} : U \to V$ such that $J_\alpha > 0$ on V. In this case, φ and ψ are said to be *equivalent.* ◊

We shall usually drop the qualifier "smooth" when referring to parameterized surfaces. Note that the parameter set U is a m-parameterized surface in $\mathbb{R}^m$. Here, we take φ to be the identity map $\iota : U \to U$.

Tangent Spaces of a Parameterized Surface

Let $\varphi : U \to \mathbb{R}^n$ be a parameterized m-surface and $\boldsymbol{u} \in U$. For small $|t|$ the line segment $\boldsymbol{u} + t\boldsymbol{e}^j$ is contained in U and is mapped by φ onto a curve in $S := \varphi(U)$ with tangent vector

$$\left.\frac{d}{dt}\right|_{t=0} \varphi(\boldsymbol{u} + t\boldsymbol{e}^j) = d\varphi_{\boldsymbol{u}}(\boldsymbol{e}^j) = \left(\frac{\partial \varphi_1}{\partial u_j}(\boldsymbol{u}), \ldots, \frac{\partial \varphi_n}{\partial u_j}(\boldsymbol{u})\right) =: \partial_j \varphi(\boldsymbol{u}),$$

where $\boldsymbol{e}^1, \ldots, \boldsymbol{e}^m$ are the standard basis vectors in $\mathbb{R}^m$. Note that $\partial_j \varphi(\boldsymbol{u})$ is just the jth column of $\varphi'(\boldsymbol{u})$. Since $\varphi'(\boldsymbol{u})$ has rank m, the vectors $d\varphi_{\boldsymbol{u}}(\boldsymbol{e}^j)$ are linearly independent and hence form a basis for an m-dimensional subspace $\mathcal{T}_{\varphi(\boldsymbol{u})}$ of $\mathbb{R}^n$, called the *tangent space of* φ *at* $\boldsymbol{u}$. Thus $d\varphi_{\boldsymbol{u}}$ is a linear isomorphism from $\mathbb{R}^m$ onto $\mathcal{T}_{\varphi(\boldsymbol{u})}$ mapping the frame $(\boldsymbol{e}^1, \ldots, \boldsymbol{e}^m)$ onto the frame $(\partial_1 \varphi(\boldsymbol{u}), \ldots, \partial_m \varphi(\boldsymbol{u}))$.[1] Note that φ is not assumed to be one-to-one, and $\varphi(\boldsymbol{u}) = \varphi(\boldsymbol{v})$ does not necessarily imply that $\mathcal{T}_{\varphi(\boldsymbol{u})} = \mathcal{T}_{\varphi(\boldsymbol{v})}$. (See Figure 12.7.)

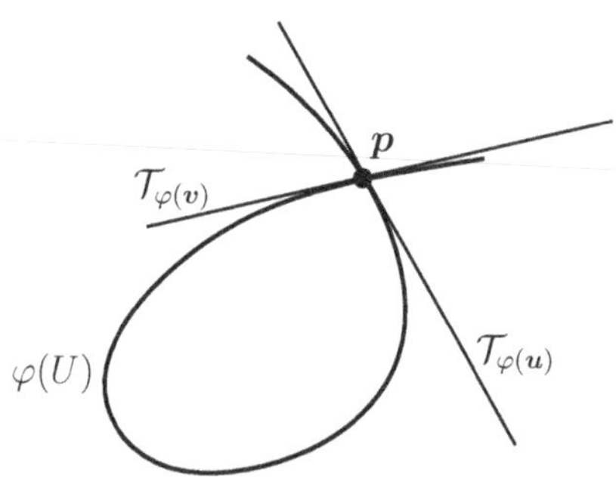

FIGURE 12.7: Tangent spaces at $\boldsymbol{p} = \varphi(\boldsymbol{u}) = \varphi(\boldsymbol{v})$.

[1]A *frame* in a finite-dimensional vector space is simply an ordered basis—see Appendix B.

Orientation of a Parameterized m-Surface

Tangent spaces may be used to assign an *orientation* to a parameterized m-surface, a notion that will be needed later to construct the integral of a differential form on a surface. First, we define orientation for the space $\mathbb{R}^m$.

Two frames $(\boldsymbol{v}^1, \ldots, \boldsymbol{v}^m)$ and $(\boldsymbol{w}^1, \ldots, \boldsymbol{w}^m)$ in $\mathbb{R}^m$ are said to be *orientation equivalent* if the determinants of the matrices

$$\begin{bmatrix} \boldsymbol{v}^1 & \cdots & \boldsymbol{v}^m \end{bmatrix} \quad \text{and} \quad \begin{bmatrix} \boldsymbol{w}^1 & \cdots & \boldsymbol{w}^m \end{bmatrix}$$

(where $\boldsymbol{v}_j$ and $\boldsymbol{w}_j$ are written as column vectors) have the same sign. Orientation equivalence is easily seen to be an equivalence relation. The collection of frames of $\mathbb{R}^m$ is therefore partitioned into two classes, one containing $(\boldsymbol{e}^1, \boldsymbol{e}^2, \ldots, \boldsymbol{e}^m)$ and the other containing $(-\boldsymbol{e}^1, \boldsymbol{e}^2, \ldots, \boldsymbol{e}^m)$. An *orientation* is assigned to $\mathbb{R}^m$ by designating one of these equivalence classes to be *positive* and the other *negative.* Any frame in the former class is then said to have *positive orientation,* while a frame in the latter class is said to have *negative orientation.* For example, if $m = 3$ and $(\boldsymbol{v}^1, \boldsymbol{v}^2, \boldsymbol{v}^3)$ has positive orientation, then so does $(\boldsymbol{v}^2, \boldsymbol{v}^3, \boldsymbol{v}^1)$, while $(\boldsymbol{v}^2, \boldsymbol{v}^1, \boldsymbol{v}^3)$ has negative orientation. By convention, the *standard* or *positive orientation of* $\mathbb{R}^m$ is the orientation obtained by designating the frame $(\boldsymbol{e}^1, \ldots, \boldsymbol{e}^m)$ to be positive. For example, in the standard orientation, the sign of the frame $(\boldsymbol{e}^m, \boldsymbol{e}^1, \ldots, \boldsymbol{e}^{m-1})$ is $(-1)^{m-1}$. We shall always assume that the spaces $\mathbb{R}^m$ have the standard orientation.

A parameterized m-surface $\varphi : U \to \mathbb{R}^n$ is said to be *orientable* if, whenever $\varphi(\boldsymbol{u}) = \varphi(\boldsymbol{v})$, $\mathcal{T}_{\varphi(\boldsymbol{u})} = \mathcal{T}_{\varphi(\boldsymbol{v})}$ and the matrix of the linear transformation

$$T_{\boldsymbol{uv}} = (d\varphi_{\boldsymbol{v}})^{-1} \circ d\varphi_{\boldsymbol{u}} : \mathbb{R}^m \to \mathbb{R}^m \tag{12.4}$$

has positive determinant. Frames $(\boldsymbol{\xi}^1, \ldots, \boldsymbol{\xi}^m)$ and $(\boldsymbol{\zeta}^1, \ldots, \boldsymbol{\zeta}^m)$ in $\mathcal{T}_{\varphi(\boldsymbol{u})}$ are then declared to be *orientation equivalent* if the frames

$$d\varphi_{\boldsymbol{u}}^{-1}(\boldsymbol{\xi}^1, \ldots, \boldsymbol{\xi}^m) := \left(d\varphi_{\boldsymbol{u}}^{-1}(\boldsymbol{\xi}^1), \ldots, d\varphi_{\boldsymbol{u}}^{-1}(\boldsymbol{\xi}^m)\right)$$

and

$$d\varphi_{\boldsymbol{u}}^{-1}(\boldsymbol{\zeta}^1, \ldots, \boldsymbol{\zeta}^m) := \left(d\varphi_{\boldsymbol{u}}^{-1}(\boldsymbol{\zeta}^1), \ldots, d\varphi_{\boldsymbol{u}}^{-1}(\boldsymbol{\zeta}^m)\right)$$

are orientation equivalent in $\mathbb{R}^m$. Since

$$T_{\boldsymbol{uv}} \circ (d\varphi_{\boldsymbol{u}})^{-1}(\boldsymbol{\xi}^1, \ldots, \boldsymbol{\xi}^m) = (d\varphi_{\boldsymbol{v}})^{-1}(\boldsymbol{\xi}^1, \ldots, \boldsymbol{\xi}^m)$$

and $\det T_{\boldsymbol{uv}} > 0$, the frames $d\varphi_{\boldsymbol{u}}^{-1}(\boldsymbol{\xi}^1, \ldots, \boldsymbol{\xi}^m)$ and $d\varphi_{\boldsymbol{v}}^{-1}(\boldsymbol{\xi}^1, \ldots, \boldsymbol{\xi}^m)$ have the same sign, hence the notion of orientation equivalence in the common tangent space $\mathcal{T}_{\varphi(\boldsymbol{u})} = \mathcal{T}_{\varphi(\boldsymbol{v})}$ is well-defined. As with the vector space $\mathbb{R}^m$, orientation equivalence on $\mathcal{T}_{\varphi(\boldsymbol{u})}$ is an equivalence relation with two equivalence classes, one containing $d\varphi_{\boldsymbol{u}}(\boldsymbol{e}^1, \boldsymbol{e}^2, \ldots, \boldsymbol{e}^m)$, the other containing $d\varphi_{\boldsymbol{u}}(-\boldsymbol{e}^1, \boldsymbol{e}^2, \ldots, \boldsymbol{e}^m)$. The *positive orientation of* φ is obtained by designating the equivalence class containing $d\varphi_{\boldsymbol{u}}(\boldsymbol{e}^1, \boldsymbol{e}^2, \ldots, \boldsymbol{e}^m)$ to be positive for every $\boldsymbol{u} \in U$. We define the *sign of* φ by

$$\operatorname{sign}(\varphi) = \begin{cases} +1 & \text{if } \varphi \text{ is positively oriented,} \\ -1 & \varphi \text{ is negatively oriented.} \end{cases}$$

Obviously, if φ is one-to-one, then it is orientable. For example, a simple smooth curve $\varphi : I \to \mathbb{R}^n$ is orientable. Since

$$d(\varphi_t)(\boldsymbol{e}^1) = \varphi'(t) = \lim_{\Delta t \to 0^+} \frac{\varphi(t + \Delta t) - \varphi(t)}{\Delta t},$$

the positive orientation is the one for which the tangent vector $d\varphi_t(\boldsymbol{e}^1)$ is in the direction of increasing t. By contrast, the curve in Figure (12.7) is not orientable.

12.3.2 Example. Let $\boldsymbol{a}^1, \ldots, \boldsymbol{a}^m$ be linearly independent vectors in $\mathbb{R}^n$ and let $\boldsymbol{b} \in \mathbb{R}^n$. Define *$m$-dimensional parameterized affine space* $\varphi : \mathbb{R}^m \to \mathbb{R}^n$ by

$$\varphi(\boldsymbol{u}) = \varphi(u_1, \ldots, u_m) = \boldsymbol{b} + \sum_{i=1}^{m} u_i \boldsymbol{a}^i.$$

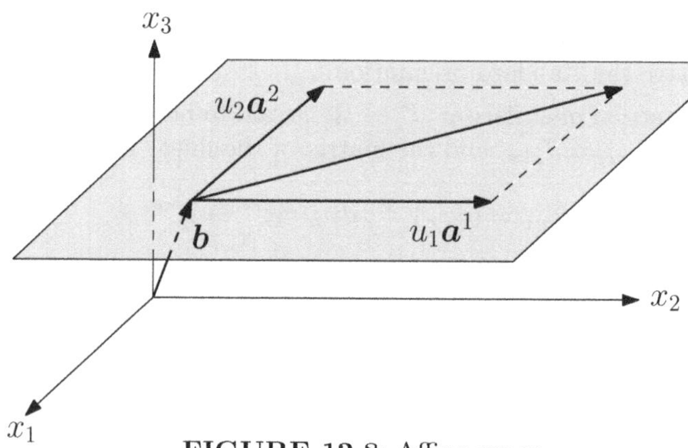

FIGURE 12.8: Affine space.

Since φ is one-to-one, it is orientable. Since $\partial_i \varphi = \boldsymbol{a}^i$, the tangent space at each point is the subspace of $\mathbb{R}^n$ with frame $(\boldsymbol{a}^1, \ldots, \boldsymbol{a}^m)$. ◊

12.3.3 Example. The *Cartesian product of circles*

$$\varphi(\theta_1, \ldots, \theta_m) = \big(r_1 \cos\theta_1, r_1 \sin\theta_1, \ldots, r_m \cos\theta_m, r_m \sin\theta_m\big), \quad r_i > 0,$$

is a parameterized m-surface in $\mathbb{R}^{2m}$. Orientability follows from the periodicity of the sine and cosine functions. ◊

Orientation of a Parameterized $(n-1)$-Surface

For $m = n-1$, the notion of orientability may be formulated more concretely in terms of a *normal vector field.*

12.3.4 Lemma. *Let* $\varphi : U \to \mathbb{R}^n$ *be a parameterized* $(n-1)$*-surface. Define* $\partial\varphi^{\perp} : U \to \mathbb{R}^n$ *by*

$$\partial\varphi^{\perp} := \sum_{i=1}^{n} (-1)^{i+n} \frac{\partial(\varphi_1, \ldots, \widehat{\varphi}_i, \ldots, \varphi_n)}{\partial(u_1, \ldots, u_{n-1})} \mathbf{e}^i,$$

where the hat indicates that φ_i *is omitted in the calculation, and let*

$$A := \begin{bmatrix} \partial_1\varphi(\boldsymbol{u}) \\ \vdots \\ \partial_{n-1}\varphi(\boldsymbol{u}) \\ \partial\varphi^{\perp}(\boldsymbol{u}) \end{bmatrix}_{n\times n}.$$

Then $d\varphi^{\perp}(\boldsymbol{u})$ *is perpendicular to the tangent space* $\mathcal{T}_{\varphi(\boldsymbol{u})}$*, and*

$$|A| = \|\partial\varphi^{\perp}(\boldsymbol{u})\|^2 = \det\left[\varphi'(\boldsymbol{u})^t\varphi'(\boldsymbol{u})\right] > 0. \tag{12.5}$$

Proof. Let $m = n-1$. For each j, the determinant

$$D_j(\boldsymbol{u}) := \begin{vmatrix} \partial_j\varphi_1(\boldsymbol{u}) & \cdots & \partial_j\varphi_n(\boldsymbol{u}) \\ \partial_1\varphi_1(\boldsymbol{u}) & \cdots & \partial_1\varphi_n(\boldsymbol{u}) \\ \vdots & & \vdots \\ \partial_m\varphi_1(\boldsymbol{u}) & \cdots & \partial_m\varphi_n(\boldsymbol{u}) \end{vmatrix}$$

has two identical rows and hence is zero. Expanding $D_j(\boldsymbol{u})$ along the first row and multiplying by $(-1)^m$ yields

$$D_j(\boldsymbol{u}) = (-1)^m \sum_{i=1}^{n} (-1)^{i+1} \partial_j\varphi_i \frac{\partial(\varphi_1, \ldots, \widehat{\varphi}_i, \ldots, \varphi_n)}{\partial(u_1, \ldots, u_{n-1})} = \partial_j\varphi(\boldsymbol{u}) \cdot \partial\varphi^{\perp}(\boldsymbol{u}).$$

Therefore, $d\partial_j(\boldsymbol{u}) \cdot \varphi^{\perp}(\boldsymbol{u}) = 0$, so $\varphi^{\perp}(\boldsymbol{u})$ is perpendicular to $\mathcal{T}_{\varphi(\boldsymbol{u})}$.

To prove the first equality in (12.5), expand $|A|$ along the last row to obtain

$$|A| = \sum_{i=1}^{n} \left[\frac{\partial(\varphi_1, \ldots, \widehat{\varphi}_i, \ldots, \varphi_n)}{\partial(u_1, \ldots, u_m)}\right]^2 = \|\varphi^{\perp}(\boldsymbol{u})\|^2 > 0,$$

the positive inequality because φ' has rank m.

For the second equality in (12.5), using what has already been established we calculate

$$\begin{aligned}\|\partial\varphi^{\perp}(\boldsymbol{u})\|^4 &= |A|^2 = |AA^t| \\ &= \begin{vmatrix} \partial_1\varphi(\boldsymbol{u})\cdot\partial_1\varphi(\boldsymbol{u}) & \cdots & \partial_1\varphi(\boldsymbol{u})\cdot\partial_m\varphi(\boldsymbol{u}) & 0 \\ \vdots & & \vdots & \vdots \\ \partial_m\varphi(\boldsymbol{u})\cdot\partial_1\varphi(\boldsymbol{u}) & \cdots & \partial_m\varphi(\boldsymbol{u})\cdot\partial_m\varphi(\boldsymbol{u}) & 0 \\ 0 & \cdots & 0 & \|\partial\varphi^{\perp}(\boldsymbol{u})\|^2 \end{vmatrix} \\ &= \|\partial\varphi^{\perp}(\boldsymbol{u})\|^2 \det\left[\varphi'(\boldsymbol{u})^t\varphi'(\boldsymbol{u})\right]. \qquad \square\end{aligned}$$

12.3.5 Corollary. *The frame* $\big(d\varphi_{\boldsymbol{u}}(\mathbf{e}^1),\ldots,d\varphi_{\boldsymbol{u}}(\mathbf{e}^{n-1}),\partial\varphi^{\perp}(\boldsymbol{u})\big)$ *is positively oriented in* $\mathbb{R}^n$.

12.3.6 Theorem. *Let* $\varphi : U \to \mathbb{R}^n$ *be a parameterized* $(n-1)$*-surface. The following statements are equivalent:*

(a) φ *is orientable.*

(b) $\varphi(\boldsymbol{u}) = \varphi(\boldsymbol{v}) \Rightarrow \partial\varphi^{\perp}(\boldsymbol{u}) = c\,\partial\varphi^{\perp}(\boldsymbol{v})$ *for some* $c > 0$.

(c) *There exists a function* $\vec{N}_{\varphi} : \varphi(U) \to \mathbb{R}^n$ *(necessarily unique) such that*

$$\begin{aligned}\vec{N}_{\varphi}(\varphi(\boldsymbol{u})) &= \|\partial\varphi^{\perp}(\boldsymbol{u})\|^{-1}\partial\varphi^{\perp}(\boldsymbol{u}) \qquad (12.6)\\ &= \frac{1}{\sqrt{\det\left[\varphi'(\boldsymbol{u})^t\varphi'(\boldsymbol{u})\right]}}\sum_{i=1}^{n}(-1)^{i+n}\frac{\partial(\varphi_1,\ldots,\widehat{\varphi}_i,\ldots,\varphi_n)}{\partial(u_1,\ldots,u_{n-1})}\mathbf{e}^i.\end{aligned}$$

Proof. For $\boldsymbol{u} \in U$, let $T_{\boldsymbol{u}} : \mathbb{R}^n \to \mathbb{R}^n$ denote the unique linear isomorphism such that

$$T_{\boldsymbol{u}}(\mathbf{e}^j) = d\varphi_{\boldsymbol{u}}(\mathbf{e}^j),\ 1 \le j \le n-1, \ \text{ and }\ T_{\boldsymbol{u}}(\mathbf{e}^n) = \partial\varphi^{\perp}(\boldsymbol{u}).$$

By Lemma 12.3.4, $\det T_{\boldsymbol{u}} = \|\varphi^{\perp}(\boldsymbol{u})\|^2 > 0$. Suppose that $\varphi(\boldsymbol{u}) = \varphi(\boldsymbol{v})$. Since $\partial\varphi^{\perp}(\boldsymbol{u}) \perp \mathcal{T}_{\varphi(\boldsymbol{u})}$ and $\partial\varphi^{\perp}(\boldsymbol{v}) \perp \mathcal{T}_{\varphi(\boldsymbol{v})}$,

$$\mathcal{T}_{\varphi(\boldsymbol{v})} = \mathcal{T}_{\varphi(\boldsymbol{u})} \ \text{ iff }\ \partial\varphi^{\perp}(\boldsymbol{u}) = c\,\partial\varphi^{\perp}(\boldsymbol{v}) \text{ for some } c \ne 0.$$

In this case, by (12.4),

$$\begin{aligned}&(T_{\boldsymbol{v}}^{-1}T_{\boldsymbol{u}})(\mathbf{e}^j) = \big(d\varphi_{\boldsymbol{v}}\big)^{-1}\big(d\varphi_{\boldsymbol{u}}(\mathbf{e}^j)\big) = T_{\boldsymbol{uv}}(\mathbf{e}^j),\ 1 \le j \le n-1, \ \text{ and}\\ &(T_{\boldsymbol{v}}^{-1}T_{\boldsymbol{u}})(\mathbf{e}^n) = T_{\boldsymbol{v}}^{-1}\big(c\,\partial\varphi^{\perp}(\boldsymbol{v})\big) = c\mathbf{e}^n.\end{aligned}$$

Thus the matrix of $T_{\boldsymbol{v}}^{-1}T_{\boldsymbol{u}}$ has columns $T_{\boldsymbol{uv}}(\mathbf{e}^1)$, …, $T_{\boldsymbol{uv}}(\mathbf{e}^{n-1})$, $c\mathbf{e}^n$. It follows that

$$0 < \det T_{\boldsymbol{u}}/\det T_{\boldsymbol{v}} = \det(T_{\boldsymbol{v}}^{-1}T_{\boldsymbol{u}}) = c\det T_{\boldsymbol{uv}}. \qquad (12.7)$$

With these preliminaries out of the way, assume that $\vec{N}_\varphi$ exists and let $\varphi(\boldsymbol{u}) = \varphi(\boldsymbol{v})$. Then

$$\|\partial\varphi^\perp(\boldsymbol{u})\|^{-1}\partial\varphi^\perp(\boldsymbol{u}) = \vec{N}_\varphi(\varphi(\boldsymbol{u})) = \vec{N}_\varphi(\varphi(\boldsymbol{v})) = \|\partial\varphi^\perp(\boldsymbol{v})\|^{-1}\partial\varphi^\perp(\boldsymbol{v}),$$

hence $\partial\varphi^\perp(\boldsymbol{u}) = c\partial\varphi^\perp(\boldsymbol{v})$ for some $c > 0$. By the first paragraph, $\mathcal{T}_{\varphi(\boldsymbol{u})} = \mathcal{T}_{\varphi(\boldsymbol{v})}$ and $\det T_{\boldsymbol{uv}} > 0$. Therefore, φ is orientable.

Conversely, assume that φ is orientable and let $\varphi(\boldsymbol{u}) = \varphi(\boldsymbol{v})$. Then $\mathcal{T}_{\varphi(\boldsymbol{u})} = \mathcal{T}_{\varphi(\boldsymbol{v})}$, hence $\partial\varphi^\perp(\boldsymbol{u}) = c\partial\varphi^\perp(\boldsymbol{v})$ for some $c \neq 0$. Since $\det[T_{\boldsymbol{uv}}] > 0$, $c > 0$ by (12.7). Therefore,

$$\|\partial\varphi^\perp(\boldsymbol{u})\|^{-1}\partial\varphi^\perp(\boldsymbol{u}) = \|\partial\varphi^\perp(\boldsymbol{v})\|^{-1}\partial\varphi^\perp(\boldsymbol{v}),$$

so $\vec{N}_\varphi$ may be unambiguously defined by (12.6). □

12.3.7 Special Cases.

(a) $n = 2$: Then $\varphi^\perp = (-\varphi_2', \varphi_1')$, the *inward normal*. (Figure 12.9.)

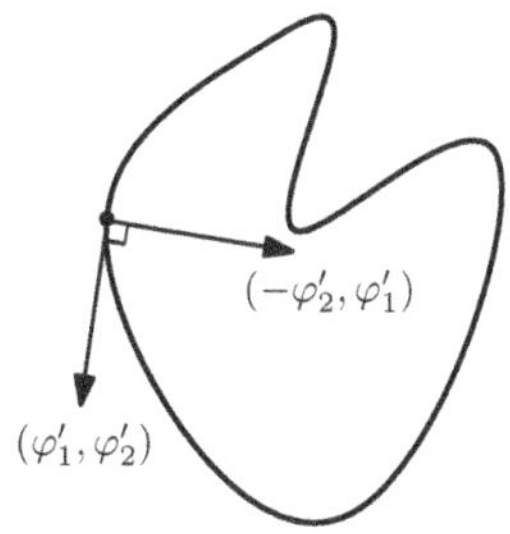

FIGURE 12.9: The inward unit normal.

(b) $n = 3$: Then

$$\partial\varphi^\perp = \left(\begin{vmatrix}\partial_1\varphi_2 & \partial_1\varphi_3\\ \partial_2\varphi_2 & \partial_2\varphi_3\end{vmatrix}, -\begin{vmatrix}\partial_1\varphi_1 & \partial_1\varphi_3\\ \partial_2\varphi_1 & \partial_2\varphi_3\end{vmatrix}, \begin{vmatrix}\partial_1\varphi_1 & \partial_1\varphi_2\\ \partial_2\varphi_1 & \partial_2\varphi_2\end{vmatrix}\right) = \partial_1\varphi \times \partial_2\varphi,$$

the familiar *cross product* of $\partial_1\varphi$ and $\partial_2\varphi$.

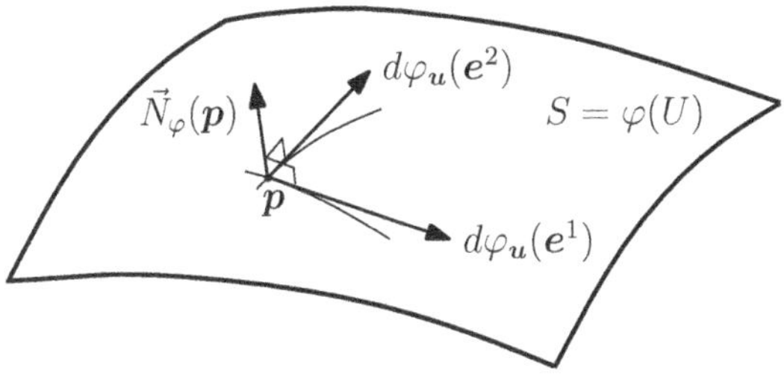

FIGURE 12.10: Normal vector to S at $\boldsymbol{p}$.

Thus the positively oriented frame $\big(d\varphi_{\boldsymbol{u}}(\boldsymbol{e}^1), d\varphi_{\boldsymbol{u}}(\boldsymbol{e}^2), \vec{N}_\varphi(\boldsymbol{p})\big)$ is a right-handed system, as shown in Figure 12.10.

(c) Let $U \subseteq \mathbb{R}^{n-1}$ be open and let $g : U \to \mathbb{R}$ be C^1. Define

$$\varphi(u_1, \ldots, u_{n-1}) = \big(u_1, \ldots, u_{n-1}, g(u_1, \ldots, u_{n-1})\big).$$

Then $\varphi(U)$ is the graph of g. Since φ is one-to-one, it is orientable. Also,

$$\partial_j \varphi = \big(0, \cdots, 0, \underset{j}{1}, 0, \cdots, 0, \partial_j g\big) \perp \big(-\partial_1 g, \cdots, -\partial_j g, \cdots, -\partial_{n-1} g, 1\big)$$

and, by elementary row operations,

$$\begin{vmatrix} 1 & \cdots & 0 & \partial_1 g \\ 0 & \cdots & 0 & \partial_2 g \\ \vdots & & \vdots & \vdots \\ 0 & \cdots & 1 & \partial_{n-1} g \\ -\partial_1 g & \cdots & -\partial_{n-1} g & 1 \end{vmatrix} = (\partial_1 g)^2 + \cdots + (\partial_{n-1} g)^2 + 1.$$

Since this is positive, by uniqueness,

$$\vec{N}_\varphi \circ \varphi = \frac{(-\partial_1 g, \cdots, -\partial_{n-1} g, 1)}{\sqrt{(\partial_1 g)^2 + \cdots + (\partial_{n-1} g)^2 + 1}} = \frac{(-\nabla g, 1)}{\sqrt{\|\nabla g\|^2 + 1}}. \qquad \diamond$$

12.3.8 Example. Let $r > 0$ and define

$$\varphi(\theta_1, \theta_2) = (r \sin\theta_1 \cos\theta_2, r \sin\theta_1 \sin\theta_2, r\cos\theta_1), \quad \theta_1 \in (0, \pi),\ \theta_2 \in (0, 2\pi).$$

The image of φ is the sphere in $\mathbb{R}^3$ with radius r and center $(0,0,0)$ and with the great circle $(r\sin\theta_1, 0, r\cos\theta_1)$ (that is, $\theta_2 = 0$) through the poles $(0, 0, \pm r)$ missing. Since

$$\begin{aligned} \partial_1 \varphi(\theta_1, \theta_2) &= r(\cos\theta_1 \cos\theta_2, \cos\theta_1 \sin\theta_2, -\sin\theta_1) \quad \text{and} \\ \partial_2 \varphi(\theta_1, \theta_2) &= r(-\sin\theta_1 \sin\theta_2, \sin\theta_1 \cos\theta_2, 0), \end{aligned}$$

by 12.3.7

$$\begin{aligned} \partial\varphi^{\perp}(\theta_1, \theta_2) &= \partial_1\varphi(\theta_1, \theta_2) \times \partial_2 \varphi(\theta_1, \theta_2) \\ &= r\sin\theta_1 \big(r\sin\theta_1\cos\theta_2, r\sin\theta_1 \sin\theta_2, r\cos\theta_1\big) \\ &= (r\sin\theta_1)\varphi(\theta_1, \theta_2). \end{aligned}$$

Therefore,

$$(\vec{N}_\varphi \circ \varphi)(\theta_1, \theta_2) = \frac{\varphi(\theta_1, \theta_2)}{\|\varphi(\theta_1, \theta_2)\|} = r^{-1}\varphi(\theta_1, \theta_2),$$

that is,

$$\vec{N}_\varphi(\boldsymbol{p}) = \frac{\boldsymbol{p}}{\|\boldsymbol{p}\|}, \quad \boldsymbol{p} \in S. \qquad \diamond$$

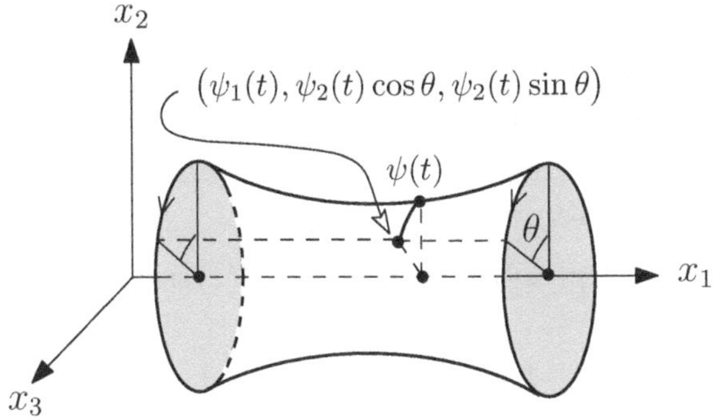

FIGURE 12.11: Surface of revolution.

12.3.9 Example. Let I be an open interval and $\psi : I \to \mathbb{R}^2$ a smooth curve with $\psi_2(t) > 0$ for $t \in I$. The *parameterized surface of revolution* in $\mathbb{R}^3$ is defined by

$$\varphi(t, \theta) = (\psi_1(t), \psi_2(t)\cos\theta, \psi_2(t)\sin\theta), \quad t \in I, \ \theta \in \mathbb{R}.$$

From (12.3.7) and the calculations

$$\begin{aligned} \partial_1\varphi(t, \theta) &= \big(\psi_1'(t), \psi_2'(t)\cos\theta, \psi_2'(t)\sin\theta\big), \\ \partial_2\varphi(t, \theta) &= \big(0, -\psi_2(t)\sin\theta, \psi_2(t)\cos\theta\big), \end{aligned}$$

we have

$$\partial\varphi^\perp(t, \theta) = \psi_2(t)\big(\psi_2'(t), -\psi_1'(t)\cos\theta, -\psi_1'(t)\sin\theta\big) \tag{12.8}$$

and

$$\partial\psi^\perp(t) = (-\psi_2'(t), \psi_1'(t)).$$

Now suppose that ψ is orientable. We claim that φ is then orientable. To see this, suppose that $\varphi(t_1, \theta_1) = \varphi(t_2, \theta_2)$. Then $\psi_1(t_1) = \psi_1(t_2)$, and because $\psi_2(t) > 0$, $\psi_2(t_1) = \psi_2(t_2)$ and hence $\theta_2 = \theta_1 + 2k\pi$. By orientability of ψ,

$$(-\psi_2'(t_2), \psi_1'(t_2)) = \partial\psi^\perp(t_2) = c\partial\psi^\perp(t_1) = c(-\psi_2'(t_1), \psi_1'(t_1))$$

for some $c > 0$. It follows from (12.8) that

$$\begin{aligned} \partial\varphi^\perp(t_2, \theta_2) &= \psi_2(t_2)\big(\psi_2'(t_2), -\psi_1'(t_2)\cos\theta_2, -\psi_1'(t_2)\sin\theta_2\big) \\ &= c\psi_2(t_1)\big(\psi_2'(t_1), -\psi_1'(t_1)\cos\theta_1, -\psi_1'(t_1)\sin\theta_1\big) \\ &= c\partial\varphi^\perp(t_1, \theta_1), \end{aligned}$$

which shows that φ is orientable. Moreover, from (12.8),

$$\|\partial\varphi^\perp(t, \theta)\| = \psi_2(t)\|\psi'(t)\|,$$

hence

$$\vec{N}_\varphi(\varphi(t,\theta)) = \frac{\partial\varphi^\perp(t,\theta)}{\|\partial\varphi^\perp(t,\theta)\|} = \|\psi'(t)\|^{-1}\big(\psi_2'(t), -\psi_1'(t)\cos\theta, -\psi_1'(t)\sin\theta\big),$$

which is the rotation of the unit normal vector $-N_\psi$ about the x_1 axis.

For the special case $\psi(x) = \big(x, f(x)\big)$,

$$\varphi(x,\theta) = (x, f(x)\cos\theta, f(x)\sin\theta)$$

and

$$\vec{N}_\varphi(\varphi(x,\theta)) = \big([f'(x)]^2+1\big)^{-1/2}\big(f'(x), -\cos\theta, -\sin\theta\big).$$

A point (x,y,z) on the surface $S = \varphi(U)$ and not on the graph of f may be written uniquely as

$$\big(x, f(x)\cos(\theta(y,z)), f(x)\sin\big(\theta(y,z)\big)\big)$$

where $0 < \theta(y,z) < 2\pi$ is the (continuous) argument of (y,z) determined by $\theta_0 = 0$ (see 9.4.6). Therefore,

$$\vec{N}_\varphi(x,y,z) = \big([f'(x)]^2+1\big)^{-1/2}\left(f'(x), -\cos\big(\theta(y,z)\big), -\sin\big(\theta(y,z)\big)\right),$$

which is continuous on S by the periodicity of sine and cosine. ◇

12.3.10 Example. The *parameterized Möbius strip* is defined by

$$\varphi(t,\theta) = \left(\left[2 + t\cos\left(\tfrac{1}{2}\theta\right)\right]\cos\theta, \left[2 + t\cos\left(\tfrac{1}{2}\theta\right)\right]\sin\theta, t\sin\left(\tfrac{1}{2}\theta\right)\right),$$

where $-1 < t < 1$ and $\theta \in \mathbb{R}$. The surface may be concretely realized by taking one end of a long strip of paper, giving it a half-twist, and gluing it to the other end.

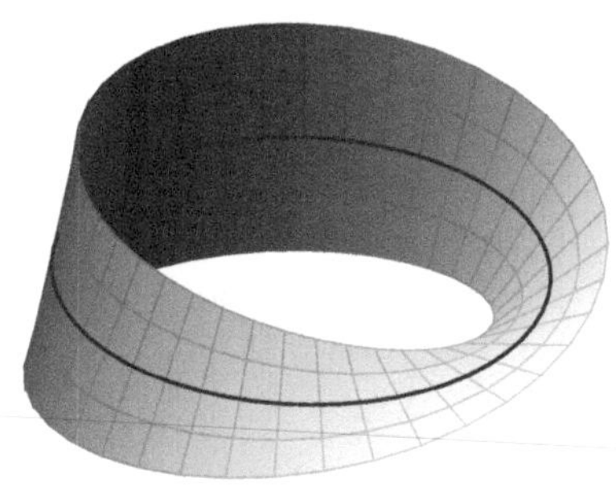

FIGURE 12.12: Möbius strip.

The Möbius strip is not orientable. Indeed, $\varphi(0,0) = \varphi(0,2\pi)$, but since

$$\partial_1\varphi(0,0) = -\partial_1\varphi(0,2\pi) = (1,0,0) \quad\text{and}\quad \partial_2\varphi(0,0) = \partial_2\varphi(0,2\pi) = (0,1,0),$$

we see that

$$\partial_1\varphi(0,0)\times\partial_2\varphi(0,0) = (0,0,1) = -\partial_1\varphi(0,2\pi)\times\partial_2\varphi(0,2\pi).$$

Therefore, $\vec{N}_\varphi$ cannot exist. ◇

Exercises

1. Assuming that $\mathbb{R}^3$ has the standard orientation, find the sign of the frames

   (a) $(\mathbf{e}^1 + \mathbf{e}^2, \mathbf{e}^2 + \mathbf{e}^3, \mathbf{e}^3 + \mathbf{e}^1)$

   (b) $(-\mathbf{e}^1 + \mathbf{e}^2 + \mathbf{e}^3, \mathbf{e}^1 - \mathbf{e}^2 + \mathbf{e}^3, \mathbf{e}^1 + \mathbf{e}^2 - \mathbf{e}^3)$.

2. Show that the frames

$$(\mathbf{e}^1 + \mathbf{e}^2 + \mathbf{e}^3, 2\mathbf{e}^1 + \mathbf{e}^2 + 3\mathbf{e}^3) \quad \text{and} \quad (\mathbf{e}^1 + 3\mathbf{e}^2 - \mathbf{e}^3, \mathbf{e}^1 + 4\mathbf{e}^2 - 2\mathbf{e}^3)$$

   in $\mathbb{R}^3$ span the same subspace but have opposite orientations.

3. Let $\varphi : U \to \mathbb{R}^n$ be a parameterized m-surface and $\psi = \varphi \circ \alpha : V \to \mathbb{R}^n$ a reparametrization of φ. Show that φ is orientable iff ψ is orientable.

4. Let $\varphi : U \to \mathbb{R}^n$ be an orientable parameterized $(n-1)$-surface and let $\psi = \varphi \circ \alpha : V \to \mathbb{R}^n$ be a reparametrization of φ. Find $\partial\psi^\perp$ in terms of $\partial\varphi^\perp$. Use the result to show that $N_\psi = N_\varphi$ on $S := \varphi(U) = \psi(V)$.

5. Use 12.3.9 to find $\vec{N}_\varphi(x, y, z)$ for the *torus*

$$\varphi(\phi, \theta) = \big(a\cos\phi, (b + a\sin\phi)\cos\theta, (b + a\sin\phi)\sin\theta\big), \quad 0 < \theta,\ \phi < 2\pi,$$

   where $0 < a < b$.

6. Find $\vec{N}_\varphi(x, y, z)$ for the following orientable 2-surfaces in $\mathbb{R}^3$:

   (a) $\varphi(t, \theta) = (t\cos\theta, t\sin\theta, t)$, $t > 0$, $\theta \in \mathbb{R}$.

   (b) $\varphi(t, \theta) = (\sinh t, \cosh t\cos\theta, \cosh t\sin\theta)$, t, $\theta \in \mathbb{R}$ (*hyperboloid of one sheet*)

   (c) $\varphi(t, \theta) = (\cosh t, \sinh t\cos\theta, \sinh t\sin\theta)$, t, $\theta \in \mathbb{R}$ (*one sheet of a hyperboloid of two sheets*)

   (d) $\varphi(t, \theta) = (t\cos\theta, t\sin\theta, \theta)$, $t > 0$, $\theta \in \mathbb{R}$ (*helicoid*)

   (e) $\varphi(t, \theta) = (t\cos\theta, t\sin\theta, \theta^2)$, $t > 0$, $\theta > 0$.

   (f) $\varphi(t, s) = (1 - s)\big(a\cos t, a\sin t, 0\big) + s\big(b\cos t, b\sin t, 1\big)$, $0 < s < 1$, where $0 < a < b$.

7. Let $V \subseteq \mathbb{R}^{n-2}$ be open and let $\psi : V \to \mathbb{R}^{n-1}$ be an $(n-2)$-parameterized surface in $\mathbb{R}^{n-1}$. Define the *cylinder* φ *over* ψ by

$$\varphi(\boldsymbol{v}, s) = \big(\psi(\boldsymbol{v}), s\big), \quad \boldsymbol{v} \in V,\ s \in (a, b).$$

   Show that

(a) φ is a parameterized $(n-1)$-surface in $\mathbb{R}^n$.

(b) $\partial\varphi^\perp(u_1,\ldots,u_{n-1}) = \big(\partial\psi^\perp(u_1,\ldots,u_{n-2}),0\big)$.

(c) φ is orientable iff ψ is orientable, in which case

$$N_\varphi(x_1,\ldots,x_n) = \big(N_\psi(x_1,\ldots,x_{n-1}),0\big).$$

8. Let $V \subseteq \mathbb{R}^{n-2}$ be open and let $\psi : V \to \mathbb{R}^{n-1}$ be an $(n-2)$-parameterized surface in $\mathbb{R}^{n-1}$. Define the *cone over* ψ by

$$\varphi(\boldsymbol{v},s) = \big((1-s)\psi(\boldsymbol{v}),s\big), \quad \boldsymbol{v}\in V,\ 0<s<1.$$

Show that

(a) φ is a parameterized $(n-1)$-surface in $\mathbb{R}^n$.

(b) $\partial\varphi^\perp(\boldsymbol{v},s) = \big((1-s)^{n-2}\partial\psi^\perp(\boldsymbol{v}),D(\boldsymbol{v},s)\big)$, where

$$D(\boldsymbol{v},s) = \begin{bmatrix} (1-s)a_{1,1} & \cdots & (1-s)a_{1,n-2} & -\psi_1(\boldsymbol{v}) \\ (1-s)a_{2,1} & \cdots & (1-s)a_{2,n-2} & -\psi_2(\boldsymbol{v}) \\ \vdots & & \vdots & \vdots \\ (1-s)a_{n-1,1} & \cdots & (1-s)a_{n-1,n-2} & -\psi_{n-1}(\boldsymbol{v}) \end{bmatrix}$$

and $[a_{i,j}]_{(n-1)\times(n-2)} = \psi'(\boldsymbol{v})$.

12.4 m-Dimensional Surfaces

Let $1 \le m < n$ and let $V \subseteq \mathbb{R}^n$ be open. Suppose that the function

$$F = (F_1,\ldots,F_{n-m}) : V \to \mathbb{R}^{n-m}$$

is C^1 on V such that the $(n-m)\times n$ matrix $F'(\boldsymbol{x})$ has rank $n-m$ at each point $\boldsymbol{x}\in V$. A set of the form

$$S = \{\boldsymbol{x}\in V \,:\, F(\boldsymbol{x}) = \boldsymbol{c}\},$$

where $\boldsymbol{c}\in\mathbb{R}^{n-m}$, is called an *$m$-dimensional level surface of F* or simply an *m-surface in $\mathbb{R}^n$*. By replacing F by $F-\boldsymbol{c}$, we may (and hereafter shall) take $\boldsymbol{c}=\boldsymbol{0}$.

Local Parametrization of an m-Surface

The following theorem shows that an m-surface may be "patched together" from a collection of one-to-one parameterized m-surfaces. This will be an important tool in the development of a theory of integration on m-surfaces.

12.4.1 Theorem. *Let* $S = \{\boldsymbol{x} \in V : F(\boldsymbol{x}) = \boldsymbol{0}\}$ *be an* m*-surface in* $\mathbb{R}^n$.

(a) *For each* $\boldsymbol{a} \in S$ *there exist open sets* $U_{\boldsymbol{a}} \subseteq \mathbb{R}^m$ *and* $V_{\boldsymbol{a}} \subseteq \mathbb{R}^n$ *with* $\boldsymbol{a} \in V_{\boldsymbol{a}}$*, and a one-to-one parameterized* m*-surface* $\varphi_{\boldsymbol{a}}$ *from* $U_{\boldsymbol{a}}$ *onto* $S_{\boldsymbol{a}} := S \cap V_{\boldsymbol{a}}$.

(b) *Each* $\varphi_{\boldsymbol{a}}^{-1}$ *is the restriction to* $S_{\boldsymbol{a}}$ *of a* C^1 *map on* $V_{\boldsymbol{a}}$.

(c) *If* $S_{\boldsymbol{a}} \cap S_{\boldsymbol{b}} \neq \emptyset$*, then the mapping*

$$\varphi_{\boldsymbol{ab}} := \varphi_{\boldsymbol{b}}^{-1} \circ \varphi_{\boldsymbol{a}} : \varphi_{\boldsymbol{a}}^{-1}(S_{\boldsymbol{a}} \cap S_{\boldsymbol{b}}) \to \varphi_{\boldsymbol{b}}^{-1}(S_{\boldsymbol{a}} \cap S_{\boldsymbol{b}})$$

is C^1 *with inverse* $\varphi_{\boldsymbol{ba}}$.

(d) *The mappings* $\varphi_{\boldsymbol{a}}$ *may be chosen so that* $\boldsymbol{0} \in U_{\boldsymbol{a}}$ *and* $\varphi_{\boldsymbol{a}}(\boldsymbol{0}) = \boldsymbol{a}$.

Proof. If (a)–(c) of the theorem hold and $\boldsymbol{a} = \varphi_{\boldsymbol{a}}(\boldsymbol{u}_0)$, then (d) may be achieved by replacing $U_{\boldsymbol{a}}$ by $U_{\boldsymbol{a}} - \boldsymbol{u}_0$ and $\varphi_{\boldsymbol{a}}$ by $\varphi_{\boldsymbol{a}}(\boldsymbol{u} + \boldsymbol{u}_0)$, $\boldsymbol{u} \in U_{\boldsymbol{a}} - \boldsymbol{u}_0$.

We prove (a)–(c) for the case $m = n - 1$, that is, for F real-valued, and then outline the proof for the general case. Since F has rank 1, $\partial_i F(\boldsymbol{a}) \neq 0$ for some index i (which typically depends on $\boldsymbol{a}$). Define a C^1 map $G_{\boldsymbol{a}} : V \to \mathbb{R}^n$ by

$$G_{\boldsymbol{a}}(x_1, \ldots, x_n) = \big(x_1, \ldots, x_{i-1}, F(x_1, \ldots, x_n), x_{i+1}, \ldots x_n\big).$$

Thus $G_{\boldsymbol{a}}$ simply replaces the ith coordinate of its argument $\boldsymbol{x}$ by $F(\boldsymbol{x})$. Note that $G'_{\boldsymbol{a}}(\boldsymbol{x})$ is the identity matrix with row i replaced by $\nabla F(\boldsymbol{x})$. A standard row reduction shows that $J_{G_{\boldsymbol{a}}}(\boldsymbol{a}) = \partial_i F(\boldsymbol{a})$. Since this is nonzero, by the inverse function theorem there exist open sets $V_{\boldsymbol{a}} \subseteq V$ and $W_{\boldsymbol{a}} = G_{\boldsymbol{a}}(V_{\boldsymbol{a}})$ in $\mathbb{R}^n$ with $\boldsymbol{a} \in V_{\boldsymbol{a}}$ such that $G_{\boldsymbol{a}}$ is one-to-one on $V_{\boldsymbol{a}}$ and $G_{\boldsymbol{a}}^{-1} : W_{\boldsymbol{a}} \to V_{\boldsymbol{a}}$ is C^1. Taking smaller $W_{\boldsymbol{a}}$ and $V_{\boldsymbol{a}}$ if necessary, we may suppose that

$$W_{\boldsymbol{a}} = (\alpha_1, \beta_1) \times \cdots \times (\alpha_n, \beta_n).$$

Note that $0 \in (\alpha_i, \beta_i)$, since $\big(a_1, \ldots, a_{i-1}, 0, a_{i+1}, \ldots, a_n\big) = G_{\boldsymbol{a}}(\boldsymbol{a}) \in W_{\boldsymbol{a}}$. Now let $(u_1, \ldots, u_n) \in W_{\boldsymbol{a}}$ and set $(v_1, \ldots, v_n) = G_{\boldsymbol{a}}^{-1}(u_1, \ldots, u_n)$. Then

$$\begin{aligned}(u_1, \ldots, u_i, \ldots, u_n) &= G_{\boldsymbol{a}}(v_1, \ldots, v_n)\\ &= \big(v_1, \ldots, v_{i-1}, F(v_1, \ldots, v_n), v_{i+1}, \ldots, v_n\big)\\ &= \big(u_1, \ldots, u_{i-1}, (F \circ G_{\boldsymbol{a}}^{-1})(u_1, \ldots, u_n), u_{i+1}, \ldots, u_n\big),\end{aligned}$$

hence

$$(F \circ G_{\boldsymbol{a}}^{-1})(u_1, \ldots, u_n) = u_i \tag{12.9}$$

and, in particular,

$$(F \circ G_{\boldsymbol{a}}^{-1})(u_1, \ldots, u_{i-1}, 0, u_{i+1}, \ldots, u_n) = 0. \tag{12.10}$$

Now set

$$U_{\boldsymbol{a}} := (\alpha_1, \beta_1) \times \cdots \times (\alpha_{i-1}, \beta_{i-1}) \times (\alpha_{i+1}, \beta_{i+1}) \times \cdots \times (\alpha_n, \beta_n)$$

and define $\varphi_{\boldsymbol{a}} : U_{\boldsymbol{a}} \to \mathbb{R}^n$ by

$$\varphi_{\boldsymbol{a}}(u_1, \ldots, u_{n-1}) = G_{\boldsymbol{a}}^{-1}(u_1, \ldots, u_{i-1}, 0, u_i, \ldots, u_{n-1}).$$

By (12.10), $F\big(\varphi_{\boldsymbol{a}}(u_1, \ldots, u_{n-1})\big) = 0$, hence $\varphi_{\boldsymbol{a}}(U_{\boldsymbol{a}}) \subseteq S_{\boldsymbol{a}}$. Conversely, by (12.9),

$$\begin{aligned}(v_1, \ldots, v_n) \in S_{\boldsymbol{a}} &\Rightarrow u_i = (F \circ G_{\boldsymbol{a}}^{-1})(u_1, \ldots, u_n) = F(v_1, \ldots, v_n) = 0\\ \Rightarrow (v_1, \ldots, v_n) &= G_{\boldsymbol{a}}^{-1}(u_1, \ldots, u_{i-1}, 0, u_{i+1}, \ldots u_{n-1}) = \varphi_{\boldsymbol{a}}(u_1, \ldots, u_{n-1}).\end{aligned}$$

Therefore, $\varphi_{\boldsymbol{a}}(U_{\boldsymbol{a}}) = S_{\boldsymbol{a}}$.

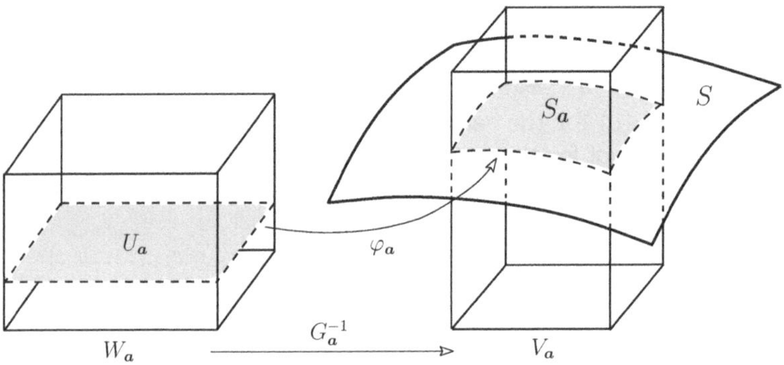

FIGURE 12.13: The mapping $G_{\boldsymbol{a}}^{-1}$.

Define the *injection mapping* $\iota_{\boldsymbol{a}} : U_{\boldsymbol{a}} \to W_{\boldsymbol{a}}$ and the *projection mapping* $\pi_{\boldsymbol{a}} : V_{\boldsymbol{a}} \to \mathbb{R}^{n-1}$, respectively, by

$$\begin{aligned}\iota_{\boldsymbol{a}}(u_1, \ldots, u_{n-1}) &= (u_1, \ldots, u_{i-1}, 0, u_i, \ldots, u_{n-1}) \text{ and}\\ \pi_{\boldsymbol{a}}(v_1, \ldots, v_n) &= (v_1, \ldots, v_{i-1}, v_{i+1}, \ldots, v_n).\end{aligned}$$

Then $\pi_{\boldsymbol{a}} \circ \iota_{\boldsymbol{a}} : U_{\boldsymbol{a}} \to U_{\boldsymbol{a}}$ is the identity function and $\varphi_{\boldsymbol{a}} = G_{\boldsymbol{a}}^{-1} \circ \iota_{\boldsymbol{a}}$. Since $G_{\boldsymbol{a}}^{-1}$ has rank n and $\iota_{\boldsymbol{a}}$ has rank $n-1$, $\varphi_{\boldsymbol{a}}$ has rank $n-1$. Also, if $\boldsymbol{v} = \varphi_{\boldsymbol{a}}(\boldsymbol{u})$, then

$$(\pi_{\boldsymbol{a}} \circ G_{\boldsymbol{a}})(\boldsymbol{v}) = (\pi_{\boldsymbol{a}} \circ G_{\boldsymbol{a}} \circ \varphi_{\boldsymbol{a}})(\boldsymbol{u}) = \pi_{\boldsymbol{a}} \circ \iota_{\boldsymbol{a}}(\boldsymbol{u}) = \boldsymbol{u} = \varphi_{\boldsymbol{a}}^{-1}(\boldsymbol{v}),$$

which shows that $\varphi_a^{-1} : S_{\boldsymbol{a}} \to U_{\boldsymbol{a}}$ is the restriction to S_a of the C^1 function $\pi_{\boldsymbol{a}} \circ G_{\boldsymbol{a}} : V_{\boldsymbol{a}} \to U_{\boldsymbol{a}}$.

Now let $\boldsymbol{b} \in S$ and $S_{\boldsymbol{a}} \cap S_{\boldsymbol{b}} \neq \emptyset$. Then $G_{\boldsymbol{b}} \circ G_{\boldsymbol{a}}^{-1}$ maps the open set $G_{\boldsymbol{a}}(V_{\boldsymbol{a}} \cap V_{\boldsymbol{b}})$ onto the open set $G_{\boldsymbol{b}}(V_{\boldsymbol{a}} \cap V_{\boldsymbol{b}})$. Also, in the preceding notation, $\varphi_{\boldsymbol{a}} = G_{\boldsymbol{a}}^{-1} \circ \iota_{\boldsymbol{a}}$ on $U_{\boldsymbol{a}}$ and $\varphi_{\boldsymbol{b}}^{-1} = \pi_{\boldsymbol{b}} \circ G_{\boldsymbol{b}}$ on $S_{\boldsymbol{b}}$, hence

$$\varphi_{\boldsymbol{b}}^{-1} \circ \varphi_{\boldsymbol{a}} = \pi_{\boldsymbol{b}} \circ G_{\boldsymbol{b}} \circ G_{\boldsymbol{a}}^{-1} \circ \iota_{\boldsymbol{a}},$$

which maps the open set $\varphi_{\boldsymbol{a}}^{-1}(S_{\boldsymbol{b}} \cap S_{\boldsymbol{a}}) \subseteq U_{\boldsymbol{a}}$ onto the open set $\varphi_{\boldsymbol{b}}^{-1}(S_{\boldsymbol{b}} \cap S_{\boldsymbol{a}}) \subseteq$

$U_{\boldsymbol{b}}$ and is C^1 with C^1 inverse $\varphi_{\boldsymbol{a}}^{-1} \circ \varphi_{\boldsymbol{b}}$. This verifies the theorem for the case $m = n - 1$.

In the general case, there exist indices $i_1 < \cdots < i_k$ in $\{1, \ldots, n\}$ such that

$$\frac{\partial(F_1, \ldots F_k)}{\partial(u_{i_1}, \ldots, u_{i_k})}(\boldsymbol{a}) \neq 0,$$

where $k := n - m$. Let $i_1' < i_2' < \cdots < i_m'$ denote the complementary indices. (In the above case, these were the indices $1, \ldots, i-1, i+1, \ldots, n$.) Define $G_{\boldsymbol{a}}(x_1, \ldots, x_n)$ to be the n-tuple $(x_1, \ldots, x_n)$, with the coordinates $x_{i_1}, \ldots, x_{i_k}$ replaced by $F_1(\boldsymbol{x})$, ..., $F_k(\boldsymbol{x})$. Then $J_{G_{\boldsymbol{a}}}(\boldsymbol{a}) \neq 0$, so the sets $V_{\boldsymbol{a}}$ and $W_{\boldsymbol{a}}$ may be obtained as before. Define

$$U_{\boldsymbol{a}} = (\alpha_{i_1'}, \beta_{i_1'}) \times \cdots \times (\alpha_{i_m'}, \beta_{i_m'}) \to \mathbb{R}^n$$

and the injection mapping $\iota_{\boldsymbol{a}} : U_{\boldsymbol{a}} \to W_{\boldsymbol{a}}$ by

$$\iota_{\boldsymbol{a}}(u_1, u_2, \ldots, u_m) = (v_1, v_2, \ldots, v_n),$$

where $v_{i_j} = 0$, $1 \leq j \leq k$, and $v_{i_j'} = u_j$, $1 \leq j \leq m$. Thus $\iota_{\boldsymbol{a}}$ places zeros in the coordinate positions $i_1 < \cdots < i_k$ and fills the complementary positions by $u_1, \ldots, u_m$. Finally, define the projection mapping $\pi_{\boldsymbol{a}} : V_{\boldsymbol{a}} \to \mathbb{R}^{n-1}$ by

$$\pi_{\boldsymbol{a}}(v_1, \ldots, v_n) = (v_{i_1'}, \ldots, v_{i_m'}).$$

The proof then proceeds as before. □

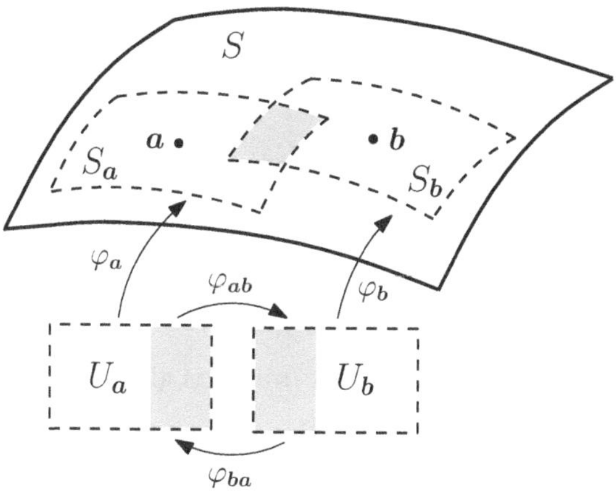

FIGURE 12.14: Transition mappings.

The functions $\varphi_a : U_{\boldsymbol{a}} \to S_{\boldsymbol{a}}$ in the theorem are called *local parametrizations of* S, and the C^1 functions $\varphi_{\boldsymbol{ab}}$ are called *transition mappings*. The sets $S_{\boldsymbol{a}}$ are called *surface elements*. A collection of local parameterizations of S whose surface elements cover S is called an *atlas* for S. Note that if F is C^r then, as an examination of the proof reveals, the local parameterizations and the transition maps are C^r as well.

12.4.2 Example. Consider the $(n-1)$-sphere $S := \{\boldsymbol{y} \in \mathbb{R}^n : \|\boldsymbol{y}\| = 1\}$ with north and south poles $\boldsymbol{p} := (0, \ldots, 0, 1)$ and $\boldsymbol{q} := (0, \ldots, 0, -1)$. Let the points $\boldsymbol{y} = (y_1, \ldots, y_n)$ and $\boldsymbol{x} = (x_1, \ldots, x_{n-1})$ be related as in Figure 12.15.

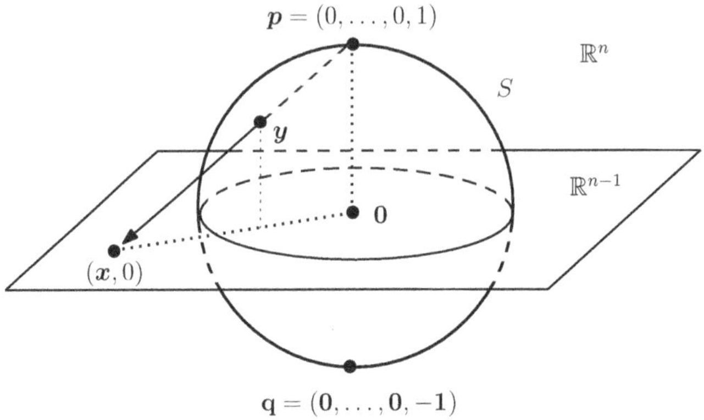

FIGURE 12.15: Stereographic projection from $\boldsymbol{p}$.

Then for some t,

$$(x_1, \ldots, x_{n-1}, -1) = (\boldsymbol{x}, 0) - \boldsymbol{p} = t(\boldsymbol{y} - \boldsymbol{p}) = \big(ty_1, \ldots ty_{n-1}, t(y_n - 1)\big),$$

hence

$$(x_1, \ldots, x_{n-1}) = \frac{1}{1 - y_n}(y_1, \ldots, y_{n-1}), \quad -1 \le y_n < 1.$$

The mapping

$$\boldsymbol{x} = \varphi^{-1}(\boldsymbol{y}) = \frac{1}{1 - y_n}(y_1, \ldots, y_{n-1}), \quad y_n < 1,$$

from $S \setminus \{\boldsymbol{p}\}$ onto $\mathbb{R}^{n-1}$, is called the *stereographic projection* from $\boldsymbol{p}$ onto the equatorial hyperplane $x_n = 0$. One readily checks that the inverse of this mapping is given by

$$\boldsymbol{y} = \varphi(\boldsymbol{x}) = \frac{1}{1 + \|\boldsymbol{x}\|^2}\big(2x_1, \ldots, 2x_{n-1}, \|\boldsymbol{x}\|^2 - 1\big), \quad \boldsymbol{x} \in \mathbb{R}^{n-1}.$$

Similarly, the *stereographic projection from* $\boldsymbol{q}$ is given by

$$\boldsymbol{x} = \tilde{\varphi}^{-1}(\boldsymbol{y}) = \frac{1}{1 + y_n}(y_1, \ldots, y_{n-1}), \quad y_n > -1$$

with inverse

$$\boldsymbol{y} = \tilde{\varphi}(\boldsymbol{x}) = \frac{1}{1 + \|\boldsymbol{x}\|^2}\big(2x_1, \ldots, 2x_{n-1}, 1 - \|\boldsymbol{x}\|^2\big).$$

The set $\{\varphi, \tilde{\varphi}\}$ is an atlas for S. The transition mapping from $\mathbb{R}^{n-1} \setminus \{\boldsymbol{0}\}$ to $\mathbb{R}^{n-1} \setminus \{\boldsymbol{0}\}$ is the self-inverse mapping

$$(\varphi^{-1} \circ \tilde{\varphi})(\boldsymbol{x}) = \frac{\boldsymbol{x}}{\|\boldsymbol{x}\|^2}. \qquad \lozenge$$

Tangent Space of an m-Surface

The local parameterizations $\varphi_{\boldsymbol{a}}$ of an m-surface $S = \{\boldsymbol{x} : F(\boldsymbol{x}) = \boldsymbol{0}\}$ may be used to construct a *tangent space* at each point $\boldsymbol{a} \in S$. Let

$$\varphi_{\boldsymbol{a}}(\boldsymbol{u}) = \varphi_{\boldsymbol{b}}(\boldsymbol{v}) \in S_{\boldsymbol{a}} \cap S_{\boldsymbol{b}}.$$

Then $\boldsymbol{v} := \varphi_{\boldsymbol{ab}}(\boldsymbol{u})$ and, by the chain rule applied to $\varphi_{\boldsymbol{a}} = \varphi_{\boldsymbol{b}} \circ \varphi_{\boldsymbol{ab}}$,

$$d(\varphi_{\boldsymbol{a}})_{\boldsymbol{u}} = d(\varphi_{\boldsymbol{b}})_{\boldsymbol{v}} \circ d(\varphi_{\boldsymbol{ab}})_{\boldsymbol{u}}.$$

Since $d(\varphi_{\boldsymbol{ab}})_{\boldsymbol{u}} : \mathbb{R}^m \to \mathbb{R}^m$ is an isomorphism, the vectors $\boldsymbol{d}^j := d\varphi_{\boldsymbol{ab}})_{\boldsymbol{u}}(\boldsymbol{e}^j)$ form a basis of $\mathbb{R}^m$. Therefore, we have the mapping of $\mathbb{R}^m$-frames

$$d(\varphi_{\boldsymbol{a}})_{\boldsymbol{u}}(\boldsymbol{e}^1, \ldots, \boldsymbol{e}^m) = d(\varphi_{\boldsymbol{b}})_{\boldsymbol{v}}(\boldsymbol{d}^1, \ldots, \boldsymbol{d}^m), \tag{12.11}$$

which shows that $\mathcal{T}_{\varphi_{\boldsymbol{b}}(\boldsymbol{v})} = \mathcal{T}_{\varphi_{\boldsymbol{a}}(\boldsymbol{u})}$ and hence makes the following definition meaningful.

12.4.3 Definition. The *tangent space* $\mathcal{T}_{\boldsymbol{x}}$ *to* S *at a point* $\boldsymbol{x} \in S$ is defined as $\mathcal{T}_{\varphi_{\boldsymbol{a}}(\boldsymbol{u})}$, where $\varphi_{\boldsymbol{a}}$ is any local parametrization of S with $\varphi_{\boldsymbol{a}}(\boldsymbol{u}) = \boldsymbol{x}$. ◊

The next proposition gives an intrinsic characterization of tangent space.

12.4.4 Proposition. *For* $\boldsymbol{x} \in S$ *let* $\Lambda_{\boldsymbol{x}}$ *denote the set of all vectors in* $\mathbb{R}^n$ *of the form* $\alpha'(0)$, *where* $\alpha : (-r, r) \to S$ *is a* C^1 *curve with* $\alpha(0) = \boldsymbol{x}$. *Then*

$$\mathcal{T}_{\boldsymbol{x}} = \Lambda_{\boldsymbol{x}} = \{\boldsymbol{z} \in \mathbb{R}^n : dF_{\boldsymbol{x}}(\boldsymbol{z}) = \boldsymbol{0}\} = \bigcap_{i=1}^{n-m} \{\boldsymbol{z} \in \mathbb{R}^n : \nabla F_i(\boldsymbol{x}) \cdot \boldsymbol{z} = 0\}.$$

Proof. Let φ be a local parametrization of S with $\varphi(\boldsymbol{u}) = \boldsymbol{x}$. A member of $\mathcal{T}_{\boldsymbol{x}}$ is of the form $\boldsymbol{z} = \sum_{i=1}^{m} a_i d\varphi_{\boldsymbol{u}}(\boldsymbol{e}^i)$. For small $|t|$, the curve

$$\alpha(t) = \varphi\left(\boldsymbol{u} + t\sum_{i=1}^{m} a_i \boldsymbol{e}^i\right)$$

lies in S, $\alpha(0) = \boldsymbol{x}$, and, by the chain rule,

$$\alpha'(0) = d\varphi_{\boldsymbol{u}}\left(\sum_{i=1}^{m} a_i \boldsymbol{e}^i\right) = \sum_{i=1}^{m} a_i d\varphi)_{\boldsymbol{u}}(\boldsymbol{e}^i) = \boldsymbol{z}.$$

Therefore, $\boldsymbol{z} \in \Lambda_{\boldsymbol{x}}$. On the other hand, if $\alpha'(0) \in \Lambda_{\boldsymbol{x}}$, then differentiating the identity $(F \circ \alpha)(t) = 0$ at $t = 0$ yields $dF_{\boldsymbol{x}}\big(\alpha'(0)\big) = \boldsymbol{0}$. We have shown that

$$\mathcal{T}_{\boldsymbol{x}} \subseteq \Lambda_{\boldsymbol{x}} \subseteq \{\boldsymbol{z} : dF_{\boldsymbol{x}}(\boldsymbol{z}) = \boldsymbol{0}\}.$$

Since $dF_{\boldsymbol{x}}(\boldsymbol{z}) = \boldsymbol{0}$ has dimension m, the three spaces must be equal. □

12.4.5 Remark. The proposition shows that if S_1 is an m_1-surface, S_2 is an m_2-surface, and $\psi : S_1 \to S_2$ is C^1, then for $\boldsymbol{x} \in S_1$ and $\boldsymbol{y} = \psi(\boldsymbol{x})$ the function $d\psi_{\boldsymbol{x}}$ maps $\mathcal{T}_{\boldsymbol{x}}$ into $\mathcal{T}_{\boldsymbol{y}}$. Indeed, if $\boldsymbol{v} \in \mathcal{T}_{\boldsymbol{x}}$, then there exists a smooth curve $\alpha_1 : (-1, 1) \to S_1$ with $\alpha_1(0) = \boldsymbol{x}$ and $\alpha_1'(0) = \boldsymbol{v}$. Then $\alpha_2 =: \psi \circ \alpha_1$ is a smooth curve in S_2 and $d\psi_{\boldsymbol{x}}(\boldsymbol{v}) = (\psi \circ \alpha_1)'(0) = \alpha_2'(0) \in \mathcal{T}_{\boldsymbol{y}}$. ◊

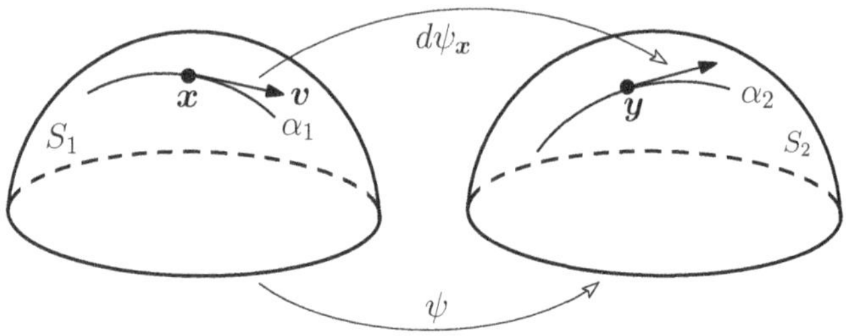

FIGURE 12.16: The mapping $d\psi_{\boldsymbol{x}} : \mathcal{T}_{\boldsymbol{x}} \to \mathcal{T}_{\boldsymbol{y}}$.

Orientation of an m-Surface

Let S be an m-surface with local parameterizations $\varphi_{\boldsymbol{a}} : U_{\boldsymbol{a}} \to \mathbb{R}^n$. Since $\varphi_{\boldsymbol{a}}$ is one-to-one, it is orientable. Suppose the parameterizations have the same orientation, that is, $\text{sign}(\varphi_{\boldsymbol{a}}) = \text{sign}(\varphi_{\boldsymbol{b}})$ for all $\boldsymbol{a}$ and $\boldsymbol{b}$. If $\boldsymbol{u} \in U_{\boldsymbol{a}}$, $\boldsymbol{v} \in U_{\boldsymbol{b}}$, and $\varphi_{\boldsymbol{a}}(\boldsymbol{u}) = \varphi_{\boldsymbol{b}}(\boldsymbol{v})$, then (12.11) shows that the orientation of $\mathcal{T}_{\varphi_{\boldsymbol{b}}(\boldsymbol{v})}$ agrees with that of $\mathcal{T}_{\varphi_{\boldsymbol{a}}(\boldsymbol{u})}$ iff $J_{\varphi_{\boldsymbol{ab}}}(\boldsymbol{v}) > 0$. Thus if $J_{\varphi_{\boldsymbol{ab}}} > 0$ whenever $S_{\boldsymbol{a}} \cap S_{\boldsymbol{b}} \neq \emptyset$, then S may be given a well-defined orientation via the orientations of the local parameterizations. In this case, S is said to be *orientable*. The *positive orientation* is obtained if each local parametrization is positively oriented.

Orientation of an $(n-1)$-Surface

Orientability of an $(n-1)$-surface may be characterized in terms of the normal vector fields $\vec{N}_{\varphi_{\boldsymbol{a}}}$. For this we need the following lemma, which relates $\vec{N}_{\varphi_{\boldsymbol{a}}}$ and $\vec{N}_{\varphi_{\boldsymbol{b}}}$ on overlapping surface elements.

12.4.6 Lemma. *Let* $\boldsymbol{x} := \varphi_{\boldsymbol{a}}(\boldsymbol{u}) = \varphi_{\boldsymbol{b}}(\boldsymbol{v}) \in S_{\boldsymbol{a}} \cap S_{\boldsymbol{b}}$, *where* $\boldsymbol{u} \in U_{\boldsymbol{a}}$, $\boldsymbol{v} \in U_{\boldsymbol{b}}$. *Then*

$$\vec{N}_{\varphi_{\boldsymbol{a}}}(\boldsymbol{x}) = |J_{\varphi_{\boldsymbol{ab}}}(\boldsymbol{u})|^{-1} J_{\varphi_{\boldsymbol{ab}}}(\boldsymbol{u}) \vec{N}_{\varphi_{\boldsymbol{b}}}(\boldsymbol{x}) = \text{sign}\big(J_{\varphi_{\boldsymbol{ab}}}(\boldsymbol{u})\big) \vec{N}_{\varphi_{\boldsymbol{b}}}(\boldsymbol{x}).$$

Proof. Since $\varphi_{\boldsymbol{a}} = \varphi_{\boldsymbol{b}} \circ \varphi_{\boldsymbol{ab}}$ and $\boldsymbol{v} = \varphi_{\boldsymbol{ab}}(\boldsymbol{u})$, the chain rule implies that

$$\varphi_{\boldsymbol{a}}'(\boldsymbol{u}) = \varphi_{\boldsymbol{b}}'(\boldsymbol{v}) \varphi_{\boldsymbol{ab}}'(\boldsymbol{u})$$

and

$$\frac{\partial(\varphi_{\boldsymbol{a},1}, \ldots, \widehat{\varphi_{\boldsymbol{a},i}}, \ldots, \varphi_{\boldsymbol{a},n})}{\partial(u_1, \ldots, u_{n-1})}(\boldsymbol{u}) = \frac{\partial(\varphi_{\boldsymbol{b},1}, \ldots, \widehat{\varphi_{\boldsymbol{b},i}}, \ldots, \varphi_{\boldsymbol{b},n})}{\partial(v_1, \ldots, v_{n-1})}(\boldsymbol{v}) J_{\varphi_{\boldsymbol{ab}}}(\boldsymbol{u}).$$

From the first equation,

$$\begin{aligned}\sqrt{\det\left(\varphi'_{\boldsymbol a}(\boldsymbol u)^t\varphi'_{\boldsymbol a}(\boldsymbol u)\right)} &= \sqrt{\det\left(\varphi'_{\boldsymbol{ab}}(\boldsymbol u)^t\varphi'_{\boldsymbol b}(\boldsymbol v)^t\varphi'_{\boldsymbol b}(\boldsymbol v)\varphi'_{\boldsymbol{ab}}(\boldsymbol u)\right)}\\ &= |J_{\varphi_{ab}}(\boldsymbol u)|\sqrt{\det\left(\varphi'_{\boldsymbol b}(\boldsymbol v)^t\varphi'_{\boldsymbol b}(\boldsymbol v)\right)}.\end{aligned}$$

The assertion now follows by recalling that

$$\vec N_{\varphi_a}(\boldsymbol x) = \frac{1}{\sqrt{\det\left(\varphi'_{\boldsymbol a}(\boldsymbol u)^t\varphi'_{\boldsymbol a}(\boldsymbol u)\right)}}\sum_{i=1}^n(-1)^{i+n}\frac{\partial(\varphi_{\boldsymbol a,1},\ldots,\widehat{\varphi_{\boldsymbol a,i}},\ldots,\varphi_{\boldsymbol a,n})}{\partial(u_1,\ldots,u_{n-1})}(\boldsymbol u)$$

and

$$\vec N_{\varphi_b}(\boldsymbol x) = \frac{1}{\sqrt{\det\left(\varphi'_{\boldsymbol b}(\boldsymbol v)^t\varphi'_{\boldsymbol b}(\boldsymbol v)\right)}}\sum_{i=1}^n(-1)^{i+n}\frac{\partial(\varphi_{\boldsymbol b,1},\ldots,\widehat{\varphi_{\boldsymbol b,i}},\ldots,\varphi_{\boldsymbol b,n})}{\partial(v_1,\ldots,v_{n-1})}(\boldsymbol v).\ \square$$

12.4.7 Theorem. *An $(n-1)$-surface S is orientable iff there exists a continuous vector field $\vec N$ on S such that*

$$\vec N\big|_{S_a} = \vec N_{\varphi_a}\quad \textit{for each } \boldsymbol a \in S. \tag{12.12}$$

Proof. If S is orientable, then $J_{\varphi_{ab}} > 0$, hence, by 12.4.6, $\vec N_{\varphi_a} = \vec N_{\varphi_b}$ on $S_{\boldsymbol a} \cap S_{\boldsymbol b}$. Therefore, (12.12) defines $\vec N$ unambiguously. Since $\vec N_{\varphi_a}$ is easily seen to be continuous on $S_{\boldsymbol a}$ and $S_{\boldsymbol a}$ is relatively open in S, $\vec N$ is continuous on S.

Conversely, assume there exists a continuous vector field $\vec N$ on S that satisfies (12.12). If $\boldsymbol x = \varphi_{\boldsymbol a}(\boldsymbol u) \in S_{\boldsymbol a} \cap S_{\boldsymbol b}$, then

$$\vec N_{\varphi_a}(\boldsymbol x) = \vec N(\boldsymbol x) = \vec N_{\varphi_b}(\boldsymbol x),$$

hence, by 12.4.6, $J_{\varphi_{ab}}(\boldsymbol u) > 0$. Therefore, S is orientable. $\square$

Let S be orientable with positive orientation. Then, by definition, the frame $(d(\varphi_{\boldsymbol a})_{\boldsymbol u}(\boldsymbol e^1),\ldots,d(\varphi_{\boldsymbol a})_{\boldsymbol u}(\boldsymbol e^{n-1}))$ in $\mathcal{T}_{\boldsymbol a}$ is designated as positive ($\operatorname{sign}(\varphi_{\boldsymbol a}) > 0$) for each $\boldsymbol a \in S$. Since the frame

$$(d(\varphi_{\boldsymbol a})_{\boldsymbol u}(\boldsymbol e^1),\ldots,d(\varphi_{\boldsymbol a})_{\boldsymbol u}(\boldsymbol e^{n-1}),\vec N(\boldsymbol a))$$

in $\mathbb{R}^n$ is positive (12.3.5), we say in this case that S *is oriented by* $\vec N$. The notion of *orientation by* $-\vec N$ is defined analogously. For example, the sphere $S = \{(x_1,\ldots,x_n) : \|\boldsymbol x\| = r\}$ is locally parameterized by the mappings φ and $\tilde\varphi$ of 12.3.8. The positive orientation is given by the unit normal vector field $\vec N(\boldsymbol p) = \|\boldsymbol p\|^{-1}\boldsymbol p$, called the *outward unit normal.*

12.4.8 Corollary. *If $S = \{\boldsymbol x : F(\boldsymbol x) = \boldsymbol 0\}$ is connected, then S is orientable and*

$$\vec N = \|\nabla F\|^{-1}\nabla F \quad \textit{or} \quad \vec N = -\|\nabla F\|^{-1}\nabla F.$$

Proof. Since $\nabla F(\boldsymbol{x})$ is perpendicular to S at $\boldsymbol{x}$, the uniqueness of $\vec{N}$ implies that

$$\vec{N}(\boldsymbol{x}) = s(\boldsymbol{x})\frac{\nabla F(\boldsymbol{x})}{\|\nabla F(\boldsymbol{x})\|}, \quad \boldsymbol{x} \in S,$$

where $s(\boldsymbol{x}) = \pm 1$ is constant on each surface element. Since the surface elements are open in S, $s(\boldsymbol{x})$ is continuous. Since S is connected, $s(\boldsymbol{x})$ must be constant on S. □

$(n-1)$-Surfaces-with-Boundary

To discuss surfaces-with-boundary, we shall need the following notation:

$$\begin{aligned} \mathbb{R}^{n-1}_{+} &:= \left\{\boldsymbol{y} \in \mathbb{R}^{n-1} : y_{n-1} > 0\right\}. \\ \mathbb{H}^{n-1} &:= \left\{\boldsymbol{y} \in \mathbb{R}^{n-1} : y_{n-1} \geq 0\right\}. \\ \partial\mathbb{H}^{n-1} &:= \left\{\boldsymbol{y} \in \mathbb{R}^{n-1} : y_{n-1} = 0\right\}. \end{aligned}$$

12.4.9 Definition. An $(n-1)$-*surface-with-boundary* is a subset of $\mathbb{R}^n$ of the form

$$S = \{\boldsymbol{x} \in W : F(\boldsymbol{x}) = 0 \text{ and } g_i(\boldsymbol{x}) \geq 0,\ i = 1, \ldots, k\},$$

where $W \subseteq \mathbb{R}^n$ is open and $F : W \to \mathbb{R}$ and $g_i : W \to \mathbb{R}$ are C^1 and satisfy the following conditions:

(a) $\nabla F(\boldsymbol{x}) \neq 0$ for all $\boldsymbol{x} \in S$.

(b) The sets $B_i := \{\boldsymbol{x} \in S : g_i(\boldsymbol{x}) = 0\}$ are pairwise disjoint.

(c) For each i and $\boldsymbol{x} \in B_i$, the vectors $\nabla F(\boldsymbol{x})$ and $\nabla g_i(\boldsymbol{x})$ are linearly independent.

The set $\partial S := \bigcup_{i=1}^{k} B_i$ is called the *boundary of* S and $S \setminus \partial S$ is the *interior*.◊

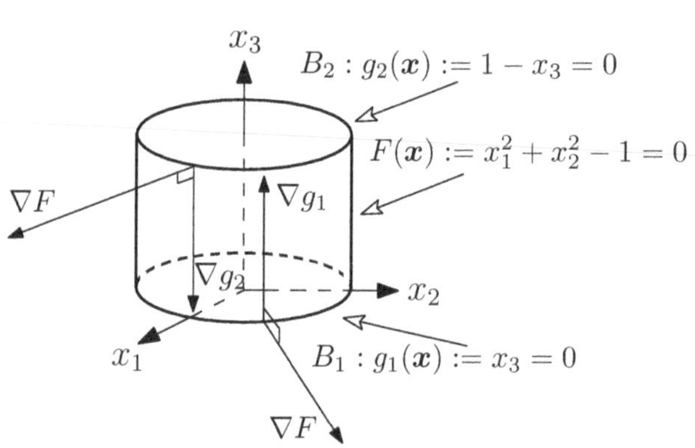

FIGURE 12.17: Cylinder-with-boundary: $x_1^2 + x_2^2 = 1$, $0 \leq x_3 \leq 1$.

If V denotes the open set $\{\boldsymbol{x} \in W : g_i(\boldsymbol{x}) > 0,\ i = 1, \ldots, k\}$, then

$$S \setminus \partial S = \{\boldsymbol{x} \in V : F(\boldsymbol{x}) = 0\}.$$

Therefore, condition (a) implies that the interior of S is an $(n-1)$-surface. Conditions (b) and (c) assert that the boundary of S is made up of disjoint $(n-2)$-surfaces. Indeed, if $F_i := (F, g_i)$, then $B_i = \{\boldsymbol{x} \in W : F_i = \boldsymbol{0}\}$ and

$$F_i' = \begin{bmatrix} \nabla F \\ \nabla g_i \end{bmatrix}$$

has rank 2. Also, because the $(n-2)$-surfaces B_i are pairwise disjoint, a local parametrization of B_i may be chosen to be disjoint from a local parametrization of B_j.

The following theorem shows that, as in the case of an $(n-1)$-surface, an $(n-1)$-surface-with-boundary may be described by a collection of local parameterizations.

12.4.10 Theorem. *Let S be an $(n-1)$-surface-with-boundary.*

(a) *If $\boldsymbol{a} \in S \setminus \partial S$, then there exists a local parametrization $\varphi_{\boldsymbol{a}} : U_{\boldsymbol{a}} \to \mathbb{R}^n$ of $S \setminus \partial S$ at $\boldsymbol{a}$ with $\varphi_{\boldsymbol{a}}(\boldsymbol{0}) = \boldsymbol{a}$.*

(b) *If $\boldsymbol{a} \in \partial S$, then there exists an open set $\tilde{U}_{\boldsymbol{a}} \subseteq \mathbb{R}^{n-1}$ and a one-to-one parameterized $(n-1)$-surface $\tilde{\varphi}_{\boldsymbol{a}} : \tilde{U}_{\boldsymbol{a}} \to \mathbb{R}^{n-1}$ with $\tilde{\varphi}_{\boldsymbol{a}}(\boldsymbol{0}) = \boldsymbol{a}$ such that if $U_{\boldsymbol{a}} := \tilde{U}_{\boldsymbol{a}} \cap \mathbb{H}^{n-1}$ and $\varphi_{\boldsymbol{a}} := \tilde{\varphi}_{\boldsymbol{a}}\big|_{U_{\boldsymbol{a}}}$, then*

  (i) $\varphi_{\boldsymbol{a}}(U_{\boldsymbol{a}})$ is open in S,
  (ii) $\varphi_{\boldsymbol{a}}(U_{\boldsymbol{a}} \cap \mathbb{R}^{n-1}_+)$ is open in $S \setminus \partial S$, and
  (iii) $\varphi_{\boldsymbol{a}}(U_{\boldsymbol{a}} \cap \partial \mathbb{H}^{n-1})$ is open in ∂S.

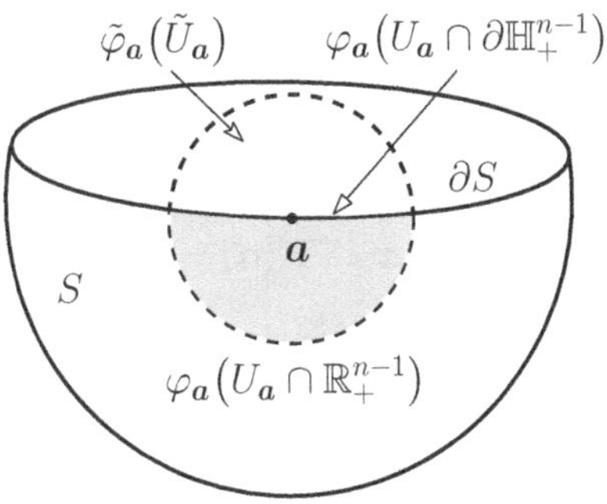

FIGURE 12.18: Surface element $S_{\boldsymbol{a}} = \varphi_{\boldsymbol{a}}(\tilde{U}_{\boldsymbol{a}} \cap \mathbb{H}^{n-1})$.

Proof. Part (a) follows from 12.4.1, since $S \setminus \partial S$ is an $(n-1)$-surface without boundary.

For part (b), we may assume without loss of generality that $\partial S = \{\boldsymbol{x} \in S : g(\boldsymbol{x}) = 0\}$. Choose a local parametrization $\psi_{\boldsymbol{a}} : W_{\boldsymbol{a}} \to \mathbb{R}^n$ of

$S' := \{\boldsymbol{x} \in W : F(\boldsymbol{x}) = 0\}$ such that $\psi_{\boldsymbol{a}}(\mathbf{0}) = \boldsymbol{a}$. Since $\psi_{\boldsymbol{a}}$ has rank $n-1$ and g has rank 1, $\partial_i(g \circ \psi_{\boldsymbol{a}})(\mathbf{0})) \neq 0$ for some i. Define $H_{\boldsymbol{a}} : W_{\boldsymbol{a}} \to \mathbb{R}^{n-1}$ by

$$H_{\boldsymbol{a}}(w^1, \ldots, w^{n-1}) = \big(w^1, \ldots, w^{i-1}, w^{i+1}, \ldots, w^{n-1}, g \circ \psi_{\boldsymbol{a}}(w^1, \ldots, w^{n-1})\big).$$

Then $H_{\boldsymbol{a}}$ has rank $n-1$ at $\mathbf{0}$, hence, by the inverse function theorem, there exist open sets $\tilde{W}_{\boldsymbol{a}} \subseteq W_{\boldsymbol{a}}$ and $\tilde{U}_{\boldsymbol{a}} = H_{\boldsymbol{a}}(\tilde{W}_{\boldsymbol{a}})$ in $\mathbb{R}^{n-1}$ with $\mathbf{0} \in \tilde{W}_{\boldsymbol{a}}$ such that $H_{\boldsymbol{a}}$ is one-to-one on $\tilde{W}_{\boldsymbol{a}}$ and $H_{\boldsymbol{a}}^{-1} : \tilde{U}_{\boldsymbol{a}} \to \tilde{W}_{\boldsymbol{a}}$ is C^1. Set

$$\tilde{\varphi}_{\boldsymbol{a}} = \psi_{\boldsymbol{a}} \circ H_{\boldsymbol{a}}^{-1} : \tilde{U}_{\boldsymbol{a}} \to S'.$$

If $\boldsymbol{u} = H_{\boldsymbol{a}}(\boldsymbol{w}) \in \tilde{U}_{\boldsymbol{a}}$, then $g \circ \psi_{\boldsymbol{a}}(\boldsymbol{w}) = g \circ \psi_{\boldsymbol{a}} \circ H_{\boldsymbol{a}}^{-1}(\boldsymbol{u}) = g \circ \tilde{\varphi}_{\boldsymbol{a}}(\boldsymbol{u})$, hence, by definition of $H_{\boldsymbol{a}}$,

$$(u_1, \ldots, u_{n-1}) = \big(w^1, \ldots, w^{i-1}, w^{i+1}, \ldots, w^{n-1}, g \circ \tilde{\varphi}_{\boldsymbol{a}}(\boldsymbol{u})\big).$$

Therefore, $u_{n-1} = g \circ \tilde{\varphi}_{\boldsymbol{a}}(\boldsymbol{u})$, so

$$g \circ \tilde{\varphi}_{\boldsymbol{a}}(\boldsymbol{u}) > 0 \text{ iff } \boldsymbol{u} \in \tilde{U}_{\boldsymbol{a}} \cap \mathbb{R}^{n-1}_+ \quad \text{and} \quad g \circ \tilde{\varphi}_{\boldsymbol{a}}(\boldsymbol{u}) = 0 \text{ iff } \boldsymbol{u} \in \tilde{U}_{\boldsymbol{a}} \cap \partial\, \mathbb{H}^{n-1}.$$

It follows that

$$\tilde{\varphi}_{\boldsymbol{a}}\big(\tilde{U}_{\boldsymbol{a}} \cap \mathbb{R}^{n-1}_+\big) = (S \setminus \partial S) \cap \psi\big(\tilde{W}_{\boldsymbol{a}}\big) \quad \text{and} \quad \tilde{\varphi}_{\boldsymbol{a}}\big(\tilde{U}_{\boldsymbol{a}} \cap \partial \mathbb{H}^{n-1}\big) = \partial S \cap \psi\big(\tilde{W}_{\boldsymbol{a}}\big).$$

Since $\psi\big(\tilde{W}_{\boldsymbol{a}}\big)$ is open in S' and $S' \supseteq S$, (i)–(iii) follow. □

Oriented $(n-1)$-Surfaces-with-Boundary

As in the non-boundary case, orientation of an $(n-1)$-surface-with-boundary S may be defined in terms of local parameterizations. By 12.4.4, the $(n-1)$-dimensional tangent space at $\boldsymbol{a} \in S$ is

$$\mathcal{T}_{\boldsymbol{a}}^S = \{\boldsymbol{z} \in \mathbb{R}^n : \boldsymbol{z} \cdot \nabla F(\boldsymbol{a}) = 0\}.$$

The new feature here is that if $\boldsymbol{a} \in \partial S$, say $\boldsymbol{a} \in B_i$, then there is also an $(n-2)$-dimensional tangent space to ∂S at $\boldsymbol{a}$, namely,

$$\mathcal{T}_{\boldsymbol{a}}^{\partial S} = \{\boldsymbol{z} \in \mathbb{R}^n : \boldsymbol{z} \cdot \nabla F(\boldsymbol{a}) = \boldsymbol{z} \cdot \nabla g_i(\boldsymbol{a}) = 0\}.$$

The connection between $\mathcal{T}_{\boldsymbol{a}}^S$ and $\mathcal{T}_{\boldsymbol{a}}^{\partial S}$ is described as follows: Let $\varphi_{\boldsymbol{a}}$ be a local parametrization of S as described in part (b) of 12.4.10, where $\varphi_{\boldsymbol{a}}(\mathbf{0}) = \boldsymbol{a}$. Since $\tilde{\varphi}_{\boldsymbol{a}}(U_{\boldsymbol{a}} \cap \partial\mathbb{H}^{n-1}) \subseteq \partial S$ and $(\boldsymbol{e}^1, \ldots, \boldsymbol{e}^{n-2})$ is a frame for $\partial\mathbb{H}_{n-1}$, $d(\tilde{\varphi}_{\boldsymbol{a}})_{\mathbf{0}}(\boldsymbol{e}^1, \ldots, \boldsymbol{e}^{n-2})$ is a frame for $\mathcal{T}_{\boldsymbol{a}}^{\partial S}$. Since the vector $d(\tilde{\varphi}_{\boldsymbol{a}})_{\mathbf{0}}(-\boldsymbol{e}^{n-1})$ is not in the subspace $\mathcal{T}_{\boldsymbol{a}}^{\partial S}$,

$$d(\tilde{\varphi}_{\boldsymbol{a}})_{\mathbf{0}}(-\boldsymbol{e}^{n-1}, \boldsymbol{e}^1, \ldots, \boldsymbol{e}^{n-2}) \tag{12.13}$$

is a frame for $\mathcal{T}_{\boldsymbol{a}}^S$. The *induced orientation* of ∂S is obtained by declaring the frame $d(\tilde{\varphi}_{\boldsymbol{a}})_{\mathbf{0}}(\boldsymbol{e}^1, \ldots, \boldsymbol{e}^{n-2})$ of $\mathcal{T}_{\boldsymbol{a}}^{\partial S}$ to have the sign of the frame (12.13). If S is positively oriented, then this sign is $(-1)^{n-1}$.

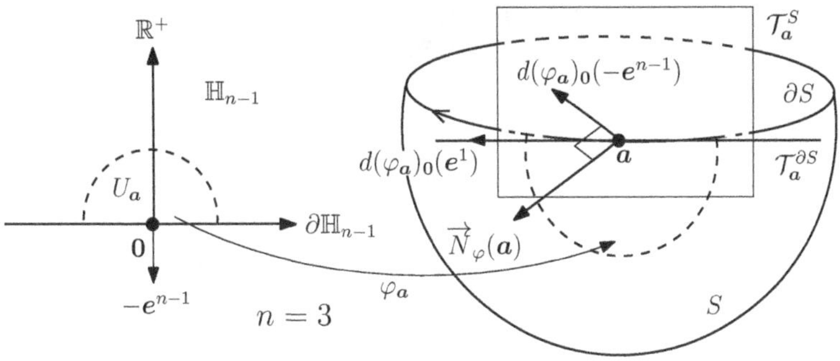

FIGURE 12.19: Induced orientation of $\mathcal{T}_{\boldsymbol{a}}^{\partial S}$.

Figure 12.19 depicts the case $n = 3$. Here, S is oriented by the normal $\vec{N}$ (pointing outward). Therefore, by definition, the frame $d(\varphi_{\boldsymbol{a}})_{\boldsymbol{0}}(\boldsymbol{e}^1, \boldsymbol{e}^2)$ is positive in $\mathcal{T}_{\boldsymbol{a}}^S$, hence so is the frame $d(\tilde{\varphi}_{\boldsymbol{a}})_{\boldsymbol{0}}(-\boldsymbol{e}^2, \boldsymbol{e}^1)$. Thus, again by definition, the frame $d(\tilde{\varphi}_{\boldsymbol{a}})_{\boldsymbol{0}}(\boldsymbol{e}^1)$ of $\mathcal{T}_{\boldsymbol{a}}^{\partial S}$ is positive in the induced orientation. Note that because the frame $(d(\varphi_{\boldsymbol{a}})_{\boldsymbol{0}}(\boldsymbol{e}^1), d(\varphi_{\boldsymbol{a}})_{\boldsymbol{0}}(\boldsymbol{e}^2), \vec{N}_\varphi)$ in $\mathbb{R}^3$ is positive (12.3.5), so is the frame $(d(\varphi_{\boldsymbol{a}})_{\boldsymbol{0}}(-\boldsymbol{e}^2), d(\varphi_{\boldsymbol{a}})_{\boldsymbol{0}}(\boldsymbol{e}^1), \vec{N}_\varphi)$. The latter therefore forms a right-handed system in $\mathbb{R}^3$. Thus if $d(\tilde{\varphi}_{\boldsymbol{a}})_{\boldsymbol{0}}(-\boldsymbol{e}^2)$ points upward, then $d(\tilde{\varphi}_{\boldsymbol{a}})_{\boldsymbol{0}}(\boldsymbol{e}^1)$ must point in the direction shown. Therefore, the induced orientation of ∂S is the one for which the surface S is on the left when ∂S is traversed in the direction of the tangent vectors $d(\tilde{\varphi}_{\boldsymbol{a}})_{\boldsymbol{0}}(\boldsymbol{e}^1)$.

Exercises

1. Let $0 < a < b$. Show that the mapping

$$\varphi(\phi, \theta) = \big(a\cos\phi, (b + a\sin\phi)\cos\theta, (b + a\sin\phi)\sin\theta\big), \quad 0 < \theta,\ \phi < 2\pi,$$

is a local parametrization of the torus $x^2 + \big(\sqrt{y^2 + z^2} - b\big)^2 = a^2$ with two circles missing.

2. Let $U = \big\{\boldsymbol{x} \in \mathbb{R}^{n-1} : \|\boldsymbol{x}\| < 1\big\}$ and define a local parametrization $\psi : U \to S^{n-1} = \{\boldsymbol{y} \in \mathbb{R}^n : \|\boldsymbol{y}\| = 1\}$ by

$$\psi(\boldsymbol{x}) = \big(\boldsymbol{x}, \sqrt{1 - \|\boldsymbol{x}\|^2}\big), \quad \boldsymbol{x} \in \mathbb{R}^{n-1}$$

Give a geometric description of ψ. Referring to 12.4.2, find the transition mapping $\tilde{\varphi}^{-1} \circ \psi$.

3. Consider the stereographic projection $\varphi_1^{-1}(\boldsymbol{y}) = \boldsymbol{x}$ from $\boldsymbol{p}$ onto the hyperplane $x_n = -1$ shown in Figure 12.20, where $\boldsymbol{y} = (y_1, \ldots, y_n)$ and $\boldsymbol{x} = (x_1, \ldots, x_{n-1})$. Calculate $\varphi_1(\boldsymbol{x})$ and $\varphi_1^{-1}(\boldsymbol{y})$ and find the transition mapping $\varphi^{-1} \circ \varphi_1$, where φ is the mapping of 12.4.2.

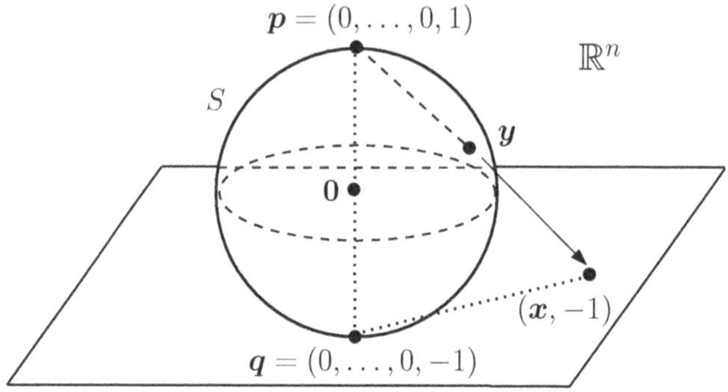

FIGURE 12.20: Stereographic projection $\varphi_1^{-1}(\boldsymbol{y})$ from $\boldsymbol{p}$.

4. Replace the sphere in 12.4.2 by the elliptic paraboloid

$$S = \left\{ (y_1, y_2, y_3) : y_3 = \left(\frac{y_1}{a_1}\right)^2 + \left(\frac{y_2}{a_2}\right)^2, \quad y_3 < 1 \right\}$$

(with $\boldsymbol{p} = (0, 0, 1)$) and find the corresponding maps φ and φ^{-1}.

5. Repeat Exercise 4 using the elliptic cone

$$S = \left\{ (y_1, y_2, y_3) : y_3^2 = \left(\frac{y_1}{a_1}\right)^2 + \left(\frac{y_2}{a_2}\right)^2, \quad 0 < y_3 < 1 \right\}.$$

6. Repeat Exercise 4 using the ellipsoid

$$S = \left\{ (y_1, y_2, y_3) : \left(\frac{y_1}{a_1}\right)^2 + \left(\frac{y_2}{a_2}\right)^2 + y_3^2 = 1 \right\}.$$

7. Find the equation of the tangent plane $\mathcal{T}_{\boldsymbol{a}}$ at $\boldsymbol{a} = (1, 1, 1)$ for each of the following surfaces:

   (a) $x_1^2 + 2x_2^2 + 3x_3^2 = 6$. (b) $x_1^2 + x_2^2 - 2x_3^2 = 0$. (c) $x_1^2 - x_2^2 + x_3 = 1$.

8. An $n \times n$ matrix A is said to be *orthogonal* if $A^t A$ is the identity matrix. Identifying a 2×2 matrix $\left[\begin{smallmatrix} x_1 & x_2 \\ x_3 & x_4 \end{smallmatrix}\right]$ with the point (x_1, x_2, x_3, x_4), show that the collection of all 2×2 orthogonal matrices is a 1-surface S in $\mathbb{R}^4$. Characterize the matrices in the tangent space to S at each of the following points:

   (a) $\begin{bmatrix} 1 & 0 \\ 0 & 1 \end{bmatrix}$ (b) $\begin{bmatrix} -1 & 0 \\ 0 & 1 \end{bmatrix}$ (c) $\begin{bmatrix} 0 & 1 \\ 1 & 0 \end{bmatrix}$ (d) $\begin{bmatrix} 1/\sqrt{2} & -1/\sqrt{2} \\ 1/\sqrt{2} & 1/\sqrt{2} \end{bmatrix}$.

   The matrices in the tangent space at the point in part (a) are the so-called 2×2 *skew-symmetric* matrices.

9. Referring to 12.4.2, let $\boldsymbol{y} \in S$ and set $T := d(\varphi^{-1})_{\boldsymbol{y}} : \mathcal{T}_{\boldsymbol{y}} \to \mathbb{R}^{n-1}$.

   (a) Prove that $\|T(\boldsymbol{v})\| = \dfrac{1}{(1-y_n)}\|\boldsymbol{v}\|$ for all $\boldsymbol{v} \in \mathcal{T}_{\boldsymbol{y}}$.

   (b) Use (a), the bilinearity of $\boldsymbol{v} \cdot \boldsymbol{w}$ and $T(\boldsymbol{v}) \cdot T(\boldsymbol{w})$, and the identity

$$2\boldsymbol{v} \cdot \boldsymbol{w} = \|\boldsymbol{v} + \boldsymbol{w}\|^2 - \|\boldsymbol{v}\|^2 - \|\boldsymbol{w}\|^2$$

   to prove that

$$\frac{\boldsymbol{v} \cdot \boldsymbol{w}}{\|\boldsymbol{v}\|\|\boldsymbol{w}\|} = \frac{T(\boldsymbol{v}) \cdot T(\boldsymbol{w})}{\|T(\boldsymbol{v})\|\|T(\boldsymbol{w})\|}, \quad \boldsymbol{v}, \boldsymbol{w} \in \mathcal{T}_{\boldsymbol{y}}.$$

   Thus, by 12.4.4, the stereographic projection preserves the angle at the intersection of a pair of simple smooth curves on S.

10. Let each of the following 2-surfaces-with-boundary be positively oriented. Find parametrizations of the boundary curves that are compatible with the induced orientation on the boundary.

    (a) $S = \{(x_1, x_2, x_3) : x_1^2 + x_2^2 = 1,\ 0 \le x_3 \le 2 - x_2\}$

    (b) $S = \{(x_1, x_2, x_3) : x_3 = x_1^2 + x_2^2,\ 0 \le x_3 \le 1 - x_1 - x_2\}$

    (c) $S = \{(x_1, x_2, x_3) : x_1^2 + x_2^2 + x_3^2 = 4,\ -2 \le x_3 \le 3 - x_1 - x_2\}$.

    Hint. For (c) the boundary is a circle on the plane $x_1 + x_2 + x_3 = 3$. Translate and rotate that plane into the plane $x_3 = 0$, find a parametric equation of the rotated circle with center $\mathbf{0}$, then reverse the procedure to find the parametrization of the original circle with appropriate orientation.

11. Let $S = \{\boldsymbol{x} : F(\boldsymbol{x}) = 0\}$ be an oriented 2-surface in $\mathbb{R}^3$, where F is C^2.

    (a) The *tangent bundle* of S is the set

$$\mathcal{T}_S = \bigcup_{\boldsymbol{x} \in S} \{\boldsymbol{x}\} \times \mathcal{T}_{\boldsymbol{x}}.$$

    Show that

$$\mathcal{T}_S = \{(\boldsymbol{x}, \boldsymbol{v}) \in \mathbb{R}^6 : F(\boldsymbol{x}) = 0 \text{ and } \boldsymbol{v} \cdot \nabla F(\boldsymbol{x}) = 0\}$$

    and that $\mathcal{T}_S$ is a 4-surface in $\mathbb{R}^6$.

    (b) The *sphere bundle* of S is the subset

$$\mathcal{T}_S^1 := \{(\boldsymbol{x}, \boldsymbol{v}) \in \mathcal{T}_S : \|\boldsymbol{v}\| = 1\}.$$

    Show that $\mathcal{T}_S^1$ is a 3-surface in $\mathbb{R}^6$.

    (c) Let $S = \{\boldsymbol{x} \in \mathbb{R}^3 : \|\boldsymbol{x}\|^2 = 3\}$. Show that the tangent space to the sphere bundle $\mathcal{T}_S^1$ at the point $(1, 1, 1, 1/\sqrt{6}, 1/\sqrt{6}, -2/\sqrt{6})$ consists of all vectors $\boldsymbol{w} \in \mathbb{R}^6$ satisfying the system

$$\begin{array}{ccccccccc} w_1 & + & w_2 & + & w_3 & & & = 0 \\ -3\sqrt{6}w_3 & + & w_4 & + & w_5 & + & w_6 & = 0 \\ w_4 & + & w_5 & - & 2w_6 & & & = 0 \end{array}$$

Chapter 13

Integration on Surfaces

Throughout the chapter m and n are fixed positive integers with $1 \le m \le n$.

In this chapter we construct the integral of a differential m-form on an m-surface in $\mathbb{R}^n$, a generalization of the line integral of a 1-form on a curve. This will provide the context for the divergence theorem and the theorems of Green and Stokes, generalizations of the fundamental theorem of calculus.

13.1 Differential Forms

Alternating Multilinear Functionals

An *m-multilinear functional on* $\mathbb{R}^n$ is a real-valued function

$$M(\boldsymbol{a}^1, \ldots, \boldsymbol{a}^m), \quad \boldsymbol{a}^1, \ldots, \boldsymbol{a}^m \in \mathbb{R}^n,$$

that is linear in each variable $\boldsymbol{a}^i$ separately. (See Section 9.7.) Such a function is said to be *alternating* if interchanging two vectors changes the sign of M:

$$M(\boldsymbol{a}^1, \ldots, \boldsymbol{a}^i, \ldots, \boldsymbol{a}^j, \ldots, \boldsymbol{a}^m) = -M(\boldsymbol{a}^1, \ldots, \boldsymbol{a}^j, \ldots, \boldsymbol{a}^i, \ldots, \boldsymbol{a}^m).$$

Thus if $\boldsymbol{a}^i = \boldsymbol{a}^j$, then $M(\boldsymbol{a}^1, \ldots, \boldsymbol{a}^m) = 0$. Note that a linear combination of alternating m-multilinear functionals is an alternating m-multilinear functional.

A *permutation of* $(1, \ldots, m)$ is a one-to-one function σ mapping $\{1, \ldots, m\}$ onto itself, frequently denoted by $(i_1, \ldots, i_m)$, where $i_k = \sigma(k)$. The *sign* $(-1)^\sigma$ *of* σ is positive (negative) if an even (odd) number of adjacent interchanges are required to transform $(i_1, \ldots, i_m)$ back to $(1, \ldots, m)$ (see Appendix B). It follows that if M is an alternating m-multilinear functional, then

$$M(\boldsymbol{a}^{\sigma(1)}, \ldots, \boldsymbol{a}^{\sigma(m)}) = (-1)^\sigma M(\boldsymbol{a}^1, \ldots, \boldsymbol{a}^m).$$

An important example is the determinant of an $n \times n$ matrix, which is multilinear and alternating on its rows as well as its columns. To build on this,

DOI: 10.1201/9781003629382-13

we introduce the following notation. Define

$$\begin{aligned}\mathbf{J}_m &= \{\mathbf{j} := (j_1, \ldots, j_m) \,:\, 1 \le j_k \le n\}, \quad \text{and}\\ \mathbf{I}_m &= \{\mathbf{i} := (i_1, \ldots, i_m) \,:\, 1 \le i_1 < i_2 < \cdots < i_m \le n\}.\end{aligned}$$

Thus $\mathbf{J}_m$ is the set of all m-tuples of (possibly repeated) indices in $\{1, \ldots, n\}$ and $\mathbf{I}_m$ the set of all strictly increasing m-tuples in $\mathbf{J}_m$. In particular, $\mathbf{I}_n = \{(1, \ldots, n)\}$. Now let A be an $n \times m$ matrix with columns $\boldsymbol{a}^1, \ldots, \boldsymbol{a}^m \in \mathbb{R}^n$ and B an $m \times n$ matrix with rows $\boldsymbol{b}_1, \ldots, \boldsymbol{b}_m \in \mathbb{R}^n$. For any member $\mathbf{j} = (j_1, \ldots, j_m)$ of $\mathbf{J}_m$ define $A_\mathbf{j}$ to be the $m \times m$ matrix whose rth row is row j_r of A and define $B^\mathbf{j}$ to be the $m \times m$ matrix whose cth column is column j_c of B, that is,

$$A_\mathbf{j} = \left[\boldsymbol{a}^1 \cdots \boldsymbol{a}^m\right]_\mathbf{j} = \begin{bmatrix} a_1^1 & a_1^2 & \cdots & a_1^m \\ a_2^1 & a_2^2 & \cdots & a_2^m \\ \vdots & \vdots & & \vdots \\ a_n^1 & a_n^2 & \cdots & a_n^m \end{bmatrix}_\mathbf{j} = \begin{bmatrix} a_{j_1}^1 & a_{j_1}^2 & \cdots & a_{j_1}^m \\ a_{j_2}^1 & a_{j_2}^2 & \cdots & a_{j_2}^m \\ \vdots & \vdots & & \vdots \\ a_{j_m}^1 & a_{j_m}^2 & \cdots & a_{j_m}^m \end{bmatrix}$$

and

$$B^\mathbf{j} = \begin{bmatrix} \boldsymbol{b}_1 \\ \vdots \\ \boldsymbol{b}_m \end{bmatrix}^\mathbf{j} = \begin{bmatrix} b_1^1 & b_1^2 & \cdots & b_1^n \\ b_2^1 & b_2^2 & \cdots & b_2^n \\ \vdots & \vdots & & \vdots \\ b_m^1 & b_m^2 & \cdots & b_m^n \end{bmatrix}^\mathbf{j} = \begin{bmatrix} b_1^{j_1} & b_1^{j_2} & \cdots & b_1^{j_m} \\ b_2^{j_1} & b_2^{j_2} & \cdots & b_2^{j_m} \\ \vdots & \vdots & & \vdots \\ b_m^{j_1} & b_m^{j_2} & \cdots & b_m^{j_m} \end{bmatrix}$$

Thus $\mathbf{j}$ selects rows from A and columns from B. For example, for $n = 4$ and $m = 3$,

$$\begin{bmatrix} 1 & 2 & 3 \\ 4 & 5 & 6 \\ 7 & 8 & 9 \\ 10 & 11 & 12 \end{bmatrix}_{(4,2,1)} = \begin{bmatrix} 10 & 11 & 12 \\ 4 & 5 & 6 \\ 1 & 2 & 3 \end{bmatrix}$$

and

$$\begin{bmatrix} 1 & 2 & 3 & 4 \\ 5 & 6 & 7 & 8 \\ 9 & 10 & 11 & 12 \end{bmatrix}^{(4,4,1)} = \begin{bmatrix} 4 & 4 & 1 \\ 8 & 8 & 5 \\ 12 & 12 & 9 \end{bmatrix}.$$

Finally, define the alternating m-multilinear functional $dx_\mathbf{j} = dx_{j_1,\ldots,j_m}$ on $\mathbb{R}^n$ by

$$dx_\mathbf{j}(\boldsymbol{a}^1, \ldots, \boldsymbol{a}^m) = \det[\boldsymbol{a}^1 \cdots \boldsymbol{a}^m]_\mathbf{j}.$$

Note that if $m = 1$, the definition reduces to $dx_j(\boldsymbol{a}) = a_j$, as defined in Section 9.7.

13.1.1 Lemma. *If* $\mathbf{i} = (i_1, \cdots, i_m)$ *and* $\mathbf{j} = (j_1, \cdots, j_m) \in \mathbf{I}_m$*, then*

$$dx_\mathbf{i}(\mathbf{e}^{j_1}, \ldots, \mathbf{e}^{j_m}) = \begin{cases} 1 & \text{if } \mathbf{i} = \mathbf{j}, \\ 0 & \text{otherwise,} \end{cases}$$

where $\mathbf{e}^1, \ldots, \mathbf{e}^n$ *are the standard basis vectors in* $\mathbb{R}^n$*.*

Proof. By definition,

$$dx_{\mathbf{i}}(\boldsymbol{e}^{j_1},\ldots,\boldsymbol{e}^{j_m}) = \det\begin{bmatrix} e_1^{j_1} & \cdots & e_1^{j_m} \\ \vdots & & \vdots \\ e_n^{j_1} & \cdots & e_n^{j_m}\end{bmatrix}_{\mathbf{i}} = \begin{vmatrix} e_{i_1}^{j_1} & \cdots & e_{i_1}^{j_m} \\ \vdots & & \vdots \\ e_{i_m}^{j_1} & \cdots & e_{i_m}^{j_m}\end{vmatrix},$$

where $e_i^j = 1$ if $i = j$ and 0 otherwise. If $j_1 < i_1$, then $j_1 < i_\ell$ for every ℓ, hence the first column is zero and the determinant is zero. Similarly, if $j_1 > i_1$, then the first row is zero and, again, the determinant is zero. If $j_1 = i_1$, then the determinant reduces to

$$\begin{vmatrix} e_{i_2}^{j_2} & \cdots & e_{i_2}^{j_m} \\ \vdots & & \vdots \\ e_{i_m}^{j_2} & \cdots & e_{i_m}^{j_m}\end{vmatrix},$$

and an induction argument completes the proof. □

13.1.2 Lemma. *Let M and M' be alternating m-multilinear functionals on $\mathbb{R}^n$. If*

$$M(\boldsymbol{e}^{i_1},\ldots,\boldsymbol{e}^{i_m}) = M'(\boldsymbol{e}^{i_1},\ldots,\boldsymbol{e}^{i_m}) \tag{13.1}$$

for all $(i_1,\ldots,i_m) \in \mathbf{I}_m$, then $M = M'$.

Proof. For $j = 1,\ldots,m$, let $\boldsymbol{a}^j = (a_1^j,\ldots,a_n^j) = \sum_{i=1}^n a_i^j \boldsymbol{e}^i$. By multilinearity,

$$\begin{aligned} M(\boldsymbol{a}^1,\ldots,\boldsymbol{a}^m) &= M\left(\sum_{i=1}^n a_i^1\boldsymbol{e}^i,\ldots,\sum_{i=1}^n a_i^m\boldsymbol{e}^i\right) \\ &= \sum_{i_1=1}^n\cdots\sum_{i_m=1}^n a_{i_1}^1\cdots a_{i_m}^m M(\boldsymbol{e}^{i_1},\ldots,\boldsymbol{e}^{i_m}), \end{aligned}$$

with the analogous equality holding for M'. It therefore suffices to show that

$$M(\boldsymbol{e}^{i_1},\ldots,\boldsymbol{e}^{i_m}) = M'(\boldsymbol{e}^{i_1},\ldots,\boldsymbol{e}^{i_m}).$$

This is clear if two of the indices i_k are equal, since then both sides are zero. If the indices are distinct, then, by permuting the vectors $\boldsymbol{e}^{i_1},\ldots,\boldsymbol{e}^{i_m}$ and attaching the appropriate signs, the indices may be brought into increasing order, and the desired equality then follows from the hypothesis. □

13.1.3 Theorem. *If M is an alternating m-multilinear functional on $\mathbb{R}^n$, then*

$$M = \sum_{(i_1,\ldots,i_m)\in\mathbf{I}_m} M(\boldsymbol{e}^{i_1},\ldots,\boldsymbol{e}^{i_m})\, dx_{i_1,\cdots,i_m}.$$

Proof. Let M' denote the alternating m-multilinear functional on the right. If $(j_1, \ldots, j_m) \in \mathbf{I}_m$, then

$$\begin{aligned} M'(\boldsymbol{e}^{j_1}, \ldots, \boldsymbol{e}^{j_m}) &= \sum_{(i_1,\ldots,i_m)\in\mathbf{I}_m} M(\boldsymbol{e}^{i_1}, \ldots, \boldsymbol{e}^{i_m})\, dx_{i_1,\ldots,i_m}(\boldsymbol{e}^{j_1}, \ldots, \boldsymbol{e}^{j_m}) \\ &= M(\boldsymbol{e}^{j_1}, \ldots, \boldsymbol{e}^{j_m}), \end{aligned}$$

the second equality from 13.1.1. By 13.1.2, $M = M'$. □

The following application of 13.1.3 will be needed later in connection with integration on surfaces.

13.1.4 Binet–Cauchy Product. *Let C be an $m \times n$ matrix and D an $n \times m$ matrix. Then*

$$\det(CD) = \sum_{\mathbf{i}\in\mathbf{I}_m} \big(\det C^{\mathbf{i}}\big)\big(\det D_{\mathbf{i}}\big).$$

Proof. Let $\boldsymbol{c}_1, \ldots, \boldsymbol{c}_m \in \mathbb{R}^n$ denote the rows of C and $\boldsymbol{d}^1, \ldots, \boldsymbol{d}^m \in \mathbb{R}^n$ the columns of D, the latter considered as variables. Define

$$M\big(\boldsymbol{d}^1, \ldots, \boldsymbol{d}^m\big) = \det(CD) = \det\big(\boldsymbol{c}_i \cdot \boldsymbol{d}^j\big)_{m\times m}.$$

Then M is an alternating m-multilinear form and, by 13.1.3,

$$M\big(\boldsymbol{d}^1, \ldots, \boldsymbol{d}^m\big) = \sum_{(i_1,\ldots,i_m)\in\mathbf{I}_m} M(\boldsymbol{e}^{i_1}, \ldots, \boldsymbol{e}^{i_m})\, dx_{i_1,\ldots,i_m}\big(\boldsymbol{d}^1, \ldots, \boldsymbol{d}^m\big).$$

Since $M\big(\boldsymbol{e}^{i_1}, \ldots, \boldsymbol{e}^{i_m}\big) = \det C^{\mathbf{i}}$ and $dx_{\mathbf{i}}\big(\boldsymbol{d}^1, \ldots, \boldsymbol{d}^m\big) = \det D_{\mathbf{i}}$, the conclusion follows. □

13.1.5 Corollary. *If C and D are $n \times n$ matrices, then*

$$\det(CD) = (\det C)(\det D).$$

13.1.6 Corollary. *If A is an $n \times m$ matrix, then*

$$\det(A^t A) = \sum_{\mathbf{i}\in\mathbf{I}_m} [\det(A_{\mathbf{i}})]^2.$$

Proof. Take $C = A^t$ and $D = A$ in the theorem and note that $C^{\mathbf{i}} = (A_{\mathbf{i}})^t$, so $\big(\det C^{\mathbf{i}}\big)\big(\det D_{\mathbf{i}}\big) = \big(\det {A_{\mathbf{i}}}^t\big)\big(\det A_{\mathbf{i}}\big) = [\det A_{\mathbf{i}}]^2$. □

From 13.1.6, we have

13.1.7 Corollary. *Let A be an $n \times m$ matrix. Then A has rank m iff $\det(A^t A) \neq 0$.*

Definition of a Differential Form

A *differential m-form on a set* $S \subseteq \mathbb{R}^n$ is a function ω that assigns to each $\boldsymbol{x} \in S$ an alternating m-multilinear functional $\omega_{\boldsymbol{x}}$ on $\mathbb{R}^n$. We shall usually drop the qualifier "differential" when referring to forms. The integer m is called the *degree* of the form. A 0-*form* is simply a real-valued function on S.

By 13.1.3, if ω is an m-form, then for each $\mathbf{i} \in \mathbf{I}_m$ there exists a unique function $g_{\mathbf{i}}$ on S such that

$$\omega_{\boldsymbol{x}} = \sum_{\mathbf{i} \in \mathbf{I}_m} g_{\mathbf{i}}(\boldsymbol{x})\, dx_{\mathbf{i}}, \quad \boldsymbol{x} \in S.$$

Conversely, if $f_{\mathbf{j}}$ is a real-valued function on S, then

$$\omega_{\boldsymbol{x}} := \sum_{\mathbf{j} \in \mathbf{J}_m} f_{\mathbf{j}}(\boldsymbol{x})\, dx_{\mathbf{j}}, \quad \boldsymbol{x} \in S, \tag{13.2}$$

defines an m-form on S. If each $f_{\mathbf{j}}$ is of class C^r on S (that is, on an open set containing S), then ω is called a *differential form of class* C^r or simply a C^r *form*, where $r \in \mathbb{Z}^+ \cup \{+\infty\}$.

The Algebra of Differential Forms

For $a \in \mathbb{R}$ and m-forms

$$\omega = \sum_{\mathbf{j} \in \mathbf{J}_m} f_{\mathbf{j}}\, dx_{\mathbf{j}} \quad \text{and} \quad \eta = \sum_{\mathbf{j} \in \mathbf{J}_m} g_{\mathbf{j}}\, dx_{\mathbf{j}}$$

on S, define m-forms $a\omega$ and $\omega + \eta$ on S by

$$a\omega := \sum_{\mathbf{j} \in \mathbf{J}_m} a f_{\mathbf{j}}\, dx_{\mathbf{j}} \quad \text{and} \quad \omega + \eta := \sum_{\mathbf{j} \in \mathbf{J}_m} (f_{\mathbf{j}} + g_{\mathbf{j}})\, dx_{\mathbf{j}}.$$

The collection of m-forms on S is easily seen to be a vector space under these operations.

It is also possibly to multiply forms. For this, the notation

$$dx_{j_1,\ldots,j_m} = dx_{j_1} \wedge \cdots \wedge dx_{j_m} \tag{13.3}$$

will be useful. The right side may be interpreted as a product of differentials, called a *wedge product* and made precise below. Because $dx_{j_1,\ldots,j_m}(\boldsymbol{a}^1, \ldots, \boldsymbol{a}^m)$ is a determinant, interchanging a pair of differentials in (13.3) changes the sign of the product. Furthermore, if there are duplicate indices, then the product is zero. Thus we have the "rules"

$$dx_j \wedge dx_i = -dx_i \wedge dx_j \quad \text{and} \quad dx_i \wedge dx_i = 0. \tag{13.4}$$

Using these rules, one can reduce any m-form to its unique *canonical representation*

$$\omega = \sum_{(i_1,\ldots,i_m) \in \mathbf{I}_m} g_{i_1,\ldots,i_m} dx_{i_1} \wedge \cdots \wedge dx_{i_m}.$$

For example, the 3-form in $\mathbb{R}^4$

$$\omega = f\,dx_2 \wedge dx_1 \wedge dx_2 + g\,dx_3 \wedge dx_2 \wedge dx_1 + h\,dx_2 \wedge dx_4 \wedge dx_1$$

has canonical representation

$$\omega = -g\,dx_1 \wedge dx_2 \wedge dx_3 + h\,dx_1 \wedge dx_2 \wedge dx_4.$$

13.1.8 Definition. Let $1 \le p,\ q \le n$. The *wedge product* or *exterior product* of the forms

$$\omega = \sum_{(j_1,\ldots,j_p)\in \mathbf{J}_p} f_{j_1,\ldots,j_p}\,dx_{j_1} \wedge \cdots \wedge dx_{j_p} \quad\text{and}\quad \eta = \sum_{(k_1,\ldots,k_q)\in \mathbf{J}_q} g_{k_1,\ldots,k_q}\,dx_{k_1} \wedge \cdots \wedge dx_{k_q}$$

is the form

$$\omega \wedge \eta := \sum_{\substack{(j_1,\ldots,j_p)\in \mathbf{J}_p \\ (k_1,\ldots,k_q)\in \mathbf{J}_q}} f_{j_1,\ldots,j_p} g_{k_1,\ldots,k_q}\,dx_{j_1} \wedge \cdots \wedge dx_{j_p} \wedge dx_{k_1} \wedge \cdots \wedge dx_{k_q}. \tag{13.5}$$

If f is a 0-form on S, then the p-form $f\omega = f \wedge \omega$ is defined by

$$f \wedge \omega := \sum_{(j_1,\ldots,j_p)\in \mathbf{J}} f f_{j_1,\ldots,j_p} dx_{j_1} \wedge \cdots \wedge dx_{j_p}. \qquad \Diamond$$

Note that the right side of (13.5) may be obtained by formally multiplying the sums defining ω and η, where the product of forms $dx_{i_1} \wedge \cdots \wedge dx_{i_p}$ and $dx_{j_1} \wedge \cdots \wedge dx_{j_q}$ is defined as

$$dx_{j_1} \wedge \cdots \wedge dx_{j_p} \wedge dx_{k_1} \wedge \cdots \wedge dx_{k_q}.$$

The rules in (13.4) may then be used to obtain the canonical representation of $\omega \wedge \eta$. The resulting form has degree $\le n$, in compliance with our definition.

13.1.9 Example. In $\mathbb{R}^4$,

(a) $$\begin{aligned}&(f_1\,dx_1 + f_2\,dx_2 + f_3\,dx_3 + f_4\,dx_4) \wedge (g_1\,dx_1 + g_2\,dx_2)\\ &\quad= (f_1g_2 - f_2g_1)\,dx_1 \wedge dx_2 - f_3g_1\,dx_1 \wedge dx_3 - f_3g_2\,dx_2 \wedge dx_3\\ &\qquad - f_4g_2\,dx_2 \wedge dx_4 - f_4g_1\,dx_1 \wedge dx_4.\end{aligned}$$

(b) $$\begin{aligned}&(f_1\,dx_1 + f_2\,dx_2 + f_3\,dx_3 + f_4\,dx_4) \wedge (h_1\,dx_1 \wedge dx_3 + h_2\,dx_2 \wedge dx_4)\\ &\quad= f_1h_2\,dx_1 \wedge dx_2 \wedge dx_4 - f_2h_1\,dx_1 \wedge dx_2 \wedge dx_3\\ &\qquad - f_3h_2\,dx_2 \wedge dx_3 \wedge dx_4 + f_4h_1\,dx_1 \wedge dx_3 \wedge dx_4.\end{aligned} \qquad \Diamond$$

It must still be shown that the definition of $\omega \wedge \eta$ in (13.5) is independent of the particular representations of ω and η. To see this, apply the rules in (13.4), first on the indices j_p and then on the indices k_q, to reduce the right side of (13.5) to

$$\sum_{\substack{(i_1,\ldots,i_p)\in \mathbf{I}_p \\ (i'_1,\ldots,i'_q)\in \mathbf{I}_q}} \tilde{f}_{i_1,\ldots,i_p} \tilde{g}_{i'_1,\ldots,i'_p}\,dx_{i_1} \wedge \cdots \wedge dx_{i_p} \wedge dx_{i'_1} \wedge \cdots \wedge dx_{i'_q}$$

where

$$\sum_{(i_1,\ldots,i_p)\in\mathbf{I}_p} \tilde{f}_{i_1,\ldots,i_p}\, dx_{i_1}\wedge\cdots\wedge dx_{i_p} \quad\text{and}\quad \sum_{(i'_1,\ldots,i'_q)\in\mathbf{I}_q} \tilde{g}_{i'_1,\ldots,i'_q}\, dx_{i'_1}\wedge\cdots\wedge dx_{i'_q}$$

are the canonical representations of ω and η. Since the latter are unique, every version of $\omega\wedge\eta$ may be reduced to the same form, hence $\omega\wedge\eta$ is well-defined.

13.1.10 Proposition. *Let ω be a p-form, η a q-form, and ν an r-form, where $1\le p,\ q,\ r\le n$. Then*

(a) *$\omega\wedge\eta$ is linear in each variable separately;*

(b) *$(\omega\wedge\eta)\wedge\nu=\omega\wedge(\eta\wedge\nu)$;*

(c) *$\eta\wedge\omega=(-1)^{pq}\omega\wedge\eta$.*

Proof. The straightforward proofs of (a) and (b) are left to the reader. For the proof of (c), let ω and η be as in 13.1.8. Then

$$\begin{aligned}\eta\wedge\omega &= \sum_{\substack{(k_1,\ldots,k_q)\in\mathbf{J}_q\\(j_1,\ldots,j_p)\in\mathbf{J}_p}} g_{k_1,\ldots,k_q} f_{j_1,\ldots,j_p}\, dx_{k_1}\wedge\cdots\wedge dx_{k_q}\wedge dx_{j_1}\wedge\cdots\wedge dx_{j_p}\\ &= \sum_{\substack{(k_1,\ldots,k_q)\in\mathbf{J}_q\\(j_1,\ldots,j_p)\in\mathbf{J}_p}} g_{k_1,\ldots,k_q} f_{j_1,\ldots,j_p}\,(-1)^{pq} dx_{j_1}\wedge\cdots\wedge dx_{j_p}\wedge dx_{k_1}\wedge\cdots\wedge dx_{k_q}\\ &= (-1)^{pq}\omega\wedge\eta,\end{aligned}$$

the last equality because pq adjacent interchanges are required. □

13.1.11 Proposition. *Let $\boldsymbol{a}^j=\sum_{i=1}^n a_i^j \boldsymbol{e}^j$, $j=1,\ldots,n$. Then*

$$\left(\sum_{i=1}^n a_i^1 dx_i\right)\wedge\cdots\wedge\left(\sum_{i=1}^n a_i^n dx_i\right) = \det[\boldsymbol{a}^1\cdots\boldsymbol{a}^n]dx_1\wedge\cdots\wedge dx_n.$$

Proof. By properties of the wedge product, the left side of the equation is

$$\sum_{i_1=1}^n\cdots\sum_{i_n=1}^n a_{i_1}^1\cdots a_{i_n}^n\, dx_{i_1}\wedge\cdots\wedge dx_{i_n} = \sum_{i_1,\ldots,i_n \text{ distinct}} a_{i_1}^1\cdots a_{i_n}^n\, dx_{i_1}\wedge\cdots\wedge dx_{i_n}.$$

If $\sigma=(i_1,\ldots,i_n)$, then $dx_{i_1}\wedge\cdots\wedge dx_{i_n}=(-1)^\sigma dx_1\wedge\cdots\wedge dx_n$, and the assertion follows from the definition of determinant. □

The proposition provides an alternate method for evaluating determinants.

13.1.12 Example. Let $A = \begin{bmatrix} 1 & 3 & 5 \\ 2 & 4 & 6 \\ 3 & -2 & 1 \end{bmatrix}$. By wedge product rules applied to the forms constructed from the columns,

$$\begin{aligned}
&(1\,dx_1 + 2\,dx_2 + 3\,dx_3) \wedge (3\,dx_1 + 4\,dx_2 - 2\,dx_3) \wedge (5\,dx_1 + 6\,dx_2 + 1\,dx_3) \\
&\quad = (-2\,dx_{1,2} - 11\,dx_{1,3} - 16\,dx_{2,3}) \wedge (5\,dx_1 + 6\,dx_2 + dx_3) \\
&\quad = (-2\,dx_{1,2,3} + 66\,dx_{1,2,3} - 80\,dx_{1,2,3},) \\
&\quad = -16\,dx_{1,2,3},
\end{aligned}$$

hence $\det(A) = -16$. ◊

The Differential of a Form

13.1.13 Definition. The *differential of a* 0*-form* f of class C^1 on $S \subseteq \mathbb{R}^n$ is its differential as a C^1 function, namely, the 1-form

$$df = \sum_{j=1}^{n} (\partial_j f) dx_j.$$

The *differential of an* m*-form*

$$\omega = \sum_{(j_1,\ldots,j_m) \in \mathbf{J}} f_{j_1,\ldots,j_m}\, dx_{j_1} \wedge \cdots \wedge dx_{j_m}$$

of class C^1 on S is the $(m+1)$-form $d\omega$ defined by

$$\begin{aligned}
d\omega &= \sum_{(j_1,\ldots,j_m) \in \mathbf{J}_m} (df_{j_1,\ldots,j_m}) \wedge dx_{j_1} \wedge \cdots \wedge dx_{j_m} && (13.6) \\
&= \sum_{(j_1,\ldots,j_m) \in \mathbf{J}} \sum_{j=1}^{n} (\partial_j f_{j_1,\ldots,j_m})\, dx_j \wedge dx_{j_1} \wedge \cdots \wedge dx_{j_m}.
\end{aligned}$$

◊

Note that if $m = n$, then $d\omega = 0$, since in the last expression every dx_j is a dx_{j_i} for some i.

As in the case of wedge products, it must be verified that the definition of $d\omega$ does not depend on the particular representation of ω. For this we use the rules in (13.4) to express ω canonically as

$$\omega = \sum_{(i_1,\ldots,i_m) \in \mathbf{I}_m} g_{i_1,\ldots,i_m}\, dx_{i_1} \wedge \cdots \wedge dx_{i_m}.$$

Here, each $g_{i_1,\ldots,i_m}$ is a linear combination the functions $f_{j_1,\ldots,j_m}$ produced by combining these functions during the reduction process. Applying the *same* sequence of operations to the sum on the right in (13.6) results in

$$\sum_{(i_1,\ldots,i_m) \in \mathbf{I}_m} \eta_{i_1,\ldots,i_m} \wedge dx_{i_1} \wedge dx_{i_2} \wedge \cdots \wedge dx_{i_m},$$

where $\eta_{i_1,\ldots,i_m}$ is precisely the same linear combination of the forms $df_{j_1,\ldots,j_m}$. Since the differential is linear on 0-forms, $\eta_{i_1,\ldots,i_m} = dg_{i_1,\ldots,i_m}$. Therefore, all versions of $d\omega$ may be reduced to the same form and hence are equal.

For the next example, we introduce the following notation and terminology from classical vector analysis.

13.1.14 Definition. The *curl* of a C^1 vector field $\vec{F} = (f_1, f_2, f_3)$ on an open subset of $\mathbb{R}^3$ is the vector

$$\operatorname{curl} \vec{F} = (\partial_2 f_3 - \partial_3 f_2)\, \mathbf{e}^1 + (\partial_3 f_1 - \partial_1 f_3)\, \mathbf{e}^2 + (\partial_1 f_2 - \partial_2 f_1)\, \mathbf{e}^3.$$

The *divergence* of a C^1 vector field $\vec{F} = (f_1, \ldots, f_n)$ on an open subset of $\mathbb{R}^n$ is defined by

$$\operatorname{div} \vec{F} = \sum_{i=1}^{n} \partial_i f_i.$$

If $\omega = \sum_{j=1}^{n} f_j\, dx_j$ we define $\operatorname{div} \omega = \operatorname{div} \vec{F}$. ◊

13.1.15 Example. In $\mathbb{R}^3$,

(a)
$$\begin{aligned} &d\big(f_1\, dx_1 + f_2\, dx_2 + f_3\, dx_3\big) \\ &= (\partial_1 f_1\, dx_1 + \partial_2 f_1\, dx_2 + \partial_3 f_1\, dx_3) \wedge dx_1 \\ &\qquad + (\partial_1 f_2\, dx_1 + \partial_2 f_2\, dx_2 + \partial_3 f_2\, dx_3) \wedge dx_2 \\ &\qquad + (\partial_1 f_3\, dx_1 + \partial_2 f_3\, dx_2 + \partial_3 f_3\, dx_3) \wedge dx_3 \\ &= (\partial_2 f_3 - \partial_3 f_2)\ dx_{2,3} + (\partial_3 f_1 - \partial_1 f_3)\ dx_{3,1} + (\partial_1 f_2 - \partial_2 f_1)\ dx_{1,2} \\ &= \mathbf{e}^1 \cdot \operatorname{curl} \vec{F}\, dx_{2,3} + \mathbf{e}^2 \cdot \operatorname{curl} \vec{F}\, dx_{3,1} + \mathbf{e}^3 \cdot \operatorname{curl} \vec{F}\, dx_{1,2}. \end{aligned}$$

(b)
$$\begin{aligned} &d\big(f_3\, dx_1 \wedge dx_2 + f_1\, dx_2 \wedge dx_3 + f_2\, dx_3 \wedge dx_1\big) \\ &= (\partial_1 f_3\, dx_1 + \partial_2 f_3\, dx_2 + \partial_3 f_3\, dx_3) \wedge dx_1 \wedge dx_2 \\ &\qquad + (\partial_1 f_1\, dx_1 + \partial_2 f_1\, dx_2 + \partial_3 f_1\ \ dx_3) \wedge dx_2 \wedge dx_3 \\ &\qquad + (\partial_1 f_2\, dx_1 + \partial_2 f_2\, dx_2 + \partial_3 f_2 dx_3) \wedge dx_3 \wedge dx_1 \\ &= (\partial_1 f_1 + \partial_2 f_2 + \partial_3 f_3)\ dx_1 \wedge dx_2 \wedge dx_3 \\ &= \operatorname{div} \vec{F}\, dx_1 \wedge dx_2 \wedge dx_3. \end{aligned}$$

◊

13.1.16 Theorem. *Let f be a 0-form, let ω and η be p-forms, and let ν be a q form, all of class C^1 on $S \subseteq \mathbb{R}^n$. Then*

(a) $d(a\omega + b\eta) = a\, d\omega + b\, d\eta,\ a,\ b \in \mathbb{R}$;

(b) $d^2\omega := d(d\omega) = 0$;

(c) $d(\omega \wedge \nu) = (d\omega) \wedge \nu + (-1)^p \omega \wedge (d\nu)$;

(d) $d(f\nu) = (df) \wedge \nu + f d\nu$.

Proof. Part (a) is clear from the definition of addition and scalar multiplication of m-forms and the linearity of the differential operator on 0-forms.

For (b), it suffices by linearity to prove that

$$d\big[(df)\,dx_{j_1}\wedge dx_{j_2}\wedge\cdots\wedge dx_{j_p}\big]=0.$$

The left side of this equation is

$$d\left[\sum_{k=1}^{n}(\partial_k f)dx_k\wedge dx_{j_1}\wedge\cdots\wedge dx_{j_p}\right]$$
$$=\left[\Big(\sum_{j=1}^{n}\sum_{k=1}^{n}\partial_j\partial_k f\Big)dx_j\wedge dx_k\right]\wedge dx_{j_1}\wedge\cdots\wedge dx_{j_p}.$$

Since $dx_k\wedge dx_j=-dx_j\wedge dx_k$ and $\partial_j\partial_k f=\partial_k\partial_j f$, the terms in the square brackets on the right cancel pairwise, producing zero, as required.

To prove (c), let

$$\omega=\sum_{\mathbf{j}\in\mathbf{J}_p} f_{\mathbf{j}}\,dx_{\mathbf{j}}\quad\text{and}\quad \nu=\sum_{\mathbf{k}\in\mathbf{J}_q} g_{\mathbf{k}}\,dx_{\mathbf{k}}.$$

By the product rule for differentials of 0-forms,

$$\begin{aligned}
d(\omega\wedge\nu)&=\sum_{\mathbf{j}\in\mathbf{J}_p,\,\mathbf{k}\in\mathbf{J}_q} d(f_{\mathbf{j}}g_{\mathbf{k}})\wedge dx_{\mathbf{j}}\wedge dx_{\mathbf{k}}\\
&=\sum_{\mathbf{j}\in\mathbf{J}_p,\mathbf{k}\in\mathbf{J}_q} g_{\mathbf{k}}(df_{\mathbf{j}})\wedge dx_{\mathbf{j}}\wedge dx_{\mathbf{k}}+\sum_{\mathbf{j}\in\mathbf{J}_p,\mathbf{k}\in\mathbf{J}_q} f_{\mathbf{j}}(dg_{\mathbf{k}})\wedge dx_{\mathbf{j}}\wedge dx_{\mathbf{k}}\\
&=(d\omega)\wedge\nu+(-1)^{-p}\omega\wedge(d\nu),
\end{aligned}$$

the last equality because p adjacent interchanges are needed to place the form $dg_{\mathbf{k}}$ in the second sum to the immediate left of $dx_{\mathbf{k}}$.

Part (d) follows from (c) with $p=0$. □

The Pullback of a Form

> *Throughout this subsection, $U\subseteq\mathbb{R}^m$ and $W\subseteq\mathbb{R}^n$ are open and $\varphi:U\to W$ is a C^1 map.*

13.1.17 Definition. The *pullback by* φ of a C^1 function (0-form) f on W is the 0-form $\varphi^*(f)$ on U defined by

$$\varphi^*(f)(\boldsymbol{u}):=f(\varphi(\boldsymbol{u})),\quad \boldsymbol{u}\in U.$$

The *pullback by* φ of the 1-form dx_j on W is the 1-form $\varphi^*(dx_j)$ on U defined by

$$\varphi^*(dx_j):=\sum_{i=1}^{m}\frac{\partial\varphi_j}{\partial u_i}\,du_i=d\varphi_j,\quad j=1,\ldots,n.$$

The *pullback by* φ of the C^1 p-form

$$\omega = \sum_{(j_1,\ldots,j_p)\in \mathbf{J}_p} f_{j_1,\ldots,j_p}\, dx_{j_1} \wedge \cdots \wedge dx_{j_p}$$

on W is the C^1 p-form $\varphi^*\omega$ on U defined by

$$\varphi^*\omega := \sum_{(j_1,\ldots,j_p)\in \mathbf{J}_p} \varphi^*(f_{j_1,\ldots,j_p})\varphi^*(dx_{j_1}) \wedge \cdots \wedge \varphi^*(dx_{j_p}). \qquad \Diamond$$

Arguments similar to those used earlier show that the definition of $\varphi^*\omega$ is independent of the representation of ω.

13.1.18 Example. Let $\varphi = (\varphi_1, \varphi_2, \varphi_3) : \mathbb{R}^2 \to \mathbb{R}^3$ be C^1. Then

(a) $$\begin{aligned}\varphi^*(f\,dx_1 \wedge dx_2) &= \varphi^*(f)\varphi^*(dx_1) \wedge \varphi^*(dx_2)\\ &= (f\circ\varphi)\left(\frac{\partial\varphi_1}{\partial u_1}du_1 + \frac{\partial\varphi_1}{\partial u_2}du_2\right) \wedge \left(\frac{\partial\varphi_2}{\partial u_1}du_1 + \frac{\partial\varphi_2}{\partial u_2}du_2\right)\\ &= (f\circ\varphi)\left(\frac{\partial\varphi_1}{\partial u_1}\frac{\partial\varphi_2}{\partial u_2} - \frac{\partial\varphi_2}{\partial u_1}\frac{\partial\varphi_1}{\partial u_2}\right) du_1 \wedge du_2.\end{aligned}$$

(b) $$\begin{aligned}&\varphi^*(f_1\,dx_1 + f_2\,dx_2 + f_3\,dx_3)\\ &= \varphi^*(f_1)\varphi^*(dx_1) + \varphi^*(f_2)\varphi^*(dx_2) + \varphi^*(f_3)\varphi^*(dx_3)\\ &= (f_1\circ\varphi)\left(\frac{\partial\varphi_1}{\partial u_1}du_1 + \frac{\partial\varphi_1}{\partial u_2}du_2\right) + (f_2\circ\varphi)\left(\frac{\partial\varphi_2}{\partial u_1}du_1 + \frac{\partial\varphi_2}{\partial u_2}du_2\right)\\ &\qquad + (f_3\circ\varphi)\left(\frac{\partial\varphi_3}{\partial u_1}du_1 + \frac{\partial\varphi_3}{\partial u_2}du_2\right)\\ &= \left[(f_1\circ\varphi)\frac{\partial\varphi_1}{\partial u_1} + (f_2\circ\varphi)\frac{\partial\varphi_2}{\partial u_1} + (f_3\circ\varphi)\frac{\partial\varphi_3}{\partial u_1}\right] du_1\\ &\qquad + \left[(f_1\circ\varphi)\frac{\partial\varphi_1}{\partial u_2} + (f_2\circ\varphi)\frac{\partial\varphi_2}{\partial u_2} + (f_3\circ\varphi)\frac{\partial\varphi_3}{\partial u_2}\right] du_2. \qquad \Diamond\end{aligned}$$

13.1.19 Theorem. *If* ω *and* η *are* C^1 *p-forms and* ν *is a* C^1 *q-form, then*

(a) $\varphi^*(a\omega + b\eta) = a\varphi^*(\omega) + b\varphi^*(\eta)$, $a, b \in \mathbb{R}$;

(b) $\varphi^*(\omega \wedge \nu) = \varphi^*(\omega) \wedge \varphi^*(\nu)$;

(c) $\varphi^*(d\omega) = d\varphi^*(\omega)$;

(d) $(\varphi^*\omega)_{\boldsymbol{u}}(\boldsymbol{a}^1, \ldots, \boldsymbol{a}^p) = \omega_{\varphi(\boldsymbol{u})}(d\varphi_{\boldsymbol{u}}(\boldsymbol{a}^1), \ldots, d\varphi_{\boldsymbol{u}}(\boldsymbol{a}^p))$.

Proof. Part (a) follows directly from the definition of pullback. Part (b) is easily established for $\omega = f\,dx_{i_1} \wedge \cdots \wedge dx_{i_p}$ and $\nu = g\,dx_{j_1} \wedge \cdots \wedge dx_{j_q}$; bilinearity of the wedge product and linearity of φ^* then imply that (b) holds generally.

For (c) it suffices, by linearity of the differential and pullback, to verify that

$$\varphi^*\big(d(f\,dx_{j_1} \wedge \cdots \wedge dx_{j_p})\big) = d\varphi^*(f\,dx_{j_1} \wedge \cdots \wedge dx_{j_p}),$$

that is,

$$\sum_{j=1}^{n}[(\partial_j f)\circ\varphi]\,\varphi^*(dx_j)\wedge\varphi^*(dx_{j_1})\wedge\cdots\wedge\varphi^*(dx_{j_p})$$
$$=\left(\sum_{i=1}^{m}\partial_i(f\circ\varphi)du_i\right)\varphi^*(dx_{j_1})\wedge\cdots\wedge\varphi^*(dx_{j_p}) \quad (13.7)$$

By the chain rule,

$$\partial_i(f\circ\varphi)=\sum_{j=1}^{n}[(\partial_j f)\circ\varphi]\frac{\partial\varphi_j}{\partial u_i},$$

hence the right side of (13.7) is

$$\sum_{j=1}^{n}[(\partial_j f)\circ\varphi]\left(\sum_{i=1}^{m}\frac{\partial\varphi_j}{\partial u_i}\,du_i\right)\wedge\varphi^*(dx_{j_1})\wedge\cdots\wedge\varphi^*(dx_{j_p}).$$

Recalling the definition of $\varphi^*(dx_j)$, we see that the last expression is precisely the left side of (13.7).

To prove (d), let ω have canonical representation

$$\omega=\sum_{(i_1,\ldots,i_p)\in\mathbf{I}_p} f_{i_1,\ldots,i_p}\,dx_{i_1}\wedge\cdots\wedge dx_{i_p}.$$

By 13.1.2, it suffices to show that

$$(\varphi^*\omega)_{\boldsymbol{u}}(\mathbf{e}^{\ell_1},\ldots,\mathbf{e}^{\ell_p})=\omega_{\varphi(\boldsymbol{u})}(d\varphi_{\boldsymbol{u}}(\mathbf{e}^{\ell_1}),\ldots,d\varphi_{\boldsymbol{u}}(\mathbf{e}^{\ell_p}))$$

for any $(\ell_1,\ldots,\ell_p)\in\mathbf{I}_p$. The left side of this equation is

$$\sum_{\mathbf{i}\in\mathbf{I}_p}\varphi^*(f_{\mathbf{i}})(\boldsymbol{u})(\varphi^*(dx_{i_1})\wedge\cdots\wedge\varphi^*(dx_{i_p}))(\mathbf{e}^{\ell_1},\ldots,\mathbf{e}^{\ell_p})$$

and the right side is

$$\sum_{\mathbf{i}\in\mathbf{I}_p} f_{\mathbf{i}}(\varphi(\boldsymbol{u}))\big(d_{x_{i_1}}\wedge\cdots\wedge d_{x_{i_p}}\big)\big(d\varphi_{\boldsymbol{u}}(\mathbf{e}^{\ell_1}),\ldots,d\varphi_{\boldsymbol{u}}(\mathbf{e}^{\ell_p})\big)$$

Hence it suffices to prove that

$$\big(\varphi^*(dx_{i_1})\wedge\cdots\wedge\varphi^*(dx_{i_p})\big)(\mathbf{e}^{\ell_1},\ldots,\mathbf{e}^{\ell_p})$$
$$=\big(d_{x_{i_1}}\wedge\cdots\wedge d_{x_{i_p}}\big)\big(d\varphi_{\boldsymbol{u}}(\mathbf{e}^{\ell_1}),\ldots,d\varphi_{\boldsymbol{u}}(\mathbf{e}^{\ell_p})\big) \qquad (13.8)$$

By multilinearity,

$$\varphi^*(dx_{i_1})\wedge\cdots\wedge\varphi^*(dx_{i_p})=\sum_{(j_1,\ldots,j_p)\in\mathbf{J}_p}\frac{\partial\varphi_{i_1}}{\partial u_{j_1}}\cdots\frac{\partial\varphi_{i_p}}{\partial u_{j_p}}\,du_{j_1}\wedge\cdots\wedge du_{j_p}.$$

Now, $du_{j_1} \wedge \cdots \wedge du_{j_p}(\mathbf{e}^{\ell_1}, \ldots, \mathbf{e}^{\ell_p}) \neq 0$ only if the p-tuple $(j_1, \ldots, j_p)$ is a permutation of $(\ell_1, \ldots, \ell_p)$. For each such p-tuple, define a permutation σ of $(1, \ldots, p)$ such that $\ell_k = j_{\sigma(k)}$. Then

$$du_{j_1} \wedge \cdots \wedge du_{j_p}(\mathbf{e}^{\ell_1}, \ldots, \mathbf{e}^{\ell_p}) = (-1)^\sigma\, du_{\ell_1} \wedge \cdots \wedge du_{\ell_p}(\mathbf{e}^{\ell_1}, \ldots, \mathbf{e}^{\ell_p}) = (-1)^\sigma$$

and

$$\frac{\partial \varphi_{i_1}}{\partial u_{j_1}} \cdots \frac{\partial \varphi_{i_p}}{\partial u_{j_p}} = \frac{\partial \varphi_{i_1}}{\partial u_{\ell_{\tau(1)}}} \cdots \frac{\partial \varphi_{i_p}}{\partial u_{\ell_{\tau(p)}}} = \frac{\partial \varphi_{i_{\sigma(1)}}}{\partial u_{\ell_1}} \cdots \frac{\partial \varphi_{i_{\sigma(p)}}}{\partial u_{\ell_p}},$$

where $\tau = \sigma^{-1}$. Thus the left side of (13.8) is

$$\left[\varphi^*(dx_{i_1}) \wedge \cdots \wedge \varphi^*(dx_{i_p})\right](\mathbf{e}^{\ell_1}, \ldots, \mathbf{e}^{\ell_p}) = \sum_\sigma (-1)^\sigma \frac{\partial \varphi_{i_{\sigma(1)}}}{\partial u_{\ell_1}} \cdots \frac{\partial \varphi_{i_{\sigma(p)}}}{\partial u_{\ell_p}}, \quad (13.9)$$

where the sum is taken over all permutations σ of $(1, \ldots, p)$.

On the other hand, since

$$d\varphi_{\boldsymbol{u}}(\mathbf{e}^{\ell_j}) = \partial_{\ell_j} \varphi(\boldsymbol{u}) = \sum_{i=1}^{p} \frac{\partial \varphi_i(\boldsymbol{u})}{\partial u_{\ell_j}} \mathbf{e}^i,$$

the right side of (13.8) is

$$\begin{aligned} &d_{x_{i_1}} \wedge \cdots \wedge d_{x_{i_p}} \left(\sum_{j=1}^{p} \frac{\partial \varphi_j}{\partial u_{\ell_1}} \mathbf{e}^j, \ldots, \sum_{j=1}^{p} \frac{\partial \varphi_j}{\partial u_{\ell_p}} \mathbf{e}^j \right) \\ &\qquad = \sum_{(j_1, \ldots, j_p) \in \mathbf{J}_p} \frac{\partial \varphi_{j_1}}{\partial u_{\ell_1}} \cdots \frac{\partial \varphi_{j_p}}{\partial u_{\ell_p}} d_{x_{i_1}} \wedge \cdots \wedge d_{x_{i_p}}(\mathbf{e}^{j_1}, \ldots, \mathbf{e}^{j_p}). \quad (13.10) \end{aligned}$$

As above, $dx_{i_1} \wedge \cdots \wedge dx_{i_p}(\mathbf{e}^{j_1}, \ldots, \mathbf{e}^{j_p}) \neq 0$ only if the p-tuple $(j_1, \ldots, j_p)$ is a permutation of $(i_1, \ldots, i_p)$. For each such p-tuple, define a permutation σ of $(1, \ldots, p)$ such that $j_k = i_{\sigma(k)}$. Then

$$d_{x_{i_1}} \wedge \cdots \wedge d_{x_{i_p}}(\mathbf{e}^{j_1}, \ldots, \mathbf{e}^{j_p}) = d_{x_{i_1}} \wedge \cdots \wedge d_{x_{i_p}}\left(\mathbf{e}^{i_{\sigma(1)}}, \ldots, \mathbf{e}^{i_{\sigma(k)}}\right) = (-1)^\sigma$$

and

$$\frac{\partial \varphi_{j_1}}{\partial u_{\ell_1}} \cdots \frac{\partial \varphi_{j_p}}{\partial u_{\ell_p}} = \frac{\partial \varphi_{i_{\sigma(1)}}}{\partial u_{\ell_1}} \cdots \frac{\partial \varphi_{i_{\sigma(p)}}}{\partial u_{\ell_p}}$$

so (13.10) reduces to

$$\sum_\sigma (-1)^\sigma \frac{\partial \varphi_{i_{\sigma(1)}}}{\partial u_{\ell_1}} \cdots \frac{\partial \varphi_{i_{\sigma(p)}}}{\partial u_{\ell_p}},$$

where the sum is taken over all permutations of $(1, \ldots, p)$. As this is precisely (13.9), the proof is complete. □

Exercises

1. Let $T_j \in \mathcal{L}(\mathbb{R}^n, \mathbb{R})$, $j = 1, \ldots, m$. Which of the following functions is multilinear on $\mathbb{R}^n$?

   (a) $M(\boldsymbol{x}^1, \ldots, \boldsymbol{x}^m) := \sum_{i=1}^m T_i(\boldsymbol{x}^i)$ (b) $M(\boldsymbol{x}^1, \ldots, \boldsymbol{x}^m) := \prod_{i=1}^m T_i(\boldsymbol{x}^i)$.

2. For fixed $\boldsymbol{c} = (c_1, c_2)$, $\boldsymbol{d} = (d_1, d_2) \in \mathbb{R}^2$ define

$$M(\boldsymbol{x}, \boldsymbol{y}) := (\boldsymbol{c} \cdot \boldsymbol{x})(\boldsymbol{d} \cdot \boldsymbol{y}) - (\boldsymbol{c} \cdot \boldsymbol{y})(\boldsymbol{d} \cdot \boldsymbol{x}), \quad \boldsymbol{x}, \boldsymbol{y} \in \mathbb{R}^2.$$

   (a) Show that M is an alternating multilinear functional on $\mathbb{R}^2$.

   (b) Express M in terms of differentials, as in 13.1.3

3. Let $M(\boldsymbol{a}^1, \ldots, \boldsymbol{a}^m)$ be a multilinear functional on $\mathbb{R}^n$ with the property that $M(\boldsymbol{a}^1, \ldots, \boldsymbol{a}^m) = 0$ whenever two of the vectors $\boldsymbol{a}^j$ are equal. Prove that M is alternating.

4. Let M be an alternating m-multilinear functional on $\mathbb{R}^n$. Show that if the vectors $\boldsymbol{a}^1, \ldots, \boldsymbol{a}_m$ are linearly dependent, then $M(\boldsymbol{a}^1, \ldots, \boldsymbol{a}^m) = 0$.

5. Let $M(\boldsymbol{a}^1, \ldots, \boldsymbol{a}^m)$ be an m-multilinear functional on $\mathbb{R}^n$. Define

$$\operatorname{Alt}(M)(\boldsymbol{a}^1, \ldots, \boldsymbol{a}^m) = \frac{1}{m!} \sum_{\sigma} (-1)^\sigma M\big(\boldsymbol{a}^{\sigma(1)}, \ldots, \boldsymbol{a}^{\sigma(m)}\big),$$

   where the sum is taken over all permutations σ of $(1, \ldots, m)$. Show that $\operatorname{Alt}(M)$ is an alternating m-multilinear functional on $\mathbb{R}^n$ and that $\operatorname{Alt}(M) = M$ iff M is alternating.

6. Prove that the vector space of m-forms on S has dimension $\binom{n}{m}$.

7. Find the canonical representation of the following forms in $\mathbb{R}^3$:

   (a) $(f_1\, dx_1 + f_2\, dx_2 + f_3\, dx_3) \wedge (g_1\, dx_1 + g_2\, dx_2 + g_3\, dx_3)$

   (b) $(f_1\, dx_1 + f_2\, dx_2 + f_3\, dx_3) \wedge (g_1\, dx_1 + g_2\, dx_2 + g_3\, dx_3)$
   $\wedge (h_1\, dx_1 + h_2\, dx_2 + h_3\, dx_3)$.

8. Find the canonical representation of the following forms in $\mathbb{R}^5$:

   (a) $(-dx_1 + dx_2 + dx_3) \wedge (dx_1 - 2dx_2 + 3dx_3)$

   (b) $(dx_1 + dx_2) \wedge (dx_1 - dx_3) \wedge (dx_2 + 2dx_3)$

   (c) $dx_1 \wedge (dx_1 \wedge dx_3 + 3dx_5 \wedge dx_4)$

   (d) $\big(dx_1 \wedge dx_2 + dx_1 \wedge dx_3\big) \wedge \big(dx_4 \wedge dx_3 + dx_2 \wedge dx_5\big)$
   $\wedge \big(dx_3 \wedge dx_1 + dx_4 \wedge dx_1\big)$.

9. Find the canonical representation of the following forms in $\mathbb{R}^n$:

   (a) $\big(dx_2 \wedge dx_4 \wedge \cdots \wedge dx_{2k}\big) \wedge \big(dx_1 \wedge dx_3 \wedge \cdots \wedge dx_{2k-1}\big)$, $2k \le n$

   (b) $\big(dx_1 \wedge dx_5 \wedge \cdots \wedge dx_{4k-3}\big) \wedge \big(dx_3 \wedge dx_7 \wedge \cdots \wedge dx_{4k-1}\big)$
   $\wedge \big(dx_2 \wedge dx_6 \wedge \cdots \wedge dx_{4k-2}\big) \wedge \big(dx_4 \wedge dx_8 \wedge \cdots \wedge dx_{4k}\big)$, $4k \le n$.

10. Show that if ω is an m-form and m is odd, then $\omega \wedge \omega = 0$. Find an example of a 2-form ω in $\mathbb{R}^4$ such that $\omega \wedge \omega \neq 0$.

11. Use the method of 13.1.12 to verify the determinants

$$\text{(a)} \begin{vmatrix} 1 & -1 & 1 \\ -1 & 1 & 1 \\ 1 & 1 & -1 \end{vmatrix} = \text{-4.} \quad \text{(b)} \begin{vmatrix} -1 & 0 & 2 \\ 2 & -1 & 0 \\ 0 & 2 & 1 \end{vmatrix} = 9. \quad \text{(c)} \begin{vmatrix} 3 & 1 & 2 \\ 0 & -1 & 1 \\ -1 & 1 & 0 \end{vmatrix} = \text{-6.}$$

12. Show directly that in $\mathbb{R}^n$, $d\big[f(dx_1 \wedge \cdots \wedge dx_n)\big] = 0$.

13. Let $f : \mathbb{R} \to \mathbb{R}$ be C^1 and define $g_j(\boldsymbol{x}) = f(x_j)$. Find the canonical representation of

(a) $d\sum_{j=1}^{n} g_j \, dx_j$ (b) $d\sum_{j=1}^{n} g_{n-j+1} \, dx_j$.

14. Find $d(f\,dg)$, where f is C^1 and g is C^2 on W.

15. A form η on W is *exact* if $\eta = d\omega$ for some form ω on W. Prove that if η is exact and $d\nu = 0$, then $\eta \wedge \nu$ is exact.

16. Let f and $\omega := \sum_{i=1}^n f_i \, dx_i$ be C^1 on an open set $W \subseteq \mathbb{R}^n$. Show that if $d(f\omega) = 0$, then $f\omega \wedge d\omega = (df) \wedge \omega \wedge \omega$.

17. Let $U \subseteq \mathbb{R}^k$, $V \subseteq \mathbb{R}^\ell$, and $W \subseteq \mathbb{R}^n$ be open and let $\varphi : U \to V$ and $\psi : V \to W$ be C^1. If ω is an m-form on W, prove that $(\psi \circ \varphi)^*\omega = \varphi^*(\psi^*\omega)$. *Hint.* Use 13.1.19(d).

18. Let $U, W \subseteq \mathbb{R}^n$ be open and let $\varphi : U \to W$ and $f : W \to \mathbb{R}^n$ be C^1. Show that $\varphi^*(dx_1 \wedge \cdots \wedge dx_n) = \det(\varphi')\, du_1 \wedge \cdots \wedge du_n$.

19. Let $F = (f_1, f_2, f_3)$ be C^1 on $\mathbb{R}^3$ and homogeneous of degree $k \in \mathbb{N}$. (See Exercise 9.3.15.) Let $\omega = f_1 \, dx_1 + f_2 \, dx_2 + f_3 \, dx_3$. Show that if $d\omega = 0$, then $\omega = df$ where $f(\boldsymbol{x}) = (k+1)^{-1} F(\boldsymbol{x}) \cdot \boldsymbol{x}$.

13.2 Integrals on Parameterized Surfaces

Recall that the length of a parameterized curve C in $\mathbb{R}^n$ is, by definition, a limit of lengths of inscribed polygonal lines. The proof of 12.2.4 shows that if the curve C is C^1, then its length may be also be approximated by tangent line segments. This idea may be extended to higher dimensions, using tangent parallelepipeds to approximate surface area. This leads ultimately to the definition of the integral of a function or a form on a surface.

Area of a Parallelepiped

13.2.1 Definition. *The parallelepiped spanned by vectors* $\boldsymbol{a}^1, \ldots, \boldsymbol{a}^m \in \mathbb{R}^n$ is the set

$$P = P(\boldsymbol{a}^1, \ldots, \boldsymbol{a}^m) := \left\{ \sum_{i=1}^m t_i \boldsymbol{a}^i \, : \, 0 \le t_i \le 1 \right\}.$$

The *volume* vol(P) *of* P is its n-dimensional Lebesgue measure. ◊

For $m = n$, there is a simple formula for the volume:

13.2.2 Lemma. $\operatorname{vol}(P) = \left|\det\begin{bmatrix}\boldsymbol{a}^1 & \ldots & \boldsymbol{a}^n\end{bmatrix}\right|$.

Proof. Denote by $T \in \mathcal{L}(\mathbb{R}^n, \mathbb{R}^n)$ the linear mapping with matrix $A := \begin{bmatrix}\boldsymbol{a}^1 \cdots \boldsymbol{a}^n\end{bmatrix}$. Since $T(\boldsymbol{e}^j) = \boldsymbol{a}^j$, a typical member of $P := P(\boldsymbol{a}^1, \ldots, \boldsymbol{a}^n)$ may be expressed as

$$\sum_{i=1}^n t_i \boldsymbol{a}^i = T\left(\sum_{i=1}^n t_i \boldsymbol{e}^i\right) = T(t_1, \ldots, t_n) \in T\left([0,1]^n\right).$$

By 11.6.3 and 11.6.8, $\lambda_n(P) = \lambda_n(T([0,1]^n)) = |\det A|\lambda_n([0,1]^n) = |\det A|$. □

If $m < n$, then $\lambda_n(P) = 0$, but P may still have positive m-dimensional Lebesgue measure, as defined in 11.6.8. Specifically, let $\mathcal{V}$ denote the linear span of the vectors $\boldsymbol{a}_1, \ldots, \boldsymbol{a}_m$ and choose an orthonormal basis $\boldsymbol{v}_1, \ldots, \boldsymbol{v}_n$ of $\mathbb{R}^n$ such that $\boldsymbol{v}_1, \ldots, \boldsymbol{v}_m$ is a basis for $\mathcal{V}$. Define $T \in \mathcal{L}(\mathcal{V}, \mathbb{R}^m)$ so that $T(\boldsymbol{v}_j) = \boldsymbol{e}_j$, $1 \le j \le m$. Thus T "rotates" and/or "reflects" $\mathcal{V}$ onto $\mathbb{R}^m \times \{\boldsymbol{0}\}$. The *area of* P is then defined by

$$\operatorname{area}\big(P(\boldsymbol{a}^1, \ldots, \boldsymbol{a}^m)\big) = \lambda_m\Big(T\big(P(\boldsymbol{a}^1, \ldots, \boldsymbol{a}^m)\big)\Big).$$

A concrete value for this area is given in the following theorem.

13.2.3 Theorem. *Let* $m < n$, $\boldsymbol{a}^1, \ldots, \boldsymbol{a}^m \in \mathbb{R}^n$, *and* $A = \begin{bmatrix}\boldsymbol{a}^1 \cdots \boldsymbol{a}^m\end{bmatrix}$. *Then*

$$\operatorname{area}\big(P(\boldsymbol{a}^1, \ldots, \boldsymbol{a}^m)\big) = \sqrt{\det(A^t A)} = \left[\sum_{\mathbf{i} \in \mathbf{I}_m} \big(\det A_{\mathbf{i}}\big)^2\right]^{1/2}.$$

Proof. Set $\boldsymbol{b}^j = T(\boldsymbol{a}^j)$ and $B = \begin{bmatrix}\boldsymbol{b}^1 & \cdots & \boldsymbol{b}^m\end{bmatrix}$. By linearity of T,

$$T\big(P(\boldsymbol{a}^1, \ldots, \boldsymbol{a}^m)\big) = P(\boldsymbol{b}^1, \ldots, \boldsymbol{b}^m) \subseteq \mathbb{R}^m,$$

hence, by 13.2.2,

$$\operatorname{area}\big(P(\boldsymbol{a}^1, \ldots, \boldsymbol{a}^m)\big) = \lambda_m\big(P(\boldsymbol{b}^1, \ldots, \boldsymbol{b}^m)\big) = |\det B|.$$

Now, the (i,j)th entry of $B^t B$ is $\boldsymbol{b}^i \cdot \boldsymbol{b}^j$, and because T preserves inner products, this is the same as $\boldsymbol{a}^i \cdot \boldsymbol{a}^j$. Therefore, $B^t B = A^t A$, hence

$$|\det B| = \sqrt{(\det B^t)(\det B)} = \sqrt{\det(B^t B)} = \sqrt{\det(A^t A)}.$$

This proves the first equality in the theorem. The second equality is from 13.1.6. □

Area of a Parameterized Surface

Let $\varphi : U \to \mathbb{R}^n$ be a parameterized m-surface in $\mathbb{R}^n$ with image S and let $\boldsymbol{u} = (u_1, \ldots, u_m) \in U$ and $\boldsymbol{a} = \varphi(\boldsymbol{u}) \in S$. Choose a small m-dimensional interval

$$Q = [u_1, u_1 + \Delta u_1] \times \cdots \times [u_m, u_m + \Delta u_m] \subseteq U, \quad \Delta u_j > 0.$$

As noted in Chapter 12, the line segments $\boldsymbol{u} + t\boldsymbol{e}_j$ in U map onto curves in S with tangent vectors

$$d\varphi_{\boldsymbol{u}}(\boldsymbol{e}^j) = \partial_j \varphi(\boldsymbol{u}), \quad 1 \le j \le m,$$

at $\varphi(\boldsymbol{u})$. The matrix with columns $\partial_j \varphi(\boldsymbol{u})$ is $\varphi'(\boldsymbol{u})$, the Jacobian matrix of φ at $\boldsymbol{u}$. By 13.2.3, the parallelepiped spanned by the vectors $\Delta u_j \partial_j \varphi(\boldsymbol{u})$ therefore

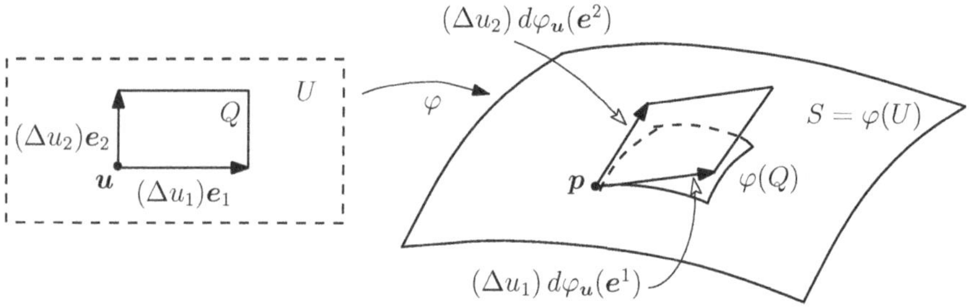

FIGURE 13.1: Parallelogram approximation to $\varphi(Q)$.

has area

$$\sqrt{\det\left(\varphi'(\boldsymbol{u})^t \varphi'(\boldsymbol{u})\right)}\, \Delta u_1\, \Delta u_2 \cdots \Delta u_m,$$

which is taken as an approximation of the area of the surface element $\varphi(Q)$. Partitioning U into a grid $\mathcal{Q}$ of intervals Q and summing these expressions, we obtain the Riemann sums

$$\sum_{\mathcal{Q}} \sqrt{\det\left(\varphi'(\boldsymbol{u})^t \varphi'(\boldsymbol{u})\right)}\, \Delta u_1\, \Delta u_2 \cdots \Delta u_m.$$

It is reasonable then to define the *area of* S as the limit of these sums as the diameters of the intervals Q tend to zero, that is,

$$\text{area}(\varphi) := \int_U \sqrt{\det\left(\varphi'(\boldsymbol{u})^t \varphi'(\boldsymbol{u})\right)}\, d\boldsymbol{u}. \tag{13.11}$$

Integral of a Function on a Parameterized Surface

Let f be a continuous, real-valued function on $S = \varphi(U)$. Motivated by (13.11) we define the *surface integral of* f *over* φ by

$$\int_\varphi f\, dS = \int_U (f \circ \varphi)(\boldsymbol{u}) \sqrt{\det\left(\varphi'(\boldsymbol{u})^t \varphi'(\boldsymbol{u})\right)}\, d\boldsymbol{u} \tag{13.12}$$

whenever the right side exists. In particular,

$$\operatorname{area}(S) = \int_\varphi 1\, dS.$$

The integral on the right in (13.12) may be interpreted as a Lebesgue integral or (if φ has compact support) as a Riemann integral. In the latter case, it is a limit of Riemann sums

$$\sum_{\mathcal{Q}} (f \circ \varphi)(\boldsymbol{u}) \sqrt{\det\left(\varphi'(\boldsymbol{u})^t \varphi'(\boldsymbol{u})\right)}\, \Delta u_1 \cdots \Delta u_m. \tag{13.13}$$

This interpretation has important physical applications. For example, if f is the density in mass per unit area of a curved sheet S in $\mathbb{R}^3$, then (13.13) approximates the mass of the surface element

$$\varphi\Big(\{\boldsymbol{u} + \sum_j t_j \boldsymbol{e}^j \ : 0 \le t_j \le \Delta u_j\}\Big),$$

hence $\int_\varphi f$ gives the mass of S. For another example, let $f(\boldsymbol{x})$ denote the temperature of the sheet at point $\boldsymbol{x} \in S$. Then $[\operatorname{area}(S)]^{-1} \int_\varphi f\, dS$ gives the average temperature of the sheet.

To evaluate (13.12), it is useful to note that since $\varphi' = \begin{bmatrix}\partial_1 \varphi & \cdots & \partial_n \varphi\end{bmatrix}$, by 13.1.6

$$\det\left(\varphi'(\boldsymbol{u})^t \varphi'(\boldsymbol{u})\right) = \sum_{(i_1,\ldots,i_m)\in \mathbf{I}_m} \left[\frac{\partial(\varphi_{i_1}, \ldots \varphi_{i_m})}{\partial(u_1, \ldots, u_m)}(\boldsymbol{u})\right]^2 \tag{13.14}$$

The following instances of 13.12 are of particular interest.

13.2.4 Special Cases.

(a) $m = 1$: Then $\det\left(\varphi'(\boldsymbol{u})^t \varphi'(\boldsymbol{u})\right) = \|\varphi'(u)\|^2$, hence

$$\int_\varphi f\, dS = \int_U (f \circ \varphi)(u) \|\varphi'(u)\|\, du = \int_\varphi f\, ds,$$

which is the line integral of Section 12.2.

(b) $m = 2$: In this case

$$\det\left(\varphi'(\boldsymbol{u})^t \varphi'(\boldsymbol{u})\right) = \det\left\{\begin{bmatrix}\partial_1\varphi \\ \partial_2\varphi\end{bmatrix} \begin{bmatrix}\partial_1\varphi & \partial_2\varphi\end{bmatrix}\right\} = \begin{vmatrix}\partial_1\varphi \cdot \partial_1\varphi & \partial_1\varphi \cdot \partial_1\varphi \\ \partial_1\varphi \cdot \partial_2\varphi & \partial_2\varphi \cdot \partial_2\varphi\end{vmatrix}$$

hence

$$\int_\varphi f\, dS = \int_U (f \circ \varphi) \sqrt{\|\partial_1\varphi\|^2 \|\partial_2\varphi\|^2 - \left[\partial_1\varphi \cdot \partial_2\varphi\right]^2}\, d\boldsymbol{u}.$$

(c) $m = n - 1$: Here

$$\det\big(\varphi'(\boldsymbol{u})^t\varphi'(\boldsymbol{u})\big) = \sum_{i=1}^{n}\left[\frac{\partial(\varphi_1,\ldots,\widehat{\varphi}_i,\ldots\varphi_n)}{\partial(u_1,\ldots,u_{n-1})}(\boldsymbol{u})\right]^2 = \|\partial\varphi^{\perp}(\boldsymbol{u})\|^2,$$

hence

$$\int_{\varphi} f\,dS = \int_U (f\circ\varphi)(\boldsymbol{u})\|\partial\varphi^{\perp}(\boldsymbol{u})\|\,d\boldsymbol{u}.$$

(d) $\varphi(u_1,\ldots,u_{n-1}) = \big(u_1,\ldots,u_{n-1},g(u_1,\ldots,u_{n-1})\big)$ (the graph of g): Let $\mathbf{i} = (1,\ldots,i-1,i+1,\ldots,n)$. Then

$$\frac{\partial(\varphi_1,\ldots,\widehat{\varphi}_i,\ldots\varphi_n)}{\partial(u_1,\ldots,u_{n-1})} = \begin{vmatrix} 1 & 0 & \cdots & 0\\ 0 & 1 & \cdots & 0\\ \vdots & \vdots & & \vdots\\ 0 & 0 & \cdots & 1\\ \partial_1 g & \partial_2 g & \cdots & \partial_{n-1}g \end{vmatrix}_{\mathbf{i}}$$

$$= \begin{cases} (-1)^{n-1+i}\partial_i g, & i < n,\\ 1, & i = n, \end{cases} \tag{13.15}$$

hence

$$\det\big(\varphi'(\boldsymbol{u})^t\varphi'(\boldsymbol{u})\big) = 1 + \|\nabla g(\boldsymbol{u})\|^2$$

and

$$\int_{\varphi} f\,dS = \int_U (f\circ\varphi)(\boldsymbol{u})\sqrt{1+\|\nabla g(\boldsymbol{u})\|^2}\,d\boldsymbol{u}. \qquad\Diamond$$

13.2.5 Example. Let S be the following portion of an n-dimensional cone:

$$S = \Big\{(x_1,\ldots,x_{n+1}) : x_{n+1}^2 = \sum_{i=1}^{n} x_i^2,\ 0 < x_{n+1} < 1\Big\}.$$

Then S is parameterized by

$$\varphi(\boldsymbol{x}) = \big(\boldsymbol{x},g(\boldsymbol{x})\big),\quad g(\boldsymbol{x}) := \|\boldsymbol{x}\|,\quad \boldsymbol{x} := (x_1,\ldots,x_n),$$

where $\nabla g(\boldsymbol{x}) = \boldsymbol{x}/\|\boldsymbol{x}\|$. If f is of the form $f(\boldsymbol{x}) = h(\|\boldsymbol{x}\|)$, then, by Exercise 11.6.3,

$$\int_{\varphi} f\,dS = \sqrt{2}\,n\,\alpha_n\int_0^1 h(r)r^{n-1}\,dr.$$

In particular, taking $h = 1$,

$$\text{area}(S) = \sqrt{2}\,\alpha_n, \qquad\Diamond$$

The following result will be needed later to construct the integral of a function on a general m-surface. It asserts that the integral over a parameterized surface φ is invariant under a change of parameter and hence may be viewed as a construct intrinsic to the image of φ.

13.2.6 Proposition. *Let U and V be open subsets of $\mathbb{R}^m$, $\alpha : V \to U$ a C^1 function with C^1 inverse, and $\varphi : U \to \mathbb{R}^n$ a parameterized m-surface. Then $\psi := \varphi \circ \alpha$ is a parameterized m-surface and $\int_\varphi f\, dS = \int_\psi f\, dS$.*

Proof. By the chain rule, $\psi'(\boldsymbol{v}) = \varphi'(\boldsymbol{u})\alpha'(\boldsymbol{v})$, where $\boldsymbol{u} = \alpha(\boldsymbol{v})$, hence

$$\begin{aligned}\det\left(\psi'(\boldsymbol{v})^t\psi'(\boldsymbol{v})\right) &= \det\left(\alpha'(\boldsymbol{v})^t\varphi'(\boldsymbol{u})^t\varphi'(\boldsymbol{u})\alpha'(\boldsymbol{v})\right)\\ &= \left[J_\alpha(\boldsymbol{v})\right]^2\det\left(\varphi'(\boldsymbol{u})^t\varphi'(\boldsymbol{u})\right).\end{aligned}$$

Therefore, by the change of variables theorem,

$$\begin{aligned}\int_\psi f\, dS &= \int_V (f\circ\psi)(\boldsymbol{v})\sqrt{\det\left(\psi'(\boldsymbol{v})^t\psi'(\boldsymbol{v})\right)}\, d\boldsymbol{v}\\ &= \int_V (f\circ\varphi)(\alpha(\boldsymbol{v}))\sqrt{\det\left(\varphi'(\alpha(\boldsymbol{v}))^t\varphi'(\alpha(\boldsymbol{v}))\right)}\,|J_\alpha(\boldsymbol{v})|\, d\boldsymbol{v}\\ &= \int_U (f\circ\varphi)(\boldsymbol{u})\sqrt{\det\left(\varphi'(\boldsymbol{u})^t\varphi'(\boldsymbol{u})\right)}\, d\boldsymbol{u}\\ &= \int_\varphi f\, dS.\end{aligned}$$

□

13.2.7 Remark. The material in this section holds in particular for a local parametrization of an m-surface as well as a local parametrization of an $(n-1)$-surface-with-boundary. In the latter case, the domain of the parametrization at a boundary point is an open set in $\mathbb{H}^{n-1}$. ◊

Integration of a Form on a Parameterized m-Surface

13.2.8 Definition. Let $\varphi : U \to \mathbb{R}^n$ be a parameterized orientable m-surface in $\mathbb{R}^n$ and let

$$\omega = \sum_{(j_1,\cdots,j_m)\in\mathbf{J}_m} f_{j_1,\cdots,j_m}\, dx_{j_1}\wedge\cdots\wedge dx_{j_m}$$

be a continuous m-form on $S := \varphi(U)$. The *integral of ω over φ* is defined by

$$\int_\varphi \omega = \int_S \omega = \operatorname{sign}(\varphi)\int_U \omega_{\varphi(\boldsymbol{u})}\left(d\varphi_{\boldsymbol{u}}(\mathbf{e}^1),\ldots,d\varphi_{\boldsymbol{u}}(\mathbf{e}^m)\right) d\boldsymbol{u}.$$

◊

The inclusion of $\operatorname{sign}(\varphi)$ corresponds to the familiar convention

$$\int_b^a f(t)\, dt = -\int_a^b f(t)\, dt$$

for Riemann integrals, which reflects the fact that the process of Riemann integration respects the natural orientation (ordering) of the interval $[a, b]$. Recalling that $d\varphi_{\boldsymbol{u}}(\mathbf{e}^j) = \partial_j\varphi(\boldsymbol{u})$ and

$$dx_{j_1}\wedge\cdots\wedge dx_{j_m}\left(\partial_1\varphi(\boldsymbol{u}),\ldots,\partial_m\varphi(\boldsymbol{u})\right) = \frac{\partial(\varphi_{j_1},\ldots,\varphi_{j_m})}{\partial(u_1,\ldots,u_m)}(\boldsymbol{u}),$$

we obtain the formula

$$\int_{\varphi} \omega = \operatorname{sign}(\varphi) \int_U \sum_{(j_1,\ldots,j_m)\in \mathbf{J}_m} (f_{j_1,\ldots,j_m} \circ \varphi) \frac{\partial(\varphi_{j_1},\ldots,\varphi_{j_m})}{\partial(u_1,\ldots,u_m)} \, d\boldsymbol{u}. \tag{13.16}$$

The following instances of (13.16) are of particular importance.

13.2.9 Special Cases. Let φ be positively oriented.

(a) $m = 1$:

$$\int_{\varphi} \sum_{i=1}^{n} f_i \, dx_i = \sum_{i=1}^{n} \int_I f_i(\varphi(t))\varphi_i'(t) \, dt,$$

which is the integral of Section 12.2.

(b) $m = n - 1$:

$$\int_{\varphi} \sum_{i=1}^{n} f_i \, dx_1 \wedge \cdots \wedge \widehat{dx_i} \wedge \cdots \wedge dx_n = \int_U \sum_{i=1}^{n} (f_i \circ \varphi) \frac{\partial(\varphi_1,\ldots,\widehat{\varphi_i},\ldots\varphi_n)}{\partial(u_1,\ldots,u_{n-1})} \, d\boldsymbol{u}.$$

In particular, for the graph $\varphi(u_1,\ldots,u_{n-1}) = \big(u_1,\ldots,u_{n-1}, g(u_1,\ldots,u_{n-1})\big)$, we have from (13.15)

$$\int_{\varphi} \sum_{i=1}^{n} f_i \, dx_1 \wedge \cdots \wedge \widehat{dx_i} \wedge \cdots \wedge dx_n = \int_U \left[f_n \circ \varphi + \sum_{i=1}^{n-1} (-1)^{n-1+i} (f_i \circ \varphi) \partial_i g \right] d\boldsymbol{u}.$$

(c) $m = 2$, $n = 3$: Let $D_{ij}(\boldsymbol{u}) := \dfrac{\partial(\varphi_i,\varphi_j)}{\partial(u_1,u_2)}$. Then

$$\int_{\varphi} \left[f_1 \, dx_2 \wedge dx_3 + f_2 \, dx_1 \wedge dx_3 + f_3 \, dx_1 \wedge dx_2 \right]$$
$$= \int_U \left[(f_1 \circ \varphi)(\boldsymbol{u}) D_{23}(\boldsymbol{u}) + (f_2 \circ \varphi)(\boldsymbol{u}) D_{13}(\boldsymbol{u}) + (f_3 \circ \varphi)(\boldsymbol{u}) D_{12}(\boldsymbol{u}) \right] d\boldsymbol{u}.$$

(d) m-form on parameterized surface $\iota : U \to U$:

$$\int_{\iota} \sum_{(j_1,\cdots,j_m)\in \mathbf{J}_m} g_{j_1,\cdots,j_k} \, du_{j_1} \wedge \cdots \wedge du_{j_m} = \int_U \sum_{\mathbf{j}\in \mathbf{J}_m} g_{\mathbf{j}}(\boldsymbol{u}) \, d\boldsymbol{u}. \qquad \Diamond$$

13.2.10 Notation. *For the integral on the left in* **(d)**, *we write* $\int_U$ *instead of* $\int_\iota$. *In particular,*

$$\int_{\iota} g \, du_{j_1} \wedge \cdots \wedge du_{j_m} = \int_U g(\boldsymbol{u}) \, d\boldsymbol{u}. \qquad \Diamond$$

13.2.11 Example. Let S be the following portion of a paraboloid:

$$S = \{(x_1, x_2, x_3) : x_1 = x_2^2 + x_3^2,\ 0 < x_1 < 1\}.$$

For purposes of integration, we may consider S to be the image of the parameterized 2-surface

$$\varphi(t, \theta) = (t, \sqrt{t}\cos\theta, \sqrt{t}\sin\theta),\quad 0 < t < 1,\ 0 < \theta < 2\pi,$$

since there are no contributions to an integral on the set where $\theta = 0$. By 13.2.9(**c**),

$$\begin{aligned}\int_S x_2^2 x_3\, dx_1 \wedge dx_2 &= \int_0^1 \int_0^{2\pi} [t^{3/2}\cos^2\theta \sin\theta] \frac{\partial(\varphi_1, \varphi_2)}{\partial(t, \theta)}\, d\theta\, dt \\ &= -\int_0^1 \int_0^{2\pi} t^2 \cos^2\theta \sin^2\theta\, d\theta\, dt \\ &= -\frac{\pi}{12}.\end{aligned}$$

◊

The following proposition, the analog of 13.2.6 for differential forms, shows that the definition of integral of a form is invariant under reparametrizations.

13.2.12 Proposition. *Let U, V be open connected subsets of $\mathbb{R}^m$, $\alpha : V \to U$ a C^1 function with C^1 inverse and positive Jacobian, and $\varphi : U \to \mathbb{R}^n$ a parameterized orientable m-surface. If ω is a continuous m-form on $\varphi(U)$, then*

$$\int_\varphi \omega = \int_{\varphi\circ\alpha} \omega.$$

Proof. Note first that $\text{sign}(J_\alpha)$ is constant since α is C^1 and V is connected. Let $\psi = \varphi \circ \alpha$. By the chain rule and the change of variables theorem,

$$\begin{aligned}&\int_V (f_{j_1,\ldots,j_m} \circ \psi) \frac{\partial(\psi_{j_1}, \ldots, \psi_{j_m})}{\partial(v_1, \ldots, v_m)}\, d\boldsymbol{v} \\ &\quad = \int_V (f_{j_1,\ldots,j_m} \circ \varphi \circ \alpha)(\boldsymbol{v}) \frac{\partial(\varphi_{j_1}, \ldots, \varphi_{j_m})}{\partial(u_1, \ldots, u_m)}(\alpha(\boldsymbol{v})) J_\alpha(\boldsymbol{v})\, d\boldsymbol{v} \\ &\quad = \int_U (f_{j_1,\ldots,j_m} \circ \varphi) \frac{\partial(\varphi_{j_1}, \ldots, \varphi_{j_m})}{\partial(u_1, \ldots, u_m)}\, d\boldsymbol{u}.\end{aligned}$$

The conclusion now follows from (13.16) and linearity of the integral. □

The final result of this section expresses $\int_\varphi \omega$ as an integral of a form on U. It will be needed in the proof of Stokes's theorem.

13.2.13 Theorem. *Let $U \subseteq \mathbb{R}^m$ be open and let $\varphi : U \to \mathbb{R}^n$ be an oriented parameterized surface. If ω is a C^1 m-form on $\varphi(U)$, then*

$$\int_\varphi \omega = \text{sign}(\varphi) \int_U \varphi^*\omega.$$

Proof. By (d) of 13.1.19, if $\iota : U \to U$ denotes the identity map then

$$\begin{aligned}\omega_{\varphi(\boldsymbol{u})}(d\varphi_{\boldsymbol{u}}(\mathbf{e}^1), \ldots, d\varphi_{\boldsymbol{u}}(\mathbf{e}^m)) &= (\varphi^*\omega)_{\boldsymbol{u}}(\mathbf{e}^1, \ldots, \mathbf{e}^m) \\ &= (\varphi^*\omega)_{\boldsymbol{u}}(d\,\iota_{\boldsymbol{u}}(\mathbf{e}^1), \ldots, d\,\iota_{\boldsymbol{u}}(\mathbf{e}^m)).\end{aligned}$$

The result now follows directly from the definition of the integral of a form (13.2.8) and 13.2.10. □

Exercises

1. Find the area of the following 2-surfaces in $\mathbb{R}^3$.

   (a) $\varphi(t, \theta) = (t\cos\theta, t\sin\theta, t)$, $t \in (0, 1)$, $\theta \in (0, 2\pi)$

   (b) $\varphi(t, \theta) = (t\cos\theta, t\sin\theta, \theta)$, $0 < t < 1$, $0 < \theta < 2\pi$

   (c) $\varphi(\theta, s) = (1 - s)\big(a\cos\theta, a\sin\theta, 0\big) + s\big(b\cos\theta, b\sin\theta, 1\big)$, $0 < s < 1$, $0 < \theta < 2\pi$, $0 < a < b$.

2. Let $\boldsymbol{a}^1, \ldots, \boldsymbol{a}^m \in \mathbb{R}^n$ be linearly independent and let $\boldsymbol{b} \in \mathbb{R}^n$. Define $\varphi : \mathbb{R}^m \to \mathbb{R}^n$ by

   $$\varphi(u_1, \ldots, u_m) = \boldsymbol{b} + \sum_{i=1}^{m} u_i \boldsymbol{a}^i.$$

   (See 12.3.2.) For a continuous function f on $\mathbb{R}^n$, prove that

   $$\int_{\varphi} f = \sqrt{\det(A^t A)} \int_{\mathbb{R}^n} (f \circ \varphi)(\boldsymbol{u})\, d\boldsymbol{u},$$

   where $A = \big[\boldsymbol{a}^1 \cdots \boldsymbol{a}^m\big]_{n\times m}$.

3. Let φ be as in Exercise 2. Show that

   $$\int_{\varphi} \sum_{\mathfrak{i}\in\mathbf{I}_m} f_{\mathfrak{i}}\, dx_{\mathfrak{i}} = \sum_{\mathfrak{i}\in\mathbf{I}_m} \det(A_{\mathfrak{i}}) \int_U f_{\mathfrak{i}} \circ \varphi\, d\boldsymbol{u}.$$

4. Show that the area of the Cartesian product of circles

   $$\varphi(\theta_1, \ldots, \theta_m) = \big(r_1\cos\theta_1, r_1\sin\theta_1, \ldots, r_m\cos\theta_m, r_m\sin\theta_m\big), \quad r_i > 0,$$

   is $(2\pi r_1)(2\pi r_2)\cdots(2\pi r_m)$.

5. Let φ be the product of two circles:

   $$\varphi(\theta_1, \theta_2) = \big(r_1\cos\theta_1, r_1\sin\theta_1, r_2\cos\theta_2, r_2\sin\theta_2\big), \quad r_i > 0,$$

   and let

   $$\begin{aligned}\omega = f_{12}\, dx_1 \wedge dx_2 + f_{13}\, dx_1 \wedge dx_3 + f_{14}\, dx_1 \wedge dx_4 \\ + f_{23}\, dx_2 \wedge dx_3 + f_{24}\, dx_2 \wedge dx_4 + f_{34}\, dx_3 \wedge dx_4.\end{aligned}$$

Show that

$$\int_{\varphi} \omega = r_1 r_2 \int_0^{2\pi} \int_0^{2\pi} \big[(f_{13} \circ \varphi) \sin\theta_1 \sin\theta_2 - (f_{14} \circ \varphi) \sin\theta_1 \cos\theta_2 \\ - (f_{23} \circ \varphi) \cos\theta_1 \sin\theta_2 + (f_{24} \circ \varphi] \cos\theta_1 \cos\theta_2\big] \, d\theta \, d\phi.$$

6. (Area of an n-dimensional simplex in $\mathbb{R}^{n+1}$). Use Example 11.5.5 to find the surface area of

$$S = \Big\{(x_1, \ldots x_{n+1}) : \sum_{j=1}^{n+1} x_j = 1 \text{ and } x_j \ge 0\Big\}.$$

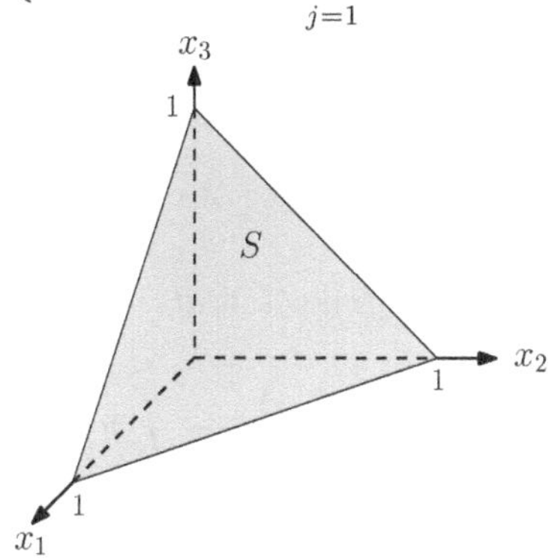

FIGURE 13.2: Two-dimensional simplex S in $\mathbb{R}^3$.

7. Let $U \subseteq \mathbb{R}^{n-2}$ be open and let $\psi : U \to \mathbb{R}^{n-1}$ be a parameterized $(n-2)$-surface in $\mathbb{R}^{n-1}$. Let $\varphi : U \times [0, h] \to \mathbb{R}^n$ be the *cylinder*

$$\varphi(\boldsymbol{u}, s) = \big(\psi(\boldsymbol{u}), s\big), \quad \boldsymbol{u} \in U, \quad 0 \le s \le h.$$

Show that $\text{area}(\varphi) = h \cdot \text{area}(\psi)$.

8. Let φ be the cylinder of Exercise 7 for $n = 3$ and $h = 1$. Show that

$$\int_{\varphi} (f_1 \, dx_2 \wedge dx_3 + f_2 \, dx_1 \wedge dx_3 + f_3 \, dx_1 \wedge dx_2) \\ = \int_0^1 \big[(f_1 \circ \varphi)\psi_1' + (f_2 \circ \varphi)\psi_2'\big] \, dt.$$

9. Let $\psi : [a, b] \to \mathbb{R}^2$ be a C^1 curve in $\mathbb{R}^2$ and let $\varphi : [a, b] \times (0, h) \to \mathbb{R}^3$ be the *cone*

$$\varphi(t, s) = \big((1 - s/h)\psi(t), s\big), \quad a \le t \le b, \quad 0 < s < h.$$

Show that the area of φ is

$$\frac{h}{2} \int_a^b \sqrt{\big[\psi_1'(t)\big]^2 + \big[\psi_2'(t)\big]^2 + h^{-2}[\psi_1(t)\psi_2'(t) - \psi_2(t)\psi_1'(t)\big]^2} \, dt.$$

Use this to show that the surface area of a right circular cone with radius r and axis length h is $\pi r\sqrt{r^2+h^2}$.

10. Let φ be the cone of Exercise 9 with $h = 1$. Show that

$$\int_\varphi (f_1\,dx_2 \wedge dx_3 + f_2\,dx_1 \wedge dx_3 + f_3\,dx_1 \wedge dx_2)$$
$$= \frac{1}{2}\int_0^1 \Big\{(f_1\circ\varphi)\psi_1' + (f_2\circ\varphi)\psi_2' + [\psi_1(t)\psi_2'(t) - \psi_2(t)\psi_1'(t)]\Big\}\,dt.$$

11. Let $\psi : [a,b] \to \mathbb{R}^2$ a parameterized C^1 curve with $\psi_2(t) > 0$ for all t. Define

$$\varphi(t,\theta) = \big(\psi_1(t), \psi_2(t)\cos\theta, \psi_2(t)\sin\theta\big), \quad t \in I,\ \theta \in (0, 2\pi),$$

which is the parameterized surface of revolution of 12.3.9. Show that

$$\text{area}(\varphi) = 2\pi\int_a^b \psi_2(t)\|\psi'(t)\|\,dt = (2\pi\overline{y})\text{length}(\psi), \tag{13.17}$$

where $(x,y) = \psi$ and $\overline{y} := \dfrac{1}{\text{length}(\psi)}\displaystyle\int_\psi y\,ds$, the y-coordinate of the centroid of ψ. Use the first part of (13.17) to find the surface area of the torus

$$\varphi(t,\theta) = \big(a\cos\theta, (b + a\sin t)\cos\theta, (b + a\sin t)\sin\theta\big), \quad 0 < \theta, t < 2\pi,$$

where $0 < a < b$. Show also that the area of the cone in Exercise 9 may be found from (13.17).

12. Let φ be the parameterized surface of revolution in Exercise 11 and let $\omega := f_1\,dx_2 \wedge dx_3 + f_2\,dx_1 \wedge dx_3 + f_3\,dx_1 \wedge dx_2$. Show that

$$\int_\varphi \omega = \int_a^b\int_0^{2\pi} (f_1\circ\varphi)\psi_2(t)\psi_2'(t)\,d\theta\,dt + \int_a^b\int_0^{2\pi} (f_2\circ\varphi)\psi_1'(t)\psi_2(t)\cos\theta\,d\theta\,dt$$
$$- \int_a^b\int_0^{2\pi} (f_3\circ\varphi)\psi_1'(t)\psi_2(t)\sin\theta\,d\theta\,dt.$$

Show also that if $\psi(t) = (t, g(t))$ (the graph of g), then this reduces to

$$\int_a^b\int_0^{2\pi} g(t)\big[(f_1\circ\varphi)g'(t) + (f_2\circ\varphi)\cos\theta - (f_3\circ\varphi)\sin\theta\big]\,d\theta\,dt.$$

13. Use Exercise 12 to evaluate

(a) $\displaystyle\int_S x_1x_3\,dx_1 \wedge dx_2$, (b) $\displaystyle\int_S x_2x_3\,dx_1 \wedge dx_2$, (c) $\displaystyle\int_S x_1^2x_2^2\,dx_2 \wedge dx_3$,

where S is the cone

$$S = \big\{(x_1, x_2, x_3) : x_1^2 = x_2^2 + x_3^2,\ 0 < x_1 < 1\big\}.$$

14. Repeat Exercise 13 using the portion of the hyperboloid

$$S = \left\{(x_1, x_2, x_3) \, : \, x_1^2 - x_2^2 - x_3^2 = 1, \, 1 < x_1 < \sqrt{2}\right\}.$$

15. Let S be the torus given by $x_1^2 + \left(\sqrt{x_2^2 + x_3^2} - b\right)^2 = a^2$, where $0 < a < b$. Use Exercise 12 to evaluate

(a) $\displaystyle\int_S x_2 \, dx_2 \wedge dx_3$ (b) $\displaystyle\int_S x_1 \, dx_2 \wedge dx_3$ (c) $\displaystyle\int_S x_2 \, dx_1 \wedge dx_3$.

13.3 Partitions of Unity

The theorem proved in this section will be used to extend the definition of the integral to functions and forms on m-surfaces. It will also be needed later in the proofs of Stokes's theorem and the divergence theorem.

13.3.1 Definition. The *support* of a continuous function $\psi : \mathbb{R}^n \to \mathbb{R}$ is defined by

$$\text{supp}(\psi) = \text{cl}\big(\{\boldsymbol{x} \, : \, \psi(\boldsymbol{x}) \neq 0\}\big). \qquad \Diamond$$

Thus, by definition of closure, $\text{supp}(\psi)$ is the smallest closed set outside of which ψ is zero.

13.3.2 Partition of Unity. *Let K be a compact subset of $\mathbb{R}^n$ and let $\{U_\mathfrak{i} : \mathfrak{i} \in \mathfrak{I}\}$ be an open cover of K. Then there exists a finite subcover $\{U_1, \ldots, U_p\}$ of K and C^∞ functions $\chi_i : \mathbb{R}^n \to [0, +\infty)$, $i = 1, \ldots, p$, such that* $\text{supp}(\chi_i)$ *is compact and contained in U_i and $\sum_{i=1}^p \chi_i = 1$ on K.*

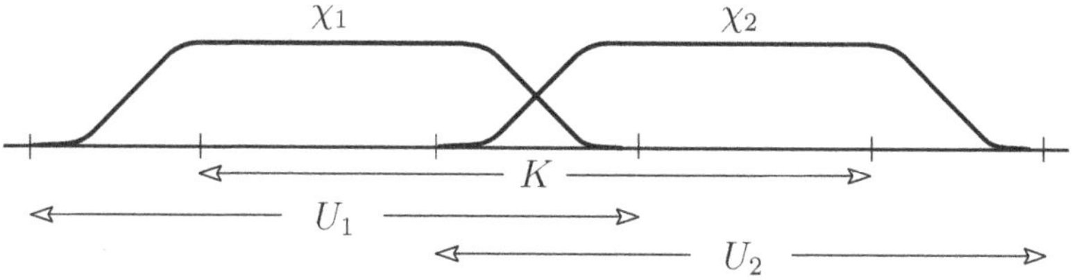

FIGURE 13.3: A partition of unity subordinate to U_1 and U_2.

The functions χ_i are said to form a *partition of unity subordinate to the open sets U_i*. They are typically used to patch together local data to form a global construct such as a surface integral, or to reduce a global problem to a local one, as in the case of the proof of Stokes's theorem.

The proof of 13.3.2 requires several lemmas which are of intrinsic interest.

13.3.3 Lemma. *Let $a < b$. Then there exists a C^∞ function $h : \mathbb{R} \to [0, +\infty)$ such that $h > 0$ on (a, b), and $h = 0$ on $(a, b)^c$.*

Proof. Define h by

$$h(x) = \begin{cases} \exp\left[(x-a)^{-1}(x-b)^{-1}\right] & \text{if } a < x < b, \\ 0 & \text{otherwise.} \end{cases}$$

Clearly, $h^{(m)} = 0$ on $[a,b]^c$ for all $m \geq 0$. Moreover, if $x \in (a,b)$, then $h^{(m)}(x)$ is a sum of terms of the form

$$\frac{\pm h(x)}{(x-a)^p(x-b)^q}, \quad p,\, q \in \mathbb{Z}^+.$$

Since the exponent $(x-a)^{-1}(x-b)^{-1}$ is negative on (a,b), by l'Hospital's rule,

$$\lim_{x \to a^+} \frac{h(x)}{(x-a)^p(x-b)^q} = 0.$$

Therefore, $\lim_{x\to a} h^{(m)}(x) = 0$, and an induction argument then shows that $h^{(m)}(a) = 0$ for all m. A similar argument holds at the point b. Thus h is C^∞ on $\mathbb{R}$. □

FIGURE 13.4: The functions h and g.

13.3.4 Lemma. *Let $a < b$. Then there exists a C^∞ function $g : \mathbb{R} \to \mathbb{R}$ such that $0 \leq g \leq 1$, $g = 0$ on $(-\infty, a]$, and $g = 1$ on $[b, +\infty)$.*

Proof. Let h be the function in 13.3.3. Then

$$g(x) := \left[\int_a^b h\right]^{-1} \int_a^x h$$

has the required properties. □

13.3.5 Lemma. *Let $I = (a_1, b_1) \times \cdots \times (a_n, b_n)$. Then there exists a C^∞ function $f : \mathbb{R}^n \to \mathbb{R}$ such that $f > 0$ on I and $f = 0$ on I^c.*

Proof. For each j, let $h_j : \mathbb{R} \to [0, +\infty)$ be a C^∞ function such that $h_j > 0$ on (a_j, b_j) and $h_j = 0$ on $(a_j, b_j)^c$. The function

$$f(x_1, \ldots, x_n) := h_1(x_1) \cdots h_n(x_n)$$

then satisfies the requirements. □

For the next lemma we define the *open cube with center* $\boldsymbol{x} \in \mathbb{R}^n$ *and edge* $2r$ by

$$\{\boldsymbol{y} \in \mathbb{R}^n : x_j - r < y_j < x_j + r,\ j = 1, \ldots, n\}.$$

13.3.6 Lemma. *Let* $K \subseteq U \subseteq \mathbb{R}^n$, *where* K *is compact and* U *is open. Then there exists a* C^∞ *function* $\psi : \mathbb{R}^n \to [0,1]$ *such that* $\operatorname{supp}(\psi) \subseteq U$ *and* $\psi = 1$ *on* K.

Proof. For each $\boldsymbol{x} \in K$, let $V_{\boldsymbol{x}}$ be an open cube with center $\boldsymbol{x}$ and edge $2r$ such that $\operatorname{cl}(V_{\boldsymbol{x}}) \subseteq U$ and let $W_{\boldsymbol{x}} \subseteq V_{\boldsymbol{x}}$ denote the concentric open cube with center $\boldsymbol{x}$ and edge r. Since K is compact, there exist finitely many cubes $W_{\boldsymbol{x}}$ whose union contains K. Denote these cubes by $W_1, \ldots, W_m$ and denote the corresponding cubes $V_{\boldsymbol{x}}$ by $V_1, \ldots, V_m$ (See Figure 13.5) By 13.3.5, for each i

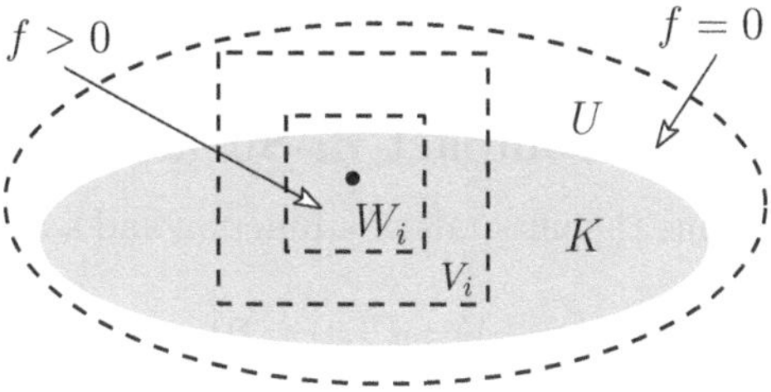

FIGURE 13.5: The cubes W_i and V_i.

there exists a C^∞ function $f_i : \mathbb{R}^n \to \mathbb{R}$ such that $f_i > 0$ on W_i and $f_i = 0$ on W_i^c. Set

$$f = \sum_{i=1}^m f_i, \quad V = \bigcup_{i=1}^m V_i, \quad \text{and} \quad W = \bigcup_{i=1}^m W_i.$$

Then f is nonnegative and C^∞ on $\mathbb{R}^n$, $f > 0$ on $W \supseteq K$, and $\operatorname{supp}(f) \subseteq \operatorname{cl}(V) \subseteq U$. Now let $a = \min_{\boldsymbol{x} \in K} f(\boldsymbol{x})$. Since $a > 0$, there exists a C^∞ function $g : \mathbb{R} \to [0,1]$ such that $g = 0$ on $(-\infty, 0]$ and $g = 1$ on $[a, +\infty)$ (13.3.4). The function $\psi := g \circ f$ then has the required properties. □

Proof of the partition of unity theorem.

For each $\boldsymbol{x} \in K$, let $\mathrm{i}(\boldsymbol{x})$ be an index such that $\boldsymbol{x} \in U_{\mathrm{i}(\boldsymbol{x})}$. Choose a bounded open set $V_{\boldsymbol{x}}$ containing $\boldsymbol{x}$ such that $\operatorname{cl}(V_{\boldsymbol{x}}) \subseteq U_{\mathrm{i}(\boldsymbol{x})}$. Since K is compact, finitely many of the sets $V_{\boldsymbol{x}}$ cover K. Denote these by $V_1, \ldots V_p$ and denote the corresponding sets $U_{\mathrm{i}(\boldsymbol{x})}$ by $U_1, \ldots, U_p$. Since $V_i \subseteq K_i := \operatorname{cl}(V_i) \subseteq U_i$, by 13.3.6 there exists a C^∞ function $\psi_i : \mathbb{R}^n \to [0,1]$ such that $\psi_i = 1$ on K_i and $\operatorname{supp}(\psi_i) \subseteq U_i$. Now set

$$\chi_1 = \psi_1 \quad \text{and} \quad \chi_i = (1 - \psi_1)(1 - \psi_2) \cdots (1 - \psi_{i-1})\psi_i, \quad i > 1.$$

Then χ_i is C^∞, $0 \le \chi_i \le 1$, and $\operatorname{supp}(\chi_i) \subseteq \operatorname{supp}(\psi_i) \subseteq U_i$. Finally, let

$$\eta_i = (1 - \psi_1)(1 - \psi_2) \cdots (1 - \psi_i).$$

For $i > 1$,

$$\eta_{i-1} - \eta_i = (1-\psi_1)(1-\psi_2)\cdots(1-\psi_{i-1})\big[1-(1-\psi_i)\big] = \chi_i,$$

hence

$$\sum_{i=1}^{p} \chi_i = \chi_1 + \sum_{i=2}^{p}(\eta_{i-1} - \eta_i) = \chi_1 + \eta_1 - \eta_p = 1 - \eta_p.$$

Since $K \subseteq \bigcup_i V_i \subseteq \bigcup_i K_i$ and $\psi_i = 1$ on K_i, $\eta_p = 0$ on K, hence $\sum_{i=1}^{p} \chi_i = 1$ on K, completing the proof. □

13.4 Integration on Compact m-Surfaces

In this section we define the integrals of a function and a form on a compact m-surface

$$S = \{\boldsymbol{x} \in V : F(\boldsymbol{x}) = \boldsymbol{0}\},$$

where $V \subseteq \mathbb{R}^n$ is open, $F : V \to \mathbb{R}^{n-m}$ is C^1, and $F'(\boldsymbol{x})$ has rank $n - m$ for all $\boldsymbol{x} \in V$.

To set the stage, let $\{(U_{\boldsymbol{a}}, \varphi_{\boldsymbol{a}}) : \boldsymbol{a} \in S\}$ be an atlas for S. By the partition of unity theorem, there exist finitely many charts $(U_i, \varphi_i) := \big(U_{\boldsymbol{a}_i}, \varphi_{\boldsymbol{a}_i}\big)$ and C^1 functions $\chi_i : \mathbb{R}^n \to \mathbb{R}$ such that the sets $S_i := \varphi_i(U_i)$ cover S, $\operatorname{supp}(\chi_i) \subseteq S_i$, and $\sum_i \chi_i = 1$ on S.

Integral of a Function

The (*surface*) *integral* of a continuous function f on S is defined by

$$\int_S f\,dS = \sum_i \int_{\varphi_i} \chi_i f = \sum_i \int_{U_i} \big[(\chi_i f)\circ\varphi_i\big](\boldsymbol{u})\sqrt{\det\big(\varphi_i'(\boldsymbol{u})^t\varphi_i'(\boldsymbol{u})\big)}\,d\boldsymbol{u}.$$

To see that the integral is independent of the system $\{(U_i, \varphi_i, \chi_i)\}_i$ and hence is well-defined, consider another such system $\{(\tilde{U}_j, \tilde{\varphi}_j, \tilde{\chi}_j)\}_j$. Since

$$\chi_i = \sum_j \chi_i\tilde{\chi}_j \quad\text{and}\quad \tilde{\chi}_j = \sum_i \tilde{\chi}_j\chi_i \quad\text{on } S,$$

we see that

$$\sum_i \int_{\varphi_i} f\chi_i = \sum_{i,j}\int_{\varphi_i} f\chi_i\tilde{\chi}_j \quad\text{and}\quad \sum_j \int_{\tilde{\varphi}_j} f\tilde{\chi}_j = \sum_{i,j}\int_{\tilde{\varphi}_j} f\tilde{\chi}_j\chi_i.$$

Set

$$\alpha_{ij} = \tilde{\varphi}_j^{-1} \circ \varphi_i : \varphi_i^{-1}(S_i \cap \tilde{S}_j) \to \varphi_j^{-1}(S_i \cap \tilde{S}_j).$$

Since $f\chi_i\tilde{\chi}_j = 0$ outside $S_i \cap \tilde{S}_j$ and $\varphi_i = \tilde{\varphi}_j \circ \alpha_{ij}$ on $\varphi_i^{-1}(S_i \cap \tilde{S}_j)$, by 13.2.6 $\int_{\varphi_i} f\chi_i\tilde{\chi}_j = \int_{\tilde{\varphi}_j} f\chi_i\tilde{\chi}_j$. Therefore,

$$\sum_i \int_{\varphi_i} f\chi_i = \sum_j \int_{\tilde{\varphi}_j} f\tilde{\chi}_j,$$

as required.

The definition of the integral is extended to a finite union S of compact m-surfaces $S_1, \ldots, S_p$ by defining

$$\int_S f\, dS = \sum_i \int_{S_i} f\, dS.$$

13.4.1 Definition. The *area of* S is defined as

$$\operatorname{area}(S) = \int_S 1\, dS. \qquad \Diamond$$

13.4.2 Example. In 11.5.6 we found that the volume of the closed ball $C_r^n(\mathbf{0}) = \{\boldsymbol{x} \in \mathbb{R}^n : ||\boldsymbol{x}|| \le r\}$ is $r^n\alpha_n$, where

$$\alpha_n = \begin{cases} \dfrac{2(2\pi)^{(n-1)/2}}{n(n-2)\cdots 3\cdot 1} & \text{if } n \text{ is odd,} \\[2ex] \dfrac{(2\pi)^{n/2}}{n(n-2)\cdots 4\cdot 2} & \text{if } n \text{ is even.} \end{cases}$$

We now show that for the sphere $S := S_r^{n-1}(\mathbf{0}) = \{\boldsymbol{x} \in \mathbb{R}^n : ||\boldsymbol{x}|| = r\}$,

$$\operatorname{area}(S) = nr^{n-1}\alpha_n = \frac{n}{r}\lambda_n\big(C_r^n(\mathbf{0})\big). \tag{13.18}$$

To this end, note that the upper hemisphere H^u of S is the graph of the function

$$g(x_1, \ldots, x_{n-1}) = \sqrt{r^2 - (x_1^2 + \cdots + x_{n-1}^2)} = \sqrt{r^2 - ||\boldsymbol{x}||^2}, \quad ||\boldsymbol{x}|| \le r.$$

Let $0 < t < 1$ and consider the part of the hemisphere H_t^u for which $||\boldsymbol{x}|| < rt$. Since

$$1 + ||\nabla g(\boldsymbol{x})||^2 = 1 + \frac{||\boldsymbol{x}||^2}{r^2 - ||\boldsymbol{x}||^2} = \frac{r^2}{r^2 - ||\boldsymbol{x}||^2},$$

by 13.2.4(**c**)

$$\operatorname{area}(H_t^u) = r\int_{||\boldsymbol{x}||<rt} \big(r^2 - ||\boldsymbol{x}||^2\big)^{-1/2}\, d\boldsymbol{x} = r(n-1)\alpha_{n-1}\int_0^{rt} \frac{s^{n-2}}{\sqrt{r^2 - s^2}}\, ds,$$

where the second equality comes from Exercise 11.6.3. The substitution $s = xr$ produces

$$\text{area}(H_t^u) = (n-1)r^{n-1}\alpha_{n-1}\int_0^t \frac{x^{n-2}}{\sqrt{1-x^2}}\,dx.$$

The lower hemisphere counterpart H_t^ℓ has the same area. Since

$$\mathbf{1}_S = \lim_{t\to 1^-} \mathbf{1}_{H_t^u} + \lim_{t\to 1^-} \mathbf{1}_{H_t^\ell},$$

it follows from Exercise 6 below that

$$\text{area}(S) = 2\lim_{t\to 1^-}\text{area}(H_t^u) = 2(n-1)r^{n-1}\alpha_{n-1}\int_0^1 \frac{x^{n-2}}{\sqrt{1-x^2}}\,dx. \tag{13.19}$$

By Exercise 5.3.7,

$$\int_0^1 \frac{x^{n-2}}{\sqrt{1-x^2}}\,dx = \begin{cases} \dfrac{(n-3)(n-5)\cdots 4\cdot 2}{(n-2)(n-4)\cdots 3\cdot 1} & \text{if } n \text{ is odd} \\ \dfrac{\pi}{2}\,\dfrac{(n-3)(n-5)\cdots 3\cdot 1}{(n-2)(n-2)\cdots 4\cdot 2} & \text{if } n \text{ is even.} \end{cases} \tag{13.20}$$

Now use (13.19) and (13.20), to obtain (13.18). ◊

Integral of an m-Form

The definition of the (*surface*) *integral of an m-form* on a (positively) oriented compact m-surface S is analogous to the case of a function on S:

$$\int_S \omega = \sum_i \int_{\varphi_i} \chi_i\omega = \sum_i \int_{U_i} (\chi_i\omega)_{\varphi_i(\boldsymbol{u})}\big(\partial_1\varphi_i(\boldsymbol{u}),\ldots,\partial_m\varphi_i(\boldsymbol{u})\big)\,d\boldsymbol{u}.$$

The argument that the integral is well-defined proceeds as above.

As in the case of functions, the integral of a form on a finite union S of oriented compact m-surfaces $S_1,\ldots,S_p$ is defined by

$$\int_S f\,dS = \sum_i \int_{S_i} f\,dS.$$

Exercises

> *For these exercises, declare a function $f : S \to \mathbb{R}$ to be* Borel measurable *if $f\circ\varphi_{\boldsymbol{a}} : U_{\boldsymbol{a}} \to \mathbb{R}$ is Borel measurable on $U_{\boldsymbol{a}}$ for each local parametrization $\varphi_{\boldsymbol{a}}$ of S.*

1. Show that the collection of Borel measurable functions on S is closed under the operations described in 10.5.3.

2. Define $\int_S f\,dS$ for nonnegative Borel measurable functions f on S.

3. Call a Borel measurable function f *integrable* if $\int_S f^+\,dS$ and $\int_S f^-\,dS$ are finite. In this case, define

$$\int_S f\,dS = \int_S f^+\,dS - \int_S f^-\,dS.$$

Prove that the resulting integral is linear and $\left|\int_S f\,dS\right| \le \int_S |f|\,dS$.

4. Formulate and prove versions of the monotone convergence theorem, Fatou's lemma, and the dominated convergence theorem for Borel measurable functions on S.

5. Define the σ-field $\mathcal{B}(S) := \{S \cap B : B \in \mathcal{B}(\mathbb{R}^n)\}$. Show that $\mathbf{1}_E$ is Borel measurable on S for every $E \in \mathcal{B}(S)$.

6. Show that

$$\mu_S(E) := \int_S \mathbf{1}_E\,dS, \quad E \in \mathcal{B}(S),$$

defines a measure on $\mathcal{B}(S)$. (μ_S is called *surface measure on* S. It provides a way of calculating the area of Borel subsets of S.)

7. Use Exercise 6 to verify the first equality in (13.19).

13.5 The Fundamental Theorems of Calculus

The fundamental theorem of single-variable calculus expresses the integral of a continuous function on an interval $[a, b]$ as a function of the boundary $\{a, b\}$. The theorem tacitly assumes that integration occurs from "left to right," which is to say that the interval $[a, b]$ is *positively oriented.* The fundamental theorem of calculus may then be stated as follows: For any primitive F of a continuous function f on $[a, b]$, the integral of f is $F(b) - F(a)$ if $[a, b]$ is positively oriented and $F(a) - F(b)$ if $[a, b]$ is negatively oriented. The theorems proved in this section are higher-dimensional versions of this formulation.

Stokes's Theorem

While the proof of the following theorem is based on many of the intricate constructs developed earlier, the conclusion of the theorem is remarkably easy to state.

13.5.1 Stokes's Theorem. *Let S be a compact oriented $(n-1)$-surface-with-boundary in $\mathbb{R}^n$ and let ω be a C^1 $(n-2)$-form on S. If ∂S has the induced orientation, then*

$$\int_{\partial S} \omega = \int_S d\omega.$$

Proof. We prove the theorem first under the assumption that ω has compact support contained $\varphi(I)$, where $\varphi : U \to \mathbb{R}^n$ is a local parametrization of S and I is a bounded $(n-1)$-dimensional interval with $\text{cl}(I) \subseteq U$. For definiteness, we assume that $\text{sign}(\varphi) = 1$, that is, the frame $\big(d\varphi_{\boldsymbol{u}}(\mathbf{e}^1), \ldots, d\varphi_{\boldsymbol{u}}(\mathbf{e}^{n-1})\big)$ is positive in $\mathcal{T}_{\varphi(\boldsymbol{u})}$. Thus φ is oriented by $\vec{N}_\varphi$.

Suppose first that U is open in $\mathbb{R}^{n-1}$ and $\varphi(U) \subseteq S \setminus \partial S$. We may then take $I = (a_1, b_1) \times \cdots \times (a_{n-1}, b_{n-1})$. Since $\omega = 0$ on ∂S, $\int_{\partial S} \omega = 0$. Set $\eta := \varphi^*(\omega)$. By 13.1.19, $d\eta := \varphi^*(d\omega)$, hence, by 13.2.13,

$$\int_S d\omega = \int_U d\eta.$$

Since η is an $(n-2)$ form on $U \subseteq \mathbb{R}^{n-1}$, it may be expressed as

$$\eta = \sum_{i=1}^{n-1} (-1)^{i+1} f_i \, du_1 \wedge \cdots \wedge \widehat{du_i} \wedge \cdots \wedge du_{n-1},$$

where f_i is C^1 on U and $\text{supp}(f_i) \subseteq I$. Then

$$\begin{aligned}
\int_U d\eta &= \int_U \sum_{i=1}^{n-1} (-1)^{i+1} \left(\sum_{j=1}^{n-1} \frac{\partial f_i}{\partial u_j} du_j \right) \wedge du_1 \wedge \cdots \wedge \widehat{du_i} \wedge \cdots \wedge du_{n-1} \\
&= \int_U \sum_{i=1}^{n-1} (-1)^{i+1} \frac{\partial f_i}{\partial u_i} du_i \wedge du_1 \wedge \cdots \wedge \widehat{du_i} \wedge \cdots \wedge du_{n-1} \\
&= \sum_{i=1}^{n-1} \int_U (\partial_i f_i) \, du_1 \wedge \cdots \wedge du_{n-1}. \qquad (13.21)
\end{aligned}$$

Since $f_i = 0$ outside I, by the Fubini–Tonelli theorem, recalling 13.2.10, we have

$$\int_U (\partial_i f_i) \, du_1 \wedge \cdots \wedge du_{n-1} = \int_{a_1}^{b_1} \cdots \int_{a_{n-1}}^{b_{n-1}} \int_{a_i}^{b_i} (\partial_i f_i) \, du_i \, du_{n-1} \cdots \widehat{du_i} \cdots du_1.$$

By the fundamental theorem of calculus, the innermost integral evaluates to

$$f_i(u_1, \ldots, b_i, \ldots, u_{n-1}) - f_i(u_1, \ldots, a_i, \ldots, u_{n-1}) = 0 - 0 = 0.$$

Therefore,

$$\int_S d\omega = 0 = \int_{\partial S} \omega.$$

Now suppose that $\varphi(U) \cap \partial S \neq \emptyset$. Then U is an open subset in $\mathbb{H}^{n-1}$ that intersects $\partial\,\mathbb{H}^{n-1}$ and I is of the form $(a_1, b_1) \times \cdots \times (a_{n-2}, b_{n-2}) \times [0, b_{n-1})$. The above argument works for every term in the sum (13.21) except the last. It is still the case that $f_{n-1}(u_1, \ldots, u_{n-2}, b_{n-1}) = 0$, but now there is no guarantee that $f_{n-1}(u_1, \ldots, u_{n-2}, 0) = 0$. Thus all we may conclude is that

$$\int_S d\omega = \int_U d\eta = -\int_{a_1}^{b_1} \cdots \int_{a_{n-2}}^{b_{n-2}} f_{n-1}(u_1, \ldots, u_{n-2}, 0)\, du_{n-2} \cdots du_1. \tag{13.22}$$

Set $V = U \cap \partial\mathbb{H}^{n-1}$. By definition of induced orientation, $\operatorname{sign}(\varphi|_V) = (-1)^{n-1}$, hence, by 13.2.13 again,

$$\int_{\partial S} \omega = (-1)^{n-1} \int_V \eta = \sum_{i=1}^{n-1} (-1)^{n+i} \int_V f_i\, du_1 \wedge \cdots \wedge \widehat{du_i} \wedge \cdots \wedge du_{n-1}.$$

Since $u_{n-1} = 0$ on V, du_{n-1} must be zero on V. Therefore, the first $n-2$ terms in the above sum vanish and we are left with

$$\int_{\partial S} \omega = -\int_V f_{n-1}\, du_1 \wedge \cdots \wedge du_{n-2},$$

which is (13.22). This verifies the theorem if ω has compact support.

In the general case, let $\{\varphi_i : U_i \to S : i = 1, \ldots, m\}$ be an atlas of local parameterizations of S and let $\{\chi_i : i = 1, \ldots, m\}$ be a partition of unity subordinate to the open sets U_i. By the first part of the proof,

$$\int_{\partial S} \chi_i \omega_i = \int_S d(\chi_i \omega), \quad i = 1, \ldots, m.$$

By the product rule, $d(\chi_i\omega) = \chi_i\, d\omega + (d\,\chi_i)\omega$. Since $\sum_i \chi_i = 1$,

$$\sum_i d(\chi_i\omega) = \Big[\sum_i \chi_i\Big]\, d\omega + \Big[d \sum_i \chi_i\Big]\omega = d\omega.$$

Therefore,

$$\int_{\partial S} \omega = \sum_i \int_{\partial S} \chi_i \omega = \sum_i \int_S d(\chi_i\omega) = \int_S d\omega. \qquad \square$$

From the first part of the proof we have

13.5.2 Corollary. *If S is a compact oriented $(n-1)$-surface-without-boundary and if ω is an $(n-2)$-form on S, then*

$$\int_S d\omega = 0.$$

13.5.3 Remark. **(a)** For the case $n = 3$, Stokes's formula takes the form

$$\int_{\partial S} [f_1\,dx_1 + f_2\,dx_2 + f_3\,dx_3] = \int_S d[f_1\,dx_1 + f_2\,dx_2 + f_3\,dx_3]. \tag{13.23}$$

The left side of (13.23) may be written

$$\int_{\partial S} \vec{F} \cdot d\vec{r}, \quad \text{where } d\vec{r} := (dx_1, dx_2, dx_3).$$

Let $\vec{F} := (f_1, f_2, f_3)$ and $(g_1, g_2, g_3) := \operatorname{curl} \vec{F}$. By 13.1.15(a),

$$d\big(f_1\,dx_1 + f_2\,dx_2 + f_3\,dx_3\big) = g_1\,dx_2 \wedge dx_3 + g_2\,dx_3 \wedge dx_1 + g_3\,dx_1 \wedge dx_2.$$

For a local parametrization φ,

$$\begin{aligned}
&\int_\varphi [g_1\,dx_2 \wedge dx_3 + g_2\,dx_3 \wedge dx_1 + g_3\,dx_1 \wedge dx_2] \\
&= \int_U \left[(g_1 \circ \varphi)\frac{\partial(\varphi_2, \varphi_3)}{\partial(u_1, u_2)} + (g_2 \circ \varphi)\frac{\partial(\varphi_3, \varphi_1)}{\partial(u_1, u_2)} + (g_3 \circ \varphi)\frac{\partial(\varphi_1, \varphi_2)}{\partial(u_1, u_2)}\right] d\boldsymbol{u} \\
&= \int_U (\operatorname{curl} \vec{F} \circ \varphi) \cdot \big(\vec{N}_\varphi \circ \varphi\big)\sqrt{\det\left[\varphi'(\boldsymbol{u})^t \varphi'(\boldsymbol{u})\right]}\, d\boldsymbol{u} \\
&= \int_{\varphi(U)} \operatorname{curl} \vec{F} \cdot \vec{N}_\varphi\, dS
\end{aligned}$$

Using a partition of unity, we see that the right side of (13.23) may then be written $\int_S \operatorname{curl} \vec{F} \cdot \vec{N}\, dS$. Thus we obtain the classical version of Stokes's formula

$$\int_{\partial S} \vec{F} \cdot d\vec{r} = \int_S \operatorname{curl} \vec{F} \cdot \vec{N}\, dS \tag{13.24}$$

(b) If S is the graph of a function $g(x, y)$ on a set $D \subseteq \mathbb{R}^2$, then S may be parameterized by $\varphi(x, y) = (x, y, g(x, y))$. Hence, if S is oriented by the upward normal

$$\frac{(-\nabla g, 1)}{\sqrt{\|\nabla g\|^2 + 1}}$$

(see (12.3.7)**(c)**), then (13.24) becomes

$$\int_{\partial S} \vec{F} \cdot d\vec{r} = \int_D \operatorname{curl} \vec{F} \cdot (-g_x, -g_y, 1)\, dx\, dy. \qquad \Diamond$$

13.5.4 Example. We verify (13.24) for

$$F(x_1, x_2, x_3) = (x_2 x_3, x_1, x_2)$$

on the cylinder-with-boundary

$$S := \big\{(x_1, x_2, x_3) : x_1^2 + x_2^2 = 1, \quad 0 \le x_3 \le 1\big\}$$

(Figure 12.17) oriented by the outward normal $\vec{N}(x_1, x_2, x_3) = (x_1, x_2, 0)$. The bottom boundary may be parameterized by $\varphi_1(t) = (\cos t, \sin t, 0)$ and the top by $\varphi_2(t) = (\cos t, -\sin t, 1)$, $0 \le t \le 2\pi$. Therefore,

$$\begin{aligned}\int_{\partial S} \vec{F} \cdot d\vec{r} &= \int_0^{2\pi} \left[-f_1(\cos t, \sin t, 0)\sin t + f_2(\cos t, \sin t, 0)\cos t\right] dt \\ &\quad + \int_0^{2\pi} \left[-f_1(\cos t, -\sin t, 1)\sin t - f_2(\cos t, -\sin t, 1)\cos t\right] dt \\ &= \int_0^{2\pi} \left[\cos^2 t + \sin^2 t - \cos^2 t\right] dt = \pi.\end{aligned}$$

To find $\int_S \operatorname{curl} \vec{F} \cdot \vec{N}\, dS$ we parameterize S by $\varphi(t, x_3) = (\cos t, \sin t, x_3)$. Then

$$\det\left[\varphi'(t, x_3)^t \varphi'(t, x_3)\right] = \det \begin{bmatrix} -\sin t & \cos t & 0 \\ 0 & 0 & 1 \end{bmatrix} \begin{bmatrix} -\sin t & 0 \\ \cos t & 0 \\ 0 & 1 \end{bmatrix} = 1,$$

and since $\operatorname{curl} F = (1, x_2, 1 - x_3)$,

$$\int_S \operatorname{curl} \vec{F} \cdot \vec{N}\, dS = \int_0^{2\pi} (\cos t + \sin^2 t)\, dt = \pi. \qquad \Diamond$$

Divergence Theorem

The divergence theorem is a variation of Stokes's theorem. In the latter theorem, integration occurs on an $(n-1)$-dimensional surface in $\mathbb{R}^n$ with $(n-2)$-dimensional boundary. In the former, the domain of integration is a *regular region*, which may be viewed as an n-dimensional surface in $\mathbb{R}^n$ with $(n-1)$-dimensional boundary.

13.5.5 Definition. A bounded open subset E of $\mathbb{R}^n$ is called a *regular region* if for each point $\boldsymbol{a} \in \operatorname{bd}(E)$ there exists a neighborhood $U_{\boldsymbol{a}}$ of $\boldsymbol{a}$ and a C^1 function $F_{\boldsymbol{a}} : U_{\boldsymbol{a}} \to \mathbb{R}$ with $\nabla F_{\boldsymbol{a}} \ne \mathbf{0}$ such that

(i) $S_{\boldsymbol{a}} := \operatorname{bd}(E) \cap U_{\boldsymbol{a}} = \{\boldsymbol{x} : F_{\boldsymbol{a}}(\boldsymbol{x}) = 0\}$,

(ii) $E \cap U_{\boldsymbol{a}} = \{\boldsymbol{x} : F_{\boldsymbol{a}}(\boldsymbol{x}) < 0\}$, and

(iii) $\operatorname{cl}(E)^c \cap U_{\boldsymbol{a}} = \{\boldsymbol{x} : F_{\boldsymbol{a}}(\boldsymbol{x}) > 0\}$. $\Diamond$

Note that $S_{\boldsymbol{a}}$ is an $(n-1)$-surface in $\mathbb{R}^n$ and hence has a local parametrization at each $\boldsymbol{x} \in S_{\boldsymbol{a}}$. Let

$$\vec{\mathfrak{n}}_{\boldsymbol{a}} = \|\nabla F_{\boldsymbol{a}}\|^{-1} \nabla F_{\boldsymbol{a}},$$

and let $\boldsymbol{x} \in S_{\boldsymbol{a}}$. For sufficiently small $|t|$, $h(t) := \boldsymbol{x} + t\vec{\mathfrak{n}}_{\boldsymbol{a}}(\boldsymbol{x}) \in U_{\boldsymbol{a}}$. Since

$$(F_{\boldsymbol{a}} \circ h)'(0) = \nabla F_{\boldsymbol{a}}(\boldsymbol{x}) \cdot \vec{\mathfrak{n}}_{\boldsymbol{a}}(\boldsymbol{x}) = \|\nabla F_{\boldsymbol{a}}(\boldsymbol{x})\| > 0,$$

$(F_{\boldsymbol{a}} \circ h)$ is strictly increasing. Because $(F_{\boldsymbol{a}} \circ h)(0) = 0$ we therefore have

$$F_{\boldsymbol{a}}(\boldsymbol{x} + t\vec{\mathfrak{n}}_{\boldsymbol{a}}(\boldsymbol{x})) \begin{cases} < 0 & \text{if } t < 0, \\ > 0 & \text{if } t > 0. \end{cases}$$

It follows from (ii) and (iii) that the normal vector $t\vec{\mathfrak{n}}_{\boldsymbol{a}}(\boldsymbol{x})$ to $S_{\boldsymbol{a}}$ at $\boldsymbol{x}$ points into E if $t < 0$ and away from E (that is, toward E^c) if $t > 0$. The *exterior unit normal vector* on $\text{bd}(E)$ is then defined by

$$\vec{\mathfrak{n}}(\boldsymbol{x}) = \vec{\mathfrak{n}}_{\boldsymbol{a}}(\boldsymbol{x}), \quad \boldsymbol{x} \in S_{\boldsymbol{a}}.$$

Uniqueness and continuity of $\vec{\mathfrak{n}}_{\boldsymbol{a}}$ shows that $\vec{\mathfrak{n}}$ is well-defined and continuous on $\text{bd}(E)$ (See Figure 13.6)

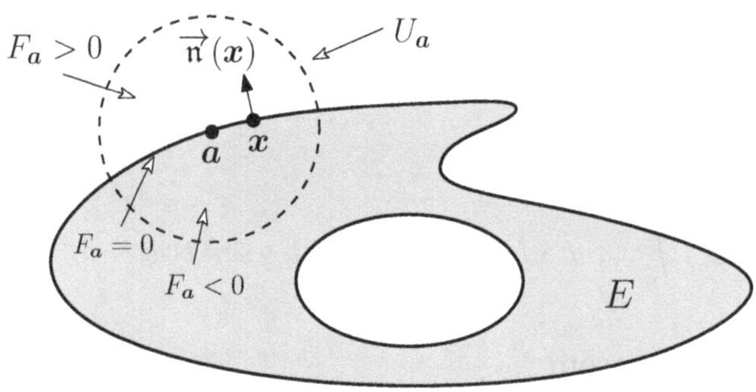

FIGURE 13.6: Regular region E.

13.5.6 Example. The *n-dimensional annulus*

$$E = \{\boldsymbol{x} \in \mathbb{R}^n : r_1 < \|\boldsymbol{x}\| < r_2\}$$

is a regular region in $\mathbb{R}^n$. Here, $\text{bd}(E)$ has the components

$$S_i = \{\boldsymbol{x} \in \mathbb{R}^n : \|\boldsymbol{x}\| = r_i\}, \quad i = 1, 2.$$

The conditions of regularity are met by defining

$$F_{\boldsymbol{a}}(\boldsymbol{x}) = \begin{cases} r_1 - \|\boldsymbol{x}\| & \text{on } \{\boldsymbol{x} \in \mathbb{R}^n : \|\boldsymbol{x}\| < (r_1 + r_2)/2\} \text{ if } \boldsymbol{a} \in S_1, \\ \|\boldsymbol{x}\| - r_2 & \text{on } \{\boldsymbol{x} \in \mathbb{R}^n : \|\boldsymbol{x}\| > (r_1 + r_2)/2\} \text{ if } \boldsymbol{a} \in S_2. \end{cases}$$

Figure 13.7 depicts the case $n = 2$. ◊

13.5.7 Divergence Theorem. *If E is a regular region in $\mathbb{R}^n$ and ω is a C^1 1-form on $\text{cl}(E)$, then*

$$\int_{\text{bd}(E)} \omega \cdot \vec{\mathfrak{n}} \, dS = \int_E \text{div}\, \omega_{\boldsymbol{x}} \, d\boldsymbol{x}. \tag{13.25}$$

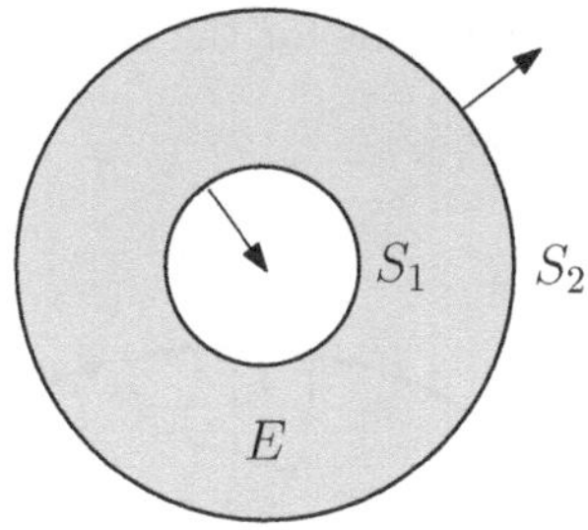

FIGURE 13.7: Annulus in $\mathbb{R}^2$ with exterior normal.

Proof. The proof uses ideas similar to those used in the proof of Stokes's theorem. By hypothesis, ω is C^1 on an open set containing $\mathrm{cl}(E)$, which we may assume also contains the sets $U_{\boldsymbol{a}}$ in 13.5.5. Since $\mathrm{cl}(E)$ is compact, by using a partition of unity as in the proof of Stokes's theorem, we may assume that for any $\boldsymbol{a} = (a_1, \ldots, a_n) \in \mathrm{cl}(E)$ and the neighborhoods W of $\boldsymbol{a}$ constructed in the proof,

$$K := \bigcup_{i=1}^{n} \mathrm{supp}(f_i) \subseteq W.$$

Suppose first that $\boldsymbol{a} \in E$. Choose an n-dimensional interval W containing $\boldsymbol{a}$ such that $\mathrm{cl}(W) \subseteq E$. If $K \subseteq W$, then $\omega = 0$ on $W^c \supseteq \mathrm{bd}(E)$, hence

$$\int_{\mathrm{bd}(E)} \omega \cdot \vec{\mathsf{n}}\, dS = 0 \quad \text{and} \quad \int_E \mathrm{div}\, \omega_{\boldsymbol{x}}\, d\boldsymbol{x} = \int_W \sum_{i=1}^{n} \partial_i f_i(\boldsymbol{x})\, d\boldsymbol{x} = 0,$$

the last equality by the Fubini–Tonelli theorem and the fundamental theorem of calculus. Therefore, (13.25) holds in this case.

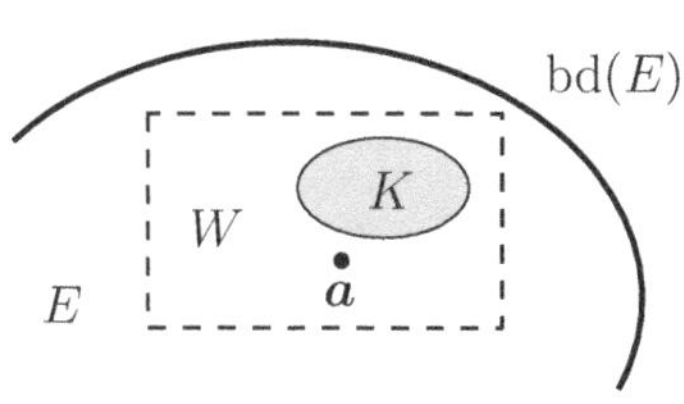

FIGURE 13.8: The case $\boldsymbol{a} \in E$.

Now let $\boldsymbol{a} \in \mathrm{bd}(E)$ and let $U_{\boldsymbol{a}}$ and $F_{\boldsymbol{a}}$ be as in 13.5.5. We may assume that the components a_i of $\boldsymbol{a}$ and $\mathsf{n}_i(\boldsymbol{a})$ of $\vec{\mathsf{n}}(\boldsymbol{a})$ are positive, otherwise apply a rotation and translation; the change of variables theorem implies that (13.25) is invariant under such transformations. (See Exercise 11 below for a special case of this.) We show that for each $i = 1, \ldots, n$, there exists a neighborhood W_i of $\boldsymbol{a}$ such that if $K \subseteq W_i$ then

$$\int_S f_i\, \mathsf{n}_i\, dS = \int_E \partial_i f_i(\boldsymbol{x})\, d\boldsymbol{x}. \tag{13.26}$$

For notational simplicity, we do this for the case $i = n$.

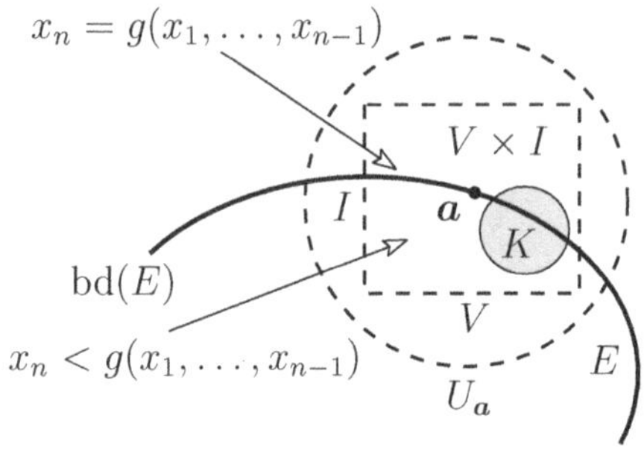

FIGURE 13.9: The case $\boldsymbol{a} \in \text{bd}(E)$.

Since $\partial_n F_{\boldsymbol{a}}(\boldsymbol{a}) \neq \mathbf{0}$, by the implicit function theorem there exists a neighborhood V of $(a_1, \ldots, a_{n-1})$, an open interval I containing a_n, and a C^1 function $g : V \to \mathbb{R}$ such that $V \times I \subseteq U_{\boldsymbol{a}}$, $a_n = g(a_1, \ldots, a_{n-1})$, and

$$F_{\boldsymbol{a}}\big(x_1, \ldots, x_{n-1}, g(x_1, \ldots, x_{n-1})\big) = 0 \text{ on } V.$$

By continuity, we may choose V and I sufficiently small so that

$$g(x_1, \ldots, x_{n-1}) > 0 \text{ and } \partial_n F_{\boldsymbol{a}}(\boldsymbol{x}) > 0 \text{ for all } \boldsymbol{x} \in V \times I.$$

Now let $\boldsymbol{x} = (x_1, \ldots, x_n) \in (V \times I) \cap E$. Since $F_{\boldsymbol{a}}(\boldsymbol{x})$ is a strictly increasing function of $x_n \in I$ when the other coordinates are fixed and since $F_{\boldsymbol{a}}(\boldsymbol{x}) < 0$, it must be the case that $0 < x_n < g(x_1, \ldots, x_{n-1})$. Thus

$$\begin{aligned} (V \times I) \cap E &= \{\boldsymbol{x} \in V \times I : 0 < x_n < g(x_1, \ldots, x_{n-1})\} \text{ and} \\ (V \times I) \cap S_{\boldsymbol{a}} &= \{\boldsymbol{x} \in V \times I : x_n = g(x_1, \ldots, x_{n-1})\}. \end{aligned}$$

(See Figure 13.9.) Note that the function φ defined by

$$\varphi(\boldsymbol{v}) := (\boldsymbol{v}, g(\boldsymbol{v})), \quad \boldsymbol{v} = (v_1, \ldots, v_{n-1}) \in V,$$

is a local parametrization of $S_{\boldsymbol{a}}$ with unit normal

$$(1 + \|\nabla g\|^2)^{-1/2}\big(- \nabla g, 1\big).$$

Since this points outward it coincides with $\vec{\mathsf{n}}$. In particular, the nth component of $\vec{\mathsf{n}}$ is

$$\mathsf{n}_n = (1 + \|\nabla g\|^2)^{-1/2}$$

on $(V \times I) \cap S_{\boldsymbol{a}}$. Therefore, if $K \subseteq V \times I$ then, by 13.2.4(d),

$$\int_{S_a} f_n \, \mathsf{n}_n(\boldsymbol{a}) \, dS = \int_{(V \times I) \cap S_{\boldsymbol{a}}} \frac{f_n}{\sqrt{1 + \|\nabla g\|^2}} \, dS = \int_V (f_n \circ \varphi)(\boldsymbol{v}) \, d\boldsymbol{v}. \quad (13.27)$$

On the other hand, since $f_n = \partial_n f_n = 0$ outside K, by the Fubini–Tonelli theorem,

$$\begin{aligned}\int_E \partial_n f_n(\boldsymbol{x})\,d\boldsymbol{x} &= \int_{(V\times I)\cap E} \partial_n f_n(\boldsymbol{x})\,d\boldsymbol{x} \\ &= \int_V \int_0^{g(v_1,\ldots,v_{n-1})} (\partial_n f_n)(v_1,\ldots,v_{n-1},x_n)\,dx_n\,dv_1\ldots dv_{n-1} \\ &= \int_V (f_n\circ\varphi)(\boldsymbol{v})\,d\boldsymbol{v}, \qquad (13.28)\end{aligned}$$

the last equality by the fundamental theorem of calculus. Setting $W_n = V\times I$ and comparing (13.27) and (13.28), we see that (13.26) holds for $i=n$. A similar proof works for $i<n$. Thus if $K\subseteq W_1\cap\cdots\cap W_n$, then (13.26) holds for all i. Summing from 1 to n we obtain (13.25). □

Connection with Stokes's Theorem

Let E be a regular region in $\mathbb{R}^n$ whose boundary is a finite union of compact connected $(n-1)$-surfaces of the form $S=\{\boldsymbol{x}: F(\boldsymbol{x})=0\}$, where $F: U\to\mathbb{R}$ is a C^1 function with $\nabla F\neq\mathbf{0}$ such that $U_{\boldsymbol{a}} = U$ and $F_{\boldsymbol{a}} = F$ for all $\boldsymbol{a}\in S$. A ball or annulus in $\mathbb{R}^n$ are simple examples. By 12.4.8, S is oriented and, for each local parametrization $\varphi: V\to S$,

$$\vec{\mathfrak{n}}(\varphi(\boldsymbol{v})) = \frac{\pm 1}{\sqrt{\det\left(\varphi'(\boldsymbol{v})^t\varphi'(\boldsymbol{v})\right)}}\sum_{i=1}^n (-1)^{i-1}\frac{\partial(\varphi_1,\ldots,\widehat{\varphi_{i-1}},\ldots,\varphi_n)}{\partial(v_1,\ldots,v_{n-1})}(\boldsymbol{v}).$$

where the sign is chosen to be the same for all $\boldsymbol{v}$. Let each S have the orientation for which the sign is $(+)$. We shall call the resulting orientation of $\mathrm{bd}(E)$ *positive.* In this setting we have the following consequence of the divergence theorem.

13.5.8 Theorem. *Let E be as described above and let*

$$\omega = \sum_{i=1}^n f_i\,dx_1\wedge\cdots\wedge\widehat{dx_i}\wedge\cdots\wedge dx_n$$

be an $(n-1)$-form on $\mathrm{cl}(E)$. If $\mathrm{bd}(E)$ is positively oriented, then

$$\int_{\mathrm{bd}(E)}\omega = \int_E d\omega.$$

Proof. Recalling the additive definition of $\int_{\mathrm{bd}(E)}\omega$, we may assume that $\mathrm{bd}(E)$ consists of a single compact connected $(n-1)$-surface S. Let

$$\eta := \sum_{i=1}^n (-1)^{i-1} f_i\,dx_i.$$

By the above,

$$(\vec{\mathfrak{n}} \cdot \eta) \circ \varphi(\boldsymbol{v}) = \frac{1}{\sqrt{\det\left(\varphi'(\boldsymbol{v})^t \varphi'(\boldsymbol{v})\right)}} \sum_{i=1}^{n} (f_i \circ \varphi)(\boldsymbol{v}) \frac{\partial(\varphi_1, \ldots, \widehat{\varphi}_i, \ldots, \varphi_n)}{\partial(v_1, \ldots, v_{n-1})}(\boldsymbol{v}),$$

hence

$$\int_\varphi \vec{\mathfrak{n}} \cdot \eta \, dS = \sum_{i=1}^{n} \int_V (f_i \circ \varphi) \frac{\partial(\varphi_1, \ldots, \widehat{\varphi}_i, \ldots, \varphi_n)}{\partial(v_1, \ldots, v_{n-1})} \, d\boldsymbol{v} = \int_\varphi \omega.$$

Using a partition of unity we obtain

$$\int_S \omega = \int_S \vec{\mathfrak{n}} \cdot \eta \, dS. \tag{13.29}$$

On the other hand,

$$\begin{aligned} d\omega &= \sum_{i=1}^{n} \left(\sum_{j=1}^{n} (\partial_j f_i) \, dx_j \right) \wedge dx_1 \wedge \cdots \wedge \widehat{dx_i} \wedge \cdots \wedge dx_n \\ &= \left(\sum_{i=1}^{n} (-1)^{i-1} \partial_i f_i \right) dx_1 \wedge \cdots \wedge dx_n \\ &= \operatorname{div} \eta \; dx_1 \wedge \cdots \wedge dx_n, \end{aligned}$$

hence, recalling 13.2.10,

$$\int_E d\omega = \int_E \operatorname{div} \eta_{\boldsymbol{x}} \; dx_1 \wedge \cdots \wedge dx_n = \int_E \operatorname{div} \eta_{\boldsymbol{x}} \; d\boldsymbol{x} \tag{13.30}$$

The conclusion now follows from (13.29), (13.30), and the divergence theorem. □

13.5.9 Remark. The divergence theorem has an interesting application to fluid dynamics. Consider a fluid moving in space. Let $\rho(\boldsymbol{x}, t)$ denote the density of the fluid in mass per unit volume at time t and point $\boldsymbol{x}$, and let $\vec{v}(\boldsymbol{x}, t)$ denote its velocity. If $\vec{\mathfrak{n}}$ is normal to a small surface element of area ΔS, then $(\rho\vec{v} \cdot \vec{\mathfrak{n}})(\Delta S)(\Delta t)$ is approximately the mass of the fluid flowing across that surface element during a small time period Δt. The rate of flow is then $(\rho\vec{v} \cdot \vec{\mathfrak{n}})\Delta S$. Adding these quantities and taking limits, we see that the rate of flow of the fluid across a surface S in the direction of the normal is given by the integral

$$\int_S \rho\vec{v} \cdot \vec{\mathfrak{n}} \, dS$$

Now let E be a regular region with smooth boundary S. Applying the foregoing to a ball B_ε in E with boundary S_ε, center $\boldsymbol{y}$, and outer normal $\vec{\mathfrak{n}}$, we see that the integral

$$\int_{S_\varepsilon} \rho\vec{v} \cdot \vec{\mathfrak{n}} \, dS$$

represents the rate of flow of the fluid out of the ball, that is, the negative of the rate of change of fluid in the ball. Since the amount of fluid in the ball at time t is $\int_{B_\varepsilon} \rho(\boldsymbol{x}, t)\, d\boldsymbol{x}$,

$$\frac{d}{dt}\int_{B_\varepsilon} \rho(\boldsymbol{x}, t)\, d\boldsymbol{x} = -\int_{S_\varepsilon} \rho\vec{v}\cdot\vec{\mathfrak{n}}\, dS = -\int_{B_\varepsilon} \operatorname{div}(\rho\vec{v})\, d\boldsymbol{x},$$

the last equality by the divergence theorem. Differentiating under the integral sign and dividing by $\operatorname{vol}(B_\varepsilon)$, we obtain

$$\frac{1}{\operatorname{vol}(B_\varepsilon)}\int_{B_\varepsilon} \partial_t\rho(\boldsymbol{x}, t)\, d\boldsymbol{x} = -\frac{1}{\operatorname{vol}(B_\varepsilon)}\int_{B_\varepsilon} \operatorname{div}(\rho\vec{v})\, d\boldsymbol{x}.$$

Letting $\varepsilon \to 0$, we obtain

$$\partial_t\rho(\boldsymbol{y}, t) = -\operatorname{div}\big(\rho(\boldsymbol{y}, t)\vec{v}(\boldsymbol{y}, t)\big).$$

In particular, if ρ is constant in time, then $\operatorname{div}(\rho\vec{v})$ is zero throughout E, hence

$$\int_S \rho\vec{v}\cdot\vec{\mathfrak{n}}\, dS = \int_E \operatorname{div}(\rho\vec{v})\, d\boldsymbol{x} = 0,$$

that is, the amount of fluid flowing out of E equals the amount flowing in. ◊

Green's Theorem

Let E be a regular region in $\mathbb{R}^2$ with boundary the union of finitely many smooth simple pairwise disjoint curves $C = \varphi(I)$. The boundary $\operatorname{bd}(E)$ is said to be *positively oriented* if the vector obtained by rotating the unit tangent vector $\vec{T}$, which is in the direction of (φ_1', φ_2'), 90 degrees clockwise. This produces the exterior normal $\vec{\mathfrak{n}}$ on C, which is in the direction of $(\varphi_2', -\varphi_1')$. The region is then to the left as the boundary is traced in the direction of the tangent vector field on each curve C.

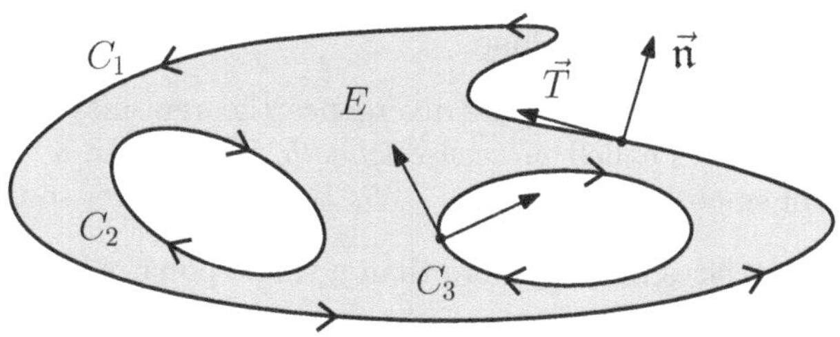

FIGURE 13.10: Regular region E in $\mathbb{R}^2$.

Now let $\omega = Q\, dx - P\, dy$. Then

$$(\omega\cdot\vec{\mathfrak{n}})\circ\varphi = (Q\circ\varphi, -P\circ\varphi)\cdot(\varphi_2', -\varphi_1')\|\varphi'\|^{-1} = \big[(P\circ\varphi)\varphi_1' + (Q\circ\varphi)\varphi_2'\big]\|\varphi'\|^{-1},$$

hence

$$\int_C \omega \cdot \vec{\mathfrak{n}}\, ds = \int_C (P\,dx + Q\,dy).$$

Summing over the curves C, we have

$$\int_{\mathrm{bd}(E)} \omega \cdot \vec{\mathfrak{n}}\, ds = \int_{\mathrm{bd}(E)} (P\,dx + Q\,dy).$$

Since

$$\operatorname{div} \omega = \frac{\partial Q}{\partial x} - \frac{\partial P}{\partial y},$$

we obtain the following important special case of the divergence theorem.

13.5.10 Green's Theorem. *Let E be a region in $\mathbb{R}^2$, as described above. If P, Q are C^1 functions on an open set containing E, then*

$$\int_{\mathrm{bd}(E)} (P\,dx + Q\,dy) = \iint_E \left(\frac{\partial Q}{\partial x} - \frac{\partial P}{\partial y} \right) dx\,dy. \tag{13.31}$$

13.5.11 Corollary. *The area of E is given by*

$$\mathrm{area}(S) = \frac{1}{2} \int_{\partial S} (x\,dy - y\,dx).$$

Proof. Apply Green's theorem to $P(x, y) = -y/2$, $Q(x, y) = x/2$, noting that $Q_x - P_y = 1$. □

13.5.12 Example. The ellipse $\dfrac{x^2}{a^2} + \dfrac{y^2}{b^2} = 1$ has parametrization $x = a\cos t$, $y = b\sin t$, $0 \le t \le 2\pi$. Therefore, the area inside the ellipse is

$$\frac{1}{2} \int_0^{2\pi} ab(\cos^2 t + \sin^2 t)\, dt = \pi ab.$$ ◊

The Piecewise Smooth Case

Both Stokes's theorem and the divergence theorem may be extended to more general surfaces called *piecewise smooth*. In the case $n = 3$, these are finite unions of smooth surfaces $S_1, \ldots, S_k$ that fit together so that

- no three surfaces meet in more than a single point, and
- the common boundary of two of these surfaces consists of finitely many disjoint piecewise smooth simple closed curves.

(See Figure 13.11.) If S is such a surface, then the surface integral $\int_S f\,dS$ is defined as the sum $\sum_{j=1}^k \int_{S_j} f\,dS$. The integral of a form on S has an analogous definition. These definitions are reasonable since, by cancelations, the common

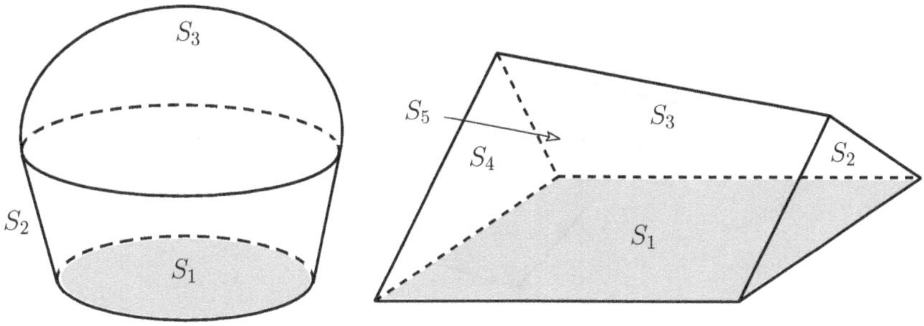

FIGURE 13.11: Piecewise smooth surfaces.

boundary of a pair of surfaces contributes nothing to the integral. We illustrate the basic idea with the simple example of a cube. Removing a face of the cube results in a surface-with-boundary Q, which we orient by the outward normal. If Stokes's theorem is applied to each of the five faces and the results are added, the integrals along the boundaries cancel and one is left with Stokes's theorem for Q:

$$\int_{\partial Q} \vec{F} \cdot \vec{dr} = \int_Q \operatorname{curl} \vec{F} \cdot \vec{N}\, dS.$$

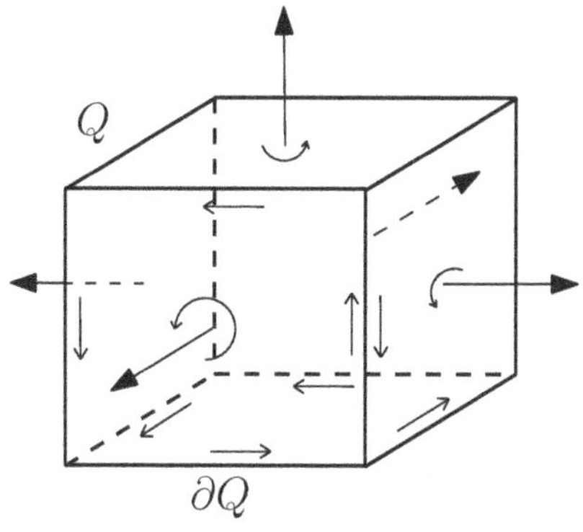

FIGURE 13.12: Oriented cube without bottom face.

Similarly, Green's theorem extends to regions in $\mathbb{R}^2$ whose boundaries are only piecewise smooth. This, of course, leads to extended versions of its corollaries. Here's an application of the extended version of 13.5.11:

13.5.13 Example. Let ∂S be a closed polygon consisting of m line segments

$$L_i := [(a_i, b_i) : (a_{i+1}, b_{i+1})], \quad i = 1, 2, \ldots, m,$$

where $(a_{m+1}, b_{m+1}) = (a_1, b_1)$ and the vertices are in counterclockwise order (See Figure 13.13)

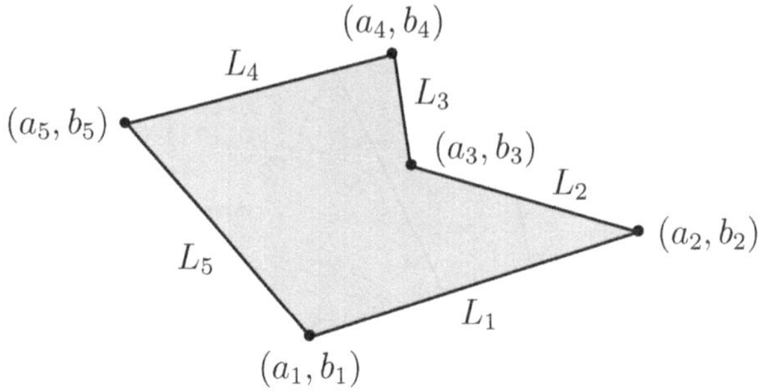

FIGURE 13.13: Closed polygon.

Then L_i has the parametrization

$$x = (1-t)a_i + ta_{i+1}, \quad y = (1-t)b_i + tb_{i+1}, \quad 0 \le t \le 1,$$

hence

$$\begin{aligned}\int_{L_i} (x\,dy - y\,dx) &= (b_{i+1} - b_i) \int_0^1 \big[(1-t)a_i + ta_{i+1}\big]\,dt \\ &\qquad - (a_{i+1} - a_i) \int_0^1 \big[(1-t)b_i + tb_{i+1}\big]\,dt \\ &= a_i b_{i+1} - a_{i+1} b_i.\end{aligned}$$

Therefore,

$$\operatorname{area}(S) = \frac{1}{2} \sum_{1=1}^{m} (a_i b_{i+1} - a_{i+1} b_i). \qquad \Diamond$$

Exercises

1. Verify directly the following version of Stokes's theorem

$$\begin{aligned}&\int_{\partial S} [f\,dx + g\,dy + h\,dz] \\ &\quad = \int_S \big[(h_y - g_z)\,dy \wedge dz + (f_z - h_x)\,dx \wedge dz + (g_x - f_y)\,dx \wedge dy\big],\end{aligned}$$

where S is the cylinder $\big\{(x, y, z) : x^2 + y^2 = 1,\ 0 \le z \le 1\big\}$.

2. For $(x, y) \ne (0, 0)$ define

$$P(x, y) = \frac{-y\,dx}{x^2 + y^2} \quad \text{and} \quad Q(x, y) = \frac{x\,dy}{x^2 + y^2}.$$

Show that

(a) $Q_x = P_y$

(b) $\int_{\varphi_r} P\,dx + Q\,dy = 2\pi$, where $\varphi_r(t) = (r\cos t, r\sin t)$, $0 \le t \le 2\pi$

(c) $\int_{\psi} P\,dx + Q\,dy = 2\pi$, where ψ is any piecewise smooth, clockwise oriented, simple closed curve enclosing $(0,0)$

(d) $$\int_0^{2\pi} \frac{\cos^{2m} t \sin^{2m} t}{a^2 \cos^{4m+2} t + b^2 \sin^{4m+2} t}\,dt = \frac{2\pi}{(2m+1)ab},\ m \in \mathbb{Z}^+,\ a, b > 0.$$

3. Let $0 < r < R$ and let $S = \{(x,y) : r^2 \le x^2 + y^2 \le R^2\}$. Verify Green's theorem on S for

$$\text{(a) } P(x,y) = \frac{-y}{\sqrt{x^2+y^2}}, \qquad Q(x,y) = \frac{x}{\sqrt{x^2+y^2}}$$

$$\text{(b) } P(x,y) = \frac{y}{\sqrt{x^2+y^2}}, \qquad Q(x,y) = \frac{x}{\sqrt{x^2+y^2}}$$

$$\text{(c) } P(x,y) = \frac{x}{x^2+y^2}, \qquad Q(x,y) = \frac{-y}{x^2+y^2}.$$

4. Use Green's theorem to evaluate the following integrals, where the curves C have counterclockwise orientation.

(a) $$\int_C \big[\sin(x-y)\,dx + \sin(x+y)\,dy\big], \quad C = \mathrm{bd}\big([0,\pi/2] \times [0,\pi/2]\big)$$

(b) $$\int_C \big[e^{-xy}\,dx + e^{xy}\,dy\big], \quad C = \mathrm{bd}\big([0,1] \times [0,1]\big)$$

(c) $$\int_C \big[\cos(xy)\,dx + \sin(xy)\,dy\big], \quad C = \mathrm{bd}\big([0,1] \times [0,1]\big)$$

(d) $\int_C \big[f(x)\,dx + g(y)\,dy\big]$, where f and g are C^1 and C is simple, closed, and piecewise C^1.

5. Use 13.5.11 to show that the area enclosed by the "elliptical astroid"

$$\left(\frac{x^2}{a^2}\right)^{1/(2m+1)} + \left(\frac{y^2}{b^2}\right)^{1/(2m+1)} = 1, \quad a > 0,\ b > 0,\ m \in \mathbb{Z}^+,$$

is given by

$$\beta \int_0^{\pi/2} \big(\cos^{2m} t + \sin^{2m} t\big)\,dt = \frac{\beta\pi}{2}\,\frac{(2m-1)(2m-3)\cdots 5 \cdot 3}{2m(2m-2)\cdots 4 \cdot 2},$$

where $\beta := 4^{-m}ab(m + \frac{1}{2})$. (See 5.3.4.)

6. Let E be a regular region in $\mathbb{R}^n$ and let f and g be C^2 on $\mathrm{cl}(E)$. Prove *Green's formulas*:

(a) $\displaystyle\int_{\text{bd}(E)} f\nabla g \cdot \vec{\mathbf{n}}\, dS = \int_E (\nabla f \cdot \nabla g + f\nabla^2 g)\, d\boldsymbol{x}$

(b) $\displaystyle\int_{\text{bd}(E)} (f\nabla g - g\nabla f) \cdot \vec{\mathbf{n}}\, dS = \int_E (f\nabla^2 g - g\nabla^2 f)\, d\boldsymbol{x}$,

where $\nabla^2 f := \displaystyle\sum_{i=1}^{n} \frac{\partial^2 f}{\partial x_i^2}$, the *Laplacian* of f.

7. A C^2 function f is said to be *harmonic* on set $S \subseteq \mathbb{R}^n$ if $\nabla^2 f = 0$ on an open set containing S.

   (a) Show that if f is harmonic on the ball $C_r(\mathbf{0})$, then $\int_{S_r(\mathbf{0})} \nabla f \cdot \vec{\mathbf{n}}\, dS = 0$.

   (b) Show that if f and g are harmonic on the region cl(E) of 13.5.6 and $\vec{\mathbf{n}}_t = \|\boldsymbol{x}\|^{-1}\boldsymbol{x}$ on $S_t := S_t(\mathbf{0})$, then

$$\int_{S_1} \nabla f \cdot \vec{\mathbf{n}}_1\, dS = \int_{S_2} \nabla f \cdot \vec{\mathbf{n}}_2\, dS$$

   and

$$\int_{S_1} (g\,\nabla f - f\,\nabla g) \cdot \vec{\mathbf{n}}_1\, dS = \int_{S_2} (g\,\nabla f - f\,\nabla g) \cdot \vec{\mathbf{n}}_2\, dS.$$

8. Let $E \subseteq \mathbb{R}^n$ be a regular region and let f be harmonic on cl(E) (Exercise 7). Show that

$$\int_E \|\nabla f\|^2\, d\boldsymbol{x} = \int_{\text{bd}(E)} f\,\nabla f \cdot \vec{\mathbf{n}}\, dS,$$

   where $\vec{\mathbf{n}}$ is the outer normal. Deduce that if $f = 0$ on bd(E) and E is connected, then $f = 0$ on E.

9. Let $E \subseteq \mathbb{R}^n$ be a regular region and let f and g be harmonic on cl(E) (Exercise 7). Show that

$$\int_{\text{bd}(E)} (f\nabla g + g\nabla f) \cdot \vec{\mathbf{n}}\, dS = 2\int_E \nabla f \cdot \nabla g\, d\boldsymbol{x},$$

   where $\vec{\mathbf{n}}$ is the outer normal.

10. Let $n > 2$. For $t > 0$, let $C_t = C_t(\mathbf{0})$, $S_t = S_t(\mathbf{0})$, and $\vec{\mathbf{n}}_t(\boldsymbol{x}) = \|\boldsymbol{x}\|^{-1}\boldsymbol{x}$, the outer normal to S_t. Suppose f is harmonic on C_r (Exercise 7). Prove the *average value property* of harmonic functions

$$f(\mathbf{0}) = \frac{1}{\text{area}(S_r)} \int_{S_r} f\, dS$$

    by verifying (a)–(f) for $0 < t \le r$. (Refer to 13.4.2.)

(a) The function $g(\boldsymbol{x}) := \|\boldsymbol{x}\|^{2-n}$, $\boldsymbol{x} \neq \mathbf{0}$, is harmonic.

(b) $\displaystyle\int_{S_t} f\,\nabla g \cdot \vec{\mathfrak{n}}_t \, dS = \frac{2-n}{t^{n-1}} \int_{S_t} f \, dS$

(c) $\displaystyle\int_{S_t} g \nabla f \cdot \vec{\mathfrak{n}}_t \, dS = 0$

(d) $\displaystyle\frac{1}{t^{n-1}} \int_{S_t} f \, dS = \frac{1}{r^{n-1}} \int_{S_r} f \, dS$

(e) $\displaystyle\frac{1}{\text{area}(S_r)} \int_{S_r} f \, dS = \frac{1}{\text{area}(S_t)} \int_{S_t} f \, dS$

(f) $\displaystyle\lim_{t \to 0} \frac{1}{\text{area}(S_t)} \int_{S_t} f \, dS = f(\mathbf{0})$.

11. Let E be a region as in the statement of Green's theorem. For the functions ψ in (a) and (b) below, prove that if the conclusion of Green's theorem holds for $\psi(E)$, then it holds for E. (This is a special case of the statement in the proof of the divergence that the region E may be rotated and translated without loss of generality.)

(a) ψ is the translation $\psi(x, y) = (x + x_0, y + y_0)$.

(b) ψ is the rotation $\psi(x, y) = \big(x \cos\theta - y \sin\theta, x \sin\theta + y \cos\theta\big)$.

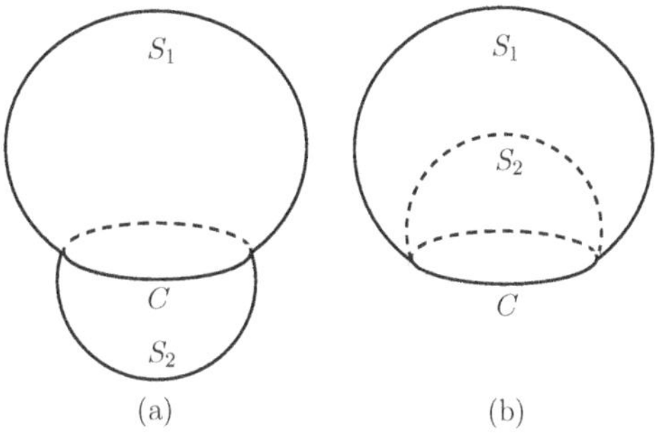

FIGURE 13.14: Surfaces S_1 and S_2 with common boundary C.

12. Orient the surfaces S_1 and S_2 in (a) and (b) of Figure 13.14 by their outer normals $\vec{\mathfrak{n}}$. Show that in

$$\text{(a)}, \int_{S_1 \cup S_2} \text{curl}\,\vec{F} \cdot \vec{\mathfrak{n}} \, dS = 0; \quad \text{(b)}, \int_{S_1} \text{curl}\,\vec{F} \cdot \vec{\mathfrak{n}} \, dS = \int_{S_2} \text{curl}\,\vec{F} \cdot \vec{\mathfrak{n}} \, dS.$$

13. Let $\boldsymbol{a} \in \mathbb{R}^n$, $n > 2$, and define an $(n-1)$ form ω on $\mathbb{R}^{n+1} \setminus \{\boldsymbol{a}\}$ by

$$\omega_{\boldsymbol{x}} = \|\boldsymbol{x} - \boldsymbol{a}\|^{-n} \sum_{i=1}^{n} (-1)^{i-1} (x_i - a_i) \, dx_1 \wedge \cdots \wedge \widehat{dx_i} \wedge \cdots \wedge dx_n.$$

Show that $d\omega = 0$. Conclude that if S is a compact, oriented n-surface-with-boundary in $\mathbb{R}^{n+1}$ and $\boldsymbol{a} \notin S$, then $\int_{\partial S} \omega = 0$.

14. Use the divergence theorem and 11.5.6 to show that the area of the sphere $S_r(\mathbf{0})$ is $nr^{n-1}\alpha_n$, derived by another method in 13.4.2.

15. Let $E \subseteq \mathbb{R}^n$ be a regular region and $\boldsymbol{a} \in E$. Define f on $\mathbb{R}^n \setminus \{\boldsymbol{a}\}$ by $f(\boldsymbol{x}) = \|\boldsymbol{x} - \boldsymbol{a}\|^{2-n}$. Show that $\operatorname{div} \nabla f = 0$. Conclude that if $C_r(\boldsymbol{a}) \subseteq E$, then

$$\int_{\mathrm{bd}(E)} (\nabla f) \cdot \vec{\mathfrak{n}}\, dS = \int_{S_r(\boldsymbol{a})} (\nabla f) \cdot \vec{\mathfrak{n}}\, dS = (2-n)n\alpha_n,$$

where $\vec{\mathfrak{n}}$ denotes the outer normals.

*13.6 Closed Forms in $\mathbb{R}^n$

13.6.1 Definition. A C^1 m-form ω on an open subset W of $\mathbb{R}^n$ is said to be *closed* if $d\omega = 0$. The form ω is *exact* if there exists a C^2 $(m-1)$-form η on W such that $d\eta = \omega$. ◊

By 13.1.16(b), an exact form is closed. The converse is false (see Exercise 13.5.2). However, there is a general class of regions on which every closed m-form is exact. We consider first the case $m = 1$.

Closed 1-Forms on Simply Connected Regions

13.6.2 Definition. An open connected subset U of $\mathbb{R}^n$ is said to be *simply connected* if for each closed C^2 curve $\varphi : [a,b] \to \mathbb{R}^n$ in U there exists a C^2 function $\Phi : [a,b] \times [0,1] \to U$ such that for all $s \in [0,1]$ and $t \in [a,b]$,

$$\Phi(t,1) = \varphi(t), \quad \Phi(t,0) = \varphi(a) = \varphi(b), \quad \text{and} \quad \Phi(a,s) = \Phi(b,s). \qquad \lozenge$$

The function Φ is called a (C^2) *homotopy between* φ *and the point* $\boldsymbol{p} : \varphi(a) = \varphi(b)$.

Note that, for each $s \in [0,1]$, $\Phi(\cdot, s)$ is a closed C^2 curve in U such that $\Phi(\cdot, 1) = \varphi$ and $\Phi(\cdot, 0)$ is a single point $\boldsymbol{p}$. Thus a simply connected region U has the property that every closed curve in U may be contracted smoothly to a point while remaining in U (see Figure 13.15). In $\mathbb{R}^2$ this means that there are no "holes" in U. In higher dimensions a simply connected set may have holes. For example, $\mathbb{R}^n \setminus C_1(\mathbf{0})$ is simply connected if $n \geq 3$. However, the holes may not be too large: the set $\mathbb{R}^3 \setminus L$, where L is a line, is not simply connected.

To prove that every closed 1-form of class C^2 on a simply connected set is exact, we follow [5].

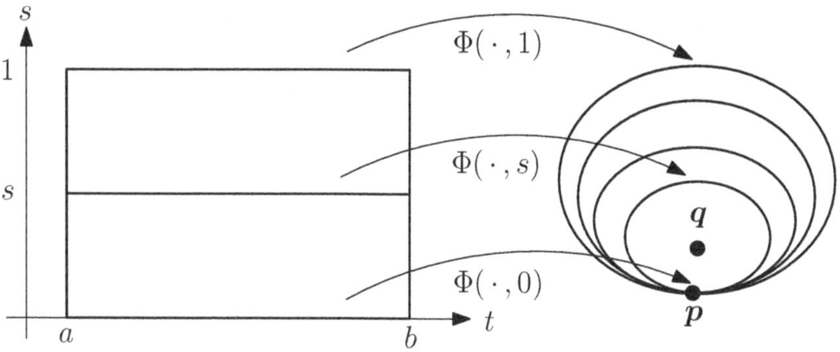

FIGURE 13.15: Curves contracting to $\boldsymbol{p}$ must pass through $\boldsymbol{q}$.

13.6.3 Lemma. *Let ω be a closed 1-form on a simply connected subset U of $\mathbb{R}^n$. Then $\int_\varphi \omega = 0$ for each closed C^2 curve φ in U.*

Proof. Let $\omega = \sum_{j=1}^n f_j\, dx_j$ and let $\Phi : [a,b] \times [0,1] \to U$ be a homotopy as in 13.6.2. By hypothesis,

$$0 = d\omega = \sum_{j=1}^n \sum_{i=1}^n \partial_i f_j dx_i \wedge dx_j = \sum_{1\le i<j\le n} (\partial_i f_j - \partial_j f_i) dx_i \wedge dx_j,$$

hence

$$\partial_i f_j = \partial_j f_i \quad \text{for all } i \text{ and } j. \tag{13.32}$$

Define C^1 functions $P(t,s)$ and $Q(t,s)$ on $S := [a,b] \times [0,1]$ by

$$P = \sum_{j=1}^n (f_j \circ \Phi)\partial_s \Phi_j \quad \text{and} \quad Q = \sum_{i=1}^n (f_i \circ \Phi)\partial_t \Phi_j,$$

where we have suppressed the variable (t,s). By the chain and product rules,

$$\begin{aligned}\partial_t P &= \sum_{j=1}^n \Big\{(\partial_s\Phi_j)[(\nabla f_j)\circ\Phi]\cdot(\partial_t\Phi) + (f_j\circ\Phi)(\partial_{ts}\Phi_j)\Big\} \quad \text{and}\\ \partial_s Q &= \sum_{i=1}^n \Big\{(\partial_t\Phi_i)[(\nabla f_i)\circ\Phi]\cdot(\partial_s\Phi) + (f_i\circ\Phi)(\partial_{st}\Phi_i)\Big\}.\end{aligned}$$

Since Φ is C^2, $\partial_{ts}\Phi_j = \partial_{st}\Phi_j$, hence

$$\begin{aligned}\partial_t P - \partial_s Q &= \sum_{j=1}^n \Big\{(\partial_s\Phi_j)[(\nabla f_j)\circ\Phi]\cdot(\partial_t\Phi) - \sum_{i=1}^n \Big\{(\partial_t\Phi_i)[(\nabla f_i)\circ\Phi]\cdot(\partial_s\Phi)\\ &= \sum_{i,j}(\partial_s\Phi_j)[(\partial_i f_j)\circ\Phi](\partial_t\Phi_i) - \sum_{i,j}(\partial_t\Phi_i)[(\partial_j f_i)\circ\Phi](\partial_s\Phi_j).\end{aligned}$$

By (13.32), $\partial_t P - \partial_s Q = 0$, hence, by Green's theorem,

$$\int_{\partial S} (P\,ds + Q\,dt) = 0.$$

Now, the positively oriented boundary of S consists of the parameterized line segments

$$\psi_1(t) = (t, 0), \quad a \le t \le b; \qquad \psi_2(s) = (b, s), \quad 0 \le s \le 1;$$
$$\psi_3(t) = (-t, 1), \quad -b \le t \le -a; \qquad \psi_4(s) = (a, -s), \quad -1 \le s \le 0.$$

(See Figure 13.6.) From the calculations

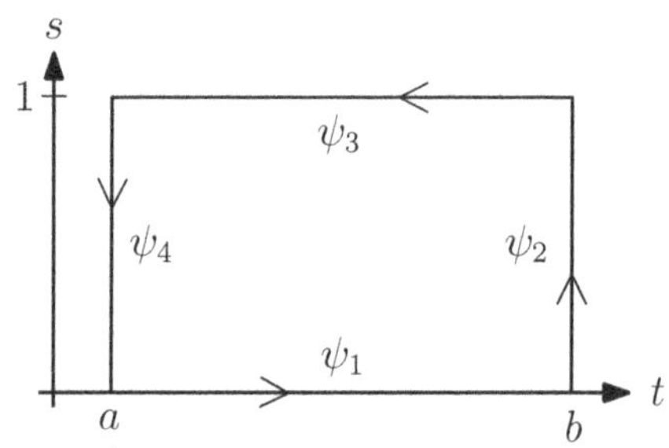

FIGURE 13.16: Boundary parametrization.

$$\int_{\psi_1} (P\,ds + Q\,dt) = \int_a^b Q(t,0)\,dt, \quad \int_{\psi_2} (P\,ds + Q\,dt) = \int_0^1 P(b,s)\,ds,$$
$$\int_{\psi_3} (P\,ds + Q\,dt) = -\int_a^b Q(t,1)\,dt, \quad \int_{\psi_4} (P\,ds + Q\,dt) = -\int_0^1 P(a,s)\,ds,$$

we have

$$\begin{aligned} 0 = \int_{\partial S} (P\,ds + Q\,dt) &= \int_0^1 \big[P(b,s) - P(a,s)\big]\,ds + \int_a^b \big[Q(t,0) - Q(t,1)\big]\,dt, \\ &= \int_0^1 \sum_{j=1}^n \big[f_j(\Phi(b,s)))\partial_s\Phi_j(b,s) - f_j(\Phi(a,s))\partial_s\Phi_j(a,s)\big]\,ds \\ &\quad + \int_a^b \sum_{j=1}^n \big[f_j(\varphi(a))\partial_t\Phi_j(t,0) - f_j(\varphi(t))\partial_t\Phi_j(t,1)\big]\,dt. \end{aligned}$$

Since $\Phi(b,s) = \Phi(a,s)$, the first integral is zero hence so is the second. Since $\partial_t\Phi_j(t,0) = 0$ and $\partial_t\Phi_j(t,1) = \varphi_j'(t)$, we see from the second integral that

$$\int_\varphi \omega = \int_a^b \sum_{j=1}^n f_j(\varphi(t))\varphi_j'(t)\,dt = 0. \qquad \square$$

13.6.4 Lemma. *Let $\varphi : [0,1] :\to \mathbb{R}^n$ be a piecewise C^1 curve such that $\varphi(0) = \varphi(1)$. Then there exists a sequence of C^∞ functions $\varphi_k : [0,1] \to \mathbb{R}^n$ with the following properties:*

(a) $\varphi_k(0) = \varphi_k(1) = \varphi(0)$ *for all* k.

(b) $\lim_k \varphi_k'(t) = \varphi'(t)$ *at each continuity point* t *of* φ'.

(c) $\lim_k \varphi_k = \varphi$ *uniformly on* $[0,1]$.

(d) *The sequences* $\{\varphi_k\}$ *and* $\{\varphi_k'\}$ *are uniformly bounded on* $[0,1]$.

Proof. By considering components, we may assume that $n = 1$, that is, φ is real-valued. Extend φ' periodically with period 1 to $\mathbb{R}$. Let M be a bound for $|\varphi'|$ on $\mathbb{R}$. By 13.3.3 there exists a C^∞ function $h : \mathbb{R} \to \mathbb{R}^+$ such that $h > 0$ on $(-1,1)$ and $h = 0$ on $(-1,1)^c$. Multiplying h by a positive constant, we may assume that $\int_{\mathbb{R}} h = 1$. Let $h_k(x) = kh(kx)$, $k = 1, 2, \ldots$. Then $h_k \geq 0$, $h_k(x) = 0$ for $|x| \geq 1/k$, and $\int_{\mathbb{R}} h_k = 1$. Define a C^∞ function g_k on $\mathbb{R}$ by

$$g_k(x) = \int_{-\infty}^{\infty} \varphi'(y)h_k(x-y)\,dy = \int_{-1/k}^{1/k} \varphi'(x+y)h_k(y)\,dy.$$

The sequence $\{g_k\}$ is uniformly bounded since

$$|g_k(x)| \leq \int_{-\infty}^{\infty} |\varphi'(x+y)|h_k(y)\,dy \leq M \int_{-\infty}^{\infty} h_k(y)\,dy = M.$$

By periodicity,

$$\int_0^1 \varphi'(x+y)\,dx = \int_0^1 \varphi'(x)\,dx = \varphi(1) - \varphi(0) = 0$$

(Exercise 5.3.1), hence, by Fubini's theorem,

$$\int_0^1 g_k(x)\,dx = \int_{-\infty}^{\infty} h_k(y) \int_0^1 \varphi'(x+y)\,dx\,dy = 0.$$

Now define φ_k on $\mathbb{R}$ by

$$\varphi_k(x) = \varphi(0) + \int_0^x g_k(y)\,dy.$$

Then (a) and (d) hold and (b) follows from

$$\varphi_k'(x) - \varphi'(x) = g_k(x) - \varphi'(x) = \int_{-1/k}^{1/k} \left[\varphi'(x+y) - \varphi'(x)\right]h_k(y)\,dy,$$

which tends to 0 at continuity points x as $k \to +\infty$.

Finally, (c) follows from (b), the inequality

$$|\varphi_k(t) - \varphi(t)| \le \int_0^t |\varphi_k'(x) - \varphi'(x)|\,dx \le \int_0^1 |\varphi_k'(x) - \varphi'(x)|\,dx,$$

and Lebesgue's dominated convergence theorem, noting that the set of discontinuity points of φ' is finite and hence has measure zero. □

13.6.5 Theorem. *Let ω be a closed 1-form on a simply connected subset U of $\mathbb{R}^n$. Then ω is exact.*

Proof. By 12.2.10 it suffices to show that $\int_\varphi \omega = 0$ for every piecewise C^1 closed curve $\varphi : [0,1] \to U$. Let $\{\varphi_k\}$ be as in 13.6.4. Since $\varphi_k \to \varphi$ uniformly on $[0,1]$ and $\varphi([0,1]) \subseteq U$, it follows that $\varphi_k([0,1]) \subseteq U$ for all sufficiently large k (Exercise 8.5.22). For such k, $\int_{\varphi_k} \omega = 0$ by 13.6.3. By (b) and (c) of 13.6.4, Lebesgue's dominated convergence theorem, and the definition of integral of a form (13.16), $\int_{\varphi_k} \omega \to \int_\varphi \omega$. Therefore, $\int_\varphi \omega = 0$, as required. □

Closed m-Forms on Star-Shaped Regions

13.6.6 Definition. A subset W of $\mathbb{R}^n$ is said to be *star-shaped with respect to $\mathbf{y} \in W$* if the line segment from $\mathbf{y}$ to any point $\mathbf{x} \in W$ lies in W:

$$\mathbf{y} + t(\mathbf{x} - \mathbf{y}) \in W, \quad 0 \le t \le 1. \qquad \Diamond$$

For example, a convex set is star-shaped with respect to every one of its points. In Figure 13.17, W is star-shaped with respect to $\mathbf{y}$ but not $\mathbf{z}$, and V is not star-shaped with respect to any of its points.

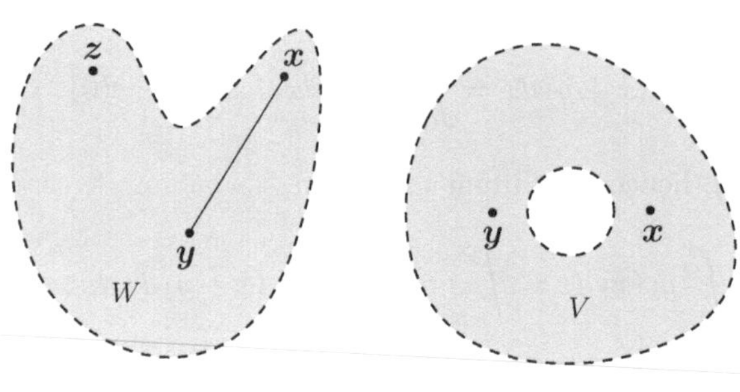

FIGURE 13.17: Star-shaped and non-star-shaped regions.

13.6.7 Poincaré's Lemma. *Let $W \subseteq \mathbb{R}^n$ be open and star-shaped with respect to some $\boldsymbol{y} \in W$. If ω is a closed C^1 m-form on W, where $1 \le m \le n$, then ω is exact.*

Proof. Define a function $\psi : [0,1] \times W \to W$ by $\psi(t, \boldsymbol{x}) = \boldsymbol{y} + t(\boldsymbol{x} - \boldsymbol{y})$. For an r-form

$$\eta = \sum_{\mathbf{j} \in \mathbf{J}_r} g_{\mathbf{j}}\, dx_{\mathbf{j}}$$

on W, define the $(r-1)$-form $\widetilde{\eta}$ on W by

$$\widetilde{\eta}_{\boldsymbol{x}} = \sum_{\mathbf{j}\in\mathbf{J}_m} \left[\int_0^1 t^{r-1}(g_{\mathbf{j}} \circ \psi)(t,\boldsymbol{x})\,dt\right] \eta_{\mathbf{j}}, \quad \text{where}$$

$$\eta_{\mathbf{j}} := \sum_{i=1}^{r} (-1)^{i-1}(x_{j_i} - y_{j_i})\,dx_{j_1} \wedge \cdots \wedge \widehat{dx_{j_i}} \wedge \cdots \wedge dx_{j_r}, \quad \mathbf{j} = (j_1, \ldots, j_r).$$

A standard argument shows that the definition of $\widetilde{\eta}$ is independent of the choice of representation of η. In particular, by putting η in canonical form we see that $\eta = 0$ implies that $\widetilde{\eta} = 0$. Furthermore,

$$d\eta_{\mathbf{j}} = \sum_{i=1}^{r} (-1)^{i-1} d(x_{j_i} - y_{j_i})\,dx_{j_1} \wedge \cdots \wedge \widehat{dx_{j_i}} \wedge \cdots \wedge dx_{j_m} = r\,dx_{\mathbf{j}}.$$

Now let

$$\omega = \sum_{\mathbf{j}\in\mathbf{J}_m} f_{\mathbf{j}}\,dx_{\mathbf{j}}.$$

Then

$$\widetilde{\omega} = \sum_{\mathbf{j}\in\mathbf{J}_m} \left[\int_0^1 t^{m-1}(f_{\mathbf{j}} \circ \psi)(t,\boldsymbol{x})\,dt\right] \omega_{\mathbf{j}},$$

and, by 13.1.16(d) (suppressing the variables $(t,\boldsymbol{x})$ in $f_{\mathbf{j}} \circ \psi(t,\boldsymbol{x})$),

$$d\,\widetilde{\omega} = \sum_{\mathbf{j}\in\mathbf{J}_m} \left\{ d\left[\int_0^1 t^{m-1} f_{\mathbf{j}} \circ \psi\,dt\right] \wedge \omega_{\mathbf{j}} + \left[\int_0^1 t^{m-1} f_{\mathbf{j}} \circ \psi\,dt\right] d\omega_{\mathbf{j}} \right\}.$$

Differentiating under the integral sign, applying the chain rule and noting that $\psi_{\boldsymbol{x}} = tI_n$ we have

$$d\left[\int_0^1 t^{m-1}(f_{\mathbf{j}} \circ \psi)\,dt\right] = \sum_{i=1}^{n} \left[\int_0^1 t^m(\partial_i(f_{\mathbf{j}}) \circ \psi)\,dt\right] dx_i.$$

Therefore, using $d\omega_{\mathbf{j}} = m\,dx_{\mathbf{j}}$,

$$d\,\widetilde{\omega} = \sum_{\mathbf{j}\in\mathbf{J}_m} \left\{ \sum_{i=1}^{n} \left[\int_0^1 t^m(\partial_i f_{\mathbf{j}}) \circ \psi\,dt\right] dx_i \wedge \omega_{\mathbf{j}} + m\left[\int_0^1 t^{m-1} f_{\mathbf{j}} \circ \psi\right] dx_{\mathbf{j}} \right\}. \tag{13.33}$$

On the other hand,

$$d\omega = \sum_{\mathbf{j}\in\mathbf{J}_m} \left(\sum_{i=1}^{n} \partial_i f_{\mathbf{j}}\,dx_i\right) \wedge dx_{\mathbf{j}} = \sum_{\mathbf{j}\in\mathbf{J}_m} \sum_{i=1}^{n} \partial_i f_{\mathbf{j}}\,dx_i \wedge dx_{\mathbf{j}},$$

hence, since $d\omega = 0$,

$$\sum_{\mathbf{j}\in\mathbf{J}_m}\sum_{i=1}^{n}\left[\int_0^1 t^m(\partial_i(f_{\mathbf{j}})\circ\psi)(t,\boldsymbol{x})\,dt\right](d\omega)_{(i,\mathbf{j})} = \widetilde{d\omega} = 0.$$

By the above definition,

$$\begin{aligned}(d\omega)_{(i,\mathbf{j})} &= \sum_{\ell=1}^{m}(-1)^\ell(x_{j_\ell}-y_{j_\ell})\,dx_i\wedge dx_{j_1}\wedge\cdots\wedge\widehat{dx_{j_\ell}}\wedge\cdots\wedge dx_{j_m}\\ &\qquad + (x_i-y_i)\,dx_{j_1}\wedge\cdots\wedge dx_{j_m}\\ &= -\,dx_i\wedge\omega_{\mathbf{j}} + (x_i-y_i)\,dx_{\mathbf{j}},\end{aligned}$$

hence

$$= \sum_{\mathbf{j}\in\mathbf{J}_m}\sum_{i=1}^{n}\left[\int_0^1 t^m(\partial_i f_{\mathbf{j}})\circ\psi\,dt\right]\left[-\,dx_i\wedge\omega_{\mathbf{j}} + (x_i-y_i)\,dx_{\mathbf{j}}\right] = 0. \tag{13.34}$$

Adding (13.33) and (13.34), we obtain

$$d\widetilde{\omega} = \sum_{\mathbf{j}\in\mathbf{J}_m}\left\{m\left[\int_0^1 t^{m-1}(f_{\mathbf{j}}\circ\psi)\right] + \sum_{i=1}^{n}\left[(x_i-y_i)\int_0^1 t^m(\partial_i f_{\mathbf{j}})\circ\psi\,dt\right]\right\}dx_{\mathbf{j}}.$$

The term in braces is simply

$$\int_0^1 \frac{d}{dt}\left[t^m f_{\mathbf{j}}\circ\psi\right]dt = t^m f_{\mathbf{j}}\circ\psi\Big|_0^1 = f_{\mathbf{j}}.$$

Therefore, $d\widetilde{\omega} = \omega$, which shows that ω is exact. □

From Poincaré's lemma we obtain the following results from classical vector analysis, where, in keeping with the spirit, we write grad f for ∇f.

13.6.8 Corollary. *Let W be an open star-shaped subset of $\mathbb{R}^3$ and let*

$$\vec{F}(x,y,z) = \big(P(x,y,z), Q(x,y,z), R(x,y,z)\big)$$

be a C^1 vector field on W. Then

(a) $\operatorname{curl}\vec{F} = \mathbf{0}$ *iff* $\vec{F} = \operatorname{grad} f$ *for some C^2 function $f: W\to\mathbb{R}$.*

(b) $\operatorname{div}\vec{F} = 0$ *iff* $\vec{F} = \operatorname{curl}\vec{G}$ *for some C^2 vector field $\vec{G}$ on W.*

Proof. (a) If $\vec{F} = \operatorname{grad} f = (f_x, f_y, f_z)$, then

$$\operatorname{curl}\vec{F} = (f_{zy}-f_{yz}, f_{xz}-f_{zx}, f_{yx}-f_{xy}),$$

which is zero because f is C^2. Conversely, assume that $\operatorname{curl}\vec{F} = \mathbf{0}$, that is,

$$R_y - Q_z = P_z - R_x = Q_x - P_y = 0.$$

Let $\omega = P\,dx + Q\,dy + R\,dz$. Then

$$\begin{aligned} d\omega &= (P_y\,dy + P_z\,dz) \wedge dx + (Q_x\,dx + Q_z\,dz) \wedge dy + (R_x\,dx + R_y\,dy) \wedge dz \\ &= (Q_x - P_y)\,dx \wedge dy + (R_x - P_z)\,dx \wedge dz + (R_y - Q_z)\,dy \wedge dz = 0 \end{aligned}$$

so ω is closed. By Poincaré's lemma, there exists a 0-form f of class C^2 on W such that $df = \omega$, that is, $\operatorname{grad} f = \vec{F}$.

(b) If $\vec{F} = \operatorname{curl} \vec{G}$, where $\vec{G} = (f, g, h)$, then

$$P = h_y - g_z, \quad Q = f_z - h_x, \quad \text{and} \quad R = g_x - f_y,$$

hence, if G is C^2,

$$\operatorname{div} \vec{F} = P_x + Q_y + R_z = (h_{yx} - g_{zx}) + (f_{zy} - h_{xy}) + (g_{xz} - f_{yz}) = 0.$$

Conversely, assume $\operatorname{div} \vec{F} = 0$ and let

$$\omega = R\,dx \wedge dy + P\,dy \wedge dz + Q\,dz \wedge dx.$$

Then $d\omega = \operatorname{div} \vec{F}\,dx \wedge dy \wedge dz$, hence ω is closed. By Poincaré's lemma,

$$\omega = d(f\,dx + g\,dy + h\,dz) = (g_x - f_y)\,dx \wedge dy + (h_x - f_z)\,dx \wedge dz + (h_y - g_z)\,dy \wedge dz$$

for some C^2 functions f, g, h on W. Therefore,

$$P = h_y - g_z, \quad Q = f_z - h_x, \quad R = g_x - f_y,$$

that is, $\vec{F} = \operatorname{curl}(f, g, h)$. □

Part III

Appendices

Appendix A

Set Theory

In this appendix we give an overview of those aspects of elementary set theory that are used throughout the book. For details the reader may wish to consult [2, 8].

Notation for a Set

A *set* is simply a collection of objects, each of which is called a *member* or *element* of the set. Sets are usually denoted by capital letters, and members of a set by small letters. If x is a member of the set A, we write $x \in A$; otherwise, we write $x \notin A$. The *empty set*, denoted by $\emptyset$, is the set with no members.

A concrete set may be described either by listing its elements or by *set-builder notation*. The latter notation is of the form $\{x : P(x)\}$, which is read "the set of all x such that $P(x)$," where $P(x)$ is a well-defined property that x must possess in order to belong to the set. For example, the set A of all odd positive integers may be described as

$$A = \{1, 3, 5, \ldots\} = \{n : n = 2m - 1 \text{ for some positive integer } m\}.$$

A set A is a *subset* of a set B, written $A \subseteq B$, if every member of A is a member of B. If $A \subseteq B$ and $A \neq B$, then A is called a *proper subset* of a set B. The empty set is a subset of every set and a proper subset of every nonempty set. Sets A and B are said to be *equal*, written $A = B$, if each is a subset of the other. If all sets under consideration are subsets of the set S, then S is called a *universal set (of discourse)*.

Set Operations

Let S be a universal set. The basic set operations are

$$\begin{aligned}
A \cup B &= \{x : x \in A \text{ or } x \in B\}, && \textit{union} \text{ of } A \text{ and } B;\\
A \cap B &= \{x : x \in A \text{ and } x \in B\}, && \textit{intersection} \text{ of } A \text{ and } B;\\
A \times B &= \{(x, y) : x \in A \text{ and } y \in B\}, && \textit{Cartesian product} \text{ of } A \text{ and } B;\\
A^c &= \{x : x \in S \text{ and } x \notin A\}, && \textit{complement} \text{ of } A \text{ in } S;\\
A \setminus B &= \{x : x \in A \text{ and } x \notin B\}, && \textit{difference} \text{ of } A \text{ and } B.
\end{aligned}$$

DOI: 10.1201/9781003629382-A

More generally, if $\{A_\mathfrak{i} : \mathfrak{i} \in \mathfrak{I}\}$ is an arbitrary collection of sets indexed by a set $\mathfrak{I}$, then the *union* and *intersection* of the collection are defined, respectively, by

$$\bigcup_{\mathfrak{i}\in\mathfrak{I}} A_\mathfrak{i} = \{x : x \in A_\mathfrak{i} \text{ for some } \mathfrak{i} \in \mathfrak{I}\},$$

$$\bigcap_{\mathfrak{i}\in\mathfrak{I}} A_\mathfrak{i} = \{x : x \in A_\mathfrak{i} \text{ for every } \mathfrak{i} \in \mathfrak{I}\}.$$

If the index set is $\{1, 2\ldots, n\}$ or $\{1, 2, \ldots, n, \ldots\}$, we use the alternate notation

$$\bigcup_{j=1}^{n} A_j = A_1 \cup A_2 \cup \cdots \cup A_n, \qquad \bigcap_{j=1}^{n} A_j = A_1 \cap A_2 \cap \cdots \cap A_n$$

and

$$\bigcup_{j=1}^{\infty} A_j = A_1 \cup A_2 \cup \ldots, \qquad \bigcap_{j=1}^{\infty} A_j = A_1 \cap A_2 \cap \ldots$$

A sequence of sets A_n is said to be *increasing* if $A_1 \subseteq A_2 \subseteq \cdots$, in which case we write $A_n \uparrow$. Similarly, the sequence is *decreasing* if $A_1 \supseteq A_2 \supseteq \cdots$, written $A_n \downarrow$. In the first case we also write $A_n \uparrow A$, where $A = A_1 \cup A_2 \cup \cdots$, and in the second $A_n \downarrow A$, where $A = A_1 \cap A_2 \cap \cdots$.

For finitely many sets we extend the definition of Cartesian product by

$$\prod_{j=1}^{n} A_j = A_1 \times \cdots \times A_n = \{(a_1, \ldots, a_n) : a_j \in A_j,\ j = 1, \ldots n\},$$

where $(a_1, \ldots, a_n)$ is an (ordered) *n-tuple*. Also, we write

$$A^n = \underbrace{A \times A \cdots \times A}_{n}.$$

In particular, for an interval $[a, b]$ and the set of all real numbers $\mathbb{R}$,

$$[a, b]^n = \underbrace{[a, b] \times \cdots \times [a, b]}_{n} \quad \text{and} \quad \mathbb{R}^n = \underbrace{\mathbb{R} \times \cdots \times \mathbb{R}}_{n}.$$

The following propositions summarize the basic properties of set operations that will be needed in the text. As with many set equalities, they may be established directly by showing that an arbitrary member of the left side of an equation is a member of the right side, and vice versa.

Proposition. *If $\{A_\mathfrak{i} : \mathfrak{i} \in \mathfrak{I}\}$ is collection of subsets of a set S, then*

(a) $\left(\bigcup_{\mathfrak{i}\in\mathfrak{I}} A_\mathfrak{i}\right)^c = \bigcap_{\mathfrak{i}\in\mathfrak{I}} A_\mathfrak{i}^c$.

(b) $A \cup \left(\bigcap_{\mathfrak{i}\in\mathfrak{I}} A_\mathfrak{i}\right) = \bigcap_{\mathfrak{i}\in\mathfrak{I}} A \cup A_\mathfrak{i}$.

(c) $\left(\bigcap_{\mathfrak{i}\in\mathfrak{I}} A_\mathfrak{i}\right)^c = \bigcup_{\mathfrak{i}\in\mathfrak{I}} A_\mathfrak{i}^c$.

(d) $A \cap \left(\bigcup_{\mathfrak{i}\in\mathfrak{I}} A_\mathfrak{i}\right) = \bigcup_{\mathfrak{i}\in\mathfrak{I}} A \cap A_\mathfrak{i}$.

Parts (a) and (c) of the above proposition are known as DeMorgan's laws. Parts (b) and (d) are called *distributive laws.*

Proposition. *The Cartesian product of sets has the following properties:*

(a) $A \times \left(A_1 \cup \cdots \cup A_n\right) = (A \times A_1) \cup \cdots \cup (A \times A_n)$.

(b) $A \times \left(A_1 \cap \cdots \cap A_n\right) = (A \times A_1) \cap \cdots \cap (A \times A_n)$.

(c) $\left(A_1 \cap \cdots \cap A_n\right) \times \left(B_1 \cap \cdots \cap B_n\right) = (A_1 \times B_1) \cap \cdots \cap (A_n \times B_n)$.

Partitions and Equivalence Relations

A collection of sets is *pairwise disjoint* if $A \cap B = \emptyset$ for each pair of distinct members A and B in the collection. A *partition* of a set S is a collection of nonempty pairwise disjoint sets whose union is S.

An *equivalence relation* on a set S is a subset R of $S \times S$ with the following properties:

- (reflexivity) xRx for every $x \in S$;
- (symmetry) $xRy \Rightarrow yRx$;
- (transitivity) xRy and $yRz \Rightarrow xRz$.

Here, as is customary, we have written xRy for $(x, y) \in R$.

There is an important duality regarding partitions and equivalence relations: If R is an equivalence relation on S, then the collection of sets of the form

$$[x] := \{y \in S : xRy\},$$

called an *equivalence class* of the relation, is a partition of S. Conversely, given a partition of S, define

$$xRy \quad \text{iff } x \text{ and } y \text{ are in the same partition member.}$$

Then R is an equivalence relation on S whose equivalence classes are precisely the members of the partition.

Functions

Let A and B be nonempty sets. A *function* or *mapping from A to B* is a rule f that assigns to each member x of A a unique member $f(x)$ of B. We then write $f : A \to B$. The set A is called the *domain of f*. The alternate notation $x \mapsto f(x) : A \to B$ is also used. If $A_0 \subseteq A$ and $B_0 \subseteq B$, then

$$f(A_0) = \{f(x) : x \in A_0\} \quad \text{and} \quad f^{-1}(B_0) = \{x \in A : f(x) \in B_0\}$$

are called, respectively, the *image of* A_0 and the *pre-image* of B_0 under f. The set $f(A)$ is called the *range* of f. A function $f : A \to B$ is said to be *onto* B if $f(A) = B$, and *one-to-one* if $x_1 \neq x_2$ implies $f(x_1) \neq f(x_2)$.

Proposition. *Let* $f : A \to B$ *be a function,* $\{A_{\mathfrak{i}} : \mathfrak{i} \in \mathfrak{I}\}$ *a collection of subsets of* A*, and* $\{B_{\mathfrak{j}} : \mathfrak{j} \in \mathfrak{J}\}$ *a collection of subsets of* B*. Then*

(a) $f^{-1}\Big(\bigcup_{\mathfrak{j}\in\mathfrak{J}} B_{\mathfrak{j}}\Big) = \bigcup_{\mathfrak{j}\in\mathfrak{J}} f^{-1}(B_{\mathfrak{j}})$.

(b) $f^{-1}\Big(\bigcap_{\mathfrak{j}\in\mathfrak{J}} B_{\mathfrak{j}}\Big) = \bigcap_{\mathfrak{j}\in\mathfrak{J}} f^{-1}(B_{\mathfrak{j}})$.

(c) $f\Big(\bigcup_{\mathfrak{i}\in\mathfrak{I}} A_{\mathfrak{i}}\Big) = \bigcup_{\mathfrak{i}\in\mathfrak{I}} f(A_{\mathfrak{i}})$.

(d) $f\Big(\bigcap_{\mathfrak{i}\in\mathfrak{I}} A_{\mathfrak{i}}\Big) \subseteq \bigcap_{\mathfrak{i}\in\mathfrak{I}} f(A_{\mathfrak{i}})$, *where equality holds if* f *is one-to-one.*

(e) $f^{-1}(B_{\mathfrak{j}}^c) = \big[f^{-1}(B_{\mathfrak{j}})\big]^c$.

(f) $f(A_{\mathfrak{i}}^c) \subseteq \big[f(A_{\mathfrak{i}})\big]^c$, *where equality holds if* f *is onto* B.

(g) $f\big(f^{-1}(B_{\mathfrak{j}})\big) \subseteq B_{\mathfrak{j}}$, *where equality holds if* f *is onto* B.

(h) $A_{\mathfrak{i}} \subseteq f^{-1}\big(f(A_{\mathfrak{i}})\big)$, *where equality holds if* f *is one-to-one.*

If $f : A \to B$ and $g : C \to D$ are functions with $B \subseteq C$, then the *composition of* g *and* f is the function $g \circ f : A \to D$ defined by

$$(g \circ f)(x) = g\big(f(x)\big), \quad x \in A.$$

If $D_0 \subseteq D$, then

$$(g \circ f)^{-1}(D_0) = f^{-1}\big(g^{-1}(D_0)\big).$$

If $f : A \to B$ is one-to-one and onto B, then the *inverse* $f^{-1} : B \to A$ is defined by the rule $x = f^{-1}(y)$ iff $y = f(x)$. One then has the identities

$$(f^{-1} \circ f)(x) = x \text{ and } (f \circ f^{-1})(y) = y, \quad x \in A,\ y \in B.$$

Thus $f^{-1} \circ f$ and $f \circ f^{-1}$ are the *identity functions* on A and B, respectively.

Cardinality

Two sets A and B are said to have the same *cardinality* if there exists a one-to-one function from A onto B. A set A is *finite* if either A is the empty set or A has the same cardinality as $\{1, 2, \ldots, n\}$ for some positive integer n. In the latter case, the members of A may be labeled with the numbers $1, 2, \ldots, n$, so A may be written $\{a_1, a_2, \ldots, a_n\}$. A set A is *countably infinite* if it has the same cardinality as the set of natural numbers. In this case the members of A may be labeled with the positive integers $1, 2, 3, \ldots$ A set is *countable* if it is either finite or countably infinite; otherwise it is said to be *uncountable.* The set of all integers is countably infinite, as is the set of rational numbers. The set $\mathbb{R}$ of all real numbers is uncountable, as is any (nondegenerate) interval of real numbers.

Appendix B

Linear Algebra

This appendix contains a brief review of the main ideas of linear algebra that will be needed in Part II of the text. For details and proofs the reader is referred to [9].

Vector Spaces. Bases

A *vector space* is a set $\mathcal{V}$ containing at least one member $\mathbf{0}$, called the *zero vector*, together with two operations $\boldsymbol{u}+\boldsymbol{v}$ and $a\boldsymbol{u}$ ($\boldsymbol{u}$, $\boldsymbol{v} \in \mathcal{V}$, $a \in \mathbb{R}$), called *vector addition* and *scalar multiplication*, respectively, such that for all $\boldsymbol{u}$, $\boldsymbol{v}$, $\boldsymbol{w} \in \mathcal{V}$ and a, $b \in \mathbb{R}$ the following axioms hold:

- Associativity of addition: $(\boldsymbol{u}+\boldsymbol{v})+\boldsymbol{w} = \boldsymbol{u}+(\boldsymbol{v}+\boldsymbol{w})$.
- Commutativity of addition: $\boldsymbol{u}+\boldsymbol{v} = \boldsymbol{v}+\boldsymbol{u}$.
- Additive identity: $\boldsymbol{v}+\mathbf{0} = \boldsymbol{v}$.
- Existence of additive inverse: $\boldsymbol{u}+(-\boldsymbol{u}) = \mathbf{0}$.
- Associativity of scalar multiplication: $(ab)\boldsymbol{u} = a(b\boldsymbol{u})$.
- Scalar distributivity: $a(\boldsymbol{u}+\boldsymbol{v}) = a\boldsymbol{u}+a\boldsymbol{v}$.
- Vector distributivity: $(a+b)\boldsymbol{u} = a\boldsymbol{u}+b\boldsymbol{u}$.
- Scalar multiplicative identity: $1\boldsymbol{u} = \boldsymbol{u}$.

A subset $\mathcal{W}$ of $\mathcal{V}$ containing the zero vector and closed under the operations of vector addition and scalar multiplication is called a *subspace of* $\mathcal{V}$. The set $\mathcal{W}$ is then a vector space under the operations it inherits from $\mathcal{V}$. A *linear combination* of vectors $\boldsymbol{v}_1, \ldots, \boldsymbol{v}_n \in \mathcal{V}$ is an expression of the form

$$c_1\boldsymbol{v}_1 + \cdots + c_n\boldsymbol{v}_n, \quad c_j \in \mathbb{R}.$$

The set of all linear combinations of $\boldsymbol{v}_1, \ldots, \boldsymbol{v}_n$ is called the *linear span of* $\boldsymbol{v}_1, \ldots, \boldsymbol{v}_n$ or the *subspace spanned by* $\boldsymbol{v}_1, \ldots, \boldsymbol{v}_n$. The vectors $\boldsymbol{v}_1, \ldots, \boldsymbol{v}_n$ are then said to *span* $\mathcal{V}$.

Vectors $\boldsymbol{v}_1, \ldots, \boldsymbol{v}_n \in \mathcal{V}$ are *linearly independent* if an equation of the form

$$c_1\boldsymbol{v}_1 + \cdots + c_n\boldsymbol{v}_n = \mathbf{0}$$

DOI: 10.1201/9781003629382-B

can hold only if $c_1 = \cdots = c_n = 0$. A (finite) *basis* for $\mathcal{V}$ is a finite set of linearly independent vectors that span $\mathcal{V}$. It follows that each member of $\mathcal{V}$ is uniquely expressible as a linear combination of the basis vectors. A vector space that has a basis is said to be *finite dimensional*; otherwise it is *infinite dimensional*. All bases in a finite-dimensional vector space $\mathcal{V}$ have the same number of vectors. This number is called the *dimension* of the vector space and is denoted by $\dim \mathcal{V}$. A *frame* for a finite dimensional vector space is an ordered basis.

If $\mathcal{V}$ is finite dimensional, then every set of linearly independent vectors may be extended to a basis, and every finite set of vectors that span $\mathcal{V}$ may be reduced to a basis.

An important example of a finite-dimensional vector space is Euclidean space $\mathbb{R}^n$ (Section 1.5). The *standard basis* in $\mathbb{R}^n$ consists of the n vectors

$$\boldsymbol{e}^1 = (1, 0, \ldots, 0), \quad \boldsymbol{e}^2 = (0, 1, 0, \ldots, 0), \quad \boldsymbol{e}^n = (0, 0, \ldots, 0, 1).$$

An example of an infinite-dimensional vector space is the set of all Riemann integrable functions on $[a, b]$ with the operations of pointwise addition and scalar multiplication.

A basis $\{\boldsymbol{w}_1, \ldots, \boldsymbol{w}_m\}$ for a subspace $\mathcal{W}$ of $\mathbb{R}^n$ is *orthonormal* if

$$\boldsymbol{w}_i \cdot \boldsymbol{w}_j = \begin{cases} 0 & \text{if } i \neq j, \\ 1 & \text{if } i = j, \end{cases}$$

where $(\cdot)$ is the usual inner (= dot) product on $\mathbb{R}^n$. For example, the standard basis is orthonormal. Every subspace of $\mathbb{R}^n$ has an orthonormal basis.

Linear Transformations

Let $\mathcal{U}$ and $\mathcal{V}$ be vector spaces. A *linear transformation* from $\mathcal{U}$ to $\mathcal{V}$ is a function $T : \mathcal{U} \to \mathcal{V}$ with the properties

$$T(\boldsymbol{u} + \boldsymbol{v}) = T\boldsymbol{u} + T\boldsymbol{v} \quad \text{and} \quad T(c\boldsymbol{u}) = cT\boldsymbol{u}, \quad \boldsymbol{u}, \boldsymbol{v} \in \mathcal{U}, \quad c \in \mathbb{R}.$$

Here, we have used the convention for linear transformations of dropping the parentheses in the notation $T(\boldsymbol{u})$ when there is no danger of ambiguity. The collection of all linear transformations from $\mathcal{U}$ to $\mathcal{V}$ is denoted by $\mathcal{L}(\mathcal{U}, \mathcal{V})$. It is a vector space under the operations $T_1 + T_2$ and cT defined by

$$(T_1 + T_2)(\boldsymbol{u}) = T_1\boldsymbol{u} + T_2\boldsymbol{u}, \quad (cT)\boldsymbol{u} = c(T\boldsymbol{u}), \quad \boldsymbol{u} \in \mathcal{U}, \ c \in \mathbb{R}.$$

If $T \in \mathcal{L}(\mathcal{U}, \mathcal{V})$ and $S \in \mathcal{L}(\mathcal{V}, \mathcal{W})$, then $ST := S \circ T$ is a member of $\mathcal{L}(\mathcal{U}, \mathcal{W})$. Also, the subspace $\mathcal{N}(T) := T^{-1}(\{\boldsymbol{0}\})$ of $\mathcal{U}$ is called the *nullspace* of T. The range of T, which is a subspace of $\mathcal{V}$, is denoted by $\mathcal{R}(T)$. If $\mathcal{U}$ and $\mathcal{R}(T)$ are finite dimensional, then

$$\dim \mathcal{N}(T) + \dim \mathcal{R}(T) = \dim \mathcal{U}.$$

If $T \in \mathcal{L}(\mathcal{U}, \mathcal{V})$ is one-to-one and onto $\mathcal{V}$, then $T^{-1} \in \mathcal{L}(\mathcal{V}, \mathcal{U})$. In this case T is said to be *invertible.* If $\mathcal{U}$ and $\mathcal{V}$ are finite dimensional, then T is invertible iff $\mathcal{N}(T) = \{0\}$ iff $\mathcal{R}(T) = \mathcal{V}$. In this case T maps a frame $(\boldsymbol{u}_1, \ldots, \boldsymbol{u}_n)$ in $\mathcal{U}$ onto a frame $(\boldsymbol{v}_1, \ldots, \boldsymbol{v}_n)$ in $\mathcal{V}$, where $\boldsymbol{v}_j = T\boldsymbol{u}_j$. We indicate this by writing

$$T(\boldsymbol{u}_1, \ldots, \boldsymbol{u}_n) = (\boldsymbol{v}_1, \ldots, \boldsymbol{v}_n).$$

Matrices

An $m \times n$ *matrix* is a rectangular array of real numbers with m rows and n columns. It is written variously as

$$A = [a_i^j]_{m\times n} = \begin{bmatrix} a_1^1 & a_1^2 & \cdots & a_1^n \\ a_2^1 & a_2^2 & \cdots & a_2^n \\ \vdots & \vdots & \cdots & \vdots \\ a_m^1 & a_m^2 & \cdots & a_m^n \end{bmatrix} = \begin{bmatrix} \boldsymbol{a}_1 \\ \boldsymbol{a}_2 \\ \vdots \\ \boldsymbol{a}_m \end{bmatrix} = \begin{bmatrix} \boldsymbol{a}^1 & \boldsymbol{a}^2 & \cdots & \boldsymbol{a}^n \end{bmatrix},$$

where $\boldsymbol{a}_i = (a_i^1, \cdots, a_i^n)$ is the *ith row of A* and $\boldsymbol{a}^j = (a_1^j, \cdots, a_m^j)$ is the *jth column of A* (written, of course, as a column). The number a_i^j located in row i and column j of the matrix is also written a_{ij} and is called the *(i, j)th entry of A.*

For $a \in \mathbb{R}$ and matrices $A = [a_i^j]_{m\times n}$, $B = [b_i^j]_{m\times n}$, and $C = [a_i^j]_{n\times p}$, the *sum* $A + B$, *scalar multiple* aA, and *product* AC are defined, respectively, by

$$A+B = [x_i^j],\ x_i^j := a_i^j + b_i^j,\ \ aA = [y_i^j],\ y_i^j := aa_i^j,\ \ AC = [z_i^j],\ \ z_i^j := \sum_{k=1}^n a_i^k c_k^j.$$

The product AC may also be written as

$$\begin{bmatrix} \boldsymbol{a}_1 \\ \boldsymbol{a}_2 \\ \vdots \\ \boldsymbol{a}_m \end{bmatrix}_{m\times n} \begin{bmatrix} \boldsymbol{c}^1 & \boldsymbol{c}^2 & \cdots & \boldsymbol{c}^p \end{bmatrix}_{n\times p} = [\boldsymbol{a}_i \cdot \boldsymbol{c}^j]_{m\times p}.$$

The $m \times n$ matrix $O_{m\times n}$ with all entries equal to 0 is called a *zero matrix.* It has the property that $A + O_{m\times n} = A$ for all $m \times n$ matrices A. The collection of $m \times n$ matrices is a vector space under the operations $A + B$ and aA and with zero $O_{m\times n}$.

The *transpose* of an $m \times n$ matrix A is the $n \times m$ matrix $A^t := [x_i^j]$, where $x_i^j = a_j^i$. For example

$$\begin{bmatrix} 1 & 2 & 3 \\ 4 & 5 & 6 \end{bmatrix}^t = \begin{bmatrix} 1 & 4 \\ 2 & 5 \\ 3 & 6 \end{bmatrix}.$$

The transpose operation has the following properties:

$$(A + B)^t = A^t + B^t,\ \ (aA)^t = aA^t,\ \ (AC)^t = C^t A^t.$$

For each n, the matrix

$$I_n := \begin{bmatrix} 1 & 0 & \cdots & 0 \\ 0 & 1 & \cdots & 0 \\ \vdots & \vdots & \cdots & \vdots \\ 0 & 0 & \cdots & 1 \end{bmatrix}$$

is called the *nth-order identity matrix.* It has the property that

$$AI_n = A \quad \text{and} \quad I_n B = B$$

for all $m \times n$ matrices A and all $n \times p$ matrices B.

An $n \times n$ matrix A is said to be *nonsingular* if there exists a matrix, denoted by A^{-1} and called the *inverse of A*, such that

$$AA^{-1} = A^{-1}A = I_n.$$

The inverse operation has the property

$$(AB)^{-1} = B^{-1}A^{-1}$$

for all nonsingular $n \times n$ matrices A and B.

An $m \times n$ matrix A is said to be in *reduced row echelon form* if the following conditions hold:

- Any nonzero row has its first entry equal to 1. This entry is then called the *leading entry of the row.*
- If rows i and k are nonzero and $i < k$, then the leading entry of row i is to the left of the leading entry of row k.
- Entries above and below a leading entry are zero.
- Any zero row is below all nonzero rows.

For example, the following matrix is in reduced row echelon form:

$$\begin{bmatrix} 0 & 1 & 0 & 3 & 0 \\ 0 & 0 & 1 & 7 & 0 \\ 0 & 0 & 0 & 0 & 1 \\ 0 & 0 & 0 & 0 & 0 \end{bmatrix}.$$

For a given n, I_n is the only $n \times n$ matrix in reduced row echelon form without any zero rows.

An *elementary row operation* on an $m \times n$ matrix A is one of the following:

- Interchange a pair of rows.
- Multiply a row by a nonzero scalar.
- Add to one row a scalar multiple of another.

An *elementary matrix* is a matrix obtained from the identity matrix by an elementary row operation. Each elementary row operation on A may be achieved by multiplying A on the left by a suitable elementary matrix. For example, the multiplication

$$\begin{bmatrix} 0 & 1 & 0 \\ 1 & 0 & 0 \\ 0 & 0 & 1 \end{bmatrix} \begin{bmatrix} 1 & 2 & 3 \\ 4 & 5 & 6 \\ 7 & 8 & 9 \end{bmatrix} = \begin{bmatrix} 4 & 5 & 6 \\ 1 & 2 & 3 \\ 7 & 8 & 9 \end{bmatrix}$$

switches the first and second rows of A, and the multiplication

$$\begin{bmatrix} 1 & 0 & 0 \\ 2 & 1 & 0 \\ 0 & 0 & 1 \end{bmatrix} \begin{bmatrix} 1 & 2 & 3 \\ 4 & 5 & 6 \\ 7 & 8 & 9 \end{bmatrix} = \begin{bmatrix} 1 & 2 & 3 \\ 6 & 9 & 12 \\ 7 & 8 & 9 \end{bmatrix}$$

adds twice row one to row two. Using elementary operations, one may transform any $m \times n$ matrix A into reduced row echelon form R. It follows that there exists a sequence of elementary matrices E_j such that

$$R = E_p E_{p-1} \cdots E_1 A.$$

The *row rank* (*column rank*) of a matrix A is the maximum number of linearly independent rows (columns) of A. The row rank of a matrix is always equal to the column rank. (This is clear for the reduced row echelon form.) The *rank* of a matrix is its row (= column) rank.

The Matrix of a Linear Transformation

Let $T \in \mathcal{L}(\mathbb{R}^n, \mathbb{R}^m)$. The *matrix of* T is defined by

$$[T] = \begin{bmatrix} T\boldsymbol{e}^1 & T\boldsymbol{e}^2 & \cdots & T\boldsymbol{e}^n \end{bmatrix}$$

(where $T\boldsymbol{e}^j$ is written as a column). If $T\boldsymbol{e}^j = (a^j_1, a^j_2, \cdots, a^j_m)$ and $\boldsymbol{x} = (x_1, \ldots, x_n) = \sum_{j=1}^n x_j \boldsymbol{e}^j$, then, by linearity of T,

$$\begin{aligned} T(x_1, x_2, \ldots, x_n) &= \sum_{j=1}^n x_j T\boldsymbol{e}^j = \sum_{j=1}^n (a^j_1 x_j, a^j_2 x_j, \cdots, a^j_m x_j) \\ &= \Big(\sum_{j=1}^n a^j_1 x_j, \sum_{j=1}^n a^j_2 x_j, \cdots, \sum_{j=1}^n a^j_m x_j \Big), \end{aligned}$$

which may be written in column matrix form as

$$[T]\boldsymbol{x}^t = \begin{bmatrix} a^1_1 & a^2_1 & \cdots & a^n_1 \\ a^1_2 & a^2_2 & \cdots & a^n_2 \\ \vdots & \vdots & \cdots & \vdots \\ a^1_m & a^2_m & \cdots & a^n_m \end{bmatrix} \begin{bmatrix} x_1 \\ x_2 \\ \vdots \\ x_n \end{bmatrix}.$$

Note that a_i^j may be expressed as $(T\boldsymbol{e}^j)\cdot\boldsymbol{e}^i$.

The operations of addition, scalar multiplication, and composition of linear transformations correspond to addition, scalar multiplication, and multiplication of matrices in the following way: If T, $T' \in \mathcal{L}(\mathbb{R}^n, \mathbb{R}^m)$ and $S \in \mathcal{L}(\mathbb{R}^m, \mathbb{R}^p)$, then

$$[T + T'] = [T] + [T'], \quad [tT] = t[T] \quad [ST] = [S][T].$$

In particular, if $T \in \mathcal{L}(\mathbb{R}^n, \mathbb{R}^n)$, then T is invertible iff $[T]$ is nonsingular.

An $n \times n$ matrix A is *orthogonal* if $AA^t = I_n$, that is, if $A^t = A^{-1}$ or, equivalently, $\det A = \pm 1$. (See below.) A linear transformation $T \in \mathcal{L}(\mathbb{R}^n, \mathbb{R}^m)$ is said to be *orthogonal* if $[T]$ is orthogonal.

Determinants

A *permutation of the n-tuple* $(1,\ldots,n)$ is a one-to-one function σ mapping $\{1,\ldots,n\}$ onto itself. It is frequently denoted by $(i_1,\ldots,i_n)$, where $i_k = \sigma(k)$. The permutation is said to be *even* or *odd* according as an even or odd number of adjacent interchanges are required to transform $(i_1,\ldots,i_n)$ to $(1,\ldots,n)$ (or vice versa). For example, $(3,2,1)$ is odd and $(4,3,2,1)$ is even. The *sign* of a permutation σ is defined by

$$(-1)^\sigma = \begin{cases} 1 & \text{if } \sigma \text{ is even,} \\ -1 & \text{if } \sigma \text{ is odd.} \end{cases}$$

We then have

$$(-1)^{\sigma\tau} = (-1)^\sigma(-1)^\tau \quad \text{and} \quad (-1)^{\sigma^{-1}} = (-1)^\sigma,$$

where, as is customary, $\tau\sigma$ stands for $\tau \circ \sigma$.

The *determinant* of an $n \times n$ matrix $A = [a_i^j]$ is defined by

$$\det A = \begin{vmatrix} a_1^1 & a_1^2 & \cdots & a_1^n \\ a_2^1 & a_2^2 & \cdots & a_2^n \\ \vdots & \vdots & \cdots & \vdots \\ a_m^1 & a_m^2 & \cdots & a_m^n \end{vmatrix} := \sum_\sigma (-1)^\sigma a_1^{\sigma(1)} \cdots a_n^{\sigma(n)},$$

where the sum is taken over all permutations σ of $(1,\ldots,n)$. For example,

$$\begin{vmatrix} a & b \\ c & d \end{vmatrix} = ad - bc,$$

since $(-1)^{(1,2)} = 1$ and $(-1)^{(2,1)} = -1$. If $T \in \mathcal{L}(\mathbb{R}^n, \mathbb{R}^n)$ we denote the determinant of the matrix of T by $\det T$ rather than by the more cumbersome $\det[T]$.

The following theorem summarizes the main properties of determinants. Parts (a)–(f) follow directly from the above definition; part (g) is proved in Chapter 13.

Theorem. *Let* $A = \begin{bmatrix} \boldsymbol{a}^1 & \cdots & \boldsymbol{a}^n \end{bmatrix}$ *be an* $n \times n$ *matrix and* $t \in \mathbb{R}$. *Then*

(a) $\det \begin{bmatrix} \boldsymbol{a}^1 \cdots t\boldsymbol{a}^j \cdots \boldsymbol{a}^n \end{bmatrix} = t \det \begin{bmatrix} \boldsymbol{a}^1 \cdots \boldsymbol{a}^j \cdots \boldsymbol{a}^n \end{bmatrix}$.

(b) $\det \begin{bmatrix} \boldsymbol{a}^1 \cdots \boldsymbol{a}^j + \boldsymbol{b} \cdots \boldsymbol{a}^n \end{bmatrix} = \det \begin{bmatrix} \boldsymbol{a}^1 \cdots \boldsymbol{a}^j \cdots \boldsymbol{a}^n \end{bmatrix} + \det \begin{bmatrix} \boldsymbol{a}^1 \cdots \boldsymbol{b} \cdots \boldsymbol{a}^n \end{bmatrix}$.

(c) *Interchanging two rows of* A *changes the sign of the determinant.*

(d) *If* A *has a pair of duplicate rows, then* $\det A = 0$.

(e) *Adding a multiple of one row to another does not change the value of the determinant.*

(f) $\det A^t = \det A$. *Thus any "row property" has a corresponding "column property."*

(g) *If* B *is an* $n \times n$ *matrix, then* $\det(AB) = (\det A)(\det B)$.

The following theorem is frequently useful in evaluating determinants.

Theorem. *Let* $A = [a_{ij}]$ *be an* $n \times n$ *matrix, and for each* (i, j), *let* A_{ij} *denote the matrix obtained by removing row* i *and column* j *from* A. *Then for each fixed* i *and* j,

$$\det A = \sum_{k=1}^{n} (-1)^{i+k} a_{ik} \det A_{ik} = \sum_{k=1}^{n} (-1)^{k+j} a_{kj} \det A_{kj}.$$

The first equality is called *expansion along row* i and the second *expansion along column* j. For example, expanding along row 1,

$$\begin{vmatrix} a_{11} & a_{12} & a_{13} \\ a_{21} & a_{22} & a_{23} \\ a_{31} & a_{32} & a_{33} \end{vmatrix} = a_{11} \begin{vmatrix} a_{22} & a_{23} \\ a_{32} & a_{33} \end{vmatrix} - a_{12} \begin{vmatrix} a_{21} & a_{23} \\ a_{31} & a_{33} \end{vmatrix} + a_{13} \begin{vmatrix} a_{21} & a_{22} \\ a_{31} & a_{32} \end{vmatrix}.$$

The formula $\begin{vmatrix} a & b \\ c & d \end{vmatrix} = ad - bc$ may then be used to complete the evaluation.

For another example, consider

$$\begin{vmatrix} I_p & C_{p \times q} \\ O_{q \times p} & D_{q \times q} \end{vmatrix} = \det D,$$

obtained by successive expansion along the first column.

The preceding theorem may be used to prove the following result.

Theorem. *Let* $A = [a_{ij}]$ *be an* $n \times n$ *matrix. Then* A^{-1} *exists iff* $\det A \neq 0$. *In this case the* (i, j) *entry of* A^{-1} *is*

$$(-1)^{i+j} \frac{\det A_{ji}}{\det A}.$$

The last theorem may be used to prove *Cramer's Rule*: Consider a system of n equations in n unknowns, written in matrix form as $A\boldsymbol{x} = \boldsymbol{b}$ or explicitly as

$$\begin{bmatrix} a_{11} & a_{12} & \cdots & a_{1n} \\ a_{21} & a_{22} & \cdots & a_{2n} \\ \vdots & \vdots & \cdots & \vdots \\ a_{n1} & a_{n2} & \cdots & a_{nn} \end{bmatrix} \begin{bmatrix} x_1 \\ x_2 \\ \vdots \\ x_n \end{bmatrix} = \begin{bmatrix} b_1 \\ b_2 \\ \vdots \\ b_n \end{bmatrix}.$$

If A is nonsingular, then the solution to the system is

$$x_j = \frac{1}{\det A} \begin{vmatrix} a_{11} & \cdots & a_{1,j-1} & b_1 & a_{1,j+1} & \cdots & a_{1n} \\ a_{21} & \cdots & a_{2,j-1} & b_2 & a_{2,j+1} & \cdots & a_{2n} \\ \vdots & \cdots & \vdots & \vdots & \vdots & & \vdots \\ a_{n1} & \cdots & a_{n,j-1} & b_n & a_{n,j+1} & \cdots & a_{nn} \end{vmatrix}.$$

Appendix C

Solutions to Selected Problems

Section 1.1

1. (b) $[(ab)+(-a)b] = [a+(-a)]b = 0 \cdot b = 0$, so uniqueness of the additive inverse implies $-(ab) = (-a)b$. A similar argument works for the second equality.

   (d) By (b), $(-1)a = 1(-a) = -a$.

   (f) Using commutativity and associativity of multiplication and the distributive law and 1.1.1(i),

$$\begin{aligned} a/b + c/d &= ab^{-1}(dd^{-1}) + cd^{-1}(bb^{-1}) = ad(b^{-1}d^{-1}) + bc(b^{-1}d^{-1}) \\ &= ad(bd)^{-1} + bc(bd)^{-1} = (ad+bc)/(bd). \end{aligned}$$

3. If $s := r/x \in \mathbb{Q}$, then, by Exercise 2, $x = r/s \in \mathbb{Q}$, a contradiction. Therefore, $r/x \in \mathbb{I}$. The remaining parts have similar proofs.

5. The left side of (a) is $\dfrac{n-1}{n}\dfrac{n-2}{n}\cdots\dfrac{1}{n} = \dfrac{n!}{n^n}$. For (b),

$$\begin{aligned} (2n)! &= \big[2n(2n-2)(2n-4)\cdots 4\cdot 2\big]\big[(2n-1)(2n-3)\cdots 3\cdot 1\big] \\ &= 2^n\big[n(n-1)(n-2)\cdots 2\cdot 1\big]\big[(2n-1)(2n-3)\cdots 3\cdot 1\big]. \end{aligned}$$

8. $f(k) = k^3 - (k-1)^3 = 3k^2 - 3k + 1$.

Section 1.2

1. (c) Follows from $a/b - c/d = (ad - bc)/bd$.

4. If $0 < x < y$, then multiplying the inequality by $1/(xy)$ and using (d) of 1.2.2 shows that $1/y < 1/x$. If $x < y < 0$, then $0 < -y < -x$, hence, by the first part, $1/(-x) < 1/(-y)$ so $1/x > 1/y$.

6. (a) By Exercise 1.1.4, $y^n - x^n = (y-x)\displaystyle\sum_{j=1}^{n} y^{n-j}x^{j-1}$. Each term of the sum is positive and less than $y^{n-j}y^{j-1} = y^{n-1}$. Since there are n terms, part (a) follows.

8. $a = ta + (1-t)a < tb + (1-t)b = b$.

DOI: 10.1201/9781003629382-C

10. If $a > b$, then $x := (a - b)/2 > 0$ and $a > b + x$, contradicting the hypothesis.

13. (b) $0 \le (x-y)^2 + (y-z)^2 + (z-x)^2 = 2(x^2+y^2+z^2) - 2(xy+yz+xz)$.

14. Expand $(x - a)^2 \ge 0$ and divide by x.

18. If $a \le x \le b$, then $x \le |b|$ and $-x \le -a \le |a|$, hence $|x| \le \max\{|a|, |b|\}$.

21. Assume without loss of generality that $S_1 = S\setminus\{a_1, \ldots, a_k\}$, so $\min S_1 = a_{k+1}$. Each of the remaining sets S_j contains at least one of $a_1, \ldots, a_k$, hence $\min S_j \le a_k < a_{k+1}$, verifying the assertion.

Section 1.3

2. (a) $\sup = 12$, $\inf = -12$. (b) $\sup = 1$, $\inf = -1$.

3. (c) $\sup = 10/3$, $\inf = 3$; (d) $\sup = \frac{3+\sqrt{5}}{2}$, $\inf = -\infty$;
   (e) $\sup = +\infty$, $\inf = -\infty$. (h) $\sup = 3$, $\inf = 0$;
   (i) $\sup = \frac{1}{2} + \frac{\sqrt{2}}{4}$, $\inf = \frac{1}{2} - \frac{\sqrt{2}}{4}$; (m) $\sup = 4/3$, $\inf = -1$.

5. Let $x, y \in A$. Then $\pm(x-y) \le \sup A - \inf A$, hence $|x-y| \le \sup A - \inf A$. Since $|x|-|y| \le |x-y|$, $|x|-|y| \le \sup A - \inf A$ so $|x| \le \sup A - \inf A + |y|$. Since x was arbitrary, we have $\sup|A| \le \sup A - \inf A + |y|$, hence $\sup|A| - \sup A + \inf A \le |y|$. Since y was arbitrary, it follows that $\sup|A| - \sup A + \inf A \le \inf|A|$.

6. (b) Since $x > 0$, $xa \le x \sup A$ for all $a \in A$, hence $\sup(xA) \le x \sup A$. Replacing x by $1/x$ proves the inequality in the other direction. The infimum case is similar.

9. Let $a < b$ and choose a rational r in $(a - \sqrt{2}, b - \sqrt{2})$. Then $r + \sqrt{2}$ is irrational and in (a, b).

12. (b) If $n := \lfloor x \rfloor = -\lfloor -x \rfloor$, then $x - 1 < n \le x$ and $x \le n < x + 1$. This is possible only if $x = n$. The converse is trivial.
    (c) By definition $-x - 1 < \lfloor -x \rfloor \le -x$.

14. Let $x := (b^m)^{1/n}$ and $y := \left(b^{1/n}\right)^m$. By definition, x is the unique positive solution of $x^n = b^m$. Since $y^n = \left[\left(b^{1/n}\right)^m\right]^n = \left[\left(b^{1/n}\right)^n\right]^m = b^m$, $x = y$.

17. Let $\ell \le x \le u$ for all $x \in A$. By the Archimedean principle, there exist positive integers m and n such that $-m < \ell \le u < n$. Set $N = \max\{m, n\}$.

20. For any $a \in \mathbb{N}$, if $r := \sqrt{n+a} + \sqrt{n} \in \mathbb{Q}$, then squaring both sides of $\sqrt{n+a} = r - \sqrt{n}$ shows that $\sqrt{n} \in \mathbb{Q}$ and hence that $n = j^2$ for some $j \in \mathbb{N}$ (1.3.11). Then $\sqrt{n+a} \in \mathbb{Q}$, hence $n = k^2$ for some $k \in \mathbb{N}$. Therefore, $a = k^2 - j^2 = (k-j)(k+j)$. If $a = 11$, then $k - j = 1$ and $j + k = 11$ so $n = 25$. If $a = 21$, then either $k - j = 1$ and $j + k = 21$ or $k - j = 3$ and $j + k = 7$. The first choice leads to $j = 10$ and $n = 100$ and the second to $j = 2$ and $n = 4$.

Section 1.4

3. Let $f(n)$ denote the sum on the left side of the equation and $g(n)$ the sum on the right. Then $f(1) = 1/2 = g(1)$. Now let $n \geq 1$. Then

$$f(n+1) - f(n) = \sum_{k=1}^{2n+2} \frac{(-1)^{k+1}}{k} - \sum_{k=1}^{2n} \frac{(-1)^{k+1}}{k} = \frac{1}{2n+1} - \frac{1}{2n+2}$$

$$g(n+1) - g(n) = \sum_{k=n+2}^{2n+2} \frac{1}{k} - \sum_{k=n+1}^{2n} \frac{1}{k} = \frac{1}{2n+2} + \frac{1}{2n+1} - \frac{1}{n+1}.$$

Since the right sides are equal, $f(n) = g(n) \Rightarrow f(n+1) = g(n+1)$.

5. $\frac{25}{3}n^3 - \frac{15}{2}n^2 + \frac{1}{6}n$.

6. (b) $\displaystyle\sum_{k=1}^{500}(4k^2 - 1) = 4\,\frac{500 \cdot 501 \cdot 1001}{6} - 500 = 167,166,500.$

7. For $n \geq 1$, let $Q(n)$ be the statement $P(n - 1 + n_0)$. Then $Q(1) = P(n_0)$ is true. Assume $Q(n) = P(n-1+n_0)$ is true. Then $Q(n+1) = P(n+n_0)$ is true. By mathematical induction, $Q(n) = P(n - 1 + n_0)$ is true for all $n \geq 1$, that is, $P(n)$ is true for every $n \geq n_0$.

8. In each case, let $f(n)$ be the left side of the inequality and $g(n)$ the right side, and let $P(n) : f(n) < g(n)$. Let n_0 be the base value of n for which $P(n)$ is true. It is straightforward to check that $f(n_0) < g(n_0)$. Assume $P(n)$ holds for some $n \geq n_0$, so that $f(n)/g(n) < 1$. Then

(a) $\dfrac{f(n+1)}{g(n+1)} = \dfrac{2n+3}{2^{n+1}} = \dfrac{f(n)}{2g(n)} + \dfrac{1}{2^n} < 1.$

(e) $\dfrac{f(n+1)}{g(n+1)} = \dfrac{2^{n+1}(n+1)!}{(n+1)^{n+1}} = \dfrac{f(n)}{g(n)} \dfrac{2}{(1+1/n)^n} < 1.$

9. Check that $6 < \ln(6!)$. For the induction step, use $(n+1)! = (n+1)n!$.

13. Let g_n denote the expression on the right in the assertion. One checks directly that $g_0 = g_1 = 1$. Let $n \geq 2$ and assume that $f_j = g_j$ for all

$2 \le j \le n$. Then

$$\begin{aligned} g_{n+1} - f_{n+1} &= g_{n+1} - f_n - f_{n-1} = g_{n+1} - g_n - g_{n-1} \\ &= \frac{1}{\sqrt{5}}\left(a^{n+2} - a^{n+1} - a^n\right) + \frac{1}{\sqrt{5}}\left(b^{n+2} - b^{n+1} - b^n\right) \\ &= \frac{a^n}{\sqrt{5}}(a^2 - a - 1) + \frac{b^n}{\sqrt{5}}(b^2 - b - 1) = 0. \end{aligned}$$

22. The set of all nonnegative integers of the form $m - qn$, $q \in \mathbb{Z}$, is nonempty (Archimedean principle), hence has a smallest member $r = m - qn$ (well-ordering principle). If $r \ge n$, then $0 \le r - n = m - (q+1)n < r$, contradicting the minimal property of r. Therefore, $m = qn + r$ has the required form. If also $m = q'n + r'$, $q' \in \mathbb{Z}$, and $r' \in \{0, \dots n-1\}$, then $|q - q'|n = |r - r'| < n$, hence $q' = q$ and $r' = r$.

Section 1.5

1. $\boldsymbol{x} = \boldsymbol{c} - \dfrac{\boldsymbol{d}\cdot\boldsymbol{e} - (\boldsymbol{b}\cdot\boldsymbol{c})(\boldsymbol{b}\cdot\boldsymbol{d})}{1 - (\boldsymbol{a}\cdot\boldsymbol{b})(\boldsymbol{b}\cdot\boldsymbol{d})}\,\boldsymbol{a}, \quad \boldsymbol{y} = \boldsymbol{e} - \dfrac{\boldsymbol{b}\cdot\boldsymbol{c} - (\boldsymbol{a}\cdot\boldsymbol{b})(\boldsymbol{d}\cdot\boldsymbol{e})}{1 - (\boldsymbol{a}\cdot\boldsymbol{b})(\boldsymbol{b}\cdot\boldsymbol{d})}\,\boldsymbol{d}.$

2. (c) By the triangle inequality,

$$||\boldsymbol{x}||_2 = ||\boldsymbol{x} - \boldsymbol{y} + \boldsymbol{y}||_2 \le ||\boldsymbol{x} - \boldsymbol{y}||_2 + ||\boldsymbol{y}||_2,$$

hence $||\boldsymbol{x}||_2 - ||\boldsymbol{y}||_2 \le ||\boldsymbol{x} - \boldsymbol{y}||_2$. Similarly, $||\boldsymbol{y}||_2 - ||\boldsymbol{x}||_2 \le ||\boldsymbol{x} - \boldsymbol{y}||_2$.

3. By 1.5.3, $||\boldsymbol{x}_1 + \boldsymbol{x}_2 + \cdots + \boldsymbol{x}_k||_2^2 = \displaystyle\sum_{i,j=1}^{n} \boldsymbol{x}_i \cdot \boldsymbol{x}_j = \sum_{j=1}^{k} \boldsymbol{x}_j \cdot \boldsymbol{x}_j.$

7. The hypotheses imply that

$$\sum_{j=1}^{n} x_j^2 = \sum_{j=1}^{n} y_j^2 = 1 \text{ and } \sum_{j=1}^{n} (x_j + y_j)^2 = 4.$$

It follows that $\sum_{j=1}^{n} x_j y_j = 1$ and $\sum_{j=1}^{n} (x_j - y_j)^2 = 0$. The same does not hold for $||\cdot||_\infty$ (take $\boldsymbol{x} = (-1, 1)$ and $\boldsymbol{y} = (1, 1)$) or for $||\cdot||_1$ (take $\boldsymbol{x} = (1, 0)$ and $\boldsymbol{y} = (0, 1)$).

Section 2.1

1. (a) $a_n = [a + b + (-1)^n(b - a)]/2$.

3. (b) If $n \ge 6$, $|(2n^2 - n)/(n^2 + 3) - 2| = |n + 6|/(n^2 + 3) \le 2n/n^2 = 2/n$. Therefore, choose $N \ge \max\{6, 2/\varepsilon\}$.

(e) $|(2 + 1/n)^3 - 8| = \left[(2 + 1/n)^2 + 2(2 + 1/n) + 4\right]/n \le 19/n$, so choose any integer $N > 19/\varepsilon$.

5. Let $r = pq^{-1}$, $p, q \in \mathbb{Z}$, $q > 0$. For all $n \geq q$, $n!r \in \mathbb{Z}$, hence $\sin(n!r\pi) = 0$.

7. Let $A = \{x_1, \ldots, x_p\}$ and $A_j = \{n : a_n = x_j\}$. One of these sets, say A_1, must have infinitely many members. Since $|x_1 - a| \leq |x_1 - a_n| + |a_n - a|$ and $a_n \to a$, letting $n \to +\infty$ through A_1 shows that $x_1 = a$. We may therefore choose $\varepsilon > 0$ so that $I := (a - \varepsilon, a + \varepsilon)$ contains no x_j for $j \geq 2$. Let $N \in \mathbb{N}$ such that $a_n \in I$ for all $n \geq N$. For such n, $a_n = a$.

8. (a) $b_n = (3a_n + 2b_n - 3a_n)/2 \to (c - 3a)/2$.

9. (a) 2 (d) $b/2\sqrt{a}$ (g) $-ka^{k-1}$ (k) 1/2.

11. Use $-r \leq a_n - b_n \leq r$ and 2.1.4.

14. (a) Suppose first that $r > 1$. Set $h_n = r^{1/n} - 1$. By the binomial theorem, $r = (1 + h_n)^n > nh_n$, hence, by the squeeze principle, $h_n \to 0$. If $r < 1$, consider $1/r$.

17. $a_n < ra_{n-1} < r^2 a_{n-2} < \cdots < r^{n-1}a_1 \to 0$. For the example, take $a_n = 2^{1/n}$.

19. Choose N such that $a_n - a < \varepsilon$ for all $n \geq N$. For such n,

$$0 \leq \min\{a_1, \ldots, a_n\} - a \leq a_n - a < \varepsilon.$$

Therefore, $\min\{a_1, \ldots, a_n\} \to a$. The converse is false: consider $a_n = 1 + (-1)^n$.

22. Suppose that $c \leq f(x) - x \leq d$ for all x, so $c + jx \leq f(jx) \leq d + jx$. Summing and using Exercise 1.4.4,

$$nc + xn(n+1)/2 \leq \sum_{j=1}^{n} f(jx) \leq nd + xn(n+1)/2,$$

hence

$$c/n + x(1 + 1/n)/2 \leq (1/n^2) \sum_{j=1}^{n} f(jx) \leq d/n + x(1 + 1/n)/2.$$

Letting $n \to +\infty$, we obtain (a). Part (b) is proved similarly.

Section 2.2

1. Since

$$\frac{a^{1/n}}{a^{1/(n+1)}} = a^{1/n(n+1)} < 1 < b^{1/n(n+1)} = \frac{b^{1/n}}{b^{1/(n+1)}},$$

$a^{1/n}$ is increasing and $b^{1/n}$ is decreasing. Each tends to 1 by Exercise 2.1.14.

3. By results of Section 2.1,

$$a_n = a(1/n + nb)^{-1} \to 0 \quad \text{and} \quad na_n = a(1/n^2 + b)^{-1} \to ab^{-1}.$$

The condition $a_{n+1} < a_n$ is equivalent to $(n^2 + n)b > 1$, which holds eventually. The condition $(n+1)a_{n+1} > na_n$ is equivalent to the inequality $(n+1)^2 > n^2$.

7. Let $f(x) = 1 + \dfrac{1}{2 + (1+x)^{-1}} = \dfrac{3x+4}{2x+3}$. Then $f : [1,2] \to [1,2]$, f is increasing and $f(a_m) = a_{m+2}$. Since a_1, $a_2 \in [1,2]$, $a_n \in [1,2]$ for all n. Since $a_1 = 1$, $a_2 = 3/2$, $a_3 = 7/5$ and $a_4 = 17/12$, the inequalities

$$a_{2n+2} < a_{2n} \quad \text{and} \quad a_{2n+1} > a_{2n-1}$$

hold for $n = 1$. Assume they hold for $n = k$. Then

$$\begin{aligned} a_{2k+4} &= f(a_{2k+2}) < f(a_{2k}) = a_{2k+2} \quad \text{and} \\ a_{2k+3} &= f(a_{2k+1}) > f(a_{2k-1}) = a_{2k+1}, \end{aligned}$$

hence the inequalities hold for $n = k + 1$.

Since the sequences $\{a_{2n}\}$ and $\{a_{2n+1}\}$ are bounded and monotone, the monotone convergence theorem implies that $a_{2n} \to a$ and $a_{2n+1} \to b$ for some a, $b \in \mathbb{R}$. Letting $n \to +\infty$ in $f(a_{2n}) = a_{2n+2}$ gives $f(a) = a$. Therefore, $a = \sqrt{2}$. Similarly, $b = \sqrt{2}$.

9. For $x > 0$, $x^2 + r \geq 2x\sqrt{r}$, hence $(x + r/x)/2 \geq \sqrt{r}$. Therefore, $a_n \geq \sqrt{r}$. For $x \geq \sqrt{r}$, $x^2 + r \leq 2x^2$, hence $(x + r/x)/2 \leq x$. Therefore, $a_n \geq a_{n+1}$. By the monotone convergence theorem, $a_n \to a$ for some $a \geq \sqrt{r}$. Letting $n \to +\infty$ in $a_n = (a_{n-1} + r/a_{n-1})/2$, yields $a = (a + r/a)/2$, which has positive solution $a = \sqrt{r}$.

Section 2.3

1. (a) 0, $\pm 3/8$ (c) ± 4, ± 6, ± 12, ± 14.

3. (d) $a_n^{2/k} = \left(1 + \dfrac{1}{2n+k}\right)^{2n+k} \left(1 + \dfrac{1}{2n+k}\right)^{-k} \to e.$

5. If $\{a_n\}$ lies in the set $\{x_1, \ldots, x_n\}$, then one of the sets $\{n : a_n = x_j\}$ must have infinitely many members and a subsequence may be constructed from these.

8. Given $\varepsilon > 0$, choose N so that $\sum_{n=N}^{\infty} |a_{n+k} - a_n| < \varepsilon$. For $m > n \geq N$,

$$|a_{mk} - a_{nk}| \leq |a_{mk} - a_{(m-1)k}| + \cdots + |a_{(n+1)k} - a_{nk}| < \varepsilon.$$

Therefore, $\{a_{nk}\}_{n=1}^{\infty}$ is Cauchy.

10. Clearly $a_n \to 0$ implies $b_n \to 0$. For the converse, suppose $a_n \not\to 0$. Choose $\varepsilon > 0$ and a subsequence such that $a_{n_k} \geq \varepsilon > 0$ for all k. Then

$$1 = b_{n_k}\left(\frac{1}{a_{n_k}^q} + \frac{1}{a_{n_k}^{q-p}}\right) \leq b_{n_k}\left(\frac{1}{\varepsilon^q} + \frac{1}{\varepsilon^{q-p}}\right),$$

hence $b_n \not\to 0$. If $0 < q < p$, then the sufficiency is false: Take $a_n = n$, $q = 1/2$ and $p = 1$. Then $b_n = \sqrt{n}/(n+1) \to 0$ but $a_n \to +\infty$.

Section 2.4

1. (a) $\liminf = -5/3$, $\limsup = 5/3$ (c) $\liminf = -14$, $\limsup = 14$
   (h) $\liminf = -\infty$, $\limsup = +\infty$.

3. Follows from Exercise 1.3.6.

5. Follows from $\{a_{n_k} : k \geq n\} \subseteq \{a_k : k \geq n\}$.

7. $0 < b - \varepsilon < b_n < b + \varepsilon \Rightarrow a_n(b-\varepsilon) < a_n b_n < a_n(b+\varepsilon)$
   $\Rightarrow (b-\varepsilon)\limsup_{n\to\infty} a_n \leq \limsup_{n\to\infty} a_n b_n \leq (b+\varepsilon)\limsup_{n\to\infty} a_n$.
   Now let $\varepsilon \to 0$.

10. Choose r so that $\liminf_{n\to\infty} b_n > r > 0$. Then, given $\varepsilon > 0$, there exists N such that $a_n > a/2$ and $b_n > r$, and

$$c_n := (b_n - 3a_n)(b_n + 2a_n) = b_n^2 - a_n b_n - 6a_n^2 < \varepsilon$$

for every $n > N$. Then $b_n - 3a_n = c_n/(b_n + 2a_n) < \varepsilon/(r+a)$, so $\limsup_{n\to\infty} b_n \leq 3a$.

12. Suppose that $\liminf_n a_n^{1/n} < \liminf_n \dfrac{a_{n+1}}{a_n}$. Choose r strictly between these numbers and then choose N such that $a_n/a_{n-1} > r$ for all $n > N$. For such n,

$$a_n > a_{n-1}r > a_{n-2}r^2 > \cdots > a_N r^{n-N},$$

hence $\liminf_n a_n^{1/n} \geq \liminf_n (a_N^{1/n} r^{1-N/n}) = r$, a contradiction. To evaluate $\lim_n n/(n!)^{1/n}$ take $a_n = n^n/n!$ and calculate

$$\frac{a_{n+1}}{a_n} = \left(\frac{n+1}{n}\right)^n \to e.$$

Section 3.1

1. Let $x_1 < \cdots < x_n$ denote the points of E and let

$$\delta = \frac{1}{2}\min\{x_j - x_i : 1 \leq i < j \leq n\}.$$

Then for each j, $(x_j - \delta, x_j + \delta) \cap E = \{x_j\}$.

4. Let ε, $M > 0$.

(b) The limit is 1. If $|x-1| < 1$, then $x > 0$, hence

$$\left|\frac{x+3}{3x+1} - 1\right| = \frac{2|x-1|}{3x+1} < 2|x-1|.$$

Therefore, choose $\delta = \min\{1, \varepsilon/2\}$.

(d) The limit is $+\infty$: $x < -\sqrt{M} - 1 \Rightarrow -x > \sqrt{M}$ and $-x-1 > \sqrt{M} \Rightarrow x^2 + x = (-x)(-x-1) > M$.

6. (a) $2/3$ (d) $+\infty$. (g) $9/25$.

7. (b) $-1/2$ (f) $\dfrac{\sqrt{b+x}-\sqrt{b-x}}{\sqrt{c+x}-\sqrt{c-x}} = \dfrac{\sqrt{c+x}+\sqrt{c-x}}{\sqrt{b+x}+\sqrt{b-x}} \to \sqrt{\dfrac{c}{b}}$

(h) $(a\sqrt{d})/(c\sqrt{b})$.

9. The limit exists at a iff $\lim_{\{x\to a,\, x\in\mathbb{Q}\}} f(x) = \lim_{\{x\to a,\, x\in\mathbb{I}\}} f(x)$. By continuity of polynomials, this is equivalent to $4a^2+2a-11 = 3a^2+a-5$. Thus $a = -3$, 2.

11. (a) a. (e) $(c\sqrt{a})/(2\sqrt{b})$.

13. Proof for the case f increasing and $L := \lim_n f(a_n) \in \mathbb{R}$: Given $\varepsilon > 0$, choose N such that $L-\varepsilon < f(a_n) < L+\varepsilon$ for all $n \geq N$. Let $x > a_N$ and let n be the least integer $> N$ such that $x < a_n$. Then $a_{n-1} \leq x < a_n$ so $L - \varepsilon < f(a_{n-1}) \leq f(x) \leq f(a_n) < L + \varepsilon$.

Section 3.2

1. (a) -1, 1 (c) $-2/3$, $2/3$ (e) -1, 1 (i) -3, 1.

3. lim sup case: Assume $a \in \mathbb{R}$. Set

$$L = \limsup_{\{x\to a,\, x\in E\}} f(x) \text{ and } L_j = \limsup_{\{x\to a,\, x\in E_j\}} f(x),\ \ j = 1, 2.$$

By 3.2.1, there exists a sequence $a_n \in E_1$ such that $f(a_n) \to L_1$. Since $a_n \in E$, by the same theorem, $L_1 \leq L$. Similarly, $L_2 \leq L$. Now let $b_n \in E$ such that $f(b_n) \to L$. Then one of the sets, say E_1, contains infinitely many terms of the sequence. Therefore, $L \leq L_1$, hence $L = \max\{L_1, L_2\}$.

4. Let $g(x) = 1/f(x)$. Then $\overline{g}(r) = 1/\underline{f}(r)$.

Section 3.3

1. By continuity, $f(2) = \lim_{x\to 2^-}(mx+3) = \lim_{x\to 2^+}(3x^2+7)$, that is, $2m + 3 = 19$. Therefore, $m = 8$.

4. This follows from

$$\lim_{\{x\to a,\, x\in\mathbb{Q}\}} d(x)g(x) = g(a) \quad\text{and}\quad \lim_{\{x\to a,\, x\in\mathbb{I}\}} d(x)g(x) = 0.$$

8. The identity implies that $f(nx) = nf(x)$, $n \in \mathbb{N}$. Also, $f(0)+f(0) = f(0)$ so $f(0) = 0$. Since $f(-x) + f(x) = f(0)$, we see that $f(-x) = -f(x)$, hence $f(nx) = nf(x)$ for all $n \in \mathbb{Z}$. Let $m, n \in \mathbb{N}$. Then $f(x) = f(nx/n) = nf(x/n)$. Replacing x by xm gives $mf(x) = f(mx) = nf(mx/n)$. Thus, $f(tx) = tf(x)$ for all $x \in \mathbb{R}$ and $t \in \mathbb{Q}$. Since f is continuous at zero and $f(x-y) = f(x) + f(-y) = f(x) - f(y)$, f is continuous on $\mathbb{R}$. Thus, $f(tx) = tf(x)$ for all $x, t \in \mathbb{R}$. Setting $x = 1$ gives the desired result.

9. (c) Let $a \in \mathbb{R}$ and $\varepsilon > 0$. Choose N so that $\sum_{n>N} 2^{-n} < \varepsilon$ and then choose $\delta > 0$ so that $(a, a+\delta)$ contains none of the numbers $c_1, c_2, \ldots, c_N$. If $a < x < a+\delta$, then

$$0 \le f(x) - f(a) = \sum_{n:a<c_n\le x} 2^{-n} \le \sum_{n>N} 2^{-n} < \varepsilon.$$

Therefore, f is right continuous at a.

If $a \notin \{c_n\}$, then we may choose δ so that $(a-\delta, a]$ contains none of the numbers $c_1, c_2, \ldots, c_N$. If $a - \delta < x < a$, then, as before,

$$0 \le f(a) - f(x) = \sum_{n:x<c_n\le a} 2^{-n} < \varepsilon.$$

Therefore, f is left continuous at a.

If $a = c_k$ and $a - \delta < x < a$, then

$$f(c_k) - f(x) = \sum_{n:x<c_n\le c_k} 2^{-n} \ge 2^{-k}$$

so f is not left continuous at c_k.

11. (a) Let $x_n \to x$ in $[0,1]$ and let $\varepsilon > 0$. By hypothesis, for each n we may choose $t_n \in (0,1) \setminus \{x\}$ such that $|t_n - x_n| < 1/n$ and $|f(t_n) - g(x_n)| < \varepsilon$. Then $t_n \to x$, hence $f(t_n) \to g(x)$. From

$$|g(x_n) - g(x)| \le |g(x_n) - f(t_n)| + |f(t_n) - g(x)|$$

we then have

$$\limsup_n |g(x_n) - g(x)| \le \varepsilon.$$

Since ε was arbitrary, $g(x_n) \to g(x)$.

(b) Note that f is continuous at x iff $f(x) = g(x)$. Let $\varepsilon > 0$. We show that the set $D_\varepsilon := \{x \in [0,1] : |f(x) - g(x)| \ge \varepsilon\}$ is finite. The desired conclusion will follow on observing that the set of discontinuities of f is precisely $\bigcup_{n=1}^{\infty} D_{1/n}$.

Suppose D_ε is infinite. Then there exists a sequence of distinct terms such that $|f(x_n) - g(x_n)| \geq \varepsilon$ for all n. By the Bolzano–Weierstrass theorem, $\{x_n\}$ has a convergent subsequence, say $x_{n_k} \to x$. Because the terms of $\{x_n\}$ are distinct, $x_{n_k} \neq x$ for all large k, hence $f(x_{n_k}) \to g(x)$. Also, by continuity, $g(x_{n_k}) \to g(x)$. But this contradicts the inequality $|f(x_{n_k}) - g(x_{n_k})| \geq \varepsilon$.

Section 3.4

2. Let $x_0 > 0$ and choose $r > x_0$ such that $f(x) < f(x_0)$ for all x with $|x| > r$. Then the maximum M of f on $[-r, r]$ is $\geq f(x_0)$, hence M must be the maximum of f on $\mathbb{R}$.

4. (e) Suppose f is not upper semicontinuous at x_0. Choose r such that $f(x_0) < r < \limsup_{x\to x_0} f(x)$ and then i such that $f_i(x_0) < r$. For each $\delta > 0$, $r < \sup_{0<|x-x_0|<\delta} f(x)$, hence there exists x_δ with $0 < |x_\delta - x_0| < \delta$ such that $f(x_\delta) > r$. Thus we have

$$f_i(x_0) \leq f(x_0) < r < f(x_\delta) \leq f_i(x_\delta) \leq \sup_{0<|x-x_0|<\delta} f_i(x).$$

Letting $\delta \to 0$ produces the contradiction

$$f_i(x_0) < r \leq \limsup_{x\to x_0} f_i(x) \leq f_i(x_0).$$

If $f_n(x) := x^n$ on [0,1], then $f = \inf_n f_n$ is discontinuous at $x = 1$.

5. (b) Define

$$f(x) = \begin{cases} (1-2x)(1-d(x)) & \text{if } 0 \leq x \leq 1/2, \\ (2x-1)d(x) & \text{if } 1/2 < x \leq 1, \end{cases}$$

where $d(x)$ is the Dirichlet function on $[0, 1]$.

7. (c) $f(x) = \tan x - x$. Since $\lim_{x\to(n+1/2)\pi-} \tan x = +\infty$, choose $b \in (n\pi, (n+1/2)\pi)$ such that $f(n\pi) = -n\pi < 0 < f(b)$.

(f) Let $f(x)$ denote the left side minus the right side of the equation. Then

$$\lim_{x\to 0+} f(x) = +\infty, \qquad \lim_{x\to\pi/2-} f(x) = -\infty,$$

hence there exist $0 < a < b < \pi/2$ such that $f(a) > 0 > f(b)$.

9. Set $g(x) = f(x) - x$. Then $g(b) \leq 0 \leq g(a)$ and the result follows from the intermediate value theorem.

Section 3.5

1. Take $f(x) = 1/x$ and $g(x) = x^2$ on $(0, 1)$.

3. (a) Use the inequality

$$\frac{|ax^2 - ay^2|}{\sqrt{ax^2+b} + \sqrt{ay^2+b}} \le \frac{\sqrt{a}\,|x-y|(|x|+|y|)}{\sqrt{x^2+b/a} + \sqrt{y^2+b/a}} \le \sqrt{a}\,|x-y|.$$

7. Given $\varepsilon > 0$, choose $\delta > 0$ such that $|f(x) - f(y)| < \varepsilon$ for all $x, y \in E$ with $|x - y| < \delta$. Then choose N such that $|a_n - a_m| < \delta$ for all $m, n \ge N$. For such m, n, $|f(a_n) - f(a_m)| < \varepsilon$.

9. The inequality $\big|\,|x| - |y|\,\big| \le |x - y|$ shows that $|x|$ is uniformly continuous. The given functions are therefore compositions of uniformly continuous functions.

13. If $0 < p \le 1$, then

$$\lim_{x\to+\infty} \frac{\sin x}{x^p} = 0, \quad \text{and} \quad \lim_{x\to 0^+} \frac{\sin x}{x^p} = \lim_{x\to 0^+} x^{1-p}\frac{\sin x}{x} = 0 \text{ or } 1.$$

Therefore, $(\sin x)/x^p$ has a uniformly continuous extension to $[0, +\infty)$. If $p > 1$, $(\sin x)/x^p$ is continuous on $(0, +\infty)$ but has no continuous extension to $[0, +\infty)$.

15. Since f may be extended continuously to $[a, b]$, it is bounded. The examples $f(x) = x$ on $(0, +\infty)$ and $f(x) = 1/x$ on $(0, 1)$ show that the assumptions cannot be relaxed.

18. $f(x)$ has unequal one-sided limits at 0 while those of $g(x)$ are equal. Hence 0 is a removable discontinuity of g but not of f.

Section 4.1

1. If f denotes the given function, $\dfrac{f(x+h) - f(x)}{h} =$

(b) $$\frac{2}{\sqrt{2x+2h+1} + \sqrt{2x+1}} \to \frac{1}{\sqrt{2x+1}}$$

(d) $$\frac{-3}{\sqrt{3x+3h+2}\sqrt{3x+2}\left(\sqrt{3x+2} + \sqrt{3x+3h+2}\right)} \to \frac{-3}{2(3x+2)^{3/2}}.$$

3. (a) $\dfrac{3(5x+7)^{1/3}}{5(3x+2)^{4/5}} + \dfrac{5(3x+2)^{1/5}}{3(5x+7)^{2/3}}$ (c) $\dfrac{4x}{(x^2+1)^2}\cos\left(\dfrac{x^2-1}{x^2+1}\right).$

4. (b) $-\dfrac{1}{y\cos xy^2} - \dfrac{y}{2x}.$

7. f is continuous at 1 iff $2a + b = 1$. For such a and b, f is differentiable iff $a + b = 3$. Therefore, $a = -2$ and $b = 5$.

11. (b) The difference quotient is

$$\frac{f(a+h^2)-f(a)}{h^2}h + \frac{f(a-h)-f(a)}{-h} \to f'(a).$$

14. For all $h \neq 0$, $[f(x+h) - f(x)]/h \geq 0$, hence $f'(x) \geq 0$.

16. Clear for $n = 2$. Suppose the assertion holds for $n \geq 2$. Then

$$\begin{aligned} D^{n+1}(fg) &= \sum_{k=0}^{n} \binom{n}{k} D[(D^k f)(D^{n-k} g)] \\ &= \sum_{k=0}^{n} \binom{n}{k} [(D^{k+1} f)(D^{n-k} g) + (D^k f)(D^{n+1-k} g)] \\ &= \sum_{k=1}^{n} \left[\binom{n}{k-1} + \binom{n}{k}\right] (D^k f)(D^{n+1-k} g) + gD^{n+1} f + fD^{n+1} g \\ &= \sum_{k=0}^{n+1} \binom{n+1}{k} (D^k f)(D^{n+1-k} g). \end{aligned}$$

18. (c) $(f' \circ g)g'' + (f'' \circ g)(g')^2$.

19. (a) $\dfrac{(-1)^n n!}{x^{n+1}}$.

21. If $x \neq 0$, $f'(x) = x^{m+n-1}\left(n \cos x^n + m\dfrac{\sin x^n}{x^n}\right)$. Also,

$$f'(0) = \lim_{x \to 0} \frac{\sin x^n}{x^n} x^{n+m-1} \begin{cases} \text{does not exist} & \text{if } n+m < 1, \\ = 0 & \text{if } n+m > 1, \\ = 1 & \text{if } n+m = 1. \end{cases}$$

Therefore, f' is continuous at 0 if $n + m \geq 1$.

23. For the second-order determinant, use the expansion

$$\begin{vmatrix} f_1 & f_2 \\ g_1 & g_2 \end{vmatrix} = f_1 g_2 - f_2 g_1.$$

For the third-order determinant, expand along a row or column and use the formula for the second-order case. The same idea may be applied to nth-order determinants.

Section 4.2

1. Set $f(x) = \cos x - \sqrt{x} + 1$. Since $f(0) > 0 > f(\pi/2)$, f has at least one zero in $(0, \pi/2)$, by the intermediate value theorem. Since $f' < 0$ on $(0, \pi/2)$, f is strictly decreasing so the zero is unique.

3. Since $f'(x) = 4x(x-1)(x-2) < 0$ on $(1, 2)$, f has at exactly one zero in the interval $(1, 2)$ iff $f(1)(= 1 + c)$ and $f(2)(= c)$ have opposite signs, that is, iff $c < 0 < c + 1$, or $-1 < c < 0$.

7. The assertion is clear if $n = 0$. Suppose it holds for all polynomials with degree $\le n$. Let $P(x)$ have degree $n + 1$ and suppose that the equation $\sin(ax) = P(x)$ has more than $n + 2$ solutions. Then $f(x) := \sin(ax) - P(x)$ has more than $n + 2$ zeros, hence, by Rolle's theorem, $f''(x) = -a^2 \sin(ax) - P''(x)$ has more than n zeros. But this means that $\sin(ax) = -P''(x)/a^2$ has more than n solutions, contradicting the induction hypothesis.

9. By the Cauchy mean value theorem,

$$|f(x) - f(y)|\,|g'(c)| = |g(x) - g(y)|\,|f'(c)| \le |g(x) - g(y)|\,|g'(c)|.$$

11. The derivative of $x^{-1} \sin x$ is negative since $\tan x > x$, $0 < x < \pi/2$.

17. Let $c_1 < \cdots < c_m$ be the distinct zeros of P'. By the intermediate value theorem, P' has a constant sign on (c_j, c_{j+1}). Therefore, $P(x)$ is strictly monotone on these intervals.

19. Let $|f'| \le c < r$. Then $g'(x) = r + f'(x) \ge r - c > 0$, so g is strictly increasing, hence one-to-one. By the mean value theorem, $|f(x) - f(0)| \le c|x|$ or $f(0) - c|x| \ge f(x) \le f(0) + c|x|$. Therefore,

$$f(0) + rx - c|x| \le g(x) \le f(0) + rx + c|x|.$$

Thus $x > 0 \Rightarrow g(x) \ge f(0) + (r - c)x \Rightarrow \lim_{x\to+\infty} g(x) = +\infty$, and $x < 0 \Rightarrow g(x) \le f(0) + (r - c)x \Rightarrow \lim_{x\to-\infty} g(x) = -\infty$. By the intermediate value theorem, $g(\mathbb{R}) = \mathbb{R}$.

22. $g'(0) = 0$, hence $f'(0) > 0$. Since $f(\pm 1/n\pi) = \pm 1/n\pi$ for all $n \in \mathbb{N}$, f is not monotone on any neighborhood of 0.

25. Let a, $b \in I$ with $a < b$ and suppose that $f'(a) < y_0 < f'(b)$, so $g'(a) < 0 < g'(b)$. Then $\big(g(x) - g(a)\big)/(x - a) < 0$ for $x \in (a, a + \delta)$, so the minimum of g cannot occur at a. Similarly, the minimum of g cannot occur at b. Thus, by the local extremum theorem, $g'(x_0) = 0$, that is, $f'(x_0) = y_0$, for some $x_0 \in (a, b)$.

28. Set $q(x, y) = [f(x) - f(y)](x - y)$, $x \neq y$. If f is uniformly differentiable on I, then

$$|f'(x) - f'(y)| \leq |f'(x) - q(x, y)| + |f'(y) - q(x, y)|$$

shows that f' is uniformly continuous. Conversely, assume that f' is uniformly continuous. By the mean value theorem, for each $x < y$ there exists a $z \in (x, y)$ such that

$$|q(x, y) - f'(y)| = |f'(z) - f'(y)|.$$

It follows that f is uniformly differentiable.

31. If such a function exists, then

$$\lim_{x \to y} \frac{f(x) - f(y)}{x - y} = \varphi(y, y)$$

so $f'(y) = \varphi(y, y)$, which is continuous in y.

Conversely, assume f is continuously differentiable on an open interval I and define

$$\varphi(x, y) = \begin{cases} \dfrac{f(x) - f(y)}{x - y} & \text{if } x \neq y, \\ f'(x) & \text{if } x = y. \end{cases}$$

Clearly φ is continuous on $\{(x, y) \in I \times I : x \neq y\}$. By the mean value theorem, $\varphi(x, y) = f'(\xi_{xy})(x - y)$, where ξ_{xy} is between x and y. The continuity of φ on $I \times I$ now follows from the continuity of f'.

Section 4.4

1. (b) $(2 - 3x)/(2x - 3)$, $x \neq 3/2$ (f) $\cos^{-1}\left(\dfrac{3x - 2}{1 - x}\right)$, $1/2 < x < 3/4$.

5. (b) Fix $y > 0$ and let $f(x) = \ln(xy) - \ln x - \ln y$. Since $f'(x) = 0$, $f(x) = f(1) = 0$ for all $x > 0$.

7. (b) $a^{x+y} = \exp((x + y)\ln a) = \exp(x \ln a)\exp(y \ln a) = a^x a^y$.

9. $x^a = \exp(a \ln x)$, hence $(x^a)' = \exp(a \ln x)(a/x) = ax^{a-1}$.

13. The derivative of the left side of (c) is

$$\frac{2}{x^2 + 1} - \frac{4x}{(x^2 + 1)^2\sqrt{1 - y^2}}, \quad y := \left(\frac{x^2 - 1}{x^2 + 1}\right)^2,$$

which reduces to 0. Therefore, the left side is constant.

14. Set $c = f'(0)$. Since

$$\frac{f(x+h)-f(x)}{h} = f(x)\,\frac{f(h)-1}{h},$$

$f'(x)$ exists and equals $cf(x)$. Therefore, $e^{-cx}f(x)$ has zero derivative, hence $e^{-cx}f(x) = f(0)$.

18. $(f^{-1})''(x) = -\dfrac{f''(f^{-1}(x))}{\left[f'(f^{-1}(x))\right]^3}$.

Section 4.5

1. (a) $p-q$ (d) -1 (g) -2 (j) 0 (m) $-\infty$ (p) 0
   (s) $+\infty$ (v) 1.

2. (c) $f(0) = \lim_{x\to 0+} f(x) = 5/3$.

3. (a) $\ln a_n = n^{-1}\ln\sin(1/n)$ is of the form $\frac{-\infty}{+\infty}$, hence has the same limit as

$$\frac{1}{n^2}\frac{\cos(1/n)}{\sin(1/n)} = -\frac{\cos(1/n)}{n}\,\frac{1}{n^{-1}\sin(1/n)} \to 0.$$

Therefore, $a_n \to 1$.

6. By logarithmic differentiation,

$$f'(x) = \left(1+\frac{1}{x}\right)x^{1/x} - \frac{\ln x}{x}\,x^{1/x}.$$

By l'Hospital's rule, $x^{1/x}\to 1$, hence $\lim_{x\to+\infty} f'(x) = 1$. Applying the mean value theorem to f on each of the intervals $[n, n+1]$ shows that $f(n+1)-f(n)\to 1$.

9. Let $L := \lim_{x\to+\infty} f'(x)/g'(x)$. By l'Hospital's rule, $\lim_{x\to+\infty} g(x)/f(x)$ exists and equals $1/L$. Another application of l'Hospital's rule yields

$$\lim_{x\to+\infty}\frac{\ln f(x)}{\ln g(x)} = \lim_{x\to+\infty}\frac{f'(x)\,g(x)}{g'(x)\,f(x)} = 1.$$

10. (a) By l'Hospital's rule, the quotient has the same limit as

$$\begin{aligned}&\frac{\alpha\beta}{2}\,\frac{f'(a+\alpha h)-f'(a+\beta h)}{h}\\ &\qquad= \frac{\alpha\beta}{2}\left[\alpha\frac{f'(a+\alpha h)-f'(a)}{\alpha h} - \beta\frac{f'(a+\beta h)-f'(a)}{\beta h}\right],\end{aligned}$$

which is $\alpha\beta(\alpha-\beta)f''(a)/2$.

12. Apply l'Hospital's rule n times to $f(x)/x^{-n}$ to obtain

$$\lim_{x\to 0+} x^n f(x) = \lim_{x\to 0+} \frac{(-1)^n f^{(n)}(x)}{n(n+1)\dots(2n-1)x^{-2n}} = \lim_{x\to 0+} ax^{2n} f^{(n)}(x),$$

where $a = \dfrac{(-1)^n(n-1)!}{(2n-1)!}$. Therefore, $\lim_{x\to 0+} x^n f(x)$ exists and equals aL.

16. By l'Hospital's rule,

$$\lim_{x\to+\infty} \frac{f(g(x))}{g(x)} = \lim_{x\to+\infty} \frac{f'(g(x))g'(x)}{g'(x)} = L.$$

For examples, take $f(x) = \sqrt{x}$, $\ln x$, or $x + 1/x$, and $g(x) = x^n$, e^x, or $\ln x$.

18. By l'Hospital's rule,

$$\begin{aligned}\lim_{x\to+\infty} f(x) &= \lim_{x\to+\infty} \frac{xf(x)}{x} = \lim_{x\to+\infty} \left[xf'(x) + f(x)\right] \\ &= \lim_{x\to+\infty} xf'(x) + \lim_{x\to+\infty} f(x).\end{aligned}$$

For the second part consider, $f(x) = \ln x$.

Section 4.6

2. Apply Taylor's theorem to the function between the inequalities to produce the number $c \in (0, x)$ in the remainder term:

(b) $f^{(k)}(x) = (-1)^k e^{-x}$, hence $e^{-x} = \sum_{k=0}^{2n-1} \dfrac{(-1)^k}{k!} x^k + \dfrac{e^{-c}}{(2n)!} x^{2n}$, where $e^{-c} \in (0, 1)$.

3. Let $I_n := \dfrac{1}{n!} \displaystyle\int_a^x (x-t)^n f^{(n+1)}(t)\,dt = -\dfrac{f^{(n)}(a)}{n!}(x-a)^n + I_{n-1}$. Iterating,

$$I_n = -\sum_{k=1}^{n} \frac{f^{(k)}(a)}{k!}(x-a)^k + I_0 = -T_n(x, a) + f(x).$$

5. By Taylor's theorem,

$$b_k = \frac{P^{(k)}(b)}{k!} = \frac{1}{k!} \sum_{j=0}^{n-k} (j+1)(j+2)\cdots(j+k)(b-a)^j a_{k+j}.$$

Section 4.7

1. (a) -1.52137970 (d) -1.42360584 (g) 1.220924381

2. (a) 0.87672621 (c) 1.55714559

4. 7.937253933

Section 5.1

3. Since $M_j(-f) = -m_j(f)$, $\overline{S}(-f, \mathcal{P}) = -\underline{S}(f, \mathcal{P})$, hence

$$\overline{\int_a^b}(-f) = \inf_{\mathcal{P}} \overline{S}(-f, \mathcal{P}) = \inf_{\mathcal{P}}(-\underline{S}(f, \mathcal{P})) = -\sup_{\mathcal{P}} \underline{S}(f, \mathcal{P}) = -\underline{\int_a^b} f.$$

Replacing f by $-f$ shows that $\underline{\int_a^b}(-f) = -\overline{\int_a^b} f$.

5. Since g may be obtained from f by changing one point at a time, we may assume that $f = g$ except at a single point $c \in (a, b)$. Let $\varepsilon > 0$ and let M be a bound for both $|f|$ and $|g|$. The point c is in at most two intervals of any partition $\mathcal{P}$, and each of these has width $\leq \|\mathcal{P}\|$. Since $f = g$ on the remaining intervals,

$$|\overline{S}(f, \mathcal{P}) - \overline{S}(g, \mathcal{P})| \leq 2M\|\mathcal{P}\|.$$

It follows from 5.1.15 that $\overline{\int_a^b} f = \overline{\int_a^b} g$. Similarly $\underline{\int_a^b} f = \underline{\int_a^b} g$.

6. (c) Let $g = \sin f$, $\varepsilon > 0$, and let $\mathcal{P}$ be any partition of $[a, b]$ such that $\overline{S}(f, \mathcal{P}) - \underline{S}(f, \mathcal{P}) < \varepsilon$. For fixed j, choose sequences a_n, $b_n \in [x_{j-1}, x_j]$ such that $g(a_n) \to M_j(g)$ and $g(b_n) \to m_j(g)$. Then

$$g(a_n) - g(b_n) \leq |f(a_n) - f(b_n)| \leq M_j(f) - m_j(f),$$

hence $M_j(g) - m_j(g) \leq M_j(f) - m_j(f)$. Therefore,

$$\overline{S}(g, \mathcal{P}) - \underline{S}(g, \mathcal{P}) \leq \overline{S}(f, \mathcal{P}) - \underline{S}(f, \mathcal{P}) < \varepsilon.$$

7. (a) Let $L = \lim_{\mathcal{P}} F(\mathcal{P})$ and $M = \lim_{\mathcal{P}} G(\mathcal{P})$. Given $\varepsilon > 0$, choose $\mathcal{P}'_\varepsilon$ and $\mathcal{P}''_\varepsilon$ such that $|F(\mathcal{P}) - L| < \eta$ for all partitions $\mathcal{P}$ refining $\mathcal{P}'_\varepsilon$ and $|G(\mathcal{P}) - M| < \eta$ for all partitions $\mathcal{P}$ refining $\mathcal{P}''_\varepsilon$, where $\eta = \varepsilon/(2|\alpha| + 2|\beta| + 2)$. Let $\mathcal{P}_\varepsilon$ denote the common refinement of $\mathcal{P}'_\varepsilon$ and $\mathcal{P}''_\varepsilon$. Then both inequalities hold for any partition $\mathcal{P}$ refining $\mathcal{P}_\varepsilon$, hence

$$|(\alpha F(\mathcal{P}) + \beta G) - (\alpha L + \beta M)| \leq |\alpha||F(\mathcal{P}) - L| + |\beta||G(\mathcal{P}) - M| < \varepsilon.$$

(b) Given $\varepsilon > 0$, choose $\mathcal{P}_\varepsilon$ such that $\overline{\int_a^b} f - \varepsilon < \overline{S}(f, \mathcal{P}_\varepsilon) \leq \overline{\int_a^b} f$. The inequality still holds if $\mathcal{P}_\varepsilon$ is replaced by a refinement. Therefore, $\overline{\int_a^b} f = \lim_{\mathcal{P}} \overline{S}(f, \mathcal{P})$.

Section 5.2

1. Assume $c_n \to c \in (a, b)$. Choose $\delta > 0$ so that $a < c - \delta < c + \delta < b$ and choose N so that $c_n \in (c - \delta, c + \delta)$ for all $n > N$. Since f has only finitely many discontinuities on $[a, c - \delta] \cup [c + \delta, b]$, f is integrable on

these intervals and the integrals are zero. Thus, given $\varepsilon > 0$, there exist partitions $\mathcal{P}_1$ of $[a, c-\delta]$ and $\mathcal{P}_2$ of $[c+\delta, b]$ such that

$$\overline{S}(f, \mathcal{P}_1) - \underline{S}(f, \mathcal{P}_1) < \varepsilon/3 \text{ and } \overline{S}(f, \mathcal{P}_2) - \underline{S}(f, \mathcal{P}_2) < \varepsilon/3.$$

Define a partition $\mathcal{P}$ on $[a, b]$ by $\mathcal{P} = \mathcal{P}_1 \cup \mathcal{P}_2$ and let $|f| \le M$ on $[a, b]$. If $\delta < \varepsilon/6M$, then

$$\overline{S}(f, \mathcal{P}) - \underline{S}(f, \mathcal{P}) \le \overline{S}(f, \mathcal{P}_1) - \underline{S}(f, \mathcal{P}_1) + \overline{S}(f, \mathcal{P}_2) - \underline{S}(f, \mathcal{P}_2) + 2M\delta < \varepsilon.$$

Therefore, $f \in \mathcal{R}_a^b$. Moreover,

$$\int_a^b f = \int_a^{c-\delta} f + \int_{c-\delta}^{c+\delta} f + \int_{c+\delta}^b f = \int_{c-\delta}^{c+\delta} f,$$

hence

$$\left| \int_a^b f \right| \le \int_{c-\delta}^{c+\delta} |f| \le 2M\delta.$$

Since δ may be made arbitrarily small, $\int_a^b f = 0$.

5. Set $M_n = \max\{f_1, \ldots, f_n\}$. Then $M_2 = \big(f_1 + f_2 + |f_1 - f_2|\big)/2 \in \mathcal{R}_a^b$. Since $M_n = \max\{M_{n-1}, f_n\}$, the general result follows by induction. A similar argument holds for min.

6. Choose x_0 such that $f(x_0) = \sup_{a \le x \le b} f(x)$. Then $\int_a^b f \le f(x_0)(b-a) < M(b-a)$.

9. Let $|f| \le M$ on $[a, b]$. Then $|F(x, y) - F(x, y_0)| \le M(y - y_0)$, hence $\lim_{y \to y_0} F(x, y) = F(x, y_0)$.

12. (a) By the approximation property, choose x_0 such that $|f(x_0)| > M - \varepsilon$. By continuity, we may take $x_0 \in (a, b)$ and we may choose $\delta > 0$ such that $|f(x)| > M - \varepsilon$ for all $x \in (x_0 - \delta, x_0 + \delta)$. Then

$$M(b-a) \ge \int_a^b |f| \ge \int_{x_0-\delta/2}^{x_0+\delta/2} |f| \ge \delta(M - \varepsilon).$$

(b) By (a), $|f(x)|^p > (M - \varepsilon)^p$ on $(x_0 - \delta, x_0 + \delta)$, hence, as in (a),

$$\delta(M - \varepsilon)^p \le \int_a^b |f|^p \le M^p(b-a).$$

Therefore,

$$\delta^{1/p}(M - \varepsilon) \le \left(\int_a^b |f|^p \right)^{1/p} \le M(b-a)^{1/p},$$

hence

$$M - \varepsilon \le \liminf_{p\to+\infty} \left(\int_a^b |f|^p \right)^{1/p} \le \limsup_{p\to+\infty} \left(\int_a^b |f|^p \right)^{1/p} \le M.$$

Since ε was arbitrary,

$$\liminf_{p\to+\infty} \left(\int_a^b |f|^p \right)^{1/p} = \limsup_{p\to+\infty} \left(\int_a^b |f|^p \right)^{1/p} = M.$$

Section 5.3

1. By a change of variables and periodicity,

$$\begin{aligned} \int_0^p f(x+y)\,dx &= \int_y^{p+y} f(x)\,dx \\ &= \int_y^p f(x)\,dx + \int_p^{p+y} f(x)\,dx \\ &= \int_y^p f(x)\,dx + \int_p^{p+y} f(x-p)\,dx \\ &= \int_y^p f(x)\,dx + \int_0^y f(x\,dx = \int_0^p f(x)\,dx. \end{aligned}$$

3. (a) On $[0,1]$, $2x/\pi \le \sin x \le x$. Since $\displaystyle\int_0^1 \frac{x}{\sqrt{x^2+1}}\,dx = \sqrt{2}-1$, the inequalities follow.

5. (a) Substituting $y = x^{1/n}$ and integrating by parts $n-1$ times yields

$$\int_0^1 \exp\left(x^{1/n}\right) dx = n \int_0^1 y^{n-1} e^y\,dy = F(1) - F(0),$$

where

$$F(y) = (-1)^{n+1} n! e^y \sum_{j=0}^{n-1} \frac{(-1)^j}{j!}\, y^j.$$

7. Use trig. substitution and Exercise 6.

9. Substituting $s = f(t)$ and integrating by parts yields

$$\int_0^y f^{-1}(s)\,ds = \int_0^{f^{-1}(y)} t f'(t)\,dt = y f^{-1}(y) - \int_0^{f^{-1}(y)} f,$$

hence

$$\int_0^x f + \int_0^y f^{-1} = y f^{-1}(y) + \int_0^x f - \int_0^{f^{-1}(y)} f = y f^{-1}(y) + \int_{f^{-1}(y)}^x f.$$

If $f^{-1}(y) \le x$, then $f(t) \ge y$ for all $t \in [f^{-1}(y), x]$, hence

$$\int_{f^{-1}(y)}^{x} f(t)\,dt + yf^{-1}(y) \ge \int_{f^{-1}(y)}^{x} y\,dt + yf^{-1}(y) = xy.$$

On the other hand, if $f^{-1}(y) \ge x$ then $f(t) \le y$ for all $t \in [x, f^{-1}(y)]$, hence

$$\int_{f^{-1}(y)}^{x} f(t)\,dt + yf^{-1}(y) \ge -\int_{x}^{f^{-1}(y)} y\,dt + yf^{-1}(y) = xy.$$

10. (b) Take $f(t) = \ln(t+1)$, $0 \le x \le 1$, and $0 \le y \le \ln 2$ in Young's inequality to obtain

$$(x+1)\ln(x+1) - x + e^y - y - 1 = \int_0^x \ln(t+1)\,dt + \int_0^y (e^s - 1)\,ds \ge xy.$$

Replace $x+1$ by x, $1 \le x \le 2$.

13. Integrate by parts to obtain

$$\int_a^b f(x)\sin(nx)\,dx = f(x)\cos(nx)\Big|_b^a + \frac{1}{n}\int_a^b f'(x)\cos(nx)\,dx.$$

17. If F is a primitive of f, then $\displaystyle\int_{u(x)}^{v(x)} f = F\big(v(x)\big) - F\big(u(x)\big)$. Now use the chain rule.

19. By l'Hospital's rule,

$$\lim_{x\to a} \frac{g(x)}{x-a}\int_a^x f = \lim_{x\to a}\left[g(x)f(x) + g'(x)\int_a^x f\right] = g(a)f(a).$$

20. (a) s_n is a Riemann sum for $\displaystyle\int_0^1 x^p\,dx$, hence $\lim_{n\to+\infty} s_n = 1/(p+1)$.

21. By the mean value theorem,

$$|f(x) - f(x_{k-1})| \le M|x - x_{k-1}| \le M(x_k - x_{k-1}) = Mh, \quad x \in [x_{k-1}, x_k],$$

hence

$$\left|\int_a^b f - \sum_{k=1}^{n} f(x_{k-1})h\right| = \left|\sum_{k=1}^{n}\int_{x_{k-1}}^{x_k} [f(x) - f(x_{k-1})]\,dx\right| \le Mnh^2.$$

Section 5.5

3. The substitution $t = \sin x$ yields

$$\int_{\pi/6}^{\pi/3} f(\sin x)\, dx = \int_{1/2}^{\sqrt{3}/2} \frac{f(t)}{\sqrt{1-t^2}}\, dt.$$

Now apply 5.5.3 with $g(t) = (1-t^2)^{-1/2}$.

7. $G(b) \le \int_a^b fg \le G(a)$. Now apply the intermediate value theorem to G.

9. Apply 5.5.3 to obtain $c \in [0,1]$ such that

$$\int_0^\pi g(x)\sin x\, dx = g(0)\int_0^c \sin x\, dx + g(1)\int_c^\pi \sin x\, dx = \cos c + 1.$$

Section 5.7

1. Let $f(x)$ denote the integrand and $0 < \varepsilon < 1$. Then $\int_0^1 f = \int_0^\varepsilon f + \int_\varepsilon^1 f$. On $(0, \varepsilon]$, $2x/\pi \le \sin x \le x$ and $1 - \varepsilon \le 1 - x < 1$, hence

$$\frac{1}{|x|^p} \le f(x) \le \frac{(\pi/2)^p}{(1-\varepsilon)^q |x|^p}.$$

On $[\varepsilon, 1)$,

$$\frac{1}{(1-x)^q \sin^p 1} \le f(x) \le \frac{1}{(1-x)^q \sin^p \varepsilon}.$$

Therefore, $\int_0^1 f$ converges iff $p,\ q > 1$.

5. Only (b) and (d) diverge.

8. Choose $r > 0$ so that $1/2 < \left(\dfrac{\sin x}{x}\right)^p < 1$ for $0 < x < r$. Then

$$\frac{1}{2}\int_0^\varepsilon x^{q-p}\, dx \le \int_0^\varepsilon \frac{\sin^p x}{x^q} \le \int_0^\varepsilon x^{q-p}\, dx.$$

Now apply 5.7.3(a).

9. (a) all p (c) all p (h) $p > -2$ (k) $p > -1$.

11. Let $g(x) = x(1+x^2)^{-1}$, $h(x) = \sin x$ and $f := gh$. Then $|\int_1^x h|$ is bounded and $g' < 0$ so, by 5.7.17, f is improperly integrable on $[1, +\infty)$. For every n,

$$\int_0^\infty |f|\, dx \ge \sum_{j=2}^n \int_{(j-1)\pi}^{j\pi} \frac{x|\sin x|}{1+x^2}\, dx \ge \sum_{j=2}^n \int_{(j-1)\pi}^{j\pi} \frac{\pi(j-1)|\sin x|}{1+\pi^2 j^2}\, dx$$
$$= M\sum_{j=2}^n \frac{j-1}{1+\pi^2 j^2},$$

where M is a positive constant. The sums in the last equality are unbounded, hence h is not improperly absolutely integrable in this case.

13. (a) Converges for all $p > 0$ if $0 < q < 1$; diverges for all $p > 0$ if $q \geq 1$.
    (b) Converges for all $p > 0$ if $0 < q < 1$; diverges for all $p > 0$ if $q \geq 1$.
    (c) Converges if $p > 2$ or $q > 2$ and diverges otherwise.
    (d) Converges if $p < 2$ or $q < 2$ and diverges otherwise.
    (e) Converges iff $q < 1$.
    (f) Converges iff $pq < 1$.

15. Integrate by parts:

$$I_n := \int_{-\infty}^{\infty} x^{2n} e^{-x^2/2}\,dx = (2n-1)\int_{-\infty}^{\infty} x^{2n-2} e^{-x^2/2}\,dx = (2n-1)I_{n-1}.$$

20. Both integrals converge. The root test is inconclusive.

24. By the Cauchy–Schwarz inequality,

$$\int_1^\infty \frac{\sqrt{f(x)}}{x} \leq \left(\int_1^\infty f(x)\,dx\right)^{1/2}\left(\int_1^\infty \frac{dx}{x^2}\right)^{1/2} < +\infty.$$

26. Let $F(x) = \int_a^x fg$, $a \leq x < b$, and let $b_n \uparrow b$. By the weighted mean value theorem,

$$F(b_m) - F(b_n) = f(c_{m,n})[G(b_m) - G(b_n)]$$

for some $c_{m,n}$ between b_m and b_n. Since G is bounded and $f(c_{m,n}) \to 0$, $\{F(b_n)\}$ is a Cauchy sequence and hence converges. Since $\{b_n\}$ was arbitrary, $\int_a^b fg$ converges.

28. Let $F(t) = \int_0^t f\,dx$. Then

$$\int_0^t f(x+c)\,dx = \int_c^{c+t} f(x)\,dx = \int_c^t f(x)\,dx + F(t+c) - F(t),$$

hence

$$\int_0^\infty f(x+c)\,dx = \int_c^\infty f(x)\,dx.$$

Similarly, $\displaystyle\int_{-\infty}^0 f(x+c)\,dx = \int_{-\infty}^c f(x)\,dx.$

Section 5.8

2. Given $\varepsilon > 0$, let A_n be covered by intervals $I_{n,k}$, $k = 1, 2, \ldots$, with total length $< \varepsilon/2^n$. Then the union is covered by intervals $I_{n,k}$, n, $k = 1, 2, \ldots$, with total length $< \varepsilon$.

6. The discontinuity set is countable, hence the integral exists. Since all lower sums are zero, the integral must be zero.

Section 6.1

1. (a) $\dfrac{m^3(m+1)}{2m+1}$ (c) $\ln(3/2)$ (e) $\dfrac{1}{m}\displaystyle\sum_{k=1}^{m}\frac{(-1)^k}{k}$ (g) $\dfrac{23}{480}$

   (i) $-\ln(m+1)$ (n) $\displaystyle\sum_{k=1}^{m}\frac{(-1)^k}{k}$.

2. (a) $\dfrac{1}{1+r^2}$.

3. (a) $193e$. (c) $(e-1/e)/2$.

5. Let $s_n = \displaystyle\sum_{k=1}^{n}\frac{1}{k}$, $u_n = \displaystyle\sum_{k=1}^{n}\frac{1}{2k-1}$, and $v_n = \displaystyle\sum_{k=1}^{n}\frac{4}{(2k-1)(2k+1)}$.

   (a) $s_{2n} = s_n/2 + u_n$, hence, by 6.1.9,
$$u_n - \tfrac{1}{2}\ln n = [s_{2n} - \ln(2n)] - \tfrac{1}{2}[s_n - \ln n] + \ln 2 \to \tfrac{1}{2}\gamma + \ln 2.$$

8. Given $\varepsilon > 0$, choose N such that $L - \varepsilon < a_k/b_k < L + \varepsilon$ for all $k \geq N$. Multiplying by b_k and summing,
$$(L-\varepsilon)\sum_{k=n}^{m} b_k < \sum_{k=n}^{m} a_k < (L+\varepsilon)\sum_{k=n}^{m} b_k, \quad m > n \geq N.$$
   Letting $m \to +\infty$ and dividing,
$$L - \varepsilon < \frac{\sum_{k=n}^{\infty} a_k}{\sum_{k=n}^{\infty} b_k} < L + \varepsilon, \quad n \geq N.$$

12. Let s_n and t_n denote the nth partial sums of $\sum a_n$ and $\sum b_n$, respectively. Then $t_k = s_{n_k}$ so $\{t_k\}$ is a subsequence of $\{s_n\}$. Therefore, if $\sum_n a_n$ converges, so does $\sum_k b_k$. If the terms a_n are nonnegative, then, for each n, $s_n \leq t_k$ for $k \geq n$, hence if $\sum_k b_k$ converges, then so does $\sum_n a_n$. The series $\sum_{n=0}^{\infty}(-1)^n$ shows that the latter assertion fails in general.

15. By summing a geometric series, a real number x with representation $b_N b_{N-1} \cdots b_0.a_1 a_2 \cdots a_n 999 \cdots$, where $a_n \neq 9$ may be written as

$$b_N b_{N-1} \cdots b_0.a_1 a_2 \cdots a_n + 10^{-n} = b_N b_{N-1} \cdots b_0.a_1 a_2 \cdots a_{n-1} a_n',$$

where $a_n' := a_n + 1$. Therefore, a real number has at least one standard representation.

Suppose that $b_N b_{N-1} \cdots b_0.a_1 a_2 \cdots = c_M c_{M-1} \cdots c_0.d_1 d_2 \cdots$ are standard representations. Then

$$\begin{aligned} |b_N b_{N-1} \cdots b_0 - c_M c_{M-1} \cdots c_0| &= |(.d_1 d_2 \cdots) - (.a_1 a_2 \cdots)| \\ &\leq \sum_{j=1}^{\infty} \frac{|d_j - a_j|}{10^j}. \end{aligned}$$

Since the representations are standard, $|d_j - a_j|$ cannot eventually equal 9, hence the right side is < 1. Therefore, since the left side is an integer, it must be zero. It follows from Exercise 1.4.15 that $M = N$ and $b_j = c_j$, $0 \leq j \leq N$. Then $a_1.a_2 a_3 \cdots = d_1.d_2 d_3 \cdots$, hence $a_1 = d_1$. An induction argument shows that $a_n = d_n$ for all n.

18 For $n \in \mathbb{N}$, $n = qm_n + r_n$, where $r_n, m_n \in \mathbb{N}$ and $0 \leq r_n \leq q-1$. Since $s_n - s_{qm_n}$ is a sum of terms of the form a_{qm_n+j}, $j = 1, \ldots q-1$, each of which $\to 0$, $s_n - s_{qm_n} \to 0$. Therefore, $s_n \to s$.

Section 6.2

1. By the ratio test, (a), (b), (e), and (f) converge; (c) and (d) diverge.

2. (a) Converges by ratio test.

   (d) Converges by ratio test.

   (g) Converges by integral test iff $p > 1$.

   (j) Diverges by limit comparison with $\sum 1/n$.

   (m) Converges by limit comparison with $\sum 1/n^2$.

   (p) Diverges by ratio test.

   (s) Diverges since $2^{\ln n} = n^p$, $p = \ln 2 < 1$.

   (v) Converges by limit comparison with $\sum 1/2^n$.

5. For all sufficiently large n, $a^n < a_n n^{1/n} < 2a_n$.

6. (a) Diverges for all p.(e) Converges iff $q > p$.(g) Converges iff $q > 1 + p$.

8. (a) Since $a_n \to 0$, $a_n^2 < a_n$ for all large n. Therefore, $\sum b_n$ converges by comparison test.

   (d) Converges by comparison test: $b_n \leq a_n$.

(h) Converges: For n sufficiently large, say $n \geq N$, $a_n < 1$, hence $b_n = Ma_N \cdots a_n < Ma_n$, where $M = a_1 \cdots a_{N-1}$.

(l) Converges by the Cauchy–Schwarz inequality.

11. The inequality implies that $\{a_n/b_n\}$ is a decreasing sequence and hence converges to $L < +\infty$. Now use the comparison test.

14. Since $\lim_{x\to\infty} f(g(x)) = \lim_{x\to\infty} g(x) = 0$, l'Hospital's rule implies that $\lim_{x\to\infty} f(g(x))/g(x) = \lim_{x\to\infty} f'(g(x)) = f'(0)$. Now apply the limit comparison test to $\sum_n f(g(n))$ and $\sum_n g(n)$.

15. (a) If $\sum f(1/n^p)$ converges, then $f(0) = \lim_n f(1/n^p) = 0$. Suppose $f'(0) \neq 0$. Then, by l'Hospital's rule, $\lim_{x\to 0} \dfrac{f(x^p)}{x^{2p}} = \infty$. Therefore, eventually $f(1/n^p) > 1/n^{2p}$ so the series diverges by the comparison test.

17. (a) $n!(e - s_n) = m(n-1)! - \displaystyle\sum_{k=1}^{n} \frac{n!}{k!} \in \mathbb{N}$.

(b)
$$\begin{aligned} e - s_n &= \sum_{k=1}^{\infty} \frac{1}{(n+k)!} \\ &= \frac{1}{(n+1)!}\left[1 + \frac{1}{n+2} + \frac{1}{(n+2)(n+3)} + \ldots\right] \\ &< \frac{1}{(n+1)!}\left[1 + \frac{1}{n+1} + \frac{1}{(n+1)^2} + \ldots\right] \\ &= \frac{1}{(n+1)!}\,\frac{n+1}{n}, \end{aligned}$$

hence $n!(e - s_n) < 1/n$. By (a) and (b), $n!(e - s_n)$ is a positive integer $< 1/n$, which is impossible.

Section 6.3

3. (a) and (c) diverge: $d_n \to 2/3$; (b) converges: $d_n \to 3/2$.

5. (a) By ratio test: series converges if $p < e$ and diverges if $p > e$. If $p = e$, series diverges by Raabe's test since then $d_n \to -1/2$.

6. Ratio test fails. Raabe: $d_n \to (1+p)/2$, hence converges if $p > 1$ and diverges if $p < 1$. Also diverges if $p = 1$, since then $a_n = 1/(2n+1)$.

10. (a) Diverges. (d) Converges iff $r > 1$.

13. $-\ln a_n / \ln n \to \ln b$.

16. (a) Converges iff $q > p$.

18. Let $\bar{c} > 1$ and choose $r \in (1, \bar{c})$. Then, for sufficiently large n, $c_n > r$, hence

$$\ln a_n^{-1} = \ln n + c_n \ln \ln n > \ln n + r \ln \ln n = \ln \left[n(\ln n)^r \right]$$

and therefore $a_n < \dfrac{1}{n(\ln n)^r}$. Since $\int_2^\infty 1/x(\ln x)^r \, dx < +\infty$, the integral and comparison tests complete the proof in this case. The case $\underline{c} < 1$ is similar.

The given series diverges.

21. Take $b_n = n \ln n$ in Kummer's test. Then

$$c_n = \left[1 + \frac{1}{n} + \frac{\beta_n}{n \ln n}\right] n \ln n - (n+1) \ln(n+1) = (n+1) \ln \left(\frac{n}{n+1}\right) + \beta_n.$$

Since the first term on the right side tends to -1, $\liminf_{n\to\infty} \beta_n > 1$ implies $\liminf_{n\to\infty} c_n > 0$, and $\limsup_{n\to\infty} \beta_n < 1$ implies $\liminf_{n\to\infty} c_n < 0$.

Section 6.4

2. Choose $r > 1$ and $N \in \mathbb{N}$ such that $|a_{n+1}|/|a_n| > r$ for all $n \geq N$. Then $|a_{N+k}| > r^k |a_N|$ for all k, hence $a_n \not\to 0$. Therefore, series diverges.

4. (a) Diverges. (b) Converges conditionally.

   (c) Converges absolutely if $p > 1$, conditionally if $p \leq 1$.

   (i) Converges absolutely if $p > 1/2$, conditionally if $p \leq 1/2$.

   (m) Converges absolutely if $p > 1$, diverges if $p < 1$.

9. Let $b_n = \dfrac{n - 1/2}{n^p + (-1)^n}$. If $p \leq 1$, then $b_n \sin n\theta$ need not tend to zero (see Example 8.3.10). For $p > 1$, it suffices by Dirichlet's test to show that $\sum |b_{n+1} - b_n| < +\infty$. This follows by limit comparison with $\sum 1/n^p$.

13. For the first series use the alternating series test. Let t_n be the nth partial sum of the second series. Then

$$\begin{aligned} t_{5n} &= 1 + \frac{1}{2} + \frac{1}{3} - \frac{1}{4} - \frac{1}{5} + \frac{1}{6} + \frac{1}{7} + \frac{1}{8} - \frac{1}{9} - \frac{1}{10} + \cdots - \frac{1}{5n} \\ &\geq \frac{1}{3} + \frac{1}{3} + \frac{1}{3} - \frac{1}{3} - \frac{1}{3} + \frac{1}{8} + \frac{1}{8} + \frac{1}{8} - \frac{1}{8} - \frac{1}{8} + \cdots \\ &= \frac{1}{3} + \frac{1}{8} + \frac{1}{13} + \cdots + \frac{1}{5n-2}. \end{aligned}$$

Thus $t_{5n} \to +\infty$, so the series diverges.

Section 6.5

3. (a), (b), (c): Double limit does not exist; only one iterated limit exists.

   (d), (g), (l): Iterated limits exist and are unequal. Hence double limit does not exist.

   (e), (h): Iterated limits exist and are equal. Double limit exists.

   (f), (i), (k): Iterated limits exist and are equal. Double limit does not exist.

   (j) If $a = b$, iterated limits exist and are equal, double limit exists. If $a \neq b$, iterated limits exist and are unequal.

9. Let $s_{m,n} = \sum_{j=1}^{m}\sum_{k=1}^{n} a_{j,k}$ and $s_n = \sum_{k=1}^{n} b_k$. Then for $m \geq n$,

$$s_n \leq s_{n,n} \leq s_{m,n} \leq s_{m,m} \leq s_{2m-1},$$

   hence the result follows from the squeeze principle.

10. (b) Let $b_n = \sum_{j=1}^{n} a_{j,n+1-j} = \sum_{j=1}^{n} \dfrac{1}{[j^2 + (n+1-j)^2]^{p/2}}$ and let $s_n = \sum_{k=1}^{n} b_k$. The minimum of $x^2+(n+1-x)^2$ on $[1,n]$ occurs at $x = (n+1)/2$ and the maximum at $x = 1$ and $x = n$, hence

$$(n+1)^2/2 \leq j^2 + (n+1-j)^2 \leq n^2 + 1, \quad 1 \leq j \leq n,$$

   and therefore

$$\frac{n}{(n^2+1)^{p/2}} \leq b_n \leq \frac{2^{p/2} n}{(n+1)^p},$$

   so the double series converges iff $p > 2$.

11. If $|r| \geq 1$, then $a_{m,n} \not\to 0$, hence the double series diverges. Let $|r| < 1$ and set $c_m = |r|^m/(1-|r|^m)$. Choose M such that $|r|^m < 1/2$ for $m > M$. Then

$$\sum_{m=1}^{\infty}\sum_{n=1}^{\infty} |r|^{mn} = \sum_{m=1}^{M} c_m + \sum_{m=M+1}^{\infty} c_m \leq \sum_{m=1}^{M} c_m + 2\sum_{m=1}^{\infty} |r|^m < +\infty.$$

   Therefore, the iterated series, and hence the double series, converges absolutely.

12. Let $L < 1$. Choose $r \in (L,1)$ and then N such that $a_{m,n}^{1/mn} < r$ for all m, $n \geq N$. For such m, n, $a_{m,n} < r^{mn}$, hence $\sum a_{m,n}$ converges by Exercises 6 and 11. If $L > 1$, choose $r \in (1, L)$ and then N such that $a_{m,n}^{1/mn} > r$ for all m, $n \geq N$. For such m, n, $a_{m,n} > r^{mn} > 1$, hence $a_{m,n} \not\to 0$, so the series diverges.

Section 7.1

1. (b) Pointwise to 0 on $(-1,1]$ for all $p \geq 0$, uniformly on intervals $[a,1]$ for $a > -1$ and $p < 1$. Uniformly on $[-1,1]$ if $p < 0$.

   (d) Pointwise to 0 on $\mathbb{R}$, uniformly on $|x| \geq a > 0$.

   (g) Uniformly to 0 on $\mathbb{R}$.

   (j) Pointwise on $\mathbb{R}$, uniformly on the sets $|x| \geq r > 1$ and $|x| \leq s < 1$.

2. (a) Pointwise but not uniformly. (b) Uniformly.

6. For example, $f_n(x) = x + 1/n$, $f(x) = g_n(x) = g(x) = x$ on $[1, +\infty]$.

10. Given $\varepsilon > 0$, choose $\delta > 0$ such that $|f(x) - f(y)| < \varepsilon$ for all $x, y \in \mathbb{R}$ with $|x - y| < \delta$. Then choose N such that $|a_n - a| < \delta$ for all $n \geq N$. For such n and for all x,

$$|f_n(x) - f(x+a)| = |f(x+a_n) - f(x+a)| < \varepsilon.$$

13. If $x \in \mathbb{Q}$ has reduced form $x = k/m$, then $f_n(x) = 1$ for all $n \geq m$. Therefore, f_n converges pointwise to the Dirichlet function $d(x)$. Suppose the convergence were uniform on $[0,1]$. Then we could find n such that $|f_n(x) - d(x)| < 1$ for all $x \in [0,1]$. In particular, $|f_n(1/m) - 1| < 1$ for all $m > n$, which is impossible since $f_n(1/m) = 0$.

17. Let $M > |f_0(x)| + 1$ for all $x \in S$. Then

$$\begin{aligned} |f_{n+1}(x) - f_n(x)| &= |\sin\left(rf_n(x)\right) - \sin\left(rf_{n-1}(x)\right)| \\ &\leq r|f_n(x)) - f_{n-1}(x)| \leq \cdots \leq r^n|f_1(x)) - f_0(x)| \\ &\leq Mr^n. \end{aligned}$$

Since $r < 1$, $\{f_n\}$ is uniformly Cauchy. Therefore, $f_n \to$ some f, uniformly on S. The generalization is proved in a similar manner, using the mean value theorem.

Section 7.2

4. Let $x > 0$. By l'Hospital's rule, n^2xe^{-nx} has the same limit as $2ne^{-nx}$, namely, 0. The convergence is not uniform on $(0,1)$, however, as may be seen by taking $b_n = 1/n$ in 7.1.5. An integration by parts shows that $\int_0^1 n^2xe^{-nx}\,dx = 1 - e^{-n}(1+n) \to 1$.

5. (d) Let $L := \lim_n \int_0^1 f_n$. By the mean value theorem, $e^{-x/n} - 1 = (-x/n)e^{-\xi/n}$, hence

$$\left|\frac{\sqrt{n}\left(e^{-x/n} - 1\right)}{x}\right| \leq \frac{e^{-\xi/n}}{\sqrt{n}} \leq \frac{1}{\sqrt{n}}$$

so $\sqrt{n}\left(e^{-x/n} - 1\right)/x$ converges uniformly to zero. Therefore, $L = 0$.

6. If $x \geq r > 0$, then $f_n(x) = \sqrt{n}/(1+n^2x^2) \leq \sqrt{n}/(1+n^2r^2)$, hence $f_n \to 0$ uniformly on $[r, +\infty)$. The convergence is not uniform on $(0,1)$, as can be seen by taking $b_n = 1/n$ in 7.1.5. A substitution shows that $\int_0^1 f_n = n^{-1/2} \arctan n \to 0$.

8. (a) $n \sin f_n \to f'/f$ uniformly $\Rightarrow \int_a^b n \sin f_n \to \ln f(b) - \ln f(a)$.

9. This follows from the inequality

$$\left| \int_a^x f_n(t)\,dt - \int_a^x f(t)\,dt \right| \leq \int_a^x |f_n(t) - f(t)|\,dt \leq \int_a^b |f_n - f|.$$

Section 7.3

1. (a) Pointwise on $(1, +\infty)$, uniformly on $[r, +\infty)$, $r > 1$.

   (d) Uniformly on $[0, +\infty)$.

   (g) Pointwise on $(0, +\infty)$, uniformly on $[r, +\infty)$, $r > 0$.

   (i) If $p > 1$, pointwise on $[0, +\infty)$, uniformly on $[0, r]$;
   If $p = 1$, converges only at $x = 0$.

2. (b) $s(x) = \dfrac{1}{1 + \ln x}$. Pointwise on $(1/e, e)$, uniformly on closed subintervals.

4. Both $s(x)$ and $c(x)$ converge uniformly on $\mathbb{R}$ by the M-test. Therefore, term by term integration is justified so

$$\int_x^{\pi/2} s(t)\,dt = \sum_n \frac{a_n}{2n+1} \cos\left[(2n+1)x\right], \quad \int_0^x c(t)\,dt = \sum_n \frac{a_n}{n} \sin(nx).$$

6. (a) Let $p \leq 1/2$ and $x \neq 0$. By l'Hospital's rule, $n^{-1}[1 - \cos(x/n^p)]$ has the same limit as $n \to +\infty$ as

$$\frac{-pxn^{-p-1}\sin(x/n^p)}{-1/n^2} = px^2 n^{1-2p} \frac{\sin(x/n^p)}{x/n^p}.$$

   Since this limit is positive, (a) follows from the limit comparison test.

   (b) Since cosine is an even function, to show uniform convergence on intervals $[a, b]$ we may assume $a = 0$. By the mean value theorem, for each $n \in \mathbb{N}$ and $x \in [0, b]$ there exists $x_n \in [0, b]$ such that

$$|1 - \cos(x/n^p)| = (x/n^p)|\sin(x_n/n^p)| \leq b^2/n^{2p}.$$

   Therefore, uniform convergence on $[0, b]$ follows from the M-test. Since $1 - \cos(x/n^p)$ does not converge uniformly to 0 on any unbounded interval, $s(x)$ does not converge uniformly on $\mathbb{R}$.

9. Let $|f'| \le M$ on I. By the mean value theorem, for each $x \in I$ and $n \in \mathbb{N}$ there exists ξ between $x/(n+1)$ and 0 such that

$$\frac{1}{n}\left|f\left(\frac{x}{n+1}\right)\right| = \frac{|xf'(\xi)|}{n(n+1)} \le \frac{rM}{n(n+1)}.$$

Therefore, $s(x)$ converges uniformly on I by the Weierstrass M-test. Since f' is bounded, the derived series

$$s'(x) = \sum_{n=1}^{\infty} \frac{1}{n(n+1)} f'\left(\frac{x}{n+1}\right)$$

converges uniformly on I and $s'(0) = f'(0)$.

11. Since $f_n \ge 0$, the partial sums of the series increase, so the conclusion follows from Dini's theorem (7.1.12).

13. For $x \in [a,b]$, either $f_n(a) \le f_n(x) \le f_n(b)$ or $f_n(b) \le f_n(x) \le f_n(a)$, hence $|f_n(x)| \le M_n := \max\{|f_n(a)|, |f_n(b)|\} \le |f_n(a)| + |f_n(b)|$. Since $\sum M_n < +\infty$, $s = \sum_n f_n$ converges uniformly on $[a,b]$. Since each $f_n \in \mathcal{R}([a,b])$, $s \in \mathcal{R}([a,b])$ and $\int_a^b s = \sum \int_a^b f_n$.

15. By Dini's theorem, the convergence of $\{g_n\}$ is uniform. Therefore, the result follows from 7.3.9.

18. Since g is continuous and $n^{-2}[g+n] \downarrow 0$, the convergence is uniform on closed bounded intervals I. By 7.3.9, $s(x)$ converges uniformly on I. The convergence is not absolute for any x (compare with $\sum_n 1/n$).

Section 7.4

1. (a) $(-1,3)$ (d) $(-1,1]$ (g) $(-1/4,1/4)$ (i) $(-1,1)$.

2. (b) $\displaystyle\sum_{n=3}^{\infty} \frac{3^{n-3}}{2^{n-2}} x^n$, $-2/3 < x < 2/3$.

3. (a) Replace x by $x-1$ in (7.12), where $|x-1|<1$, to obtain

$$\begin{aligned} x\ln x &= (x-1)\ln x + \ln x \\ &= \sum_{n=1}^{\infty}\frac{(-1)^{n+1}}{n}(x-1)^{n+1} + \sum_{n=1}^{\infty}\frac{(-1)^{n+1}}{n}(x-1)^{n} \\ &= \sum_{n=2}^{\infty}\frac{(-1)^{n}}{n-1}(x-1)^{n} + \sum_{n=1}^{\infty}\frac{(-1)^{n+1}}{n}(x-1)^{n} \\ &= (x-1) + \sum_{n=2}^{\infty}\left[\frac{(-1)^{n}}{n-1} + \frac{(-1)^{n+1}}{n}\right](x-1)^{n} \\ &= (x-1) + \sum_{n=2}^{\infty}\frac{(-1)^{n}}{n(n-1)}(x-1)^{n}. \end{aligned}$$

4. (a) $\displaystyle\sum_{n=1}^{\infty}\frac{(-1)^{n+1}2^n+3^n}{n}x^n,\ |x|<1/3$ (e) $\displaystyle\sum_{n=0}^{\infty}\frac{(-1)^n}{(2n+1)!}x^n,\ x>0$

(g) $\displaystyle\sum_{n=0}^{\infty}\frac{(-1)^n 4^n}{(2n+1)!}x^{2n+1},\ x\in\mathbb{R}.$

5. Use $\arccos x = \pi/2 - \arcsin x$ and (7.20).

9. (a) $\displaystyle\sum_{n=1}^{\infty}(-1)^n\frac{x^{2n-1}}{(2n-1)(2n+1)!}.$

10. (b) $\displaystyle\frac{x(1-x^2)}{(1+x^2)^2}.$

11. 27/4.

12. (a) $\displaystyle\sum_{n=1}^{\infty}(-1)^{n+1}c_n x^n,\ |x|<1,\quad c_n := \sum_{k=1}^{n}\frac{(-1)^k}{k}.$

(d) $\displaystyle\sum_{n=0}^{\infty}c_n x^{2n+1},\ x\in\mathbb{R},\quad c_n := \sum_{k=0}^{n}\frac{(-1)^k}{(2k+1)!(n-k)!}.$

16. For $|x|<(\sqrt{5}-1)/2$,

$$\begin{aligned} (1-x-x^2)s(x) &= \sum_{n=0}^{\infty}c_n x^n - \sum_{n=0}^{\infty}c_n x^{n+1} - \sum_{n=0}^{\infty}c_n x^{n+2} \\ &= c_0 + c_1 x - c_0 x + \sum_{n=2}^{\infty}(c_n - c_{n-1} - c_{n-2})x^n \\ &= 1. \end{aligned}$$

18. Replace x by $-t^2$ in (7.19) to obtain

$$\frac{1}{\sqrt{1+t^2}} = \sum_{n=0}^{\infty} \frac{(-1)^n(2n)!}{(n!)^2 4^n} t^{2n}, \quad |t| < 1.$$

Integrating from 0 to x yields the desired representation.

21. (a)) Choose r such that $R_s^{-1} = \limsup_n |c_n|^{1/n} < r < 1$. Then

$$|c_{n^2}|^{1/n} = \left(|c_{n^2}|^{1/n^2}\right)^n < r^n \to 0,$$

hence $R_t = +\infty$.

(b) If $c_n = (1 + a/n^p)^n$, $p > 0$, then $R_s = 1$ and

$$R_t = \lim_n \left(1 + \frac{a}{n^{2p}}\right)^{-n} = \begin{cases} e^{-a} & \text{if } p = 1/2, \\ 0 & \text{if } p < 1/2 \text{ and } a > 0 \\ +\infty & \text{if } p < 1/2 \text{ and } a < 0 \\ 1 & \text{if } p > 1/2. \end{cases}$$

22. (a) If $0 < R_s < +\infty$, choose N such that $|c_n|^{1/n} < 2R_s^{-1}$ for all $n \geq N$. For such n, $|c_n|^{1/n^2} < (2R_s^{-1})^{1/n} \to 1$, hence $R_t \geq 1$. Similarly, $|c_n|^{1/n^2} > (R_s^{-1}/2)^{1/n}$ for infinitely many n, hence $R_t \leq 1$.

27. By the alternating series test, $\sum_{n=0}^{\infty} c_n x^n$ converges at $x = -1$, hence the result follows from Abel's continuity theorem.

28. (a) Let $s_n(x) = \displaystyle\sum_{k=1}^{n} kx^k$ and $s(x) = \dfrac{x}{(1-x)^2}$, $x \in [0, 1)$. By 7.4.6 and the boundedness of f, $s_n(x) \to s(x)$ uniformly on $[0, r]$, $0 < r < 1$.

30. Define h on $I \cup J$ by

$$h(x) = \begin{cases} f(x) & \text{if } x \in I, \\ g(x) & \text{if } x \in J. \end{cases}$$

By 7.4.19, $f = g$ on $I \cap J$, hence h is well-defined and analytic on $I \cup J$.

33. (a) By 7.4.13, if the series $g(x)$ converges for $|x - a| < r_1$, then $f(x)g(x) = \sum c_n(x-a)^n$, where

$$c_0 = a_0 b_0 = a_0 = 1 \text{ and } c_n = \sum_{k=0}^{n} a_k b_{n-k} = a_0 b_n - b_n = 0, \quad n \geq 1.$$

Therefore, $f(x)g(x) = 1$ for $|x - a| < r_1$.

(b) Suppose $|a_n| \le M^n$ for all n. If $|b_j| \le (2M)^j$ for $1 \le j \le n-1$, then

$$|b_n| \le \sum_{k=1}^{n} |a_k||b_{n-k}| \le \sum_{k=1}^{n} 2^{n-k} M^k M^{n-k} < (2M)^n.$$

By induction, $|b_n| \le (2M)^n$ for all n.

(c) By 7.4.16, there exists a constant $M > 0$ such that $|a_n| \le M^n$ for all n, hence (b) holds. By 7.4.16, g is analytic at a.

Section 8.1

1. Only (b) and (d) are not metrics.

3. Symmetry and coincidence are clear. To verify the triangle inequality $d(x,y) \le d(x,z) + d(y,z)$ simply note that if $x_j \ne y_j$ then either $x_j \ne z_j$ or $y_j \ne z_j$ so that every index j contributing to $d(x,y)$ also contributes to $d(x,z) + d(y,z)$.

5. By the triangle inequality,

$$d(x,y) \le d(x,a) + d(a,y) \le d(x,a) + d(a,b) + d(b,y),$$

hence $d(x,y) - d(a,b) \le d(x,a) + d(b,y)$. Similarly $d(a,b) - d(x,y) \le d(x,a) + d(b,y)$.

10. Let $\{x_n\}$ be a Cauchy sequence in E. Some E_j must contain a subsequence of $\{x_n\}$, and since E_j is complete, the subsequence converges to a member of E_j. By Exercise 9, $\{x_n\}$ converges.

 The assertion is false for infinitely sets. For example, let $\{r_1, r_2, \ldots\}$ be an enumeration of the rationals, and take $E_n = \{r_n\}$ (or $\{r_1, \ldots, r_n\}$).

13. The proof of (a) is straightforward. For the necessity in (b), let $\{(x_n, y_n)\}$ be Cauchy in Z. Since $d(x_n, x_m) \le \eta\big((x_n, y_n), (x_m, y_m)\big)$, $\{x_n\}$ is Cauchy in X. Similarly, $\{y_n\}$ is Cauchy in Y. The converse is clear. Part (c) is proved in a similar manner, and (d) follows from (b) and (c).

15. Part (a) is straightforward. For example, if $\rho(x,y) = 0$, then $\rho(x,y) = d(x,y)$, hence $x = y$. Parts (b) and (c) follow from the observation that $\rho(x,y) = d(x,y)$ if either term is less than a. Part (d) follows from (b) and (c).

 The metrics need not be metrically equivalent: Take d to be the usual metric on $\mathbb{R}$.

 The function σ does not define a metric on X since $\sigma(x,x) = a > 0$.

18. (a) The triangle inequality follows from the observation that the function $t(1+t)^{-1}$ is increasing on $[0, +\infty)$. The remaining properties of a metric

are easily established. Parts (b) and (c) follow from the definition of ρ and the equation

$$d(x, y) = \frac{\rho(x, y)}{1 - \rho(x, y)},$$

noting that $\rho < 1$.

The metrics $|x-y|$ and $|x-y|/(1+|x-y|)$ are not metrically equivalent.

20. By Exercise 18, each ρ_k is a metric on X. It follows easily that ρ is a metric on X. For (b), suppose $\rho(x_n, x) \to 0$. Since $\rho_k \le 2^k \rho$, $\rho_k(x_n, x) \to 0$. By Exercise 18, $d_k(x_n, x) \to 0$. Conversely, suppose $d_k(x_n, x) \to 0$, hence $\rho_k(x_n, x) \to 0$, for each k. Given $\varepsilon > 0$, choose $M \in \mathbb{N}$ such that $\sum_{n>M} 2^{-n} < \varepsilon/2$ and choose $N > M$ so that

$$\rho_1(x_n, x) + \rho_2(x_n, x) + \cdots + \rho_M(x_n, x) < \varepsilon/2$$

for all $n \ge N$. For such n, $\rho(x_n, x) < \varepsilon$.

23. For x, $y \in [1, b]$,

$$\begin{aligned} |f_n(x,y) - f(x,y)| &= \left| \frac{y(1+x^n)^{1/n} - x(1+y^n)^{1/n}}{y(1+y^n)^{1/n}} \right| \\ &= \left| \frac{(1+x^n)^{1/n} - x(1+y^{-n})^{1/n}}{(1+y^n)^{1/n}} \right| \\ &\le \frac{|(1+x^n)^{1/n} - x| + |x - x(1+y^{-n})^{1/n}|}{(1+y^n)^{1/n}} \\ &\le x\left[(1+x^{-n})^{1/n} - 1\right] + x\left[(1+y^{-n})^{1/n} - 1\right] \\ &\le b\left[(2^{1/n} - 1) + (2^{1/n} - 1)\right] \to 0. \end{aligned}$$

Section 8.2

1.

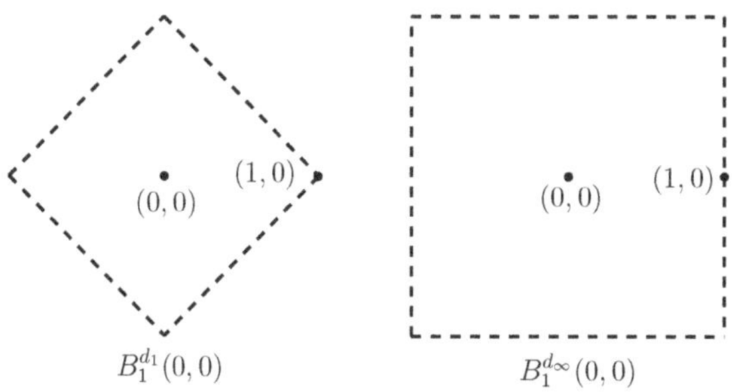

FIGURE C.1: Open balls for Exercise 1.

3. $r = d(x, y)/2$.

5. If $\boldsymbol{x}$, $\boldsymbol{y} \in B_r(\boldsymbol{a})$ and $0 < t < 1$, then

$$\begin{aligned}\|t\boldsymbol{x} + (1-t)\boldsymbol{y} - \boldsymbol{a}\| &= \|t(\boldsymbol{x}-\boldsymbol{a}) + (1-t)(\boldsymbol{y}-\boldsymbol{a})\| \\ &\le \|t(\boldsymbol{x}-\boldsymbol{a})\| + \|(1-t)(\boldsymbol{y}-\boldsymbol{a})\| \\ &< tr + (1-t)r = r.\end{aligned}$$

In general, spheres are not convex. (Consider $(\mathbb{R}^2, d_2)$.)

8. By Exercise 8.1.6, ρ is a metric. Since e^x is a continuous function on $\mathbb{R}$ with continuous inverse, $\rho(x_n, x) \to 0$ iff $|x_n - x| \to 0$. Therefore, ρ is topologically equivalent to the usual metric of $\mathbb{R}$. $(\mathbb{R}, \rho)$ is not complete in this metric. For example, $\{-n\}_{n=1}^{\infty}$ is a Cauchy sequence in $(\mathbb{R}, \rho)$ with no limit. Therefore, ρ cannot be metrically equivalent to the usual metric of $\mathbb{R}$.

12. Let $\{f_n\}$ be a sequence in C converging uniformly to f. Then $f_n(x) = f_n(1-x)$ for all n and x. Taking limits yields $f(x) = f(1-x)$ for all x. To see that C is not closed in the metric of Exercise 8.1.22, define $f_n \in C$ by $f_n(1/2) = 1$, $f_n(x) = 0$ if $x \in [0, 1/2 - 1/n] \cup [1/2 + 1/n, 1]$ and linear on $[1/2 - 1/n, 1/2 + 1/n]$.

Section 8.3

1. (a) $\mathrm{cl}(A) \cup \mathrm{cl}(B)$ is closed and $\supseteq A \cup B$, so $\mathrm{cl}(A) \cup \mathrm{cl}(B) \supseteq \mathrm{cl}(A \cup B)$. Similarly, $\mathrm{cl}(A \cup B) \supseteq \mathrm{cl}(A)$ and $\mathrm{cl}(A \cup B) \supseteq \mathrm{cl}(B)$.

   (d) $\mathrm{int}(A) \cup \mathrm{int}(B)$ is open and $\subseteq A \cup B$, hence $\mathrm{int}(A) \cup \mathrm{int}(B) \subseteq \mathrm{int}(A \cup B)$. The example $A = (0, 1]$, $B = (1, 2)$ in $\mathbb{R}$ produces strict inclusion.

   (f) $\mathrm{bd}(\mathrm{cl}(A)) = \mathrm{cl}(\mathrm{cl}(A)) \setminus \mathrm{int}(\mathrm{cl}(A)) \subseteq \mathrm{cl}(A) \setminus \mathrm{int}(A) = \mathrm{bd}(A)$. The example $A = \mathbb{Q}$ in $\mathbb{R}$ produces strict inclusion.

3. (b) $\big\{(x, y, 0) : x^2 + y^2 = 1\big\}$ together with the point $(1, 0, 1)$.

   (e) $\{(1, 0), (0, 0)\}$.

   (f) The circle $\big\{(x, y) : x^2 + y^2 = 1\big\}$ together with the point $(0, 0)$.

6. (a) By 8.3.6, $y \in \mathrm{cl}_Y(A)$ iff for any sequence $\{a_n\}$ in A with $a_n \to y$, $y \in A$. The same characterization can be given for $y \in \mathrm{cl}_X(A) \cap Y$.

8. The sequence $\{f_n\}$ has no cluster points in $\big(\mathcal{C}([0, 1]), \|\cdot\|_\infty\big)$, hence the set $\{f_1, f_2, \ldots\}$ is closed. The identically zero function is a cluster point of the sequence in $\big(\mathcal{C}([0, 1]), \|\cdot\|_1\big)$, hence the set is not closed in this space.

9. (a) B is open and $B \subseteq C$, hence $B \subseteq \mathrm{int}(C)$. The example $B_1(x) = \{x\}$ and $C_1(x) = X$ in a nontrivial discrete space gives strict inclusion.

12. (b) By 8.3.9, for any $y \in \mathbb{R}$ there exist integers $n_k > 0$, m_k such that $n_k/(2\pi) + m_k \to y - x/(2\pi)$ hence

$$\sin(n_k + x) = \sin\left[2\pi\left((n_k + x)/(2\pi) + m_k\right)\right] \to \sin(2\pi y).$$

Therefore, the set is dense in $[-1, 1]$.

16. Let $u \in U$ and choose $\varepsilon > 0$ such that $B_\varepsilon(u) \subseteq U$. Since Y is dense in X, $B_\varepsilon(u) \cap U \cap Y = B_\varepsilon(u) \cap Y \neq \emptyset$.

    If U is not open, then the assertion may not hold. For example, take $X = [0, 1]$, $Y = (0, 1]$, and $U = \{0\}$.

20. (a) Let $u, v \in I := \bigcup_{\mathfrak{i} \in \mathfrak{I}} I_\mathfrak{i}$ and $t \in (0, 1)$. Then $u \in I_\mathfrak{i}$ and $v \in I_\mathfrak{j}$ for some $\mathfrak{i}, \mathfrak{j} \in \mathfrak{I}$. Since $I_\mathfrak{i} \cap I_\mathfrak{j} \neq \emptyset$, $I_\mathfrak{i} \cup I_\mathfrak{j}$ is an interval. Therefore, $tu + (1 - t)v \in I_\mathfrak{i} \cup I_\mathfrak{j} \subseteq I$, hence I is an interval. Since each $I_\mathfrak{i}$ is open, I is open.

Section 8.4

1. (b), (k) (o), (r) Limit and iterated limits are 0.

   (e) Limit does not exist. One iterated limit is 0, the other is 1.

   (i) Limit and iterated limits exist and $= 1/2$.

2. (a) The limit is 1 since

$$\left|\frac{x^2 - 5y^2}{x^2 + 3y^2} - 1\right| = \frac{8y^2}{x^2 + 3y^2} < \frac{8y^2}{(|y|/a)^{2/p} + 3y^2} \le 8a^{-2/p}|y|^{2(1-1/p)} \to 0.$$

   (b) The limit does not exist, as may be seen by converting to polar coordinates.

3. By the Cauchy mean value theorem, for each $x \neq y$ there exists a number $\theta = \theta(x, y)$ between x and y such that

$$g(x, y) = \frac{f'(\theta)}{\cos\theta}.$$

   Since $\lim_{y \to x} \theta(x, y) = x$, define $g(x, x) = f'(x)/\cos x$.

6. This follows from Exercise 8.1.5

7. Given $\varepsilon > 0$, choose $\delta > 0$ such that $|f(x) - f(a)| < \varepsilon$ for all x, a with

$|x-a|<\delta$. Let $\sqrt{(x-a)^2+(y-b)^2}<\delta/2$. Then

$$\begin{aligned}\left|\sqrt{x^2+y^2}-\sqrt{a^2+b^2}\right| &= \frac{|x^2+y^2-a^2-b^2|}{\sqrt{x^2+y^2}+\sqrt{a^2+b^2}}\\ &\le \frac{|x-a|(|x|+|a|)+|y-b|(|y|+|b|)}{\sqrt{x^2+y^2}+\sqrt{a^2+b^2}}\\ &\le |x-a|+|y-b|\\ &\le 2\sqrt{(x-a)^2+(y-b)^2}<\delta,\end{aligned}$$

hence $|g(x,y)-g(a,b)|<\varepsilon$.

8. For a proof using the sequential criterion for uniform continuity, let $x_n - a_n,\ y_n - b_n \to 0$. Then $\alpha x_n + \beta y_n - (\alpha a_n + \beta b_n) \to 0$, hence $g(x_n, y_n) - g(a_n, b_n) = f(\alpha x_n + \beta y_n) - f(\alpha a_n + \beta a_n) \to 0$.

   The functions xy and $\sin(xy)$ are not uniformly continuous on $\mathbb{R}^2$. (For the former take $x_n = y_n = n + 1/n$ and $a_n = b_n = n$. For the latter take $x_n = y_n = \sqrt{2\pi}\,[n + 1/(3n)]$ and $a_n = b_n = \sqrt{2\pi}\, n$.)

11. This follows from the inequalities

$$|f_j(x)-f_j(a)| \le \|f(x)-f(a)\| \le \sum_{j=1}^{n} |f_j(x)-f_j(a)|.$$

12. We prove the uniform continuity part. Given $\varepsilon > 0$, choose a fixed n such that $\rho(f_n(x), f(x)) < \varepsilon/3$ for all $x \in X$. Then choose $\delta > 0$ such that $\rho(f_n(x), f_n(a)) < \varepsilon/3$ for all $x,\ a \in X$ with $d(x,a) < \delta$. The triangle inequality then shows that $\rho(f(x), f(a)) < \varepsilon/3$ for all $x,\ a \in X$ with $d(x,a) < \delta$.

Section 8.5

1. (a) compact (b) closed, not bounded

   (f) bounded, not closed (h) neither bounded nor closed.

3. Compact case: Let $\{U_\mathfrak{i} : \mathfrak{i} \in \mathfrak{I}\}$ be an open cover of $C := C_1 \cup \cdots \cup C_k$, where each C_j is compact. For each j there exists a finite set $\mathfrak{I}_j \subseteq \mathfrak{I}$ such that $\{U_\mathfrak{i} : \mathfrak{i} \in \mathfrak{I}_j\}$ covers C_j. If $\mathfrak{I}_0$ is the union of the $\mathfrak{I}_j$, then $\{U_\mathfrak{i} : \mathfrak{i} \in \mathfrak{I}_0\}$ is a finite subcover of C.

4. Such an intersection is closed and contained in a compact set and is therefore compact.

7. If E is totally bounded, then $\mathrm{cl}(E)$ is totally bounded. Since X is complete, $\mathrm{cl}(E)$ is complete. Therefore, by 8.5.8, $\mathrm{cl}(E)$ is sequentially compact. In particular, every sequence in E has a cluster point in X.

Conversely, assume every sequence in E has a subsequence that converges in X. Let $\{y_n\}$ be a sequence in $\mathrm{cl}(E)$. For each n, choose $x_n \in E$ such that $d(x_n, y_n) < 1/n$. By hypothesis, a subsequence x_{n_k} converges to some $x \in X$, hence $y_{n_k} \to x$. Therefore, $\mathrm{cl}(E)$ is sequentially compact hence totally bounded.

11. Suppose $\bigcap_{n=1}^{\infty} C_n = \emptyset$. Then $\bigcup_{n=1}^{\infty} C_n^c = X$, hence $\{C_n^c : n \in \mathbb{N}\}$ is an open cover of X and therefore also of C_1. Choose $n \in \mathbb{N}$ such that $C_1 \subseteq C_1^c \cup \cdots \cup C_n^c$. Taking complements, $C_n = C_1 \cap \cdots \cap C_n \subseteq C_1^c \subseteq C_n^c$, which is impossible.

13. By the approximation property of suprema, there exist sequences $\{a_n\}$ and $\{b_n\}$ in A such that $d(a_n, b_n) \to d(A)$. Since A is compact, there exists a subsequence $\{a_n'\}$ of $\{a_n\}$ converging to some $a \in A$. Similarly, there exists a subsequence $\{b_n''\}$ of the corresponding subsequence $\{b_n'\}$ that converges to some $b \in A$. It follows that $d(a, b) = \lim_n d(a_n'', b_n'') = d(A)$.

    For the example, take $A = \{f_n\}$ in $\mathcal{C}([0,1])$ with the sup metric, where $f_n(x) = x^n$. Then $d(A) = 1 > d(f_n, f_m)$ for all m, n.

15. (a) For any $a \in A$, $d(A, x) \le d(a, x) \le d(a, y) + d(y, x)$, hence $d(A, x) - d(y, x) \le d(a, y)$. Taking the infimum over a yields $d(A, x) - d(y, x) \le d(A, y)$ or $d(A, x) - d(A, y) \le d(y, x)$. Interchanging x and y yields (a).

    (b) If $x \notin \mathrm{cl}(A)$ there exists $r > 0$ such that $B_r(x) \cap \mathrm{cl}(A) = \emptyset$. Then $d(a, x) \ge r$ for all $a \in A$, hence $d(A, x) > 0$. Conversely, assume $x \in \mathrm{cl}(A)$ and let $a_n \in A$ with $a_n \to x$. Since $d(A, x) \le d(a_n, x) \to 0$, $d(A, x) = 0$.

    (c) By (b), the denominator of $F_{AB}(x)$ is positive, hence F_{AB} is well-defined. Continuity follows from (a), and clearly $0 \le F_{AB} \le 1$. The last assertions follow from (b).

    (d) $U = \{x \in X : F_{AB}(x) < 1/2\}$, $V = \{x \in X : F_{AB}(x) > 1/2\}$.

19. Let $x_n := f(1/n)$ and $y_n := f(2\pi - 1/n)$. Then $\lim_n x_n = \lim_n y_n = (1, 0)$ but $f^{-1}(x_n) = 1/n \to 0$ and $f^{-1}(y_n) = 2\pi - 1/n \to 2\pi$.

21. Each set is a continuous image of the compact set $A \times B$.

Section 8.6

3. Suppose that $\mathcal{F}$ is equicontinuous at a. Given $\varepsilon > 0$, choose $\delta > 0$ such that $\rho(f(x), f(a)) < \varepsilon$ for all $x \in X$ with $d(x, a) < \delta$ and all $f \in \mathcal{F}$. Given sequences $\{f_n\}$ in $\mathcal{F}$ and $\{x_n\}$ in E with $x_n \to a$, choose N such that $d(x_n, a) < \delta$ for all $n \ge N$. For such n, $\rho(f_n(x_n), f_n(a)) < \varepsilon$.

    Conversely, suppose that $\mathcal{F}$ is not equicontinuous at a. Then there exist $\varepsilon > 0$ and members x_n of E and f_n of $\mathcal{F}$ such that $d(x_n, a) < 1/n$ but $\rho(f_n(x_n), f_n(a)) \ge \varepsilon$. Therefore, the sequential condition does not hold.

7. Let $x > a \geq c$. By the mean value theorem applied to the function $f(z) = z^{-p}$ on (na, nx),

$$\left| \frac{1}{(nx)^p} - \frac{1}{(na)^p} \right| = \frac{pn|x-a|}{y_n^{p+1}} \quad \text{for some } y_n \in (na, nx).$$

Since $y_n^{p+1} \geq (nc)^{p+1} \geq c^{p+1}$, $|(nx)^{-p} - (na)^{-p}| \leq p|x-a|c^{-(p+1)}$, which shows equicontinuity.

9. Take $x_n = a + \pi/n$ in Exercise 3. Then $x_n \to a$ but

$$\sin(nx_n) - \sin(na) = -2\sin(na),$$

which has no limit if a is a nonzero rational number.

11. By the mean value theorem, $|f(x) - f(y)| \leq M|x - y|$.

14. Let $\|f_\mathfrak{i}\|_\infty \leq M$ for all i. Then $|F_\mathfrak{i}(x) - F_\mathfrak{i}(y)| \leq M|x - y|$, hence $\mathcal{F}$ is uniformly equicontinuous on $[a, b]$. It follows that the uniform closure $\mathcal{G}$ of $\mathcal{F}$ in $C([a, b])$ is uniformly equicontinuous on $[a, b]$ (Exercise 6). Since $\mathcal{G}$ is also closed and bounded, it is compact (Arzelà–Ascoli Theorem), hence totally bounded.

Section 8.7

1. (c) not connected. (d) path connected, hence connected.
(e) connected iff $-1 \leq a \leq 1$.

5. Then $f(u)$ and $f(v)$ have opposite signs, say $f(u) < 0 < f(v)$. Since the range of f is connected, it contains the interval $(f(u), f(v))$.

7. Let $f = (g, h) : X \to \mathbb{R}^2$ and $L := \{(x, x) : x \in \mathbb{R}\}$. Then L separates $\mathbb{R}^2$ into two open half-planes H_1 and H_2. Choose any $x_0 \in X$ and suppose $f(x_0) \in H_1$. Then $E := f^{-1}(H_1^c) = f^{-1}(H_2)$ is both open and closed. Since X is connected, $E = \emptyset$. Therefore, $f(X) \subseteq H_1$.

9. Consider the case $B := B_1(\mathbf{0})$. Any point in B^c may be connected to the sphere $S := S_2(\mathbf{0})$ by a radial line segment. Since S is path connected (8.7.10), B^c is path connected.

12. Denote the union by A. Let $f : A \to \{0, 1\}$ be continuous. Since A_n is connected, $f(A_n)$ is a single point. Since $A_n \cap A_{n+1} \neq \emptyset$, an induction argument shows that f is constant.

16. Suppose that $f : L \to C$ is such a function. Then $f^{-1} : C \to L$ is continuous (8.5.11). Remove a point p from the interior of L. Then f^{-1} maps the connected set $C \setminus f(p)$ onto the disconnected set $L \setminus p$.

The function $f(t) = (\cos t, \sin t)$ maps $[0, 2\pi]$ continuously onto the circle $x^2 + y^2 = 1$.

20. Let $\boldsymbol{x} \in \text{bd}(A)$ and $\varepsilon > 0$. Then there exist $\boldsymbol{u}, \boldsymbol{v} \in B_\varepsilon(\boldsymbol{x})$ such that $f(\boldsymbol{u}) \geq c$ and $f(\boldsymbol{v}) < c$. Since $B_\varepsilon(\boldsymbol{x})$ is convex, it is connected, hence $f\big(B_\varepsilon(\boldsymbol{x})\big)$ is an interval and so must contain c. Taking $\varepsilon = 1/n$, we may construct a sequence $\boldsymbol{x}_n \to \boldsymbol{x}$ with $f(\boldsymbol{x}_n) = c$ for each n. Therefore, $f(\boldsymbol{x}) = c$. This shows that $\text{bd}(A) \subseteq B$.

    The example $f(x) = x^2$ on $\mathbb{R}$ with $c = 0$ shows that the inclusion may be strict.

22. (a) $C_{\boldsymbol{x}}$ is connected by Exercise 13. Let $\boldsymbol{u} \in C_{\boldsymbol{x}}$ and choose $\varepsilon > 0$ such that $B_\varepsilon(\boldsymbol{u}) \subseteq U$. Since $B_\varepsilon(\boldsymbol{u})$ is connected, $B_\varepsilon(\boldsymbol{u}) \cup C_{\boldsymbol{x}}$ is connected, hence $B_\varepsilon(\boldsymbol{u}) \subseteq C_{\boldsymbol{x}}$. Therefore, $C_{\boldsymbol{x}}$ is open. If $C_{\boldsymbol{x}} \cap C_{\boldsymbol{y}} \neq$, then $C_{\boldsymbol{x}} \cup C_{\boldsymbol{y}}$ is connected hence $C_{\boldsymbol{x}} = C_{\boldsymbol{y}}$. Therefore, U is a union of pairwise disjoint components.

    (b) Choose a point with rational coordinates in each component in (a). Since these points form a countable set, the union is countable.

Section 8.8

3. Choose a sequence of polynomials P_n converging uniformly to f on $[a, b]$. By hypothesis, $\int_a^b fP_n = 0$ for all n, hence $\int_a^b f^2 = 0$. Since f is continuous, $f = 0$. If $a \geq 0$, then the polynomials with even powers form a separating algebra, hence the result follows as before.

6. By the Stone–Weierstrass theorem, there exists a sequence of functions g_n in $\mathcal{A}$ converging uniformly to f. Set $f_n = g_n - g_n(x_0)$. Then $f_n \in \mathcal{A}$ and $g_n(x_0) \to 0$, hence

$$\|f_n - f\|_\infty \leq \|f_n - g_n\|_\infty + \|g_n - f\|_\infty = |g_n(x_0)| + \|g_n - f\|_\infty \to 0.$$

9. By 8.8.8, there exists a sequence $\{T_n\}$ of trigonometric polynomials converging uniformly to f on $[0, 2\pi]$. For any j, $\sin(jx)$ and $\cos(jx)$ are linear combinations of products $\sin^m x \cos^n x$, hence, by hypothesis, $\int_0^{2\pi} f(x)T_n(x)\,dx = 0$ for all n. Therefore, $\int_0^{2\pi} f^2 = 0$ so $f = 0$.

11. The set of all functions of the form $T(x) := b_0 + \sum_{j=1}^m b_j \sin(jx)$ on $[-\pi/2, \pi/2]$ is an algebra $\mathcal{A}$ containing the constant functions. Since $\sin x$ separates points, so does $\mathcal{A}$. Therefore, given $\varepsilon > 0$, $\|f - T\|_\infty < \varepsilon/2$ for some T. Since $f(0) = 0$, $|b_0| < \varepsilon/2$. Therefore, $\|f - (T - b_0)\|_\infty < \varepsilon$.

15. The functions $\sum_{i=1}^n g_i(x)h_i(y)$ form an algebra and separate points of $X \times Y$.

Section 8.9

1. Assume that X has the decreasing sequence property and let $\{x_n\}$ be a Cauchy sequence in X. Take $C_n = \text{cl}\big(\{x_n, x_{n+1}, \ldots\}\big)$. Then C_n is closed,

$C_{n+1} \subseteq C_n$ and $d(C_n) \to 0$ (because $\{x_n\}$ is Cauchy). By assumption, there exists $x \in X$ such that $x \in C_n$ for all n. It follows that some subsequence of $\{x_n\}$ converges to x. Therefore $x_n \to x$ (Exercise 8.1.9).

3. Let $\{r_1, r_2 \ldots\}$ be an enumeration of $\mathbb{Q}$. Then $U_n := \{r_n, r_{n+1}, \ldots\}$ is open and dense in $\mathbb{Q}$ but $\bigcap_n U_n = \emptyset$.

Section 9.1

1. (a) $\dfrac{2y\,dx - 2x\,dy}{(x+y)^2}$ (e) $\cos(x^2y)(2xy\,dx + x^2\,dy)$

   (h) $e^{xy^2}(y^2\,dx + 2xy\,dy)$.

2. (b) $\begin{bmatrix} e^x \sin y & e^x \cos y \\ e^y \cos x & e^y \sin x \end{bmatrix}$. (c) $\dfrac{1}{(x^2+y^2)^2}\begin{bmatrix} y^3 - x^2y & x^3 - xy^2 \\ 4xy^2 & -4y^3 \end{bmatrix}$.

3. Let $\Delta = \{(x, x) : x \in \mathbb{R}\}$.

   (a) Differentiable on $\mathbb{R}^2$ iff p, $q > 3$, in which case partials are continuous.

   (d) Differentiable on $\mathbb{R}^2$ iff p, $q > 1$. Partials are continuous iff $p > 2$.

4. (a) Differentiable and partials continuous iff $p + q > 1$.

   (d) Differentiable and partials continuous iff $p + q > 2s + 1$.

8. $\boldsymbol{x} \cdot \nabla f(\boldsymbol{x}) = \boldsymbol{a} \cdot f(\boldsymbol{x})$, $\boldsymbol{x} \cdot \nabla g(\boldsymbol{x}) = g(\boldsymbol{x})$.

10. $e^{-f(\boldsymbol{x})}(e^{x_1}, e^{x_2}, \ldots, e^{x_n})$.

12. (a) $\dfrac{x_i}{\|\boldsymbol{x}\|}$ (c) $\dfrac{\|\boldsymbol{x}\|^2 - x_i^2}{\|\boldsymbol{x}\|^3}$.

Section 9.2

1. Let α denote the right side of the inequality. Clearly, $\|T\| \le \alpha$. If $\|\boldsymbol{x}\| \le 1$, then $\|T\boldsymbol{x}\| \le \|T\|\|\boldsymbol{x}\| \le \|T\|$, hence $\alpha \le \|T\|$.

3. Since $\nabla(\psi^{-1}) = -\psi^{-2}\nabla\psi$, the assertion follows from the scalar product rule.

4. (a) Let $f(\boldsymbol{x}) = \boldsymbol{x}$ and $\psi(\boldsymbol{x}) = \|\boldsymbol{x}\|$ in the product rule (9.2.6). Since $df_{\boldsymbol{x}}$ is the identity transformation and $\nabla\psi(\boldsymbol{x}) = \boldsymbol{x}/\|\boldsymbol{x}\|$,

$$dg_{\boldsymbol{x}}(\boldsymbol{h}) = \|\boldsymbol{x}\|\boldsymbol{h} + \|\boldsymbol{x}\|^{-1}(\boldsymbol{x} \cdot \boldsymbol{h})\boldsymbol{x}.$$

   Therefore,

$$dg_{\boldsymbol{x}}(\boldsymbol{x}) = \|\boldsymbol{x}\|\boldsymbol{x} + \|\boldsymbol{x}\|^{-1}(\boldsymbol{x} \cdot \boldsymbol{x})\boldsymbol{x} = 2\|\boldsymbol{x}\|\boldsymbol{x}.$$

6. Let $\eta(\boldsymbol{h})$, $\mu(\boldsymbol{k})$ be such that

$$f(\boldsymbol{a}+\boldsymbol{h}) = f(\boldsymbol{a})+df_{\boldsymbol{a}}(\boldsymbol{h})+\|\boldsymbol{h}\|\eta(\boldsymbol{h}), \quad g(\boldsymbol{b}+\boldsymbol{k}) = g(\boldsymbol{b})+dg_{\boldsymbol{b}}(\boldsymbol{k})+\|\boldsymbol{k}\|\mu(\boldsymbol{k})$$

for all $\boldsymbol{h} \in \mathbb{R}^p$, $\boldsymbol{k} \in \mathbb{R}^q$ with $\|\boldsymbol{h}\|, \|\boldsymbol{k}\|$ sufficiently small, and

$$\lim_{\boldsymbol{h}\to\boldsymbol{0}} \eta(\boldsymbol{h}) = \lim_{\boldsymbol{k}\to\boldsymbol{0}} \mu(\boldsymbol{k}) = \boldsymbol{0}.$$

Let $T(\boldsymbol{h}, \boldsymbol{k}) = \alpha df_{\boldsymbol{a}}(\boldsymbol{h}) + \beta dg_{\boldsymbol{b}}(\boldsymbol{k})$. Then T is linear in $(\boldsymbol{h}, \boldsymbol{k})$ and

$$\varepsilon(\boldsymbol{h}, \boldsymbol{k}) := F(\boldsymbol{a}+\boldsymbol{h}, \boldsymbol{b}+\boldsymbol{k}) - F(\boldsymbol{a}, \boldsymbol{b}) - T(\boldsymbol{h}, \boldsymbol{k}) = \alpha\|\boldsymbol{h}\|\eta(\boldsymbol{h}) + \beta\|\boldsymbol{k}\|\mu(\boldsymbol{k}).$$

Since $\|(\boldsymbol{h}, \boldsymbol{k})\| = \sqrt{\|(\boldsymbol{h}\|^2 + \|\boldsymbol{k}\|^2} \geq \|\boldsymbol{h}\|, \|\boldsymbol{k}\|$,

$$\frac{\|\varepsilon(\boldsymbol{h}, \boldsymbol{k})\|}{\|(\boldsymbol{h}, \boldsymbol{k})\|} \leq \frac{|\alpha|\|\boldsymbol{h}\|\|\eta(\boldsymbol{h})\| + |\beta|\|\boldsymbol{k}\|\|\mu(\boldsymbol{k})\|}{\|(\boldsymbol{h}, \boldsymbol{k})\|} \leq |\alpha|\|\eta(\boldsymbol{h})\| + |\beta|\|\mu(\boldsymbol{k})\|.$$

10. Part (a) follows from Exercise 8.5.14. For (b), set $g(t) = \|f(t) - \boldsymbol{v}\|^2$. Then $g'(t) = 2(f(t) - \boldsymbol{v}) \cdot f'(t)$, and since $g(t_0)$ is the minimum value of g, $g'(t_0) = 0$.

Section 9.3

1. $g'\big(\varphi(x)\psi(y)\big)\big(\varphi'(x)\psi(y), \varphi(x)\psi'(y)\big)$.

3. $g_x\big(\boldsymbol{a}\cdot\boldsymbol{x}, \boldsymbol{b}\cdot\boldsymbol{x}\big)\boldsymbol{a} + g_y\big(\boldsymbol{a}\cdot\boldsymbol{x}, \boldsymbol{b}\cdot\boldsymbol{x}\big)\boldsymbol{b}$.

7. $Tf'(x)$.

10. (a) Let $g(t) = f(\boldsymbol{a} + t\boldsymbol{u})$. By definition, $D_{\boldsymbol{u}}f(\boldsymbol{a}) = g'(0)$. On the other hand, by the chain rule, $g'(t) = \boldsymbol{u}\cdot\nabla f(\boldsymbol{a} + t\boldsymbol{u})$. Setting $t = 0$ yields (a).
(c) $\lim_{t\to 0} \dfrac{f(t\boldsymbol{u}) - f(\boldsymbol{0})}{t} = \lim_{t\to 0} \dfrac{ab^2}{a^2 + b^4t^2}$ exists for all $\boldsymbol{u} = (a, b)$. f is not continuous at $(0, 0)$, since $f \to 0$ along $y = 0$ but $f = 1/2$ along $y = \sqrt{x}$, $x > 0$.

12. Let $F(x) = \int_a^b f(t, x)\,dt$. By the mean value theorem,

$$\frac{F(x+h) - F(x)}{h} = \int_a^b f_x(t, x + rh)\,dt, \quad \text{for some } r = r(t, x, h) \in (0, 1).$$

Since f_x is uniformly continuous, $f_x(t, x + rh) \to f_x(t, x)$ uniformly in t on $[a, b]$ as $h \to 0$. Therefore, $F'(x) = \int_a^b f_x(t, x)\,dt$.

15. Let $\varphi(t) = t^{-p}f(t\boldsymbol{x})$. By the product rule and the chain rule,

$$\varphi'(t) = \frac{-p}{t^{p+1}}f(t\boldsymbol{x}) + \frac{1}{t^p}\nabla f(t\boldsymbol{x})\cdot\boldsymbol{x}.$$

If f is homogeneous of degree p, then φ is a constant function, hence

$$\frac{p}{t^{p+1}} f(t\boldsymbol{x}) = \frac{1}{t^p} \nabla f(t\boldsymbol{x}) \cdot \boldsymbol{x}.$$

Setting $t = 1$ produces the desired identity. On the other hand, if the identity holds, then $t\boldsymbol{x} \cdot \nabla f(t\boldsymbol{x}) = pf(t\boldsymbol{x})$ for all t and $\boldsymbol{x}$, hence $\varphi'(t) = 0$. Therefore, $\varphi(t) = \varphi(1)$, which shows that f is homogeneous of degree p.

17. Fix $\boldsymbol{y} \in C$ and define g on U by $g(\boldsymbol{x}) = f(\boldsymbol{x}) - f(\boldsymbol{y}) - df_{\boldsymbol{y}}(\boldsymbol{x})$. Then $g(\boldsymbol{x}) - g(\boldsymbol{y}) = f(\boldsymbol{x}) - f(\boldsymbol{y}) - df_{\boldsymbol{y}}(\boldsymbol{x} - \boldsymbol{y})$ and $dg_{\boldsymbol{z}} = df_{\boldsymbol{z}} - df_{\boldsymbol{y}}$ (9.1.7), hence the result follows from 9.3.6 applied to g.

Section 9.4

1. (a) $\{(x, y) : x \neq y\}$ (b) $\{(x, y) : x + y \neq (2n + 1)\pi/2,\ n \in \mathbb{Z}\}$
   (e) $\{(x, y) : xy \neq 0\}$ (f) $\{(x, y) : x,\ y > 0,\ y \neq x\}$
   (i) $\{(x, y) : y \neq \pm x\}$ (j) $\{(x, y, z) : xyz \neq 0\}$
2. (i) $x = \frac{1}{2}\big(u + \sqrt{u^2 - 4v}\big),\ \ y = \frac{1}{2}\big(u - \sqrt{u^2 + 4v}\big)$.
   (v) $x = \frac{1}{\sqrt{2}}\left(u - \sqrt{u^2 - 4v^2}\right)^{1/2},\ \ y = \frac{1}{\sqrt{2}}\left(u + \sqrt{u^2 - 4v^2}\right)^{1/2}$.
4. Set $u = x(x^2 + y^2)^{-1}$ and $v = y(x^2 + y^2)^{-1}$. Square and add. $J_f = -1$.

Section 9.5

1. Let $f(x, y) = x + y^2 + e^{xy} - 1$. Then $f_x(0, 0) = 1$ and $f_y(0, 0) = 0$, so the implicit function theorem guarantees a local solution $x = x(y)$ but says nothing about a solution $y = y(x)$.

5. Let

$$\begin{aligned} F(x, y, z) &= \sin(x + z) + \ln(y + z) - \sqrt{2}/2 \ \text{ and} \\ G(x, y, z) &= e^{xz} + \sin(\pi y + z) - 1. \end{aligned}$$

Then, at $(\pi/4, 1, 0)$, $F = G = 0$ and

$$\frac{\partial(F, G)}{\partial(x, y)} \frac{\partial(F, G)}{\partial(y, z)} \frac{\partial(F, G)}{\partial(x, z)} \neq 0.$$

8. Let $F = x - y + z + u^2 - 2$, $G = -x + 2z + u^3 - 2$, $H = -y + 3z + u^4 - 3$. Then, at $(1, 1, 1, 1)$, $F = G = H = 0$ and

$$\frac{\partial(F, G, H)}{\partial(x, y, u)} \frac{\partial(F, G, H)}{\partial(y, z, u)} \frac{\partial(F, G, H)}{\partial(x, z, u)} \neq 0.$$

9. (b) Let $a := f_x(0,0)$ and $b := f_y(0,0)$. The condition is $b(a+1) \neq 0$. The derivative is $\dfrac{-f_x(x,y)f_x\big(f(x,y),y\big)}{f_y(x,y)f_x\big(f(x,y),y\big) + f_y\big(f(x,y),y\big)}$.

11. (a) The condition is $a(a^3 - ab^2 - b^3) \neq 0$ where $a := f_x(0,0)$, $b := f_y(0,0)$.

13. $f'(1) + g'(1) + h'(1) \neq 0$.

15. Let $y = F(x_1, \ldots, x_n)$. If x_1 is a function of $x_2, \ldots, x_n$, then, assuming the necessary differentiability,

$$0 = \frac{\partial y}{\partial x_n} = F_{x_1}\frac{\partial x_1}{\partial x_n} + F_{x_n},$$

hence $\dfrac{\partial x_1}{\partial x_n} = -\dfrac{F_{x_n}}{F_{x_1}}$. In this manner we obtain

$$\frac{\partial x_2}{\partial x_1}\frac{\partial x_3}{\partial x_2}\cdots\frac{\partial x_n}{\partial x_{n-1}}\frac{\partial x_1}{\partial x_n} = (-1)^n\frac{F_{x_1}}{F_{x_2}}\frac{F_{x_2}}{F_{x_3}}\cdots\frac{F_{x_{n-1}}}{F_{x_n}}\frac{F_{x_n}}{F_{x_1}} = (-1)^n.$$

Section 9.6

1. (b) $z_{rr} = t^2 z_{xx} + 2t z_{xy} + z_{yy}$, $\quad z_{tt} = r^2 z_{xx} + 2r z_{xy} + z_{yy}$.

(e)
$$\begin{aligned} z_{rr} &= (e^{2r}\sin^2 t)z_{xx} + (e^{2r}\cos^2 t)z_{yy} + (2e^{2r}\sin t\cos t)z_{xy} \\ &\quad + (e^r \sin t)z_x + (e^r\cos t)z_x, \\ z_{tt} &= (e^{2r}\cos^2 t)z_{xx} + (e^{2r}\sin^2 t)z_{yy} - (2e^{2r}\sin t\cos t)z_{xy} \\ &\quad - (e^r\sin t)z_x - (e^r\cos t)z_x. \end{aligned}$$

(f) $z_r = axz_x$, $z_t = byz_y$, $z_{rr} = a^2x^2z_{xx} + a^2xz_x$, $z_{tt} = b^2y^2z_{yy} + b^2yz_y$.

4. $F_x + z_xF_z = 0$, hence

$$0 = F_{xx} + 2z_xF_{xz} + z_{xx}^2F_{zz} + z_{xx}F_z = F_{xx} - 2\frac{F_x}{F_z}F_{xz} + \frac{F_x^2}{F_z^2}F_{zz} + z_{xx}F_z$$

and so

$$z_{xx} = -\frac{1}{F_z}F_{xx} + 2\frac{F_x}{F_z^2}F_{xz} - \frac{F_x^2}{F_z^3}F_{zz}.$$

5. (a) $u_t = -k^2u$, $u_{xx} = -u$.

(b) By logarithmic differentiation,

$$u_t = \left(\frac{x^2}{4k^2t^2} - \frac{1}{2t}\right)u, \quad u_{xx} = \left(\frac{x^2}{4k^4t^2} - \frac{2}{4k^2t}\right)u.$$

7. The second-order partial derivatives are

$$\begin{aligned}
w_{\rho\rho} &= (\sin\phi\cos\theta)^2 w_{xx} + (\sin\phi\sin\theta)^2 w_{yy} + (\cos\theta)^2 w_{zz} \\
&\quad + (2\sin\phi)\big[(\sin\phi\sin\theta\cos\theta)w_{xy} + (\cos\phi\cos\theta)w_{xz} + (\cos\phi\sin\theta)w_{yz}\big], \\
w_{\theta\theta} &= (\rho\sin\phi)^2\big[(\sin^2\theta)w_{xx} + (\cos^2\theta)w_{yy} - 2(\sin\theta\cos\theta)w_{xy}\big] \\
&\quad - (\rho\sin\phi)[(\cos\theta)w_x - (\sin\theta)w_y], \\
w_{\phi\phi} &= \rho^2\big[(\cos\phi\cos\theta)^2 w_{xx} + (\cos\phi\sin\theta)^2 w_{yy} + (\sin\phi)^2 w_{zz}\big] \\
&\quad + 2\rho^2\big[(\cos^2\phi\sin\theta\cos\theta)w_{xy} - (\cos\phi\sin\phi\cos\theta)w_{xz} - (\cos\phi\sin\phi\sin\theta)w_{yz}\big] \\
&\quad - \rho\big[(\sin\phi\cos\theta)w_x + (\sin\phi\sin\theta)w_y + (\cos\phi)w_z\big].
\end{aligned}$$

9. $f_{x_i} = px_i\|\boldsymbol{x}\|^{p-2}g'\left(\|\boldsymbol{x}\|^p\right)$, hence

$$\begin{aligned}
f_{x_i x_i} &= p\left[\|\boldsymbol{x}\|^{p-2} + (p-2)x_i^2\|\boldsymbol{x}\|^{p-4}\right]g'\left(\|\boldsymbol{x}\|^p\right) + p^2x_i^2\|\boldsymbol{x}\|^{2(p-2)}g''\left(\|\boldsymbol{x}\|^p\right), \\
f_{x_i x_j} &= px_ix_j\left[(p-2)\|\boldsymbol{x}\|^{p-4}g'\left(\|\boldsymbol{x}\|^p\right) + p\|\boldsymbol{x}\|^{2(p-2)}g''\left(\|\boldsymbol{x}\|^p\right)\right] \quad (i \neq j).
\end{aligned}$$

Section 9.7

1. (b) $\dfrac{\partial^3 f}{\partial x^3}(dx)^3 + 3\dfrac{\partial^3 f}{\partial x^2\partial y}(dx)^2\,dy + 3\dfrac{\partial^3 f}{\partial x\partial y^2}\,dx\,(dy)^2 + \dfrac{\partial^3 f}{\partial y^3}(dy)^3.$

2. (a) $2y^2(3x+y)\,(dx)^2 + 12xy(x+y)\,dx\,dy + 2x^2(x+3y)\,(dy)^2$

   (b) $\dfrac{6}{x^4y}(dx)^2 + \dfrac{4}{x^3y^2}\,dx\,dy + \dfrac{2}{x^2y^3}(dy)^2$

   (c) $-y^2\sin(xy)\,(dx)^2 + 2\big[\cos(xy) - xy\sin(xy)\big]\,dx\,dy - x^2\sin(xy)\,(dy)^2$

   (d) $2f(x,y)\big[(2x^2+1)\,(dx)^2 + 4xy\,dx\,dy + (2y^2+1)\,(dy)^2$

   (e) $\dfrac{1}{(x^2+y)^2}\left[2(y-x^2)\,(dx)^2 - 4x\,dx\,dy - (dy)^2\right].$

3. zero.

5. (a) $f + h_1\dfrac{\partial f}{\partial x_1} + h_2\dfrac{\partial f}{\partial x_2} + h_3\dfrac{\partial f}{\partial x_3}$. The terms are evaluated at $\boldsymbol{a}$.

8. By induction,

$$\frac{\partial^p f(\boldsymbol{x})}{\partial x_1^{p_1}\partial x_2^{p_2}\ldots\partial x_n^{p_n}} = b_1^{p_1}b_2^{p_2}\ldots b_n^{p_n}\varphi^{(p)}(\boldsymbol{b}\cdot\boldsymbol{x}),$$

hence

$$\begin{aligned}
(\boldsymbol{x}\cdot\nabla)^p f(\mathbf{0}) &= \varphi^{(p)}(0)\sum\binom{p}{p_1,p_2,\ldots,p_n}(b_1x_1)^{p_1}(b_2x_2)^{p_2}\ldots(b_nx_n)^{p_n} \\
&= \varphi^{(p)}(0)(\boldsymbol{b}\cdot\boldsymbol{x})^p,
\end{aligned}$$

where the second equality follows from the multinomial theorem.

11. (a) $x + y - \frac{1}{6}(x+y)^3$ (d) $x + y - \frac{1}{3}(x+y)^3$.

Section 9.8

2. $x^2+2y^2+3z^2-xy-yz-xz = \frac{1}{2}\left[(x-y)^2+(y-z)^2+(x-z)^2\right]+y^2+2z^2 \geq 0.$

3. (a) $(0,0)$: local min; $(-4/3, 4/3)$: saddle

   (d) $(1,1)$, $(-1,-1)$: local max; $(0,0)$: saddle

   (f) $(2,-2)$: saddle

   (i) $(1/3,1/3)$: local max; $(0,0)$, $(0,1)$, $(1,0)$: saddle.

4. (a) Use polar coordinates to optimize the resulting single-variable function $g(\theta) = \cos\theta + \sin\theta$, $g'(\theta) = -\sin\theta + \cos\theta$, $0 \leq \theta \leq 2\pi$. The critical points of g occur at values of θ that satisfy $\sin\theta = \cos\theta = \pm\sqrt{2}/2$. At these values, $g(\theta) = \pm\sqrt{2}$. Also, $g(0) = g(2\pi) = 1$. Therefore, the maximum and minimum values of f are $\pm\sqrt{2}$.

6. (b) The only critical point is $(2/3,-1/3)$. On bd(D), $f = x^2 - x + 2$, $-1 \leq x \leq 1$, which has critical point $(\pm1/\sqrt{2}, \pm1/\sqrt{2})$. Checking the values of f at these points and at $(\pm1,0)$ shows that the maximum of f is $f(-1,0) = f(-1/\sqrt{2},-1/\sqrt{2}) = 2$ and the minimum is $f(2/3,-1/3) = -1/3$.

   (d) The only critical point is $(0,0)$. On bd(D), $f = \pm\sin\left(x\sqrt{1-x^2}\right)$, $-1 \leq x \leq 1$, which has critical points $(\pm1/\sqrt{2}, \pm1/\sqrt{2})$. Checking the values of f at these points and at $(\pm1,0)$ shows that the extreme values of f are $\pm\sin(1/2)$.

10. Since

$$\lim_{(x,y)\to(0^+,0^+)} f(x,y) = \lim_{(x,y)\to(+\infty,+\infty)} f(x,y) = +\infty,$$

   f has a minimum on $(0,c)\times(0,d)$ for suitable c, $d > 0$, and the minimum must occur at a critical point. The unique critical point is $(a^{2/3}b^{-1/3}, a^{-1/3}b^{2/3})$, which gives the minimum $3(ab)^{1/3}$.

11. Let $f(m,b) = \sum_{i=1}^{n}\left(y_i - mx_i - b\right)^2$. Since not all x coordinates are the same, m must be bounded. Since the data is bounded, b must be bounded. Therefore, the minimum exists and must occur at the unique critical point (m,b) of f, which is determined by the system

$$\sum_{i=1}^{n}(y_i - mx_i - b)(-x_i) = \sum_{i=1}^{n}(y_i - mx_i - b)(-1) = 0.$$

   It follows that $\boldsymbol{x}\cdot\boldsymbol{y} - m\|\boldsymbol{x}\|^2 - nb\overline{x} = m\overline{x} - \overline{y} + b = 0$.

15. Let $f(x,y) = ax^2 + 2bxy + y^2$ and $g(x,y) = x^2 + y^2 - c^2$. The equation $\nabla f = \lambda\nabla g$ yields $ax + by = \lambda x$ and $bx + y = \lambda y$. Multiplying the first equation by x and the second by y and then adding yields

$$f(x,y) = \lambda(x^2 + y^2) = \lambda c^2.$$

Since the system $(a-\lambda)x + by = bx + (1-\lambda)y = 0$ has a nontrivial solution iff the determinant of the coefficient matrix is zero, we obtain $\lambda^2 - (a+1)\lambda + a - b^2 = 0$. Solving for λ we see that the maximum and minimum values of f on the circle are

$$\lambda c^2 = \left[a + 1 \pm \sqrt{(a+1)^2 + 4(a-b^2)}\right](c^2/2).$$

17. We minimize $f(x,y) := (x-1)^2 + (y-2)^2 + (z-3)^2$ subject to the constraint $g(x,y,z) := x^2 + y^2 - z = 0$. From $\nabla f = \lambda \nabla g$ we have

$$x - 1 = \lambda x, \quad y - 2 = \lambda y, \quad z - 3 = -\lambda/2,$$

from which it follows that $y = 2x$ and $z = 3-(x-1)/2x$. From $z = x^2+y^2$ we then have $3 - (x-1)/2x = 5x^2$, or $10x^3 - 5x - 1 = 0$.

19. We minimize $f(x,y) := (x-1)^2 + (y-2)^2 + (z-3)^2$ subject to the constraint $g(x,y,z) := z^2 - x^2 - y^2 - 1 = 0$. From $\nabla f = \lambda \nabla g$ we have

$$x = \frac{1}{1+\lambda}, \quad y = \frac{2}{1+\lambda}x, \quad z = \frac{3}{1-\lambda},$$

hence $y = 2x$ and $z = 3x/(2x-1)$. Substituting into $z^2 - x^2 - y^2 = 1$ yields the desired polynomial.

22. Let $f(x,y,z) = x + 2y + 3z$, $g_1(x,y,z) = x + y + z - 1$, and $g_2(x,y,z) = x^2 + y^2 + z^2 - 1$. From $\nabla f = \lambda_1 \nabla g_1 + \lambda_2 \nabla g_2$,

$$1 = \lambda_1 + 2\lambda_2 x, \quad 2 = \lambda_1 + 2\lambda_2 y, \quad 3 = \lambda_1 + 2\lambda_2 z.$$

Subtracting yields $1 = 2\lambda_2(y-x) = 2\lambda_2(z-y)$ so $y - x = z - y$ or $x - 2y + z = 0$. Combining this with the constraint $x + y + z = 1$ yields $y = 1/3$ and $z = 2/3 - x$. From the constraint $x^2 + y^2 + z^2 = 1$ we obtain $x^2 - 2x/3 - 2/9 = 0$ so $x = (1 \pm \sqrt{3})/3$. The maximum value of f (≈ 3.154694) occurs when $x = (1-\sqrt{3})/3$, the minimum (≈ 0.845293) when $x = (1+\sqrt{3})/3$.

24. We minimize $f(\boldsymbol{x}) := \sum_{i=1}^n (x_i - b_i)^2$ subject to the constraint $g(\boldsymbol{x}) := \boldsymbol{a} \cdot \boldsymbol{x} - c = 0$. From $\nabla f = \lambda \nabla g$ we have $(x_j - b_j) = \lambda a_j/2$, hence

$$(x_j - b_j)^2 = \frac{\lambda^2 a_j^2}{4} = \frac{\lambda(a_j x_j - a_j b_j)}{2}, \quad 1 \le j \le n.$$

Adding and using the constraint,

$$f(\boldsymbol{x}) = \frac{\lambda^2}{4}\|\boldsymbol{a}\|^2 \text{ and } f(\boldsymbol{x}) = \frac{\lambda}{2}(\boldsymbol{a} \cdot \boldsymbol{x} - \boldsymbol{a} \cdot \boldsymbol{b}) = \frac{\lambda}{2}(c - \boldsymbol{a} \cdot \boldsymbol{b})$$

Therefore, $\lambda = 2(c - \boldsymbol{a} \cdot \boldsymbol{b})\|\boldsymbol{a}\|^{-2}$, which gives the desired conclusion.

26. Let $f(\boldsymbol{x}) = \|\boldsymbol{x}-\boldsymbol{a}\|^2$ and $g(\boldsymbol{x}) = \|\boldsymbol{x}\|^2 - 1$. The equation $\nabla f = \lambda \nabla g$ leads to the system $x_j - a_j = \lambda x_j$, or $x_j(1-\lambda) = a_j$, $j = 1, \ldots, n$. Therefore, $x_j = a_j/(1-\lambda)$, so by the constraint $\|\boldsymbol{a}\|^2 = \sum_{j=1}^{n} a_j^2 = (1-\lambda)^2$, hence $\boldsymbol{x} = \pm\boldsymbol{a}/\|\boldsymbol{a}\|$. The distance to the sphere is then the smaller of

$$\left\| \pm \boldsymbol{a}\|\boldsymbol{a}\|^{-1} - \boldsymbol{a}\right\| = \left|1 \pm \|\boldsymbol{a}\|^{-1}\right| \|\boldsymbol{a}\|,$$

namely, $\left|1 - \|\boldsymbol{a}\|^{-1}\right| \|\boldsymbol{a}\|$.

27. (a) Let $f(\boldsymbol{x}) = \boldsymbol{a} \cdot \boldsymbol{x}$ and $g(\boldsymbol{x}) = \sum_{i=1}^n b_i/x_i - 1$. From $\nabla f = \lambda \nabla g$ we have $a_i = -\lambda b_i/x_i^2$, hence

$$x_i = \mu\sqrt{b_i/a_i} \quad \text{and} \quad b_i x_i^{-1} = \sqrt{a_i b_i}/\mu, \quad \mu := \sqrt{-\lambda}.$$

The constraint implies that $\mu = \sum_{i=1}^n \sqrt{a_i b_i}$. Since $a_i x_i = \mu\sqrt{a_i b_i}$, the minimum is $\left(\sum_{i=1}^{n} \sqrt{a_i b_i}\right)^2$.

That the value is indeed the minimum may be argued as follows. If $\boldsymbol{x}$ is any point satisfying the constraint, then

$$f(\boldsymbol{x}) = a_1 x_1 + a_2 x_2 + \cdots + a_{n-1} x_{n-1} + \frac{a_n b_n}{1 - \sum_{i=1}^{n-1} b_i/x_i},$$

where $x_i > b_i$. Thus $|f| \to +\infty$ as the variables $x_1, x_2, \ldots, x_{n-1}$ become large or as $\sum_{i=1}^n b_i/x_i$ nears 1. Therefore, the minimum occurs in the interior of a compact set, hence at the point obtained above.

30. Since $\mathrm{cl}(U)$ is compact, there exist points $\boldsymbol{u}, \boldsymbol{v} \in \mathrm{cl}(U)$ such that

$$f(\boldsymbol{u}) \le f(\boldsymbol{x}) \le f(\boldsymbol{v}) \quad \text{for all } \boldsymbol{x} \in \mathrm{cl}(U).$$

If $f(\boldsymbol{u}) = f(\boldsymbol{v})$, then f is a constant function and the result follows. If $f(\boldsymbol{u}) < f(\boldsymbol{v})$, then one of the points, say $\boldsymbol{u}$, must lie in U. By 9.8.2, $f'(\boldsymbol{u}) = \boldsymbol{0}$.

Section 10.1

1. Let μ be as in 10.1.5 with $p_k = 1/k$, or let μ be as in ??, and take $A_k = \{k, k+1, \ldots\}$.

3. By the inclusion–exclusion principle and additivity,

$$\mu(A \cup B) = \mu(A) + \mu(B) - \mu(A \cap B) = \mu(A), \quad \text{and}$$
$$\mu(A) = \mu(A \setminus B) + \mu(A \cap B) = \mu(A \setminus B).$$

5. Let $B = A_1 \cup \cdots \cup A_n$. By 10.1.6(c),

$$\mu(A_1 \cup \cdots \cup A_{n+1}) = \mu(B \cup A_{n+1}) = \mu(B) + \mu(A_{n+1}) - \mu(B \cap A_{n+1}).$$

By the induction hypothesis,

$$\begin{aligned}&\mu(B) + \mu(A_{n+1})\\ &= \sum_{i=1}^{n+1} \mu(A_i) - \sum_{1 \le i < j \le n}^{n} \mu(A_i \cap A_j) + \cdots + (-1)^{n-1}\mu(A_1 \cap \cdots \cap A_n)\end{aligned}$$

and

$$\begin{aligned}\mu(B \cap A_{n+1}) = \sum_{i=1}^{n} \mu(A_i \cap A_{n+1}) &- \sum_{1 \le i < j \le n}^{n} \mu(A_i \cap A_j \cap A_{n+1})\\ &+ \cdots + (-1)^{n-1}\mu(A_1 \cap \cdots \cap A_n \cap A_{n+1}).\end{aligned}$$

Subtracting produces the desired formula.

7. For any finite subset S of $\mathbb{N}$ with at least m members, define

$$E_S = \bigcap_{j \in S} E_j \cap \bigcap_{j \in S^c} E_j^c.$$

There are countably many such sets, they are pairwise disjoint, and

$$\bigcup_{k=1}^{\infty} E_k \supseteq \bigcup_S E_S = A.$$

Moreover, $E_k \cap E_S \ne \emptyset$ iff $k \in S$, in which case $E_k \supseteq E_S$. Therefore, by additivity,

$$\begin{aligned}\sum_{k=1}^{\infty} \mu(E_k \cap A) &= \sum_{k=1}^{\infty} \sum_S \mu(E_k \cap E_S) = \sum_S \sum_{k=1}^{\infty} \mu(E_S \cap E_k)\\ &= \sum_S \sum_{k \in S} \mu(E_S \cap E_k) \ge \sum_S m\mu(E_S) = m\mu(A).\end{aligned}$$

Section 10.2

1. Clearly, $\lambda^*(A) \le \alpha(A)$. To show the reverse inequality, let $\{I_k\}$ be a sequence in $\mathcal{I}$ that covers A. Given $\varepsilon > 0$ choose $J_k \in \mathcal{J}$ containing I_k such that $|J_k| < |I_k| + \varepsilon/2^n$. Then $\{J_k\}$ covers A and $\sum_k |I_k| \ge \sum_k |J_k| - \varepsilon \ge \alpha(A) - \varepsilon$. Since ε was arbitrary, $\sum_k |I_k| \ge \alpha(A)$. Therefore, $\lambda^*(A) \ge \alpha(A)$.

4. Let $\{I_k\}$ be any sequence of intervals that covers A. Then the intervals $\boldsymbol{x} + I_k$ cover $\boldsymbol{x} + A$, hence

$$\lambda^*(\boldsymbol{x} + A) \le \sum_{k=1}^{\infty} |\boldsymbol{x} + I_k| = \sum_{k=1}^{\infty} |I_k|.$$

Since $\{I_k\}$ was arbitrary, $\lambda^*(\boldsymbol{x} + A) \le \lambda^*(A)$. Since $A = (A + \boldsymbol{x}) - \boldsymbol{x}$, the reverse inequality also holds.

Section 10.3

2. For any $C \subseteq \mathbb{R}^n$,

$$C \cap (-E) = -[(-C) \cap E] \quad \text{and} \quad C \cap (-E)^c = -[(-C) \cap E^c],$$

hence, using $D := -C$ as a test set for E,

$$\lambda^*(C \cap (-E)) + \lambda^*(C \cap (-E)^c) = \lambda^*(D \cap E) + \lambda^*(D \cap E^c) = \lambda^*(D).$$

By Exercise 10.2.5, $\lambda^*(D) = \lambda^*(C)$. Therefore, $-E \in \mathcal{M}$ and $\lambda(-E) = \lambda(E)$.

4. Let $D = \bigcup_{k=1}^{\infty}(r_k - \varepsilon/2^{k+1}, r_k + \varepsilon/2^{k+1})$, where $\{r_1, r_2, \ldots\}$ is an enumeration of $\mathbb{Q}$.

Section 10.4

1. Let $\{r_1, r_2, \ldots,\}$ be an enumeration of the rationals in $[0,1]$ and set $A = \bigcup_{k=1}^{\infty}(r_j - \varepsilon/2^{k+1}, r_j + \varepsilon/2^{k+1})$ and $C = [0,1] \cap A^c$. Then C is compact and $\lambda(A) < \varepsilon$, hence

$$\lambda(E \setminus C) = \lambda([0,1] \setminus C) = \lambda([0,1] \cap A) < \varepsilon.$$

4. Let $\mathcal{F}$ denote the collection of all Borel sets B such that rB is a Borel set. Then $\mathcal{F}$ is a σ-field containing all intervals, hence $\mathcal{F} = \mathcal{B}$. A similar argument works for the other sets.

Section 10.5

1. For $t \in \mathbb{R}$,

$$\begin{aligned} \{x : f(x) < t\} &= \bigcup_{k=1}^{\infty} \{x \in A_k : f(x) < t\} \\ &= \bigcup_{n=1}^{\infty} \{x : (f\mathbf{1}_{A_k})(x) < t\} \cap A_k \in \mathcal{F}. \end{aligned}$$

5. (a) $\{x : g(x) < h(x)\} = \bigcup_{r\in\mathbb{Q}} \{x : g(x) < r < h(x)\} \in \mathcal{F}$.

   (d) Since gh is measurable the assertion follows from

$$\{x \in S : g(x)h(x) \leq 1\} \cap \{x \in S : g(x)h(x) \geq 1\}.$$

8. If $\mathbf{1}_E$ is measurable, then $E = \{x : \mathbf{1}_E(x) > 0\} \in \mathcal{F}$. Conversely, if $E \in \mathcal{F}$ and $t \in \mathbb{R}$, then

$$\{x : \mathbf{1}_E(x) \leq t\} = \begin{cases} \emptyset & \text{if } t < 0, \\ E^c & \text{if } 0 \leq t < 1, \text{ and} \\ S & \text{if } t \geq 1. \end{cases}$$

   In each case, $\{x : \mathbf{1}_E(x) \leq t\} \in \mathcal{F}$, hence $\mathbf{1}_E$ is measurable.

10. $\mathbf{1}_{A\Delta B}(x) = 1$ iff $\mathbf{1}_A(x) - \mathbf{1}_B(x) = 1$ or $\mathbf{1}_B(x) - \mathbf{1}_A(x) = 1$ iff $x \in A \setminus B$ or $x \in B \setminus A$.

14. The range of f is $\{1/k : k \in \mathbb{N}\}$. Since $f(x) = 1/k$ iff $1/x - 1 < k \leq 1/x$ iff $1/(k+1) < x \leq 1/k$, the assertion follows from Exercise 7.

17. Let $\varepsilon > 0$ and choose $N \in \mathbb{N}$ such that $2^N > 1/\varepsilon$ and $f \leq N$ on S. Let $k > N$, so $0 \leq f \leq k$. Then, in the notation of the proof of 10.5.8,

$$f_k = \sum_{j=1}^{k2^k} \frac{j-1}{2^k}\mathbf{1}_{A_{k,j}}, \quad \text{where}$$
$$A_{k,j} = \left\{x \in S : (j-1)2^{-k} \leq f(x) < j2^{-k}\right\}, \quad j = 1, 2, \ldots, k2^k.$$

   For any $x \in S$ there exists $j \in \{1, 2, \ldots, k2^k\}$ such that $x \in A_{k,j}$, hence $0 \leq f(x) - f_k(x) = f(x) - (j-1)2^{-k} \leq 1/2^k < \varepsilon$.

19. (a) That $\mathcal{F}$ is a σ-field follows from properties of preimages.

   (b) Since $f^{-1}(I_1 \times \cdots \times I_m) = \bigcap_{j=1}^m f_j^{-1}(I_j)$, $\mathcal{F}$ contains all intervals, hence, by minimality, $\mathcal{F} = \mathcal{B}(\mathbb{R}^m)$.

   (c) If $A \in \mathcal{B}(\mathbb{R})$ and $B := F^{-1}(A)$, then $B \in \mathcal{B}(\mathbb{R}^m)$, hence, from (b), $g^{-1}(A) = f^{-1}(B) \in \mathcal{B}(\mathbb{R}^n)$.

Section 11.2

3. Since $f^{-1}(\{d^2\}) = \displaystyle\bigcup_{k=1}^{\infty} (d/10^k, (d+1)/10^k) \cap \mathbb{I}$,

$$\int_{[0,1]} f\,d\lambda = \sum_{d=1}^{9} d^2\lambda\left(f^{-1}(\{d^2\})\right) = \frac{1}{9}\sum_{d=1}^{9} d^2.$$

5. (a) If $\int_E |g|\,d\lambda = 0$, then $g = 0$ a.e., hence both integrals in (a) are zero. Suppose $\int_E |g|\,d\lambda \neq 0$. Since $m\int_E |g| \le \int_E f|g| \le M\int_E |g|$ on E, $a := \left(\int_E |g|\,d\lambda\right)^{-1}\int_E f|g|\,d\lambda$ satisfies the requirement.
   (b) For example, take $E = (-1,1)$ and $f = g = \mathbf{1}_{(-1,0)} - \mathbf{1}_{(0,1)}$, so
$$\int_E fg = \int_E \left[\mathbf{1}_{(-1,0)} + \mathbf{1}_{(0,1)}\right] = 2, \quad \int_E g = 0.$$
   (c) Given $\varepsilon > 0$, choose $\delta > 0$ such that $-\varepsilon < f(t) - f(x) < \varepsilon$ for all $t \in (x-\delta, x+\delta)$. If $y \in (x, x+\delta)$, then, by (a),
$$\begin{aligned}\int_{[a,y]} f\,d\lambda - \int_{[a,x]} f\,d\lambda - f(x)(y-x) &= \int [f(t) - f(x)]\mathbf{1}_{[x,y]}(t)\,dt \\ &= a_y \int \mathbf{1}_{[x,y]}\,d\lambda = a_y(y-x),\end{aligned}$$
   where $|a_y| \le \varepsilon$. Dividing by $y - x$ proves the assertion for right-hand limits.

7. Use 11.2.17 and the Stone–Weierstrass theorem.

9. By the approximation property of integrals, given $\varepsilon > 0$, there exists a simple function g such that $||f-g||_1 < \varepsilon$ and $g = 0$ outside an interval $[a,b]$. If $k > b$, then $\int_{[k,k+1]} g\,d\lambda = 0$, hence
$$\left|\int_{[k,k+1]} f\,d\lambda\right| = \left|\int_{[k,k+1]} (f-g)\,d\lambda\right| \le \int_{[k,k+1]} |f-g|\,d\lambda < \varepsilon.$$

11. Let $A = \{x \in I : |f(x)| > \varepsilon\}$. Then
$$\int_I f^2\,d\lambda \ge \int_A f^2\,d\lambda \ge \varepsilon^2\lambda(A).$$

15. By linearity of the integral, $\int_0^1 P(x)f(x)\,dx = 0$ for every polynomial $P(x)$ whose terms have even powers. Let $g : [0,1]$ be continuous. Since x^2 separates points of $[0,1]$, by the Stone–Weierstrass theorem there exists a sequence of such polynomials P_n such that $||g - P_n||_\infty \to 0$. Then
$$\left|\int_0^1 fg\right| = \left|\int_0^1 f(g - P_n)\right| \le ||g - P_n||_\infty \int_0^1 |f| \to 0,$$
    so $\int_0^1 gf = 0$. Since the continuous functions are dense in $\mathcal{L}^1[0,1]$, there exists a sequence of continuous functions g_n such that $\int_0^1 |f - g_n| \to 0$. Then
$$\int_0^1 f^2 = \int_0^1 f(f - g_n) \le ||f||_\infty \int_0^1 |f - g_n| \to 0.$$
    Therefore, $\int_0^1 f^2 = 0$, hence $f = 0$ a.e.

Section 11.3

1. In each case the integrand $f_k \to 0$. Moreover, in (a) the functions are bounded, and in (b) $|f_k(x)| \le 1/x^2$ on $[1, +\infty)$, and $|f_k(x)| \le kx^{3/2}/(1 + k^2x^2) \le 1/\sqrt{x}$ on $[0, 1]$. Thus in each case the dominated convergence theorem may be used.

2. (b) $k \ln(1 + f/k^2) \le f/k$.

   (c) $k \sin\left(f/k\right) = \dfrac{f \sin\left(f/k\right)}{f/k} \to f$ and $k \sin(f/k) \le f$.

   In each case apply the dominated convergence theorem.

3. $g(1 + f/k)^n e^{-f} \uparrow g$.

5. $|f(x)\sin(tx)/x| \le |tf(x)|$, hence is integrable in x for each t. If $t_k \to t$, then $\{t_k\}$ is bounded hence $|t_k f(x)| \le C|f(x)|$. Now apply the dominated convergence theorem.

8. Assume first that $f \ge 0$. Since $\mathbb{R}^n = \bigcup_{k=1}^{\infty} A_k$, 11.3.4 implies that

$$\int f\, d\lambda = \sum_{k=1}^{\infty} \int_{A_k} f\, d\lambda = \sum_{k=1}^{\infty} a_k \lambda(A_k).$$

   The general result follows by considering $f^{\pm}$.

12. By 11.2.6, the series $\sum_{k=1}^{\infty} |f_k(\boldsymbol{x})|$ converges a.e.

15. By Fatou's lemma, $\int \liminf (f_k - g)\, d\lambda \le \liminf \int (f_k - g)\, d\lambda$.

19. Let $\mu_+(E) = \int_E f^+\, d\lambda$ and $\mu_-(E) = \int_E f^-\, d\lambda$, $E \in \mathcal{M}(\mathbb{R}^n)$. Then $\mu_\pm$ are measures that agree on compact intervals. Since every open interval is an increasing union of compact intervals, by continuity from below, the measures $\mu_\pm$ agree on open intervals. Since every open set is a countable disjoint union of open intervals, they agree on open sets. By 10.4.4, for any bounded $E \in \mathcal{M}$ there exists a decreasing sequence of bounded open sets $U_k \supseteq E$ such that $\lambda(U_k) \to \lambda(E)$. By Exercise 11.2.8, $\mu_\pm(U_k) \to \mu_\pm(E)$. Therefore, $\mu_\pm$ agree on bounded Lebesgue measurable sets. In particular, if $A_k = \{x \in [-k, k] : f(x) > 0\}$, then $\int_{A_k} f\, d\lambda = 0$. Since $\mathbf{1}_{A_k} f \ge 0$, it follows that $\mathbf{1}_{A_k} f = 0$ a.e., hence $\lambda(A_k) = 0$. Since $A_k \uparrow \{x \in \mathbb{R} : f(x) > 0\}$, $\lambda\big(\{x \in \mathbb{R} : f(x) > 0\}\big) = 0$. A similar argument shows that $\lambda\big(\{x \in \mathbb{R} : f(x) < 0\}\big) = 0$.

Section 11.5

1. The given set is the union of the sets

$$E_{j,k} = [-k, k] \times \cdots \times \overset{j}{\big([-k, k] \cap \mathbb{Q}\big)} \times \cdots \times [-k, k], \quad k \in \mathbb{N},\ 1 \le j \le k,$$

   over all k and j, and $\lambda(E_{j,k}) = (2k)^{n-1}\lambda(\mathbb{Q} \cap [-k, k]) = 0$.

2. (a) By the Fubini–Tonelli theorem and a substitution,

$$\begin{aligned}\int_I f\,d\lambda &= \int_0^\infty \cdots \int_0^\infty x_1\cdots x_n e^{-\|\boldsymbol{x}\|^2}\,dx_1\cdots dx_n \\ &= \int_0^\infty \cdots \int_0^\infty x_2\cdots x_n e^{-(x_2^2+\cdots x_n^2)}\,dx_2\cdots dx_n \int_0^\infty x_1 e^{-x_1^2}\,dx_1 \\ &= \frac{1}{2}\int_0^\infty \cdots \int_0^{+\infty} x_2\cdots x_n e^{-(x_2^2+\cdots x_n^2)}\,dx_2\cdots dx_n \\ &= \cdots = \frac{1}{2^n}.\end{aligned}$$

5. By the Fubini–Tonelli theorem, the integral, denote it by I, equals

$$\int_{0\le x\le x_1\le\cdots\le x_{m-1}\le 1}\int_{x_{m-1}}^1 x\,dx_m\,d\lambda(x,x_1,x_2,\ldots,x_{m-1}).$$

Performing the inner integration, we have

$$I = \int_{0\le x\le x_1\le\cdots\le x_{m-1}\le 1}(1-x_{m-1})x\,d\lambda(x,x_1,x_2,\ldots,x_{m-1}).$$

Integrating with respect to x_{m-1},

$$I = \frac{1}{2}\int_{0\le x\le x_1\le\cdots\le x_{m-2}\le 1}(1-x_{m-2})^2 x\,d\lambda(x,x_1,x_2,\ldots,x_{m-2}).$$

Continuing we obtain

$$I = \frac{1}{m!}\int_0^1 (1-x)^m x\,dx = \frac{1}{(m+2)!},$$

the last equality by 5.3.4.

7. The function

$$F(t,x) := t^{-p}f(t)\mathbf{1}_{[x^{1/p},1]}(t) = t^{-p}f(t)\mathbf{1}_{(0,t^p]}(x)$$

is Borel measurable on $(0,1)\times(0,1)$ and is integrable since

$$\begin{aligned}\int_{(0,1)}\int_{(0,1)}|F(t,x)|\,d\lambda(t,x) &= \int_{(0,1)} t^{-p}|f(t)|\int_{(0,1)}\mathbf{1}_{(0,t^p]}(x)\,dx\,dt \\ &= \int_{(0,1)} t^{-p}|f(t)|\,t^p\,dt \\ &= \int_{(0,1)}|f|\,d\lambda < +\infty.\end{aligned}$$

Repeating the calculation with $F(t,x)$ instead of $|F(t,x)|$,

$$\int_{(0,1)} g(x)\,dx = \int_{(0,1)}\int_{(0,1)} F(t,x)\,d\lambda(t,x) = \int_{(0,1)} f\,d\lambda.$$

9. (a) Let $I_c(t) = \int_0^c e^{-xt}\sin x\,dx$. Using the given identity, we have

$$\int_0^c \frac{\sin x}{x}\,dx = \int_0^c \int_0^\infty e^{-xt}\sin x\,dt\,dx = \int_0^\infty I_c(t)\,dt, \qquad (\dagger)$$

the second equality by the Fubini–Tonelli theorem, valid since

$$\int_0^c \int_0^\infty e^{-xt}|\sin x|\,dt\,dx = \int_0^c \frac{|\sin x|}{x}\,dx < +\infty.$$

Integrating by parts twice yields

$$I_c(t) = \frac{1}{1+t^2}\big[1 - e^{-ct}(\cos c + t\sin c)\big]$$

so $\lim_{c\to+\infty} I_c(t) = (1+t^2)^{-1}$. Moreover, if $c \geq 1$ then

$$|I_c(t)| \leq \frac{1+e^{-ct}(1+t)}{1+t^2} \leq \frac{2}{1+t^2}.$$

Since the last function is integrable on $[0,+\infty)$, $(\dagger)$ and Lebesgue's dominated convergence theorem imply that

$$\int_0^\infty \frac{\sin x}{x}\,dx = \lim_{c\to+\infty}\int_0^\infty I_c(t)\,dt = \int_0^\infty \frac{1}{1+t^2}\,dt = \frac{\pi}{2}.$$

11. (a) By translation invariance and symmetry,

$$\begin{aligned} f * g(\boldsymbol{x}) &= \int_{\mathbb{R}^n} f(\boldsymbol{x}-\boldsymbol{y})g(\boldsymbol{y})\,d\boldsymbol{y} = \int_{\mathbb{R}^n} f(-\boldsymbol{y})g(\boldsymbol{x}+\boldsymbol{y})\,d\boldsymbol{y} \\ &= \int_{\mathbb{R}^n} f(\boldsymbol{y})g(\boldsymbol{x}-\boldsymbol{y})\,d\boldsymbol{y} = g * f(\boldsymbol{x}). \end{aligned}$$

14. Let $\alpha = \big[G(b) - G(a)\big]$ and $\beta = \big[F(b) - F(a)\big]$. By the Fubini–Tonelli theorem,

$$\begin{aligned} \int_{[a,b]} F(x)g(x)\,dx - \alpha F(a) &= \int_{[a,b]} [F(x) - F(a)]g(x)\,dx \\ &= \int_{[a,b]}\int_{[a,b]} \mathbf{1}_{[a,x]}(t)f(t)g(x)\,dt\,dx \\ &= \int_{[a,b]}\int_{[a,b]} \mathbf{1}_{[t,b]}(x)f(t)g(x)\,dx\,dt, \end{aligned}$$

and by a notation switch,

$$\begin{aligned} \int_{[a,b]} G(x)f(x)\,dx - \beta G(a) &= \int_{[a,b]} [G(x) - G(a)]f(x)\,dx \\ &= \int_{[a,b]}\int_{[a,b]} \mathbf{1}_{[a,x]}(t)g(t)f(x)\,dt\,dx \\ &= \int_{[a,b]}\int_{[a,b]} \mathbf{1}_{[a,t]}(x)g(x)f(t)\,dx\,dt. \end{aligned}$$

Adding, we have

$$\int_{[a,b]} F(x)g(x)\,dx - \alpha F(a) + \int_{[a,b]} G(x)f(x)\,dx - \beta G(a)$$
$$= \int_{[a,b]}\int_{[a,b]} \big[\mathbf{1}_{[a,t]}(x) + \mathbf{1}_{[t,b]}(x)\big] f(t)g(x)\,dx\,dt$$
$$= \int_{[a,b]}\int_{[a,b]} f(t)g(x)\,dx\,dt = \alpha\beta,$$

which implies the desired conclusion.

Section 11.6

3. Take $f(\boldsymbol{x}) = h(||\boldsymbol{x}||)$ in (11.16).

5. Let $\boldsymbol{x} = (x_1, \ldots, x_n)$. The hole H may be described by

$$H = \Big\{(\boldsymbol{x}, x_{n+1}) \,:\, -\sqrt{1 - ||\boldsymbol{x}||^2} \le x_{n+1} \le \sqrt{1 - ||\boldsymbol{x}||^2},\ \ ||\boldsymbol{x}|| \le R\Big\}.$$

By Exercise 3,

$$\lambda(H) = 2\int_{||\boldsymbol{x}||\le R} \sqrt{1 - |||\boldsymbol{x}||^2}\,d\lambda(\boldsymbol{x}) = 2n\alpha_n \int_0^R \sqrt{1 - r^2}\,dr$$
$$= n\alpha_n\big[R\sqrt{1 - R^2} - \arcsin\sqrt{1 - R^2} + \pi/2\big].$$

6. Define a C^∞ map $\varphi : \mathbb{R}^n \times (0, 2\pi) \to \mathbb{R}^{n+1}$ by

$$\varphi(x_1, \ldots, x_n, \theta) = (x_1, \ldots, x_{n-1}, x_n\cos\theta, x_n\sin\theta).$$

The condition $x_n > 0$ implies that φ is one-to-one. Since $E_r = \varphi(E \times (0, 2\pi))$ and

$$J_\varphi = \begin{vmatrix} I_{n-1} & 0 & 0 \\ 0 & \cos\theta & -x_n\sin\theta \\ 0 & \sin\theta & x_n\cos\theta \end{vmatrix} = x_n,$$

the change of variables theorem implies that

$$\lambda_{n+1}(E_r) = \int_{\varphi(E\times(0,2\pi))} 1\,d\lambda_{n+1} = \int_0^{2\pi}\int_E x_n\,d\lambda_n(x_1, \ldots, x_n)\,d\theta$$
$$= 2\pi\int_E x_n\,d\lambda_n(x_1, \ldots, x_n).$$

Section 12.1

1. If $\psi = \varphi\circ\alpha$ and $\phi = \psi\circ\beta$, where α and β are C^1 with C^1 inverses, then $\phi = \varphi\circ\gamma$, where $\gamma = \alpha\circ\beta$ is C^1 with C^1 inverse $\beta^{-1}\circ\alpha^{-1}$.

3. There are two unit tangent vectors at the point where $t = \pm 1$, namely, $(\pm e^1 + e^2)/\sqrt{2}$.

4. (a) The trace is the parabola $(x, 1 - 2x^2)$. The tangent vector field is $(\cos t)\boldsymbol{e}^1 - 2(\sin(2t))\boldsymbol{e}^2$.

5. (b) $x = a\cos t,\ y = b\sin t,\ z = ab\sin(2t),\ 0 \le t \le 2\pi$.

6. If $\varphi(t) = \boldsymbol{x}$ for infinitely many t, then, by the Bolzano–Weierstrass theorem, there would exist a point $t \in [a, b]$ and a sequence $t_n \to t$ with $t_n \neq t$ for all n such that $\varphi(t_n) = \boldsymbol{x}$ for all n. Then

$$\varphi'(t) = \lim_n \frac{\varphi(t_n) - \varphi(t)}{t_n - t} = \lim_n \frac{0 - 0}{t_n - t} = 0,$$

hence φ is not smooth.

Section 12.2

1. Only (b) and (c) are rectifiable.

2. (b) $-\sqrt{5}/4$.

4. (b) $x = 1 + 2\cos t,\ y = 2 + 3\sin t,\ 0 \le t \le 2\pi$.

$$\text{length} = \int_0^{2\pi} \sqrt{4\sin^2 t + 9\cos^2 t}\, dt.$$

5. (b) 6.

7. $x' = r'\cos\theta - r\theta'\sin\theta,\ y' = r'\sin\theta + r\theta'\cos\theta,\ (x')^2 + (y')^2 = (r\theta')^2 + (r')^2$.

9. (a) $W = -\displaystyle\int_a^b \Big(\nabla P(\varphi(t))\Big) \cdot \alpha'(t)\, dt = \int_b^a \frac{d}{dt}(P \circ \varphi)(t)\, dt.$

Section 12.3

1. (a) If $T(\boldsymbol{e}^1, \boldsymbol{e}^2, \boldsymbol{e}^3) = (\boldsymbol{e}^1 + \boldsymbol{e}^2, \boldsymbol{e}^2 + \boldsymbol{e}^3, \boldsymbol{e}^3 + \boldsymbol{e}^1)$, then

$$[T] = \begin{bmatrix} 1 & 0 & 1 \\ 1 & 1 & 0 \\ 0 & 1 & 1 \end{bmatrix},$$

which has positive determinant. Therefore, the sign of the frame is positive.

5. Take $\psi(\phi) = (a\cos\phi, b + a\sin\phi),\ 0 < \phi < 2\pi$. By 12.3.9,

$$\vec{N}_\varphi(\varphi(\phi, \theta)) = a(\cos\phi, \sin\phi\cos\theta, \sin\phi\sin\theta).$$

Setting $x = a\cos\phi$, $y = (b + a\sin\phi)\cos\theta$ and $z = (b + a\sin\phi)\sin\theta$, we have

$$a\sin\phi = \sqrt{y^2+z^2} - b, \quad \cos\theta = \frac{y}{\sqrt{y^2+z^2}}, \quad \sin\theta = \frac{z}{\sqrt{y^2+z^2}},$$

hence

$$\vec{N}_\varphi(x,y,z) = x\boldsymbol{e}^1 + \frac{y(\sqrt{y^2+z^2}-b)}{\sqrt{y^2+z^2}}\boldsymbol{e}^2 + \frac{z(\sqrt{y^2+z^2}-b)}{\sqrt{y^2+z^2}}\boldsymbol{e}^3.$$

6. Let $(x,y,z) = \varphi(t,\theta)$ and use $\partial\varphi^\perp = (x_t, y_t, z_t) \times (x_\theta, y_\theta, z_\theta)$.
(a) $\partial\varphi^\perp(t,\theta) = t\big(-\cos\theta, \sin\theta, 1\big)$, so

$$\vec{N}_\varphi(t\cos\theta, t\sin\theta, t) = \frac{1}{\sqrt{2}}\big(-\cos\theta, -\sin\theta, 1\big).$$

Therefore, $\vec{N}_\varphi(x,y,z) = \dfrac{1}{\sqrt{2}}\left(\dfrac{-x}{z}, \dfrac{-y}{z}, 1\right)$.
(f) Let $f(s) = a + s(b-a)$. Then $\varphi(t,s) = (f(s)\cos t, f(s)\sin t, s)$ and $\partial\varphi^\perp(t,s) = f(s)(\cos t, \sin t, a-b)$. Therefore,

$$\vec{N}_\varphi(x,y,z) = \frac{1}{\sqrt{1+(a-b)^2}}\left(\cos t, \sin t, a-b\right),$$

where $x = f(s)\cos t$, $y = f(s)\sin t$, and $z = s$, so

$$\vec{N}_\varphi(x,y,z) = \frac{1}{\sqrt{1+(a-b)^2}}\left(\frac{x}{f(z)}, \frac{y}{f(z)}, a-b\right).$$

7. If $\boldsymbol{v} \in V$ and $\boldsymbol{u} = (\boldsymbol{v}, s)$, then

$$\det\left(\varphi'(\boldsymbol{u})^t\varphi'(\boldsymbol{u})\right) = \det\left(\psi'(\boldsymbol{v})^t\psi'(\boldsymbol{v})\right) \neq 0,$$

hence φ is a parameterized $(n-2)$-surface. Moreover, since the determinant $\dfrac{\partial(\varphi_1,\ldots,\varphi_{n-1})}{\partial(u_1,\ldots,u_{n-1})}$ has a column of zeros, setting $\boldsymbol{u} = (\boldsymbol{v}, s)$, we have

$$\begin{aligned}
\partial\varphi^\perp(\boldsymbol{u}) &= \sum_{i=1}^{n}(-1)^{i+n}\frac{\partial(\varphi_1,\ldots,\widehat{\varphi}_i,\ldots,\varphi_n)}{\partial(u_1,\ldots,u_{n-1})}(\boldsymbol{v}, u_{n-1})\boldsymbol{e}^i \\
&= \sum_{i=1}^{n-1}(-1)^{i+n}\frac{\partial(\psi_1,\ldots,\widehat{\psi}_i,\ldots,\psi_{n-1})}{\partial(u_1,\ldots,u_{n-1})}(\boldsymbol{v})\boldsymbol{e}^i \\
&= \big(\partial\varphi^\perp(\boldsymbol{v}), 0\big),
\end{aligned}$$

proving (b). Part (c) follows from (b) and 12.3.6.

Section 12.4

3. The transition mapping is $\varphi^{-1} \circ \varphi_1(\boldsymbol{x}) = \frac{1}{2}\boldsymbol{x}$, as may be seen from

$$\boldsymbol{x} = \varphi_1^{-1}(\boldsymbol{y}) = 2(1-y_n)^{-1}(y_1, \ldots, y_{n-1}), \quad y_n < 1,$$

hence

$$\varphi_1(\boldsymbol{x}) = (4 + \|\boldsymbol{x}\|^2)^{-1}(4x_1, \ldots, 4x_{n-1}, \|\boldsymbol{x}\|^2 - 4).$$

5. For (y_1, y_2, y_2) on the cone,

$$(x_1, x_2) = \varphi^{-1}(y_1, y_2, y_2) = \frac{1}{1-y_3}(y_1, y_2), \quad 0 < y_3 < 1,$$

hence $\varphi(x_1, x_2) = ((1-y_3)x_1, (1-y_3)x_2, y_3)$, where

$$y_3 = \frac{\alpha}{1+\alpha} \quad \alpha = \left(\frac{x_1}{a_1}\right)^2 + \left(\frac{x_2}{a_2}\right)^2.$$

7. (a) $\{\boldsymbol{v} : v_1 + 2v_2 + 3v_3 = 0\}$.

9. The transpose of $T\boldsymbol{v}$ is

$$\frac{1}{1-y_n}\begin{bmatrix} 1 & 0 & \cdots & 0 & \dfrac{y_1}{1-y_n} \\ 0 & 1 & \cdots & 0 & \dfrac{y_2}{1-y_n} \\ \vdots & \vdots & & \vdots & \vdots \\ 0 & 0 & \cdots & 1 & \dfrac{y_{n-1}}{1-y_n} \end{bmatrix}\begin{bmatrix} v_1 \\ v_2 \\ \vdots \\ v_n \end{bmatrix} = \frac{1}{1-y_n}\begin{bmatrix} v_1 + \dfrac{y_1 v_n}{1-y_n} \\ v_2 + \dfrac{y_2 v_n}{1-y_n} \\ \vdots \\ v_n + \dfrac{y_{n-1} v_n}{1-y_n} \end{bmatrix}.$$

For part (a), use the identities

$$\sum_{j=1}^{n-1} y_j^2 = 1 - y_n^2 \quad \text{and} \quad \sum_{j=1}^{n-1} v_j y_j = -v^n y_n,$$

a consequence of $\|\boldsymbol{y}\| = 1$ and $\boldsymbol{y} \cdot \boldsymbol{v} = 0$.

11. (a) The conditions are

$$F(\boldsymbol{x}) = 0 \quad \text{and} \quad G(\boldsymbol{x}, \boldsymbol{v}) := v_1\partial_1 F(\boldsymbol{x}) + v_2\partial_2 F(\boldsymbol{x}) + v_3\partial_3 F(\boldsymbol{x}) = 0.$$

The Jacobian matrix of (F, G) (suppressing $\boldsymbol{x}$ and $\boldsymbol{v}$) is

$$\begin{bmatrix} \partial_1 F & \partial_2 F & \partial_3 F & 0 & 0 & 0 \\ \partial_1 G & \partial_2 G & \partial_3 G & \partial_1 F & \partial_2 F & \partial_3 F \end{bmatrix},$$

which has rank 2 since for each $\boldsymbol{x}$, $\partial_i F(\boldsymbol{x}) \neq 0$ for some i.

Section 13.1

3. Fix $\boldsymbol{a}^i$ for $i \geq 3$ and set $B(\boldsymbol{a}, \boldsymbol{b}) := M(\boldsymbol{a}, \boldsymbol{b}, \boldsymbol{a}^3, \ldots, \boldsymbol{a}^m)$. Then

$$B(\boldsymbol{a}+\boldsymbol{b}, \boldsymbol{a}+\boldsymbol{b}) = B(\boldsymbol{a}, \boldsymbol{a}) + B(\boldsymbol{b}, \boldsymbol{a}) + B(\boldsymbol{a}, \boldsymbol{b}) + B(\boldsymbol{b}, \boldsymbol{b}),$$

hence $B(\boldsymbol{b}, \boldsymbol{a}) + B(\boldsymbol{a}, \boldsymbol{b}) = 0$. This shows that M changes sign if the first two arguments are interchanged. The other pairs are treated similarly.

7. (a) $(f_1g_2 - g_1f_2)\,dx_{1,2} + (f_1g_3 - g_1f_3)\,dx_{1,3} + (f_2g_3 - g_2f_3)\,dx_{2,3}$.

8. (b) $-dx_{1,2,3}$.

9. (a) $(-1)^{k(k+1)/2}\,dx_1 \wedge \cdots \wedge dx_{2k}$.

11. (b) $(-\,dx_1 + 2\,dx_2) \wedge (-\,dx_2 + 2\,dx_3) \wedge (2\,dx_1 + \,dx_3) = 9\,dx_{1,2,3}$.

13. (a) $\displaystyle\sum_{j=1}^{n}\left(\sum_{i=1}^{n}\frac{\partial g_j}{\partial x_i}\,dx_i\right) \wedge dx_j = \sum_{j=1}^{n} f'(x_j)\,dx_j \wedge dx_j = 0.$

15. By 13.1.16, $d(\omega \wedge \nu) = (d\omega) \wedge \nu + (-1)^p \omega \wedge (d\nu) = \eta \wedge \nu$.

17. By 13.1.19(d),

$$\begin{aligned}
[\varphi^*(\psi^*\omega)](\boldsymbol{a}^1, \ldots, \boldsymbol{a}^m) &= (\psi^*\omega)_{\varphi(\boldsymbol{u})}(d\varphi_{\boldsymbol{u}}(\boldsymbol{a}^1), \ldots, d\varphi_{\boldsymbol{u}}(\boldsymbol{a}^m)) \\
&= \omega_{\psi(\varphi(\boldsymbol{u}))}\big((d\psi)_{\varphi(\boldsymbol{u})}(d\varphi_{\boldsymbol{u}}(\boldsymbol{a}^1)), \ldots, (d\psi)_{\varphi(\boldsymbol{u})}(d\varphi_{\boldsymbol{u}}(\boldsymbol{a}^m))\big) \\
&= \omega_{\psi\circ\varphi(\boldsymbol{u})}\big(d(\psi\circ\varphi)_{\boldsymbol{u}}(\boldsymbol{a}^1), \ldots, d(\psi\circ\varphi)_{\boldsymbol{u}}(\boldsymbol{a}^m)\big),
\end{aligned}$$

the last equality by the chain rule.

19. By Exercise 9.3.15, $\boldsymbol{x} \cdot \nabla f_j(\boldsymbol{x}) = kf_j(\boldsymbol{x})$. Since $d\omega = 0$, $\partial_1 f_2 = \partial_2 f_1$, $\partial_3 f_1 = \partial_1 f_3$ and $\partial_2 f_3 = \partial_3 f_2$ (13.1.15). Therefore,

$$\begin{aligned}
\partial_1 f &= \frac{1}{k+1}\,(x_1\partial_1 f_1 + x_2\partial_1 f_2 + x_3\partial_1 f_3 + f_1) \\
&= \frac{1}{k+1}\,(x_1\partial_1 f_1 + x_2\partial_2 f_1 + x_3\partial_3 f_1 + f_1) \\
&= f_1.
\end{aligned}$$

Similarly, $\partial_2 f = f_2$ and $\partial_3 f = f_3$.

Section 13.2

1. (b) Since $\partial\varphi^{\perp}(t,\theta) = (\sin\theta, -\cos\theta, t)$, $\|\partial\varphi^{\perp}(t,\theta)\|^2 = 1 + t^2$, hence

$$\operatorname{area}(\varphi) = 2\pi \int_0^1 \sqrt{1+t^2}\,dt = \tfrac{1}{2}\big[\sqrt{2} + \ln(1+\sqrt{2})\big].$$

4. For the case $m = 2$:
$$\varphi'(\theta_1, \theta_2) = \begin{bmatrix} -r_1 \sin\theta_1 & 0 \\ r_1 \cos\theta_1 & 0 \\ 0 & -r_2 \sin\theta_2 \\ 0 & r_2 \cos\theta_2 \end{bmatrix},$$
hence $\det\left[\varphi'(\theta_1, \theta_2)^t \varphi'(\theta_1, \theta_2)\right] = (r_1 r_2)^2$. Therefore,
$$\operatorname{area}(\varphi) = \int_0^{2\pi}\int_0^{2\pi} r_1 r_2\, d\theta_1\, d\theta_2 = (2\pi r_1)(2\pi r_2).$$

6. In the notation of Example 11.5.5, the surface is the graph of
$$g(x_1, \ldots x_n) := 1 - \sum_{j=1}^{n} x_j, \quad (x_1, \ldots, x_n) \in S(1, n).$$
Therefore, the surface area is
$$\int_{S(1,n)} \sqrt{1 + \|\nabla g\|^2}\, d\lambda_n = \sqrt{1+n}\,\lambda_n\big(S(1,n)\big) = \frac{\sqrt{n+1}}{n!}.$$

7. Let $u_{n-1} = s$. Since
$$\varphi(u_1, \ldots, u_{n-1}) = \big(\psi_1(u_1, \ldots, u_{n-2}), \ldots, \psi_{n-1}(u_1, \ldots, u_{n-2}), u_{n-1}\big),$$
$$\frac{\partial(\varphi_1, \ldots, \widehat{\varphi}_i, \ldots, \varphi_n)}{\partial(u_1, \ldots, u_{n-1})} = \begin{cases} \dfrac{\partial(\psi_1, \ldots, \widehat{\psi}_i, \ldots, \psi_{n-1})}{\partial(u_1, \ldots, u_{n-2})} & \text{if } i \le n-1 \\ 0 & \text{if } i = n. \end{cases}$$
Therefore, $\det\big(\varphi'(\boldsymbol{u})^t\varphi'(\boldsymbol{u})\big) = \det\big(\psi'(\boldsymbol{u})^t\psi'(\boldsymbol{u})\big)$, so
$$\operatorname{area}(\varphi) = \int_0^h \int_U \sqrt{\det\big(\psi'(\boldsymbol{u})^t\psi'(\boldsymbol{u})\big)}\, d\boldsymbol{u} = h \cdot \operatorname{area}(\psi).$$

8. From $\varphi(t, s) = (\psi_1(t), \psi_2(t), s)$, we have
$$\frac{\partial(\varphi_2, \varphi_3)}{\partial(t, s)} = \begin{vmatrix} \psi_2'(t) & 0 \\ 0 & 1 \end{vmatrix}, \quad \frac{\partial(\varphi_1, \varphi_3)}{\partial(t, s)} = \begin{vmatrix} \psi_1'(t) & 0 \\ 0 & 1 \end{vmatrix}, \quad \frac{\partial(\varphi_1, \varphi_2)}{\partial(t, s)} = \begin{vmatrix} \psi_1'(t) & 0 \\ \psi_2'(t) & 0 \end{vmatrix}.$$

11. By 12.3.7(b), $\partial\varphi^{\perp}(t, \theta) = \psi_2(t)\big(\psi_2'(t), -\psi_1'(t)\cos\theta, -\psi_1'(t)\sin\theta\big)$, hence
$$\|\partial\varphi^{\perp}(t, \theta)\| = \psi_2(t)\|\psi'(t)\|.$$
For the torus, take $\psi(t) = (a\cos t, b + a\sin t)$, $0 < t < 2\pi$. Then
$$\operatorname{area}(\varphi) = 2\pi \int_0^{2\pi} \psi_2(t)\|\psi'(t)\|\, dt = 2\pi a \int_0^{2\pi} (b + a\sin t)\, dt = 4\pi^2 ab.$$
For the cone, take $\psi(t) = (t, rt/h)$, $0 \le t \le h$, so
$$\operatorname{area}(\varphi) = 2\pi \int_0^h (rt/h)\sqrt{(r/h)^2 + 1}\, dt = \pi r\sqrt{r^2 + h^2}.$$

13. (a) Take $g(t) = t$, $0 < t < 1$, so $(x_1, x_2, x_3) := \varphi(t) = \big(t, t\cos\theta, t\sin\theta\big)$ and

$$\int_\varphi \omega = \int_0^1 \int_0^{2\pi} t\big[(f_1 \circ \varphi) + (f_2 \circ \varphi)\cos\theta - (f_3 \circ \varphi)\sin\theta\big]\, d\theta\, dt.$$

Since $f_1 = f_2 = 0$ and $f_3 = x_1 x_3 = t^2 \sin\theta$,

$$\int_\varphi \omega = -\int_0^1 \int_0^{2\pi} t^3 \sin^2\theta\, d\theta\, dt = -\frac{\pi}{4}.$$

15. (b) A local parametrization of S is

$$(x_1, x_2, x_3) := \varphi(t, \theta) = \big(\psi_1(t), \psi_2(t)\cos\theta, \psi_2(t)\sin\theta\big), \quad 0 < \theta, t < 2\pi,$$

where

$$\psi_1(t) = a\cos t \ \text{ and } \ \psi_2(t) = b + a\sin t.$$

Therefore, by Exercise 12,

$$\begin{aligned}\int_\varphi &\big(f_1\, dx_2 \wedge dx_3 + f_2\, dx_1 \wedge dx_3 + f_3\, dx_1 \wedge dx_2\big)\\ &= a\int_0^{2\pi}\int_0^{2\pi} (f_1 \circ \varphi)(b + a\sin t)\cos t\, d\theta\, dt\\ &- a\int_0^{2\pi}\int_0^{2\pi} (f_2 \circ \varphi)(b + a\sin t)\sin t\cos\theta\, d\theta\, dt\\ &- a\int_0^{2\pi}\int_0^{2\pi} (f_3 \circ \varphi)(b + a\sin t)\sin t\sin\theta\, d\theta\, dt.\end{aligned}$$

Since $f_2 = f_3 = 0$ and $f_1 = x_1 = a\cos t$,

$$\int_\varphi \omega = a^2 \int_0^{2\pi}\int_0^{2\pi} (b + a\sin t)\cos^2 t\, d\theta\, dt = 2a^2 b\pi^2.$$

Section 13.4

4. For example, let f_n and g be Borel measurable and integrable on S with $|f_n| \le g$ for all n. If $f_n \to f$ on S, then $\int_S f_n\, dS \to \int_S f_n\, dS$. This follows directly from the dominated convergence theorem applied to $f_n \circ \varphi_{\mathbf{a}}$.

7. This follows from Exercise 6 and $H^u_{1/n} \cup H^\ell_{1/n} \uparrow S$.

Section 13.5

1. The parameterizations are

   - $S : (\cos t, \sin t, z)$,

- bottom boundary: $(\cos t, \sin t, 0)$,
- top boundary: $(\cos t, -\sin t, 1)$,

where, $0 \le t \le 2\pi$ and $0 \le z \le 1$. The left side of Stokes's formula then reduces to

$$-\int_0^{2\pi} [f(\cos t, \sin t, 0) + f(\cos t, -\sin t, 1)] \sin t\, dt + \int_0^{2\pi} [g(\cos t, \sin t, 0) - g(\cos t, -\sin t, 1)] \cos t\, dt \quad (\dagger)$$

and the right side to

$$\int_0^1 \int_0^{2\pi} [(h_y - g_z)(\cos t, \sin t, z) \cos t + (f_z - h_x)(\cos t, \sin t, z) \sin t]\; dt\, dz$$
$$= \int_0^{2\pi} \int_0^1 [-g_z(\cos t, \sin t, z) \cos t + f_z(\cos t, \sin t, z) \cos t]\; dz\, dt,$$

where we have used

$$h_y(\cos t, \sin t, z) \cos t - h_x(\cos t, \sin t, z) \sin t = \frac{d}{dt} h(\cos t, \sin t, z).$$

Make the substitution $t = 2\pi - s$ in $(\dagger)$ to complete the argument.

3. (a) Both sides equal.

4. (d) 0.

5. Use the parametrization $x = a\cos^{2m+1} t$, $y = b\sin^{2m+1} t$, $0 \le t \le 2\pi$, to first obtain

$$\frac{1}{2}\int_{\varphi} (x\, dy - y\, dx) = ab\left(m + \tfrac{1}{2}\right) \int_0^{2\pi} \cos^{2m} t\, \sin^{2m} t\, dt.$$

Then use $\sin(2t) = 2\sin t \cos t$ and 5.3.4.

9. Apply the divergence theorem using $d(fg) = f\,dy + g\,df$ and $\operatorname{div} d(fg) = f\nabla^2 g + 2\nabla f \cdot \nabla g + g\nabla^2 f$.

12. In (a), the induced orientation of C from S_1 is the opposite of the induced orientation from S_2. By Stoke's theorem,

$$\int_{S_1} \operatorname{curl} \vec{F} \cdot \vec{n}\, dS = \int_C \vec{F} \cdot d\vec{r} = -\int_{S_2} \operatorname{curl} \vec{F} \cdot \vec{n}\, dS,$$

hence

$$\int_S \operatorname{curl} \vec{F} \cdot \vec{n}\, dS = \int_{S_1} \operatorname{curl} \vec{F} \cdot \vec{n}\, dS + \int_{S_2} \operatorname{curl} \vec{F} \cdot \vec{n}\, dS = 0.$$

In (b), the induced orientations of C from S_1 and S_2 are the same, so

$$\int_{S_1} \operatorname{curl} \vec{F} \cdot \vec{n}\, dS = \int_C \vec{F} \cdot d\vec{r} = \int_{S_2} \operatorname{curl} \vec{F} \cdot \vec{n}\, dS.$$

14. Let $F(\boldsymbol{x}) = \boldsymbol{x}$, $\boldsymbol{x} \in \mathbb{R}^n$. On $S := S_r(\mathbf{0})$, $\vec{\mathfrak{n}} = \boldsymbol{x}/r$, hence

$$\int_S \vec{F} \cdot \vec{\mathfrak{n}}\; dS = \int_S r\, dS = r \cdot \operatorname{area}(S).$$

Since $\operatorname{div} \vec{F} = n$,

$$\int_{\|\boldsymbol{x}\|<r} \operatorname{div} \vec{F}\, d\boldsymbol{x} = \int_{\|\boldsymbol{x}\|<r} n\, d\boldsymbol{x} = nr^n \alpha_n.$$

Bibliography

[1] Apostol, Thom, *Mathematical Analysis, 2nd ed.* Addison-Wesley, 2000.

[2] Bloch, Ethan, *Proofs and Fundamentals*, Birkhaüser, 2000.

[3] Burckill, J.C. and H. Burkill, *A Second Course in Mathematical Analysis*, Cambridge Press, 1970.

[4] Dudley, Richard, *Real Analysis and Probability*, Cambridge University Press, 2002.

[5] Fleming, Wendell, *Functions of Several Variables, 2nd ed.* Springer Verlag, 1977.

[6] Mendelson, Elliot, *Number Systems and the Foundations of Analysis*, Dover, 2008.

[7] Rudin, Walter, *Principles of Mathematical Analysis, 3rd ed.* McGraw-Hill, 1976.

[8] Sibley, Thomas, *Foundations of Mathematics*, John Wiley & Sons 2009.

[9] Strang, Gilbert, *Linear Algebra and Its Application, 4th ed.* Thomson Brooks/Cole, 2006.

[10] Wilder, Raymond, *Introduction to the Foundations of Mathematics, 2nd ed.* John Wiley & Sons, 1965.

Index

For Product Safety Concerns and Information please contact our EU representative GPSR@taylorandfrancis.com Taylor & Francis Verlag GmbH, Kaufingerstraße 24, 80331 München, Germany